SUSTAINABLE AGRICULTURAL DEVELOPMENT: THE ROLE OF INTERNATIONAL COOPERATION

SUSTAINABLE AGRICULTURAL DEVELOPMENT:
THE ROLE OF INTERNATIONAL COOPERATION

PROCEEDINGS
OF THE
TWENTY-FIRST
INTERNATIONAL CONFERENCE
OF AGRICULTURAL ECONOMISTS

Held at Tokyo, Japan
22–29 August 1991

Edited by
G.H. Peters, Agricultural Economics Unit, Queen Elizabeth House,
University of Oxford, England
and
B.F. Stanton, Cornell University, USA
Assisted by
G.J. Tyler
University of Oxford

INTERNATIONAL ASSOCIATION OF
AGRICULTURAL ECONOMISTS
QUEEN ELIZABETH HOUSE
UNIVERSITY OF OXFORD

1992

Dartmouth

Published by
Dartmouth Publishing Company Limited
Gower House
Croft Road
Aldershot
Hants GU11 3HR
England

Dartmouth Publishing Company
Old Post Road
Brookfield
Vermont 05036
USA

A CIP catalogue record for this book is available from the British Library and the US Library of Congress

ISBN 1 85521 272 2

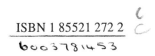

Printed and bound in Great Britain by
Biddles Ltd, Guildford and King's Lynn

CONTENTS

Preface   x
John W. Longworth
Introduction   xiii
George H. Peters and Bernard F. Stanton

OPENING SESSION

Elmhirst Memorial Lecture: Agrarian structure, environmental
    concerns and rural poverty   3
Vijay S. Vyas

Presidential Address: Human capital formation for sustainable
    agricultural development   18
John W. Longworth

SECTION I – GLOBAL STRATEGIES FOR THE DEVELOPMENT
OF AGRICULTURAL AND RURAL RESOURCES

Food security: issues and options   29
Kirit S. Parikh

World agriculture in the next century: challenges for production
    and distribution   44
Nikos Alexandratos

The efficiency of agricultural markets in directing agricultural
    development   66
Ulrich Koester

Achieving a reasonable balance between the public and private
    sectors in agriculture   81
Lawrence D. Smith and Anne M. Thomson

Sustainability: the paradigmatic challenge to agricultural
    economists   92
Richard B. Norgaard

The green revolution, dryland agriculture and sustainability:
    insights from India   102
K.N. Ninan and H. Chandrashekar

Section summary   115
David Colman

SECTION II – INTERNATIONAL TRADE AND ITS IMPACT ON
DOMESTIC POLICIES

Implications of the GATT negotiations: the process of reaching an
    agreement on agriculture   119
Aart de Zeeuw

Agricultural and trade policy reform: implications for agricultural
    trade   129
D. McClatchy and T.K. Warley

The Soviet Union's agrarian sector on the way to a market
    economy   152
A. Emel'ianov

Transition to market economies in Eastern Europe: the case of
    Hungary   158
Ferenc Fekete and Csaba Forgåcs

Evaluating policy choices in developing countries: the policy
    analysis matrix   166
Eric Monke and Scott Pearson

General equilibrium effects of trade liberalization in the presence
    of imperfect competition   181
Thomas Hertel, Kent Lanclos and Marie Thursby

Section summary   192
David Colman

SECTION III – STRUCTURAL ADJUSTMENT POLICIES AND
AGRICULTURAL DEVELOPMENT

Government and agricultural development   197
Eliseu Alves, Clovis de Faro and Elisio Contini

Adjustment policy and agricultural development   211
Patrick Guillaumont

Inflation and agriculture: ten years of high inflation and
    government debt in Brazil   227
Guilherme L.S. Dias

Impact of policy reforms on the agricultural sector in Chile   241
Eugenia Muchnik de Rubinstein

Decollectivization in East and Central Europe 254
Karen Brooks and Avishay Braverman

Agricultural policy and structural adjustment in Cameroon 265
Joseph Ntangsi

Structural adjustment policies and agricultural development in
    Morocco 281
Mokhtar Bouanani and Wallace E. Tyner

Agricultural trade and pricing policies in developing countries:
    implications for policy reform 292
Alberto Valdés

Section summary 306
Antonio Brandão

SECTION IV – THE POTENTIAL OF BIOTECHNOLOGY FOR AGRICULTURE AND THE FOOD INDUSTRY

Key elements of modern biotechnology of relevance to agriculture 311
W. James Peacock

Comments from various perspectives 317
*C.H. Hanumantha Rao, Cornelis L.J. Van der Meer and
    Randolph Barker*

Plant breeders' rights legislation, enforcement and R & D: lessons
    for developing countries 330
Carl E. Pray

Financing agricultural research in the presence of international
    benefit spillovers: the need for institutional coordination and
    innovation 343
David B. Schweikhardt and James T. Bonnen

Biotechnology and the International Agriculture Research Centres
    of the CGIAR 355
M.P. Collinson and K. Wright Platais

The role of the private and public sectors in the development and
    diffusion of biotechnology in agriculture 371
R.K. Lindner

Section summary 386
Herbert Stoevener

SECTION V – ENVIRONMENTAL ENHANCEMENT
STRATEGIES AND POLICY LINKAGES

Sustainable development: concepts and strategies   391
Sandra S. Batie

Sustainable development strategies in less favoured and marginal
   production areas   405
R. P. Yadav

Management of fisheries as a common property resource   422
Trond Bjørndal

Management of common-pool forest resources   435
Randall A. Kramer and Vishwa Ballabh

Climate change, agriculture and the environment: some economic
   issues   450
Richard M. Adams

Air pollutants and options for their control: experiences from the
   European scene   462
Sten Nilsson

Section summary   474
Herbert Stoevener

SECTION VI – FARM HOUSEHOLDS AS THE DOMINANT
INSTITUTIONAL UNITS IN AGRICULTURE

Household time allocation – the ultimate determinant of improved
   agricultural technology adoption in Nigeria: an empirical
   activity interphase impact model   481
Anthony Ikpi

The theory of resource allocation by farm households: the role of
   off-farm employment, household production and transaction
   costs   502
Günther H. Schmitt

Family farms, cooperatives and collectives   520
Yoav Kislev

Contributions of women and household members to the economy
   in rural areas   533
Helen H. Jensen

Agriculture, rural labour markets and the evolution of the rural
    non-farm economy   542
Steven Haggblade and Carl Liedholm

Capital formation of the farm household and resource allocation
    in agricultural – rural sectors: an analysis of sustainability of
    Japanese agriculture   558
Ryohei Kada

Section summary   568
Mahabub Hossain

SECTION VII – SMALLHOLDER AGRICULTURE AND
CHANGES IN DEVELOPMENT

Will biotechnology alleviate poverty?   573
Iftikar Ahmed

Agricultural biotechnology and the poor in developing countries   592
Robert W. Herdt

Green revolution, agrarian structure and income distribution in
    Asia   607
Keijiro Otsuka

'Small farmers' in China and their development   619
Niu Ruofeng and Chen Jiyuan

An assessment of the role of informal finance in the development
    process   644
Richard L. Meyer and Geetha Nagarajan

Informal credit markets in Bangladesh agriculture: bane or boon?   657
K.A.S. Murshid

Section summary   670
Mahabub Hossain

SYNOPTIC VIEW

Csaba Csáki – President Elect   673

INDEX   689

PREFACE

The title of this volume was the theme for the 21st International Conference of Agricultural Economists (XXI ICAE) held in Tokyo, 22–29 August 1991. In 1989, when the Executive of the International Association of Agricultural Economists (IAAE) agreed to adopt this theme for the 1991 conference, the concept of sustainability was being widely discussed. By the time of the conference, the world-wide debate was in full swing. It would have been difficult to have chosen a more appropriate theme for an international professional meeting of agricultural economists in Japan in 1991. The question of sustainability in relation to modern, high-technology agricultural production systems is nowhere more evident than in Japan. Furthermore, with a growing internationalization of agricultural production systems, the issues surrounding the sustainability of these systems are becoming of world-wide importance both in poor countries as well as in the advanced world.

The appropriateness of the theme, the professional quality of the programme and the attractiveness of the venue drew a record number of participants from 62 different countries. Once at the conference, many participants became involved both by formally presenting, or commenting on, papers and by taking part in the vigorous discussions which followed most presentations. All of the participants are to be thanked for their contribution to the various aspects of the programme and the general success of the conference.

The smooth running of the conference was largely due to the enormous efforts of the Japanese Organizing Committee (JOC) chaired by Professor Kawano. Formally, the JOC consisted of 26 people who were primarily responsible for the various major activities. However, each member of the JOC was supported by one or more sub-committees. In total the JOC represented a voluntary workforce of over 340 Japanese agricultural economists. The involvement of such a large number of local agricultural economists in the organization of the conference and the associated professional activities not only greatly assisted the smooth running of the meetings but also facilitated the very considerable involvement of the Japanese profession in the conference. In return, the XXI ICAE gave Japanese agricultural economists a platform on which to demonstrate their professional competence and enhance their profile both internationally and within the Japanese professional hierarchy. Well before the event, the Japanese Science Council nominated the XXI ICAE as one of the six nationally recognized international conferences to be held in Japan in 1991. Not only did this provide considerable incentive for sponsors in Japan to assist the JOC by making donations, but it also raised the standing of the agricultural economics profession in Japan.

The extraordinary thoroughness with which the JOC approached their task became obvious to me during the five six-monthly visits I made to Japan commencing in May 1989 and finishing in May 1991. In addition, Professor Bernard Stanton, the Vice-President (Programme) visited Japan on two occasions in the run-up to the conference. Communications between the IAAE Executive, the JOC and intending participants were greatly facilitated by the availability of facsimile machines. Indeed, the organization of such a large professional conference, a task undertaken largely on a voluntary basis by members of the JOC and the Executive of the IAAE, would have been extremely difficult without facsimile machines.

This volume makes available all the major papers presented at the conference. In addition, it provides a summary of the discussions which followed. The task of building the formal programme for the conference was undertaken by Professor Stanton. He invited seven internationally well-known colleagues to assist with developing the seven major sections of the programme. Professor Stanton and the seven sectional organizers (their names are listed elsewhere in this volume) produced a nicely balanced but intellectually challenging general programme. The IAAE editor (Professor George Peters) has had principal responsibility for editing these major papers and the sectional summaries. He has, of course, been assisted in this work by Bud Stanton and Godfrey Tyler. To these people, in particular, the profession owes a great debt of gratitude. They have made available an excellent volume in a most timely manner.

A companion volume (*IAAE Occasional Paper No. 6*), which will contain the contributed papers, is being edited by Bruce Greenshields and Margot Bellamy. Bruce Greenshields did his usual excellent job in organizing the Contributed Papers Competition for the XXI ICAE. With the publication of the contributed papers volume for this conference, he will complete many years of service to the IAAE in connection with an important component of our conferences. Bruce has now accepted the editorship of the IAAE journal, *Agricultural Economics*. I would like to take this opportunity to wish him well with this new task. Another person who contributed significantly to the success of the Tokyo Conference was Charles Caudill. As Chairperson of the Discussion Groups and Mini-symposia, Charles (ably assisted by Larry Sivers) played a major part in bringing people together in smaller forums where the free interchange of ideas was possible. The poster sessions at the Tokyo Conference were another highlight. These were organized by Kenji Ozawa, the member of the JOC assigned to take charge. The availability of ample, high-quality space for the display of posters, together with the high quality of the posters themselves, and the enterprising way in which the poster sessions sub-committee of the JOC made various Excellence Awards, all contributed to a most successful poster session component.

The excellent facilities for the display of poster papers was only part of the advantages enjoyed by the IAAE in having access to the Kogakuin University facilities opposite the Keio Plaza Hotel. The Association is most grateful to the Council of Kogakuin University for making their facilities available for the conference. These most spacious modern offices and lecture theatres greatly enhanced the environment of the conference.

A large number of other donors made significant contributions to the conference. In particular, a number of Japanese agencies made generous donations to assist with bringing participants from soft currency countries to the meetings. These agencies, together with several non-Japanese donor institutions, have all been thanked and listed elsewhere. However, it is important for readers of this volume to recognize that many of their colleagues from the Third World and the former Eastern Bloc countries who had the opportunity of participating in the XXI ICAE were only able to attend the meetings because of the generosity of international aid agencies. In this respect, the XXI ICAE itself was an excellent example of the second half of the conference theme.

Conferences like the Tokyo gathering always require a great deal of behind-the-scenes clerical and secretarial support. I know that the JOC was well served in this regard, as was the IAAE Executive. In particular, I would like to thank my Secretary, Mrs Cherelle Mungomery, for her dedication to the tasks associated with preparing for the conference over the preceding two years. In addition, she attended the conference and took charge of the *Cowbell*, the daily newspaper, distributed to all participants. For this particular conference, the *Cowbell* was a major source of information not only about the conference and its daily activities but also about the IAAE. The daily preparation of this newspaper was a burdensome and demanding task. It is also important to recognize the efficiency with which the printing and distribution, not only of the English-language *Cowbell* but also of a companion Japanese *Cowbell*, was undertaken by a sub-committee of the JOC. These arrangements, like everything else associated with the conference, were exceedingly well done.

No doubt, readers of this volume who had the pleasure of attending the XXI ICAE will enjoy the opportunity to reflect once more upon the high-quality papers presented. Furthermore, I am sure that not only these people but all members of the international agricultural economics profession will find this volume of great reference value for many years to come.

JOHN W. LONGWORTH
Immediate Past President

INTRODUCTION

The theme of the IAAE 21st International Conference, held in Tokyo, Japan was 'Sustainable Agricultural Development: The Role of International Cooperation'. The organization of the papers provided the usual difficult challenge to those bearing particular responsibility for it. On this occasion the traditional Elmhirst Memorial Lecture (V.S.Vyas) and Presidential Address (John Longworth) were followed by seven sections of plenary and invited papers, for each of which a session organizer assisted the Vice-President Programme. These, with the named organizer, were:

I   *Global Strategies for the Development of Agricultural and Rural Resources*
      David Colman, University of Manchester, United Kingdom

II   *International Trade and its Impact on Domestic Policies* Karl Meilke, University of Guelph, Canada

III   *Structural Adjustment Policies and Agricultural Development*
      Antonio Brandao, World Bank, Washington

IV   *The Potential of Biotechnology for Agriculture and the Food Industry*
      J.G.Ryan, Australian Centre for International Agricultural Research

V   *Environmental Enhancement Strategies and Policy Linkages*
      Herbert Stoevener, Virginia Polytechnic Institute and State University, USA.

VI   *Farm Households as the Dominant Institutional Units in Agriculture*
      Günther Schmitt, University of Göttingen, Germany

VII   *Smallholder Agriculture and Changes in Development*
      Mahabub Hossain, Bangladesh Institute of Development Studies

Within each section the first two papers printed were normally those delivered to plenary sessions of the Conference. This does not apply in Section IV, where there was one plenary paper followed by three extended comments. In accordance with precedent, the papers within each section are followed immediately by comments of invited discussion openers (except in very few cases where the exigencies of operating a large and complex programme involved unexpected absence). There were few alterations in the advertised programme of papers. The major difficulty appeared in Section II where Alexei Emel'ianov was unable to be present. His paper was presented by Dr Karen Brooks, and we have been able to include a translation of the original address. It is one of very few cases in which there is a full citation of an author's affiliations. As a general point, readers will note that in August 1991 the Soviet Union existed as a single entity. That is particularly relevant to Professor Emel'ianov's paper. No attempt has been made, throughout the volume, to alter references to the country as it was. The IAAE is strictly an organization without any form of political affiliation and the use of geographical descriptions, in that and all other cases, is simply a matter of convenience.

Within Section II the absence of another speaker resulted in delivery of a swiftly prepared paper by Professors Fekete and Forgacs. This also occurred in Section VII, where our debt is to Richard L. Meyer and Geetha Nagarajan. It must also be emphasized that the smooth operation of the Conference owed much to the work of the Japanese Organizing Committee in general, and to Professors Keizo Tsuchiya and Fumio Egaitsu in particular.

Once again, rapporteur reports summarizing the conference floor discussion are excluded, though the names of rapporteurs (who are thanked for their valuable assistance), chairpersons and participants are recorded within the section summaries. The latter were first prepared for delivery on the Conference floor by four of the organizers present, with three of them, David Colman, Herbert Stoevener and Mahabub Hossain summarizing both their own sessions and one other. They each have an attribution, though in all cases the Editor has made some additions, based on information contained in rapporteur summaries. Though the main summaries are naturally placed at the end of each section, they could equally well serve as introductions to the relevant blocks of the programme, a point which readers may wish to bear in mind. A quick overview of the wealth of material presented in Tokyo, and an impression of the atmosphere of the conference, appear in the traditional synoptic view of proceedings by our President Elect.

A theme can never be a rigid guideline. Those who devote their attention to the study of agricultural economics have many, diverse, interests and it is impossible to choose a single strand of thought which expresses both those interests and the current issues which may be of pressing policy significance. The general aim on this occasion was neatly expressed by Professor Shigeto Kawano, as Chairman of the Japanese Organizing Committee, in his invitation note to prospective conference delegates:

> Whether in a developed or developing nation, progress in agriculture, really the principal industry of any country, is a basic condition for economic advancement. All countries have their own particular agricultural problems, and seeking solutions through cooperation among nations is a vital task in the world's economic community.

Readers of this volume will, however, naturally expect discussion of the theme to be the element within the Conference which differentiates it from earlier meetings. In fact, it appears in many guises; first as a direct concern with 'sustainability', highlighted in the Presidential Address, mentioned in so many papers, and most heavily represented in Section V, where it is married with discussion of the managerial issues surrounding common resources and of the controversial issue of climatic change. The second strand of 'international cooperation' is very much in evidence in Section IV, which discusses research in biotechnology, both in terms of its possible effects and in respect of the international organization of research and the diffusion of its results. Some of that discussion continues in Section VII. Needless to say, it is also a key element in the many papers which mention the international organization of agricultural markets and its interface with national policies. Readers will note that the GATT round of negotiations was mentioned often. At the time of going to press (February 1992) it was clear that difficulties were still being experienced in bringing the negotiations to a successful conclusion.

The Conference Proceedings were to have been edited once again by our colleague, Allen Maunder. Sadly, he died, in retirement from his Oxford University post, in February 1991. It is fitting to record the contributions which he made, over so many years, to our activities. Coupled with his name must go that of his wife, Pearl, who had attended many conferences and who acted in an invaluable secretarial capacity. The Editorship remains in Oxford (G.H.Peters was assisted by Godfrey Tyler). Mrs Maunder was not directly involved in the work, but it must be recorded that she did provide much valuable advice relating to the task of editorship itself, and of the running of the Oxford office. Without that it would have been a much more difficult task to complete the work.

We owe a particular debt to Debbie Noon (University of Queensland), who was present in Tokyo and handled so much of the initial word processing. Within the Conference office we also received help from Cherelle Mungomery (University of Queensland) and from Lona Christoffers and Charlene Carsrud (Iowa State University). It was the last appearance as Secretary/Treasurer of Jim Hildreth. Characteristically, he was able to provide guidance based on long experience. It was also obvious to those who were in the office that a major burden was, and had been, borne by John Longworth, as President. In Oxford, much of the secretarial work was undertaken by Denise Watt. It became quite clear that she and Debbie Noon understood a language which can be beyond the scope of uninitiated mortals. Help was also received from Owen Jones, Roger Crawford, Julia Knight and (not least) Judith Peters. The staff of Dartmouth Publishing Company provided invaluable assistance once again. We are particulary grateful to Sonia Bridgman, Editor, Dartmouth Publishing Company.

Apart from this volume, other activities of the Conference are reported elsewhere. Bruce Greenshields and Margot Bellamy are editing *IAAE Occasional Papers 6* and reports of discussion groups are being distributed as a *Members' Bulletin*.

The views expressed in this book are not necessarily those of the IAAE, nor of the institutions with which the various authors are affiliated.

G.H. PETERS
Editor, International Association of Agricultural Economists

B.F. STANTON
Vice President Programme

OPENING SESSION

ELMHIRST MEMORIAL LECTURE

VIJAY S. VYAS*

Agrarian Structure, Environmental Concerns and Rural Poverty

I feel deeply honoured by the opportunity to address my professional colleagues on this singular occasion of honouring the founder and for a long time
the moving spirit of our Association, Dr Leonard Elmhirst. When Professor
Longworth invited me to deliver the sixth Elmhirst Lecture, I could not help
but remind myself that every one of my five distinguished predecessors had
made important contributions to our understanding not only of the rural economy
but also of the broader socio-economic system of which it is a part. I cannot
claim either the erudition or the sweep or the vision of my illustrious predecessors. But circumstances have enabled me to observe carefully the changes
which were taking place in the rural areas of the developing world from a
number of vantage points: from the perspective of a village level researcher in
poverty-stricken villages in Western India; as an academic in universities and
research institutions; as a policy adviser to the state and the national governments and as senior adviser in the World Bank's Washington establishment.
These observations, filtered through my early economic training, which was
in the true humanistic traditions of the Bombay School of Economics, have
led me to arrive at some tentative conclusions on the nature, causes and
consequences of rural poverty. Some of these I venture now to share with you.

Poverty eradication and sustainable development are now being recognized
as important dimensions of development. Public pronouncements at the national level as well as the policy documents issued by international agencies
reflect these overriding concerns; so also does the work of scholars and the
stand taken by social activists. Various factors have helped in highlighting
these two issues. The 'lost decade of the eighties' had brought to the fore the
plight of the poor, particularly in the countries of Africa and Latin America.
The development of means of communication has heightened the awareness
of the poor regarding their miserable plight. When juxtaposed with the immense
possibilities of improving their economic well-being, this awareness has been
further sharpened. With the spread of democratic institutions, the importance
of the poor as a constituency has been enhanced, and one finds a large number

*Institute of Development Studies, Jaipur, India. The author is grateful to his colleagues at IDS
and elsewhere who contributed substantially to the content and logic of this presentation. In
particular, the contributions of Deepak Gianchandani, Anil Gupta, Vidya Sagar and Davendra
Tyagi are gratefully acknowledged.

3

of spokesmen for poverty alleviation. Concern for environmental degradation, particularly for unsustainable agriculture, has become urgent with the alarming speed of deforestation and desertification, spread of soil erosion, ingress of salinity, water shortages, increase in pests and crop diseases and other symptoms.

While there is seldom any dissent on the importance of both these issues, there is a discernible lack of clarity on the nature of the interrelationship between rural poverty and environmental constraints which inhibit sustainable agriculture, and a good deal of confusion on how to tackle them simultaneously. This lecture will be a modest effort to discuss some aspects of poverty eradication and (economically and environmentally) sustainable agriculture and the critical role which agrarian structure plays in reaching these objectives. The first section deals with the nature and extent of rural poverty in the developing countries, the impact on poverty during the last few years and the approaches to poverty eradication. In the second section environmental aspects of sustainable agriculture are outlined, pressure on natural resources, particularly land, is highlighted and the question of poverty and over-exploitation of resources is examined. The third section underscores the economic aspects of sustainable agriculture and elicits the importance of structural change in agriculture as reflected in the access to land, an issue which has been generally side tracked in the discussion on sustainable agriculture. The last section points out the scope for conflict as well as convergence between the twin goals of poverty eradication and sustainable agriculture through the mediation of structural change in agriculture which ensures more equitable access to land for the poor.

For empirical support to various premises I have relied heavily on the experience of the developing countries of South Asia, particularly India. This reliance is partly because I am more familiar with this part of the world and partly because the largest number of the poor inhabit the countries of this region. I may point out one more limitation of my presentation. The time-horizon I have selected for the prognosis is medium-term, that is, the decade of the 1990s. I am aware that we are on the threshold of some momentous changes in technology, such as advances in biotechnology, and maybe also in social organization. However, these changes will take time to unfold fully and make their impact felt.

EXTENT AND NATURE OF RURAL POVERTY

Over the years an awareness of the poverty problem has resulted in a deepened interest among policy makers and researchers alike in the concepts and measurement of poverty. So much so that, in the words of a researcher, it has become an 'industry'. The cause of poverty alleviation in recent years suffered as the definitions of poverty adopted were either too wide or too narrow. With too broad a definition poverty alleviation could hardly be distinguished from socio-economic development as such, with too narrow a definition the purpose of poverty alleviation, enabling the poor to lead a decent life, is defeated. Thus a UN Report, true to its tradition, defined minimum subsistence to include physiological, legal and social necessities. It did, however, suggest that these were roughly hierarchical: each employing a higher quality of, or additional, goods and services than the *preceding* one (UN, 1969). At the other extreme is the narrow, single index, definition of poverty using per capita income or per capita consumption.

A definition of poverty currently in vogue is one which translates a calorie norm into monetary value and adds expenditure on other non-food items generally consumed by the households in that income bracket which satisfies the minimum calorie intake.[1] The latter, that is expenditure on non-food items, is generally arrived at on the basis of budget studies. A serious lacuna in this definition is the total reliance on *private income*. It is important to remember that in one of the pioneering efforts in measurement of poverty, that by a group of experts set up by the Planning Commission of India in the 1960s, it was explicitly mentioned that, while defining poverty in terms of minimum household income, the group had assumed that *public services* like elementary education and health were available to all (Sushma, 1988). In later discussions this vital aspect – provision of and access to public goods – was ignored. In any event an income-based poverty line can only serve as a first approximation. For a poverty alleviation strategy, access to public goods, like elementary health and education, needs to be taken into account. A partial correction to the income/calorie-based definition is provided by the recent efforts of UNDP to construct a Human Development Index (HDI) for different countries.[2]

The poor households subsist at various levels of deprivation and their poverty could be transitory or chronic. We are here concerned with the chronically poor, although it should be understood that transient poverty, say due to a natural calamity, can soon degenerate into chronic poverty in the absence of some devices – household, social or public – to absorb adverse shocks. The chronic poor are those who lack the skills, assets and stamina. The intensity of poverty of different sections of the poor households could be measured by the 'Sen Index of Poverty'.[3] Less sophisticated but equally remarkable is Lipton's categorization of poor into the poor and the ultra poor on the basis of whether the household is able to meet its calorie needs (Lipton, 1983). These distinctions are important to evaluate the effects of various poverty alleviation measures.

Extent of poverty

For international comparisons, frequent use is made of the World Bank's country-specific statistics on income levels and income distribution. The World Bank also makes use of global norms to arrive at approximate figures of the poor. Thus, according to the *World Development Report, 1990* (World Bank 1990), in 1985 an estimated 1.1 billion people in the developing world lived in poverty. Of these nearly 633 million were absolutely poor. These estimates were based on an income of $375 per person per year for the poor and $275 per person per year for absolutely poor (amounts being in constant 1985 *PPP* prices). The World Bank's estimates are not universally accepted, but they are more comprehensive in their coverage than other global estimates (Saith, 1990).

The geographical distribution of the world's poor as estimated by the World Bank suggests that one-third of the developing world's total population, of which 18 per cent were extremely poor, was living in poverty. Major concentration of poverty was in South Asia, which accounted for nearly half of the developing world's poor, and nearly half of the extremely poor. Sub-Saharan Africa reported an equally high proportion of the poor in relation to the region's overall population, and Latin America and the Caribbean region was not far behind.

Recent developments

In recent years, on the basis of some welfare indicators, such as life expectancy, infant mortality and literacy, one finds that there has been a distinct improvement in living standards in practically all developing regions. Per capita income in developing countries has increased by nearly 3 per cent per year and more people have access to health services, drinking water and better sanitary facilities. The daily supply of calories has improved significantly (UN, 1990). The improvement has not been uniform over the regions or among countries in any particular region. Also the 'average' figures at the country level mask severe deprivation suffered by various groups of people. Yet the fact remains that a remarkable improvement in the living standards of the people in the developing world has taken place, according to these welfare measures.

The same cannot be said about the performance on the basis of an income or expenditure based poverty criterion. The East Asian countries continued their commendable progress towards poverty eradication, the Philippines being the main exception, and some of the poor countries in South Asia, for example India, did improve their performance. There was a stagnation in Sub-Saharan Africa and a retrogression in countries of Latin America, mainly because of the burden of external debt and the falling commodity prices. Putting these two facts together – better performance on the basis of some of the welfare norms and persistence of poverty in terms of entitlement for goods and services – it is clear that rigours of poverty have eased, even in the countries where widespread poverty prevails (Jodha, 1988).

Characteristics of the poor

A large number of studies the world over have brought out the main characteristics of the poor. While the situations differ from country to country, certain common features stand out. Most of the poor in the developing world are concentrated in rural areas, and a large number of them depend directly or indirectly on agriculture; they own or control few physical productive assets.

They face a fragmented and exploitative market for their main asset: labour; women are more discriminated against in this market. Those among the poor households who have a land base have marginal holdings and/or are tenant-at-will. They are generally over-represented in the marginal regions.

Their capacity to take risks is probably weak. They have a larger dependence on common property resources.

The poor are handicapped by social disabilities as much as by economic constraints. They are powerless and marginal, in spite of their large numbers in several countries, as producers as well as consumers. They have a larger number of dependants, though not necessarily large-sized households. Mostly they do not possess saleable skills, are malnourished and physically weak. In other words, not only do they not have adequate claim on current production, but also their future prospects are not very encouraging.[4]

Approach to poverty alleviation

Basically, two approaches have been adopted to resolve the poverty problem. These are not exclusive, yet the emphasis on one or the other is quite clear. In one approach a large measure of reliance is placed on the 'trickle-down' effect of rapid growth, while the other relies more on public intervention.

There cannot be any doubt about the fact that without growth poverty eradication will not be a practical proposition. However, a heavy reliance on 'trickle-down' or 'spread effect' of growth raises several questions and is not supported by historical experience. Historically 'trickle-down' has benefited the poor when at least one of the following two conditions had been present: (a) the rate of growth of the economy had been very high, say 8 to 10 per cent per annum, or (b) the poor had some asset base. Neither of the two conditions can be taken for granted.

There are not many examples of countries recording a consistently high rate of growth in recent times. Experience suggests that, even if a spurt in growth rate is achieved, it becomes difficult to maintain the tempo without incurring inflationary pressures or running into balance of payment difficulties, usually both.[5] I will come back to the transient nature of such growth swings in inequitable societies a little later. Similarly, if the asset base of the poor is absent or depleted – and the assets could comprise not only physical assets, like land, but also assets such as skills and physical stamina – the poor households will not be able to respond to the growth stimuli. That is why even those who put much faith in 'trickle-down' also worry about the *composition* of growth rather than growth *per se*. For example, a plea for labour-intensive patterns of growth is advocated in order to ensure a larger participation of the poor in the growth process (World Bank, 1990).

The interventionist policies by themselves cannot be counted upon to eradicate mass poverty in the developing countries. This is not only because of the faulty project designs and wrong implementation strategies of the poverty alleviation projects, on which much has been written, but is also due to the diversion of scarce resources which could enhance the asset base of the poor households benefiting from these projects. The case of India's Integrated Rural Development Programme may be used to illustrate these points. In scale of operations and clarity of objectives it could be considered as an exemplary programme, yet its cost-effectiveness can be seriously questioned. There has been a massive investment of resources and political support at all levels, but the impact on eradication of poverty has not been commensurate. A careful researcher has estimated that over a period of ten years this programme has helped nearly 9.3 per cent of the poor households in rural areas to cross the poverty line. Even this small impact was selective in terms of the households and the areas. For example, the households who were nearer to the poverty line received more benefits and the programme fared better in developed regions as compared to backward and remote regions (Hirway, 1990).

Over a period of time there have been some improvements in performance: bureaucracy has become more sensitive, local institutions are getting involved, some forms of participatory management are being experimented with. However, participatory management, which is the king-pin in the delivery of these programmes, has to go beyond tokenism. A prerequisite to genuine participation by the people is an institutional underpinning. With gross inequities in the ownership of assets, particularly with a large section being assetless, it is doubtful whether participatory institutions capable of delivering economic goods and services in an equitable manner would emerge.

The experiences of India and other countries suggest that the 'trickle-down' effect of growth depends on the recipients having access to assets, possession of skills or physical stamina. The public programmes of poverty eradication have helped the category of the poor households so endowed. It will be instructive in this context to review the quality of the main productive asset in the rural areas of the developing countries, that is land, and then look into the question of access to this resource by the poor. The next section addresses the first question and in the following section the issues pertaining to agrarian structure are examined.

THE LAND RESOURCES

In the context of poverty eradication there are questions on environmental sustainability as well as economic sustainability of the present mode of agriculture. Let me briefly comment on these two aspects. Current anxiety about the sustainability of the present mode of agriculture,[6] from the environmental angle, can be ascribed to two related factors: the growing pressure of population on land resources and deteriorating quality of earth resources (land, water, air) at least partly due to intensification of agriculture.

For the poor countries raising agricultural productivity without endangering its sustainability poses a serious challenge. In the past, the 'open frontiers' provided an outlet for the poor. Most of the developed countries of today could circumvent the initial crowding of the people on land by using spaces available within the countries or because migration to other land-abundant countries was easy. This is no longer true for the developing countries whose surplus rural labour cannot find any outlet in the countries where the land is still abundant and labour relatively scarce. The poor in the developing countries have to find the solution to the poverty problem within their own country. In this situation the main source of livelihood in the rural areas, agriculture, plays a pivotal role.

The possibility of raising agricultural production has now to be examined only in the context of intensive, and not extensive, agriculture. This is true globally, and with a few exceptions for a large number of countries. According to Food and Agricultural Organization (FAO) estimates, the amount of arable land in developing countries will increase by less than 10 per cent to 850 m. hectares, by the year 2000. Amount of land per inhabitant in the developing world will fall from 0.85 hectares at the beginning of the 1980s to 0.6 hectares by 2000. New land which will be brought under cultivation will be of poor quality.[7] In any case, by 2025 no high-quality land will be available for arable purposes. Already the area under severe risk of erosion and salinity is increasing, as the lands on slopes of hills, fringes of forests and borders of deserts are brought under the plough.

Another way the arable area can be extended is through irrigation. Among several possible benefits, such as increase in productivity, or lowering of risks, a major benefit of irrigation is the multi-season cropping. Particularly in areas where concentration of rainfall to a few months makes crop production in non-rainy seasons very risky, if not an impossible enterprise, perennial irrigation contributes substantially to increase in arable areas. In several parts of the developing world, especially Asia and Africa, the land base was significantly augmented with the introduction of irrigation. The area of irrigated lands has trebled since 1950, but almost all easy sites for irrigation development have already been exhausted. Now the cost of new irrigation development is becoming prohibitive. Also ecological damage done by surface as well as ground water irrigation to the soils and environment, coupled with the increasing threat of water-borne diseases, is making the irrigation option much less attractive. As it is, nearly 10 per cent of the irrigated hectares have already become saline. Another 25 per cent show early signs of salinity.

In brief, there is very little scope for extensive farming, either by extension of cultivated areas on fallow lands or by an extension of irrigation. Whatever solutions to poverty and environment degradation we seek should be in the context of fixed, rather shrinking, land surface and water resources. This would require greater ingenuity and greater care to conserve the land and water resources. Eminent scientists such as M.S. Swaminathan have been warning us for more than two decades of ecological disasters if proper methods of conservation are not adopted.[8]

The poor and environmental degradation

It is sometimes presumed that the poor are responsible for the depletion of natural resources, not because of any perversity, but because of the compulsions of poverty. How far are these fears genuine? There is a great dearth of usable data to establish any proposition very firmly. However, a few tentative conclusions can be advanced. In the first place, in the resource-rich regions there is no evidence to suggest that lands cultivated by the poor, in this case the holdings of the small and marginal farmers, are in any way inferior to the holdings of the large or medium-sized farmers. At the same time, in ecologically degraded tracts or the resource-poor regions there is a relatively larger concentration of the poor holdings. The latter fact by itself does not suggest that the poor are 'responsible' for environmental degradation in these areas.

The poor could be charged with over-exploitation and consequent degradation of natural resources on two possible counts. Because of a faster rate of growth of population and because of their larger dependence on the natural resources (say, for fuel and fodder) coupled with high subjective discount rates, they may be responsible for fast depletion of some of the resources (mainly forests) and degradation of other resources (such as soils). Of these possibilities the argument based on higher population growth is more serious and has a wider currency. If this is justified, the expanding numbers in poor households would appear to have been pushing cultivation to the fragile or marginal lands, with serious consequences.[9]

But this view has been seriously challenged by some scholars in India and other South Asian countries. In a recent publication results of case studies from four countries – Pakistan, India, Bangladesh and Nepal – and from Bihar state of India, were reported. The authors brought out the complexity of the economic and demographic relationships. Some of their major findings, summarized in an overview by the editor Garry Rogers, may be noted: (a) there is no evidence in the studies that the poor have relatively high fertility, and there is a hint that their fertility may be lower than middle-income groups at least. Moreover, because child survival rates are lower among the poor, the achieved family size appears to be particularly low among the poorer groups; and more explicitly (b) there is a powerful association between household size and wealth – poor households are small (Rogers 1989). The effect of poverty on the exploitation of resources does not arise from the large size of the households of the poor but, as the authors of the study quoted above have put it, is due to 'pre-existing patterns of inequality and exploitation'.

The poor are also blamed for 'over-exploiting' the resources, as they depend heavily on natural resources and also heavily discount the future value of income streams. This is particularly true of the exhaustible resources, such as forest products. However, one should not over-estimate the damage caused to the natural resources by petty pilferage by the poor. This pales into insignificance when compared to the depletion of natural resources by the rich and their wasteful life style (WCED, 1989). More demands on land and natural resources are made by the life style of the rich than by the need-based exploitation of the poor. We will not be able to talk about preservation of environment unless we seriously question, and suggest alternatives to, the present style of living.

The fact that the poor depend heavily on natural resources gives them greater motivation to conserve these resources, provided their interests are not overlooked. No plans of environmental protection can succeed unless they involve the poor as participants *and* beneficiaries, whether those plans be for regeneration of forests or control of water pollution or soil erosion. Unless the poor cooperate the rot cannot be stopped, and in order to ensure their cooperation some tangible benefits should accrue to them. The arguments that environmental protection measures generate positive externalities do not persuade poor households to sacrifice their immediate income for the sake of medium to long-term gains of the community. A cardinal principle of sustainability is to ensure protection, if not an enhancement, of the incomes of the small and marginal farmers and others whose livelihoods depend on these resources.

Poor households are as much interested in conserving natural resources as any other section of society.[10] The agricultural and forestry practices improvised over generations bear testimony to the ingenuity of the poor to eke out a living from shrinking and depleted resources. In any effort to preserve environment the poor must be stake-holders. It is important in this connection to recognize the role of collective efforts to conserve resources. With their subsistence orientation, the poor cannot be disciplined to conserve resources with reliance on the instruments of pricing and subsidies. They can be disciplined mainly by community efforts. In the past many of the conservation practices depended on community efforts, such as preservation and rejuvenation of village commons. With the disintegration of the relevant social institutions these practices are also being gradually abandoned.

The relationship between the people and their environment was based on a robust system which ensured that the basic needs of the households dependent on natural resources were met and the households in turn preserved and safeguarded the environment. This balance was disturbed by rapacious claims on these resources from the rich and the powerful, by pressure of population from the households which could not find enough alternative sources of earning a livelihood and by the erosion of institutional safeguards to protect the environment. Underlying all these causes was the deterioration of agrarian structure, basically a deteriorating relationship to land.

AGRARIAN STRUCTURE

The countries, and regions within the large countries, where least impact has been made on the eradication of poverty often suffer from an inequitable agrarian structure, or marginalized and depleted resources, or both. Inequitable agrarian structure is characterized by unequal access to land for the rural households depending on agriculture either as owners or as lease holders and a large army of landless labourers depending on wage labour in agriculture. Extreme forms of such structures exist in parts of Latin America. In a number of South Asian countries also large landowners coexist with numerous small owners, owner-cum-tenants, pure tenants and landless labourers. The rigours of small ownership holdings are eased if there is an active lease market, or when non-farm employment opportunities are present and expanding. In the absence of either of these two conditions a strong correlation exists between the inequality in ownership of land and the extent of poverty, as in South Asia, Southern Africa and large parts of Latin America (World Bank, 1990).

In the absence of a buoyant non-farm sector which, as I will discuss later, itself is predicated in the countries we are referring to upon a dynamic agriculture, two important factors could contribute to the relaxation of limitations imposed by an inequitable agrarian structure: functioning of the land and labour markets, and technology. Unfortunately, in countries with large concentrations of poverty, as in the countries of South Asia, these factors reinforce the inequities.

Land and labour markets

In a situation of grossly unequal access to land the functioning of labour and land markets does not provide the poor with an alternative of non-farm work or of augmenting their holdings with leased-in land. The labour market is fragmented, not only on the basis of skills compounded by difficulties in transport and communication, but also because of the differences in gender, castes, tribes and other social characteristics. All these impose restrictions on the mobility of labour. The macro policies of the governments, especially those pertaining to pricing and subsidies, inhibit whatever limited opportunities may arise to expand non-farm employment. An abundant supply of rural labour with limited and more or less fixed demand in agriculture and related activities results in low wage rates, except during the time-constrained seasonal operations.

In any case, as is now well documented, the employment elasticity of labour in agriculture is quite low. Even in the intensively cultivated areas it will not at the margin exceed 0.5, compared to addition to the workforce in the farm sector which exceeds 2 per cent per year or more in the countries of South Asia (Asian Development Bank, 1978). Further, most of the additional employment even in the good agricultural years gets translated into fuller employment for under-employed family labour rather than additional work for wage-labour. The dependence of labour on the landlords in other markets, for example for credit, compounds the handicaps, which in extreme circumstances lead to what are known as semi-feudal conditions (Bhaduri, 1973).

Following the Marshallian formulation many commentators on the rural scene have concluded that the organization of production on leased-in lands, especially when these are under sharecropping arrangements, may not be as efficient as on the owner-operated holdings. These findings have recently been questioned (Otsuka and Hayami, 1988). There is a large body of empirical evidence, and plausible logic to explain it, which suggests that, even in circumstances of overcrowded agriculture, crop-sharing provides for the landless rural labourer and marginal farmers a better alternative to limited opportunities for wage-labour. That is why, despite the numerous lacunae in lease contracts, the poor households do take recourse to sharecropping. For example, in India the area under cultivated holdings is larger than the area under owned holdings in the lower-size groups of land holdings. The difference is obviously accounted for by land leased in by the small and marginal landowners.

However, this option of augmenting land is not available to many poor households when the large holdings are cultivated extensively, or in some cases left uncultivated. Another situation, where access of the poor to land is circumscribed, arises when, because of capital-intensive technology or other socio-economic considerations, an 'inverse' tenancy system prevails, with medium and large farmers leasing-in land rather than the small farmers having access to it.[11] The imperfections in the lease market are compounded because of misplaced policy interventions (for example an insistence on owner operation in much land legislation), which have only resulted in 'concealed' tenancy and made the terms of lease more onerous for small tenants.

Finally, I touch upon one other alternative which, in the past and at least in some countries, was favoured as a panacea for increasing the incomes of small farmers: joint or collective farming. For quite some time this solution had an ideological appeal. However, even in those days perceptive writers had unravelled all the inconsistencies in the logic on which a case for joint farming was based.[12] The subsequent historical developments, first in China and now in Eastern Europe and the Soviet Union, have clearly demonstrated, if such evidence was at all necessary, that technological and social correlates of agriculture militate against the principle of collectivism at the production stage.

Technological developments

Technological development in agriculture in the developing countries in recent years have been mostly aimed at yield augmentation to ensure food security at the national level. The major vehicle for yield augmentation was high yielding varieties (HYV) of seeds, mainly of wheat and rice and to some extent maize, which with the complement of assured water supply and chemical fertilizers and pesticides promised a remarkable increase in yield per hectare. In theory this technology was divisible and neutral to scale, but in practice it favoured medium to large farmers who could have recourse to their own savings, or had easy access to financial institutions to obtain credit, to purchase inputs. It is true that over a period of time state interventions in the credit market did ease the conditions of access for small farmers in a number of countries. Thus those who had smallholdings in the well-watered areas could take advantage of the advances of HYV technology, though only after a time-lag. The implicit bias of technological development was clearly in favour of the resource-rich farmers in the resource-rich regions. In regions with uncertain or inadequate water supply, that is in the large, rain-fed tracts, no yield-augmenting technology of the type of HYV technology was available. There the access to land and poverty were more clearly and inversely correlated.

POVERTY, AGRARIAN STRUCTURE AND SUSTAINABILITY

Let me at this stage pose the basic question: can one by-pass a defective agrarian structure and lift the poor households in rural areas from the mire of poverty? The relationship between poverty and access to land could be subjected to several qualifications. Population pressure, quality of land, nature of technology and development of non-farm activities make this relationship rather complex (Mellor and Desai, 1986). However, if we take into account the medium to large countries, where a large proportion of the rural workforce is dependent on agriculture, the links between the agrarian structure and the extent of poverty are fairly general. Contemporary history hardly provides an example of a relatively *swift and sustainable* transformation taking place in the context of an inegalitarian agrarian structure in such societies.

The underlying logic is not difficult to follow. Under special circumstances a section of the farm sector can develop, irrespective of the overall stagnation in agriculture, and can forge strong linkages with the non-farm sector, for example, with the modernization of agriculture on large holdings or export-oriented developments on these holdings. Thus a farm sub-sector catering to a sizable urban demand or export markets can coexist with a largely stagnant rural sector. In such cases, really, two worlds, of the poor and the non-poor, exist without close interaction. These cases, although rare, are not unknown. Even in these cases sustainable development is questionable in the absence of some mechanism which ensures an expanding rural demand. And expansion in the rural demand, especially if the sector is dominated by a large number of poor households, will require a more dynamic small-farm sector or a substantial rise in real wages. An inequitable agrarian structure mitigates against both these developments. Even if the overall growth rate is high, in the absence of an equitable agrarian structure the sustainability of such growth is always doubtful. Brazil in Latin America and the Philippines in Asia provide good examples of the need for an equitable agrarian structure as a precondition for sustainable growth.

However, the reverse is not automatically proven. What we can say now – without revisiting the now famous size productivity debate – is that there is enough evidence to suggest that size by itself is not a constraint in raising productivity. A more important question is whether this type of small farm-dominated system would be environmentally sound, and whether and how it would make agriculture sustainable. By itself a small farm-dominated and economically viable agriculture is no guarantee of sustainable agriculture. But it could pave the way for it. By generating surplus it enables poor households to invest in skills and diversifying enterprises, thus relieving pressure from land. Even then the tasks of facilitating skill formation by education and training, and diversifying enterprises with the provision of infrastructure and development of markets, falls on public investment policies. More than investment policies, the policies on science and technology are of critical importance.

Technological developments

Even in the past, technological developments have yielded several land-augmenting measures. From perennial irrigation to non-photo-sensitive and short maturing varieties, a number of technological interventions can be identified which have increased the usable land surface. Also the small farmers, especially those who were potentially viable, could be made viable by generating and extending a technology which enabled them to generate higher values of products from their limited land base. In other words, where quality of land could, to some extent at least, substitute for quantity. This could mean high-value crops but it could also mean a different enterprise mix. The crops grown by the small farmers and the resource base available to them generally do not attract the attention of the mainstream scientific establishments to improve their economic prospects.

A small farm-dominated agriculture will by definition be in nature an intensive agriculture. With the present-day technology this would mean high-yielding varieties of seeds, chemical fertilizers and chemical pesticides. These are not the ingredients for sustainable agriculture. If there is a threat to sustainability it comes largely from these sources. The only way to cope with this problem is to harness technology to solve the problems of soil nutrients, pests and diseases in a way that does least harm to the environment. There are several promising leads in different regions. But today these efforts do not constitute the mainstream technological research. It is only by according priority to these efforts and providing them with resources that a credibility will be built up and the technology generation in these directions given a fillip.

The other constraints which have to be faced by the small farmers are their limited access to cash resources and hence inability to use high capital-intensive techniques. Coupled to that is their inability to take risks. Both these aspects take us from the field of technology to that of macro policies. The critical role of macro policies in rural poverty alleviation is now fully appreciated. The role of pricing, subsidies, exchange rates and trade policies are well documented. What is largely missing in this discussion is (a) how the poor can benefit from the free markets and (b) how they can be insured against the attendant risks. The poor face two difficulties *vis-à-vis* markets. In the first place, they find it difficult to 'enter'. This may not be as much of a handicap in the markets for agricultural products, because of the competitive structure of these markets. However, it may not be so in the case of the labour and lease markets, as noted above. These markets are fragmented and extra-economic considerations do play their part. The second major handicap is the return to small farmers, again, in the lease and labour markets. Because of uneven bargaining power it is not easy for them to secure a 'fair' wage or a 'fair' share from gross produce (in the case of leased-in land).

The role of macro policies, pricing, trade and subsidies, will become more important once technologically feasible solutions are available. They will fine-tune the basic thrusts provided by institutional change and technology. This has been amply demonstrated by the experience of the High Yielding Varieties Programme in some countries, such as India, where the institution of price and credit policy helped once a superior technology was available for extension (Vyas, 1989). The role of the state in this respect is important. State policies should facilitate an easy entry and ensure fair returns. The other aspect is the regulatory aspect. However, in both these cases more than the state efforts the organizations of the poor can play a critical role. The implementation of the legislative provisions for minimum wages or the stipulated tenant's share in the case of leased-in land illustrate the regulatory role of the state as much as the critical part played by peasants' movements.

If we are not yet fully equipped to use the market institutions in favour of the poor, we are at a very rudimentary stage in our understanding of the risks faced by them, their coping strategies and the institutions and macro policies which can either help them to minimize the incidence of risk or equip them to face the risk or compensate them in the event of an adverse outcome. Here again there is a role for researchers, for social activists and the state to devise appropriate 'safety-nets'.

The task of poverty alleviation is as important as the move towards sustainable agriculture. In fact, properly conceived poverty alleviation could be a step in the direction of an environmentally safe world. For poverty alleviation, an access to assets is important. The earlier we realize the interdependence of agrarian structure, environmental sustenance and poverty eradication in the over-populated developing countries, the more realistic will be our approach to attaining sustainable agricultural development.

NOTES

[1]It seems that in this respect the concept of poverty is going in circles. In the early part of this century, English social commentators, notably B.S.Rowntree (Rowntree, 1901, quoted by Scott, 1981), defined primary poverty as the 'minimum necessary expenditure for maintenance of merely physical health'. Note that this concept was not much different from the current definition of absolute poverty.

[2]The methodology and justification for the use of the HDI is provided in a recent study by the United Nations Development Programme (UN, 1990).

[3]Fully described in Sen (1973).

[4]In an as yet unpublished document, Subodh C. Mathur has presented a review of literature bearing on the characteristics, economic environment and behaviour of the rural poor in India. My account draws heavily on his work.

[5]A discussion of the failure of different types of models of economic growth to sustain a 'trickle down' effect is available (Saith, 1990).

[6]A satisfactory definition of sustainable development was adopted by the FAO Council in 1988: 'The management and conservation of the natural resource base, and the orientation of technological and institutional change, in such a manner as to ensure the attainment and continued satisfaction of human needs for present and future generations.'

[7]In fact the new land often proves not to be of much help. An estimated 20 per cent of new ranches in the Amazon basin 'failed within a few years. The loss of existing agricultural land through erosion is estimated at between 6 and 7 million hectares per year, with an additional 1.5 million hectares being lost through waterlogging, salinity and excessive alkali' (World Resources Institute, 1990).

[8]This was the theme of Swaminathan's Presidential Address to the Agricultural Science Section of the Indian Science Congress at Varansi in January 1968.

[9]Subodh C. Mathur (see note 4) quotes a number of studies in India suggesting that as a general rule poverty and large households are closely linked.

[10]Professor Anil Gupta of IIM, Ahmedabad, India, in the Newsletter *Honey Bee,* published on behalf of an international network for documentation of local innovations, has drawn attention to numerous examples of eco-friendly techniques, and also suggests that small peasants have a logical basis for environmental concern.

[11]I have drawn attention elsewhere to the phenomenon of 'reverse tenancy' in the early years of the green revolution in India (Vyas, 1970).

[12]A critical review of issues in collectivization of agriculture in South Asia is provided by Deshpande (1977).

REFERENCES

Asian Development Bank, 1978, *Rural Asia, Challenge and Opportunities*, Asian Development Bank, New York.

Bhaduri, A., 1973, 'A Study in Agricultural Backwardness under Semi-Feudalism', *Economic Journal*, March.

Deshpande, S.H., 1977, *Some Problems of Cooperative Farming*, Asia Publishing House, Bombay.

Hirway, I., 1990, 'Poverty Alleviation Programme in India: Past Experience and Future Direc-
    tion', in *ISS Proceedings and Papers of a Workshop*, The Hague.
Jodha, N.S., 1988, 'Poverty Debate in India, A Minority View', *Economic and Political Weekly*,
    Special Number, November.
Lipton, M., 1983, 'Poverty, Undernutrition and Hunger', *World Bank Staff Working Paper No.597*,
    World Bank, Washington.
Mellor, J. and Desai, G.M., 1986, 'Agricultural Change and Rural Poverty: A Synthesis', in J.
    Mellor and Desai, G.M. (eds), *Agricultural Change and Rural Poverty*, Delhi University Press,
    Delhi.
Otsuka, K. and Hayami, Y., 1988, 'Theories of Share Tenancy: A Critical Review', *Economic
    Development and Cultural Change*, 37 (1).
Rogers, G., 1989, *Population Growth and Poverty in Rural Asia*, Delhi University Press, Delhi.
Saith, A., 1990, 'Development Strategy and the Rural Poor', *Journal of Peasant Studies*, 17 (2).
Scott, W., 1981, *Concept and Measurement of Policy*, UNRISD, Geneva.
Sen, A.K., 1973, 'Poverty, Inequality and Unemployment', *Economic and Political Weekly*,
    Special Number, August.
Sushma, S., 1988, '*Poverty Measurement: Some Issues*, Institute of Development Studies, Jaipur.
United Nations, 1969, *Social Policy and the Distribution of Income in the Nation*, United Na-
    tions, New York.
United Nations, 1990, *Human Development Report, 1990*, United Nations, New York.
Vyas, V.S., 1970, 'Tenancy in a Dynamic Setting', *Economic and Political Weekly*, Review of
    Agriculture, June.
Vyas, V.S., 1989, 'India's Rural Development Strategies: Lessons in Agricultural Growth and
    Poverty Alleviation', in Longworth, J. (ed.), *China's Rural Development Miracle*, Queens-
    land University Press, St. Lucia.
World Bank, 1990, *World Development Report, 1990*, Oxford University Press, Oxford.
World Commission on Environment and Development, 1989, *Our Common Future*, World
    Commission on Environment and Development, New York.
World Resources Institute, 1990, *World Resources, 1990–91*, World Resources Institute, New
    York.

PRESIDENTIAL ADDRESS

JOHN W. LONGWORTH*

Human Capital Formation for Sustainable Agricultural Development

It has been a great honour and privilege, and it has given me great pleasure, to have served as President of the IAAE which, as we all know, is one of the oldest international professional associations in the world. Previous Presidents have chosen a wide variety of topics for their Presidential Addresses. Most have discussed research issues of interest to the profession and usually, of course, the subjects considered have been closely related to the theme of the Conference.

Today I will be talking not about research itself but about the training of researchers and research administrators. I want to discuss 'Human Capital Formation for Sustainable Agricultural Development'. Although, the idea of 'sustainable development' had its origins in United Nations debates of the 1960s, it gained increasing credibility in the first half of the 1980s, reaching a new level of significance with the release in 1987 of the World Commission on Environment and Development report, generally known as the Brundtland Report. Thus, in 1988, when the theme for this Conference was being considered, 'sustainable development' was very much in the ascendancy as a concept.

In recent years, there have been a large number of papers and books published; some important conferences, including the current conference, have been organized; and there have even been some previous Presidential Addresses on the topic – notably, in the context of today's meeting, Sandra Batie's very thought-provoking presentation to the American Agricultural Economics Association in 1989.

My purpose today is not to provide a definitive statement on sustainable development, nor to attempt to set a research agenda. Instead, I accept that there has been a permanent shift in public attitudes: it is not just a swing of the pendulum but a permanent shift, not only in the developed world, but also in Third World countries in favour of sustainable development. In particular, people are demanding sustainable agricultural production systems. These demands have generated enormous challenges for agricultural scientists. At the same time, university programmes and research administrators have been slow to respond. The agricultural economists' profession has the potential, in fact it has a responsibility, to play a major role in making agricultural educa-

*University of Queensland, Australia.

tion in the universities and colleges of the world more relevant to solving the problems of sustainable agricultural development.

I do *not* for a moment advocate disciplinary arrogance. I do *not* want to argue for a kind of disciplinary imperialism. Indeed, as Batie and others have pointed out, sustainable development concepts should force us to re-examine closely some of our most basic methodology. But in doing so, there is a real risk that we could 'throw the baby out with the bath water'. As one of the major social science professions closely involved with agricultural education and administration, we must not abdicate our professional responsibility in the face of the challenges posed by sustainable development issues. We have unique contributions to make to public debates in regard to sustainable agricultural development. In particular, we can contribute a great deal to the training of the next generation of agricultural scientists, administrators and agribusiness people so that they are better equipped to cope with questions of sustainability.

On a world-wide basis, the number of university and college courses in agricultural science, and related disciplines such as agricultural economics and farm management, increased dramatically in the 1950 to 1990 period. One of the most important factors influencing this growth in tertiary training opportunities was the perceived need to enlist the assistance of modern science to solve the world food problem. That is, for the last four decades, agricultural education at the tertiary level has been primarily oriented to increasing food production. Consequently, a large proportion of the human capital (researchers, extension workers, agricultural administrators and agribusiness people) created by these educational programmes has been used to address production-related problems. The result has been a massive increase in agricultural productivity. But these gains have not been achieved without putting great pressure on the natural environment.

Gradually, the negative impacts of the gains in agricultural productivity (soil degradation, salination, species extinctions and so on) have become increasingly obvious and important. The recent growing public awareness of the need to develop sustainable agricultural production systems has given respectability to ideas which previously were dismissed by many production-oriented educators, administrators and scientists as counterproductive.

University programmes have been slow to adjust to the new reality. While the intellectual challenges associated with 'making two ears of corn grow where one grew before' inspired agricultural scientists in the mid-twentieth century, the challenge for the twenty-first century is how to ensure that the hard won gains of the last 40 years can be maintained and even developed further within sustainable farming systems.

THE CHALLENGE FOR AGRICULTURAL EDUCATORS

Two fundamental changes need to be widely implemented if mainstream tertiary agricultural education and research is to answer the sustainability challenge. First, undergraduate and postgraduate curricula in agricultural science must provide a greater awareness of the long-term costs and benefits of technological and social change. Sustainability refers not only to physical environments but to social (and economic) environments as well. Indeed, it is the conflict between these two aspects of sustainability which creates most of the fundamental problems facing Third World agriculture today.

Secondly, researchers need to be taught how to identify the real problems and to be rewarded for tackling these issues. A great deal of the agricultural research effort over the last 40 years has been misdirected. Most decisions about *precisely* what research will be undertaken are in the hands of the researchers themselves. Their concepts of 'what counts' towards their own professional advancement greatly influences *exactly* what research is undertaken. We need to question whether the traditional personal reward structures for agricultural scientists are consistent with the social goal of working towards long-term sustainable agricultural systems.

THE KNOWLEDGE EXPLOSION:
THE NEED FOR A NEW STRATEGY

The amount of scientific information relevant to agriculture has expanded greatly in the last 40 years. No longer is it possible to 'cover everything', even in relatively high specialized university programmes. A new strategy for training agricultural scientists is required.

Some would advocate the holistic agricultural systems approach. While there is great merit in a systems approach to research (see, for example, Nagy and Sanders, 1990), it is not the answer in regard to agricultural education. Students still need a rigorous disciplinary base on which to build. The question is, which disciplines and what degree of depth in each is required?

The traditional agricultural science programme has been built on the basic physical and biological science disciplines. Exposure to the social and behavioural sciences has been minimal. These traditional programmes, for the most part, do not place sufficient emphasis on the social science concepts relevant to analysing sustainable development issues. A new educational strategy is required which retains a strong disciplinary basis but, at the same time, inculcates a different philosophy and set of values. The next generation of agricultural scientists, extension workers and administrators must place a much higher value on the need for agricultural technology which can increase productivity within a production system which is sustainable in social and economic terms as well as in a technological sense.

SUSTAINABLE DEVELOPMENT: SOME OF THE BROAD ISSUES

Sustainable development is a more complicated concept than many advocates of the idea acknowledge. In an excellent brief review of the issues associated with the concept of sustainable development, Veeman (1989) suggests there are three interwoven aspects to be considered: a growth component, a distributional component and an environmental component. The following three sub-sections draw heavily upon Veeman's ideas.

Growth component

Early theories of economic growth placed great emphasis on the accumulation of physical capital and the need for a high marginal rate of savings to finance capital accumulation. Gradually, the emphasis shifted to acknowledge the contribution of human capital formation to the growth process. The recent emphasis on sustainable development has added two more dimensions: the need to give greater weight to the stabilization of growth over time and to the intergenerational implications of economic growth; and the need to emphasize the role of natural resources in long-term economic growth.

There are two major difficulties with the traditional approach to analysing economic growth which the recent sustainability debate has moved to centre stage. The first concerns the hypothesis that the role of natural resources in economic progress declines as economies become more industrialized. There are at least two compelling reasons why this hypothesis should be rejected. First, many of the natural resource products and services which are inputs to human well-being (such as clean air, personal space and 'green' surroundings) are not included in the conventional indicators of growth such as changes in GNP. Secondly, the income elasticities of demand for these (mostly non-market) products and service appear to be extremely high. Therefore, from both the supply side and the demand perspective, natural resources tend to become increasingly important determinants of aggregate human welfare as growth progresses.

The second major difficulty with conventional discussions about growth which the sustainability debate has highlighted is that national accounting measurement procedures do not allow for the depreciation/deterioration in natural resource assets. Improvements in national income (and hence economic growth) based on changes in such indices as GNP, therefore, may seriously overstate the true rate of improvement in the welfare of society. Economic progress, especially in the Third World but also in wealthy countries, makes sustainable development possible. Yet, paradoxically, the growth aspect of sustainable development is frequently in conflict with the other two aspects of sustainable development (that is, the distributional and the environmental aspects). Agricultural students need to be educated to appreciate the complexity of this paradox. They must be given a conceptual framework and a set of analytical tools/skills with which to resolve this conflict on a case-by-case basis. Traditional agricultural science curricula have concentrated on scientific and technological approaches to increasing productivity and hence growth. They have not developed human capital which can recognize and contribute to the solution of the growth/sustainability paradox. Hence the emergence of a plethora of environmental science courses. Agricultural scientists have become the 'bad guys' trained to exploit the natural environment in the name of agricultural development. Environmental scientists are the 'good guys' trained to protect the environment.

Agricultural educational programmes for the future must seek the middle ground. They must be designed to train people who can devise agricultural production methods which both contribute to increased productivity (growth) and satisfy the distributional and environmental aspects of sustainable development.

Distributional component

Economic growth only becomes economic development when the benefits of growth are widely distributed. A major debate has emerged in the last decade about whether agricultural research projects should be screened for distributional consequences. Research, for example, which promises to lead to significant growth (increased agricultural productivity) but which will benefit larger, wealthier landholders rather than the poorer smallholders, is seen as inappropriate research. This raises at least two questions.

First, can research administrators identify such 'inappropriate' research *ex ante* with any certainty? Research originally conceived as inappropriate on 'distributional' grounds may become most appropriate *ex post*. As in the case examined by Yee and Longworth (1985), this could occur because the advances achieved eventually prove not to be biased either towards wealthy/ larger producers or towards certain factors of production (such as capital) because conditions in the factor markets (and hence factor rewards) change during the gestation of the research. Secondly, and more fundamentally, should agricultural research be used to solve distributional problems in the rural sector? Research policy is an extremely blunt instrument with which to attack such problems. Other more direct policy measures such as tax reform and land reform are more appropriate.

This is not the place to pursue this debate. However, it illustrates the critical need for agricultural educators who are training future agricultural researchers and administrators to acquaint their students with these broader issues of research policy. While the need to develop appropriate technology in a technological sense is relatively straightforward, the broadening of the definition of appropriate technology to encompass its distributional consequences raises a more complicated set of issues. Most agricultural science students are not being trained to understand or to address these distributional aspects.

Environmental component

Economists have a long tradition of tackling environmental issues rather differently from biological scientists. The concepts of externalities, property rights, optimum rates of depletion and option values are only four of the many ideas which economists have developed to help analyse environmental issues. Perhaps the biggest difference between the economists' paradigm and that of most biological scientists is that economists do not perceive natural resources as a fixed quantum with a predetermined finite capacity to satisfy the needs of mankind. Instead, economists stress the ingenuity of man. Our capacity to adjust over time and our creation of new institutions (for example, property rights), technological change (for example, development of fusion energy) and substitution opportunities (such as alternative food sources) can all greatly change the value to society of a particular set of natural resources. In general, economists are more optimistic and positive about the environmental component of sustainable development than most biological scientists (Goeller and Weinberg, 1976).

Natural resources are often grouped into renewable or flow resources (fisheries, forests, rangelands, natural populations and so on) and non-renewable or stock resources (minerals, for example). For certain analytical purposes, this is a most convenient division. In the case of renewable resources, a number of conservation or management strategies have been suggested, such as maximum sustainable yield (MSY) and optimum stocking rate or carrying capacity, by biologically trained scientists. Unfortunately, in practice, it is usually extremely difficult to implement these strategies with any degree of precision. Furthermore, they are not usually optimal in an economic sense. Normally, for example, the economic optimum level of use for a renewable resource will be less intensive than that suggested by the MSY criteria. This is another instance where the paradigm of the economist is more constructive and positive with respect to the environmental component of sustainability than the approaches advocated by ecologists and other biological scientists.

There is a world-wide need for mankind to develop appropriate policies and management strategies for renewable resources such as soil, pastures, native forests and native terrestrial and marine animal populations. In many parts of the world, over-exploitation is causing irreversible changes. Renewable resources are becoming non-renewable. Appropriate policies and management strategies cannot be developed, nor successfully implemented, from a purely biological perspective. For example, new institutions which create appropriate economic and social incentive structures are usually required. To be relevant in the future, agricultural research will need to address these non-biological constraints to the development of sustainable systems.

This last point is illustrated on a grand scale by rangeland degradation and desertification in the half of China known as the pastoral region (Zhang *et al.*, 1991). The introduction of new physical and biological technology as part of 'The Great Leap Forward" (1958–61) had a disastrous impact on the rangelands of north and north-western China. The situation has been exacerbated by the reforms of the early 1980s (Liu *et al.*, 1991). Although these changes have led to substantial and widespread development, the production systems on which this progress has been based are not sustainable in the long term (Longworth *et al.*, 1990). While the reforms have created new incentives for the utilization of the natural pasture resources of the region, the necessary institutional arrangements to discourage or prevent exploitation by over-grazing are not in place. Further major socio-economic institutional reforms will be required in the pastoral region of China if massive and irreparable damage to the rangelands is to be avoided (Niu and Chen, 1991; Williamson and McIver, 1991; Du *et al.*, 1991).

In the case of non-renewable resources, questions about how and when they should be utilized raise such issues as intergenerational equity, option values and resource stewardship. Indeed, even such fundamental philosophical questions as the rights of man versus the rights of other living creatures, may also be raised.

None of the ideas and broad issues discussed above are new, yet traditional agricultural science education, based as it is on the basic biological and physical sciences, does not equip students to address these complex matters. Future training programmes will need to recognize and remedy these deficiencies.

THE CHALLENGE FOR
AGRICULTURAL RESEARCH ADMINISTRATORS

Agricultural scientists have made great progress in the last 40 years in terms of raising agricultural output per unit of land and per unit of labour. A great deal of 'the right' research must have been successfully undertaken. At the same time, enormous amounts of time and effort (money) have been devoted to 'research' which has had no practical pay-off. A major part of the motivation for most research activity is the personal rewards it will bring to the researcher. Research administrators and policy makers must be careful to structure the reward system so that 'appropriate' research is rewarded the most. Unfortunately, traditional reward structures for agricultural researchers, especially those employed in publicly funded research institutions and universities, do not encourage people to address directly the complex problems associated with sustainable development. The long-term, multi-disciplinary, non-scientific features of the problems involved 'frighten' young, ambitious and capable agricultural scientists.

The challenge for agricultural research administrators of the future is how to attract the best researchers to these complex areas of research. Two major barriers to progress are the conventional disciplinary divisions between research groups and conventional wisdom among agricultural scientists as to what constitutes 'good research'. Researchers, like all human beings, prefer to work with and to receive the acceptance (accolades) of their peers. Consequently, research institutes and university departments tend to develop enclaves of scientists of like training (and hence values). Cross-fertilization of ideas is actively discouraged because the 'best' journals in any field only accept research papers which maintain the traditional paradigm for that discipline. Future agricultural research administrators will need to break down these disciplinary barriers if worthwhile research on the development of sustainable systems is to be undertaken.

For most agricultural scientists, the personal need to be accepted as a scientist is in conflict with their social responsibility to tackle the major sustainability problems facing agricultural industries. Agricultural educators and research policy makers and administrators need to be fully aware of this conflict. Unless this problem is addressed, the *raison d'être* for agricultural science education, as distinct from a general scientific education, will disappear.

CONCLUSIONS

Sustainable agricultural development is like motherhood: no reasonable person is opposed to the idea in principle. Yet, in practice, much of modern agricultural output arises from production systems which appear to be unsustainable in the longer term. Gradually, the future sustainability of a large part of agricultural production has become an important issue in many countries. Agricultural educators, scientists and research administrators who have played a major role in the development of modern agriculture over the last 40 years have been slow to recognize the new challenges ahead.

There are no easy solutions. However, the problems associated with sustainability are not amenable to purely scientific solutions. The economic and social dimensions are critical if meaningful progress is to be achieved. Agricultural economics, farm management and rural sociology have been the poor cousins of the biological and physical science branches of agricultural science for the last four decades, yet these social science disciplines can make major contributions towards a better understanding of how to develop and establish sustainable agricultural systems. Sustainability issues will require agricultural educators, scientists and research administrators of the twenty-first century to place greater emphasis than has been the case in the past on social and economic aspects of agricultural production systems. The agricultural economics profession needs to adopt a proactive role to ensure that this change of emphasis occurs.

REFERENCES

Batie, Sandra S., 1989, 'Sustainable Development: Challenges to the Profession of Agricultural Economics', *American Journal of Agricultural Economics*, 71, December.

Du, Y., Williamson, G.J., and Drynan, R.G., 1991, 'Fiscal Reform and the Chinese Wool Industry,' poster paper presented at the XXI International Conference of Agricultural Economists, Tokyo, Japan, 22–29 August.

Goeller, H.E. and Weinberg, A.M., 1976, 'The age of substitutability', *Science*, 191, pp. 683–9.

Liu, Y., Chen, J., Lin, X. and Du, Y., 1991, 'Institutional Change and Economic Development: Introduction of the Household Responsibility System in the Pasture Region of Northern China', poster paper presented at the XXI International Conference of Agricultural Economists, Tokyo, Japan, 22–29 August.

Longworth, J.W., Chen, J. and Niu, R., 1990, 'Economic development in the pastoral areas of China: progress, problems and some policy suggestions', in *Proceedings of the International Symposium on Food Nutrition and Social Economic Development*, Chinese Academy of Agricultural Science, Beijing, China.

Nagy, J.G. and Sanders, J.H. 1990, 'Agricultural technology development and dissemination within a farming systems perspective', *Agricultural Systems*, 32, pp. 305–20.

Niu, R. and Chen, J., 1991, ' "Small Farmers" in China and Their Development', invited paper presented at XXI International Conference of Agricultural Economists, Tokyo, Japan, 22–29 August.

Veeman, T.S., 1989, 'Sustainable development: Its economic meaning and policy implications', *Canadian Journal of Agricultural Economics*, 37, December.

Williamson, G.J. and McIver, R., 1991, 'Implications of Constrained Factor Mobility for Sustainable Economic Development in China's Pastoral Areas', poster paper presented at the XXI International Conference of Agricultural Economists, Tokyo, Japan, 22–29 August.

Yee, Y.L. and Longworth, J.W., 1985, 'Biases in research: The case of rubber growing in Malaysia', *Journal of Agricultural Economics*, 36, January, pp. 15–29.

Zhang, C., Zhou, L. and Xu, Y., 1991, 'The Easily Forgotten Half of China: Towards Sustainable Economic Development in the Pastoral Region', poster paper presented at the XXI International Conference of Agricultural Economists, Tokyo, Japan, 22–29 August.

SECTION I

Global Strategies for the Development of Agricultural and Rural Resources

KIRIT S. PARIKH*

Food Security: Issues and Options

INADEQUATE INCOME AS THE
MAIN PROBLEM OF FOOD INSECURITY

All countries are concerned with food security in some sense or other. Yet what is meant by the concept is not always clear. Food security should mean that a country is able to provide 'adequate' food to all its citizens under all circumstances that can reasonably be expected. Whether this is accomplished through market mechanisms or through government organizations is not fundamental to the notion of food security. Moreover, the objective should be to provide food security to all as a matter of right, without inflicting any humiliation on the poor or the poor countries. The level of food energy intake at which a person can be considered adequately fed is still a point of unsettled debate among experts (Srinivasan, 1983), yet no matter how one measures hunger and poverty, one finds that large numbers (hundreds of millions) of people in the world, almost all of them in the developing countries, suffer from persistent hunger (FAO, 1986; World Bank, 1986, 1990).

In addition to those who suffer persistent hunger many others, who normally get enough to eat, live precariously on the margin of subsistence. They are vulnerable to all manner of external influences which can easily reduce their food consumption and make them join the ranks of the hungry. A major threat to the already inadequate food consumption of the poor is that of a drop in real income. The poor usually suffer a loss of income when agricultural production is reduced or disrupted owing to unusual weather or wars. Apart from such a sudden decline in real income, a creeping loss may occur if employment opportunities do not keep pace with the growth of the labour force.

The real income of the poor can also fall when food prices rise. If that occurs the poor, who are often net buyers of food, obtain less to eat. Prices can increase for many reasons. This is obvious during drought and may be compounded if the poor also lose employment opportunities. Even when drought, or some other supply difficulty, strikes in a far-off rich country, the price of food for the poor can increase. The rich country will either export less, or import more, foodgrains forcing up prices in world markets. A poor importing country will be unable to import as much as before, or get as much food aid as before, and its domestic price will increase. Food prices can also alter when a major country drastically changes its trade policy and decides to import more

*Indira Gandhi Institute of Development Research, Bombay, India.

or export less, as happened in the early 1970s when the Soviet Union suddenly imported nearly 20 million tonnes of grain. Even a booming domestic economy can aggravate the hunger of some poor people if their incomes do not rise as rapidly as increases in food prices because incomes of others in the economy rise even faster.

In an economic, but not a moral sense, the world food system functions efficiently. It provides food at reasonable prices to those who have the money to buy it (Johnson, 1984). Yet the biological needs of those with insufficient purchasing power are not met by the system. The hungry in the world are hungry because they are poor. They are poor because they own too few resources of land, capital or skills. Hunger is primarily a problem of poverty and not of food production. Thus, if all the poor are given additional income, more food will be demanded *and produced.* However, if more food is produced because farmers are given greater price incentives, the poor whose incomes have not changed will remain hungry. Thus food security can be provided to an individual either by increasing money income or by decreasing the price at which 'adequate' food is made available. One has to recognize, therefore, that to deal with hunger is to deal with poverty and under-development.

The same is true for food security for a country. If it has enough income, it need not strive to be self-sufficient. It can import the food it needs. But if it is poor and deficient in food production it becomes more vulnerable to transient influences which reduce domestic production or increase world market prices. Lack of food security is a problem only for poor people and poor nations. While it is conceivable – for example, in some nuclear winter scenarios – that global food production could fall so far below demand that even rich nations would face difficulty, such drastic supply shortages are very unlikely. The technological food production potential of the world, even without invoking exotic technologies, is so large that inability to produce food at any cost cannot be expected to threaten food security of the rich nations (Linnemann *et al.*, 1979; FAO/UNFPA, 1980). We will therefore look at options only for those less fortunate. While famines and other transient food insecurity problems attract much media attention and academic analysis, the major challenge is from persistent food insecurity which affects hundreds of millions. This is considered first.

PERSISTENT FOOD INSECURITY: POLICY OPTIONS

Agricultural development: price policy options

A large proportion of the population, particularly the poor, in most developing countries, depend on agriculture for their income. In the short run it is often far easier to improve their position through agricultural development than to absorb them in sufficient number in alternative productive employment in non-agricultural sectors. Of course, in the long run the bulk of the labour force in most countries has to be employed outside farming. But many developing countries have been unable to increase industrial employment at a sufficiently fast pace. Moreover, even those not employed in agriculture depend on it to provide food. Thus agricultural development has to be at the centre of any food security strategy in most poor countries.

The major elements of strategies for promotion of agricultural development consist of price policy, trade policy, investment, infrastructure development, research to foster better varieties, extension to bring new technology to farmers, and institutional reforms. Appropriate policy in each of these areas is not easy to determine, for it involves complex trade-offs and feedbacks. We look at some of these in turn.

Output price policy

Developing countries are often advised to get their prices 'right'. While the importance of 'right' prices cannot be under-estimated, it is not obvious what such prices are. The majority of the world's hungry are net buyers of food. For them, lower food prices mean more food. However, low prices which help the poor also inhibit producers of food. Thus the developing countries appear to face a real dilemma, though perhaps one which is less severe than it may appear at first sight. While farmers do respond with larger output when the price of a specific crop is increased, the output expansion for food in aggregate when all agricultural prices are increased is not as large as it could be for a single crop. This follows because when the price of only one crop is increased resources such as land, labour or water can be diverted from alternative uses. For agriculture as a whole, output expansion has to be realized by general intensification and augmentation of resources, which takes time, particularly in developing countries.

Thus there is some logic in maintaining the relatively low prices for agricultural output characteristic of developing countries. High prices stimulate output only modestly and hurt many of the poor.

High output prices and long-term supply response

It can be argued that high output prices for agricultural products, maintained for some years, would attract more investment into agriculture, thus securing a supply response in the long run much larger than in the short. While this is possible, once again it must be recognized that additional investment in agriculture might well come from reduction in investment in other sectors. Thus, while agricultural GDP would increase, total GDP would barely be affected.

Simulations with a system of linked general equilibrium policy models of a number of countries have shown this in the context of trade policy analysis (Fischer *et al.*, 1988). When agricultural trade is liberalized by all market economies the relative prices affecting agriculture increase in most developing countries. The impact after 15 years of higher prices is demonstrated in Table 1. It can be seen that improved terms of trade for agriculture would indeed stimulate agricultural output in the long run, but the impact on total GDP would be negligible. Moreover, even after 15 years of sustained policy, higher

agricultural GDP would not be adequate to compensate the poor for higher
food costs and they would continue to remain worse off in terms of welfare,
compared to the corresponding year of the reference scenario.

The estimates in Table 1 should be interpreted with care, as they involve
pitfalls of aggregation and index number construction. The lack of impact of
agricultural changes on total GDP for developed market economies is to be
expected as farming is a small part of the economy. For the developing
countries, even keeping index number problems in mind, it is more surprising
since, even with large changes in relative agricultural prices and consequent
significant changes in agricultural GDP, aggregate GDP changes but little.

To assess the effects on welfare average equivalent income is basically a
better measure than per capita GDP since it accounts for consumer preferences

TABLE 1   *Impact of relative agricultural prices on GDP and agricultural GDP (percentage change over the corresponding year of the reference
scenario after 15 years of price policy change)*

Country	Pa/Pn	GDP	GDP (ag)	Average equiv. income[†]	Average calorie intake	Persons in hunger
Japan	−39	+0	−6			
EC	−12	+0	−8			
USA	−5	+0	1			
New Zealand	8	1	11			
Canada	13	−0	17			
Australia	9	+0	1			
Austria	6	+0	+0			
Turkey	−10	1	−9	2	+0	−6
Pakistan	−1	3	−1	3	1	−17
Nigeria	−9	−1	−1	+0	1	−57
Egypt	8	−3	5	−2	−0	*
Mexico	−5	−4	1	−4	+0	−3
India	3	+0	−0	1	−0	2
Argentina	48	−0	47	3	−2	31
Brazil	25	−1	7	−1	−2	50
Indonesia	17	1	6	−0	2	*
Thailand	20	+0	6	n.a.	−0	3
Kenya	15	3	10	n.a.	3	−14

Notes: Pa/Pn = price of agriculture/price of non-agriculture weighted by
domestic production.
GDPs are at constant 1970 prices.
Numbers rounded off: +0 means > 0; < 0.5 and −0 means > −0.5 but < 0.
* = no hunger in the reference scenario.
† = income needed to provide the same utility at base year prices as
provided by current consumption.
n.a. = not available.

(at least the preferences implied by the demand system of the 'average consumer'), which per capita GDP ignores. In general, the changes in equivalent income are similar to those in GDP, though for some countries the two do show different signs.

What emerges from these scenarios is that higher domestic food prices in developing countries increase domestic agricultural output, but even when average real income improves the result might, in some important cases, be lower calorie intake and an adverse impact on the chronically hungry. In general, therefore, the case for 'getting prices right' cannot be made with reference to its supposed effects on overall economic growth. We can also extrapolate from the results with reference to effects on nutrition and hunger. As the scenarios show, higher trend prices of food may hurt the poor, even when increasing the average income in a country. It is emphasized that in these scenarios we have assumed normal weather and have accounted for the macro-economic feedbacks and adjustments which might evolve after 15 years of policy change. Thus price increases of similar magnitude which occur suddenly could be expected to be even worse for the poor. This indicates the importance of stabilization of domestic food prices. In the context of variable weather and fluctuating world prices a flexible trade policy and/or a stock policy can be used to stabilize domestic prices.

Trade policy

Trade policy is closely linked to price policy. When a country allows free trade, domestic prices will equal trade prices. If a tariff is imposed on traded goods the domestic price will differ from the border price by the amount of tariff. If no trade is permitted domestic prices will be independent of world prices. Thus using trade policy as an instrument of agricultural development runs into problems similar to the ones encountered by price policy which are described above.

Yet it is often argued that an appropriate trade policy helps promote development in a number of ways. First, under free trade with equality between domestic and world relative prices, investment allocation becomes optimal. Second, free trade provides competition for domestic sectors, provoking greater efficiency. Third, an outward-oriented strategy which promotes exports provides flexibility in importing critical goods, including new technology, and in meeting unforeseen bottlenecks. These arguments for freer trade are not universally accepted. When applied to agriculture their force is quite weak. Many studies have shown that gains from increased allocative efficiency consequent to free trade in general (Whalley, 1985) and trade liberalization in agriculture in particular (Parikh *et al.,* 1988) are marginal. Tables 2 and 3 show the results of agricultural trade liberalization simulations using the IIASA/FAP system of linked models referred to earlier. The changes in GDP and in the number of persons hungry are rather small, compared with the corresponding year in the reference scenario.

The argument that rent-seeking activities in a protected economy lead to production inefficiency, and that freer trade leads to substantial gain as pro-

TABLE 2   *Impact on GDP (at 1970 world prices) of various agricultural trade liberalization scenarios (percentage change after 15 years relative to the reference scenario)*

Scenario*	World	OECD	CMEA**	Developing
F-ALLME	0.28	0.63	–0.30	–0.22
F-OECD	0.22	0.48	–0.40	–0.02
F-LDC	0.05	0.15	–0.11	–0.10

Notes: * = Agricultural free trade by ALLME – all market economies – by OECD countries and by LDC – less developed countries.
** = Council of Mutual Economic Assistance formed by the Eastern European countries.

TABLE 3   *Impact on hunger of various agricultural trade liberalization scenarios (percentage change in persons hungry relative to the reference scenario)*

Scenario	Five years after	Fifteen years after
F-ALLME	+0.8	+1.4
F-OECD	+3.3	+3.6
F-LDC	–4.7	–4.6

duction moves to the efficiency frontier, is not likely to be significant for agriculture in developing countries. Activity is in the hands of millions of private producers who act competitively and no further gains in efficiency are likely to accrue from additional competition from abroad.

Thus, while many developing countries singly, and developing countries as a group, would be better off in terms of reducing the number of hungry persons with agricultural trade liberalization than without, we can conclude that agricultural trade policy, like price policy, cannot by itself lead to better food security. In fact, price and trade policy, and other policies which rely on market mechanisms, are not very effective in bringing food to the poor. They will provide food to those who have money to buy it but not to those who lack adequate purchasing power.

PRODUCER AND CONSUMER SUBSIDIES FOR FOOD SECURITY

Adverse impacts on poor consumers of high food prices needed for agricultural growth can be redressed through various subsidy schemes linked to input prices. We examine two such options.

Input subsidy and low output price

Some developing countries maintain relatively low agricultural prices but subsidize inputs such as fertilizer, water or electricity to stimulate higher production. A subsidy on farm inputs when output prices are low is really a subsidy to consumers. Also an input subsidy accrues to all farmers, big or small, who use the inputs. It also stimulates input use by poorer farmers because their credit needs and risks are reduced. On the other hand, low output prices impose a tax on farmers in proportion to their marketable surplus, and larger farmers have more available for sale. While this may seem a way out of the dilemma of requiring high output prices to stimulate production and low output prices needed to protect poor consumers, problems remain.

The consumer subsidy implied by low output prices benefits all consumers, rich or poor. Thus, if the subsidy rate is high, as it has to be to help poor consumers, the total cost of the subsidy becomes large. Thus a successful programme, effective in stimulating farmers to use more of the input, leads to an increasing strain on the government budget. If it is impossible to finance the cost with increased taxes, other expenditure has to be cut. In many cases this is likely to affect public investment, which will slow down the growth of the economy. Furthermore, if lower public investment impinges on irrigation, electrification, power, transport and, possibly, agricultural research, agricultural output could suffer, and the net effect after some years could be lower production than would have been obtained without any input subsidy. Of course, the outcome depends on the magnitude of the various effects and feedbacks involved.

Input subsidy costs can grow rapidly. In India, fertilizer support, amounting to some 30 per cent of its cost, has grown from Rs 5 billion in 1980–1 to more than Rs 40 billion in 1990–1 and is now 5 per cent of the central government budget. Simulations with an empirical applied general equilibrium model of India have been made to explore alternatives to the fertilizer subsidy (Narayana, Parikh and Srinivasan, 1991). In one such scenario fertilizer subsidy was assumed to be withdrawn in 1989. At the same time, rural works programmes were substituted with the aim of distributing 20 kg per capita of wheat, annually, among the two poorest rural classes as wages. An irrigation subsidy was also assumed to be introduced to enhance the irrigated area annually by one million hectares. Prices and tax rates were kept fixed at reference run levels. The results (as described more fully in Parikh and Suryanarayana, 1989) shown in Table 4, when compared with the reference run, show higher GDP and higher welfare for the poorest rural class from 1990 onwards. Agricultural GDP, at constant as well as current prices, is also higher, though, of course, farm income, including fertilizer subsidy, is lower. Owing to the 30 per cent increase in fertilizer prices, foodgrain output falls by 3.3 per cent in 1990. In year 2000, however, as a result of the additional irrigation capacity created, foodgrain output is simulated to be 4.5 per cent higher, even though fertilizer use is some 3.5 per cent lower.

Thus, in designing a subsidy programme for food security, one should think of the opportunity cost of the funds involved to see if greater security could be provided through alternative policies.

TABLE 4   *Alternative to fertilizer subsidy*

	Year	Reference run (fert. subsidy (30%) continued)	No subsidy from 1989, rural works programme, additional irrigation
GDP 70 (10^9 1970 Rs.)	1990	746.01	747.28
	2000	1262.93	1337.20
GDP Agr. 70 (1970 Rs.)	1990	247.47	248.74
	2000	315.55	332.01
Fertilizer subsidy (current price)	1990	41.26	—
	2000	52.83	—
Fertilizer use (10^3 N)	1990	10007	8736
	2000	12874	12407
Total Irrigated area (10^6 hectares)	1990	56.19	57.19
	2000	77.85	91.71
GDP Agr. (10^9 current Rs)	1990	535.04	547.51
	2000	718.57	759.55
Food grains (10^6 tonnes)	1990	161.03	155.74
	2000	209.47	218.86
Rural poorest class (equiv. income)	1990	120.20	124.50
	2000	124.70	129.40

Dual pricing to direct subsidies and taxes

Whereas a subsidy through output price goes to all consumers and an input price subsidy accrues to all producers, one might want to subsidize only the poor consumers or the smaller farmers to limit subsidy costs. To reach only poor consumers selectively, the prices of those foods which are consumed mainly by them, such as millet, sorghum and other similar coarse grains, might be kept low. Yet this is not easy to accomplish. If prices of coarse grains alone were reduced farmers would shift resources out of their production. Moreover, when coarse grains are cheap, more of them may be fed to animals to produce meat for richer consumers. However, farm prices can be maintained if distribution is handled through ration shops selling coarse grains at subsidized prices. In this case, to reach the poor effectively, public food distribution has to cover urban as well as rural areas.

To be able to improve consumption of coarse grains among the poor, output needs to be increased. The difficulty in this case is that new production technology using high-yielding varieties is mainly available for wheat and rice. As yet there has not been a major research breakthrough specific to coarse grains. Whether this is due to neglect by researchers because of their urban or 'class' bias, or because of the inherent difficulties and uncertainty of success in scientific research is not clear. Whatever the explanation of this situation, a major increase in coarse grain output appears to require marked price incentives to farmers and ultimate strain on government budgets.

One way to subsidize consumers, without nationalizing the food trade and without injuring farm incomes, is to provide a limited quantity of food at a subsidized price to all consumers through ration shops. Additional purchases of food would be permitted in the open market, where prices would be higher than they would be in the absence of partial rationing. Farm incomes would thus be protected. Such dual pricing policies are adopted in a number of developing countries and do indeed provide some relief to poor consumers. The problem remains that of coverage; there are difficulties in directing the subsidy only to the poor. Usually the benefits of such rationing are limited to residents of urban areas or larger towns. The rural poor, who constitute the bulk of the problem in many countries, are frequently ignored. Covering the entire population with such schemes involves a high subsidy cost and consequent financial burdens.

Even though partial rationing and dual pricing do not provide fully satisfactory solutions, it is worth noting that, given limited financial resources for subsidizing food, it is better to provide smaller quantities with larger concessions on the ration price than to distribute more food per person at a smaller subsidy per food unit. The subsidy is then more likely to reach the poor, who would be able to buy from the ration shop all the food to which they are entitled. An extension of this argument is indeed that it is best to give direct income subsidies to poor consumers (if they can be identified) rather than using price policy as an instrument for providing them with food security.

INFRASTRUCTURE DEVELOPMENT, RESEARCH, EXTENSION AND FOOD FOR WORK

Agricultural growth that achieves low food prices is needed to improve food security. Experience in many developing countries has shown that development of land through levelling, irrigation or drainage, and provision of facilities such as roads, markets and electricity, are critical in stimulating agricultural growth without raising food prices.

Once the potential for agricultural growth through infrastructure development is realized, further improvement requires technological progress in the form of newer higher-yield varieties. This requires research, which cannot be centralized as varieties need to be adapted to local agro-climatic conditions. Moreover, sustained efforts in extension are needed to persuade farmers that any new variety is indeed better. The natural scepticism of farmers, so poor that they cannot afford to take any risk, takes time to overcome. Other support in the form of credit and assured supply of inputs such as seeds, water, fertilizers and pesticides are also vital. A comparison of agricultural development in different districts of India has confirmed the importance of infrastructure and extension (Parikh, Mahendra, and Shantanu, 1991).

For many African countries, land development efforts which expand cultivable area have a large potential in stimulating agricultural output. For land-scarce countries of Asia, irrigation and drainage works improve land productivity by raising yields as well as by increasing multiple cropping. For many Asian countries, rural work programmes (or food for work) may be the most attractive way to stimulate agricultural growth, since they generate additional income for the poor which is needed not only to alleviate hunger but also to absorb the output of agriculture.

The level of agricultural growth that can be sustained is linked with both the growth rate of the national economy and the incomes policy. Even a poor developing country can run into the paradox of hunger amidst abundance, if the growth and distribution of income are such that adequate effective demand for food is not generated. This was seen in India where, during the year May 1990 to May 1991, food prices increased by 16.5 per cent even though government stocks of foodgrains stood at 21 million tonnes in May 1991. Note that more than 200 million persons suffer from persistent hunger in India. Simulation with a general equilibrium model of Bangladesh has also shown that it can also run into the problem of surplus rice and yet have hungry people if agriculture grows rapidly. This further underlines the need to make agricultural development policy a part of the overall growth strategy.

Experience of highly motivated and skilled engineers and scientists who have worked in rural areas has shown the tremendous potential and economic profitability of labour-intensive land development schemes in different parts of the world. Such schemes can also be a vehicle for generating off-peak season employment for the rural poor and be an anti-poverty measure. However, to be effective such schemes must be well-engineered, economically attractive, relevant to the needs of the people and efficiently executed.

Of course, while such a strategy of labour-intensive land development, properly organized and implemented, can help in reducing hunger, one must recognize that it also requires the resources necessary to reach all the poor. The developing countries do not have these resources. Aid on a substantial scale, tied to such programmes, can effectively provide food security to all in a reasonable time if it is forthcoming.

TRANSIENT FOOD INSECURITY

Famines, as examples of extreme food insecurity, affect relatively very few people, but pose immediate danger of death. However, since relief efforts are needed for a short period, international action is easily mobilized against famines. We examine here some policy options to deal with transient food insecurity.

Domestic buffer stocks versus reliance on trade

To provide security against transient disturbances, a country may either oper-ate domestic buffer stocks or rely on foreign trade (that is maintain a buffer stock of foreign exchange). It may be noted that reliance on foreign trade does not imply free trade. That is a separate issue.

As already emphasized, the world market cannot be relied upon to provide food at stable prices. This is not to under-estimate the importance and the substantial amount of food aid given as famine relief, but to emphasize that, as a course of normal strategy of development and for food security, develop-ing countries must take account of the extra expenses involved in relying primarily on the world market for food in dealing with emergencies short of famines.

Though the International Monetary Fund (IMF) cereal facility is very help-ful in providing access to foreign exchange to meet unexpected import costs, the IMF has eventually to be repaid. The facility does make it possible for a country to rely more on trade and to aim for slightly lower food self-sufficiency than would otherwise be the case. Yet there are some issues which should be noted. It is often not easy to be an intermittent exporter of small quantities of food. If a country is larger and its domestic production fluctuates in a way that makes it a major exporter in some years and a major importer in others, its exports will depress world prices and imports will increase them. It will, therefore, on balance, need to export a larger quantity to pay for the import of a given quantity. This extra cost will have to be balanced against the costs of domestic storage. The relative costs depend on the variability of domestic production, the variability of prices on the world market, the costs of domestic storage and the size of the country's needs compared to the world market. If the terms of IMF cereal facility were further softened, trade could be made more attractive than domestic buffer stocking. However, a country which is small, or badly located geographically, may not be able to export small quan-tities of intermittent surplus. In such cases operation of domestic buffer stocks becomes unavoidable.

A common buffer stock for food security: would it work?

It is often suggested that developing countries should maintain a common buffer stock of food grains to enhance food security. One could conceive of an agency holding stocks in a number of countries, which could be released to countries in need on the basis of specific criteria, for example, the shortfall below an accepted calorie availability norm.

The basic idea of such a scheme would be that individual nations could maintain a smaller domestic buffer stock than otherwise for given levels of confidence and food availability. Also their dependence on an international grain market dominated by a few large private traders could be reduced. However, by and large, there are very few grain surplus developing countries. For them grain exports constitute a major source of foreign exchange earn-ings. They would have a strong incentive to operate on the free market, particularly when international prices are high. A common buffer stock scheme could, therefore, run into difficulties precisely in those years when it was most needed.

AN INSURANCE SCHEME

The IMF cereal facility provides access to foreign exchange to import cereals. However, it does not protect a country against price increases on the world market resulting from actions of other bigger, and richer, players. If insurance cover was provided against price increases then greater reliance could be placed on international markets. Countries would be encouraged to participate if the premium for cover was less than the cost of maintaining a domestic buffer stock. To ensure the availability of food when required the insurance agency might want to operate a buffer stock itself.

CONCLUSIONS

Food security is a problem of poverty and under-development. Though economic development can eventually be expected to absorb the bulk of the active population, in the medium term agricultural growth must itself play an important role in alleviating poverty and hunger. An integrated approach to agricultural growth and poverty alleviation is needed. More aid, a softer IMF cereal facility, along with an insurance scheme against higher import prices, and a better trading environment, can provide developing countries with the necessary resources, but they must themselves follow policies to provide food security for all.

REFERENCES

FAO/UNFPA, 1980, *Potential Population Supporting Capacities of Lands in the Developing World*, FAO, Rome.

Fischer, G., Frohberg, K., Keyzer, M.A. and Parikh, K.S., 1988, *Linked National Models: A Tool for International Food Policy Analysis*, Kluwer Academic, Netherlands.

Food and Agriculture Organization (FAO), 1986, The Fifth World Food Survey, FAO, Rome.

Johnson, D. Gale, 1984, 'A World Food System: Actuality or Promise?', paper presented at the 75th Anniversary Colloquium on World Food Policy,' Harvard Business School, 8–11 April 1984.

Linnemann, Hans, Jerrie De Hoogh, Keyzer, Michiel A. and Henk, D.J. Van Heemst, 1979, *MOIRA Model of International Relations in Agriculture*, North Holland, Amsterdam.

Narayana, N.S.S., Parikh, Kirit S. and Srinivasan T.N., 1991, *Agriculture, Growth and Redistribution of Income: Policy Analysis with a General Equilibrium Model of India*, Allied Publishers and North Holland, Amsterdam.

Parikh, K.S., Fischer, G., Frohberg K. and Gulbrandsen, O., 1988, *Towards Free Trade in Agriculture*, Martinus Nijhoff, Netherlands.

Parikh, K.S. and Suryanarayana, M.H., 1989, 'Food and Agricultural Subsidies: Incidence and Welfare under Alternative Schemes', Discussion Paper No. 14, Indira Gandhi Institute of Development Research, Bombay.

Parikh, Kirit S., Mahendra, Dev S. and Shantanu, Deshpande, 1991, 'The Role of Technology in Agricultural Development in India', Discussion Paper No. 42, Indira Gandhi Institute of Development Research, Bombay.

Srinivasan, T.N., 1983, 'Hunger: Defining It, Estimating Its Global Incidence and Alleviating It', in D. Gale Johnson and G. Schuh (eds), *The Role of Markets in the World Food Economy*, Westview Press, Boulder, Co.

Whalley, J., 1985, *Trade Liberalization among Major World Trading Areas*, MIT Press, Cambridge, MA.

World Bank, 1986, *Poverty and Hunger: Issues and Options for Food Security in Developing Countries*, Washington, DC.

World Bank, 1990, Handbook of Trade and Development Statistics 1985, Geneva.

DISCUSSION OPENING – TRUMAN R. PHILLIPS*

Dr Parikh focuses his analysis on options touted to benefit the poor. They are of particular concern because lack of food security is stated to be a problem only for poor people and poor nations. Within that context the central theme is persistent food insecurity which affects hundreds of millions of individuals; he does recognize that the problems caused by famine and other transient episodes have their own obvious importance, but it is persistence which he sees as the major challenge.

The main effort is devoted to appraisal of three policy response options (price policies for agricultural output, trade policies and farm input subsidies) where effectiveness is assessed using the IIASA/FAP system of linked general equilibrium models. Analysis of three further alternatives (dual pricing, infrastructure development plus research and extension, and holding of food reserves) is conducted in more general terms.

A key finding is that price and trade policies which rely on market mechanisms are not particularly effective in bringing food to the poor. Specifically, the number of hungry persons would not be greatly reduced according to the simulations after 15 years of either higher prices for agricultural output or trade liberalization. Referring to Indian fertilizer subsidies it is shown that their cost could become a substantial burden on the national budget, and it is suggested that there are alternative programmes which are more cost-effective. Dr Parikh then suggests that rationing and dual pricing could be cheaper options than general price policies or input subsidies, but despite that it would remain difficult to reach the poor, and especially those in rural areas. Under the heading of infrastructure development, research and extension, plus food for work programmes, he highlights the option of promoting labour-intensive land development, allied to keeping food prices low. It is noted, however, that a number of poor countries would need aid on a substantial scale to undertake such programmes successfully. Less is said about overcoming transient food insecurity by maintaining domestic buffer stocks or using the IMF cereal facility.

The general conclusions of the paper appear to me to be defensible, though there are several specific results which appear worthy of note and discussion. Firstly, while it may be generally true that price and trade policies are ineffective in reducing food insecurity, the simulation results do produce some startling

*Centre for Food Security, University of Guelph, Canada.

differences. Contrast the projected decline in the number of hungry in Pakistan and Nigeria (17 and 57 per cent respectively) with the increases of 31 and 50 per cent in Argentina and Brazil. Given this range of estimates it would be safer to conclude that the option of high agricultural prices as a means of influencing food security needs to be carefully examined for each country.

Secondly, the impact of the simulation of free trade by all market economies, or by OECD countries, suggests that there could be 1.4 or 3.6 per cent more hunger after a 15-year period. By contrast, when LDC liberalization is considered, the result quoted involves a 4.6 per cent reduction in the numbers affected by hunger. It is not clear from the paper if this result refers to trade liberalization by LDC countries alone, or if it is an overall effect which also includes simultaneous developed country action. If it is the latter it suggests that LDC liberalization would more than compensate for the unfavourable effect stemming from the developed world. Though all of these changes do not represent major deviations from the base projection, it is worth pointing out that LDC liberalization has some beneficial impact.

My third comment is more general; it relates to breadth in analytical approach and, in particular, to the options which Dr Parikh has considered. Specifically, one might attempt to identify key strategies which could be adopted to reduce food insecurity, and then consider the instruments or interventions to be used to realize them. Five strategies are listed in Figure 1; increase food supply, increase stability of supply, increase access to food, increase food quality and increase food intake. These goals could be reached by making investments of various types (in agriculture, human resource development (HRD), health, infrastructure or research); or through interventions in economic policy or by aid transfers. Space does not allow a full discussion of the detailed issues involved; all that can be done is to suggest that the shaded areas represent those investments or interventions which could be used to fulfil each goal. The boxes which are marked 'X' indicate the strategies and investment and intervention choices implicit in Dr Parikh's paper. Given the nature of this meeting, it is perhaps not surprising that he tended to focus on the strategy of increasing food supply, and the use of economic policy. He must be congratulated for what he has been able to accomplish, but clearly his coverage has not been comprehensive. Those of us interested in the major problem of persistent food insecurity should not, however, lose sight of all of the options which are available.

	Increase food supply	Increase stability of supply	Increase access to food	Increase food quality	Increase food intake
			STABILITY		
INVESTMENT					
Agriculture	X				
HRD					
Health					
Infrastructure					
Research	X				
INTERVENTIONS					
Economic policy	X	X	X		
Transfers (aid)					

FIGURE 1   *Food security strategies, investments and interventions*

NIKOS ALEXANDRATOS*

World Agriculture in the Next Century:
Challenges for Production and Distribution

INTRODUCTION

For reasons of space, analysis will focus only on issues of primary concern to the developing countries, concretely on the situation and prospects to the year 2000 for (a) food and nutrition, (b) agricultural production and (c) the scope for agriculture-oriented strategies. The red thread connecting these issues runs from the realization that the pessimistic assessment about resumption of economic growth in the many countries with food and nutrition problems does not augur well for the solution of such problems through rapid growth in per capita incomes. Given the generally high dependence of many developing countries on agriculture, and the increasing recognition that improved policies for agriculture can be important for the resumption of economic growth, the natural question concerns the potential scope for reversing past unfavourable trends in agricultural production (stagnant or negative in per capita terms). In turn the link between agricultural production growth and nutrition is both direct (increasing food supplies) and, more significantly, indirect through income generation in agriculture, income distribution and linkages to the rest of the economy (demand, foreign exchange). In this context we present the results of a detailed multidisciplinary evaluation of production prospects to the year 2000, focusing on potential gains in yields in different agro-ecological zones. We then briefly review some recent contributions to the literature on general equilibrium modelling which generally establish that agriculture-oriented development strategies can be instrumental in stimulating economic growth and improving income distribution and nutrition. Exploiting the potential of such strategies is perhaps the major global challenge in both production and distribution, if we are not to enter the twenty-first century with numerous developing countries in conditions of perennial crisis. Our detailed evaluation of the agricultural prospects of individual countries in a global context can be a useful input into general equilibrium analyses aimed at examining the case for such strategies.

*Global Perspective Studies Unit, FAO, Rome. The views expressed do not necessarily represent those of FAO. Useful comments from D. Colman, D. Brooks, J. Bruinsma, R. Gaiha, A. Gurkan, J. Greenfield, C. Joseph, D. Norse, K. Stamoulis, T. Kelley White, P. Yotopoulos and G. Zanias are gratefully acknowledged. Help with the data was provided by M.G. Ottaviani and I. Reyes.

To avoid overlap with other Conference papers, many issues which constitute major challenges for production and distribution in the next century are mentioned only in passing. These include the environmental dimension, agricultural policy reforms in Eastern Europe/USSR, trade liberalization and agricultural adjustment in the developed countries, technology generation and diffusion and the possible effects of climate change. In particular, we do not discuss detailed policies required for raising agricultural production performance, combating rural poverty and improving food security, which are amply covered in the companion paper by Kirit Parikh.

FOOD AND NUTRITION IN THE DEVELOPING COUNTRIES

Average food supplies for *direct* human consumption[1] in the LDCs stand now at 2470 calories per person per day, having risen from 1950 calories in the early 1960s. At the same time, 'indirect' consumption of cereals in the form of feed also more than doubled from 15 to 35 kg per head. This happened over a period of rapid LDC population growth, an increase of 1.6 billion, or 77 per cent. In the face of these developments, one is tempted to say that significant progress has been made. However, exclude China and the picture is rather more sombre: calories supplied per person increased from 2070 to 2400 over the same period. Under the circumstances, under-nutrition persisted, though its precise magnitude remains uncertain. Existing data are inadequate for a comprehensive measure of people with food intakes below any given threshold, though attempts have been made to arrive at estimates.[2] The latest World Bank study on poverty suggests considerable reduction of population shares below national poverty lines (World Bank, 1990). Since income or expenditure required to meet minimum food requirements are commonly important criteria for fixing poverty lines, it can be deduced that the incidence of under-nutrition followed a similar pattern.

Country data on the evolution of per capita food supplies permit us to obtain an idea of progress made in terms of the *LDC population living in countries with per capita food supplies below given thresholds.* In the early 1960s, 1070 million people (75 per cent of the LDC population, excluding China) lived in countries with per capita food supplies for direct human consumption under 2200 calories. The comparable number for 1986/8 is 1470 million or 54 per cent of the population. At the other extreme, some 17 per cent of the LDC population (excluding China) lives today in countries with relatively 'high' food supplies per head (over 2800 calories), up from under 1 per cent in the early 1960s. If it were possible to estimate a *true* measure of under-nutrition (proportions of the population with, rather than living in countries with, food intakes below any given threshold) it would probably also show considerable improvement following the increase in the national level averages. Unlike income, there are limits to the extent to which food can be unequally distributed when the average is high, simply because of physiological limits to the amount a person can consume directly. However, so long as many developing countries continue to have per capita food availability nearer to 2000 than to 3000 calories, caution is required in interpreting increases in the average as implying reduced under-nutrition, since at low levels of the average there is considerable scope for improvement in the mean to be offset by worsening distribution.

Such a basically optimistic evaluation is marred by several negative aspects. Firstly, population growth has meant that any given fall in the proportion of under-nourished does not necessarily translate into a reduction in the absolute numbers of persons in this category. It is thought that such numbers increased somewhat and remain in the area of some 500 million (Alexandratos, 1988). Secondly, the great majority of the countries which failed to make progress, or suffered declines, are precisely those with very low food availability levels in the first place (see Table 2). Thirdly, there has been an abrupt interruption of progress in the 1980s affecting mostly the countries which were far from having attained high consumption levels, as the following data show:

Number of countries with under 2300 calories at beginning of each decade according to growth rates of per capita food supplies

	Negative	0–1.0% p.a.	Over 1.0% p.a.	Total countries
1961–70	14	36	29	79
1970–80	14	27	23	64
1980–88	23	16	7	46

Though imperfect, the use of per capita food availability estimates will continue to hold centre stage in the analysis and discussion of historical developments in food and nutrition at the global level and in sketching out scenarios for the future. The material is simply all that is available for almost all countries on a fairly consistent basis. It is therefore justifiable to devote some effort to examining the determinants of the relative positions of individual countries with respect to per capita food supplies. To explore hypotheses, the assumption is made that the food balance sheet data are a good proxy for national average food consumption levels. This is not to say that they can accurately measure the nutritional value of food ingested, but we have to accept the assumption as a working hypothesis. This is not to deny that information for some countries is probably more accurate than for others,[3] but we have no way of establishing whether inaccuracies suspected for some countries (for example, those in conditions of war or with undeveloped statistical services) result in under- or over-estimation in a consistent manner.

An Appendix contains the results of cross-section analysis (equations 1–6). Inter-country differentials in food consumption levels are related to those in a number of other variables which are commonly thought to be important. These include per capita private consumption expenditure (PCE) corrected for purchasing power parities;[4] real food prices;[5] shares of population below the poverty line; income distribution statistics; per capita food imports; and the growth rate of per capita agricultural production. The two latter variables are hypothesized to influence consumption through effects on food prices and, for the countries with a high share of population in agriculture and a high share of the rural sector in aggregate poverty, the growth of agricultural production influences both average incomes and the incidence of poverty. The variables used do appear to have a statistically significant effect on food consumption levels. There are, however, too many outliers, and the regressions explain only 37–65 per cent of total inter-country variability. It can be noted that the calorie variable shows little sensitivity to variation in per capita PCEs in the cross-section data (there is low elasticity with respect to income of 0.09 to 0.14 at the mean calorie value in the different inter-country cross-section equations). We note, however, that the data used relate calories to PCE, while much common demand analysis, which generally reports higher elasticities, relates food expenditure to income or total expenditure.[6] Reutlinger and Selowsky (1976), faced with the same problem of implausibly low calorie income elasticities estimated from inter-country cross-section data, opted for using assumed elasticity values of 0.15 and 0.30 at the level of average calorie requirements for the different LDC regions in their subsequent work.

In the light of these findings (which use the maximum number of countries reporting relevant data), we resort to an impressionist examination of the tail-ends of the distribution of per capita food consumption revealed when countries are ranked according to the growth rate of this variable over the period 1969/71–1986/8 (Table 1). Not surprisingly, all countries in the upper group had sustained economic growth (per capita PCE grew at more than 2 per cent annually in the 17-year period) and the ones in the lower group had zero or negative growth rates. Six of the ten countries in the upper group were fuel exporters, while a high proportion of the ten countries in the lower group have experienced unsettled political conditions, including civil strife and war. The last five columns of Table 1 show developments in the key variables of cereals self-sufficiency, net per capita imports of cereals, and per capita gross agricultural production. The last two variables represent the sources of supply, though not in the sense of an exact accounting identity. As such, they largely indicate how different countries achieved (or failed to achieve) improvements through varying combinations of changes in domestic production and net imports.

In all ten countries in the bottom group decreases in per capita food supplies were associated with declining per capita production, and only modest increases in food imports. By contrast, in the top group, the unfavourable growth in per capita production was compensated by sizable increases in net food imports and diminished self-sufficiency, including some countries in which the growth of per capita production was positive and rather high by historical standards. This material, plus that for other countries not in the table, lends support to the proposition that countries enjoying rapid economic progress also experience rapid growth in their net food imports even when their own production grows relatively fast (Mellor, 1988). However, this phenomenon is not universal. In some of the larger countries (for example China, Indonesia, Brazil), spillover effects of accelerated agricultural growth into increased food imports are less straightforward. The explanation seems to be that accelerated agricultural growth has, in the first place, import substitution effects. It can, however, stimulate food imports with varying time-lags, the effects depending on initial conditions and the extent to which agricultural growth stimulates progress in other sectors and affects income distribution (de Janvry, Sadoulet and White, 1989; Anderson, 1989).

For the moment we remain with the proposition that it will be difficult, without adequate income growth, to achieve quantum jumps in the nutritional situation of the different countries, keeping in mind that the implicit calories/ PCE elasticities (Table 1) vary widely between them.[7] Of major importance for the issue at hand is the agricultural dimension of this proposition, in the sense that, in the countries with high economic dependence on agriculture, the overall economy may not be in a position to take off without sustained gains in agricultural productivity, except in the presence of special circumstances, such as well managed mineral sector booms. We shall examine the scope for increases in agricultural productivity in the next section. Taking that, along with the prospects for overall economic growth, into account we consider the situation which might emerge as we enter the twenty-first century, in terms of food consumption levels, production and food deficits in the different LDC regions.

From the standpoint of the global issue of food and nutrition there is particular interest in prospects for countries and country groups which have been falling behind. Table 2 presents an overall picture of what could emerge at the end of the century. It comes from the assessment carried out some five years ago for the FAO study 'World Agriculture: Toward 2000' (Alexandratos, 1988). With few exceptions, affecting possible outcomes for individual countries rather than regions, this assessment remains generally valid today.[8] The broad pattern is that, for the developing countries as a whole, the rate of improvement in per capita food availability (measured in calories) could be lower than in the historical period (which included the boom period of the 1970s and, for China, the early 1980s). This slow-down is the natural consequence of the fact that some country groups (Near East/North Africa, China) have reached fairly advanced consumption (calorie) levels.[9] When China is excluded, however, no slow-down is foreseen, with slower growth in the nutritionally better-off regions being more than compensated by continued growth in the country groups at middling levels and rather significant acceleration in South Asia, particularly in India.[10]

This leaves most countries in Sub-Saharan Africa, some in South Asia and a few in Latin America and the Caribbean which, having started from very low food availability levels, could see little progress over the next ten years. In particular, in Sub-Saharan Africa, the outlook for slow growth in per capita incomes (itself a considerable improvement over the past record of decline) leaves little scope for optimism. Our food projections for the region are, if anything, less pessimistic than justified by the 0.5 per cent annual growth rate in incomes per head. They reflect, above all, an evaluation based on cautious optimism that agricultural policy reforms, under way in many countries of the region, will allow the agricultural growth rate to recover to at least match that of population (3.3 per cent per annum 1986/88–2000). If this does occur, given the importance of agriculture in the majority of the countries in the region, overall economic prospects might turn out to be less pessimistic than the overall assessments indicate. In conclusion, in the year 2000 the world will most likely continue to be faced with a situation characterized by significant, though somewhat lower, proportions of LDC population (which, in the meantime, will have grown by 1050 million to 4860 million) living in countries in which food availabilities will still be near 2000 calories per person, despite the expectation that the LDC average will have continued to increase.

SCOPE FOR AGRICULTURAL PRODUCTIVITY GROWTH

The food projections of the previous section are the calorie outcomes of the projections of demand, production and trade for individual commodities and countries. For reasons of space the underlying material is presented with minimal commentary (Table 3), concentrating on the factors affecting assessment of cereal sector land productivity (details appear in Alexandratos, 1988 and, for the developed countries, Alexandratos, 1990). Again excluding China the LDC aggregate picture for cereals is as follows (5-year averages):

Production aggregates for cereals LDCs, excluding China

	1963/7	1973/7	1984/8	2000
Harvested area (m. ha)	280	308	327	381
Yield (tons/ha)	1.17	1.39	1.79	2.21
Production (m. tons)	318	428	585	843

Growth of yields has been, and will continue to be, the mainstay of production increases. The factors that may account for inter-country yield differentials are examined first as a prelude to looking into the yield projections. The cereals yield for the LDCs as a whole is the average of very widely differing country yields, ranging from under 0.5 tons/ha (Botswana, Niger) to around 5.0 tons/ha (Egypt, Korea). Several attempts have been made to explain inter-country differentials in agricultural productivity using production functions incorporating conventional inputs (labour, land, fertilizer, machinery) as well as non-conventional categories such as the quality of human capital represented by general and technical education (see, for example, Hayami and Ruttan, 1985; Lau and Yotopoulos, 1989).

In particular, Hayami and Ruttan (1985, pp. 129–33) emphasize the close link between inter-country differences in levels of industrialization and the use of off-farm inputs substituting for land and labour. Others have also drawn attention to the close association between agricultural and overall economic or industrial growth (FAO, 1990, p. 5; World Bank, 1986, p. 80; Hwa, 1988; Singer, 1984; de Janvry and Sadoulet, 1988). One may generalize and formulate the hypothesis that, when factors relating to country-specific agricultural resource characteristics (quantity and quality) and product mix are controlled for, inter-country differences in agricultural productivity will be closely associated with those affecting per capita incomes. The latter variable is a good proxy reflecting forces for change, because it stimulates demand and is associated with changes in factor prices which raise the opportunity cost of labour.[11] In relation to the latter, the higher the incomes which can be earned in the non-agricultural sector, the higher will be the incomes which must be generated per unit of land in agriculture, given the quantity and quality of land per person. Agricultural productivity is also likely to be positively correlated with overall per capita income, since countries generally increase the rate of protection of agriculture (or shift from negative to positive protection) as they move from being low-income and predominantly agricultural, to being high-income and predominantly industrial (Anderson and Hayami,1986). Inter-country regression results (equations 7–9 in the Appendix) show that per capita incomes together with indices of land quantity, land quality and crop mix (a proxy for the fact that agricultural research has benefited wheat and rice more than coarse grains) explain much of the inter-country variation in agricultural productivity per labour unit, and in cereal yields.

The projected LDC average yield of 2.21 ton/ha is a composite of very diverse agro-ecological, country and crop-mix situations. This assessment is the result of detailed multidisciplinary evaluation of the potential for individual commodities, resource characteristics and production technologies in each country, given demand and an assumption that countries would rather produce than import. Economic considerations enter the evaluation in a non-explicit manner, with particular attention to the fact that in many countries with adverse ecologies and geographical positions (for example, land-locked ones) there are severe limits to the profitable introduction of improved technologies (Matlon, 1990). The results have still to be tested in a formal economic modelling sense. Work is currently under way to integrate the results, as far as possible, for aggregate agriculture, or particular sub-sectors, with the rest of the economy as depicted in economy-wide general equilibrium models. A rough and ready test is to compare the total GDP and agricultural production projections of Table 2 (Columns 10, 12). On the agricultural productivity side, yield projections can be seen in perspective by examining the extent to which average LDC cereal yields may progress by the year 2000 to be nearer the best currently achieved in LDCs, after controlling for agro-ecological factors and crop mix. Relevant information is given in Table 4. For example, the last three columns show that today's irrigated wheat yields are in the range 1.1–4.2 ton/ha, while the average yield achieved in the top 20 per cent of LDC irrigated land is 2.9 tons/ha. The average LDC irrigated wheat yield is projected to increase from 2.2 tons/ha today to 2.7 tons/ha in 2000; that is, it would still be below that achieved by the top quintile.

Two considerations are relevant here: first, our land/water/ecology classifications are not sufficiently fine and, therefore, there is no presumption of complete resource homogeneity within each class. This is more so for rain-fed than for irrigated conditions. For rain-fed crops, therefore, ecological factors, and not only policy/management ones, will continue to act as constraints. Second, micro-level studies show persistence of inter-farm yield differentials and this phenomenon can be assumed to apply also to the inter-country universe.

In practice, farmer and farming-system heterogeneity persists even under fairly homogeneous resource characteristics. As Pingali *et al.* (1990, p. 2) report, 'the larger yield gap is between farmers, rather than between the farmer and experiment station.... Reducing these differences between farmers in knowledge and ability will be the primary source of productivity growth in the post-green revolution period'. These findings refer to irrigated rice yields in which resource heterogeneity is less important than in rain-fed conditions. In the latter it would be even more difficult to remove constraints to narrowing inter-country yield gaps by focusing primarily on upgrading farmer capabilities. Add to this the continued persistence of inter-country differences in overall development levels (as suggested by the hypothesis discussed above) and we can see why the existence of wide inter-country yield gaps for the same crop and land/water class is a poor guide to what is achievable by those countries lagging behind in the yield league.

CONCLUSIONS

The majority of countries in which the global food and nutrition problem is concentrated are the low-income ones with generally high dependence on agriculture. The prospects for them to achieve significant gains in food and nutrition through quantum jumps in per capita incomes are not bright. If this is a correct assessment, any future examination of the incidence of under-nutrition conducted in the year 2000 would probably conclude that the phenomenon was far from having been eradicated, or even significantly reduced in terms of absolute numbers. In the search for policies that would make it possible for affected countries to make progress, we first note that such prospects are commonly derived in a global context with the aid of models which very rarely recognize a well identified role for agriculture (for example, Fardoust and Dareshwar, 1990; UNCTAD, 1990). Yet there is growing recognition that without significant policy reforms in favour of agriculture the prospects for accelerated overall development will remain poor in many countries. In this spirit the World Bank assessment for Sub-Saharan Africa (World Bank, 1989, p. 8) emphasizes that, 'in contrast to the past, the future strategy sees agriculture as the primary foundation for growth'. Similar views permeate other studies (FAO, 1986, 1988). The possibility for explicitly modelling the role of agriculture in analyses addressing issues of LDC growth and development in *a global context* should be given serious consideration. An encouraging step in this direction is represented by the work of Adelman *et al.* (1989).

As noted earlier, inter-country analyses have established the close association between the growth of agriculture and that of industry or the overall economy, although attempts to establish causality have been rarer. There are, however, a number of statements or articulated hypotheses in the literature suggesting that pro-agriculture strategies may be more efficient, compared with alternatives, in meeting desirable objectives in both food and agriculture as well as in industrialization and overall development (see, for example, Mellor, 1986, 1989 and, for more general discussion, Flanders, 1969).[12] Some of these ideas have prompted empirical work, often using general equilibrium models which distinguish one or more agricultural sectors (see, for example, de Janvry and Sadoulet, 1987, 1988; Adelman, 1984; Adelman and Taylor, 1990a; Parikh *et al.*, 1988). Models can be used to investigate the effects of pro-agriculture strategies by computing counterfactual scenarios. The results, in many cases, indicate their superiority over alternatives on a number of criteria, including total GDP growth, income distribution and gains in specific food and nutrition variables. In particular, application of a SAM-based model to Mexico (Adelman and Taylor, 1990a) shows that such strategies could secure Pareto improvements within a positive sum game; that is, gains in food and agriculture and in other indicators without any of the population groups distinguished being made worse off.

These findings could be useful for re-evaluating the situation of those countries which are assessed in global analyses as having poor growth prospects in the 1990s. As noted, the majority of these countries are low-income with high dependence on agriculture. Mellor (1989) argues that one of the conditions for effectiveness of pro-agriculture strategies is that agriculture should be a major sector of the economy, a condition obviously fulfilled in most of these countries. Yet other conditions may not be fulfilled, in particular the existence of an industrial sector capable of benefiting from, and responding to, growth impulses originating from policy-induced increases in agricultural productivity. For example, Adelman *et al.* (1989) find that shifting investment from industry to agriculture would be 'more beneficial to semi-industrialized countries with their large urban sectors than to less-industrialized countries in low-income Asia and, especially, Africa', and 'in both of the lower income regions, the increase in agricultural investment tends to choke off the supply of capital to industry to an extent that may be excessive, leading to a sharp rise in the price of urban goods. This again reflects the high initial weight of agriculture in these economies.' Likewise, de Janvry and Sadoulet (1988) conclude that the very low-income countries (agriculture share in GDP 44 per cent) would gain less than the next group (low-income, agriculture share 33 per cent) from an increase in agricultural productivity.

Whether the results of the different modelling exercises can be taken as empirical evidence helping us establish criteria for categorizing countries from the standpoint of the growth-enhancing potential of the pro-agriculture strategies is an open question. Moreover, a general problem shared by some of these models (with the possible exception of Parikh *et al.*, 1988) is that the food and agriculture sector is not yet satisfactorily represented (despite a fair amount of disaggregation) in terms of specificities and constraints regarding nutrition, resources and technology and, occasionally, conditions in international agricultural commodity markets.[13]

We conclude that there is much scope for improving our insights into the relative merits and demerits of pro-agriculture strategies by closer interfacing between detailed multidisciplinary sectoral analysis, like that presented in the preceding two sections, and more aggregative general equilibrium approaches. Such insights are all the more needed today, since development prospects for many LDCs for the 1990s are little better than the dismal record of the 1980s. If such prospects materialize one of the major challenges facing agriculture on the way to the twenty-first century will be its role in conditions of protracted overall economic crisis. Some recent studies suggest that agriculture can play a stabilizing role and cushion declines in output, employment and incomes associated with economic recession (Goldin and Rezende, 1990).

TABLE 1   *Ten best and ten worst performers (out of 93 LDCs) in per capita food availabilities 1969/71 to 1986/88*

										CEREALS	
		Calories			*PCE per capita (1980 I$ prices)*		*Self-Sufficiency Ratio*		*Net Imports (Kg/cap)*		*Gr.Rate Ag.Pro/ capita*
			Gr.Rate			Gr.Rate					
Country	1969/71	1986/88	1970–88	1970	1985	1970–85	1969/71	1986/88	1969/71	1986/88	1970–88
S.Arabia	1865	2810	3.1	894	2268	9.8	54	34	90	485	4.9
Algeria	1830	2790	2.4	563	1047	4.2	74	31	37	208	−0.8
Mauritania	1950	2625	2.2	330	408	2.2	58	40	57	101	−1.0
China	1990	2640	2.0	607	1466	5.5	99	97	3	7	2.7
Egypt	2470	3340	2.0	472	802	3.6	77	49	33	168	−0.2
Iraq	2260	2950	2.0	795	1014	2.1	82	36	46	228	−1.3
Syria	2390	3140	1.9	992	2049	5.9	76	96	63	90	1.5
Indonesia	2020	2650	1.6	370	883	6.5	94	96	7	10	2.3
Myamar	2050	2550	1.6	281	384	2.4	113	98	−25	−11	1.9
Libya	2435	3480	1.5	–	5420[1]	...	25	15	175	378	1.5
Malawi	2360	2100	−0.9	192	228	0.0	101	84	7	2	−0.9
Central African Rep.	2170	2050	−0.8	350	329	0.0	87	76	8	17	0.0
Sierra Leone	2080	1840	−0.7	405	329	0.0	86	72	26	37	−1.3
Angola	2030	1805	−0.7	684	315	−5.9	110	51	−11	28	−5.4
Mozambique	1800	1620	−0.6	803	422	−4.7	88	52	11	27	−3.2
Zambia	2190	2065	−0.6	448	284	−3.4	86	91	49	19	−2.1
Chad	2140	1840	−0.6	309	184	−3.4	98	92	3	10	−1.1
Zaire	2255	2115	−0.6	229	120	−6.9	75	70	12	13	−1.1
Madagascar	2460	2255	−0.5	568	398	−2.4	99	89	3	19	−1.1
Afghanistan	2275	2110	−0.5	578	487	−1.1	93	90	16	14	0.0

Notes: Self-sufficiency ratio = production % of domestic demand or disappearance (all uses, excluding stock changes). *Growth rates* are % p.a. computed from OLS fitted exponential trends. Zero denotes actual zero or growth rate not statistically significant at 5% level. I$ stands for international dollars, i.e. corrected for purchasing power parities (see Kravis 1986).

Sources: FAO, except for per capita PCEs which are from Summers and Heston (1988). For China PCE = 0.6 GDP and the growth rate is for per caput GDP.
[1]Per capita GNP in 1988, current US$ as computed by the World Bank Atlas method, from World Bank (1990).

TABLE 2   *Developing countries – population, calories, GDP, agricultural production, cereals imports : historical values and projections*

	Population (million)			Calories (per caput/day)			GROWTH RATES (% per c..p/annum) Calories		PCE or GDP	
	1988	2000	88–2000	69/71	87/89	2000	70–89	87/89-2000	70–85	89–2000
	(1)	(2)	(3)	(4)	(5)	(6)	(7)	(8)	(9)	(10)
All Developing	3811	4861	2.0	2100	2471	2620	1.0	0.5	3.3	3.2
Sub-Saharan Africa	456	671	3.3	2083	2120	2190	0.1	0.3	−1.4	0.5
Near East/North Africa	279	381	2.6	2383	3019	3090	1.3	0.2	3.4	2.1
Asia	2651	3278	1.8	2014	2431	2605	1.2	0.6	4.2	
South Asia	1078	1398	2.2	2006	2186	2396	0.5	0.8	1.3	3.2
East Asia	1572	1879	1.5	2019	2610	2752	1.6	0.4	5.3	5.1
Latin America and Caribbean	425	532	1.9	2507	2724	2906	0.5	0.5	2.3	2.3
All Developing, excl. China	2725	3585	2.3	2153	2406	2583	0.7	0.7	2.2	
Asia, excluding China	1565	2001	2.1	2037	2290	2526	0.7	0.8	2.6	

	GROWTH RATES (per cap/annum) Ag Prod		Net Imports (kg/cap) CEREALS			SSR(%)		
	70–88	86/88-2000	69/71	86/88	2000	69/71	86/88	2000
	(11)	(12)	(13)	(14)	(15)	(16)	(17)	(18)
All Developing	1.1	1.2	7.6	18.5	20.2	97	92	92
Sub-Saharan Africa	−1.3	0.1	4.9	17.5	25.1	101	85	83
Near East/North Africa	0.2	0.6	29.0	125.9	155.3	93	68	61
Asia	1.7	1.7	7.1	6.8	6.1	96	97	98
South Asia	0.5	1.0	10.3	3.1	1.1	94	100	99
East Asia	2.4	2.0	5.0	9.3	9.7	97	9	97
Latin America and Caribbean	0.6	0.7	0.3	19.7	−1.8	105	92	101
All Developing excl. China	0.5	0.8	7.7	23.2	26.9	98	89	89
Asia excl. China	1.0	1.1	6.8	6.9	9.4	97	97	96

Notes: Country groups are defined in Alexandratos (1988) and (for col.10) in World Bank (1990). Population numbers are from the latest (1990) UN Assessment, medium variant projection; they are somewhat different (generally higher) than those underlying the projections in columns 6,8,12,15,18. The growth rates in column 9 are for PCE from Summers and Heston (1988) and refer only to countries in each group with data in the source. The growth rates in column 10 are for GDP from World Bank (1990, table 1.3) and do not correspond exactly to the country coverage used here; in particular the growth rates for the group Near East/North Africa refers to the World Bank group "Middle East, North Africa and Other Europe" which includes also Greece, Hungary, Poland, Portugal, Yugoslavia and Romania and excludes Saudi Arabia.

Columns 4,5,7,8 updated in June 1991 with latest data up to 1989.

TABLE 3   *LDCs, Selected historical data and projections: cereals and total agriculture*

		CEREALS (including milled rice)						GROWTH RATES (%per annum)				
		Domestic Use			Prod	Net Trade	SSR		Cereals		Total Agriculture	
		Total	Food						Demand	Prod	Demand	Prod
		mil tons	%Total	kg/cap	mil tons		%					
Developing Countries[1]	69/71	490	76	145	481	−16 / −20[2]	98	70–88	3.5	3.2	3.6	3.3
	86/88	868	74	171	797	−67 / −76[2]	92	88–2000	2.8	2.9	3.1	3.0
	2000	1247	66	174	1152	−95 / −112[2]	92					
Sub-Sah. Africa	69/71	37	81	112	37	−2	100	1970–88	3.1	1.9	2.7	1.8
	86/88	61	86	119	52	−7	85	86/88–2000	3.8	3.6	3.7	3.4
	2000	100	82	121	83	−17	83					
N.East/N.Africa	69/71	52	62	183	46	−6	87	1970–88	4.4	2.5	4.6	3.0
	86/88	106	57	212	72	−36	68	86/88–2000	2.8	2.0	3.3	3.0
	2000	153	52	204	93	−60	61					
Asia	69/71	338	82	150	332	−12	98	1970–88	3.4	3.5	3.7	3.7
	86/88	590	80	181	571	−17	97	86/88–2000	2.7	2.7	3.1	3.1
	2000	830	70	187	811	−19	98					
Asia (exc.China)	69/71	178	86	148	174	−9	98	1970–88	2.8	3.1	3.3	3.3
	86/88	285	83	155	276	−10	97	86/88–2000	2.6	2.5	2.8	2.7
	2000	398	82	173	380	−18	96					
Latin America Carib	69/71	63	54	120	66	3	105	1970–88	3.5	2.8	3.1	2.9
	86/88	111	51	137	102	−8	92	86/88–2000	3.0	3.7	2.8	2.8
	2000	164	46	141	165	1	101					
Developed Countries	69/71	627	26	150	646	21	103	1970–88	1.3	1.5	1.3	1.4
	86/88	786	22	143	838	80	107	86/88–2000	0.9	2.3	0.9	1.3
	2000	888	21	141	1000	112	127					
E.Europe/USSR	69/71	237	29	198	231	0	98	1970–88	1.6	0.8	1.5	1.2
	86/88	323	21	173	288	−34	89	86/88–2000	1.0	1.1	1.1	1.1
	2000	366	21	174	331	−35	90					
Market Economies	69/71	390	24	127	415	22	107	1970–88	1.1	1.9	1.2	1.6
	86/88	464	23	129	550	115	119	86/88–2000	0.9	1.5	0.7	1.3
	2000	522	21	124	669[3]	147[3]	128					

[1] 94 Developing Countries accounting for 98% of LDC population.

[2] Net Cereal Imports for all LDCs. The bulk of cereal imports by the LDCs not in the group of 94 is accounted for by Taiwan (China Province – 5.4 million tons). The latest forecast for LDC net cereal imports is 84.6 million tons in 1990/91 (July-June) (FAO, Food Outlook, 2/91), which is very close to 84 million tons obtained by interpolating for 1990/91 between 68.9 million tons for 1983/85 (the base year of the projection study) and 112 million tons projected for 2000.

[3] Projected production and net exports required for world balance. The 86/88 production is below "normal" because of the North American drought. The 5-year average 1984/88 is 580 million tons, meaning an average growth rate of production to year 2000 of 1.0% p.p.a.

Sources: Alexandratos (1988), with historical data updated. For comparison with LDC cereal deficit projections of other studies see Alexandratos (1990:78–80).

TABLE 4   *Selected cereal yields (tons/ha) by major land-water classes[1],
LDCs excluding China*

	% of Area under Crop	Average Yield		Country Range[2]	Top 20%
		mid-80's	2000	mid-80's	
WHEAT	100	1.6	2.3		
Uncertain Rainfall	8	1.0	1.6	0.5–1.8	1.7
Good Rainfall	27	1.5	2.0	0.6–2.2	2.2
Irrigated	40	2.2	2.7	1.1–4.2	2.9
RICE (Paddy)	100	2.4	3.0		
Problem Areas	10	1.3	1.6	0.7–2.1	1.9
Naturally Flooded	7	1.9	2.5	1.0–4.1	2.7
Irrigated	43	3.1	3.6	2.0–6.3	4.8
MAIZE	100	1.6	2.2		
Uncertain Rainfall	13	1.1	1.5	0.5–2.5	2.1
Problem Areas	27	1.1	1.4	0.3–2.5	1.7
Good Rainfall	46	1.7	2.5	0.5–3.7	2.5
Irrigated	11	3.1	3.9	1.8–7.0	5.7
BARLEY	100	1.1	1.7		
Low Rainfall	25	0.6	0.9	0.3–1.1	1.1
Irrigated	14	1.5	2.3	1.0–2.8	2.2
MILLET	100	0.6	0.8		
Low Rainfall	36	0.4	0.5	0.2–0.6	0.5
Uncertain Rainfall	25	0.6	0.8	0.3–1.2	0.8
Good Rainfall	12	0.9	1.3	0.4–2.0	1.4
SORGHUM	100	1.0	1.4		
Low Rainfall	27	0.4	0.6	0.2–1.1	0.5
Uncertain Rainfall	27	0.9	1.0	0.4–3.1	2.3
Good Rainfall	22	1.5	2.2	0.4–3.9	3.5

[1]The definition of the land-water classes is as follows:

Low-rainfall rainfed land (or arid/semi-arid): Rainfall providing 1–119 growing days, soil quality very suitable, suitable or marginally suitable.

Uncertain rainfall rainfed land (or dry sub-humid): Rainfall providing 120–179 growing days, soil quality very suitable or suitable.

Good rainfall rainfed land (or moist sub-humid): Rainfall providing 180–269 growing days, soil quality very suitable or suitable.

Problem lands (or humid): The term is used to designate areas with excessive moisture and/or unsuitable soils. In these areas rainfall provides more than 269 growing days, soil quality is very suitable, suitable or marginally suitable. Also included in this class is that part of the 120–269 growing days zones where soil rating is only marginally suitable.

Naturally flooded lands: Land under water for part of the year and lowland non-irrigated paddy-fields (gleysols).

Irrigated land: It comprises both fully irrigated lands which are equipped for irrigation and suitable drainage and not suffering from water shortages and partially irrigated lands which are equipped for irrigation but lacking drainage or reliable water supplies or with low quality and reliability of distribution.

[2]Excluding countries with less than 50000 ha. of the specified land class under the crop.

APPENDIX

Estimated cross-section equations, developing countries (t-ratios in parentheses)

Independent Variable	Equations and Dependent Variable								
	(1) C	(2) C	(3) C	(4) C	(5) C	(6) C	(7) ln(LABPROD)	(8) ln(CERYIELD)	(9) ln(CERYIELD)
CONSTANT	179	954	1574	1474	58	827	1.06	4.84	5.80
ln(PCE)	332.7	259.7	209.2	285.6	295.3	211.0	0.83	0.43	0.30
	(10.1)	(4.6)	(2.9)	(3.4)	(3.9)	(5.0)	(10.8)	(6.4)	(4.3)
PRICE		−240.6		−556.7					
		(−1.2)		(−2.0)					
POV			−23.5	−6.5					
			(-2.7)	(-2.2)					
POV²			0.18						
			(2.1)						
CERM						3.29			
						(4.10)			
CERM²						−0.006			
						(−3.2)			
GPROD						31.74			
						(2.4)			
DISTR					19.8				
					(2.0)				
ln(LQUANT)							0.43	−0.23	−0.20
							(6.7)	(−4.1)	(−3.9)
ln(LQUAL)							0.17	0.27	0.24
							(2.5)	(4.7)	(4.4)
ln(CRMIX)									0.15
									(3.7)
$\bar{R}^2$	0.55	0.37	0.56	0.54	0.49	0.64	0.79	0.54	0.61
NO. COUNTRIES	85	37[1]	37[2]	22[3]	18[4]	75[5]	79[6]	79	79

Variables

C = calories per capita per day, average 1980/85, direct human consumption, from the FAO Food Balance Sheets.

PCE = Private Consumption Expenditure per capita in international dollars, average 1980/85 at 1980 prices, from Summers and Heston (1988).

= Food Price Index from the ICP project, 1980, from UN/CEC 1987.

PRICE = Percent of population below poverty line, from the data base of UNDP (1990).

POV = Percent of total household income accruing to the bottom 40% of households, mostly 1980s, from World Bank (1990), table 30.

DISTR = Net cereals imports, Kg. per capita, annual average for 1980/85.

CERM = Growth rate of agricultural production per capita, 1975–85, % p.a.

GPROD = Gross output (all crop and livestock products) per person in the agricultural labour force, average for 1984/88, in I$ of 1979/81.

LABPROD = Land in agricultural use per person in the Ag. labour force, mid-1980s, in ha.

LQUANT = Ratio of "good" land to total land, all in terms of harvested area, mid-1980s; "Good" land = land in the classes: irrigated, good rainfall.

LQUAL naturally flooded; the balance of land comprises the classes low rainfall, uncertain rainfall and problem areas. For explanations see Alexandratos(1988):128–129.

=Ratio of wheat and rice output to total cereals output, 1984/88.

CRMIX

Notes: [1]Countries with food price data.

[2]Countries with poverty data.

[3]Countries with price and poverty data.

[4]Countries with income distribution data.

[5]Countries with net imports of cereals in 1980/85.

[6]The 85 countries with per capita PCE data minus those in which cereals account for less than 15% of land minus China (no land quality data).

NOTES

[1]In FAO Food Balance Sheets food supplies for direct human consumption for any commodity/country/year are computed from the following equation:

Food = Production + (Imports - Exports) + (Opening Stocks – Closing Stocks) – Feed – Seed – Industrial Non-Food Uses – Waste.

The terms 'per capita food supplies for direct human consumption' and 'consumption' are used interchangeably, unless otherwise specified.

[2]See FAO (1985), Alexandratos (1988). The existence and definition of a calorie threshold is itself a subject of controversy (FAO, 1985; Srinivasan, 1983).

[3]See Svedberg (1987) and Berry (1984) for Sub-Saharan Africa, and Tyagi (1990) for differences in estimates of per capita cereals consumption from survey data and those from national averages in India. Dowler and Seo (1985) argue that Food Balance Sheets overestimate both average national food consumption and increases with GNP per person. There are, however, strong arguments to suggest that data from household surveys are not necessarily better indicators of average national food consumption (Naiken and Becker, 1990). For more general discussion see FAO (1983).

[4]The data are in international dollars (I$) at 1980 prices as provided by the UN International Comparisons Project (ICP) complemented by Summers and Heston (1988). Use of these data should avoid the problem of non-comparability across countries of the conventional national accounts PCE data which, when converted to a common dollar numeraire on the basis of exchange rates, tend to under-state the purchasing power of the low-income countries (Kravis, 1986, Yotopoulos, 1989).

[5]1980 data for the 37 developing countries in the ICP-phase IV project. We obtain a real food price index as the ratio of each country's food price level to that for all goods and services (GDP data from UN/CEC, 1987, Table 9). It is noted that food, as well as most tradable commodities, appears as generally more expensive in the developing countries than in the developed ones, despite the fact that a dollar buys more food in the former than in the latter, as the following example shows (USA = 100).

	USA	France	Italy	India	Chile
Price of food	100	136.3	105.4	59.7	99.3
Price of all goods and services (GDP)	100	124.1	88.6	41.9	67.9
Real price of food (ratio)	100	109.8	119.0	142.5	146.5

[6]These results are, therefore, not necessarily incompatible with the findings of other food demand analyses but, if valid, they imply that even at low levels of income additional expenditure on food buys calories of increasing unit cost. Research on this issue, using more detailed income group data for individual countries, seems to confirm the prevalence of low-income elasticities of demand for food when the latter is expressed in terms of calories (Behrman and Deolalikar, 1987; Behrman, Deolalikar and Wolfe, 1988; see also Ravallion, 1990). The fact that the income elasticity of food demand appears lower when demand is expressed in calories rather than in terms of expenditure on food is the result of (a) the changing commodity composition of diets towards higher priced calories as incomes rise and (b) the rising share in total food expenditure of the cost of desirable attributes of food commodities other than calorie content, such as processing and distribution margins. In relation to this latter factor, Schiff and Valdés (1990) rightly point out that a distinction should be made between the effects of incremental income on the demand for *nutrients* and that for *nutrition*. The latter is hypothesized to be positively correlated with the non-calorie attributes of food products and also to be influenced by expenditure of incremental income on non-food goods and services having an impact on nutrition, such as sanitation, medical services and refrigeration.

[7]For the 85 countries with both calorie and PCE growth rates for 1970–88 and 1970–85, respectively, the elasticity can be computed as 0.16 from the equation:

$$(1 + GR.CAL) = a(1 + GR.PCE)^b, (R^2 = 0.41, t\text{-value of } b = 7.6).$$

[8]At the time of the assessment we were working with base year data 1982/4 for per capita food availabilities and with data up to 1985 for production and trade. We had used exogenous GDP growth assumptions which, for the country groups for which comparisons can be made, are very close to those of the most recent projections of the World Bank shown in column 10 of Table 2 (World Bank, 1990, Table 1.3; for comparisons with the FAO assumptions see FAO, 1990, p. B.10). The projected availabilities (col. 9, Table 2) are the result of separate projections of 52 individual food commodities for each country obtained using the Engel demand functions of the FAO demand model (FAO, 1971). The projections thus obtained have been subjected to a series of adjustments for nutritional consistency as well as following the evaluation of each country's production and trade possibilities and on the basis of whatever country-specific information concerning plans and so on was available at the moment. (For more details see Alexandratos, 1988.)

[9]Naturally, demand for both direct and indirect (feed) uses can be expected to grow faster than indicated in Table 2. The projections for China are very conservative: 2730 calories in 2000 compared with 2630 calories for the latest three-year average, 1987/9, in the FAO food balance sheets. These projections reflect, among other things, the conclusions of Chinese studies which project 2700 and 2800 calories for the years 2000 and 2020, respectively. An alternative 'nutritionists' scenario' projects a flat 2400 calories for both future years *(Problems of Agricultural Economy,* no. 8, August 1990, translations or excerpts in English in US, *Joint Publications Research Service – JPRS,* CAR-90-090, 7 December 1990, and *China News Analysis,* Hong Kong, 15 January 1990). The overwhelming consideration in these projections seems to be the need to promote nutritionally desirable consumption patterns while 'maintaining a correspondence between the growth of valid demand for food products and the development of production in agriculture and the food industry'. This slow growth in food consumption in the future would seem to be unlikely, given past trends, if per capita total consumption expenditure were to grow as projected by the same studies: 3.6 per cent per annum in the 1990s and 3.2 per cent per annum in 2000–20.

[10]The IIASA reference scenario for India projects increases in calorie intake to 2533 by 2000, compared with 2430 in this paper (Parikh *et al.,* 1988).

[11]Pickett (1990a) has this to say on agriculture and overall development: 'Why is Ethiopia so poor? The tempting answer is the dominance of traditional agriculture in total economic activity. To accept this, however, is to confuse effect with cause. Ethiopia is not poor because of its traditional agriculture, but its agriculture is traditional because the country is so poor.' De Janvry and Sadoulet (1988) investigate short-term causality (running from agriculture to industry) by introducing one-year lags in the regression analysis.

[12]Some 40 years ago, W.A. Lewis (1953) in advice to the Government of Ghana concluded that 'the main obstacle [to industrialization] is the fact that agricultural productivity per man is stagnant... Number one priority is therefore a concentrated attack on the system of growing food in the Gold Coast, so as to set in motion an ever increasing productivity. This is the way to provide the market, the capital and the labour to industrialization [p. 22]. ... Our concern is not with the amount of food available, but with food production per person engaged in agriculture. If the food supply were adequate, the Gold Coast should still be straining to have fewer farmers, each producing more, since this is the way to stimulate the other sectors of the economy [p. 3]. ... To increase yield per acre is usually the cheaper way [p. 2].' Picket (1990b) ascribes the economic retrogression of Ghana to the fact that the government followed policies virtually opposite to those recommended by Lewis. His arguments assume new strength today with the hindsight of development failures in countries which neglected, or discriminated against, agriculture. They are also strengthened by the fact that progress in agricultural research over the last few decades has greatly enlarged the technologies available for increasing agricultural productivity. At the same time, the current preoccupation with poverty also favours agriculture as the sector whose more rapid development is likely to produce the quickest payoff in terms of rural poverty alleviation.

[13]For example, in Adelman and Taylor (1990a) the 'unimodal' (campesino-friendly) superior strategy for Mexico leads to Pareto improvements by, *inter alia,* implying production growth in the 'basic-grains' sector of 11.5 per cent per annum for ten years while some population groups arrive at spectacular increases in calorie consumption, including one group (the 'capital-

ists') shown as consuming some 4500 calories per capita per day. In particular the basic grains growth implications (without other agriculture sub-sectors growing by less than in the baseline case) would require some very spectacular combinations of land and yield expansion rates which are highly implausible. One may also question the feasibility of Mexico turning from net importer to net exporter of grains, given international markets characterized by protectionism and export subsidies. The authors rightly warn that these results should be interpreted qualita tively and as indicative of broad orders of magnitude because of model limitations, such as fixed-price SAM model, perfectly elastic output supplies, exogenous assumptions of agricultural productivity growth (personal communication of J.E. Taylor). In a subsequent application of a CGE model to Mexico, the authors find that the introduction of price adjustments and non-linearities dampen but do not alter fundamentally their conclusions (Adelman and Taylor, 1990b).

REFERENCES

Adelman, I., 1984, 'Beyond Export-Led Growth', *World Development*, vol. 12, no.9.
Adelman I. and Taylor, J., 1990a, *Changing Comparative Advantage in Food and Agriculture: Lessons from Mexico*, OECD (Development Centre Studies), Paris.
Adelman, I. and Taylor, J., 1990b, 'Is Structural Adjustment with a Human Face Possible? The Case of Mexico', *Journal of Development Studies*, vol. 26, no. 3.
Adelman, I., Bourniaux, J. and Waelbroeck, J., 1989, 'Agricultural Development-Led Industrialisation in a Global Perspective', in J. G. Williamson and V. R. Panchamuki (eds), *The Balance between Industry and Agriculture in Economic Development, Vol. 2, Sector Proportions*, Macmillan, London, pp. 320–39.
Alexandratos, N. (ed.) 1988, *World Agriculture: Toward 2000, an FAO Study*, Belhaven Press, London and New York University Press, New York (French edn, Alexandratos, N. (sous la direction de) *L'Agriculture mondiale, Horizon 2000, étude de la FAO*, Economica, Paris, 1989).
Alexandratos, N. (ed.), 1990, *European Agriculture: Policy Issues and Options to 2000: an FAO Study*, Belhaven Press, London and Columbia University Press, New York (French edn. Alexandratos, N. (sous la direction de) *Agriculture européenne: Enjeux et options à l'Horizon 2000: Etude de la FAO*, Economica, Paris, 1991).
Anderson, K., 1989, 'Does Agricultural Growth in Poor Countries Harm Agricultural-Exporting Rich Countries?' *Agricultural Economics*, vol. 3.
Anderson, K. and Hayami, Y., 1986, *The Political Economy of Agricultural Protection, East Asia in International Perspective*, Allen & Unwin, London.
Behrman, J. and Deolalikar A., 1987, 'Will Developing Country Nutrition Improve with Income? A Case Study for Rural South India', *Journal of Political Economy*, vol. 95, no. 3.
Behrman, J., Deolalikar, A. and Wolfe, B., 1988, 'Nutrients, Impacts and Determinants', *The World Bank Economic Review*, vol. 2, no. 3.
Berry, S., 1984, 'The Food Crisis and Agrarian Change in Africa: a Review Essay', *African Studies Review*, vol. 27, no. 2.
de Janvry, A. and Sadoulet, E., 1987, 'Agricultural Price Policy in General Equilibrium Models: Results and Comparisons', *American Journal of Agricultural Economics*, vol. 69, no. 2.
de Janvry, A. and Sadoulet, E., 1988, 'The Conditions for Compatibility between Aid and Trade in Agriculture', *Economic Development and Cultural Change*, vol. 37, no. 1.
de Janvry, A., Sadoulet, E. and Kelley White, T., 1989, *Foreign Aid's Effect on US Farm Exports, Benefits or Penalties?* USDA, Foreign Agricultural Economic Report No. 238, Washington, DC.
Dowler, E. and Seo, Y. O., 1985, 'Assessment of Energy Intake: Estimates of Food Supply v. Measurement of Food Consumption', *Food Policy*, August.
FAO, 1971, *Agricultural Commodity Projections 1970–1980*, Rome.
FAO, 1983, *A Comparative Study of Food Consumption Data from Food Balance Sheets and Household Surveys*, FAO Economic and Social Development Paper No. 34, Rome.
FAO, 1985, *The Fifth World Food Survey*, Rome.
FAO, 1986, *African Agriculture: The Next 25 years*, Rome.

FAO, 1988, *Potentials for Agricultural and Rural Development in Latin America and the Caribbean*, Rome.

FAO, 1990, *Long-term Strategy for the Food and Agriculture Sector*, (document CL 98/13), Rome.

Fardoust, S. and Dareshwar, A., 1990, *A Long-term Outlook for the World Economy: Issues and Projections for the 1990s*, World Bank, Policy and Research Series No. 12, Washington, DC.

Flanders, M. June, 1969, 'Agriculture versus Industry in Development Policy: The Planner's Dilemma Re-examined', *Journal of Development Studies*, 5(3) April, pp. 171–89.

Goldin, I. and Rezende, G.C., 1990, *Agriculture and Economic Crisis: Lessons from Brazil*, OECD Development Centre Studies, Paris.

Hayami, Y. and Ruttan, V., 1985, *Agricultural Development, an International Perspective*, Johns Hopkins University Press, Baltimore and London.

Hwa, E-C., 1988, 'The Contribution of Agriculture to Economic Growth: Some Empirical Evidence', *World Development*, vol. 16, no. 11.

Kravis, I., 1986, 'The Three Faces of the International Comparison Project', *The World Bank Research Observer*, vol. 1, no. 1.

Lau, L. and Yotopoulos, P., 1989, 'The Meta-Production Function Approach to Technological Change in World Agriculture', *Journal of Development Economics*, vol. 31.

Lewis, W.A., 1953, *Report on Industrialization and the Gold Coast'*, Government Printing Office, Accra, Gold Coast.

Matlon, P., 1990, 'Improving Productivity in Sorghum and Pearl Millet in Semi-arid Africa', *Food Research Institute Studies*, vol. XXII, no.1.

Mellor, J., 1986, 'Agriculture on the Road to Industrialization', in J. Lewis and V. Kallab (eds), *Development Strategies Reconsidered*, Transaction Books, New Brunswick and Oxford.

Mellor, J., 1988, 'Food Demand in Developing Countries and the Transition of World Agriculture', *European Review of Agricultural Economics*, vol. 15, no. 4.

Mellor, J., 1989, 'Rural Employment Linkages through Agricultural Growth: Concepts, Issues, and Questions', in J. G. Williamson and V. R. Panchamuki (eds), *The Balance between Industry and Agriculture in Economic Development, Vol. 2, Sector Proportions*, Macmillan, London, pp. 305–19.

Naiken, L. and Becker, K., 1990, 'Food Consumption Statistics – The Potential Role of Data Collected in Household Income/Expenditure Surveys', *FAO Quarterly Bulletin of Statistics*, vol. 3, no. 4.

Parikh, K., Fischer, G., Frohberg, K. and Gulbrandsen, O., 1988, *Towards Free Trade in Agriculture*, M. Nijhoff Publishers, Dordrecht.

Pickett, J., 1990a, 'Comparative Advantage and the Economic Development of Ethiopia', in S. Aziz (ed.), *Agricultural Policies for the 1990s*, OECD Development Centre Studies, Paris.

Pickett, J., 1990b, 'Agriculture and Industry in the Ghanaian Economy", in S. Aziz, (ed.), *Agricultural Policies*.

Pingali, P., Hoya, P. and Velasco, L., 1990, *The Post-Green Revolution Blues in Asian Rice Production*, IRRI Paper No. 90–01, IRRI, Los Baños.

Ravallion, M., 1990, 'Income Effects on Undernutrition', *Economic Development and Cultural Change*, vol. 38, no. 3–4.

Reutlinger, S. and Selowsky, M., 1976, *Malnutrition and Poverty, Magnitude and Policy Options*, Johns Hopkins University Press for the World Bank, Baltimore and London.

Schiff, M. and Valdés A. 1990, 'Nutrition: Alternative Definitions and Policy Implications', *Economic Development and Cultural Change*, vol. 38, no. 2.

Singer, H., 1984, 'Success Stories in the 1970s: Some Correlations', *World Development*, vol. 12, no. 9.

Srinivasan, T., 1983, 'Hunger: Defining It, Estimating Its Global Incidence and Alleviating It', in D. Gale Johnson and G. Schuh (eds), *The Role of Markets in the World Food Economy*, Westview Press, Boulder, Co.

Summers, R. and Heston, A., 1988, 'A New Set of International Comparisons of Real Product and Price Levels: Estimates for 130 Countries, 1950–1985', *Review of Income and Wealth*, series 34, no. 1.

Svedberg, P., 1987, *Undernutrition in Sub-Saharan Africa: A Criticala Assessment of the Evidence*, World Institute for Development Economics Research, Working Paper 15, Helsinki.

Tyagi, D., 1990, *Managing India's Food Economy*, Sage Publications, London.
UNCTAD, 1990, *Trade and Development Report 1990*, UN, New York.
UNDP, 1990, *Human Development Report 1990*, Oxford University Press for the UNDP, New York and Oxford.
United Nations and Commission of the European Communities, 1987, *World Comparisons of Purchasing Power and Real Product for 1980*, New York.
World Bank, 1986, *World Development Report 1986*, Oxford University Press for the World Bank.
World Bank, 1989, *Sub-Saharan Africa: From Crisis to Sustainable Growth*, Washington, DC.
World Bank, 1990, *World Development Report 1990*, Oxford University Press for the World Bank.
Yotopoulos, P., 1989, 'Distributions of Real Income: Within Countries and by World Income Class', *Review of Income and Wealth*, series 35, no. 4.

DISCUSSION OPENING – ROBERT L. THOMPSON*

In scope this paper is much narrower than its title suggests. The focus is confined to prospects to the year 2000, and there is very little discussion of issues which appear to be of prospective importance beyond the turn of the century. I will first discuss the paper as presented, before moving on to the broader perspective.

The paper does make a useful addition to the literature in emphasizing the need to give priority to agriculture in Third World countries in order to boost development in general and to solve nutritional problems. However, it pays lamentably little heed to the fact that the developing countries which have performed best over the last two decades have followed an open, trade-oriented, strategy (World Bank, 1986).The author's mercantilistic assertion that 'countries would rather produce than import' flies in the face of this experience and the modern principles of comparative advantage (Abbott and Thompson, 1988; Jabara and Thompson, 1980).

The model employed in the paper to investigate inter-country differences in food consumption levels is neither a structural model nor a reduced form, and it accounts for a disappointingly small fraction of observed variability. A richer specification could draw on human capital theory and the new household economics. The opportunity cost of time in the household is an important determinant of the mix of foods demanded, because different products have varying time intensities in consumption. It is the increasing opportunity cost of time which is the very essence of economic growth. Therefore, one should expect the product mix to alter as income levels change. Moreover, differences in education and other social variables are likely to affect the efficiency of household production and, in turn, the quantities of products demanded in different countries. In cross-country studies similar to that attempted one should find sufficient variation in the wage rate and in other relevant indicators to be able to isolate their influence on the quantity of each good demanded (Thompson and Schuh, 1975).

The inter-country productivity analysis does stress the importance of human capital in accounting for differences in crop yields among farms. Spe-

*Purdue University, USA.

cifically, it makes an important observation that yield differences are often greater among farms than between experimental stations and farms. Better educated farmers also tend to be superior in marketing, thus obtaining higher prices for their output. The disappointing aspect of the inter-country yield analysis is the absence of relative prices from the set of explanatory variables. Relatively few empirical studies have found significant effects of relative product and input prices on crop yields using data from individual countries, largely because there is insufficient variation in the price ratio over time within any one country to identify the relationship. Across countries this problem is eliminated and significant positive relationships can be found (Lyons and Thompson, 1981). Omission of relative prices is particularly unfortunate in the light of the extent to which so many countries distort product and input prices through public policy; the potential impact is completely ignored.

I concur with the author's calling for more computable general equilibrium analysis to be undertaken to reveal linkages between agriculture and the rest of the economy, and with his concern about the quality of many existing models. While often powerful, it is frequently the case that the empirical content is not credible because model building has been solely in the hands of economists. They lack an input from multidisciplinary teams including experts in production technologies and in the complexity of the markets involved.

Let me now look ahead at the critical challenges facing world agriculture as it enters the twenty-first century. The first of these is that of satisfying consumer requirements. Alexandratos has dealt with this only in the context of the Third World, largely ignoring the high-income countries. There consumers are increasingly concerned about the nutritional characteristics and safety of their food. The high-income countries are also faced with demographic changes, sometimes in the age structure of their populations, sometimes in ethnic characteristics brought about by immigration. Both affect the mix of goods demanded. To add to that, women are being increasingly drawn into the labour force and the opportunity cost of housekeeping time is rising. Changes in food demand patterns are obvious from the increased frequency of eating out in restaurants and fast food establishments, and the use of the domestic microwave oven. The whole food system needs to adapt to these changes, and they are particularly challenging to the farm sector. Historically, most farmers have displayed antagonistic attitudes toward food marketers and processors. If farmers are to thrive in an evolving economy, in which their own numbers are becoming relatively small, they will have to abandon this historical animosity, begin to view themselves as part of the food system, and form political coalitions with marketers and processors who account for a far larger fraction of value added in the products which modern consumers buy.

The second major challenge will be the continuing internationalization of world agricultural markets. There is already an important trend under way in which there is faster growth in trade in processed food products, often with brand identity, than in raw commodities. Countries which equip themselves to compete in export markets for branded, value added, goods will gain significantly larger foreign exchange earnings as well as providing a more secure future for their farmers. Part of the international challenge is likely to come from continuing liberalization of markets as the political power of farmers wanes, requiring adjustment to lower price supports. It will also come as the regional trading blocs, which have emerged in the 1980s and 1990s, mature and the forces of interregional competition within each bloc begin to make themselves felt. The formerly centrally planned economies, in addition to resolving their internal adjustments, can also be expected to play an increased role on the international market scene.

Growing concern for the quality of the environment will pose a third critical challenge to farmers. Water quality is already a key worry among the non-farm population and farming practice is seen as the culprit in polluting surface and ground water. In many cases agricultural price supports have artificially raised product prices relative to input prices, and the price of land relative to the price of inputs which substitute for land. Both encourage intensification of crop and animal agriculture, leading to the overloading of the environmental capacity to absorb chemical and manure applications. In some cases farming has been falsely accused of causing damage, though whether culpable or not, agriculture will increasingly be called upon to demonstrate that it is operating in the most benign manner possible. Little needs to be said about the other major environmental issue of global warming: it is quite obviously an issue for the twenty-first century.

Fourth, agriculture is increasingly becoming a highly technological activity. We are in the golden age of the biological sciences, the results of which could transform production techniques of the next century. That, however, is not all. Developments in robotics and in machine vision will bring farm machinery to new heights of sophistication, and low-cost information transmission and processing capacity will vastly expand the power of the farmer in management and marketing. Farmers of the future will need higher levels of educational attainment to cope with applying increasingly sophisticated technologies.

The fifth challenge will be the changing political scene. In public policy priorities are already shifting away from traditional commodity price and income supports, subsidies and quotas, towards a new agenda: that of environmental policy, trade arrangements, animal welfare, food safety and science policy. Farm organizations, in isolation, will see their political power dwindle unless they form political alliances with other groups.

In summary, in the early twenty-first century the poor countries of the world are still likely to be struggling to provide an adequate food supply to the bulk of their populations, especially in Africa. There is growing recognition of several failings of past policy. Hunger is, to a great extent, attributable to poverty, and broad-based economic growth is the surest cure. Priority for agriculture in national economic development is essential. Cheap food policies which maintain artificially low prices to appease urban residents is a significant impediment to agricultural development. It is also impossible centrally to plan complex agricultural systems; the probability of success in development is vastly enhanced when a country follows a trade and market-oriented strategy.

While the developing world grapples with these challenges, many similar problems will be confronting the recently centrally planned economies now attempting to implant a market-based agricultural economy. Agriculture in the high-income countries, as we have seen, will have its own array of issues to face. There will be more than enough for the agricultural economics profession to do to help farmers and agribusiness everywhere to adapt to the future.

REFERENCES

Abbott, P.C. and Thompson, R.L., 1988, 'Changing Agricultural Comparative Advantage', *Agricultural Economics*, 1,(2), pp. 97–112.
Jabara, C.L. and Thompson, R.L., 1980, 'Agricultural Comparative Advantage Under International Price Uncertainty, *American Journal of Agricultural Economics*, 62,(2), pp. 188–98.
Lyons, D.C. and Thompson, R.L., 1981, 'The Effects of Distortions in Relative Prices on Corn Productivity and Exports: A Cross-Country Study', *Journal of Rural Development*, 4,(1), pp. 83–102.
Thompson, R.L. and Schuh, G.E., 1975, 'A Metademand Function', contributed paper at the Annual Meeting of the American Agricultural Economics Association, Ohio State University.
World Bank, 1986, *World Development Report 1986*, Oxford University Press, Oxford.

ULRICH KOESTER*

The Efficiency of Agricultural Markets in Directing Agricultural Development

INTRODUCTION

Over the last 50 years the economic performance of most developing countries and of all Eastern European countries has been disappointing. Moreover, the last five years have led to the fall of Eastern Europe's postwar political and economic system. There seems to be agreement that socialist economies have proved less efficient than market economies. Hence, most Eastern European countries (including the Soviet Union) are in the process of adopting market economies. Similar changes are occurring in many developing countries.

Of course, I do not dare challenge the widely accepted argument that markets are one of mankind's most important social inventions (Meade). Instead, I intend to consider whether specific shortcomings of agricultural markets can only be rectified by government interference. The reasoning is based on Corden's statement that 'theory does not "say" – as is often asserted by the ill-informed or the badly taught – that "free trade is best". It says that, given certain assumptions, it is "best"' (1974, p. 7). The task of this paper is to discuss whether the performance of agricultural markets can be, and has actually been, improved by government interference, in developed and developing countries, and in those in transition between socialism and markets.

CLARIFICATION OF CONCEPTS AND DEFINITIONS

Markets are institutional arrangements which facilitate the exchange of goods and services and, thus, allow for the division of labour. A free market, that is a market without government interference, is efficient if it results in a higher level of individual utility than any other means of coordinating economic decisions, such as bureaucracy (central planning), dictatorship, voting or negotiating. Economists often assume that markets are efficient if they are Pareto-efficient, that is if it is not possible to increase the welfare of one individual without impairing the welfare of others. It has to be noted that this definition of efficiency is based on value judgements as it neglects distributional aspects and relies on the individualistic paradigm. Managed markets are ones in which government interference is employed in the pursuit of certain

*Institut für Agraroekonomie der Christian-Albrechts-Universitaet, Kiel, Germany.

objectives. Managed markets are efficient if they result in a higher level of individual utility than free markets.

As with the term 'efficiency', any definition of agricultural development will necessarily be based on value judgements. It is plausible that policy makers in developed countries will have a different perception of agricultural development from those in developing countries. Both are likely to employ a broad definition of 'development' which includes more dimensions than agricultural productivity alone. The present state of government interference in agricultural markets is supposedly the consequence of a wide set of policy objectives. Hence the question is whether free markets are better than managed markets at generating allocative efficiency and satisfying other selected policy objectives.

One problem requires initial clarification. While it may be possible to model the agricultural sector and the overall economy under free market conditions it is less clear what kind of managed markets should be used for comparison. One point of reference could be markets managed in an optimal way to overcome the deficiencies of free markets. Markets managed in this way would, by definition, be more efficient than free markets. An alternative would be to take the present state of market intervention as a reference and to investigate whether free markets would be more efficient.

THE CASE OF DEVELOPED COUNTRIES

Free trade advocates generally blame agricultural policy makers for the high degree of protection in most developed countries. Some even wonder why the General Agreement on Tariffs and Trade (GATT) is needed to lower protection rates since individual countries should have an interest in reducing protection themselves. If there is any economic rationale for the present degree of protection it must lie either in failure of free agricultural markets to maximize economic efficiency or in pursuit of policy objectives which cannot be maximized by market forces alone.

Arrow and Debreu have demonstrated that market economies are only Pareto-efficient under certain circumstances, such as a complete set of markets and perfect information (Stiglitz, 1986, p. 257). As these conditions are not met in reality, there *may* exist a set of taxes, subsidies and broader government interference which *could* make everyone better off. However, there is general agreement among policy analysts that the present state of interference in developed countries is not cost-efficient and not necessarily an improvement on the situation which would otherwise prevail. Past experience will be discussed in relation to five statements.

First, free markets have most likely become more efficient in directing agricultural development than they were in the past.
— The price instability supposedly inherent in free agricultural markets has been used to justify intervention. However, prices would be less unstable now, under free trade, than they would have been in the past because the regional and intertemporal integration of markets has improved. Lower communica-

tion, transport and storage costs as well as institutions such as futures markets and insurance schemes would contribute to more stable prices.

— Market intervention is mainly pursued because agricultural incomes are supposedly too low under free market conditions. However, if labour markets are highly integrated, differentials in labour income will reflect opportunity costs. Interference on product markets, which is the predominant type of intervention, will have a greater effect on the number of people employed in the sector than on their labour income per unit of time.

Second, present policies have become increasingly inefficient over time.

— According to the previous statement, the economic environment has changed. Hence policies should also have changed to remain efficient. However, experience teaches that fundamental changes in policies are rare. Instead, policy measures are difficult to reform and the reform which does occur rarely implies changes in the set of policy instruments, but only the addition of new policy instruments (Petit, 1989).

— As the agricultural sector becomes more integrated into the general economy, old policy instruments, especially interference on product markets, become less efficient in transferring income to farmers.

Third, policy evaluation based on the pure theory of welfare economics can be misleading, because it neglects both the costs of regulation, fraud and evasion, and political economy aspects.

— Based on welfare economics, a hierarchy of policies can be set up for any given externality. For example, deficiency payment systems are supposedly superior to direct price support for correcting externalities on factor or product markets. However, experience in the EC – the tomato and olive oil scandals – illustrates that deficiency payment systems open the door to fraud. The same holds true for personal transfer payments if the group of recipients is open-ended. The attractive German social security system for farmers has enticed many non-farmers to enter agriculture on a part-time basis.

— Reasoning based on pure welfare economics ranks uniform protection rates higher than non-uniform protection rates. However, import control is much easier than export control. EC experience proves that administrative costs rise and fraud is much more prevalent if protection covers export commodities. The EC, for example, agreed to participate in the embargo of the Soviet Union in 1980. While it can be assumed that the EC administration tried to enforce this embargo, EC agricultural exports to the Soviet Union increased by roughly 500 per cent over the preceding years in 1980.

Fourth, the performance of agricultural markets and agricultural development are not mainly directed by government intervention on agricultural markets, but by macro-economic policies and the macro-economic environment.

— The importance of the exchange rate for agriculture in the United States has been analysed in detail by Schuh (1976, 1981).

— The experience gained from the liberalization of New Zealand's economy (Sandrey and Reynolds, 1990) illustrates the importance of the macroeconomic environment for agriculture.

— Experience gained in the EC supports the hypothesis that unemployment and wage rates in non-farm sectors have a greater influence on farmers' income than agricultural policies.

Fifth, economists could play a greater role in reforming policies in developed countries if they focused their research more on the effects of alternative institutions and the ex post evaluation of policies.

— There is ample evidence that the outcome of policy decisions depends on existing institutions. The EC's agricultural financial system, for example, leads to divergent national interests that make it difficult to reach decisions which are in line with overall EC welfare.

— Continuing evaluation is needed to inform policy makers *and* the public of the effects of past policies as compared to alternatives. This might encourage policy makers to act more in the public interest and less in the sole interest of the farming community. Evaluation should also measure the institutional costs of regulation, such as administrative and lobbying costs, and fraud.

THE CASE OF DEVELOPING ECONOMIES

Developing countries rely even less than developed countries on markets for directing agricultural development. Equity and food security arguments make it understandable that developing countries intervene more. However, the decline or stagnation of many developing economies has most likely been abetted by government interference. In the following discussion I draw attention first to external and then to internal trade in agricultural products.

Misleading arguments for external trade interference

Some countries intervene in external agricultural trade because they espouse a particular definition of food security, namely autarky. Such a policy *cannot* be called efficient, either in terms of pure economic efficiency or in terms of food security. In general, a country can best feed its population if it participates in the international division of labour and abstains from autarkic policies (McIntire, 1981; Valdés and Siamwalla, 1988, pp. 103).

Plausible arguments for external trade interference

Past research has illustrated the importance of macro-economic policies and especially exchange rates for agricultural development (Valdés, 1989). Free agricultural trade can hardly be recommended if exchange controls are applied and a foreign exchange allocation system is in place. Exchange controls normally imply over-valuation and depressed prices for tradables such as agricultural products. Under these conditions, agricultural markets will not be able to play their normal role in directing agricultural development.

However, even if market exchange rates do reflect the shadow prices of foreign exchange, free external trade *may not* maximize economic efficiency and contribute to food security in an efficient way. First, intervention in foreign trade may be advisable because of unstable import or export parity prices. This instability can be caused by fluctuations in world market prices for agricultural commodities and/or exchange rates, although the second cause is most likely more relevant for developing than for developed countries, as the fluctuations of the Chilean Peso against the US Dollar illustrate (Figure 1). Since developing countries have not yet established institutions such as future markets which help farmers to reduce the risk of exchange rate and commodity price fluctuations, and because farmers in developing countries are more risk-averse than those in developed countries, governments may be well advised to stabilize import and export prices within a price band, as has been done in Chile (Muchnik and Allue, 1991, pp. 67).

Second, interference in external agricultural trade may also improve internal allocation and food security in the case of land-locked countries which are characterized by a wide margin between import and export parity prices. Koester (1986) reports that export parity prices for maize at some locations in South-East Africa were negative in 1977/8, while import parity prices at most locations were twice as high as world market prices at East African ports. Some of these countries are self-sufficient in normal years but either exporters or importers in others. In such a situation, world market prices would destabilize internal markets. Moreover, since private traders and stockholders face significant risks in holding carry-over stocks, free markets will not result in a

FIGURE 1   *Monthly parity change Chilean peso against US dollar*

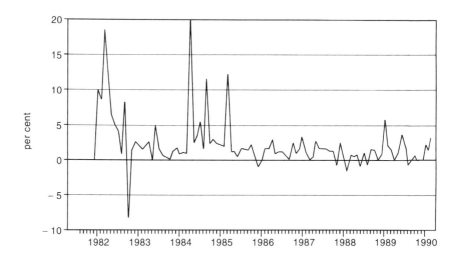

Source: IMF, International Financial Statistics, Various Issues

cost-efficient trade-off between external trade and stockpiling. Interference in the storage sector might improve welfare and food security.

Third, there is evidence that many developing countries aim at perfect self-sufficiency in food even in cases in which they could be exporters. Their export potential might not be used because export markets are not secure and planning for exports is too risky. This risk can follow from (a) variability in production which makes the potential exporter an unreliable source of supply for importers; (b) variability in production in importing countries which makes the demand for exports unreliable; and (c) variability in export prices due to unstable world market prices. If potential exporters want to mature as reliable sources of supply they have to hold large volumes of stocks that are unlikely to be held without government assistance.

Fourth, the need to produce public goods may justify other types of intervention in external trade. Taxing agricultural exports may be necessary to generate public revenue. This cause of interference is not due to market failure in agriculture, but to market failure elsewhere (Stiglitz, 1991, p. 430).

These theoretical arguments for intervention in external agricultural trade do not prove that current intervention in most countries actually increases efficiency. It may be that developing countries would be better off without trade interference because of by-product distortions and poorly designed policies. There is ample evidence that countries which have interfered less have performed better than others (Papageorgiou, Choksi and Michaely, 1990). The latter have, in many cases, merely substituted policy failure for market failure. Empirical studies in the last decade have contributed to our understanding of the political economy of state interference, lending support to the notion that it is better to accept some market failure in order to avoid the effects of policy failure (Krueger, 1990; Jones and Krueger, 1990).

There is some concern that external trade liberalization by a group of similar countries – for example neighbouring countries in Africa – could depress world market prices and make them worse off since all would seek to export the same commodities. However, empirical studies do not support this presumption (Koester, Schafer and Valdés, 1990; DeRosa and Greene, 1991).

Internal trade

Internal trade in agricultural products is heavily regulated in most developing countries. Policy makers often claim that food security is so important that it cannot be left to the vagaries of the market. There are some reasonable arguments which support the thesis of inefficient internal agricultural markets. First, functioning markets depend on institutions which define and secure property rights and which facilitate the exchange of commodities. If these institutions do not exist in the early stages of development, a rationale for state interference exists. Second, wealth and income distribution – both of which affect the outcome of the market process – are very uneven in many developing countries. Hence the market may yield results which are unacceptable for society and some interference may be justified.

However, this does not imply that all governments which rely less on markets and more on intervention are actually accelerating agricultural development. While it is difficult to identify all those policies in developing countries which actually contribute to economic efficiency, equity and/or food security, it is possible to specify which policies are most likely to be counterproductive with respect to such objectives.

Many governments, especially in Africa, mistrust private traders and support monopolistic parastatals which conduct trade in the main agricultural commodities. These parastatals are supposed to implement the governments' food policy via guaranteed prices. However, there is evidence that marketing margins in African countries are higher than in other developing countries, owing to parastatals (Ahmed, 1988). Furthermore, numerous studies prove the inefficiency of parastatal pricing policies, especially if pan-seasonal and pan-territorial prices are enforced. Under these conditions, prices on parallel markets are more unstable than under a more liberal system, and private stockpiling and transport are crowded out. The consequential reduced use of traditional storage and transport systems places an additional drain on the public sector and results in economic costs for the society.

It is probably true that market failures are more severe than elsewhere in poor and backward countries, but the same is also true of government failure (Stern, 1991, p. 429). Interference, which may be based on sound arguments, can often not be implemented adequately because of a lack of manpower and information, and the influence of interest groups. It is mainly as a result of the political economy of intervention that trade interference in domestic markets is so inefficient. The fact that many countries are reluctant to liberalize internal trade, despite evidence that current intervention conflicts with all or most of the officially stated objectives, is also due to political considerations.

TRANSFORMING A COMMAND ECONOMY INTO A MARKET ECONOMY

Most Eastern European and some developing countries are in the process of moving from a command to a market economy. It is hoped that allowing market forces to direct development will improve economic performance. However, the evidence so far is not at all overwhelming. It indicates that introducing market forces abruptly may not produce the expected results. There are grounds for believing that the state should not – at least for a transition period – withdraw from economic activities. While a government's tasks will be different during this transition period than under the command economy, it may even be that total government activity should increase. The efficiency of markets in directing agricultural development during the transition period will be discussed in relation to external and internal trade interference.

External trade

The formerly centrally planned countries have some freedom in designing their agricultural trade policies because they are not subject to most GATT regulations. Therefore, it seems worthwhile to explore whether they need specific agricultural trade policies.

It is well established in trade theory that the equalization of a nation's domestic and border prices will only maximize economic efficiency if there are no externalities in domestic production or consumption, and the nation is a price taker on world markets. External free trade will not maximize efficiency if domestic markets do not function well, as will be the case when an economy is in transition between socialism and free markets. A lack of market institutions, deficient communication systems, poorly defined property rights, an inadequate legal system and the fact that the population is not acquainted with the workings of a market economy will all contribute to high transaction costs.

A major determinant of transaction costs is the risks involved in the exchange of commodities under uncertainty with respect to future market conditions. A trader who wants to export knows neither the selling price which he will receive when he delivers his products abroad nor the future foreign exchange rate. This price uncertainty is especially high for exporters in countries which are in transition because many institutions, such as futures markets, do not exist. Inadequate communication networks and under-developed trade links enhance this uncertainty. It can be expected that these transaction costs will decline over time if trade is permitted. Since resources might be misallocated if current transaction costs are used as a point of reference, society may be well advised to intervene in foreign trade.

Trade policy could also be used to ease the speed of adjustment. It may be argued that free agricultural trade would cause 'unacceptable' social hardship for the farming population, unemployment and social unrest. This is actually a traditional argument for agricultural protection in many developed countries. However, countries in the transition phase should beware of the lessons learned in developed countries, which indicate that agricultural protectionism tends to become permanent as the farm lobby perennially advances the social hardship argument. Indeed, protectionism, by generating unfounded expectations, may be the cause of future social hardship.

A specific trade policy for agricultural products may help to exploit the trade preferences granted by some developed countries. Preferential access to developed markets may generate direct income transfers if adequate policies are instituted.

Most of these arguments for managed external trade during the transition phase are based on the hypothesis that intervention at the border is more efficient than internal policy measures. This may be the case because governments lack the administrative capacity needed to institute internal measures. However, it should be noted that while trade policy *could* contribute to the achievement of the objectives stated above, whether it does contribute depends very much on the trade situation of the economy and the instruments applied. As noted above, intervention in import trade is easier to administer than intervention in export trade. However, if a country both exports and imports agricultural products, border protection for imports alone will distort domestic price ratios. By-product distortions will be the consequence. To avoid this, trade measures will also have to include export subsidies which are costly and lead to fraud. Hence countries may be better off restricting intervention to those importables for which the costs of by-product distortions are lower than the expected benefits. It may even be best for a country to abstain from direct trade intervention altogether and concentrate instead on stabilizing import and export prices. In general, in designing agricultural trade policies, countries are well advised to build in flexibility and reversibility. Furthermore, since trade intervention is more easily justified in the transition period than it is in later periods, policies which are hard to abandon or reform should be avoided.

Internal trade

While it may be too early to draw general conclusions from the liberalization experience of the Eastern European countries, some insights can be gained already (Klaus, 1991; Koester, 1991). First, agricultural development does not result from liberalizing agricultural markets alone. More important impulses – at least during the transition period – can be generated by macro-economic stabilization policies. Of particular importance are exchange rates and monetary stability. If the exchange rate is set below the shadow rate, that is, if the currency is revalued as occurred in the case of the German Democratic Republic (GDR), sectors producing tradables, such as agriculture, will be taxed. As a result, the agricultural sector will come under stronger pressure to adjust than it would have at equilibrium exchange rates. The significance of monetary stability in the first stage of the liberalization process can be illustrated for the case of Poland. The inflation rate in Poland was roughly 1000 per cent in the latter part of 1989. Real interest rates were highly negative. Hence storage costs were much lower than expected gains from sales in future periods and there was a strong incentive to withhold supply: Poland faced a shortage of some food supplies despite having an exceptionally good harvest. This situation had completely changed by the middle of 1990. The inflation rate had dropped far more than nominal interest rates, leading to a real interest rate of some 30 to 40 per cent. This interest rate – together with uncertainty about future prices – has made stockpiling an extremely risky undertaking. Hence intertemporal price relationships are very weak, which affects agricultural prices more than others because farm production is highly seasonal. It may be more reasonable from a private point of view not to store between surplus and deficit seasons, but rather to export in the surplus season and to import in the deficit season. Clearly, this can be very costly from a macro-economic point of view. Internal prices will drop to export parity levels in the surplus season and rise to import parity levels in the deficit season. The resulting instability could affect agricultural production adversely and cause social hardship. Hence the government may decide to stabilize internal prices by interfering in the storage sector.

Second, the move to a market economy exerts pressure on collective farms to adjust. The agricultural sector can only cope with this adjustment if the necessary institutional flexibility exists. For example, if collective farm managers are not allowed to dismiss workers and if the transfer of land is inhibited, then the ability to adjust will be marginal. Hence, the liberalization of product markets must be complemented by changes in the legal and institutional framework. What is needed is not just the removal of old and discredited institutions but their replacement by new institutions which enable markets to function.

Third, liberalization is expected to result in economic benefits because individuals are allowed to respond to private incentives. However, the intensity of this response not only depends on the magnitude of the incentives, but also on human behaviour. If individuals are not used to collecting information, assessing the consequences of alternative actions and accepting risk, then private incentives will not necessarily lead to economic success. Hence the liberalization process must be supported by institutions which coordinate extension, training and the dissemination of market information.

Fourth, the farm sector in command economies was often integrated into the rural economy in a different way from that which prevails in market economies. In the case of the GDR, the entire rural economy was dominated by collective farms and most rural economic and social activities were coordinated by them. If restructuring is left to market forces, many collective farms could collapse as their most profitable branches could be sold off, leaving behind a financially unsound core. Government interference could help to avoid this outcome.

CONCLUDING REMARKS AND SUMMARY

(1) There is much support for the hypothesis that markets are not efficient in directing agricultural development. This support can be summarized as follows (Stern, 1991):

    (a) Market failure reduces the efficiency of agricultural markets. This market failure is more severe in developing countries than in developed countries and is most pronounced in economies that are in a transition phase from a command to a market economy.

    (b) Societies and governments also pursue non-economic objectives. The income objective is the most important argument for state intervention in developed countries and the food security objective often dominates debates about agricultural policy in developing countries.

    (c) Governments might be inclined in specific cases to reject the individualistic paradigm underlying a market economy. Individual preferences may be overridden by governments in developing countries that wish to improve food security, or when external effects – including the responsibility for future generations – are taken into account.

(2) However, these arguments do not necessarily imply that a society with government intervention will be better off than one without. Government or policy failure has to be balanced against the negative effects of

no intervention. There is ample evidence that policy failure is often worse than market failure (Krueger, 1990; Meier, 1990). The causes of policy failure lie not only in suboptimal instruments, timing and intensity, but also in governments that pretend to know more than they really do (Hayek, 1989). Optimistic governments fail to take into account sufficiently the indirect effects and by-product distortions of intervention and their inability to withdraw intervention and change policy tools when circumstances change.

(3)   Nevertheless, it would be an exaggeration to assert that no government intervention is justified. The final judgement depends very much on the state of the economy and on the governments' ability to implement rational policies.

REFERENCES

Ahmed, R., 1988, 'Pricing Principles and Public Intervention in Domestic Markets', in J.W. Mellor and R. Ahmed, (eds), *Agricultural Price Policy for Developing Countries*, published for the International Food Policy Research Institute, Johns Hopkins University Press, Baltimore and London.

Corden, W.M., 1974, *Trade Policy and Economic Welfare*, Clarendon Press, Oxford.

DeRosa, D. and Greene, J., 1991, 'Schaden gleichzeitige Abwertungen den Exporten aus Schwarzafrika?', *Finanzierung & Entwicklung*, March.

Hayek, F.A., 1989, 'The Pretence of Knowledge. Nobel Memorial Lecture, December 11, 1974', *The American Economic Review*, vol. LXXIX, no. 6.

Jones, R.W. and Krueger, A.O., 1990, *The Political Economy of International Trade*, Basil Blackwell, Cambridge, MA.

Klaus, V., 1991, 'A Perspective on Economic Transition in Czechoslovakia and Eastern Europe', in S. Fischer, D. de Tray and S. Shah (eds), *Proceedings of the World Bank Annual Conference on Development Economics 1990*, The World Bank, Washington, DC, p. 13.

Koester, U., 1986, *Regional Cooperation to Improve Food Security in Southern and Eastern African Countries*, Research Report 53, International Food Policy Research Institute, Washington, DC.

Koester, U., 1991, 'The Experience with Liberalization Policies – The Case of the Agricultural Sector', *European Economic Review*, vol. XXXV.

Koester, U., Schafer, H. and Valdés, A., 1990, *Demand-Side Constraints and Structural Adjustment in Sub-Saharan African Countries*, International Food Policy Research Institute, Washington, DC.

Krueger, A.O., 1990, *Perspectives on Trade and Development*, Harvester Wheatsheaf, Hemel Hempstead.

McIntire, J., 1981, *Food Security in the Sahel: Variable Import Levy, Grain Reserves, and Foreign Exchange Assistance*, Research Report 26, International Food Policy Research Institute, Washington, DC.

Meier, G.M., 1990, 'Trade Policy, Development, and the New Political Economy', in R.W. Jones and A.O. Krueger (eds), *The Political Economy of International Trade*, Basil Blackwell, Cambridge, MA. p. 179.

Muchnik E. and Allue, M., 1991, 'The Chilean experience with agricultural price bands: the case of wheat', *Food Policy*, vol. XVI, no. 1.

Papageorgiou, D., Choksi, A.M. and Michaely, M., 1990, *Liberalizing Foreign Trade in Developing Countries – The Lessons of Experience*, The World Bank, Washington, DC.

Petit, M., 1989, *Pressures on Europe's Common Agricultural Policy*, International Food Policy Research Institute, Washington, DC.

Sandrey, R. and Reynolds, R., 1990, *Farming without Subsidies, New Zealand's Recent Experience*, Wright and Carman Ltd., Upper Hutt, New Zealand.

Schuh, E., 1976, 'The New Macroeconomics of Agriculture', *American Journal of Agricultural Economics*, vol. 58, no. 5, p. 802.

Schuh, E., 1981, 'Floating Exchange Rates, International Interdependence and Agricultural Policy', in G. Johnson and A. Maunder (eds), *Rural Change – The Challenge for Agricultural Economists*, International Association of Agricultural Economics, Institute of Agricultural Economics, Oxford.

Stern, N., 1991, 'Round Table Discussion', in S. Fischer, D. de Tray and S. Shah (eds), *Proceedings of the World Bank Annual Conference on Development Economics 1990*, The World Bank, Washington, DC, p. 429.

Stiglitz, J., 1986, 'The New Development Economics', *World Development*, vol. XIV, no. 2.

Stiglitz, J., 1991, 'Round Table Discussion', in S. Fischer, D. de Tray and S. Shah (eds), *Proceedings of the World Bank Annual Conference on Development Economics 1990*, The World Bank. Washington, DC, p. 430.

Valdés, A., 1989, 'Impact of Trade and Macroeconomic Policies on Economic Growth', in J.W. Longworth (ed.), *China's Rural Development Miracle, with International Comparisons*, reprint No. 175, International Food Policy Research Institute, Washington, DC.

Valdés A. and Siamwalla, A., 1988, 'Foreign Trade Regime, Exchange Rate Policy, and the Structure of Incentive', in J.W. Mellor and R. Ahmed (eds), *Agricultural Price Policy for Developing Countries*, Johns Hopkins University Press, Baltimore and London, p. 103.

DISCUSSION OPENING – WALTER J. ARMBRUSTER*

Professor Koester is to be commended for his insightful examination of the efficiency of markets in directing agricultural development across the array of industrialized, developing and formerly centrally planned countries. His purpose was to ask whether market performance is improved by government interference in the different stages of development.

Koester defines a free market as being efficient 'if it results in a higher level of individual utility than any other means of coordinating economic decisions...'. He argues that 'managed markets are efficient if they result in a higher level of individual utility than free markets', and then poses the question as to 'whether free markets are better than managed markets at generating allocative efficiency and satisfying other selected policy objectives'.

Developed Countries

As Koester points out, most developed countries practise intervention to correct perceived failures of free market economies to maximize economic efficiency or achieve other policy objectives. One frequent rationale for intervention is to provide adequate information to help markets function better. Environmental goals also frequently trigger intervention because of the difficulties of internalizing the costs involved. Typically, of course, attempts to ensure higher, more stable, or adequate income is a major rationale for intervention. It has become increasingly clear, however, that performance of agricultural markets and agricultural development are affected more by macro-economic policies and influences than by agriculture-specific policies.

*Farm Foundation, Oak Brook, Illinois, USA.

Koester provides supporting evidence for his thesis that current levels of government intervention in developed countries are neither cost-efficient nor always better than what would exist in their absence. He then argues that economists could play a greater role in reforming developed country policies by focusing research more on the effects of alternative institutions and evaluation of policies. They could then inform policy makers and the public of the effects of past policies, leading to future decisions more in the public than in the farmer-specific interest.

His argument is an interesting hypothesis, but one that I think has received ample testing in developed countries. Certainly, within the United States there is a large body of literature analysing policy effects and comparing them with specified alternatives. Evidence suggests that economic analysis may play some role in policy decisions but probably not an overriding one. It takes more than economic analysis to change policies made through the political process. The political arena has many actors with strong self–interest. Agricultural and food system participants have proved skilled at lobbying for and achieving their particular policy goals, even though they may appear narrow, in spite of analyses of public interest impacts. Surely the debate leading up to the Food Security Act of 1985 in the USA ranks it as one of the most widely analysed and discussed agricultural policy development processes ever. Yet many question the effectiveness of all that input in shaping the final content of legislation.

I pose two questions for agricultural economists. Has an adequate amount of sound economic analysis been conducted on policy alternatives in most countries? Is it possible to increase the influence of economic analysis to achieve better subsequent policy and, if so, how?

Developing countries

Koester believes that developing countries interfere with the market even more than developed countries in directing agricultural development. He identifies plausible arguments for external trade interference which are supported by studies cited. However, he argues that 'these theoretical arguments for intervention in external agricultural trade do not prove that current intervention in most countries actually increases efficiency'. He concludes that it is better to accept some market failure in order to avoid the effects of policy failure.

While this is an interesting conclusion, its application leaves me somewhat perplexed. How much market failure is acceptable to avoid the effects of policy failure? Can economists adequately measure various degrees of market failure and determine the trade-offs with possible policy failure? Or would it be better to adopt policies to deal with perceived market failures but concentrate on designing them so that they correct the market failure and phase out when that is achieved?

Turning to internal agricultural trade, Koester argues that the market may yield results unacceptable to society, since wealth and income distribution, which integrally influence market outcomes, are so uneven in many developing countries. Particularly in early stages of development, some interference may be justified to facilitate exchange of commodities.

As Koester sees it, while interference may be justified by sound intellectual arguments, implementation is often difficult. Besides lack of manpower and information, the influence of interest groups also affects implementation and leads to inefficiency best described as the much dreaded policy failure. He then concludes that many countries hesitate to liberalize internal trade even in the face of evidence that policies fail to meet official objectives because political considerations override.

I do not find that conclusion very startling, since any intervention in a market economy is driven by the political process which determines its type and level. Once created there is a tendency for 'property rights' to be appropriated by participants in the marketing system. This increases the difficulty of reversing policies initiated by groups with enough political strength to obtain the intervention. Perhaps the best hope when creating interventions in developing countries is to have a well-defined phase-out as part of the initial political agreement built into the implementation process.

Although economists could hypothesize a number of limitations on intervention, the real question is whether we can design efficient, politically feasible policies that help rather than hinder progress in developing economies. Can we design them with built-in decision points for their removal when they have accomplished their purpose but not yet wreaked havoc on the economy? Would more research on policy implementation alternatives be helpful in achieving policy objectives?

Transition economies

Koester then turned to the case of transforming command economies to market economies, which is surely one of the most interesting current topics. He argues the need for government involvement in the transition period during which markets are not likely to operate well because of 'a lack of market institutions, deficient communication systems, poorly defined property rights, an inadequate legal system and the fact that the population is not acquainted with the workings of a market economy'. All of these features will contribute to high transaction costs.

Koester suggests that transition countries learn from developed countries 'that agricultural protectionism tends to become permanent as the farm lobby perennially advances the social hardship argument. Indeed, protectionism, by generating unfounded expectations, may be the cause of future social hardship.' He suggests that, in designing agricultural policies, transition countries need to incorporate flexibility and reversibility. Indeed, designing new institutions to allow markets to function efficiently requires attention, not only to physical infrastructure and the legal framework, but also to the extension, education and information-providing institutions. The creation of such institutions is necessary for a successful economy, even though the costs involved might appear to diminish the perceived rate of transition.

I raise the same questions as for developing economies: Can we design and implement efficient, politically feasible policies? Can we incorporate decision points? Do we need to research policy implementation alternatives?

LAWRENCE D. SMITH AND ANNE M. THOMSON

Achieving a Reasonable Balance Between
the Public and Private Sectors in Agriculture

INTRODUCTION

Over the last decade the debates on liberalization and privatization arising from the whole process of structural adjustment in the Third World and the political transformation of Eastern Europe, and the discussions surrounding the Uruguay Round of GATT negotiations, have each called for a re-evaluation of the relative roles of the public and private sector in agriculture.

Much of this debate has been couched in dogmatic terms, based on a belief that, in many countries, the economic situation could only improve as a result of government disengagement from economic activity. Where the support of economic theory is sought, the paradigm is, explicitly or implicitly, that of neo-classical micro-economics with its emphasis on increasing competitiveness and hence efficiency. This view is reinforced by stressing government failure which highlights the problems the state faces with incentives, regulations, and absence of competitive forces leading to inefficiency, lack of dynamism and rent-seeking activities.

In the process of advancing their cause, the proponents of a reduced role for the government have in general neglected the conventional welfare economics analysis of the role of government which justifies intervention to correct for market failures arising from the existence of public goods, externalities, economies of scale and natural monopolies. Moreover, there is little reference to those ideas associated with traditional development economics, such as the role of the state in extracting a surplus from agriculture to support the development of other sectors, in developing institutions which promote growth and in achieving a socially and politically acceptable income distribution. As the neo-classical paradigm has no easy way of incorporating these concerns, they are being ignored in the current debate on the relative role of the public and private sector in agriculture.

AN INSTITUTIONAL APPROACH

In this paper, we suggest that an approach drawing on the 'new institutional economics' goes some way to incorporating the concerns of both these para-

*University of Glasgow, Scotland, UK.

digms. This requires us to examine afresh the objectives of, and constraints facing, different types of economic agents, and to consider what constitutes and determines the set of regulations and conventions, or institutional framework, within which they operate and which governs their behaviour. This framework embraces factors ranging from the nature and definition of property rights to the accessibility and scope of the legal system for enforcing contracts. An appropriate institutional framework provides an environment which enables economic agents, both individually and collectively, to widen their range of options and creates an incentive structure which encourages efficient choices. We examine some of the characteristics of individuals, groups and the state in the next sections, as a prelude to considering the relative roles of the public and private sectors in agriculture.

PROBLEMS FACING INDIVIDUAL ECONOMIC AGENTS

One of the most basic lessons of economics is that the division of labour, specialization and exchange are essential for increasing productivity, encouraging economic growth and raising living standards. This requires cooperation between human individuals acting as economic agents but, as Binswanger and Rosenzweig (1985) remind us, this is conditioned by certain specific characteristics of human behaviour, of the economic environment and of the agricultural sector.

The human characteristics highlighted are self-interest, the value of consumption and dislike of effort, risk aversion and bounded rationality in decision making, given limited ability to collect, process and analyse information. Because acquiring information is costly, an optimizing person will stop when marginal costs and benefits are equated, resulting in imperfect and asymmetrically distributed information. This, in turn, means that the outcome of any economic activity cannot be predicted with certainty. Costly information is also a factor which impedes the universal definition of property rights and their proper enforcement. This generates a range of principal – agent problems, issues of incentives, the need for screening and monitoring, and increased transactions costs which eventually may make potential exchange uneconomic, resulting in incomplete markets. Externalities, including public or collective goods, arise from the inability to define and enforce property rights in a relatively costless manner. Indivisibilities, leading to economies of scale and the possibility of monopoly-type situations, are another characteristic of the economic environment.

Agricultural activity has its own peculiar features, which are most marked in Third World countries. One is that in many countries the sector is dominated by individuals engaged in relatively small-scale production and marketing activities. The spatial dispersion of agriculture means that transport, travel and information costs are high and time-intensive. The seasonality of production creates problems for the smooth functioning of many output and input markets. Moreover, agricultural production is notoriously susceptible to a variety of risks which can have profound effects at both micro and macro levels, especially in low-income, agriculturally dominated economies (Hazell, 1988).

Rather than being a rarity, we believe there are elements of 'collectiveness' in many agricultural sector activities which make effective individual provision difficult or uneconomic. It is hard to make agricultural techniques non-excludable, particularly those which are not capital-embodied, given the highly visible nature of most agricultural activities. Spatial dispersion and small size call for shared transport networks and marketing facilities for produce collection and input distribution. There are also numerous examples of externalities where individual action may not be adequate, such as pest and disease prevention and control, food safety provisions and the maintenance of minimum quality standards. Nevertheless, individuals are sometimes willing to provide goods with collective properties, for example, where private benefits exceed total cost or where the nature of the good, or the institutional framework, can be modified to allow more of the benefits to be captured by the provider. Even so, provision will be Pareto-inefficient as long as the free-rider problem persists. Indivisibilities and increasing returns are frequently significant in organized agricultural research and in the provision of irrigation and water control facilities. These may lead to local natural monopolies if not to national ones.

It is paradoxical to find agricultural markets used so frequently as a textbook example of perfect competition. In reality, the spatial dispersion and small scale of most producers and consumers, and the extent of output and price fluctuations, mean that asymmetric information is a considerable problem. Whereas small farmers and traders encounter high opportunity costs in obtaining information (other than that relating to their own operations), the benefits are circumscribed by their small or infrequent transactions. Large producers and traders generate substantially more information or can obtain it at lower unit costs and have more opportunities to benefit from it. Institutions can evolve to cope with some information problems, for example the practice of holding periodic markets at a customary time and place, or the use of commission agents to reduce search and transactions costs. Many more remain intractable at the individual level.

Small farmers, especially in less developed countries, face particular problems from risk and incomplete markets. Faced with income instability arising from production fluctuations, macro-economic causes, political disturbances or international price instability, they are rarely in a position to use capital or credit markets to smooth out their income flows, especially when poorly specified property rights in assets such as land prevent their use as collateral. Nor are well-functioning national insurance markets available, owing to severe moral hazard and adverse selection problems resulting from imperfect information. Local insurance might overcome the information problem but is limited if weather and disease factors cause a high covariance of risks across farms. Even if futures markets were available, high transactions costs would probably preclude their use by small farmers. Low-income consumers can be profoundly affected by their inability to protect themselves from real income variations arising from food price fluctuations associated with agricultural instability. There are also considerable risks involved in marketing, both from the physical deterioration or destruction of produce and from unpredicted changes in prices during the marketing process. There are numerous examples of actions taken by individuals on their own account, or through non-market institutional frameworks such as the extended family or the firm, to mitigate or minimize these various risk problems.

In low-income countries, the commercialization of agriculture is hindered by inadequate information and incomplete capital and credit markets. Combining this with weak infrastructure and transport links leads to market segmentation and the possibility of local oligopoly or monopoly power. Individuals may face problems financing production or marketing activities, particularly given their seasonal nature. Instability may reduce private investment in the agricultural sector. Thus commercialization and the achievement of an efficient, responsive agricultural sector requires the development of an appropriate institutional framework which will encourage collective response both at the group and state level to complement individual action.

VOLUNTARY GROUP ACTIVITIES

As suggested above, groups of individuals acting cooperatively can sometimes improve information provision, reduce or share risks, and provide not only private goods but collective goods as well. Olson's (1965) seminal work on pressure-group activity can be extended to a wide range of agricultural activities. Small groups have several organizational advantages over larger groups. Typically, they will be self-selected. This guards against adverse selection and makes it easier to achieve homogeneity of objectives and to sustain mutual trust without expensive monitoring to prevent shirking. There is little need for formal organization, and internal costs can be kept low. Nevertheless, the incentive to form a group may be hampered by the free-rider problem. This can be overcome by inducing, or coercing, potential members to join the group. In close-knit societies moral suasion and threats can succeed, but these techniques may fail as the group becomes increasingly heterogeneous.

In terms of inducement, Olson emphasizes the 'by-product' approach, where public-good provision is linked to that of a desirable private good that is best provided collectively, such as insurance, or goods where there are marked economies of size. Armah's (1989) description of wholesalers' market associations in Ghana exemplifies this approach. One major function is the arbitration of disputes among members. This is a partially non-rival collective good which should reduce the need for costly and lengthy litigation. The second function, representing members in dealing with local councils and government, that is lobbying, is Olson's classic example of group provision of a collective good. Acting as guarantors of credit is the third function. The association is well placed to provide this private good, insurance, as it can screen potential members to avoid adverse selection and advise on retailers with known bad-debt problems. The organization can also act as an information exchange on market conditions between members, which should improve the market's effectiveness unless wholesalers' associations increase information asymmetry, which raises barriers to entry.

For many types of collective goods there are potential advantages, such as economies of scale or positive externalities, in expanding the group size. But as size increases, both free-rider and leadership problems become more acute. Potential leaders are typically motivated by the direct or indirect benefits arising from this role. This leads to a major paradox in the cooperative movement as traders and politicians, rather than small farmers themselves, frequently instigate and lead these types of activities because the indirect benefits may embrace information about members or political allegiance. Whether these are the 'best' leaders to advance the interests of small farmers is debatable, especially where asymmetric information means that the general membership have little idea of the way in which their funds and resources are being used.

THE ROLE OF THE STATE

Where voluntary group provision of collective goods is circumscribed by leadership and free-rider problems, government intervention or provision may be the only alternative. The provision of collective goods is one of the major roles of the state, made possible by its power of legitimate coercion, which in turn is based on its comparative advantage in violence (North, 1981). To understand its other major economic role, it has to be appreciated that the state itself is composed of individuals and groups of individuals, acting separately or in coalitions to achieve their own self-interest, but constrained by the need to remain in power. To this end, resources are expropriated from the population at large, or sections of it, and redistributed to those in power, or their supporters.

The state's self-interest may be constrained by a constitution which determines the bounds of its legitimate actions, by international obligations and treaties, or by its vulnerability to being replaced. Within these constraints, however, it can formalize the institutional framework within the legal system, particularly the assignment of property rights, in its own short-term interests, and/or in such a way as to reduce the risk and costs of exchange, thus increasing economic wealth and its longer-term opportunities. This may be at as simple a level as setting a uniform system of weights and measures to be used in markets, or as complex as assigning plant breeders' intellectual property rights.

The perceived legitimacy of the state is an important factor in determining the costs of enforcing this legal structure. For example, if urban food security is a keystone of government legitimacy, its failure to provide this may lead to civil disorder, and the state will become increasingly involved in high-cost policing to maintain its position. A state with a long time horizon will be cautious as to how far self-interest can dominate its function in securing collective advantage.

The state has a number of distinctive advantages over private agents. It has the ability to change institutions by law, rather than by developing them on the basis of voluntary consent, although it may jeopardize its popularity or legitimacy in the process. Thus it can specify land tenure arrangements, or stipulate price bands within which commodities must be purchased or sold. It can raise funds through compulsory taxation to pay for the goods and services it provides, such as roads, or agricultural research institutes, thus removing the financial aspects of the free-rider problem at a stroke (though still leaving the decoupling of those who benefit and those who pay the cost). It has potentially a much greater spread of risk, both spatially and over a range of goods and services, than any private trader and can, in addition, change institutions in such a way as to create and transfer risk. Nevertheless, the state may still be risk-averse in many respects, being willing, for example, to pay substantial insurance premiums to achieve food security or income stability for its potential supporters.

It has, however, a significant number of disadvantages as well. The state typically comprises politicians, elected or otherwise, a bureaucracy of civil servants and officials, and a wide range of agencies and organizations to implement its policies and programmes. These often develop their own institutions and objectives, which may lead to significant principal – agent problems. It may be difficult for government to set clear objectives, implement an appropriate incentive structure, or monitor performance, each of which may result in organizational inefficiency.

The state also faces some of the problems which confront private agents, particularly in terms of imperfect information. However, it has different options as to how to deal with these. Unless it is an active participant in markets it will have to collect information explicitly and at a cost. The usual pattern of centralized information collation, followed by instructions being sent back out to the regions, is particularly problematic in agriculture, where climatic and production conditions can vary significantly over space and over time. At the same time, as many centrally planned economies have found to their cost, the state requires more information than any individual private agent, in order to regulate agricultural markets and provide the most appropriate level of public goods and services.

ASSESSING THE RELATIVE ROLE
OF THE PUBLIC AND PRIVATE SECTORS

If we are to assess the appropriate balance between the private and public sector, and ultimately to make informed normative statements to guide policy making on this topic, we have to develop a methodology for comparing different institutional frameworks. As yet there is little theoretical work to guide this search. One way forward is to attempt to identify and measure all of the likely types and levels of costs associated with different institutional configurations. One possible four-fold categorization is transformation costs, transaction costs, risk costs and the costs of rent-seeking activities (Thomson and Smith, 1991).

Even if precise measurement is difficult, it should be possible to make qualitative estimates of the impact of major institutional changes. Providing explicit attention is paid to issues of distribution and dynamic efficiency, this approach offers several advantages over a narrow concentration on allocative and technical efficiency alone. It moves the focus away from an abstract view of markets, towards an approach solidly based on identifying what actually occurs in specific markets, on trying to understand the functions performed by different agents and on identifying the implications, in both cost and distributional terms, of changing the institutional configuration. As more applied work is undertaken which focuses on institutions, a pragmatic approach to policy issues, based on past experience, can be developed. *Ex ante* prediction of the effects of institutional change is likely to remain difficult and imprecise, but appropriate *ex post* analysis can only improve the process.

In using this approach, it is useful to appreciate that the provision of any good or service can be categorized into four functions: providing the funding, providing the good or service itself, regulating its provision, and consuming it (Ross, 1988). Each of these functions may be provided by the state or the private sector or a mixture of the two. Even where individuals cannot easily undertake some functions, for example regulation or provision of public goods, voluntary group activity, such as producer associations, may well be able to do so.

The appropriate mix of public and private activities is also likely to vary at different stages of economic development, as a more complete set of integrated markets evolves. Some of the relevant factors are the level of existing commercialization of the agricultural sector; the degree of market integration; the level of urbanization; the level of agricultural technology and sensitivity to climatic variation; the importance of food in the average budget; the importance of agriculture in the macro-economy; the pattern of income distribution; the structure of agriculture; the overall size of the public sector and the state's revenue-raising capacity; the availability of skilled manpower and of administrative capacity; the internal cohesion of the state; the legitimacy and contestability of the state; the nature of the demand for collective goods, which will vary with income levels; and the objectives of the state.

We can use Ross's classification, in conjunction with a cost categorization, to examine past experience of what might constitute a reasonable balance between public and private sector in agriculture, taking for illustrative purposes production and output marketing activities.

Production activities

The evidence on institutions and agricultural production seems to indicate a clear advantage, in terms of transformation costs, for private-sector production over the public sector. Most countries which have tried to institute state production have eventually fallen foul of the problem of appropriate incentive structures. This can happen with large-scale private farming as well, but institutions have developed to counter this, both at the managerial and at the workforce levels. There seems little reason for the state to fund private agricultural production directly, given the possibility of regulating prices, subsidies, or deliveries to the state. There may, however, be a case for providing improved access to finance through state-backed credit systems and/or relaxing restrictions on the private financial sector in the hope of reducing transactions and risk costs. State regulation also occurs in other forms, two obvious examples being restrictions on maximum farm size for income distribution reasons, or controls on crop choice or technology to counteract negative, or engender positive, externalities.

Output marketing activities

Given its pivotal role both in coordinating production and consumption decisions and in determining the size and distribution of the agricultural surplus, the marketing system is the area where state activity is potentially most influential. We have already indicated that exchange is vital to increasing productivity and that, even in the absence of state intervention, groups will develop institutional frameworks which reduce the transactions costs of engaging in trade. Adam Smith certainly realized that orderly markets can only exist with general acceptance of a series of rules and regulations, often legitimized by the state. Where economic development is hindered by missing and incomplete markets, this can be traced back to the absence of an appropriate institutional framework which it may be impossible to provide in the short term. In the interim, direct government action may be required. For example, where financial markets are poorly developed the government may have to provide much of the finance to mobilize the marketable agricultural surplus.

Many elements of the physical infrastructure of marketing, such as roads and market-places, have substantial public-good features which may justify government funding and/or provision. A similar argument is used to justify government funding and/or provision of information, grading schemes and so on.

Manipulating prices is a major instrument in distributing the agricultural surplus and hence in income distribution. However, governments lacking the administrative machinery to monitor the implementation of these pricing decisions may have no option but to enter the marketing system themselves. The experience of state participation in marketing is mixed. As with agricultural production activities, it is often difficult to develop a suitable incentive structure for state or parastatal marketing organizations. There are two general reasons for this. First, some commodities require rapid and risky pricing, buying and selling, or arbitrage decisions, and public-sector agencies can rarely accommodate these. Second, the use of the marketing system to achieve income distribution goals sometimes leads to a financially non-viable framework which undermines motivation. On the other hand, the provision of a secure and stable food supply in situations where international trade is difficult may require inter-year storage which may be too risky for the private sector to provide or finance and, in these circumstances, public-sector funding or provision may be essential.

ACHIEVING AN APPROPRIATE BALANCE

These two examples illustrate in very broad terms the issues involved. The specific costs will vary according to the institutional setting which itself develops and changes for three major reasons: in response to changes in economic opportunities (changing relative prices and technological change); changes in the distribution of economic and political power; and changing public concerns (changes in the social welfare function). Change, or the possibility of change, may have negative effects, for example in engendering rent-seeking behaviour, shrinking productive output and reducing economic opportunity. On the positive side, it has the potential to increase productive efficiency through decreasing transactions costs. It can increase the range of types of economically viable transactions by improving information flows, reducing risk and/or reducing transactions costs. Related to this, and central to our present concern, is that it can alter the absolute or comparative advantage of different types of economic agents, such as the central government and the private sector, in performing various economic activities.

One of the paradoxes about the potential role of the state in the agriculture sector is that it has most scope for encouraging commercialization, providing information and reducing risk in low-income predominantly rural economies, where it is most likely to be weak itself, and have limited technical capacity to carry out these tasks. Conversely, the state can mobilize most funds and resources to assist the agricultural sector in situations where output and resource markets already work effectively. It is also clear that the returns to changing the relative balance between the public and private sector depend on the existing institutional framework. Where there is already an effective and active state presence, whether in marketing or research and development, the costs of building on this may be relatively low. Where state activity has been ineffective and regulation has been ignored, then the costs of developing an effective state presence will be much greater. This seems to us an area where vicious and virtuous circles abound.

CONCLUSIONS

Self-interest, including self-preservation, is the major motivating force of all economic agents, whether individuals, voluntary groups or governments. At the same time, bounded rationality, coupled with characteristics of the economic environment and agricultural production, ensures that all economic agents operate in an uncertain world with imperfect information, which imposes considerable costs on society. Economic organizations are means of reducing these costs and raising productivity, but inevitably, self-interest means that they attempt to maximize their own gains in the process.

Deciding what constitutes a reasonable balance between the public and private sectors in agriculture will always be difficult. Every institutional configuration has different distributional consequences and implications for dynamic efficiency which are difficult to predict. However, it may be possible to develop a cost-based methodology for comparing institutional frameworks which will assist this debate. What is apparent, even in the absence of empirical verification, is that there are a host of problems such as incomplete markets, asymmetric information and the 'collectiveness' of many agricultural activities, which indicate a potentially wide role for governments in the funding, provision and regulation of agricultural sector activities.

In virtually all countries which have developed a successful and responsive agricultural sector the public sector has played a significant role in encouraging commercialization, in reducing risk and in developing markets and new technologies. Undoubtedly, many other countries, both capitalist and socialist, have pursued non-viable agricultural policies, requiring more resources and technical capacity than were at the state's command. Merely reducing the level of state activity is a rather negative approach. Of equal, or more importance, is the need to identify areas where the state can withdraw immediately because it is duplicating, or impeding, private-sector operations, those where sequential withdrawal may be possible, and those where increased state intervention or support is required to develop an institutional framework encouraging the sustained effective operations of the agricultural sector.

REFERENCES

Armah, P.W., 1989, 'Post Harvest Maize Marketing Efficiency: The Ghanaian Experience', PhD thesis, University of Aberystwyth.
Binswanger, H.P. and Rosenzweig, M.R., 1985, 'Behavioural and Material Determinants of Production Relations in Agriculture', Agricultural Research Unit, Discussion Paper ARU 5, World Bank.
Hazell, P.B.R., 1988, 'Risk and Uncertainty in Domestic Production and Prices', in J.W. Mellor and R. Ahmed (eds), *Agricultural Price Policy for Developing Countries,* Johns Hopkins University Press, Baltimore.
North, D., 1981, *Structure and Change in Economic History,* W.W. Norton, New York.
Olson, M., 1965, *The Logic of Collective Action,* Harvard University Press, Boston.
Ross, R.L., 1988, *Government and the Private Sector: Who Should Do What?*, Crane Russak, New York.
Thomson, A. and Smith, L.D., 1991, 'Liberalisation of Agricultural Markets: an Institutional Approach', Centre for Development Studies Occasional Paper No. 8, University of Glasgow.

DISCUSSION OPENING – JEAN MARC BOUSSARD*

It is difficult to criticize a paper which strongly emphasizes the orthodox theoretical view that state intervention is required whenever markets fail to play the role they are supposed to play (informing producers about consumer wants and enlightening consumers as to the scarcity facing producers). I find myself in total agreement with the authors and congratulate them for reminding us of the issues at this time when the formerly socialist countries, in particular, are under-estimating the shortcomings of markets and exaggerating their virtues.

However, a few comments are needed to support and supplement the excellent paper. The authors contend that costly mistakes have been made in many centrally planned economies since the state requires more information than any other agent in order to regulate markets. Actually, in such countries, many responsible people know exactly what to do to operate the economy, either on

*Institut National de la Recherche Agronomique, Ivry, France.

a Pareto basis, or at least to improve significantly its efficiency. The mistakes which have been made have often been the outcome of the decision process, which surprises everybody and is often contrary to common sense. The results have not been due to lack of information but to the way in which it has been processed.

Information-processing methods are the key characteristics of a system. Therefore, before issuing recommendations with respect to any economic system, one should examine carefully what information is processed and how it is done. The free market is one extreme method of transmitting information between producers and consumers with prices, or their equivalents, being the principal vehicles of information. Frank Hahn periodically reminds us that such an extreme cannot possibly be met in any real-world situation. Central planning is another extreme method with many defects. There are intermediate situations. None of them is intrinsically superior to another and any diagnosis must rely on careful analysis of what is occurring. This is perhaps the main defect of the paper: it does not emphasize the 'case by case' aspect of the difficulty of assessing the role of the state in agriculture.

Assume a circular farm with uniform potentiality and uniformly spread out labourers. When supervision is from the centre it can be shown that the time spent by the supervisor in travelling between the centre and labourers visited at random, and with an equal probability of selection, will grow as the power 3/2 of the surface of the circle and, therefore, of total production. This very simple example shows how management of the economy from the centre is irrational. It also explains how in market economies very large production units, whatever their institutional environment, must fail. In addition, it helps us understand such institutions as sharecropping contracts, where the beneficiary (the share-cropper) monitors himself or herself. This kind of model is very useful, since it shows us that not only state intervention, but all sorts of institution building, must be considered in the light of economic analysis of the sector.

Sometimes free market advocates may be concerned that food producers and consumers are generally not very price responsive. Thus prices give deceptive signals concerning scarcities and wants. These prices reflect short-run situations, which are different from long-run equilibria. Thus free agricultural markets are best represented by 'deterministic chaos' models with detrimental welfare implications. In such cases there is justification for intervention. In developed countries, the state then sets prices, but often does so without regard to the level of demand. Intervention in which the government buys any quantity offered at the fixed announced prices is not sustainable in the long run. If agricultural production functions are really homogeneous of the first degree it can be shown that, in the long run, production may expand indefinitely, since the marginal cost will be constant and equal to the price for any quantity. This is broadly what has happened in the European Community and the United States for grain, or in the Ivory Coast for cocoa.

It is therefore necessary to make use of all available knowledge in analytical economics for forecasting the possible implications of recommendations relating to the organization of the agricultural sector. Otherwise there will be a risk of creating problems which are more important, and more difficult to solve, than those which would otherwise exist.

RICHARD B. NORGAARD*

Sustainability: The Paradigmatic Challenge to Agricultural Economists

INTRODUCTION

Western and westernlzed societies – whether capitalist or socialist, democratic or authoritarian – increasingly sanctioned technocrats during the nineteenth and early twentieth centuries to combine shared values, beliefs and knowledge, and act on behalf of the public. This authorization of agricultural scientists, engineers, foresters and planners was rooted in a common vision of progress and a common faith in the way Western science and technology could accelerate development (Comte, 1848). During the twentieth century the sanction was increasingly extended to economists (Pechman, 1989) and after the Second World War carried them, naively to be sure, to the head of the global pursuit for economic progress (Lasch, 1991; Nelson, 1991). How history unfolded was a product of a myriad of different factors in different places, but economists assumed the burden of trying to guide, explain and rationalize development. In the process, economics acquired a conventional wisdom.

The international discourse on sustainable development challenges the shared assumptions, understandings and rationalizations accumulated among economists during the second half of this century. Neither defensive arguments nor modest accommodations serve the profession well. The discourse is incorporating and going beyond the technological optimism and technocratic progressivism of economists and other key historic players while incorporating and transforming the environmental pessimism, preservationist inclinations, cultural survivalism and participatory approaches of new players (Colby, 1990). To re-establish a constructive role in development discourse, economists will need to assume a philosophical base adequate to encompass the broad and changing patterns of thinking and new linkages to action. Consider the following complementary directions in which economics might move.

THE RIGHTS OF FUTURE GENERATIONS

First, the international discourse on the sustainability of development is primarily concerned with (a) the rights of future generations to the services of natural and human-produced assets and (b) whether existing formal and informal institutions which affect the transfer of assets to future generations are

*Energy and Resources Programme, University of California at Berkeley, USA.

adequate to ensure the quality of life in the long run. This neo-classical framing contests implicit premises of economics as now practised. A less than generous interpretation is that existing reasoning and practice tacitly assume that current generations hold all rights to assets and *should* efficiently exploit them. A more generous interpretation might be that existing reasoning and practice assume that the mechanisms affecting the maintenance and transfer of assets to future generations are both working optimally and unaffected by current economic decisions. If this were the case, intergenerational equity need not be considered. This more generous interpretation, however, is not supported by current theoretical elaborations, conceptual discussions in the academic literature, or the reasoning employed in justifying practice. On the contrary, questions of intergenerational equity have frequently been obliquely pursued as problems of market failures. While environmental externalities are surely important, solving them could either improve or worsen distributional inequities.

A simple overlapping generations model demonstrates that efficient levels of consumption and investment and their associated prices, including the rate of interest, are a function of the way income from rights to natural and other assets is distributed across generations (Howarth and Norgaard, 1990; Norgaard and Howarth, 1991). This 'finding' is conceptually elementary yet at odds with existing theoretical elaborations and understandings of agricultural, development, environmental, forestry and resource economics. The relationship

FIGURE 1   *Allocative efficiency*

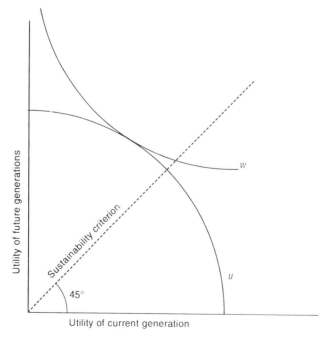

Utility of current generation

between the intertemporal allocative efficiency and the intergenerational distribution of resource and environmental rights is illustrated in Figure 1 (Bator, 1957). Every point on the utility frontier results from an efficient allocation of resources associated with different distributions of resource rights between generations. Where a society is located on the utility frontier is determined by the initial distribution of rights to productive assets, including natural assets. While efficiency is important in that it puts society on the utility frontier, the sustainability of development depends on whether a society is above the 45° line. Even inefficient solutions can be sustainable. In the absence of a social welfare function to determine the optimum point, the tenor of the development discourse indicates that sustainability is a minimum criterion on which there is broad consensus.

Since Harold Hotelling's formulation of 1931, economists have pondered how stock, exhaustible or depletable resources should be used over time. With the energy crisis of 1973/4, economists renewed their attention to the intertemporal allocation of exhaustible resources. Students of resource economics are now well aware of the 'Hotelling Rule' that the rent from a stock resource being exploited 'optimally' increases at the rate of interest. While the literature repeatedly refers to *optimal* paths of extraction, the paths explored to date have been merely the *efficient* path associated with the existing intergenerational distribution of rights to resources. If rights are redistributed across generations, the efficient path changes (Howarth and Norgaard, 1990). Similarly, the literature in environmental economics identifies the conditions under which it is 'socially' efficient to exploit a species to extinction. If, however, society *a priori* decides that future generations have a right to roughly the current diversity of species, the efficient allocation would only rarely lead to extinction.

Methods for valuing non-market environmental services are quite well developed and, when applied, frequently show that non-market goods and services have considerable value. Including non-market environmental values in benefit–cost analyses can shift the efficient path of resource use towards the future. But as a general means for ensuring resources for future generations, expanding economic analysis to incorporate how *this* generation values non-market goods and services will not necessarily result in their being saved for *future* generations. Ultimately, we are concerned with maintaining natural assets for future generations because we sense that *they* will need these assets, not because *we* somehow value them. The rights of future generations can be thought of as politically determined constraints to economic optimization and as such should not be economically valued.

No doubt there exists an economist who has never experienced the slightest moral qualm over discounting the benefits to be received and the costs to be borne by future generations. The academic literature and discussions within development agencies, however, reflect considerable unease (Markandya and Pearce, 1988). With lower discount rates, it appears that more investments in renewable resources and a larger stock of renewables would be justified. Similarly, it appears on preliminary analysis that lower rates of discount favour using stock resources more slowly. Thus many see a strong link between the rate of discount, the conservation of resources and hence the sustainability of development. But by reframing questions of the future in terms of the intergenerational distribution of rights to natural and other assets, the case for using lower discount rates to protect future generations becomes moot. When the rights to assets are redistributed between generations, the investment opportunities for and savings of current generations, and hence the interest rate, change accordingly. The interest rate may increase or decrease, but this is unimportant, for it is simply an equilibrating price. What is important is the particular ways in which consumption, savings, and investment change in order to ensure that real assets back the rights of future generations.

Several conclusions from this reframing are important. First, with redistribution there is a change in the types of investments that the current generation undertakes to meet its own commodity time preference. For these investments, the values placed on future consumption are appropriately discounted. Second, in order to meet the rights of coming generations to real assets, the current generation might invest. The returns over time from such investments facilitate attaining the objective of transferring assets to match the rights of future generations. The fact that investments can yield a return is important and helps determine the most cost-effective way of meeting the objective. But the benefits that accrue to future generations from investments undertaken to ensure their rights cannot be measured by current preferences, nor should they be discounted. For exactly the same reasons, when a development decision imposes on the rights of future generations, the costs of ensuring those rights by other means must be deducted from the benefits of the project (Mikesell, 1991). The optimal portfolio of investments (and disinvestments) to meet the rights of future generations is determined according to the cost-effectiveness of alternative combinations of ways of sustaining their rights over time. This framing resolves some of the key, long-standing ethical dilemmas of usury (Norgaard and Howarth, 1991).

Efficiency with 'trickle ahead' is no more suitable as an operating norm for development thinking than is efficiency with 'trickle-down'. Incorporating questions of intergenerational equity into the neo-classical framework opens economics up to the future and resolves many of the conflicts between earlier economic reasoning and environmental reasoning. This opening, however, also reorients economics and politics. Historically, economists have assumed the mandate of informing the political process as to which choices are efficient and of helping implement legislation efficiently. This perspective portrays politics as a black box that cannot choose without the help of economists and cannot implement without the help of economists, while never indicating what politics does do. Several economists are arguing that questions of sustainability lie beyond economics. Pearce (Markandya and Pearce, 1988; Pearce and Turner, 1990) and Daly (Daly and Cobb, 1989) argue that environmental constraints on economic optimization are necessary to achieve sustainability. Such formulations, however, beg the question of how the constraints are determined and chosen. I argue that economists need explicitly to recognize that sustainability is an equity question being debated in various moral discourses utilizing ecological reasoning and that sustainability will be chosen through politics. Economists in this framing can inform the political process of the impacts of different equity decisions and the most cost-effective ways of reaching them. Economics can interact with moral discourse, environmental lines of reasoning and the political process, but cannot 'rationalize' them. This brings us to the second point.

CONCEPTUAL PLURALISM

Second, the methodological premises on which economists base their role in society have become dysfunctional. Economics needs methodological foundations which complement the way science works in society. History has not unfolded in accordance with Auguste Comte's positivistic, deterministic, methodological vision – the Newtonian view that underlay technocratic progressivism. The methodological questions raised and answers suggested by Karl Popper in the 1920s are awash in a sea of other issues which philosophers no longer pretend they can part and walk through to the safety of a promised land (Rorty, 1979).

Two things have changed. First, we now know that different patterns of thinking are inherently different. Neo-classical market economics is an atomistic–mechanistic model which views systems as consisting of parts and relations between the parts which do not change. Evolutionary thinking looks at systems as undergoing changes in their parts and relations. One cannot have it both ways. More importantly, economic thinking inherently gives value to *things*, the more the better, while evolutionary thinking inherently values the *diversity* that sustains change. The discord between the two secular religions– environmentalism and economism–are rooted in just such irreconcilable differences.

Environmental systems can be thought of as consisting of biological systems of interacting living organisms and physical systems made up of hydrological cycles, climate systems and so on. The two broad types of systems, of course, also interact. Biological environmental scientists have multiple ways of simplifying the complexity of living systems into formal, tractable models – population dynamics, energetics, food webs, biogeochemical cycles, species coevolution, communities, hierarchy theory, succession, light patch dynamics, and many others. Similarly, physical environmental scientists have alternative ways of looking at different types of interrelated phenomena. While individual natural scientists specialize in one or two approaches for their own research and no doubt wish that other patterns of thinking would merge with their own, they are conceptual pluralists in practice, eclectically drawing on a variety of patterns of thinking to understand natural phenomena. Natural scientists, however, still suffer from conceptual monism themselves because of past philosophical traditions (Norgaard, 1989).

The understanding of complex environmental problems on which social decisions are based requires some sort of a resolution of these logically conflicting ways of understanding the parts. The resolution can be thought of as taking place through a social process of discourses within academe, between scientists in academe, public agencies, and non-governmental organizations, and among scientists and the public at large. In many cases, little consensus is reached on the key characteristics of the system that has gone astray. Rather, agreement is frequently reached on the existence and general nature of problems and on the boundaries of solution sets. Economists could contribute much more effectively to the social process of reaching environmental understanding if they were not conceptual monists, if they knew that the environmental scientists who reach conclusions contrary to economic reasoning were probably arguing logically along a pattern of thinking that does not merge with economic reasoning.

More broadly, if economists hold to the methodological belief that knowledge is accumulating to one congruent understanding, they will continue to miss the insights provided by incongruent ways of knowing. Multiple insights guard against mistaken action based on one perspective on the complexities of the world around us. This does not mean 'anything goes'. We must still insist that arguments be logical; we simply cannot insist that different logical arguments merge within the guise of a higher, more inclusive logic. The collective understanding necessary for collective action is reached through discourse, finding common ground, agreeing on critical issues, and compromise. It is a social process thoroughly intertwined with what Western culture has thought could be understood separately as politics. Economists would be more effective participants in the social process of understanding and formulating solutions if they received specific training in alternative patterns of thinking, in how they are used in the other sciences, and in how they inherently favour different values. If economists hold to the belief that knowledge consists of universal laws with universal applicability and the public keeps economists in positions of authority, we will apply our knowledge accordingly and destroy the diversity in the cultural and ecological systems we are trying to sustain (Norgaard, 1990).

DEVELOPMENT AS SOCIAL SYSTEM
AND ECOLOGICAL SYSTEM COEVOLUTION

Third, economists need a distinct alternative theoretical framework to acquire insights to which the neo-classical paradigm blinds us. With such a model, economists could deliberately maintain conceptual pluralism within the discipline. It would be advantageous if patterns of thinking were endogenous to the model. If both the way we think and our ability to transform our environment unsustainably makes us unique among species, perhaps there is a correlation.

Consider thinking of development as a process of coevolution between knowledge, values, organization, technology and the environment. Each sub-system is related to each of the others, yet each is also changing and effecting change in others. Deliberate innovations, chance discoveries and random changes occur in each sub-system, which affects the distribution and qualities of components in each of the other sub-systems Whether new components prove fit depends on the characteristics of each of the sub-systems at the time. With each sub-system putting selective pressure on each of the others, they coevolve in a manner whereby each reflects the other. Thus everything is coupled, yet everything is changing (Norgaard, 1988).

Though neo-classical models do not link technology to social organization, development economists are well aware of the connections from their own field experiences and those of others. A coevolutionary framework explains this linkage directly. A coevolutionary framework also helps emphasize that knowledge systems evolve with organizational systems, that much of what we know and even how we collectively know are a function, for example, of the way we organize society into numerous fiefdoms of resource management agencies. It suggests that the evolution of economic systems has been affected by the way we have thought about economic systems. A coevolutionary framework stresses that environmental systems have evolved along with people, including with the way people know, what they value, how they are organized, and the tools they have available to them (Figure 2).

FIGURE 2   *A coevolutionary framework*

VALUES

KNOWLEDGE   TECHNOLOGY

ENVIRONMENT   ORGANIZATION

The coevolutionary framework provides its own insights into the nature of sustainability. While most societies coevolved with their ecosystems, modern societies are coevolving around the combustion of fossil hydrocarbons. This has driven a wedge between the coevolution of social and ecological systems (Norgaard, 1984). With modernization, capturing the energy of the sun through ecosystem management became increasingly unimportant as new technologies became ever more effective at tapping the energy of coal and petroleum and using it in ever more novel ways. Social systems coevolved around the expanding means of exploiting hydrocarbons and only later adopted institutions to correct the damage this coevolution entailed for ecosystems. Hydrocarbons freed societies from immediate environmental constraints, but not from the ultimate constraints on the atmosphere and oceans to absorb greenhouse gases or on the biosphere to withstand toxins. But having coevolved to a dead end, we are stuck with ways of knowing, organizing, valuing and doing things, with tightly intertwined roots of unsustainability. It is this dilemma which drives us to look to traditional peoples for 'new' insights.

The coevolutionary framework contrasts sharply with the positivist, atomistic–mechanistic frame of neo-classical economics. A coevolutionary view assumes the nature of the parts in systems and relations between them change over time, whereas atomistic–mechanistic models assume that parts and relations stay the same, though their number and relative strengths can vary. In the coevolutionary view, how we know affects the types of social organization, technology and values which prove fit. Everything is explained by everything else since each affected the evolution of the other. At the same time, in a coevolutionary model, change is typically taking place. The coevolutionary model is useless for the sorts of predictions which Newtonian thinking does so precisely, but it helps explain why mechanistic predictions do not come true. Indeed, it helps explain why what seemed to have been key variables around which predictions were thought to be needed frequently turn out to be irrelevant. And yet the coevolutionary view does have design value. It highlights the way in which the evolutionary process will have more potential, and probably continue long into the future, the greater the diversity. Whereas the neo-classical framework almost inherently assumes substitutability and favours efficiency, the coevolutionary framework assumes interconnectedness and inherently favours diversity. The neo-classical and coevolutionary frameworks highlight each other's nature, strengths, weaknesses and appropriateness of use.

CONCLUSIONS

Railroad engineers plotted the paths of progress in North America during much of the nineteenth century, while their European counterparts connected the frontiers of Africa, Asia and Latin America to the industrial world. Their field knowledge and vision of the future made them the experts who testified in legislatures and schemed in bureaucratic agencies. Their base of understanding, however, did not evolve, as the future they envisioned and helped implement unfolded in unexpected ways. History continues to unfold in ever more surprising ways. The international discourse on sustainable development challenges agricultural economists to adopt a broader, more robust, paradigmatic base. In this paper I have argued for recasting the way in which economics relates to the moral discourse on and politics of intergenerational equity, adopting conceptual pluralism, and acquiring some facility with alternative patterns of thinking. This would allow the profession to adapt to the surprises it helped create and prevent it from joining the railroad engineers.

REFERENCES

Bator, Francis, 1957, 'The Simple Analytics of Welfare Maximization', *American Economic Review*, vol. 47, p. 22–59.
Colby, Michael, 1990, *Environmental Management in Development: The Evolution of Paradigms*, World Bank Discussion Paper 80, World Bank, Washington, DC.
Comte, Auguste, 1848, *A General View of Positivism*, translated into English by J.H. Bridges in 1865, reprinted 1971, Brown Reprints, Dubuque, Iowa.
Daly, Herman E. and Cobb, John B., 1989, *For the Common Good: Redirecting the Economy Toward Community, the Environment, and a Sustainable Future*, Beacon Press, Boston.
Howarth, Richard B. and Norgaard, Richard B., 1990, 'Intergenerational Resource Rights, Efficiency, and Social Optimality', *Land Economics*, vol. 66,(1), p. 1–11.
Lasch, Christopher, 1991, *The True and Only Heaven: Progress and Its Critics*, W.W. Norton, New York.
Markandya, Anil and Pearce, David W., 1988, 'Environmental Considerations and the Choice

of Discount Rate in Developing Countries', Environmental Department Working Paper, World Bank, Washington, DC.

Mikesell, Raymond F., 1991, 'The Multilateral Development Banks and Sustainable Development', paper prepared for presentation at the Western Economics Association Meetings, 30 June 1991 in Seattle, Washington.

Nelson, Robert H., 1991, *Reaching for Heaven on Earth: The Theological Meaning of Economics*, Rowman and Littlefield, Savage, Maryland.

Norgaard, Richard B., 1984, 'Coevolutionary Development Potential', *Land Economics*, vol. 60,(2), pp. 160–73.

Norgaard, Richard B., 1988, 'Sustainable Development: A Co-evolutionary View', *Futures*, December, p. 606–20.

Norgaard, Richard B., 1989, 'The Case for Methodological Pluralism', *Ecological Economics*, vol. 1, p. 37–57.

Norgaard, Richard B., 1990, 'Environmental Science as a Social Process', *Environmental Monitoring and Assessment*, forthcoming.

Norgaard, Richard B. and Howarth, Richard B., 1991, 'Sustainability and Discounting the Future', in Robert Costanza (ed.), *Ecological Economics: The Science and Management of Sustainability*, Columbia University Press, New York.

Pearce, David W. and Turner, R. Kerry, 1990, *Economics of Natural Resources and the Environment*, Johns Hopkins University Press, Baltimore.

Pechman, Joseph A. (ed.), 1989, *The Role of Economists in Government: An International Perspective*, New York University Press, New York.

Rorty, Richard, 1979, *Philosophy and the Mirror of Nature*, Princeton University Press, Princeton.

DISCUSSION OPENING–JAMES P. HOUCK*

My comments on Richard Norgaard's paper are divided into three parts: general remarks, my simplified interpretation of his key ideas, and some specific points of critique.

General remarks

In addition to the paper prepared for this conference, I have also examined a more detailed paper on this general topic written by Norgaard and Howarth for the World Bank. This has given me a distinct advantage in a consideration of the conference paper.

Norgaard's present paper is not a self-contained argument or a sustained line of reasoning that leads him to the conclusions and results he advances. These are contained in earlier more extended and technical pieces. Norgaard has thought long and deeply about the difficult concepts he lays before us. His ideas about sustainability and intergenerational relations are powerful and challenging. They take us into areas where economic measuring is difficult and controversial.

Norgaard urges us to lay aside neo-classical and partial equilibrium analysis in order to accommodate more normative modes of thinking. The broader system which he proposes would tie together economics, ecology and other disciplines previously linked only tangentially.

*University of Minnesota, USA.

Key ideas

In my view the core of Norgaard's theoretical base can be summarized rather succinctly. Using the standard and well-accepted assumptions about economic agents, resources, technology, markets and prices, the following three points can be advanced:

(1) Strictly efficient, dynamic paths of equilibrium solutions can lead to 'non-sustainable' results. This is true whether key resources are renewable or non-renewable.
(2) One reason why non-sustainable results can emerge is that negative externalities are created for which no good markets or appropriate public controls exist as dynamic development proceeds.
(3) Even if all the externality problems raised are overcome, non-sustainable results may occur because of the initial allocation of incomes, resources, property rights and intergenerational transfers of assets. Thus the institutional set-up and the institutional environment can create non-sustainable results in an otherwise perfect market, efficient, system.

Clearly, Norgaard is concerned with the third item in the list and the silence of standard economic theory and most economists on such issues.

Some points of critique

In my opinion, economists reading this paper will quickly want to know how some terms are to be defined: 'sustainability', 'rights', 'coevolutionary theories', 'diversity' and others. It is always tedious to define terms, but without that being done, sensible debate is stifled and confused.

Neo-classical and even partial equilibrium models are not themselves to blame for environmental myopia. Surely it is economists themselves who are at fault. I have no doubt at all that it will be traditional models that give us the toe-hold to reach for successful new ways to think about sustainability. We *evolve* from where we are to what we will become.

Richard Norgaard has given us much to consider about these new and perplexing problems. We need to take his advice and move into a serious discussion of the way we assert and account for the welfare of those who will follow us on this planet. Future generations have never done anything for us, nor can they. We cannot negotiate with them except through the intermediary of moral principle.

K.N. NINAN* AND H. CHANDRASHEKAR**

The Green Revolution,
Dryland Agriculture and Sustainability : Insights from India

INTRODUCTION

The green revolution of the 1960s, which enabled LDCs like India to over-come chronic food deficits and large food imports through a significant rise in domestic foodgrain output, continues to evoke considerable interest among researchers and policy makers regarding the nature of its impact, particularly its implications for growth, equity and sustainability. These concerns have brought to the fore a number of issues.

Firstly, why is it that, despite the impressive growth achieved by Indian agriculture following the green revolution, instability too has shown a tendency to rise, as indicated by a number of studies (Mehra, 1981; Hazell, 1982; Nadkarni and Deshpande, 1982, 1983; Rao *et al.*, 1988)? Though a similar phenomenon had been observed in the context of traditional agriculture in India (Sen, 1967), instability has worsened in the latter period. These findings raised doubts as to whether greater instability was an inevitable price to be paid for attaining higher rates of agricultural growth. Though interventionist policies, such as buffer stock operations and foreign trade, can help to cushion the effects of fluctuations in domestic output, they may be costly relative to measures to achieve agricultural growth with stability. The importance of achieving sustained agricultural growth cannot be disputed since it affects the interests of producers, consumers and the economy as a whole, and more so since the poverty ratio has been observed to move in line with the vagaries of agriculture (Ahluwalia, 1978). There has, therefore, been considerable interest in understanding the nature of the association between growth and instability and causes underlying them.

Secondly, though there is widespread acknowledgement regarding the green revolution's role in boosting foodgrain output in many LDCs, including India (at least in its initial phase), doubts have been growing regarding its cost-effectiveness and sustainability, as suggested by some studies from India (Rao, 1983; Nadkarni, 1988), which note a steady and more than proportion-ate rise in the cost of cultivation, following the green revolution. If this is so, it suggests that the strategy of growth based on the green revolution technology is not sustainable.

*Institute for Social and Economic Change, Bangalore, India.
**Dryland Development Board, Government of Karnataka, Bangalore, India.

Thirdly, the intensification and chemicalization of agriculture associated with the green revolution has also brought in its train serious ecological problems such as waterlogging, salinity and alkalinity in the irrigated regions and soil erosion, declining water-table and desertification in the dry or semi-arid regions. These focus attention on the environmental costs of various growth strategies.

Lastly, despite the scale neutrality of the new agricultural technology, institutional and economic constraints coupled with the policy bias favouring well-endowed regions, crops and farms have not only resulted in a widening of interpersonal and regional inequities but also led to sectoral imbalances within the agricultural sector, a notable example being the failure of the green revolution to benefit pulses and oilseeds in India, unlike its positive impact on wheat and rice. That growth without equity is not sustainable needs no elaboration.

Keeping these concerns in view and given the constraints of a conference, this paper addresses itself to the following issues. Apart from a fresh look at the growth/instability question with particular reference to dryland agriculture in India, it also attempts to verify the general validity of the 'increasing costs of cultivation' argument using disaggregated data. It then assesses the impact of watershed development programmes on the agriculture and economy of dry regions in India.

Equity and food security considerations apart, the prohibitive costs of future irrigation investments and remote possibilities of irrigation benefiting around 40 per cent of India's arable lands, even if potentials are fully realized, explain this focus on dryland agriculture, in addition to the green revolution. The analysis is both at national (All-India) level and for Karnataka state in South India, which reports about two-thirds of its area as drought-prone with low and uncertain rainfall. The data are from the official reports of the goverments of India and Karnataka and cover the period 1955–6 to 1988–9, unless otherwise indicated.

GROWTH AND INSTABILITY

For analysing the growth and instability question only major crops and crop-groups have been considered. For measuring growth, linear trends with slopes expressed as percentages at respective means have been computed; for instability the coefficient of variation around the trend (CVT) or mean (CV – in cases where the trends are not statistically significant or negative) have been used. These figures are presented for the pre- and post-green revolution phases of Indian agriculture. Though production variability is the outcome of area and yield variability, and their interaction effects, our interest is primarily in yield instability. This is because, while adjustments in area reflect conscious decision making by farmers (hence area is subject to endogenous factors within a farmer's control), yield variations also arise on account of exogenous factors like weather which are beyond a farmer's control. Yield rather than area variability has also been the major source of production variability in India. Moreover, land being a limiting factor in India, as in most Asian countries, efforts for sustained agricultural growth have to concentrate on accelerating and stabilizing yield rates.

In line with most Asian country experiences, foodgrain production in India – dominated by rice and wheat – has kept pace with population increases; but marginally below domestic demand. Annual growth rates in foodgrain output have been 2.6 per cent between 1955–6 to 1988–9, as against 2.2 and about 3 per cent for population and domestic demand respectively. While area and yield were the major sources of output growth in the years prior to the green revolution, it has been largely dominated by yield thereafter. A review of crop performance during the pre- and post-green revolution periods (periods I and II) reveals that at the All-India level all crops except coarse cereals and pulses, registered a significant growth in output in period II (Table 1). Wheat, rice and sugar-cane, which had the benefit of irrigation and modern inputs, fared better than other crops. For most crops, yield rather than area has been the major source of output growth in this period. However, despite a significant improvement in yields, coarse cereals lost area to wheat and rice, resulting in their slow growth. Near stagnation in area and yields contributed to slow growth of pulses too. Other crops, groundnut and cotton, registered lower growth rates in period II compared to period I.

Unlike the situation at the All-India level, where the dry crops had to bear the brunt of the backlash effects of the green revolution, Karnataka's experience stands in sharp contrast where they have shared in the gains of agricultural growth. Here coarse cereals and pulses registered significant increases in output in period II. In fact, among coarse cereals, maize and finger millet fared exceedingly well with annual growth rates in output ranging between 5 and 6 per cent (Ninan, 1991). Other dry crops, groundnut, and cotton have fared better in this period. What is significant is that in most of these cases output growth was realized through significant improvements in yield rates rather than through extension of cultivated area. Obviously, the efforts of the local agricultural scientists and extension workers to evolve and popularize location-specific dry farm technologies has had a better pay-off in Karnataka.

A perplexing feature, however, is that, along with growth, instability in yields has increased for most crops except wheat and sugar-cane, where it has reduced. The irrigated crops, wheat, rice and sugar-cane, have shown lower levels of instability compared to other crops. Rice, despite being a largely irrigated crop, reported an increase in yield instability, indicating the diverse environments and constraints under which it is grown in India, as in the rest of Asia. Access to yield-enhancing technologies has encouraged rice to move to marginal areas also. For most crops except sugar-cane (at the All-India level) fluctuations in yields have been of a higher amplitude than in area (Ninan, 1991). Despite the impressive performance of dry crops in Karnataka, instability in output and yields too have risen sharply in period II. For instance, the CV of finger millet output rose from 6 to 53 per cent over the two periods, minor millets from 7 to 37 per cent, and pearl millet from 22 to 27 per cent (Ninan, 1991).

This phenomenon of instability moving in sympathy with agricultural growth is reported from other regions as well, including South and South-east Asia (Barker *et al.*, 1981), though there are exceptions also (for example, for Bang-

TABLE 1   *Growth and instability in Indian agriculture (for All India and Karnataka State, 1955–6 to 1988–9)*

Crops/crop groups	Percentage of crop's area		Growth rates (per cent per annum)				Instability (CVT/CV, per cent)			
	to gross sown area (1986–7)	under irrigation	Pre-GR		Post-GR		Pre-GR		Post-GR	
			P	Y	P	Y	P	Y	P	Y
ALL INDIA										
Rice	23	43	4.0*	2.4*	2.6*	2.0*	5.0	5.1	8.5	7.0
Wheat	13	77	3.3*	2.3**	4.9*	3.1*	8.2	7.2	7.1	5.8
Coarse cereals	23	9	2.6*	1.6*	0.5	1.4*	3.1	3.5	9.8†	7.3
Pulses	13	10	0.4	−0.01	0.6	0.3	9.1†	8.3†	10.5†	8.5†
Foodgrains	72	33	3.0*	2.2*	2.5*	2.3*	4.4	3.8	6.7	5.4
Groundnut	4	15	3.4*	neg.	1.3**	1.2**	5.6	5.5†	8.6	12.1
Cotton	4	31	3.2*	3.00**	1.8*	2.1*	9.9	8.8	11.3	8.9
Sugar-cane	2	82	7.0*	4.1*	2.6*	1.4*	9.4	5.2	8.3	4.4
KARNATAKA STATE										
Rice	10	61	4.4*	1.4**	0.8**	0.9*	5.4	4.7	11.2	9.3
Wheat	2	28	7.0*	5.2*	0.1	1.2***	17.4	9.3	27.3†	19.2
Coarse cereals	40	8	3.1*	2.4	2.4*	2.5*	5.1	17.2†	18.7	18.7
Pulses	14	4	−0.05	0.9	1.0***	0.2	4.9†	4.9†	15.4	13.1†
Foodgrains	66	18	3.3*	2.4*	1.3	1.4*	4.2	4.2	11.6	10.9
Groundnut	9	20	−2.4	−1.7	1.4***	1.0***	16.2	13.2	20.8	14.6
Cotton	4	18	−0.04	4.8*	1.9**	4.9*	12.3†	11.5	23.1	22.7
Sugar-cane	2	99	9.3*	3.9*	3.5*	−0.4	5.5	6.1	11.0	8.7†

Notes: P-Production; Y-Yield per ha.
Pre-GR: Pre-green revolution period, 1955–6 to 1964–5.
Post-GR: Post-green revolution period, 1967–8 to 1988–9.
Growth rates are linear trends with slope expressed as per cent at respective means.
*, **, ***=significant at 1, 5 and 10 per cent levels of significance; remaining growth rates are not significant at the above levels of significance.
CVT/CV are coefficient of variation around trend and mean, respectively, expressed in percentage terms; † = CVS, the rest are CVTs.

Sources: *Agricultural Statistics at a glance*, Directorate of Economics and Statistics, Ministry of Agriculture, Govt. of India, February, 1989; and publications of the Directorate of Economics and Statistics, Government of Karnataka.

ladesh's case, see Alauddin and Tisdell, 1988). Further, though the level of relative instability in India is low by international standards, absolute variability for large low-income countries like India is high (Valdés and Konandreas, 1981); moreover with large on-farm retentions of crop produce and highly skewed income distribution patterns in LDCs, as in India, even small variations in relative instability have important economic and welfare implications (Hazell, 1982). The factors contributing to yield instability are briefly as follows:

(1)   Climatic factors like rainfall are an important factor behind yield variability. A recent study suggests that agricultural output has become more sensitive to rainfall in the post-green revolution period, with the elasticity of output with respect to rainfall increasing for most crops except wheat and some rabi (winter) crops in this period (Rao *et al.*, 1988). This is because of the strong complementary relationship between use of modern inputs and availability of moisture either through rainfall or irrigation, and also the fact that growth has taken place under diverse environments and constraints. Other empirical evidence (cf. Nadkarni and Deshpande, 1982), however, revealed that rainfall could not wholly

account for the increased variability, which arises on account of other factors too.

(2)   The extension of cultivation to marginal lands or riskier regions has contributed to higher yield instability (Sen, 1967). While this seems appropriate and valid for that particular phase of agricultural growth dominated by extensive methods, the same cannot be said about the subsequent phase (in the post-green revolution period) when intensive methods have become prominent. Particularly in the ecologically fragile areas such as semi-arid/dry regions the use of yield-enhancing technologies may have accentuated instability because of the strong complementary relationship between these modern inputs and availability of moisture, a critical factor in these areas. Despite coarse cereal area falling and yields rising, instability has increased sharply, which supports the above observations.

(3)   Factors associated with the new agricultural technology, particularly widespread use of HYVs, chemical fertilizers and greater susceptibility of new crop varieties to pests and diseases, have accentuated instability (Mehra, 1981; Hazell, 1982). For instance, the shift from the traditional crop varieties with a diversified genetic base to HYVs which have a narrow, common genetic base has contributed to greater instability. So, also, the substitution of chemical fertilizers for organic manures has had an adverse effect on soil health, aggravating instability. The yield instability-accentuating properties of HYVs and chemical fertilizers are, however, neutralized in the presence of (assured) irrigation which has a stabilizing effect on yields (Mehra, 1981; Rao et al., 1988).

(4)   The new agricultural technology, along with a changed price and policy environment, has altered the production opportunities facing farmers, which has effected changes in the traditional crop patterns and systems. The shift from risk-spreading and risk-adjusting crop systems and practices like multicropping and crop diversificaion to monocropping and crop specialization has tended to increase risks in farming. New profit-making opportunities have encouraged farmers to neglect some ecologically (apart from nutritionally) beneficial crops. For instance, pulses which had to bear the brunt of the backlash effects of the green revolution are good nitrogen fixers and help in enhancing soil fertility. Further, the mutual compensating mechanism of crops/regions (which helped in reducing overall instability) whereby a bad crop in one region can be offset by a good crop in another, has been affected owing to shifts in the cropping pattern and systems; yield correlations across crops and regions have increased following the green revolution (Hazel, 1982; Nadkarni and Deshpande, 1982, 1983; Anderson et al., 1987).

(5)   The transition from subsistence to commercialized agriculture involving greater dependence of farmers on off-farm or market inputs, market-oriented production and so on, while bringing about closer integration of farmers with the larger economy, has also made them more vulnerable to market uncertainty and market distortions (Nadkarni and Deshpande, 1983; Rao et al., 1988).

(6) Differences in infrastructure such as irrigation, institutions to provide inputs, credit, extension and other support services, have also influenced yield variability (Nadkarni and Deshpande, 1983; Rao *et al.*, 1988).

(7) Environmental degradation (shrinking common property resources, deforestation) induced by demographic and economic pressures have affected women's time allocation patterns, with more time being required for fuel and forage gathering. This has affected household and farm productivity, more so since they provide specialized skills and labour for agriculture.

(8) Access to free or subsidized inputs may also have played a role in accentuating instability. Since farmers get these inputs at prices or terms which do not reflect their real costs, they are tempted to use them intensively beyond economic or efficient levels, with adverse user costs and negative externalities. Also the advent of modern inputs has lulled farmers into complacency and a neglect of traditional environmentally sound crop practices and resource maintenance techniques evolved over centuries; modern inputs are looked upon as mere substitutes for traditional inputs and practices rather than as complementary.

COST ECONOMICS

The transition from a low- to high-cost economy in the post-green revolution period is another conspicuous attribute of Indian agriculture (Rao, 1983; Nadkarni, 1988). These observations were, however, based on highly aggregative data at country or sectoral (agricultural) level. Using disaggregated data, an attempt is made to see how far this is valid as to crop and region. Input–output ratios (that is, value of inputs as a ratio of the value of output) and the share of paid out costs in the total unit cost of production have been computed for two points of time. Two series of input–output ratios have been computed for each crop, one using paid out costs only and the other considering total costs inclusive of the imputed value of all owned inputs, including imputed wages for family labour. These data are available for the period 1970–1/1971–2 to 1982–3, covering the post-green revolution period, and hence give an opportunity to see how far the green revolution is to be blamed for the rise in the cost of cultivation. Triennial averages have been used to arrive at these ratios.

Evidence in Table 2 suggests that for both irrigated and dry crops the green revolution belt, as well as other areas, is afflicted by the malady of increasing costs of cultivation. The input–output (IO) ratios show positive signs in most cases. These increases are modest for some crops or regions, conspicuous for others. Wheat, and rice in particular – which benefited the most from the green revolution – have reported substantial increases in these ratios in a number of states. However, within the green revolution belt, while these ratios for wheat registered a conspicuous increase in Haryana, in Punjab the rise was only marginal. For rice, not only the traditional rice-growing areas (such as West Bengal) but also new areas (such as Haryana) have witnessed sizable increases. The dry crops, sorghum and maize, also reported a sharp upward swing in these ratios.

The burden of rising production costs in Indian agriculture is borne out by the fact that the share of paid out costs in the total unit cost of production has risen for most crops, in all regions, for the period under review. The increases are particularly sharp for some of the coarse cereals, sorghum, pearl millet and maize. Rice and wheat, too, reported large increases in some regions. This rise is largely on account of the growing importance of market or factory-produced inputs such as HYVs, fertilizers, pesticides and modern farm machinery in the input profile of Indian farming after the green revolution.

This emerging feature of Indian agriculture, whereby more inputs are required to produce a given level of output, indicates that it is becoming less efficient and more expensive. In fact, in some instances, inputs inclusive of all paid out costs and the imputed value of owned inputs exceed the returns (for example, with rice in Haryana, sorghum in Andhra Pradesh). Studies by the IFPRI, at Washington, confirm that aggregate input productivities of Indian agriculture are falling. The narrowing gap between (the value of) inputs and outputs also implies that smaller surpluses are available for future investment in agriculture. In fact, studies suggest that real public and private investment in agriculture for India has declined in recent years (Shetty, 1990); and so also have agricultural growth rates (Nadkarni, 1988). These are disturbing trends indeed, which could impede sustained agricultural growth in India.

The results presented above suggest that (a) Indian agriculture has reached that phase where diminishing returns are in operation; (b) it is increasingly relying on scarce, costly, external rather than local resources; (c) input prices are rising faster than farm product prices, which supports the views of the farmers' lobby that domestic policies are framed to deliberately depress farm product prices; and (d) terms of trade have turned adverse for agriculture. These tendencies are also visible in varying measure in other regions of Asia and the less developed world.

IMPACT OF WATERSHED DEVELOPMENT PROGRAMMES

The biases and weaknesses associated with the green revolution prompted efforts for suitable alternatives that combine environmental concerns with development needs. To illustrate this, the case of watershed development programmes, which are being implemented on an experimental basis in the dry belt of India, is cited here. As against the green revolution strategy which focused exclusively on croplands and ignored environmental costs, these programmes are holistic in nature, covering both arable and non-arable land development. These programmes seek to promote soil and moisture conservation as well as to enhance the productive capacity of drylands.

Karnataka has been in the forefront in experimenting with such programmes since 1984. Though a long-term impact assessment of these programmes may have to wait, available evidence (Table 3) suggests that they have had a favourable impact in terms of growing additional crops, enhancing crop yields and net returns, and generating employment. These programmes thus offer a

TABLE 2   *Cost economics of Indian agriculture (for the period between 1970–1/1971–2 to 1972–3 and 1979–80/1980–1 to 1982–3)*

Crops and states		Input–output ratios			Share of paid-out costs in total unit cost of production (%)		
		Base period	Terminal period	Per cent increase/ decrease	Base period	Terminal period	Per cent increase/ decrease
WHEAT							
Haryana	(a)	0.37	0.52	41	46	60	30
	(b)	0.67	0.88	31			
Punjab	(a)	0.47	0.52	11	51	62	22
	(b)	0.84	0.82	−2			
Uttar Pradesh	(a)	0.42	0.46	10	34	50	47
	(b)	0.75	0.74	−1			
Rajasthan	(a)	0.40	0.55	38	41	61	49
	(b)	0.71	0.88	24			
RICE							
Andhra Pradesh	(a)	0.47	0.57	21	52	60	15
	(b)	0.80	0.91	14			
West Bengal	(a)	0.32	0.53	66	30	53	77
	(b)	0.70	0.94	34			
Orissa	(a)	0.38	0.47	24	41	54	32
	(b)	0.74	0.83	12			
Haryana	(a)	0.67	0.85	27	68	63	−7
	(b)	0.74	1.01	37			
Karnataka	(a)	0.25	0.38	52	38	51	34
	(b)	0.49	0.65	33			
SORGHUM							
Maharashtra	(a)	0.35	0.55	57	20	56	180
	(b)	0.74	0.97	31			
Karnataka	(a)	0.28	0.34	21	20	39	95
	(b)	0.60	0.67	12			
Andhra Pradesh	(a)	0.44	0.71	61	36	46	28
	(b)	0.85	1.20	41			
PEARL MILLET							
Gujarat	(a)	0.42	0.47	12	25	60	140
	(b)	0.77	0.79	3			
Rajasthan	(a)	0.33	0.28	−15	20	38	90
	(b)	0.85	0.73	−14			
Haryana	(a)	0.47	0.40	−15	19	17	−11
	(b)	1.15	0.93	−19			
MAIZE							
Rajasthan	(a)	0.27	0.36	33	13	42	223
	(b)	0.67	0.86	28			
Himachal Pradesh	(a)	0.35	0.56	60	35	43	23
	(b)	0.78	1.24	59			

Note: 1. Input–Output ratios are the value of inputs expressed as a ratio of the value of output.
2. Two types of input–output ratio have been computed for each crop, denoted as (a) and (b) respectively: (a) only paid-out costs including value of farm-produced inputs and own bullock labour are used to arrive at the value of inputs here; (b) inputs here include total costs consisting of all paid-out costs plus rental value of own land, interest on own fixed capital and imputed wages for family labour.
3. The ratios/percentages are arrived at using triennium averages of the available data. Base period is based on data for 1970/1–72 to 1973–4 and terminal period on 1979–80/1980–1 to 1982–3.

Source: Computed from data published in *Indian Agriculture in Brief* (21st edn), Directorate of Economics and Statistics, Ministry of Agriculture, Govt. of India, 1987.

ray of hope for the disadvantaged dryland farmers to participate in the development process.

CONCLUSIONS

While irrigated crops and those with access to modern farm technology have dominated the growth process in India, dry crops and drought-prone regions have also shared the gains of agricultural growth in certain pockets. But this growth has been accompanied by increasing yield instability and production costs. These are obstacles to realizing sustained agricultural growth. Whether attaining higher yields involves a trade-off with greater stability or not is difficult to surmise. While evidence presented here points to instability rising with growth, experience of some developed countries and LDCs shows higher yields being combined with greater stability (Barker *et al.*, 1981; Alauddin and Tisdell, 1988). But assuming that such trade-offs do exist, how do countries respond? Generalizations are difficult since risk preferences among countries, and between farmers, could vary; and so also perceptions and policy responses. LDCs like India when faced with a severe food crisis in the 1960s gave emphasis to raising yields. This strategy complemented by stabilization policies, has enabled India to expand food supplies as well as to cope with fluctuations in domestic output; but these have not necessarily been costless.

Reducing risks and costs in farming without inhibiting growth constitute an important agenda for agricultural planners and scientists. Risk reduction or management has to centre around (a) crop, varietal and economic diversification (especially in dry regions where agriculture has to survive on a poor resource base); (b) development of crop varieties and technologies that can withstand environmental stresses and shocks; (c) development of infrastructure such as irrigation, input delivery systems and market institutions, and (d) stabilization policies (crop insurance, buffer stocks, target-oriented public distribution systems, price support) to insulate producers and consumers from weather and market-related risks. Cost-reduction strategies should focus on (a) economizing and, wherever possible, substituting external with local resources: for instance, legumes in traditional crop rotations were an inexpensive source of nitrogen (local resource) as against expensive synthetic nitrogen (external resource); so also organic manures which are less hazardous and expensive, and labour-intensive, have been increasingly discarded after the advent of chemical fertilizers; (b) research and development of cost-effective technologies; (c) resource conservation (soil and water conservation); and (d) shifting to sustainable alternatives (natural farming, watershed management, biological methods of pest control and nutrient use). The accent on 'resource exploitation' which characterized the green revolution-based growth strategy has to give place to one based on 'resource conservation' using modern science and traditional wisdom.

The strategies for promoting sustained agricultural growth will have to keep in view the diverse environments and constraints under which agricultural growth is taking place. While in the irrigated regions, or those with plentiful water, emphasis has to be on improving water-use efficiency through proper

TABLE 3   *Impact of watershed development programmes on dryland development in Karnataka State, India.*

Item	Variables			
	CROP YIELDS (Quintals per ha)			
	With watershed		*Without watershed*	
Crops	*Mittemari*	*Gonur*	*Mittemari*	*Gonur*
Finger millet	12.3	—	9.6	—
Groundnut	13.2	—	9.6	—
Sorghum	—	8.5	—	6.0
Groundnut+redgram	—	5.5+2.1	—	2.8+0.5
Pearl millet+horsegram	—	2.9+1.3	—	1.8+0.4
Minor millet+horsegram	—	2.6+0.8	—	0.2+0.2
Finger millet+horsegram	—	5.3+1.1	—	2.5+0.5
Sunflower+redgram	—	2.5+1.0	—	(not grown)
Sorghum+redgram	—	3.7+3.2	—	(not grown)
Maize+redgram	—	6.4+1.4	—	(not grown)

RELATIVE ECONOMICS OF HORTICULTURAL CROPS

	Annual		*Horticultural*	
	Sorghum	*Groundnut*	*Mango*	*Acid lime*
Benefit–cost ratio	1.2	1.5	6.9	4.9
			Sweet lime	*Cashew*
			2.9	1.3

NET RETURNS (Achalu micro-watershed)

	Year:	*1986–7 (Bench-mark)*	*1989–90*
Net returns in rupees per ha of net cropped area		–46	2575

(Joladarashi watershed)

	With watershed	*Without watershed*
	(rupees per ha)	
Crops		
Sorghum	1003	642
Coriander+safflower	1042	902
Safflower (local)	403	309

EMPLOYMENT (Achalu micro-watershed)

	Year:	*1986–87 (Benchmark)*	*1989–90*
Man-days per ha of net cropped area		67	106

Source: State Watershed Development Cell, Government of Karnataka, Bangalore.

regulation and management (relevant for South and South-east Asia), in the dry and semi-arid regions where water, apart from land are overriding constraints, the policy goals should aim at moisture and soil conservation and encouraging income-enhancing crops (such as fruit trees) and economic activities (pastoral, agro-forestry) that are less water- and land-intensive. Ultimately, an economic environment that is growth- and equity-promoting and which interacts in harmony with nature is the desired goal.

REFERENCES

Ahluwalia, Montek, S., 1978, 'Rural Poverty and Agricultural Performance in India', *Journal of Development Studies*, vol. 14.
Alauddin, M. and Tisdell, C., 1988, 'Impact of New Agricultural Technology on the Instability of Foodgrain Production and Yield', *Journal of Development Studies*, vol. 29, no. 2.
Anderson, J.R., Hazell, P.B.R. and Evans, L.T., 1987, 'Variability of Cereal Yields – Sources of Change and Implications for Agricultural Research and Policy', *Food Policy*, vol.12, no. 3, August.
Barker, R., Gabler, E.C. and Winkelmann, D., 1981, 'Long-Term Consequences of Technological Change on Crop Yield Stability:The Case for Cereal Grain', in Valdés, A. (ed.), *Food Security for Developing Countries*, Westview Press, Boulder, Colorado.
Hazell, P.B.R., 1982, *Instability in Indian Foodgrains Production*, Research Report No. 30, International Food Policy Research Institute, Washington, DC.
Mehra, S., 1981, *Instability in Indian Agriculture in the Context of the New Technology*, Research Report No. 25, International Food Policy Research Institute, Washington, DC.
Nadkarni, M.V., 1988, 'The Crisis of Increasing Costs of Cultivation – Is there a way out?', *Economic and Political Weekly*, vol.XXIII, no. 39, 24 September.
Nadkarni, M.V. and Deshpande, R.S., 1982, 'Agricultural Growth, Instability in Productivity and Rainfall – The Case of Karnataka', *Economic and Political Weekly*, vol.XVII, no. 52, December.
Nadkarni, M.V. and Deshpande, R.S., 1983, 'Growth and Instability in Crop Yields – A Case Study of Agriculture in Karnataka, South India', *Regional Studies (UK)*, vol.17, no. 1.
Ninan, K.N., 1991, 'The Green Revolution, Dryland Agriculture and Sustainability – Insights from India', ISEC, mimeo, Bangalore.
Rao, C.H.H., Ray, S K and Subbarao, K., 1988, *Unstable Agriculture and Droughts*, Vikas Publishers, New Delhi.
Rao, V.K.R.V., 1983, *India's National Income 1950–1980 – An Analysis of Economic Growth and Change*, Sage Publications, New Delhi.
Sen, S.R., 1967, 'Growth and Instability in Indian Agriculture', *Journal of Indian Society of Agricultural Statistics*, vol. 19.
Shetty, S.L., 1990, 'Investment in Agriculture: Brief Review of Recent Trends', *Economic and Political Weekly*, vol.XXV, nos. 7–8, 17–24 February.
Valdés, A. and Konandreas, P., 1981, 'Assessing Food Insecurity Based on National Aggregates in Developing Countries', in A. Valdés (ed.), *Food Security for Developing Countries*, Westview Press, Boulder, Colorado.

DISCUSSION OPENING – MAURIZIO MERLO*

The theme explored by Ninan and Chandrashekar is a vast and difficult one. In fact, the authors make clear that their objective is limited to providing some insights relating to Indian agriculture generally, but probing more deeply into the specific experience of Karnataka State.

*University of Padua, Italy.

The insights relate first to production performance which is associated with growing instability in agricultural output. It is also stressed that the green revolution has increased cultivation costs and seems to have created external costs because of environmental degradation. The widening of income disparity amongst farmers, and regional inequities and other sectoral imbalances are also said to have stemmed from the green revolution.

Coming to the key issue of the paper, namely agricultural growth and its associated instability, one can note that a very technological view is adopted, since growth is defined in terms of physical production and yield per hectare of the main crops in India. The paper attempts to compare growth before and after the green revolution, which occurred in the 1960s. The analysis therefore distinguishes the two key periods as 1955–6 and 1966–89. The dividing line, probably chosen because of data availability, seems to be hardly justifiable, given the production performances reported in Table 1. Especially when expressed in terms of yield per hectare, the growth rate does not vary significantly between the two periods. Indeed it seems to be higher, particularly in Karnataka State, before the green revolution. Certainly, some explanation should be provided to avoid this apparent anomaly. It would also be interesting to distinguish between growth specifically due to the green revolution from more general effects of irrigation expansion or watershed management.

The instability associated with growth is, however, quite clearly demonstrated in Table 1. The reasons for yield instability rising after the green revolution include rainfall failure, cultivation of ecologically fragile marginal land, sensitivity of HYVs and chemical fertilizers to moisture, specialization of agricultural systems, transition from subsistence to commercialized agriculture, environmental degradation and neglect of traditional practices and resource maintenance. The paper seems to blame the green revolution for causing instability, but it can equally well be argued that instability is intrinsic to any form of development. Social and economic turbulence has been part of the development process, and agriculture in this context, as the weakest (and residual) sector of the economy, has always been liable to experience difficulties and disarray. In other words, instability has always marked the transition from subsistence to commercialized farming. Agricultural policies have to some extent been originally conceived and implemented specifically to alleviate the socio-economic conditions of rural areas during the process of economic development.

From this point of view the paper appears to be too limited to technological aspects, overlooking any socio-economic and historical perspective, and failing to refer to basic economic theories which would have helped to explain its main thesis. The authors are right to pose the question concerning the extent to which attaining higher yields involves a trade-off with greater stability, but their answers appear to be limited.

Coming to the cost economics, Table 2 quite clearly shows a noticeable increase in input–output ratios and of the share of paid out costs in total unit costs of production. Above all it is paid out costs for fertilizers, pesticides, seeds and equipment, which have increased in weight. The trend in itself is not surprising: it can be found in many other developed and less developed countries and inevitably it assumes dramatic connotations when agriculture progresses from an essentially subsistence orientation to a commercialized structure. The main surprise, given the rather poor average production performances of India (1.6 ton/ha of wheat, 1.3 of rice and 1.1 of maize), is that a share of 50–60 per cent of paid out costs seems quite high. Perhaps it would be wise to ascertain how much the supposed scale neutrality of the new technologies is actually verifiable, or whether monopoly marketing of such inputs as seed and fertilizer is contributing to the size of costs.

The assertion that the terms of trade have become adverse for agriculture, and consequently that the farmers' lobby is displeased, seems quite reasonable, and would certainly satisfy any farmers' audience around the world.

The key question of the paper is whether modern agricultural technology is not, or at least is no longer, viable for India and other LDCs. The issue is no longer left only to agricultural economists and scientists; it has importance and momentum in the political arena, being linked to various theses advanced by radical and fundamentalist movements which question the very concept of development. These views have certainly surfaced in India. Objectively speaking, there are arguments which cannot be easily dismissed as suggested by Ninan and Chandrashekar. On balance, however, it would perhaps be better to argue that it is the old-fashioned and highly costly technology, which has characterized the green revolution, that is no longer acceptable. Technology in itself is the inescapable key to meeting production objectives, but it needs to be a new 'clean' higher-level technology originating from greater genetic improvements, but with more attention being paid to traditional practices and avoiding excessive reliance on pesticides and fertilizers. The huge investments made, in recent years, by the very same big companies responsible in the past for the old technology, are promising signals that things are moving quickly in the right direction. The real danger for India, and many other poor countries, is that the older, environmentally unfriendly, technologies will remain in use.

The general conclusions are very important; namely that stability of agricultural growth needs to be based on diversification, technologies effectively suited to Indian agriculture, substitution of external with local resources, less accent on resource exploitation which characterized the green revolution, and more emphasis on resource conservation using modern science and traditional wisdom. Technology in itself cannot be the only solution. Ninan and Chandrashekar are right to underline the fact that the development of infrastructure, such as watershed management, irrigation and services, can go a long way in aiding the expansion of agricultural production with greater stability. Nevertheless technology, of a suitable form, still has a vital role to play.

DAVID COLMAN•

Section Summary

Readers should not need reminding that the overall theme for the Conference has been *Sustainable Agricultural Development: The Role of International Cooperation*. The aim of Section I was to lay out the background for an informed debate of the main theme. It began with a broad assessment of the future of the food and agricultural system considering major priorities and elements of strategy. Kirit Parikh, with his current vantage point as Director of the Indira Gandhi Centre, and former role as Head of the Food and Agricultural Programme at the International Institute of Applied Systems Analysis (IIASA), chose the issue of food security (principally in developing countries) as the dominant and enduring challenge for the future. He argued, as have others at this Conference, that 'getting prices right' will not make a major contribution to this problem because of inelastic aggregate supply response. In that situation, and given that low food prices alleviate the problems of the poor, the meaning of the 'getting prices right' slogan is far from clear. Instead, echoing Theodore Schultz, he argued that the key strategy lies with policies to stimulate production and productivity growth.

This is the argument also put forward by Nikos Alexandratos, who, armed with the FAO projection of production and demand in developing countries up to AD2000, stresses the strategy of investment in productive capacity. The two plenary papers, and the relevant discussion openings, indicated that agriculture, particularly in developing countries, remains at the top of the agenda. There are still problems relating to sustaining the growth of output, many issues concerning equity in distribution, and a strong debate on pricing policies.

The first invited paper session was designed to give the Conference an opportunity to debate issues of privatization, open competition and the political process of reducing government interference in markets. Ulrich Koester's paper considered the efficiency of markets in agricultural development, while the paper by Lawrence Smith and Ann Thomson considered the balance between public and private sectors. Both papers, but Koester's in particular, accepted that there were good grounds for maintaining a good deal of government control over trade, but argued for progressive reduction of government interference in domestic markets. Smith and Thomson set out criteria which might help determine where government intervention might still be needed, but argue using the 'new institutional' economics for a case-by-case approach

*University of Manchester, UK.

115

rather than sweeping generalizations. As Koester argues, it is a matter of weighing 'government failure' against 'market failure'. There was surprisingly little dissent by the discussants or from the floor.

The second invited paper session was designed to begin to air the issue of sustainability in the context of global strategies. Richard Norgaard's paper provided a stimulating challenge to the profession by arguing that the paradigm of neo-classical economics and our usual tool-kit is inadequate, and when logically applied can, and does, lead to unsustainable paths of development. He emphasized the necessity of recognizing the worth of physical science-based paradigms of sustainability, and the fact that these would lead to other solutions and cannot readily be integrated with our own models. Sustainability, as he stated, is seen as a matter of intergenerational equity, with the current distribution of property rights leading to market situations of over-exploitation of resources and unsustainability. This provoked a particularly sharp and important rejoinder from the floor, emphasizing that current inter-personal, inter-country inequity is not only at the heart of the property rights issue, but that now fashionable concern for intergenerational equity may cause us to overlook the fundamental problem of current inequity.

In a complementary paper, K.N. Ninan and H. Chandrashekar considered the practical requirements of sustainable agricultural growth strategies for Indian agriculture. They argued for a return to a more integrated form of farming relying less on purchased chemical inputs, and for investment strategies to reduce production instability.

Chairpersons: David Colman, Georges Bublot, Michel Petit.
Rapporteurs: Michio Kanai, Willis Oluoch-Kosura, N.K.Mukherjee.
Floor discussion: H.Popp, D.Belshaw, J.Kola, V.I.Isaksson, G.Schmitt, M.Petit, J.C.Wetts and Dilip Shah.

SECTION II

International Trade and its Impact on Domestic Policies

AART DE ZEEUW*

Implications of the GATT negotiations:
The Process of Reaching an Agreement on Agriculture

INTRODUCTION

We find ourselves, in August 1991, near to the conclusion of the Uruguay Round where negotiations on agriculture are key issues. I think that the place of this conference is well chosen, since Japan is a country where agricultural reform is imperative. The country has to face new challenges as a result of the internationalization of the economy and the agricultural sector and the growing tension between agriculture and the environment. But not only Japan is involved; almost all developed and many developing countries are facing similar problems.

THE PRESENT SITUATION

So far the good intentions of the industrialized countries under the OECD umbrella to reduce agricultural support have come to nothing. The 24 OECD countries have seen a 12 per cent rise in the total costs of agricultural support to $300 billion in 1990. The consumer carries a large share of this burden in the form of high food prices. But the taxpayer too had to contribute more in 1990 than in the previous year. The rise was mainly due to an increase in production and a fall in world market prices, though the countries concerned have also neglected to take the necessary steps to put their agricultural affairs in order. A fundamental change in agricultural policy is clearly inevitable.

The solution to the agricultural problem must be found multilaterally, so that the different countries can share the costs involved. So far this opportunity has not been taken up, and more opportunities have been lost. An example is the Ministerial Conference which took place within the framework of the Uruguay Round in Brussels in December 1990. The chance of a successful conclusion to the Uruguay Round has not been entirely lost but speed is certainly needed if the talks are not to turn into a complete failure. The conclusion of the G7 meeting in London to agree before the end of the year gives some hope.

* Adviser to the Minister of Agriculture, Nature Management and Fisheries, Ministry of Food and Agriculture, The Netherlands.

The solution to the agricultural problem, to which agricultural economists have contributed their share, is on the table. What has been lacking so far is the political courage to strike out along a new path in agricultural policy: that of a more market-oriented policy, where supply and demand forces play a more prominent role, both in the internal and in the external market. Let us hope that politicians are not going to hide behind social and environmental issues to try to delay the reform policies that are so urgently needed. In my opinion there is a real danger that this will happen. Agriculture, like the rest of the economy, will have to be integrated more closely into the international trading system. It will have to play the rules of the game; that is, face up to more external competition and a reduction in government support on the internal market – in other words, to begin a recoupling of markets and a decoupling and reduction of support to the agricultural sector, which means that farm support moves from subsidized prices to direct payments if necessary.

That is the way agricultural policies will have to go to keep the international trading system, which has worked so well in the industrial sector, on its feet. Many countries, including Japan, have benefited from it, especially after the Second World War. Major trading partners such as the USA and many other exporting developed and developing countries have asked for the agricultural sector to be included in this trading system. Europe and Japan will have to concede. They should stop trying to make the agricultural sector an exception, arguing food security and the preservation of the environment. Trade liberalization for agricultural products can go very well with food security and a sound environmental policy. There are other policy instruments apart from border protection to realize these objectives.

THE COURSE OF THE NEGOTIATIONS

Japan, the EC and South Korea were the main countries, during the 1990 Ministerial Conference in December in Brussels, who were not prepared to take a compromise proposal put forward by the Swedish Minister Mr Hellström as a basis for further discussions. The proposal included a binding commitment on a 30 per cent reduction in support and protection over the period 1990–5. The commitments were to cover three areas: internal support, border protection and export subsidies.

The EC had always taken the stand that it was prepared to commit itself on a reduction in support and protection in a general sense, without specific commitments on export subsidies and border protection. The general support, laid down in the AMS (the aggregate measurement of support) would be reduced by 30 per cent over a ten-year period, 1986–1995. The EC argued that this would indirectly lead to a reduction in border protection and export subsidies.

At the eleventh hour the EC also proved to be willing to commit itself, after all, in the area of export subsidies, but not enough for the other big trading partners. This concession could not salvage the negotiations as the EC and other countries were not willing to make an important step in the area of border protection. The USA, the Cairns group and most of the non-exporting developing countries were very much in favour of the tariffication of all border measures. The EC only wanted to take a step in this direction under certain conditions. One such condition was that the EC be allowed to neutralize the effect of fluctuations in the exchange rate of the dollar. The EC also wanted to be able to reduce the effect on internal prices of major world market price fluctuations.

A third condition was the right to raise import duties in the future for products, notably cereal substitutes, which are now entering the EC free of import levies. These conditions, and especially the last one, were not acceptable to the USA and the Cairns group. Japan, South Korea and some European countries share the EC's reluctance to accept tariffication. These countries do not want real competition in their markets with basic products from third countries. They plead food security and the specific provisions in the General Agreement under Article XI, allowing quantitative import restrictions on the condition that the country concerned shall curb domestic production effectively.

Fewer differences existed in the area of internal support. This has to be lowered in a way comparable to that of border measures and export subsidies in so far as those support measures are trade-distorting. An agreement on a list of acceptable internal policy measures which would not be submitted to reduction was already at hand. The real cause behind the breakdown of the negotiations was the unwillingness of countries, like Japan, the EC and other European countries to commit themselves under GATT to reform their agricultural policy drastically. They were not prepared to accept the re-instrumentalization of agricultural policy as the USA and the Cairns countries had asked them to do. What these countries wanted to achieve was the full integration of the agricultural sector into the GATT trading system so that the agricultural sector would be governed by the same trade rules as other sectors. This means that, as the only instrument for border protection, import duties are to be applied and reduced and that export subsidies are to be banned. This might be compensated for by more direct income support payments.

THE FUTURE

The question now is whether, since Brussels, these positions have changed and whether as a consequence the chances for a successful conclusion to the Uruguay Round have increased. The position of the EC is of vital importance here. Since Brussels the EC has not really changed its view, but it is noteworthy that since the breakdown of talks the European Commission has, through its agricultural commissioner, Mr MacSharry, come up with new ideas for the future of the Common Agricultural Policy (CAP). The Commission now admits that the CAP has failed and is in urgent need of reform. Although the member-states agree on the necessity of reform, a consensus on the direction this reform is to take is still nowhere in sight. The outcome of the internal discussions on the coming CAP reform, therefore, is still uncertain. But it is by no means unlikely that the two processes, the Uruguay Round trade negotiations on the one hand, and the internal CAP reform discussions on the other, will influence one another.

The failed summit in Brussels and the discussions following MacSharry's proposals will mean that the other negotiating partners will have to face the facts once more: the agricultural reform policy is a very delicate business and is still an uphill struggle, not only within the EC but also in other European countries and undoubtedly also in Japan and South Korea.

I hope this will lead to a GATT agreement as a first step to a real market-oriented agricultural policy where trade barriers are reduced to a minimum and agricultural support is substantially reduced. A solution in the long term should, I think, include the following elements:

— Border protection to be replaced by fixed import duties. These are then substantially and progressively reduced according to a schedule to be established. This will bring about a genuine coupling of the internal and external markets which makes it possible to compete with other countries on the internal market.
— Export subsidies to be governed by more stringent GATT rules and to be phased out at a rate equal to that applied to import duties.
— Internal subsidies, in so far as they are product or market-related, to be phased out at a rate equal to that of border protection.
— Finally, strict rules are necessary to prevent veterinary and phytosanitary measures from becoming unnecessary trade barriers.

Is it possible for countries such as Japan, South Korea and many European countries to cope with the competition from third countries which will then arise? I think, generally speaking, that the answer is yes, provided the following conditions are met:

— First, these countries must retain the possibility to maintain certain levels of import duties to compensate for the differences in climate and agricultural structure compared to that of North and South America, Australia and New Zealand.
— Second, safeguard measures are established to provide temporary extra protection in the form of escape clauses which come into force in times of a significant surge in imports and/or in the case of a significant decline in world market prices.
— Third, temporary income support payments are to be handed out to producers in regions where agricultural restructuring is needed.
— Finally, there must be a possibility to give extra support to permanently disfavoured areas; and stringent environmental measures are to be taken into account.

The main thing to be decided on is not by how much support and protection are to be reduced over the years to come but what trade political instruments may be implemented in the agricultural sector in the coming years. The central question is whether import duties will become the main instrument for border protection and genuine tariffication will be introduced.

Genuine tariffication means that the complete isolation of internal markets from the world market will no longer be a possibility. If this road is not taken, international agricultural policies will be determined even more by government regulation. To keep prices artificially high, independent of what happens in the world market, will inexorably lead to production control. Such a policy is inextricably bound up with quantitative import and export control measures.

I hope that the latter scenario will not be followed. I am convinced that the agricultural sector will benefit greatly from a reduction of government intervention in the market. Experience has shown that sectors which are not protected by guaranteed minimum prices show a much more favourable income development in the long run than protected sectors. Countries where an agricultural policy of guaranteed and often high minimum prices has been pursued can hardly maintain their competitive edge. The lack of built-in incentives due to a lack of outside competition leads to agricultural structures to which adjustments cannot easily be made. I think, therefore, that it is also in the interest of these countries to expand their market access opportunities and to allow imports from third countries. This will force producers to look for more efficient production methods, such as expansion in scale. Right now it is difficult for these countries, including Japan, to compete on account of their labour and land productivity, which is far too low. In my opinion the government should provide financial support as a temporary measure so that the necessary structural reforms can be made. The Japanese rice culture and the Canadian dairy sector are examples of areas where such reforms are needed. A similar situation will occur in the USA for such protected commodities as dairy products, sugar, cotton and peanuts. The dairy and cereal sectors in the European Community will not escape the adaptation process either.

These reforms will, in the end, offer more promising prospects for the agricultural sector, not least through the creation of new markets in the non-food sector as a result of competitive prices. There is the potential for producing renewable energy resources or building materials. The coming months will have to provide clarity over the way agricultural policies will develop. The internal discussions in the European Community will no doubt add considerable weight. It is of course tempting for Japan and other countries to wait for the outcome of these discussions before taking a definitive stand in the Uruguay Round trade negotiations. This, however, would be a very risky undertaking, as it is by no means clear what the outcome of the internal EC discussions is going to be. The wrong outcome could put the whole Uruguay Round at risk and with it the international trading system which is so vital to both the developed and developing countries, not least for a country such as Japan. An early Japanese stand in the negotiations which would point the direction their future agricultural policy is going to take may provide an important stimulus not to let the Uruguay Round negotiations on agriculture fail.

CONCLUSIONS

During the whole five-year span of the Uruguay Round there has been a willingness at the highest possible political levels to integrate all agricultural commodities in the existing liberal trade system of the GATT. The course of the negotiations has made it crystal clear that this is only possible if the developed world is willing to bring about fundamental changes in current agricultural policy. Until now policy has been based mainly on guaranteed prices for the producer, independent of market forces. It is generally accepted that this must be replaced by a much more market-oriented policy, where market forces are allowed their free play. The negotiations have shown that the most realistic way to realize this goal is:

— to accept the tariffication of all border measures and a subsequent substantial reduction of tariffs over time (five to ten years): along with a safeguard clause to prevent prices from falling too far below a reference price, and a minimum access commitment during the transitional period. For certain basic products a temporary exception could be negotiated, using a strengthened Article XI (the possibility of quantitative import restrictions in the case of production control). Rebalancing of import restrictions can only be negotiated on the basis of Article XXVIII, and would require compensation.
— to accept for the transitional period (five to ten years) that existing export support cannot be more than the difference between internal and external market prices and cannot be more than the import tariff at the border. After the transitional period a new decision will be necessary on the question of the continuation of export support.
— to accept a list of national support measures, not falling under the reduction commitment, to make it possible to execute support programmes to restructure agriculture; to compensate for environmental measures and so on. This list cannot include product- or market-related support.
— to accept that developing countries with a backward agriculture need more time to introduce the new rules and that certain importing developing countries need extra help in case trade liberalization results in higher international market prices for some basic commodities.
— to accept stringent rules in the area of veterinary and phytosanitary regulations, to prevent them being used as non-tariff barriers.

If contracting parties are willing to accept this agenda as the basis for an agreement, I am sure that the Uruguay Round will not fail. A failure will certainly lead to alternative regional agreements between, for example, the USA and South America, Japan and other countries in the Pacific regions and Eastern Europe and Western Europe. Regional agreements need not be bad if they can be integrated into a world-wide agreement on trade. If not, many developing countries, in particular, will fail to obtain real opportunities for export to the more wealthy regions. It is for that reason, in my view, that regional agreements, instead of a world-wide agreement, are not acceptable. Furthermore, regional trade agreements will not solve the real agricultural trade problems. In fact, to bring supply and demand into balance requires a real international effort to open up markets world-wide through multilateral negotiations.

DISCUSSION OPENING – LOUIS P. MAHÉ*

Mr de Zeeuw opened his paper with a brief description of the critical situation of international relations associated with agriculture, moving on to present an interpretation of the events which led to failure of the Brussels negotiations in December 1990. He then outlined the main directions for reform in agricultural marketing policies which he regards as necessary for future negotiations to be successful.

I do not profoundly disagree with his views, particularly those concerning the need for reforming agricultural policies in industrialized countries and subjecting them to the discipline of GATT. My remarks and questions concern, first, the costs of current policies and, second, the logic underlying the positions adopted by those taking part in the international game which the current negotiations represent.

In his review of the current situation our speaker reminded us of the extent to which agriculture is supported in OECD countries, and drew attention to the fact that reform intentions have come to nothing. The measure of support cost used is the PSE – Producer Subsidy Equivalent, plus the other budget expenses in favour of the sector. However, as everyone knows, the PSE represents a transfer rather than an economic cost in terms of welfare. Even if a part of the collective waste due to lobbying is added to the classical deadweight loss represented by the welfare triangles, the total is certainly less than the 300 billion dollars mentioned, although it would remain high. My comment is not intended as an academic debating point, it simply represents the limits of the contribution which we, as agricultural economists, have made to the debate.

Mr de Zeeuw stated that agricultural economists have done their share of the work in revealing the extent of the problems posed by protectionism, and stated that it is now the politicians' turn to show courage. It cannot be denied that important analytical work has been done, but I feel that I can put all my colleagues' minds at ease! There is still much for us to do and unemployment is not just around the corner. Do we really know, at the world level, what the losses in GNP and in employment caused by national policies really are? Further, can we estimate the long-run effects on the remuneration of productive resources? We do know much about the transfers involved, and they are obviously of importance in themselves, but that is not the end of the story. There is still need for more analysis both to inform politicians (and perhaps make them more courageous) and to communicate results in a form understandable to the general public.

Against the background of more adequate information we should also question the logic of our institutions. As agricultural economists we need to examine the national and international institutional conditions which act as a brake on

*Ecole Nationale Supérieure Agronomique, Rennes, France.

reform. This is the subject now explored in the 'new political economy' which focuses on the complexity of the institutional processes through which decisions are made. Though supported by an underlying democratic system, our processes are influenced by lobbying and affected by the behaviour of bureaucracies. For example, at the national level, we must find a parry to producers' lobbies. Since their members have similar interests, they are the strongest spontaneously organized groups, and thus the most efficient in influencing politicians clearly sensitive to the votes of their electorates.The interests of taxpayers and above all of consumers, who are too scattered to organize themselves spontaneously, in matters of agricultural policy need to find an institutionalized means of expression. This is difficult because of the high costs of organization and of free-rider behaviour. In addition, ministries specializing in the administration of agriculture have an excessive role in formulating policies; they display systematic bias in favour of short-term sectoral interests, and frequently freeze reform. This is a real challenge in our societies and Mr de Zeeuw is correct in arguing that there is real risk of the true nature of the agricultural debate being eclipsed by weak arguments about social, regional, environmental and food security issues. I would like to strengthen his remarks about the latter. It is unacceptable, from an intellectual as well as a political point of view, that advanced countries in Europe, and above all Japan, should refer to the food security argument at all. The monetary resources are available to obtain food supplies without difficulty – even by imports!

At the international level we must channel external forces so that they support the processes of internal reform, and we must avoid stopping the current negotiations needlessly. It is clumsy, and probably counterproductive, to adopt the view that some countries must 'give in'. It is risky to view international negotiations as a game between countries in which there are winners and losers. This only provides nationalistic arguments to lobbies which are only too willing to denounce external pressures, and does not help public understanding of the issues involved.

In the second part of his paper, Mr de Zeeuw turned to a description of the negotiating process, stressing the extreme opposition between, to simplify, the United States and the Cairns group on the one hand, and the EC, Japan and other European countries on the other. In my view this opposition cannot be considered as one between loyal supporters of international trade, which appears to be the way in which the Cairns group views the position, and the others, who are less keen to allow the full integration of agriculture into the world trade nexus. In fact, the reluctance of countries to reform agricultural policies under the GATT system is a much more widespread attitude, and indeed one which the United States has itself adopted for some time using the waiver and Article XI. Much depends on internal interests rather than on purely doctrinal conceptions favouring liberalization. It is true that Japan does not take the same view of international exchange in food as it does in the cases of cars and electronic goods. However, there are interests at work on the other side. The United States and the Cairns group are in favour of freer trade since they might benefit as exporters. In particular the United States is mainly interested in effects on grain and oil-producing crops, and little mention is made of milk products, sugar and other commodities. Indeed, it appears unlikely that the United States could have sustained the extreme zero-option position because of internal political considerations. That may well have motivated their search for some compromise. Similar remarks can apply to other parties, notably Canada.

While the reform protagonists may discern some potential gains, the countries which have applied a brake to liberalization are in a very different position; they are traditionally importers (the Community has involuntarily become an exporter) and can expect few benefits to their agriculture. Of course they could gain overall, but again any doctrinal faith in free trade is submerged by political realities. GATT could sometimes be regarded as an intruder upsetting the status quo. However, the position in the Community has itself changed since December 1990. Driven by other internal pressure, notably budgetary cost, the Commission has begun a significant reform process – in my view it is a true revolution – which changes the whole shape of the continued negotiation. Indeed the proposals almost exceed the expectations of the USA and the Cairns group! Notice, however, that they are mainly directed towards larger-scale arable farming and have less impact on animal products and on sugar. This does not displease the United States, though it may disappoint Australia and New Zealand. The reform suggested by Mr MacSharry would lead to the near elimination of export refunds on cereals, and direct support towards payments decoupled from production. While Mr de Zeeuw seems to be rather pessimistic about the chances of this reform being successful and having an influence on trade negotiations, the link appears to be inevitable. Even if it is only partially adopted, the position of the Community can no longer be that of November 1990, even though the latter is still the official one. The probable extension of the Uruguay Round negotiation into 1992 will reinforce the link between it and CAP reform.

This evolution of the CAP appears to me to be directed along the lines of liberalization which Mr de Zeeuw favours, and seems compatible with his minimum conditions for GATT progress. Other negotiators should react by facilitating the Community reform process and not put a brake on it by making extreme demands such as the total decoupling of supports, insistence on complete internal transmission of world price movements, or disallowing some limited action affecting imports of animal feedstuffs. The process of reform could be slow, but that is itself necessary to avoid too sudden exposure of European (or Japanese) agriculture to external competition. I have criticized the CAP often enough not to be classed as one of its supporters and therefore feel that I can safely applaud the latest proposals as being along the right lines. They may not meet some strict conditions, but imperfect instruments which lead in a worthwhile direction have much in their favour. I also believe that they should be supported within Europe; there are critics (the United Kingdom and the Netherlands) who often appear to call for reform but nevertheless oppose the MacSharry scheme because of effects which it might have on larger farmers. That appears to be a case of conflict between principles and a position dictated by internal political interest.

Mr de Zeeuw does not make any forecast about the likely progress of the negotiations. He clearly indicates his views about the desirable outcome of at least partial liberalization, but in general he remains cautious about the chances of success. My own view is that the nature of the international game does not provide a favourable setting for the advance of multilateralism. The advantages of freer agricultural trade are not obvious, they are scattered among all citizens, who are badly represented internally in most countries, and hence in the negotiations. Free trade is very much a 'public good' which is not easy to secure in the context of multilateral bargaining. The most-favoured-nation clause reinforces this problem by dissipating the benefits of concessions made by any one country to all participants, thus increasing the risk of free-rider behaviour. In such circumstances, closer interactions between countries through bilateral or regional agreements become very tempting.

Despite the initial gap between negotiating positions, and the differences in the political and economic conditions in the countries affected, the combination of internal and external factors now at work could produce an agreement. However, there are so many obstacles to be overcome that we can only expect it to be one of limited significance, though perhaps one which will begin to subject agricultural policies to some international discipline.

D. McCLATCHY AND T.K. WARLEY*

Agricultural and Trade Policy Reform: Implications for Agricultural Trade

NEGOTIATING AGRICULTURE IN THE URUGUAY ROUND

As is well understood, agricultural trade problems among the developed countries stem from domestic agricultural policies and, in particular, from the production-stimulating and consumption-suppressing effects of commodity-centred agricultural price and income support policies. The resultant increase in export supplies and decrease in import demands, and the insulation of national markets, have the general effects of depressing and destabilizing international market prices and distorting trade volumes and trade patterns. Agricultural trade arrangements and practices are generally designed to support national agricultural policies and programmes. This situation has prevailed for a very long time and, whilst constantly a subject in the GATT, the Uruguay Round is the first occasion when there has been agreement that domestic policies should be fundamentally changed to reduce their adverse trade effects.

The Pauline conversion to effect agricultural policy and trade reform by bringing about 'substantial and progressive' reductions in trade-distorting subsidies and by opening import markets was driven by several factors. In the broadest view, the Uruguay Round is concerned with strengthening the multilateral trading system – and providing an alternative to managed trade, aggressive unilateralism and regionalism – with a combination of trade liberalization, rule making and institutional reform (Lawrence and Schultze, 1990; Oxley, 1990). Reform of trade in agriculture in this context is one of the 'backlog' market access issues (along with trade in textiles and safeguard measures) that must accompany the extension of the GATT into the 'new areas' of services, intellectual property and international investment and reform of the GATT institutional system (Schott, 1990). Within the narrow domain of agriculture, the re-emergence of structural surpluses of major commodities by the mid-1980s, and the mounting financial and political costs of the resultant competitive subsidization, ensured that agricultural trade would be high on the UR agenda. Other forces in play were the imperatives of the United States

*Policy Branch, Agriculture Canada, and University of Guelph, Canada, respectively. Senior authorship is not assigned. Helpful comments from several colleagues are gratefully acknowledged. The views expressed are the responsibility of the authors and the paper does not necessarily reflect the views of the Government of Canada.

129

reducing its budget deficit and the inability of the European Community to continue financing open-ended price supports at high levels once it became a major net exporter of most temperate-zone agricultural products. We would like to think that the supply from our profession of empirical knowledge about the size of the income transfers involved in national agricultural policies and their negative macro-economic, welfare and trade effects (for example, Stoeckel, 1985; BAE, 1985; World Bank, 1986; OECD, 1987; USDA, 1988) also influenced the willingness to enter into negotiations leading to mutual disarmament in contending farm programmes and agricultural trade arrangements and their concerted reform. Whatever the influence of these and related forces, since the mid-1980s there has been a coincidence of interest among those who wish to see fundamental reforms in national agricultural policies (the 'desubsidizers') and trade specialists who wish to subject agricultural trade to the authority and the disciplines of the GATT (the trade 'liberalizers').

Objectives

The United States' strategic objectives in agriculture – shared with other exporters – included ensuring that future growth in world import demand would be met from low-cost sources; the initiation of reform of the European Community's common agricultural policy before the EC deepened its relationships with other Western and Eastern European states; and curbing the tendency of less developed countries to switch from taxing farmers to subsidizing them as development takes hold (USDA, 1990). It has also been suggested that a US strategy was to pursue its internally-desired domestic agricultural policy reform through an international negotiation, and to shield the resultant agreement on reforms from domestic agricultural and congressional resistance by linking the agricultural component of the negotiations to agreement on the other elements of the negotiations and thus to the success of the Uruguay Round as a whole (Paarlberg, 1991). The Cairns group of medium-sized and smaller exporters, above all else, seek an end to the economic damage being wreaked on their agricultural export sectors by the subsidy and protection policies of the USA and the EC. Additionally, individually lacking the retaliatory power of the big trading countries, they seek, in particular, to have agricultural commerce subjected to the rule of international law. The EC, the other countries of Western Europe and Japan, while anxious to reduce the deterioration in economic and political relations caused by disputes over agricultural trade issues with the USA and the Cairns group countries, nonetheless sought to retain national agricultural policy flexibility and autonomy in the selection of a timetable for domestic farm policy reform.

The specific objectives of the major participants have reflected these strategic goals. The United States has sought primarily to reduce the European Community's negative effects on world agricultural markets through reductions in the level of its support prices, border protection and Community preference, and by eliminating its export subsidies. The Cairns group countries have shared the same objective with respect to the EC, and with the United States have also sought improved access to the markets of Japan and other developed countries. However, the subsidy and protection policies of the United States have also been a target for the Cairns group countries. The negotiating positions of the EC and Japan have been essentially defensive. The Community has sought to commit itself to only a modest level and pace of support reduction, and its proposals would allow it to retain a good deal of Community preference, to continue to insulate its producers from world markets with variable import charges (albeit to a reduced degree), to increase protection for oilseeds and feed grain substitutes, and to avoid specific commitments on export subsidy reductions. As an exporter, the Community fears the release of the full production capacity of US agriculture and needs assurance that US grain support will be reduced. Japan has sought primarily to avoid substantial cuts in farm support and protection. Indeed, it wishes to have its internal support and restrictive import regimes endorsed on food security grounds.

Modalities

As has been described elsewhere (for example, IATRC, 1990), the negotiating framework anticipated commitments being made in three areas: domestic subsidies, border protection and export competition. The EC in early 1991, confirmed its willingness to conduct negotiations to achieve specific binding commitments in each of the three areas. This significantly increased the prospects of an agreement without guaranteeing it. To the extent that the commitments in each area will include a reduction in a base period level of a particular type of support or protection, such reductions are likely to be phased in linearly over a five to ten year transition period beginning the year after an agreement is reached. Let us summarize the essence of the approaches emerging in each area, and the linkages between them.

The aggregate measure of support (AMS) is now being taken by most countries to be a measure of the extent to which internal or domestic support measures are trade distorting. However, countries still use the term differently, and its definition by certain major proponents has evolved over time. Although essentially calculated at the individual commodity level, there is no agreement yet about the level of commodity aggregation at which commitments would be taken. It does seem clear that required reductions will be in the total value of the AMS – $(P \times Q)$ rather than the per unit value (P) – to give countries the flexibility of reducing their effective support prices or the quantity of production eligible for support, or a combination. Use of the AMS beyond a limited range of (albeit major) temperate farm products is by no means assured – though there is talk of 'equivalent commitments' in other commodity areas – and the AMS does not measure subsidies to farm product processors which are not passed back to farmers. Drawing the line between what is to be reduced ('amber') and what is to be exempt ('green') is a necessary and important first step which has yet to be completed. Negotiations on domestic support measures are focused on reductions, though Canada for one would also like to get some precise definition of the rules concerning which subsidies are countervailable.

Border protection

Commitments seem likely to differ from those for domestic subsidies in two important ways. Firstly, they will apply at a relatively disaggregated (for example, tariff line) level and could extend across the whole range of raw and processed agricultural products traded. Secondly, they will include a combination of commitments on per unit value (P) and import volume (Q), but not trade values ($P \times Q$). It has yet to be decided whether reductions in agricultural tariffs will be achieved via a request and offer approach, a formula approach, or a combination. Since tariffs are currently a relatively less important form of border protection for agricultural products, the most important instrument under discussion here is 'tariffication' – the conversion of GATT- inconsistent non-tariff barriers to 'equivalent tariffs' and their subsequent reduction. Tariff rate quotas – a limited volume of imports to which a relatively low tariff rate is applied, increasing over time in some cases – are likely to be an important element of access commitments. Similarly, given the perception that tariffication is a means of moving towards a situation of reliance on tariffs as the only legitimate form of border protection for agricultural products, the outcome of the Uruguay Round (UR) with respect to rules (particularly Article XI:2(c)) about when quantitative restrictions may continue to be used is a controversial and important element of the negotiations on market access. The removal of all country-specific exceptions and the binding of all terms of access are also important goals for many countries in the border protection area.

Export subsidy reduction

Commitments may be specified in terms of expenditures ($P \times Q$), volumes (Q) or a combination of these two. The per unit (P) option has much less support. As with the AMS, the level of commodity disaggregation of export subsidy commitments is still unclear. The main target of reductions is likely to be export subsidies applied to major temperate farm products at the first level of significant trade. However, depending upon clarification of the distinction between primary and industrial products, individual commitments on each specific processed product line are conceivable. Before export subsidies are reduced, they have to be defined. 'Producer-financed exports' (levy and two-price pooling schemes), 'concessional' (as opposed to 'grant') and 'tied' food aid, and export credits are all contentious issues still awaiting resolution. Since export subsidies are unlikely to be eliminated in the UR (as Japan and Canada, among others, would have liked) the clarification of rules governing their continuing use – in particular Article XVI:3 and its economically disreputable 'equitable world market share' clause – is part of the outcome being sought under export competition. Negotiations in this area also cover rules about restrictions on exports – something of interest to major importing countries like Japan.

Linkages

There are some theoretical linkages between commitments in these three areas. The EC, in its 1990 ('pre-Brussels') offer to reduce only overall support (AMS) levels, argued that this would automatically force down internal support prices and that reductions in import charges and export subsidy levels would necessarily follow. Simple analytics suggest that, under a predominantly market price support (MPS) or two-price system, reductions in tariff levels must force down domestic price levels (and, in turn, export subsidies), while reductions in export subsidy levels (volumes or expenditures) must force down either domestic support prices or managed domestic supplies or both. In contrast, under an MPS system, increased quantitative access may have relatively little impact on domestic market prices while causing increases in subsidized exports via a displacement effect (as with EC beef, butter and sugar).

Quantitatively, it can be deduced that, under a pure MPS system, the percentage reduction in export subsidy expenditures would always be higher (and sometimes much higher) than the logically corresponding percentage reduction in the tariff or the price support gap. The same is not true for subsidized export volumes, however. When supply and demand are relatively price-inelastic and export volumes relative to production are high, the export volume reduction percentage may well be lower than the logically corresponding percentage reduction of the price support gap.

In the real world such theoretical constraints on the equivalence of relative 'depths of cut' in different areas, which anyway would vary from country to country, tend to disappear or to be too difficult to calculate, for a variety of reasons. One is the existence of 'water in the tariff' – typical of the EC situation – where duty-paid import prices are well in excess of internal market prices, and thus could be reduced without markedly affecting the latter. Another is the widespread existence of support systems for given commodities which include a combination of market price support and direct payments. A third is the frequent occurrence of some form of supply control as part of the support system.

We can, however, conclude that reduction commitments in all three areas are necessary for at least the following reasons. Internal support reductions are required because export subsidy reductions and trade barrier reductions are irrelevant for support systems based on direct payments. Export subsidy reductions are necessary in order to impose effective constraints on domestic price (or production) levels because tariff reductions will not do it while the tariff 'contains water'. And specific commitments on tariff levels themselves are the only way to ensure their reduction.

In the end, it will probably be political considerations which determine the appropriate relationships between rates of reduction in each of the three areas. The importance of each area in overall agricultural support varies greatly between countries. Border protection is by far the most important element of Japanese support. In the EC it is border protection and export subsidies. In North America, with some notable commodity exceptions, it is direct government payments. In order for the overall package to be acceptable (that is for ministers to be able to sell it to their domestic constituencies) it must be perceived as 'fair' in imposing equivalent commitments and burdens on each country. This implies a bias towards equality in the rates of reduction in each of the three main areas of commitment. The outcomes of other parts of the agricultural negotiations and of other parts of the overall negotiations having a bearing on agriculture are also uncertain at this time.[1]

OVERVIEW AND EXPECTATIONS

Through the fog of uncertainty that necessarily attends this stage of the multilateral trade negotiations, it is possible to make a number of observations on where the negotiations stand. First, there is a genuine willingness to bring about substantive and enduring agricultural and trade policy reforms, and to do this by subjecting national agricultural policies and their derivative agricultural trade arrangements and practices to binding international disciplines. The disagreements are over the extent and the speed with which trade distorting domestic and export subsidies and border protection should be reduced, not over the commitment to do it. This is a discontinuity in the history of the treatment of agriculture in the GATT. Second the initial gulf that separated the major protagonists has been narrowed. Now that the United States and the smaller net exporters have lowered their sights and abandoned their demand for a degree of desubsidization and liberalization to which it was politically impossible for their negotiating partners to accede, the task has become that of getting Europe and Japan to take a somewhat longer step down the reform path than they initially offered. Third, whereas in previous GATT rounds agricultural trade reform has been addressed in isolation, and ultimately unsuccessfully, in the Uruguay Round a durable link has been forged between progress in agriculture and the other areas of the multilateral negotiations, including the new areas of services, intellectual property and international investments. The significance of this development may be generally underestimated. The willingness, shown in Montreal and Brussels, of the majority of the Cairns group countries to scuttle the whole negotiation rather than again acquiesce in the GATT's failing to deliver its promised benefits to agricultural exporting countries is an entirely new departure in the four decades-long pursuit of agricultural trade reform. Similarly, it appears that the US Administration is not willing to take to Congress for 'fast-track' ratification a draft agreement which does not include a substantial agricultural component. Fourth, to an important degree the most difficult negotiations are now in national capitals. At various international venues, the political commitment to subject agricultural policies and trade arrangements to new GATT disciplines has been made. At home, national authorities are presiding over an intense struggle for influence between, on the one hand, manufacturing and service industry groups who do not wish the Uruguay Round to collapse for want of an agricultural agreement and some competitive agricultural groups who would rather compete with the farmers of other countries than with those countries' treasuries and, on the other, agricultural groups that prefer regulation, subsidization and protection to competition. Interestingly, the pro-reform group includes food manufacturers who require access to competitively priced farm products if they are to face freer trade in consumer food products, and the opponents of reform include some agricultural export groups who fear the loss of present support.

In forging an agricultural agreement it is necessary to keep in mind that the Uruguay Round is designed to create a stronger and more effective GATT for the twenty-first century. Accordingly, it is important that the search for an agricultural agreement in the short term should not lead to 'the planting of viruses' that will undermine and weaken the GATT in the longer term. Several examples can be given which illustrate the dangers and which explain much of the reluctance of many countries to accept some features of the agricultural negotiating proposals of the Community and Japan. First, there is a reluctance to accept ill-defined 'non-trade concerns' such as 'food security', 'rural culture', 'regional development', 'structural adjustment', 'environmental protection', 'animal welfare', or 'fourth criteria standards' as justification for continuing subsidization and protection. At a minimum, the exporters are asking that these legitimate objectives be pursued with trade-neutral policy instruments. Second, there is apprehension that the admission of exchange rate changes as a basis for adjusting international obligations (as the Community proposes as part of its 'corrective factor') would redefine the character and value of GATT tariff bindings. A similar concern attends the suggestion that obligations should be adjusted for differential rates of inflation. Third, unless they are very clearly time-limited in applicablity and/or permit only a partial and diminishing 'correction' for world market price fluctuations, the admission of arithmetically determined corrective factors to tariff commitments would take us a long step towards the generalization of variable import levies and mark a departure from the restraints of the injury tests embodied in the GATT's present safeguards procedures. Fourth, unless accompanied by an effective cap or reduction commitment on domestic support prices, a decision not to place producer-financed export programmes under export subsidy disciplines would risk these becoming, in effect, consumer-financed export-dumping programmes and distorting trade more than equivalent regular export subsidies. Fifth, a decision to permit reductions in domestic support to count as 'effective supply management' could lead to the proliferation of GATT-legal import quotas under Article XI. Finally, it could be a dangerous precedent to permit the 'rebalancing' concept to lead to increases in protection in a negotiation designed to reduce it, and without formal recourse to the compensation provisions of Article XXVIII.

Our expectation is that an agreement on agricultural policy and trade reform will be reached in the Uruguay Round. It is clear that it will contain commitments that fall well short of those proposed initially by the USA and the Cairns group, but be more ambitious than the 1990 offers of the EC and Japan. The proposals by Hathaway (1990), Miller (1990), Hellstrom (GATT 1990) and IATRC (1991) suggest the elements and orders of magnitude of a feasible agreement. Trade-distorting internal subsidies, border protection and export subsidies might each be reduced by 30–50 per cent over a five-year period from a multi-year base, possibly 1986–90. The probability is that some aggregate measure of support will be used to define and verify subsequent reductions in the internal subsidies provided by 'amber' programmes at the commodity or commodity group level. Improved market access seems likely to be accomplished through the tariffication process. However, the USA and Cairns group are unlikely to agree to increased import protection for oilseeds and feed grain substitutes or to safeguard measures that do not allow market prices and exchange rate changes to have a substantial and increasing influence on the returns to producers in all countries. Prospects for removing all or most country-specific exceptions, derogations and waivers, and for binding all or most terms of access, still look good. Export subsidies are the most politically offensive of all trade instruments and any agreement must provide for them to be cut by at least as much as domestic supports and access barriers, and probably by agreement to reduce both quantities of exports receiving subsidies and aggregate budgetary outlays. Special treatment for developing countries is likely to be reflected in slower rates of reduction commitments.

Such an agreement, though modest by the standards set by the radical initial proposals and 1990 offers of the United States and the Cairns group, would be a major accomplishment, for three reasons. First, it would mark an end to the situation wherein agricultural trade was in large measure outside the GATT and the start of an era in which the GATT's authority over the sector was acknowledged. Second, placing a ceiling on agricultural support and protection – and, more importantly, locking countries into a programme of reducing both – would reinforce policy changes that are already afoot in some countries and, beyond that, preclude a reversion to the unfettered competitive subsidization that has occurred in the past. This is the moment to ensure that, as national agricultural policies respond to future changes in economic conditions and political circumstances, they will be channelled in internationally constructive directions by legally binding quantitative limits on the levels of support and protection that can be provided and by clear guidelines on the agricultural policy instruments that are internationally acceptable. Finally, as is discussed in the next section, the economic effects of an agreement to reduce support and protection by the above orders of magnitude would be significant.

ECONOMIC IMPLICATIONS OF LIBERALIZING REFORM

The past decade has seen a range of modelling activity seeking to predict the effects of agricultural trade liberalization. In extending the results to conclusions about the potential impacts of a successful UR outcome, very little attention has been paid to the choice of the appropriate reference scenario. Some authors foresee a GATT breakdown generating a 'doomsday' or 'Götterdämmerung' situation characterized by escalation of export subsidy 'wars', heightened protectionism, trade 'bullying' by the big power triad, and increased world market distortion. Clearly, if this is the appropriate reference scenario, the quantitative results from most model analyses need to be somewhat inflated before they can be interpreted as indicative of the difference success in the UR could make! Others assume, often implicitly, that if the UR negotiations fall apart the status quo will continue. While views about the alternatives vary widely, many people (including ourselves) believe that a continuation of existing agricultural and trade policies is the least likely response to a UR failure. We are more of the persuasion that signficant changes in national agricultural policies are already in progress[2] and will continue even without an agreement on agriculture in the UR, while recognizing that a UR breakdown would inevitably escalate trade tensions at least in the short term. Under this scenario, the potential impact of the UR *per se* may be over-estimated by the various model results. The value of the results then lies in what they tell us about the potential costs and benefits of trade-liberalizing agricultural policy reform in general, rather than the UR in particular.

A full defence of our thesis – that global agricultural policy reform has a momentum of its own largely independent of, albeit reinforced by, the UR – is not possible, given the time and space constraints we face. May it suffice here to say that we would point to significant changes which have already occurred in Australia, Brazil, China, Japan, Mexico, New Zealand, Poland, Sweden and the USA or which are at present under discussion in the EC, Canada, Norway, the USSR and Eastern European countries, as current examples of unilateral agricultural liberalization. We would also point to budgetary restraints, macroeconomic costs, distributional and environmental concerns, pressures from other countries, World Bank/IMF conditionality, overhanging GATT panel decisions, bilateral and regional trade arrangements, and a general shift in focus away from 'farm policy' and towards agrifood and rural policy (in which more stakeholders have a voice) as forces underlying these reforms and common to many countries. The new policies being introduced in response to these forces are generally less distortive of global agricultural resource use, production, consumption and trade than the policies of yore. Thus the changes are for the most part consistent with GATT objectives but, rather than being GATT-driven, such unilateral reforms can be seen as the precondition that will make a multilateral agreement on agriculture achievable. For once, domestic desires and international imperatives are working together. In this light, the purposes of the UR negotiations may be those of locking governments into reshaping the objectives and reinstrumenting the programmes of their agrifood and rural policies in a manner in which they have already embarked, and of permitting them to move faster and further along the path of national policy reform because acting in concert will reduce the economic, social and political costs involved.

Model results

Useful reviews of the multicommodity agricultural trade models and what they say about the impacts of trade liberalization have already appeared in various places (for example, Meilke and Larue, 1989, Gardner, 1990, Blandford, 1990, Goldin and Knudsen, 1990). No comprehensive survey will be attempted here. However, we discuss briefly the flavour of the main results which seem to be emerging from this quantitative analysis of the effects of policies on trade, reflect on the interpretation of these results, and finish with a suggestion about model improvement in the light of the way the technical negotiations are evolving.

A common perception is that, given the differences which exist in the structure of the models, the base period used, the country and commodity scope and how 'liberalization' is defined, there is a perhaps surprising degree of commonality in the results they generate, at least with respect to the world price impacts of a full liberalization on the part of the OECD countries – arguably the most commonly analysed scenario. It would appear that collective OECD support and protection during the 1980s served to depress world dairy product prices the most and grains prices the least, with the price impacts for sugar and meats falling in between. There is a consistently small to negligible price impact on oilseed markets. Perhaps the greatest inconsistency is in the foodgrain area, where varying results reflect differences in the way livestock feed demand is modelled, and where there still appears to be some room for debate about the overall direction of the price impact of programmes: not all policies pull foodgrain prices in the same direction.

Surprisingly little reported or discussed, though obviously much analysed, given the nature of the models, are the effects of OECD liberalization on patterns and volumes of world production and trade, in total and commodity by commodity. As one example, Tyers' and Anderson's results for one scenario include an overall increase in world trade volume of 25 per cent for the products covered by their analysis, with substantial increases in rice, red meats, dairy products and sugar trade volumes and a noteworthy decline in wheat trade. The published evidence seems to point to a considerable reshuffling of trade between OECD countries, with increased exports of grains and beef from Australia and North America and of rice from the USA, decreased exports of dairy products from the USA and Europe and of sugar and wheat from the EC, and increases in Japanese imports of most products. The developing countries' trade will also be affected by OECD liberalization. The evidence seems to favour the LDCs increasing their production of livestock products, sugar, wheat and rice while possibly contracting in the soybeans and coarse grains areas. As a result, their overall cereal imports might change very little, but these countries as a group would be slower to move towards a net import position for rice, beef and sugar.[3] Given that the situation varies from one developing country to another, such generalizations are of limited usefulness. A wealth of detail on individual countries and commodities can be found in the studies collected in Goldin and Knudsen (1990).

Several analysts have addressed the question of what difference it could make if the LDCs as well as the OECD countries liberalized. One of the most comprehensive studies is the recent work by Anderson and Tyers (1990). Clearly, their concept of LDC liberalization goes well beyond the obligations to change agricultural support and protection which might derive from the UR, but not, perhaps, beyond what many developing country governments would nevertheless contemplate in a climate of global trade liberalization and economic policy rationalization. The important conclusions seem to be that, when LDCs liberalize too, the economic gains to them as a group are much greater, and individual country exceptions to the general result that LDCs are net beneficiaries are much harder to find. Because of the generally suppressive effect of the LDCs' food, exchange rate, and non-agricultural protection policies, LDC liberalization can be expected to stimulate Third World agricultural production. Consequently, if coupled with OECD liberalization, it has the potential to dampen the overall stimulus to world dairy and meat prices, to render doubtful the direction of the overall effect on wheat and sugar prices and to reduce coarse grain prices. Agricultural exports (imports) by LDCs would be correspondingly higher (lower), with the reverse applying to industrial countries. The overall volume of world agricultural trade seems likely to be stimulated significantly more under this scenario (Tyers and Anderson, 1991).[4]

Zietz and Valdés (1990) reviewed some studies projecting changes in world prices and production and trade up to the year 2000. It seems clear that, even if only OECD countries liberalize, and given that liberalization will at best be partial, any world price benefits in the grains area will only slow (but not reverse) the long-run downward trend in real world grain prices which can be expected over the next decade and beyond (Anderson, 1990). When it comes to trade volumes, Zietz and Valdés estimate the cereal import needs of the LDCs at around 175 million tonnes by the year 2000. They estimate that this figure – which would represent an increase of 125 million tonnes from the early 1980s – is likely to be only marginally lower in the event of agricultural trade liberalization. Examining a 50 per cent OECD liberalization scenario Anderson and Tyers (1991) found that the disincentive effect of reform on food production in protected countries merely slows the output expansion resulting from normal productivity growth. Even with such liberalization, they predict that Western Europe's food production in 2000 would be 20 per cent higher than in 1990. Even in Japan they foresee no decline in aggregate farm output, with contraction in grains output being offset by expansion in intensive livestock production.

It is generally recognized that even the most sophisticated economic models are, of necessity, great over-simplifications of reality. But do their results provide at least 'broad' accuracy? Despite a few suggestions that they are very inaccurate – for example, Sharples (1987) argued that the modelling exercises tend to greatly under-estimate the benefits to be derived from trade liberalization – most economists seem to have accepted that they do. Perhaps this question needs to be revisited. Many obvious avenues for model 'improvement' have been tried, like making models 'general' rather than 'partial', 'dynamic' rather than 'static', more 'disaggregated' with respect to commodity and country coverage, representing support policies in explicit rather than in proxy (for example, price wedge equivalent) fashion, and employing some 'gross trade flow' rather than a 'net trade' structure. Given the need for models to be manageable and transparent, there tends to be a trade-off involved, with improvement in one dimension often implying a loss of specificity or accuracy in other dimensions. However, the overall impression we have is that these kinds of 'improvements' do not lead to radical changes in the results generated.[5]

One recent development in the technical aspects of the agricultural negotiations seems to be suggestive of a specific improvement which could readily be made to many models in the way major support policies are represented. As a final point we wish to elaborate on this suggestion.

Support and stabilization

In our view, there are two principal types of commodity price support in the OECD countries today. These are illustrated in Figures 1 and 2. We call them 'stabilizing' support and 'constant margin' (or 'world price top-up') support. Our premise is that, while some examples of 'constant margin' support exist (for example, tariffs, Canadian grains transportation subsidies), the vast majority of support in the OECD countries today is of the stabilizing variety, involving some administered or target level of producer prices, with the market price support gap rising or falling inversely with world price levels.[6]

In Geneva there has been a growing recognition that, while less insulation of producers from market realities might be a commendable long-term goal, significant reductions of current levels of support are only achievable if that support is able to retain, at least initially, a large measure of its stabilizing element. In other words, the stabilizing feature may be able to be diluted, too, but only gradually: support can not be abruptly switched from 'stabilizing' to 'constant margin' before being gradually reduced as 'pure' tariffication and earlier versions of proposed AMS reduction commitments would imply. As a result, the 'fixed external reference price' (FERP) principle, as an approach to AMS reduction commitments, has now gained broad acceptance in the negotiations. Proposed first by the EC and more recently embraced by the USA and the Cairns group, it essentially implies a commitment to reduce the level of (effective) producer support prices rather than to reduce the margin by which they exceed world price levels. The counterpart of this principle on the market access (tariffication) side would be something like the EC's 'corrective factor' without the trappings of a separate and direct exchange rate adjustment. Here the concept is more controversial, but will, we predict, in carefully circumscribed and time-limited form, and with some misgivings, be ultimately accepted, too, in return for an appropriately enhanced EC offer. A corresponding adjustment for export subsidy reduction commitments is yet to be proposed but can be anticipated if, and to the extent that, these commitments are expressed in terms of export subsidy expenditures. To be sure, it might be desirable if over the long term national commodity policies were to offer farmers no more than 'stop loss' price guarantees where the 'floor' price was significantly less than long-run market price, and better yet if gross or net farm income were to become the target variable in agricultural 'safety net' programmes (which is the direction in which Canadian stabilization programmes are evolving). However, there is no prospect that universal agreement on such far-reaching changes in agricultural commodity programmes can be reached in the Uruguay Round: farm product prices will continue to be at least partly 'stabilized' at above world transaction prices in developed countries for some time to come.

The basic 'price gap' model, with domestic prices fully linked to world prices, essentially represents all support as being of the 'constant margin' variety. In some models the reality of 'stabilizing' support is incorporated indirectly through the use of price transmission coefficients with a value less than unity. Our concern here is whether these coefficients are generally low enough for the OECD countries, even when estimated rather than judgemen-

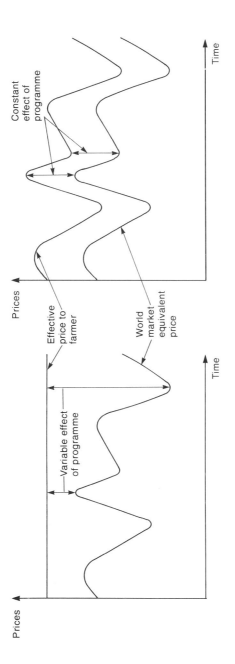

FIGURE 1   *Stabilizing support*

FIGURE 2   *Constant margin support*

tal. Our own preference is for models in which individual support policies are represented more explicitly. The existence of 'stabilizing' support would imply that the price support level should then be the exogenous policy variable and that the price support 'gap' or 'margin' should be endogenous. While recognizing that some progress with specification or respecification of some models along these lines has been made – for example, the OECD's 'MTM' model – some mis-specification still exists. To the extent that it does, the models concerned will be deficient for the purposes of simulating the impacts of a UR agreement involving incomplete reduction of support and protection based on a fixed external reference price principle.

Perhaps one reason for agricultural trade modellers having been reluctant to move far beyond the simple price gap approach is that, for purposes of comparative static analysis of full liberalization scenarios, it does not make any difference: the end-point is the same. Clearly, however, the time-path to this end-point will be different if support reductions are made according to the fixed reference price principle. And if the liberalization is only partial, the end-point reached can also be significantly different. This latter difference becomes more important the more a commodity's world price level is distorted because of existing support. The most striking example of this is probably the case of dairy products. Under a FERP-based support reduction internal support prices would be reduced by an agreed amount, independent of what happened to world prices. In contrast, under a constant margin reduction approach, internal support prices would only fall to the extent that world prices did not rise in response to the support reductions. A corollary of this is that, while under a constant margins reductions approach a 100 per cent reduction is necessary to remove all trade distortions, under a FERP-based commitment distortion-free trade may require a reduction of less or greater than 100 per cent of the support margin measured at base period price levels, depending on whether the base period world price level is distorted downward or upward. It can be shown that, if world prices are distorted downward – as seems to be the case for most products – the percentage 'depth of cut' necessary to reach a situation of distortion-free trade can be derived from the extent to which the support price in the most heavily distorting country exceeds the free trade equilibrium world price. To illustrate this with a simple numerical example: if the 'free trade' world price for product X was $200/tonne, this being twice its current level of $100/tonne (that is, a 50 per cent downward distortion), and if the price support level of $400/tonne in the most heavily supporting country was four times the current world price level (a price support gap of 75 per cent), then a 67 per cent reduction in FERP-based support would be sufficient to reach a point of distortion-free trade since this would reduce the support price in that most heavily supporting country to $200/tonne (abstracting, for purposes of simple example, from distortions in world prices of X which might derive indirectly from support for other commodities).

Another, and perhaps even more important implication of FERP-based support reductions when world prices are significantly distorted downward is that world price rises resulting from linearly phased support reductions will not occur linearly but are likely to be concentrated in the early years of the reduction period. Perhaps the best intuitive explanation of this phenomenon is that, in such a case, the 'apparent' or 'gross' level of support (measured at current world prices) is greater than the 'real' or 'net' support (measured at undistorted or free-trade world prices). If gross support in different countries is distributed across a range of values (say price gaps between 0 and 75 per cent in the example above) the linear FERP-based reductions in 'gross' support levels will result in non-linear (more 'up-front') reductions in 'net' support and correspondingly early removal of the majority of the world price distortion. In addition, expectational behaviour may result in the market 'anticipating' the structural price rises which would be implicit in a set of agreed support reduction commitments, in which case the price gains may accrue in full very soon after the ratification of the agreement.

In summary, we conclude that the adoption of the FERP principle will imply that a modest reduction in support margins – say 30–50 per cent – will suffice to remove a large part of existing production and trade distortions in cases like the dairy sector where world market prices appear to be heavily distorted downward by existing support and protection. Furthermore, rather rapid corrections of a large part of those world price distortions appear to be in prospect. In the case of products like feed grains, on the other hand, where existing world prices appear to be little out of line due to offsetting distortions on the supply and demand sides, much production, consumption and trade distortion is likely to remain after such a modest cut.

It is important to keep the magnitude of these trade liberalization results in perspective. It seems clear that huge changes in world agricultural trade patterns are going to occur over coming years as the result of differential rates of population and income growth, great differences in income elasticities of demand between commodities and between rich and poor countries for given commodities, and differential rates of productivity growth. In addition, as we have seen with China and Vietnam, and may well see with Eastern Europe and the USSR, fundamental policy changes in major producing (and consuming) regions can have important implications for world trade. When superimposed on these potential changes, the possible impacts of any conceivable trade liberalization scenario look rather modest, at best.

Finally, it is important to recall that the UR negotiations are set against a backdrop of a long-term downward trend in real world market food prices, driven largely by technological advances and productivity gains. A relatively small one-time price boost due to policy reform cannot be expected to reverse this trend for long, and may only slow it. The widely held notion that global interventions by governments in agriculture serve to reduce world prices – a notion that we agricultural economists have fostered to no small degree – is becoming debatable, at least in several important commodity areas. No doubt UR-led liberalization on the part of OECD countries over the next decade will provide a boost to world prices in many commodity areas, *ceteris paribus*. However, it may well be that it is the reforms that the Eastern European countries and the majority of the LDCs introduce in the same time period (and independent of the UR outcome) which will tip the balance as to whether the world price effect of overall policy reform in coming years is positive or negative in many product sectors.

CONCLUDING OBSERVATIONS

While we have suggested that a feasible agricultural accord may contain modest numbers for agreed support and protection reductions, particularly when compared with the endogenous reforms that are already in train, we do not wish to detract from the importance of a UR agreement on agriculture. Rather we would want to emphasize five matters that lend importance to the agricultural component of the Uruguay Round. First, in narrowly agricultural terms, commitments made now will hinder a future resurgence of agricultural subsidization and market insulation in the developed countries and discourage the emergence of a second generation of agricultural protectionism in what are now developing countries as they experience economic growth. Second, even a modest accord would provide important benefits in the efficiency of world agricultural resource use, more defensible income distributions and a reduction in international political disharmony. Third, this is the first time that an attempt has been made within the GATT to place comprehensive international disciplines on the conduct and content of national economic sectoral policies. In this sense, what is being attempted in agriculture may be pioneering a course which may have to be emulated in other basic and high technology sectors in which there is a high degree of government intervention. Fourth, whereas agriculture has been peripheral in previous GATT rounds and after preliminary skirmishes among the 'Big 3' has been left aside, in the Uruguay Round an agricultural accord is a *conditio sine qua non* to the completion of the round and hence to the future of the international economic order. Fifth, if it seems that the political and programme concessions demanded of Europe and Japan seem onerous it is not just because the external effects of their domestic farm programmes are particularly distortive of resource use and markets: it is also because they are being challenged to acknowledge their responsibilities as guardians of an open and rules-based multilateral trading system and asked to make an investment in the international stability, prosperity and harmony that only such a system can provide.

The position of the developing countries in the Uruguay Round is also a departure from the past. In so far as previous rounds dealt primarily with market access issues it was easy for the rich countries to put the developing countries' interests in this area aside. In the Uruguay Round the developing countries' insistence on satisfaction on 'backlog' access issues (agriculture, textiles, tropical products, safeguards and 'voluntary' export restraints) as a precondition for their agreement to extending the GATT into new areas and to strengthening its institutional features has put the developing countries in the position of *demandeurs*. The blocking power of the developing countries, when joined with an indigenous policy ambience in the developed countries that favours unilateral agricultural policy reform, may just prove to be the combination that succeeds in 'bringing agriculture into the GATT'.

However, the agenda for international agricultural and trade policy in the Uruguay Round is overwhelmingly concerned with placing international disciplines on the familiar instruments of agricultural support and protection – domestic subsidies, import barriers and export aids. We suspect that by the time that this Conference reconvenes our attention will be focused on the next wave of impediments to international commerce in food and agriculture. These will be the incidental international effects, or the protectionist use, of national and sub-national interventions and regulations directed at food safety, animal welfare, environmental protection, resource stewardship and labour standards (Hillman, 1991), and measures aimed at creating competitiveness. At present, the international community of nations is ill-equipped to handle these issues through the GATT or other international institutions (Runge, 1990). Accordingly, these are the matters which require the urgent attention of the international community of agricultural economists.

NOTES

[1]In agriculture, the major other elements to be settled are the articulation of a code that will prevent divergent national sanitary and phytosanitary standards being unnecessary barriers to trade and the extent of the special and differential treatment to be accorded to developing country food exporters and importers with respect to their support and protection reduction commitments. On the first, there seems to be an excellent prospect of a code being agreed that will entail harmonization of international standards, equivalence of results, and appeal to scientific knowledge in the resolution of disputes. On special and differential treatment, there seems to be an inclination to be more accommodating to the wishes of developing country importers than to exporters, but to insist that the importers' obligations be commensurate with their level of development. The perception is that the LDCs have enjoyed the benefits of the GATT system while assuming few of its obligations. For instance, the tariffs levied by many LDCs are high and unbound, and many have sought exemptions for their import controls on balance of payments grounds under Article XVIII:B. There is no disposition to allow the LDCs a free ride in reforming the agricultural trading system. To do so would be bad for the cause of agricultural reform, for the GATT system, and for the LDCs themselves. The groups negotiating on subsidy-countervail issues, market access, safeguards, tropical products, natural resource-based products and dispute settlement mechanisms are also likely to produce agreements with a bearing on the conduct of agricultural trade.

[2]The agricultural policy developments we observe include the following. The old arguments for food self-sufficiency have lost much of their force in a world where affordable abundance at falling real prices is assured to all but the penurious. The historical preoccupation of policy with farm-level prices and incomes is being diluted as governments accept a wider responsibility for encouraging agrifood sector development through: (a) the delivery of good macroeconomic policies; (b) the provision of collective services; (c) the promotion of competitiveness, and (d) the correction of such market failures as dysfunctional instability, non-competitive market behaviour and the underpricing of resource and environmental externalities. Ministers of agriculture are assuming responsibility for rural development and for enhancing the social and amenity value of the countryside. To be sure, commodity-centred farm price and income support and stabilization still account for the bulk of sectoral expenditures, but commodity policies are being transformed from an undifferential mechanism for making income transfers to farmers into a more restrained societal commitment to underwrite the adequacy and stability of factor returns in agriculture. Payment limits are replacing open-ended price guarantees. Producers are being exposed to the market value of incremental output through coresponsibility levies and 'stabilizers'. Within the limits of budgetary exigencies, decoupled income payments have increasing appeal in delivering measured benefits to selected groups. Payments are justified less for producing farm products and more for the supply of environmental goods. And commodity programmes focus less on the augmentation of farmers' incomes through price 'supports' and more on their stabilization through economic 'safety nets' that come into force only in periods of exceptional market or production stress. These movements towards more trade-friendly policy objectives (competitive development rather than insulating subsidization) and instruments (direct payments and adjustment assistance rather than commodity price supports) are at once both the goals of the UR agricultural negotiations and the means that will permit their attainment.

[3]These big models tend to exclude, at least in disaggregated form, products such as fruits and vegetables and tropical products which are of major export importance to LDCs. Moreddu *et al.* (1990) suggested that, given generally low OECD import duties, this trade would not be much affected by OECD liberalization – a result which seems to be borne out by Mabbs-Zeno and Krissoff's analysis (1990) of cocoa, tea and coffee and by a recent UNCTAD (1990) study which covered cotton, groundnuts, copra, palm oil and tobacco as well as the tropical beverages. However, the research of Islam (1990) suggests that the LDCs have much to gain from an easing of OECD tariffs and, more importantly, non-tariff barriers (largely sanitary and phytosanitary regulations) in the fruits and vegetables area.

[4]We have chosen not to say anything in this paper about the stabilizing impact of liberalizaton on world market prices, something of particular importance for LDCs, and which has been emphasized repeatedly by, among others, Anderson and Tyers in a number of papers (for example, Anderson and Tyers, 1990). This observation, of course, does not preclude other major causes of instability still remaining after such liberalization.

[5]It might be worth noting that, even if a 'super-model' could be constructed containing all the above-noted attributes, it would still have a very simple neo-classical economic structure. Does this result in misleading information about the way actual policies can affect the market? Take the case of directed export subsidies as an example. Is it possible that, because of the way it is legislated, regulated and administered, and perhaps because of concentration among international grains traders and the psychology of the market, the US Export Enhancement Program is able to apply sustained downward pressure on international grains prices to an extent which would not be possibe with a 'regular' export subsidy as represented in models where such prices are determined solely by supply and demand fundamentals? Such issues deserve further attention. In the meantime, however, the existing models provide our only guide to the quantitative impacts of policy reform.

[6]Whether this type of stabilizing support, provided more or less uniquely to the farm sector, is economically sensible or justified has been (and no doubt will continue to be) debated at length. The reality is that it exists, and continues to attract a great deal of political support. While a few farm commodity groups in a few countries prefer to manage their own market risk insurance, the majority of farm groups in the industrialized countries want the security of a floor price. The socialization of farm-level risk is a near-universal feature of the commodity policies of developed countries. Moreover, because instability will continue to be a feature of agriculture's product and factor markets even if agricultural support and protection are reduced, the provision of farm income 'safety nets' will probably emerge from the UR as an internationally acceptable form of government intervention in agriculture.

REFERENCES

Anderson, K., 1990, 'Policy Implications of Model Results', in Goldin and Knudsen, *Agricultural Trade Liberalization*.

Anderson, K. and Tyers, R., 1990, 'How Developing Countries Could Gain from Agricultural Trade Liberalization in the Uruguay Round', ch. 2, in Goldin and Knudsen, *Agricultural Trade Liberalization*.

Anderson, K. and Tyers, R., 1991, 'Effects of Gradual Food Policy Reforms in the 1990s', *Eur, Rev. Agr. Econ.* 18,(4), October.

BAE (Australian Bureau of Agricultural Economics), 1985, *Agricultural Policies in the European Community: Their Origins, Nature and Effects on Production and Trade*, Policy Monog. 2, Canberra.

Blandford, D., 1990, 'The Costs of Agricultural Protection and the Difference Free Trade Would Make', ch. 9, in Sanderson, F. (ed.) *Agricultural Protectionism in the Industrialized World*, Resources for the Future, Washington, DC.

Gardner, B., 1990, 'Recent Studies of Agricultural Trade Liberalization', in *Agricultural and Governments in an Interdependent World*, (Proceedings of 20th International Conference of Agricultural Economists), Gower, Aldershot.

General Agreement on Tariffs and Trade, 1990, 'Elements for Negotiation of a Draft Agreement on the Agricultural Reform Programme', 6 December, mimeo.

Goldin I and Knudsen O., (eds), 1990, *Agricultural Trade Liberalization: Implications for Developing Countries*, OECD, Paris.

Hathaway, D.E., 1990, 'Agriculture', in J.J. Schott (ed.), *Completing the Uruguay Round: A Results-Oriented Approach to the GATT Negotiations*, Institute for International Economics, Washington, DC.

Hillman, J.S., 1991, *Technical Barriers to Agricultural Trade*, Westview Press, Boulder, Co.

International Agricultural Trade Research Consortium, 1990, *The Comprehensive Proposals for Negotiations in Agriculture*, Commissioned Paper No. 7, University of Missouri, Columbia.

International Agricultural Trade Research Consortium, 1991, *Reviving the GATT Negotiations on Agriculture*, Commissioned Paper No. 8, University of Minnesota, St. Paul.

Islam, N., 1990, 'Horticultural Exports of Developing Countries: Past Performances, Future Prospects, and Policy Issues', Res. Rep. 80, Int. Food Policy Res. Inst., Washington, DC.

Lawrence, R.Z. and Schultze, C.L., (eds), 1990, *An American Trade Strategy*, The Brookings Institution, Washington, DC.

Mabbs-Zeno, C and Krissoff, B., 1990, 'Tropical Beverages in the GATT', ch. 6 in Goldin and Knudsen, *Agricultural Trade Liberalization*.

Meilke, K.D. and Larue, B., 1989, 'A Quantitative Assessment of the Impacts of Trade Liberalization on Canadian Agriculture', in *Agriculture in the Uruguay Round of GATT Negotiations: Implications for Canada's and Ontario's Agrifood Systems*, AEB 89/6, University of Guelph, Guelph.

Miner, W.M., 1990, *The GATT Negotiations and Canadian Agriculture: Preparing for the Brussels Ministerial Meeting*, occasional paper in International Trade Law and Policy, Centre for Trade Policy and Law, Carleton University, Ottawa, November.

Moreddu, C., Parris, K. and Huff, B., 1990, 'Agricultural Policies in Developing Countries and Agricultural Trade', ch. 4, in Goldin and Knudsen, *Agricultural Trade Liberalization*.

Organization for Economic Cooperation and Development, 1987, *National Policies and Agricultural Trade*, OECD, Paris.

Oxley, A., 1990, *The Challenge of Free Trade*, Harvester Wheatsheaf, New York.

Paarlberg, R., 1991, 'Why Agriculture Blocked the Uruguay Round: Evolving Strategies in a Two-level Game', Harvard Center for International Affairs, mimeo.

Runge, C.F., 1990, 'Trade Protectionism and Environmental Regulations: The New Nontariff Barriers', *Northwestern Journal of International Law and Business*, 11,(1).

Schott, J.J., 1990, *Completing the Uruguay Round: A Results-Oriented Approach to the GATT Trade Negotiations*, Institute for International Economics, Washington, DC.

Sharples, J.A., 1987, *Estimating the Gains from Less Distorted Agricultural Trade*, Working Paper 87–1, Int. Agric. Trade Research Consortium.

Stoeckel, A., 1985, *Intersectoral Effects of the CAP: Growth, Trade and Unemployment*, occasional paper no. 95, Bureau of Agricultural Economics, Australian Government Publishing Service, Canberra.

Tyers, R. and Anderson, K., 1991, *Disarray in World Food Markets*, Cambridge University Press, Cambridge (forthcoming).

United Nations Conference on Trade and Development, 1990, 'Agricultural Trade Liberalization in the Uruguay Round: Implications for Developing Countries', UNCTAD ITP/48, New York.

United States Department of Agriculture, 1988, *Estimates of Producer and Consumer Subsidy Equivalents: Government Intervention in Agriculture, 1982–86*, USDA Economic Research Service, Staff Report No. AGEC 880127, Washington, DC.

United States Department of Agriculture, 1990, *Multilateral Trade Reform: What the GATT Negotiations Mean to U.S. Agriculture*, Staff Briefing Paper.

World Bank, 1986, *World Development Report 1986*, Part II, Oxford University Press, New York.

Zietz, J. and Valdés A., 1990, 'International Interactions in Food and Agricultural Policies: Effects of Alternative Policies', ch. 3 in Goldin and Knudsen, *Agricultural Trade Liberalization*.

DISCUSSION OPENING – DAVID BLANDFORD*

McClatchy and Warley do an excellent job of reviewing the progress made on agriculture in the Uruguay Round negotiations, identifying what type of agreement might emerge, and assessing its implications. I share their view of the importance and technical feasibility of obtaining a meaningful agreement on agriculture, and the broad outline of its likely content. Only time will tell if our collective optimism is justified by the signing of such an agreement.

The authors rightly stress the importance of bringing agricultural policy under greater international discipline. As is widely known, the OECD has been in the vanguard of establishing the case for reforming agricultural policies, not only because of the improvements in international markets that would result from such reform, but also because of its potential domestic benefits (OECD, 1982, 1987). Despite the complexity and political sensitivity of agricultural policy in many countries, there is now far greater awareness of the implications of existing policies and their costs, partly as a result of the analysis conducted by international and national organizations and private individuals, particularly in the academic community. Greater transparency concerning the nature and implications of policies is essential for stimulating discussion of reform options, and consequently the monitoring of policies is an important part of the OECD's continuing work (for example, OECD, 1991a).

McClatchy and Warley point out that the trade problems which are the focus of the current negotiations in Geneva are created by domestic agricultural programmes. There is a broad consensus on the direction, if not the magnitude, of the effects of these programmes on international trade. There is also a broad consensus on the desirability of reforms to minimize the trade distorting effects of the programmes. There is somewhat less consensus on whether this implies eliminating existing programmes altogether, modifying them, or replacing them with alternatives. Since a major objective of the OECD is the promotion of economic growth and social welfare in member-countries, the secretariat spends a great deal of time thinking about the role of government in agriculture. Consequently, in discussing the paper, I should like to focus on some of the policy issues which may not be resolved as a result of the current negotiations, but which are likely to prove important in the future.

As correctly emphasized in the paper, a major driving force behind the GATT negotiations is recognition of the growing costs of agricultural support. The latest estimates place the level of agricultural assistance in the OECD countries as a whole at 44 per cent of the total value of output.[2] This translates into almost 300 billion US dollars of transfers from consumers and taxpayers in 1990. Although the primary focus in the media tends to be on the costs of support for government budgets, over 60 per cent of the total is actually due to the implicit tax levied on consumers through higher food prices.

The high costs of agricultural support, particularly the taxpayer costs, have been an important factor in stimulating policy reform. In so far as the reduc-

*Organization for Economic Co-operation and Development, Paris.[1]

tion of government costs leads to less trade distortions, budgetary and trade reform objectives will coincide. However, such a correspondence may not always exist. For example, the taxpayer costs of agricultural support can be reduced by shifting the burden to consumers. This may intensify trade distortions.

For this reason, the possible focus in the trade negotiations on a reduction in aggregate support ($P \times Q$), suggested by McClatchy and Warley, leaves open the possibility that international obligations can be met by countries, but at the risk of perpetuating domestic distortions. Countries may choose to operate on quantity rather than price and use production quotas to reduce total support. Recent quantitative analysis by the OECD (1991b) suggests that reduction of quotas could be an effective way to reduce both budgetary costs and trade distortions. Unfortunately, domestic consumers would then lose the potential benefits of reducing support. The negative distributional and efficiency aspects of quotas, and their limitations in achieving long-term adjustment, have been well established (OECD, 1990). The use of quantitative controls to meet international obligations on reducing support is unlikely to result in domestic markets which are more integrated into international markets.[3]

An undue focus on budgetary aspects poses an additional problem in that it has been argued that, if governments wish to provide income support to agriculture, this should be done through direct payments (OECD, 1990). The use of such payments offers the possibility of greater transparency and the reduction of demand-side distortions in the trade equation. If direct payments are de-coupled from production, supply-side distortions can also be reduced. There should be some recognition of the desirability of directing the reduction of support to those measures which distort both production and consumption, and of replacing non-transparent indirect subsidies with transparent direct subsidies, substantially delinked from production. A move to direct payments, providing that these are implemented in an appropriate manner, would seem to be a desirable way of achieving simultaneously domestic and international benefits from the reform of agricultural policies.

In addition to the budgetary question, there are a number of other issues which, if not exactly neglected in the negotiations, have not been centre stage. These are likely to be increasingly important in the future, and have the potential to create contradictions between domestic and international objectives. Probably the two most important areas are those of environment/public goods issues and stability.

The issues surrounding agriculture and the environment are complex. Agricultural externalities in the form of contamination of groundwater by chemicals, soil erosion, loss of wetlands and other wildlife habitats are increasingly recognized. Similarly, there is growing perception of the role of agriculture as a supplier of public goods, such as landscape and public amenity, as well as a sink for carbon dioxide in a world subject to possible changes in climate. The determination of appropriate policy responses to these issues is difficult. If there is widespread market failure in agriculture, should this be rectified primarily by regulation, through taxes, or through subsidies? These alternatives are unlikely to have equivalent domestic and international implications.

One particularly troubling issue is seen when the optimal policy choice implies a level of production greater than that which would be forthcoming under free trade. The achievement of an appropriate supply of public goods by agriculture may, in some cases, require a production-related subsidy. Furthermore, there is the possibility that, in the absence of coordinated international approaches to the problem of externalities, border measures may be necessary to prevent domestic measures, which seek to correct market failure, from being undermined. The problem of the environment is likely to be of increasing importance not only domestically, but also internationally. There is substantial potential for conflict between domestic and international objectives in this area.

The issue of stability is also one which has the potential to create contradictions between domestic and international policies. McClatchy and Warley indicate that many domestic policies have a stabilization component. There is room for debate on the degree to which current policy measures actually stabilize domestic markets, but nevertheless the stability issue is likely to become increasingly important if there is a general movement towards more open markets. Up to now, much of the focus in OECD countries has been on measures which might be used to guarantee a degree of producer income stability. In the future, the question of whether there is a role for governments in stabilizing consumer prices, for example, through public stockholding is likely to gain prominence.

The concluding comments to their paper made clear that McClatchy and Warley are well aware of the likely future importance of these issues. Like them, I believe that, although it may not be possible to address all agricultural issues fully in the current GATT round, the achievement of an agreement which would bring agriculture under greater international discipline is of major importance. It is to be hoped that an agreement can be reached that will lead both to a reduction in trade distortions and to the full realization of the domestic gains from a more efficient allocation of resources.

NOTES

[1]The views expressed are the responsibility of the author and do not necessarily reflect those of the OECD or its Member countries.

[2]These figures do not include estimates for Iceland and Turkey, which are full members of the OECD, nor for Yugoslavia, which takes part in some of the work of the Organization.

[3]The logical implication of the use of quotas to achieve reductions in support and export subsidies by net exporters is to move towards self-sufficiency, even if there would otherwise be net imports at free trade prices.

REFERENCES

OECD, 1982, *Problems of Agricultural Trade*, OECD, Paris.
OECD, 1987, *National Agricultural Policies and Agricultural Trade*, OECD, Paris.
OECD, 1990, *Reforming Agricultural Policies: Quantitative Restrictions on Production: Direct Income Support*, OECD, Paris.

OECD, 1991a, *Agricultural Policies, Markets and Trade: Monitoring and Outlook*, OECD, Paris.

OECD, 1991b, *Changes in Cereals and Dairy Policies in OECD Countries: A Model-based Analysis*, OECD, Paris.

A. EMEL'IANOV*

The Soviet Union's Agrarian Sector on the Way to a Market Economy

INTRODUCTION

Like ecology and energy, the food problem is global in character. It is no wonder, therefore, that national leaders and scholars are uniting in their efforts to solve it. The Soviet Union is the largest world importer of agricultural products and foodstuffs, buying 40–45 million tons of grain on average annually and many other products as well, for a total outlay of some 20–21 billion dollars per year. Such a level of imports represents a heavy burden for the Soviet economy, particularly under conditions of scarce hard currency resources. *Perestroika* has brought about a reduction in international tensions and freed huge resources that can now be devoted to peaceful development, including those in the food complex. But the fate of *perestroika* in the Soviet Union depends upon supplying food to the population. The condition of the market in food determines popular attitudes towards all policies, even the most major of them, and towards the country's leaders. Moreover, stabilization of the entire economy, whose current status threatens to bring down *perestroika* itself, depends upon the food supply. Comprehensive resolution of the economic crisis is tied to the transition to a market economy. But it must be clearly understood that practical movement towards a market economy is currently being stymied by the breakdown of the market in food.

One of the main errors of *perestroika* consists in the fact that the reorganization of the economy did not begin with the village, with a restructuring of the food complex. Had we followed this course, we might have found a sound way, from April 1985 onward, to solve the enormous problems that were strangling the country. We might have been able to avoid the importing of food that represents such a disgrace to us as a nation. We might have been able to limit substantially the export of oil, gas and other natural resources that are being sold to obtain dollars to pay for importing food. The extraction of these resources in Siberia and in the Far North and East is associated, moreover, with complex economic, social and ecological problems.

From this there follows a practical lesson for present policy: it is clear that, without fundamental improvement in matters relating to food, all other steps

*A. Emel'ianov is deputy chairman of the Committee on Agrarian Issues and Foodstuffs of the Supreme Soviet of the USSR, professor and academician of the All-Union Lenin Academy of Agricultural Sciences, and head of the department at Moscow State University. The text has been translated by Jean Laves Hellie.

152

towards developing a market mechanism and saving *perestroika* will have no effect whatsoever. That is why it is so important to give immediate priority ortiy to a radical restructuring of the food complex, and not theoretically but in practice.

Currently these issues are intimately connected to solving general problems in the transition to a market economy. What are some of the new aspects of these problems? What do people in the Soviet Union see as the solution to the food crisis that has developed? I hasten to observe that there is no simple approach to solving the problem of food. I will focus on two of the most general concepts among the great variety of such approaches. Many leaders, scholars and individuals with direct practical experience consider it essential for additional resources (technology, construction materials, fertilizers and so on) to be allotted to the agrarian sector, while others assert that, although there is no denying the need to strengthen the material and technical base, the main transformation must take place in the socio-economic structure of the village, in the forms of property and ownership.

These two views cannot be considered to be mutually exclusive. Adherents of the first do not deny the necessity of new forms of ownership, while adherents of the second do not reject the advisability of saturating the village with resources. The difference of opinion arises over what is considered primary. It is on this that the formulation of the specific measures of agrarian policy depends in each case.

The new view of the economy increasingly leads one to conclude that what is fundamental here is not saturating agriculture with resources but a thorough restructuring of agrarian policy, a restructuring that will permit increasing the return on both currently available and newly applied resources. What are the immediate measures to be instituted in restructuring fundamental policy in the agrarian sector as we move to a market economy?

(1) Of primary significance is reducing losses and ensuring proper storage and complete processing of products raised. At present we are wasting far more than we buy on the world market. Improving storage of products raised assumes not only allocating additional resources to develop the infrastructure but also changing our concepts about such development. It is important to move procurement and processing of agricultural products closer to the point of production and to overcome excessive concentration in this sphere.

(2) Of primary importance, too, is allocating resources to the development of social services in the village – to building roads, houses and other elements of the social infrastructure.

THE MAJOR ISSUE

Now I will move to the main issue upon which a comprehensive resolution of the food crisis depends. This has to do with the fundamental transformation of the socio-economic structure of the village, with the forms of property and ownership. Attention to these issues has increased as the orientation has shifted towards a market economy.

I should emphasize at the outset that many specialists have arrived at a rather facile perspective on the entry of the agro-industrial sphere into the market. They believe that the agrarian sector is better prepared than other spheres of the economy to shift to the market in terms of both timing and extent. Supposedly, in their view, the market mechanism can be introduced more rapidly in this sector, and it can also be introduced completely and in all spheres. This view, it seems to me, is far from indisputable. It reflects an insufficient knowledge of the specific features of the agrarian sector and the actual situation in the countryside and in the related branches of the agro-industrial complex. This kind of facile perspective, in my opinion, is fed by formal comparisons with Western countries and even certain countries of the former socialist bloc. But no other country, after all, had a barracks–gulag system like ours for 73 years. No other country besides the Soviet Union allowed its countryside and land to reach such a state of devastation. No other country destroyed the most elementary foundations of its market infrastructure the way we did. Great caution, therefore, should be exercised in drawing analogies between the Soviet Union and other countries as far as the agrarian sector's transition to the market is concerned.

No matter what aspect of the transition to the market one takes, its resolution in the agrarian sector involves not fewer but, indeed, greater difficulties than in other areas. I will begin with the main issue, the formation of multiple economic structures, denationalization and privatization in the agrarian economy. In recent years, new, unfamiliar economic forms have been developing, especially peasant farms and various kinds of cooperatives. A vigorous debate is taking place over the question of the socio-economic structure of our rural areas. Some maintain that famine is associated with collective and state farms (*kolkhozes* and *sovkhozes*) and that the only hope for feeding the populace lies in peasant farms. Their opponents assert with equal fervour that an orientation towards peasant farms would represent a step backwards, a movement against progress in the use of productive forces running counter to the earth's age-old rotation. But the truth, as so often happens, lies somewhere in between. Let us examine briefly the theoretical and practical aspects of this question.

In seeking a normal structure for the agrarian sector we are not starting with a *tabula rasa*. In the USSR, the question of multiple economic structures, of various types of ownership in the village, is being worked out under actual conditions where the rural situation is defined by the *kolkhozes* and *sovkhozes*.

Can individual farms become the dominant and generally prevailing form in the countryside today? I believe not. Nor is it a matter only of the fact that the entire system as it has become established is tearing out the peasant way of life like a foreign body, preventing its emergence in every possible manner. The main problem has to do with the social base of this way of life. This base is not very broad. I cannot say that there is no one who wants to obtain some land and run a family farm. Such people do exist; indeed, there are quite a number of them, and if one starts with zero, their numbers are growing. These people are true heroes who must be supported in every way possible. It is only unfortunate that the 'partocratic' system places all kinds of obstacles in their path. But from an objective standpoint, it must be said that they represent only an inconsequential minority among all the peasants, among all rural workers.

The entire history of our country since 1917 has led to the spiritual, indeed, physical degradation of the peasantry; it has stifled their proprietary initiative, and their capacity for independent action. The majority of peasants have become accustomed to taking no responsibility for anything, to taking no interest in economic matters. The pay is decent, and if it is not, it is possible to expand a private plot or operate on a black market, which goes on everywhere. But to undertake running an independent farm involves huge risk and responsibility. A whole way of life changes. By no means everyone is ready for this, even if there were comprehensive support from the farmers' movement. This is all the more so when the situation is the precise opposite. There are enormously complex problems in obtaining technology, fertilizers, construction materials and other resources. The entire agricultural industry is oriented towards serving the *kolkhozes* and *sovkhozes*. Also price and credit policies are not directed towards supporting the new system.

We should also mention another vital factor that works against peasant farms. It reflects the specific features of rural life and places the farmer in *de facto* dependence upon the *kolkhoz* and *sovkhoz*. In the city, the social infrastructure is not tied directly to a particular enterprise. The emerging private entrepreneur can avail himself of the possibilities of the city as a whole. In the village, on the other hand, all social factors are tied territorially to a particular *kolkhoz* or *sovkhoz*.

The peasant way of life is destined to play an important role in the restoration of an agrarian market economy, and not so much quantitatively as qualitatively. As should be apparent from everything we have said, it cannot have the necessary weight in quantitative terms. At the beginning of the current year, there were approximately 50 000 privately owned farms in the country. They provided less than one per cent of agricultural products in 1990. The number of peasant farms will increase. Nonetheless, we must not forget the main factor – the narrow social base and lack of preparation on the part of the overwhelming majority of the peasantry for independent ownership, risk and responsibility, and their unwillingness to undertake running a private farm.

This process can only be carried out voluntarily. It would be stupid to disband the strong *kolkhozes* and *sovkhozes* where production is going well and people's lives and the services they receive are well structured. Here, too, however, those who want to run their own farm should be given the opportunity to get a share of land and a corresponding segment of the production resources. Such *kolkhozes* and *sovkhozes*, too, must work to improve their structure on democratic principles. But they should not be broken up into individual plots for the sake of a fashionable movement towards the market.

In other words, economically strong agricultural enterprises cannot provide any noticeable base for independent farming. Many people believe that exciting prospects for extending the peasant way of life can be realized through immediate development of economically weak, unprofitable *kolkhozes* and *sovkhozes* that are essentially moribund. Such notions reflect a poor knowledge of the actual situation in the village. After all, it is in such enterprises, more than anywhere else, that the weakest stratum of the peasantry is to be found. They would represent, therefore, an extremely fragile internal base for farming.

TRANSITION TO THE MARKET

Now I will address the question of the fate of such enterprises in the transition to the market. A person can be declared bankrupt who was an owner, made all decisions himself, and therefore bears responsibility for all results. But the weak *kolkhozes* and *sovkhozes* have been brought to their present condition by state policies. Moreover, we must remember that it is old people who predominate in such enterprises, and that it was they who originally bore the burden of previous agrarian policies. Each such *kolkhoz* or *sovkhoz* must be considered individually and treated like a human being with a serious, long-neglected disease. A sober analysis of actual conditions leads one to conclude that it is the *kolkhozes* and *sovkhozes* that continue to be the mainstay of the rural socio-economic structure. Even while supporting peasant ownership in every possible way, we must be fully aware of the fact that the overwhelming share of agricultural products is produced by *kolkhozes* and *sovkhozes*. In order to arrive at a solution to the food problem, we must make greater use of this sector.

But how should this be done? After all, it is absolutely clear that, in their present state, the *kolkhozes* and *sovkhozes* as they developed historically are doomed; they have no future. Their radical restructuring is necessary in order for each worker to be transformed from a day labourer into a true owner. But this should not necessarily take place on the basis of privatization, with physical distribution of land and the other means of production, as some have proposed.

The tendency is towards farming based on primary cooperatives that possess their own land, technology, other means of production and the products raised. Under these terms, a *kolkhoz* or *sovkhoz* will become an association of primary cooperatives. An individual family can become a primary production cell. This will be a kind of farm, but for the time being within the *kolkhoz* or *sovkhoz*. A collective of a current farm, team or brigade can organize itself into a cooperative. They will delegate several common functions to the *kolkhoz* or *sovkhoz*, assigning the necessary resources for that purpose to it.

The overall process of denationalization also embraces *kolkhozes* and *sovkhozes*. It can take place in a number of ways, including conditional allotment of land and property and distribution of shares and stock. As a rule, moreover, the labour investment of each member is taken into consideration in establishing effective personal property. Those who want to leave the *kolkhoz* or *sovkhoz* receive their share directly, as they say, in kind or in cash. Those who continue working in the enterprise receive dividends for their shares. Such conditional division facilitates the creation of (family) farms and various kinds of cooperatives within the framework of the *kolkhoz* or *sovkhoz*.

CONCLUSION

I would like to emphasize my main point once again. The forms of rural ownership to be selected in each given region should be decided not in Moscow or in scholars' studies but by the peasants themselves. Displays of traditional 'concern' about them or foisting new progressive forms on them is just as repressive as collectivization.

The agrarian sector's transition to market relations assumes the restructuring of state regulation of agriculture with regard to both selling products and supplying resources. One proposal being advanced is to give *kolkhozes*, *sovkhozes* and other enterprises complete freedom in these two spheres.

However attractive such a concept may appear to be superficially, it can become a trap for the majority of agricultural enterprises. They may fall under the yoke of an even greater monopoly on the part of their suppliers. After all, there is no market infrastructure to ensure that such enterprises receive an equal opportunity for partnership in the agro-industrial complex. What is needed, therefore, is a flexible transitional mechanism for resolving this issue that will take account of actual conditions.

The agrarian sector's transition to the market is closely linked to implementation of land reform. A detailed analysis of these problems is a topic for another article. Here I will articulate briefly only the main aspects of it. In establishing the forms of land relations, regional differences must be taken into full account. Central Russia, where the countryside has been depopulated and the land has been neglected and is falling into disuse, is one thing. Quite another, for example, is Central Asia, where there is not enough land to go around in the populous villages. In such places it is a complex matter to allow land to become private property and even the object of inheritance. A considerable role is also played by particular ethnic features and traditions.

Radical laws on land reform are the basis for resolving this issue. But they are only the basis and cannot work without a specific mechanism. Therefore the mechanism for realizing land reform must include the restructuring of the party political hierarchy that still holds power. In this regard, the resolutions of the Russian Congress of Peoples' Deputies and the Russian Supreme Soviet on the impermissibility of combining leadership positions in party organs and soviets of peoples' deputies at all levels of power are extremely important.

Cooperation with other countries in a whole range of areas can also have a positive effect on the fundamental restructuring of the food complex in the Soviet Union. Of great importance to us is studying foreign experience in the technical and technological sphere, in food processing and in developing minor economic forms. We would welcome direct contributions by Western partners in the development of the rural food infrastructure, especially in food processing. There may also be other forms of mutually advantageous contacts. Such cooperation would not only be in the interests of the Soviet peasantry but would also help to reinforce the positive tendencies in world development associated with *perestroika*.

FERENC FEKETE AND CSABA FORGåCS*

Transition to Market Economies in Eastern Europe: The Case of Hungary

INTRODUCTION

A new political system began in Hungary with the free elections of March 1990. Associated economic changes involve a major transition from a 'command' economy to one based on the operation of market forces, which are also expected to affect the agriculture and food economy and to be based on a mixed ownership structure with a dominance of private property. Hungary was the Central European centrally planned economy (CMEA) with the earliest record of undertaking economic reform as well as being a leader in the more recent democratic political changes. As such her experience is that of a model case in the region.

LESSONS AND CRITICAL CONSEQUENCES OF THE PAST

For almost half a century Hungary has experienced far-reaching structural changes, some of which have resulted in failure. The unsuccessful experiments generally were those which related to the planning and control of agriculture at the macro level with the lack of the advantageous effects of market mechanisms. However, the first impact of change was in land holding with the land reform of 1945. Under the impact of wartime damage and a weakened economy, and with the further complication of political dependence, this did no more than transform ownership relationships in landed property, replacing a system which had been semi-feudal. It did not allow modernization to proceed, since it created very small-scale farms (minifundia) with weak command over means of production.

The period was also short-lived. By 1948, a communist regime was established which switched the pattern towards large-scale farming in many ways similar to the *kolkhozes* of the Soviet Union. This became dominant by 1961 and in many cases, probably as much as one-third, there was little change until very recently. They lacked capital and equipment, since the modernization of agriculture was not given a high priority by a regime which was more concerned with one-sided industrialization. Furthermore, they lacked strong proprietary motivation and suffered from the effects of unfavourable internal terms of trade. Following the death of Stalin in 1953, some efforts were made

*Budapest University of Economic Sciences, Hungary.

to encourage homestead farming by cooperative members through subsidization. It also became possible for small-scale farming to be carried on by those who were not principally engaged in cooperatives (industrial workers or pensioners, for example). Though neither of these initiatives was strengthened following the attempted revolution in 1956, there was a persisting shift in patterns of production with labour-intensive work tending to become concentrated in small-scale enterprises.

Throughout the 1960s and 1970s, the growth rate of food demand exceeded that of supply in Hungary. Furthermore, the main strategy involved attempts to increase the quantity rather than the quality of food produced, and in a general situation of scarcity there was minimal incentive for quality improvement. It should be emphasized, however, that agricultural performance, in the 1970s in particular, had some satisfactory features. For some years Hungary was one of the few countries in which incomes derived from farming for those people whose principal occupation was on the land approached parity with those of other sectors, even though their hours and working conditions were possibly less advantageous. Farming as a secondary activity was also profitable. This general position was maintained through the decade, since the price explosions on energy markets did not adversely affect the costs of farm inputs until the decade was drawing to its close.

By then, however, there were also other difficulties, in particular the decline of international demand consequent upon the effects of agricultural growth in the European Community and the greater internal integration of its food trade. By the middle of the 1980s, Hungarian agriculture, and much of the remainder of the economy, was drifting into a critical state. Indeed the centrally emphasized task of 'maintaining the living standard' could be achieved only with the restriction of agriculture to limit its resource use. There was little incentive to increase the technical efficiency of production or to provide stimulus to the labour force. Some growth did occur in the so-called 'second' and 'third' economies, but the mainstream was lagging.

In this critical situation, transition to a market economy and transformation of the structure of agricultural production to one involving a greater variety of type and scale become issues of key importance. It is complicated by the need to work within the context of international changes and by more general internal problems. Hungary has a general shortage of capital and major problems in effectively allocating what it has, following the principle of opportunity costs. The overall problem is severe. Though technologically outdated, many production processes are excessively capital-intensive and need renewal undertaken within a more effective allocation process.

APPROACHING A NEW OWNERSHIP SYSTEM IN AGRICULTURE

To create better propects for an agricultural sector which faces external constraint and internal difficulties, the key requirement appears to be that of reforming landownership relationships which can be harmonized with a market mechanism. New legislation on landownership is urgently required as soon as possible and this needs to be coupled with revised regulations concerning ownership of non-land capital assets used in farming. Cooperative law also needs attention.

Reforming ownership structures

Approximately 70 per cent of the country (9.3 million hectares) is cultivated, of which 12 per cent belongs to state farms and state combines, while 85 per cent is within the cooperative sector (including the amount farmed as household plots of members). Within the latter, two separate cases have to be considered. A large portion, of approximately 60 per cent, is both owned and used by cooperatives; of the remainder ownership of land rests with members or with individuals who are not themselves members. In considering reform of ownership there are two complicating factors:

(1)   The status of ownership in 1949 could be taken as the basis for reprivatization or compensation, with those who owned land at that time, or their descendants, having a prior claim.
(2)   While this appears to have merit it would obviously not necessarily completely reflect the current situation and the position of those who are at present engaged in farm work on particular tracts of land. A historical basis would not necessarily represent the facts of the contemporary operating pattern. Nor would either necessarily coincide with the need to ensure that future operation would be placed in the hands of those most suited to efficient farming, either on a private or a cooperative basis.

The Compensation Law passed by parliament does not imply that direct reprivatization must occur; indeed it has elements of compromise between the rights of prior ownership and those who make up the current farming peasantry.

Privatization processes in agriculture

Privatization is not only a way of handling issues of ownership; it is potentially an important tool in increasing agricultural efficiency. The need for change is widely recognized in Hungary for both reasons, though more conflicts are generated by the question of how to manage the process.

As far as the privatization of state farms is concerned, they can be placed into three groups. The first consists of those dealing with breeding of high-quality livestock of all types, or producing hybrid seeds, which act as suppliers to other enterprises. These so-called 'national farms' are not to be privatized; they will remain in the state sector into the indefinite future. A second group consists of state farms whose activity is profitable at the present time. These will eventually be passed into private ownership, but the process is not being given a high priority. That is reserved for the third group, of loss makers, which may be privatized as soon as possible. Government agencies will handle the process.

The privatization of cooperatives will be managed by the organizations themselves. The government wishes to help the process rather than intervening in it, hence decisions will largely be left to cooperative members. Where farming is efficiently conducted and where members are satisfied with their incomes, it is anticipated that no major land reorganization will be requested. However, it can be noted that in many cases the relationship between membership and management has often been more akin to an employee–employer form instead of reflecting a situation in which members were co-owners with an influence on managerial decisions. This internal issue must be addressed and discussions are taking place about new cooperative law to tackle the problem. It is a complicated issue since it involves questions both of ownership and of motivation.

In other cases, particularly where there is less efficient management, members may insist on the return of their land in order to engage in private farming. Even here, however, there is debate about the extent to which some cooperative activity should be retained and managed in order to handle the joint supply of inputs and the group marketing of products. It would be the basic activity of farm operation which would move onto a private, family, basis. A consequential problem would be that of establishing ownership in land and fixed assets.

STRUCTURAL CHANGES AS A SOURCE OF INCREASED EFFICIENCY

The policy environment

Unless real production costs can be reduced and quality improved, Hungarian agriculture cannot become competitive in world markets. In the past relatively high output prices and an increasing demand for products allowed managers to operate profitably despite basic weakensses in farm performance. The record is not consistently poor; the annual growth rate of 3–4 per cent through the 1970s was a notable achievement. Nevertheless, substantial and increasing subsidization has been needed to bridge the gap between high input costs and the relatively low world market prices of agricultural products. At the same time, strong pressure was placed on farming to export more in order to win badly-needed foreign exchange. As competitiveness weakened, the state budget could only provide a decreasing amount of food export subsidies. Further, the burden of foreign debt, and the interest payments involved, has limited the resources available for general modernization. It is a great challenge to increase efficiency in such a situation.

Inner restructuring of large farms

It has already been suggested that a key element in improving efficiency will be a change in the way in which farming is managed, and aspects of the process have been discussed. A number of points require elaboration. In the case of those state farms which remain as such it is not expected that management practices will remain static. Large organizations can be maintained to provide a broad framework, but operation can be divided into smaller decision-making units which are more sensitive to market conditions as well as being more able to supervise day-to-day farming operations. If privatization is desired, different kinds of economic decision-making units can be set up within the former large farm. These could include either cooperatives or companies as well as, possibly, family enterprises. The government agencies handling this process need to be able, within a defined legal framework, to cope with the problems of land and asset redistribution which will be involved.

Whenever cooperatives are involved, an atmosphere must be developed in which members have the right to make their own decisions. This will involve allowing them to modify the organizational structure of existing cooperatives or to break them up and found new enterprises. That might include reversion to private farming. The key element is that of allowing flexibility of choice, taking account of the need to create units of efficient scale appropriate to particular types of farming conditions. That is likely to require experiment in order for a more varied pattern to emerge, and for managerial capacity to be used which is appropriate to the scale and nature of enterprises.

Promoting viable family farms

People who believe in the comparative advantage of small farms often say 'I would do better, if it was mine.' There may be much truth in this view but other conditions, apart from ownership, are needed for successful small-scale operation. Apart from land and labour, substantial investments are required to sustain profitable businesses. Most of the people insisting on setting up on their own account tend to be aged 60 to 70, rather than 30 to 40. Apart from lacking access to finance, their knowledge of modern equipment and chemicals is limited. They may have some important skills, particularly in horticulture or animal husbandry, but this needs to be used effectively. In this context, Hungary has experience in allowing linkages to develop between small- and large-scale farming, which could be fostered. Potential does exist, in particular, for marketing the output of small-scale producers through cooperatives which would overcome the difficulty of lack of market information which is most acute among the former. This should be recognized and allowed.

FRAMEWORK AND GOALS OF A NEW AGRICULTURAL POLICY

Formulation of a new agricultural policy is approaching its final stage. The major issues have already received much discussion. Though new laws have yet to pass through parliament, the basic principles and goals which are emerging can already be outlined.

Liberalizing prices

Hungary has already had experience in applying market prices for agricultural products since a process of liberalization has been proceeding since the early 1980s. Step-by-step development occurred until the radical change early in 1990, when, except for bread-making wheat and some milk, the selling prices received by farmers became market prices, which also acted as the base for retail prices. This led to a significant increase in consumer prices in 1990, resulting in a decline in consumption. At the same time, rapid general inflation, affecting inputs to the whole food sector, has offset the increase in selling prices, creating a difficult situation. This has been prevalent for some time.

As in any open economy, Hungary periodically has had to adjust the internal price system to world market prices. There has been a specific relationship between consumer prices and producer prices because of government wage control policies. As a result of increasing input prices, there has been a gradual increase in farm producer prices. Together with a substantial reduction in government subsidies, the higher level of agricultural and food prices has become one of the major causes of the recent relatively rapid inflation. The unfavourable changes in the terms of trade have also given rise to inflationary forces.

Cutting subsidization of agricultural production

Subsidies to agriculture were significant in the 1960s and 1970s, though the level was tending to decrease. This was offset by a gradual increase in the prices of food products. In the 1980s, there were more substantial reductions in subsidies and the growth in output prices could only partially cover inflation-affected input costs. By the middle of the 1980s, the farming sector was already paying more to the budget in taxation than it received in subsidies.

Setting up the system of Agricultural Market Regulation (AMR)

Effective operation of liberalized markets requires a defined legal framework, appropriate institutions and suitable economic conditions. All of this involves different elements of policy, and it will take time and much discussion of the economic and political background to frame suitable agricultural market regulations. The basic concepts underlying current government thinking is that AMR should be a tool for intervening in demand, supply and price interlinkages when direct action is needed to reach a desired market equilibrium. It should provide stability within the market to avoid major damage through unexpected short-term adverse movements in producer prices and, against that background, allow the farm sector to increase its efficiency. This is regarded as important domestically, in order to ensure an adequate food supply, and externally to promote a satisfactory export performance. The latter is particularly important since agriculture has achieved an annual export surplus of 1–1.2 billion dollars and the foreign exchange is vital as a source of finance for debt service and reduction, and for obtaining imports needed in other sectors.

There are three major tasks for the government to undertake. First, an organizational structure, involving a Committee of Regulation and a Grain Committee, has to be established. Following from that, the second issue centres on finance for establishing and operating an intervention fund. Thirdly, some thought is being given to the issue of production control, through quotas, which may be needed to guard against potential over-supply.

SUMMARY AND CONCLUSIONS

(1)   Hungary now has 130 per cent self-sufficiency in food production. This is unique in Eastern Europe.

(2)   Substantial development was achieved in food production during the 1960s and 1970s, one of the results of that being a growing ability to supply export markets. In the late 1970s and 1980s, the increase in production costs began to exceed that of selling prices of agricultural products.

(3)   From the beginning of the past decade export achievement has been a vital element in assisting the balance of payments situation, though fewer resources have been available to farmers to help them sustain their level of performance.

(4)   Transition to a multi-party political system has already taken place following the recent election of a new parliament. Nevertheless, it will take some years to make decisive changes in the economic structure of the country.

(5)   The ultimate economic policy goal is to allow market forces to play a dominant role in the whole national economy, as well as in the food sector. Very important decisions have already been taken concerning price setting, price liberalization and agricultural subsidy reduction. The government has yet to take decisions about Agricultural Market Regulation to prevent major instability appearing during the process of moving towards a market economy. This involves very complex questions relating to appropriate ways of matching progress towards liberalization with market stabilization. It is against that background that reform of the structure of Hungarian agriculture will have to occur.

DISCUSSION OPENING – TADEUSZ HUNEK*

Fekete and Forgacs have provided a very solid background for understanding the process of agricultural reform in post-socialist countries of Central and East Europe. There is nothing which I wish to add to it directly and I will simply congratulate them for their skill in providing an overview of experience in their country at very short notice in order to overcome difficulties with

*Institute of Agricultural Economics, Warsaw, Poland.

the conference schedule. Notice that they did emphasize the importance of agriculture as an export sector in Hungary, thus providing me with the opportunity to add a little detail relating to the main topic of our session – the possibilities for increasing the agricultural trade of the European region.

The following four factors will affect the agricultural trade situation of Eastern Europe:

(1)   The food sector in that region is still one of the most important parts of national economies, with 25–35 per cent of the labour force being engaged in food production, providing some 30–35 per cent of material product. The transition to a more market-oriented economy has decreased food demand by as much as 15–25 per cent in some cases, hence the export of food surplus is a key factor in preserving agriculture.

(2)   Many agricultural and food products are competitive in world agricultural markets, mainly owing to the low cost of labour in agriculture and food processing.

(3)   There is a general expectation, strongly shared by World Bank missions, that the food sector could provide a driving force in economic recovery through rationalization of its economic performance. As much as 25–30 per cent of output might be expected to enter trade.

(4)   Foreign trade liberalization, as a part of economic reform, can also be expected to have an impact on food imports. In the case of Poland, the share of imported food in total consumption rose from 2 per cent in 1989 to 10 per cent in 1990 and there is every indication that growth will continue.

The outlook for the future is obviously still very obscure, since so many changes are taking place, or are about to take place, and events are moving with great rapidity. All that we can be certain of is that there will be strong pressure from within all of the countries concerned to increase the agricultural trade of Eastern Europe, both on the export and the import side, and that this will have important implications affecting both the wider European market and the world in general.

ERIC MONKE AND SCOTT PEARSON*

Evaluating Policy Choices in Developing Countries:
The Policy Analysis Matrix

INTRODUCTION

Developing country governments often alter substantially the incentives facing producers and consumers. Commodity markets are subject to taxes, subsidies and control of international trade through licensing or government monopolization. Prices in domestic factor markets for land, labour and capital, as well as the foreign exchange rate also receive ample attention from policy makers. A host of reasons account for these interventions including policy maker support to rent-seeking interests, promotion of non-efficiency objectives of society, provision of government revenues, control of inflation and compensation for budget deficits, compensation for the presence of market failures and stabilization of domestic markets in the face of international instability.

Whatever the rationale for government policies, many economists view the number of interventions as excessive and give this perceived glut much of the blame for stifling economic growth. But convincing policy makers of the merits of reform requires a detailed disentangling of the economic effects of policies. How do policies affect incentives among alternative commodities and technologies? How does the incidence of policy vary among regions? How would changes in various policies alter the incentive structure? The answers to such questions lie at the heart of most successful reform programmes.

The Policy Analysis Matrix (PAM) provides a framework for analysis of such questions (Monke and Pearson, 1989). A PAM portrays the pattern of incentives at the micro-economic level (producers, processors and marketing agents) and estimates of the impacts of policies on this pattern. This information is used to explore several topics of interest to policy makers: the pattern of comparative advantage and the potential for the economy to exploit this advantage; the formulation of public investment policy to support particular commodities, regions, and farm types; and the allocation of public research and development expenditures within the agricultural sector. A PAM does not provide information on all the important questions about policies which affect the agricultural sector (for example, risk and stabilization issues often require

*University of Arizona and Food Research Institute, Stanford University, USA respectively. We wish to thank Neil Conklin, Roger Fox and Mark Langworthy for helpful comments. Remaining errors are our own.

more information than that needed for the PAM), but PAM issues are often at the heart of policy debates about the most desirable course of agricultural growth and development. Further, the PAM approach is useful as a way to organize existing knowledge about agriculture. PAM results thus serve as an information baseline for monitoring and evaluation of the effects of policy and for identifying policy-relevant research needs.

The following two sections of the paper describe the analytical structure of the PAM and explain the meaning of the elements in the matrix. The construction of a PAM begins with the estimation of costs and returns at market prices, hence this calculation of private profitability shows the actual competitiveness of the agricultural activity. Subsequent analysis focuses on disentangling the effects of policies on observed (private) costs and returns. In the end, the analyst emerges with estimates of the efficiency or potential competitiveness of the activity (the costs and returns that would prevail if there were no government policies) and estimates of the magnitude of policy transfers to and from producers. Transfers are estimated for each input and output relevant to the activity, and their aggregate effect on profits are derived. Such results are communicated easily to policy makers and provide them with indicators of the quantitative importance of individual policies as well as a clear sense of the aggregate effect of policies on representative agricultural activities.

The next section of the paper considers application of PAM methods to two types of systems. The first system perspective is provided by the commodity chain–representative combinations of production, marketing and processing activities that are necessary to link farmers to consumers. Particular advantages of this perspective arise in identifying constraints at all stages of the marketing chain rather than just at the farm level and in assessing the impacts on farmers of changes in post-farm activities. The second system perspective is that of the farming system. Many agricultural observers argue that a full understanding of farmer behaviour requires an aggregated perspective on all production activities of the farm. This paper shows how 'whole farm' PAMs are constructed and used. The final section reviews the principal types of policy analysis that are aided by organizing information in the PAM framework.

THE POLICY ANALYSIS MATRIX

One of the principal motivations for the development of PAM was the need for easy communication between economic analysts and policy makers. Many decision makers often have only a limited exposure to the principles of economics and little time to digest the results of economic analyses. To be effective, therefore, presentations of the economic effects of policy ought to use perspectives understood by a larger group than just economists. At the same time, all methods need to satisfy the requirements of sound economic research, specifically an analytical framework which is theoretically rigorous and which tolerates variations in the quantity and quality of information available.

The cornerstone of PAM is the concept of economic profit. Profit is defined as the difference between revenues and costs – the value of outputs minus the costs of all inputs. When calculated at observed market prices, the result is termed 'private profit'. The definition of private profit is embodied in the first row of the PAM: A-B-C=D (Figure 1a). The letter A is used to represent the value of revenues at market prices. The costs of inputs are divided into two categories. The cost of tradable inputs – letter B in Figure 1a – is the value of inputs available in world markets. In practice, many of these inputs are produced domestically. However, these commodities are treated as tradable inputs because they are also available in international markets and represent potential imports or exports. The second category of input costs is primary domestic factor costs, denoted by the letter C in PAM. Primary domestic factors are land, labour and capital. They are treated separately from tradable inputs because they are usually available only in domestic markets. Some intermediate inputs, such as electricity or transportation services, may be similar to primary domestic factors in that they also are available only in domestic markets. In a PAM,

	Revenues	Input costs		Profits
		Tradable commodities	Primary domestic factors	
Market values	A	B	C	D
Efficiency values	E	F	G	H
Effects of divergences	I	J	K	L

(a) *The structure of the matrix*

	Revenues	Input costs		Profits
		Tradable commodities	Primary domestic factors	
Market values	$P^d Q^d$	$\sum_i p_i^d q_i^d$	$\sum_j w_j^d l_j^d$	D
Efficiency values	$P^w Q^s$	$\sum_i p_i^w q_i^s$	$\sum_j w_j^s l_j^s$	H
Effects of divergences	I	J	K	L

(b) *A disaggregated view of the matrix*

FIGURE 1   *The policy analysis matrix*

these intermediate input costs are disaggregated into tradable and primary domestic factor components, avoiding the need for a third category of inputs.

The second row of the matrix is intended to show what private costs and returns would be without domestic policies. This part of the analysis requires assessment of all policies that affect producer incentives. The list of potential policy interventions is large since policy makers have so many ways to express their dissatisfaction with efficient market outcomes. Desires to alter outcomes in commodity markets are usually pursued through commodity price policies – taxes, subsidies and quantitative controls that apply to domestic production or trade of the commodity. A second category relates to macro policies which affect incentives throughout the economy rather than just in a single commodity market. Macro policies include factor market policies that directly influence the prices for labour, capital and land; exchange rate policies that directly affect the domestic prices of internationally traded commodities relative to non-traded commodities; and macro-economic policies which influence the distribution of purchasing power between government and the private sector.

The results of policy assessments are summarized in the second and third rows of the matrix. Social profit (H in Figure 1(a)) is measured in a manner analogous to the calculation of private profitability – the value of outputs minus the costs of tradable inputs and primary domestic factors, all measured at efficiency prices ($H=E-F-G$). Because efficiency values exclude the influence of domestic government policies, social profit can be interpreted as showing the potential competitiveness, or comparative advantage, of the activity.

Calculation of the individual revenue and cost elements – E, F and G – is an exercise in efficiency pricing and borrows heavily from the logical foundations of international trade theory and social cost–benefit analysis. For example, the Little-Mirrlees method of project evaluation argues that efficiency prices for tradable outputs (E) and tradable commodity inputs (F) are represented by world prices, because these prices would prevail in the economy if there were efficient markets and no domestic government policies. A similar conclusion about the relevance of world prices as efficiency prices comes from international trade theory – setting domestic prices equal to world prices allows the economy to exhaust potential gains from trade and realize maximum national income. Trade theory also provides the theoretical basis for efficiency pricing of primary domestic factors (G). Efficiency prices of domestic factors are defined as the prices that would prevail if the factors were employed so as to maximize national income. Because maximum national income involves the production of commodities at world prices, factor prices are implicitly linked to world market prices even though primary factors are not traded internationally.

As an additional category, market failures must be considered if efficiency prices are to be consistent with the maximization of total income. Like policies, market failures alter costs and revenues and prevent the economy from realizing potential income gains. Market failures fall into three categories. Perhaps the best known type is imperfect competition, in which a small number of sellers or buyers is able to influence aggregate supply or demand and therefore exert some influence on market price. The second category of market failures includes externalities, such as pollution, and public goods, such as transport and communications infrastructure. Of particular relevance to the PAM calculations are the externalities involving producers. These arise when producers are unable to charge consumers for the full value of the things which they supply, or when producers do not pay all the costs associated with their activities.

Institutional market failures, constituting a third category, are less clearly defined than the first two categories. They include situations in which markets are inadequately developed or do not exist because of a lack of adequate rules and regulations to ensure fair behaviour in the market and to prevent cheating. For example, formal capital markets are often under-developed in rural areas, at least partly because banks lack sufficient authority to pursue repayment of defaults. Diagnosing institutional market failure is complicated because public investments may be necessary for the development of markets. In the formal credit market, for example, transport and communications infrastructure also influence the decision to establish bank branches. To break such constraints and integrate the local capital market into the national network requires an investment decision by the public sector (provision of public goods) rather than regulatory reform. The presence of institutional market failures is thus more difficult to confirm than are the other types.

The difference between private (market) values and social (efficiency) values is defined as the net effect of divergences; these values make up the third row of PAM. Divergences can be evaluated for each of the categories of revenues and costs (I, J and K). From this information, the analyst or policy maker can identify the most important distorting policies and begin to see how one distortion complements or contradicts other distortions affecting the agricultural activity. When the values of divergences are dominated by policy distortions, the final row of the matrix is sometimes represented as the effects of policies rather than the effects of divergences (policy distortions and market failures). This simplified representation is not strictly correct when market failures are significant; the analyst should disaggregate the third row of the PAM into sub-rows showing the effects of distorting policies, market failures and efficient policies that offset market failures.

EMPIRICAL ESTIMATION

Budgets of costs and returns are the principal sources of information needed to construct the first row of the PAM. Figure 1b contains a disaggregated view of the calculation and makes clear the linkage to budget data. The private market value of revenues (A) is calculated as the observed price of output, P^d, times the quantity of output produced, Q^d (Figure 1(b)). The value of entry B is calculated as the market price for each input times the quantity used of that input, summed across all inputs ($\sum p_i^{d*} q_i^d$ in Figure 1(b)). Primary factor costs (C) are calculated in an analogous manner as the market price for each primary factor times the quantity used ($\sum w_j^{d*} l_j^d$). Because the market prices used to value outputs and inputs may be very different from world market prices or opportunity costs, these potential differences are recognized in Figure 1b by attaching a superscript (^{d}) to the private market values.

Budgets for costs and returns could be constructed for every farm or firm in the market, yielding a comprehensive set of profitability estimates. Since the preparation of such estimates would stretch the availability of resources for research, and would overwhelm analysts and policy makers, it is common for empirical estimation to concentrate on a small set of budgets for representative groups of farms or firms. These budgets may be quite specific with respect to region, agro-ecological zone and technology, but they should be representative of broad groups in the market rather than exact portraits of actual farms or firms.

Budget data may be collected from surveys initiated by the researchers. More likely, secondary data will provide at least part of the information needed. If secondary information is of sufficient quality, fieldwork efforts focus on verification, updating and collection of details about input–output relationships. Even with such a seemingly straightforward exercise, however, problems arise with respect to proper calculation procedures. Common complications are the treatments of non-marketed outputs and inputs, such as farm family labour. Non-marketed items are evaluated at their market-equivalent values, implying that their value to the household or firm is the same as their value in the market. Family labour, for example, is valued at the market wage for hired labour, adjusted for sex, age and skill level. In many situations, family labour may not be able to find hired employment as an alternative to working on their home farm, and the analyst may feel that the appropriate opportunity cost is less than the market wage. Such perspectives are readily incorporated within the market-equivalent approach to pricing. When private profitability calculations turn out to be negative, the result can be interpreted as showing acceptance of rates of return (to family labour, for example) which are less than the market value.

The budget for private costs and returns can be modified to generate the second row of PAM entries, the social (efficiency) values. Some of the transformations needed to convert private values to social values are straightforward. The efficiency prices for outputs and tradable commodity inputs are world prices, indicated in Figure 1(b) as P_i^w for outputs and p_i^w for tradable commodity inputs. World prices are used as the efficiency standards, even though these prices may be distorted by policies and market failures in foreign countries. Foreign policies are usually beyond the influence of domestic politicians, and (distorted) world prices thus represent the prices that would prevail in the economy in the absence of domestic policy. Such situations may seem unfair to the domestic agricultural sector (indeed, they may provide a non-efficiency rationale for domestic policy distortions), but world prices continue to represent the opportunity cost of the commodity to the domestic economy.

More problematic are the calculations of efficiency prices for primary domestic factors, denoted as w_j^s in Figure 1(b). One approach to estimation is to make use of the linkage between world commodity prices and factor prices with the help of a general equilibrium model. Unfortunately, such models are generally unavailable or lack the necessary detail to price the primary factors used in agricultural activities. The next best approach is to exploit the double-constraint structure of PAM. When direct derivation of efficiency values is too difficult, the analyst can estimate the values indirectly by identifying the particular policies and market failures that influence factor prices. Adjustments of private market prices to their efficiency values are based on assessments of the quantitative significance of policy distortions (particularly factor price policies) and factor market failures. Sensitivity analyses are also useful procedures to evelute the impact of changes in social factor price estimates.

The most difficult information to obtain for social evaluation is the quantity data – Q^s, q_i^s, and l_j^s. A full assessment of the impacts of policy on profitability requires accounting for the effects of price divergences on output level and input use. Three categories of effects can cause social quantity measures to be different from private quantity measures (Q^d, q_i^d and l_j^d): changes in relative input prices can alter the combination of inputs used to produce a given level of output; changes in input prices can alter the amounts of inputs used and thus the level of output; and changes in output prices can encourage changes in input use that in turn change the level of output.

Measurements of these effects usually require a long time-series of detailed data that are rarely available, even for developed economies. The empirical approach to such problems is to rely on assumptions of fixed input–output coefficients (as done in social cost–benefit analysis) and thus to preclude any price response by the producer. A less restrictive approach is that used in linear programming analysis. In this approach, a set of alternative technologies (each with fixed input-output coefficients) is used to portray production alternatives. The technique with the largest social profit becomes the budget relevant for the second row of the PAM.

PAMS AS SYSTEMS

A policy analysis matrix can be estimated for any production activity which can be represented by a budget of costs and returns. These could include farm production, industrial processing or production, and marketing or other service sector activities. However, the analyst may also wish to present PAM results at a more aggregated level. One of these aggregation exercises involves representation of a commodity chain as a set of farm production, marketing and processing activities which is the essential link between producers and consumers. Consumption depends simultaneously on all of these activities, and knowledge of the complete pattern of incentives is needed to assess actual or potential competitiveness.

In a PAM, the aggregation of farming, marketing and processing activities is referred to as a commodity system (Figure 2). PAMs for individual activities (farm, farm-to-processor, processing and processor-to-market) are added together to generate measures of aggregate competitiveness and policy transfers. The measures of private profitability for the system require careful interpretation. Private profit for the commodity system is the aggregate of profits that accrue to different activities, whereas competitiveness at private market prices depends on positive profitability for each of the activities. Social profitability and total transfers have interpretations like those made at the activity level. The social profit of the system proves a particularly useful measure because some domestically produced outputs can be compared to world market coun-

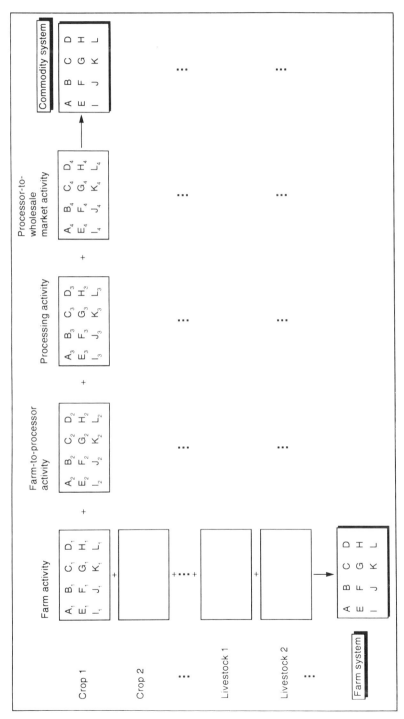

FIGURE 2   *PAMs for commodity and farm systems*

terparts only after they have been processed and delivered to a wholesale market. Milk, for example, is not traded internationally in raw form and becomes tradable only when processed.

Empirical estimation of the system values is more complicated than direct addition of the results at the activity level. Each output and input relevant to the production process can be counted only once. Double-counting can be a particular source of confusion for outputs, because the output from one activity is an input to the next activity in the commodity chain. Adding values across different activities requires a common numeraire, such as hectares or units of the final product. Conversion ratios are applied to the relevant activity budgets to convert values to this common numeraire. A third complication arises because four activities may be an inappropriate number to portray the system. For some commodities, processing is trivial and can be ignored. For others, two marketing activities (farm-to-processor and processor-to-market) may understate the number of transactions required to handle, transport and store the commodity. Generally, however, the four-activity framework has proved to be a workable starting-point for empirical analysis, and it is straightforward to expand or contract the number of activities recognized in the system.

The second aggregated perspective for PAM analysis is the farm system (Figure 2). Farmers do not think only in terms of individual commodities, but consider the farm as a composite of commodities, animals and technologies, with numerous complementarities and constraints binding together the different activities. Figure 2 shows that the PAM for the whole farm can be constructed as a composite of the farm activities in the relevant commodity systems. For the purposes of budget calculations, the whole farm can be described in terms of revenues and costs. Revenues come from the sale or home consumption of crops and livestock products. The farm's costs include inputs used in the production of crops and livestock and transportation expenditures used to service the farm's crop and livestock activities.

Some care needs to be taken in the farm system aggregation. Crop and livestock activities must be weighted to reflect their relative importance in the total area and total livestock population of the representative farm; costs and uses for inputs serving multiple production activities, such as machinery, must add up to totals that are consistent with aggregate availability; and all intra-farm transportation activities must be represented somewhere in the individual activity budgets; and a numeraire (usually land area) must be chosen to allow addition across a disparate group of commodities. Provided that all inputs and outputs are attributed to one of the commodity or livestock commodity systems, the 'whole farm' PAM will give an accurate accounting of revenues, costs and profits.

The results of whole farm analyses provide insights into aggregate farm income and the net effect of policies and market failures on income. Such calculations are particularly useful in comparisons across different farm systems. Whole farm results also provide a convenient framework in which to discuss farm-level issues, such as the total demand for farm labour and capital equipment. However, they are less useful in highlighting the relative importance of particular policies. The effect of policy distortions on total revenues, for example, is a composite of effects of commodity policies for all the outputs of the farm. Without commodity system PAMs, the analyst is unable to explain (or even calculate) the values of any of the elements in the whole farm PAM. Consequently, commodity system PAMs are necessary complements to whole farm PAMs.

PAM AND POLICY EVALUATION

Because both the rows and the columns of the matrix are based on accounting identities, the entries in the matrix satisfy a double-constraint consistency check that characterizes all successful accounting methods. The aggregate impact of divergences on the incentives facing the producer (L) can be represented in two ways: as the difference of the elements in the third row (I-J-K) or as the difference between private and social profits (D-H). These results are useful to determine the source of competitiveness: namely whether the activity is profitable because of the support of policy ($H<0$, $L>0$) or because of natural comparative advantage ($H>0$).

By considering the pattern of incentives with and without policy, the methodology can play a useful role in identifying new and efficient policy interventions. First, the results can be used to identify public interventions that assist economic growth. From a micro-economic perspective, growth opportunities are represented by excess profits. Often these profits are the consequence of changes in technologies, prices of outputs, exchange rates and domestic factor prices. When excess profits are positive, the industry has an incentive to expand production. Increased production allows for decreases in imports or increases in exports (if the good is a tradable) or a reduction in prices (if the good is a non-tradable). Perhaps most important, increased production entails increased demand for domestic factor inputs and allows increases in factor prices. Ultimately, excess profits are eliminated and the industry is ready for another round of growth.

Social profits are indicators of opportunities for economic growth. If positive, policy makers can assess the possibilities for expansion of the activity. A decision by producers to expand output requires that private profits are positive (producers have no particular interest in social profits) and may require the elimination of distorting policies in order to improve incentives. Public investments, such as improvements and expansion of road networks, may be needed if the profitable activity is to be introduced to other regions. In this circumstance, social profits can be compared with public investment costs to determine the efficiency of expanding the activity. Finally, policy makers may need to supplement PAM results to evaluate the merits of expansion. If the country has a large share of world exports, for example, the benefits of increased production will be at least partly transferred to foreign consumers through reduced export prices. If the country is a large importer, the gains from increased production will be augmented by the consequent effect on import prices.

Another use of the results in the analysis of growth is to simulate hypothetical technical changes. The appearance of negative social profitability in an existing activity need not imply that the activity should be abandoned since production of the commodity may serve well some non-efficiency objective of the economy. Policy makers can then explore the possibility of 'inducing' social profitability through the invention and dissemination of a new technology. The new technology can be described in terms of quantities of inputs and outputs, and social prices can be applied to evaluate social profitability. The impact of the technical change on social profitability is compared with the expected costs of research and development to identify efficient opportunities for public expenditures.

Other opportunities to identify efficient policy interventions arise when market failures are present. PAM results provide measures of the cost of market failure. Hence by calculating PAMs for all affected activities and aggregating, the total cost of the market failure to the economy can be estimated. This value is compared with the costs of implementing a policy to offset the market failure to distinguish between worthwhile countervailing policies and those that are uninteresting because their introduction in the economy has transactions costs that are too high. Decomposition of the final row of the matrix is essential in this context to recognize the effects of policy distortions, the effects of market failures and the effects of efficient government policies.

In addition to identification of efficient policy interventions, another potential use of the technique involves the evaluation of distorting policy. In this application, a major difference arises between it and most social cost–benefit approaches. Many methods of social cost–benefit analysis attempt to adjust efficiency values to reflect the concerns of society about income distribution and other non-efficiency objectives, such as food security. In reality, these adjustments are impossible to calculate. The flaws in attempts to value non-efficiency objectives are the presumptions that society's preferences for all aspects of the economy can be identified, compared with one another and ranked; and that some individual (or set of individuals) can be selected by the economic analyst and deemed the proper spokesman for society. Even if such difficulties were resolvable, the importance of non-efficiency objectives is neither fixed through time nor determined in advance. Instead, their importance at any point in time is the outcome of debates among policy makers.

The argument against the modification of efficiency values in social price calculations does not imply that analysts should (or can) ignore non-efficiency issues in the evaluation of economic policy. Consideration of non-efficiency objectives instead is deferred to a later stage of policy analysis, after quantitative estimates of the effects of policy on efficiency have been made. Efficiency is not the only objective of economic policy, and non-efficiency objectives provide possible explanations why divergences may be desirable. Measures of transfers (I, J, K and L) and the sacrifices of potential income (H, or profitability measured at social prices) allow PAM results to serve as inputs into policy debates about the desirability of trade-offs between efficiency and non-efficiency objectives. With the PAM approach, the analyst is not forced to make definitive statements about 'proper' and 'improper' policy.

CONCLUSIONS

The strength of the PAM approach is its simple framework, which is capable of showing non-economists critical facets of many important issues in economic policy. At the same time, the method is sufficiently general in structure to accept analytical sophistication. Both aspects are essential for successful policy analysis. Results have to be understood easily by non-economists (particularly policy makers) to have an impact on policy debates. Many economic analyses are unused for want of an understanding audience. Policy analysis frameworks should be sufficiently flexible in information requirements for analyses to be performed in data scarce environments. For example, the results of general equilibrium models can be useful for the determination of the social efficiency prices of primary domestic factors, and econometric estimates of input–output relationships and output responses can help in the estimation of the input and output quantities relevant to calculation of social efficiency revenues and costs. However, such estimates are not essential to the construction of PAMs, and researchers have other options for acquiring the necessary estimates.

Initial efforts to develop PAMs will require assumptions or guesses about appropriate values for some parameters. The results can then be enhanced by the sequential improvement in the quantity and quality of information. The important point is that, once an initial PAM baseline is prepared, the analyst can see the relative importance of various information gaps and begin to organize subsequent research efforts in an efficient manner. Such procedures allow policy makers and analysts alike to improve and deepen their understanding of the relationships between policies, agricultural competitiveness (private profitability) and efficiency (social profitability).

REFERENCE

Monke, E. and Pearson, S.R., 1989, *The Policy Analysis Matrix for Agricultural Development*, Cornell University Press, Ithaca and London.

DISCUSSION OPENING – ALEXANDER SARRIS*

A major contribution to policy debates in developing countries in recent years has been empirical estimation of direct and indirect costs of micro-economic or macro-economic policies on various sectors in the economy, and on agriculture in particular. A series of techniques has been applied, largely in the context of the general push towards market and institutional reforms necessitated by liberalization and structural adjustment programmes. As Monke and Pearson suggest, one problem with such analyses is communication of the results to policy makers in a way amenable to easy policy evaluation and

*University of Athens, Greece.

debate. Their contribution lies in proposing an organizational framework for presentation of results to make them more accessible.

The Policy Analysis Matrix (PAM) suggested as a means of easier communication between analyst and policy maker is based on analysis of profitability of specific activities within a production system. In other words, the key information communicated to the policy maker is the private profitability of the particular production or marketing activities under the current, presumably distorted, policy regime, and in an undistorted policy environment. The idea is that the policy maker can assess the cost of the policy or other distortions on private incentives and hence can be helped in forming a judgement about appropriate policies. Information is communicated by arranging the revenues, tradable input costs, non-tradable input costs, and profits, under the current policy regime, in a row; then placing the same information in a subsequent row but with valuation at efficiency prices. The difference between rows is the key information that will presumably help the policy maker obtain a better perspective on the impact of policies.

The belief underlying the suggestion seems to be that the most important information which might motivate policy reform or debate concerns divergence between undistorted and distorted prices for an activity, and the resulting impact on private profits. I suggest that such information, valuable as it might be, is only part of the story. Policies are instituted for a variety of objectives, and policy makers are often quite willing to sacrifice private profitability to promote some other objective. Unless the impact of a particular set of policies on this variety of objectives is also analysed, the information conveyed is only partial. For instance, a policy of export taxation applied to a particular agricultural product might be motivated by fiscal needs, in full recognition of the fact that the incentive to private producers is reduced. In such a case exhibiting the PAM for the commodity does not help the policy debate. What might help in such a case would be joint presentation of the disincentives to private production, along with the cost of enforcement of the policy as well as the fiscal benefit, and the contribution of various distortions to these.

The second criticism of the PAM is that it seems to rely on the comparison between distorted and completely undistorted prices in evaluation of costs. Most policy makers, and particularly those concerned with specific sectoral policies, have a limited set of instruments at their disposal. Hence the most interesting information for them would be the implications of the particular reforms within their discretionary power, leaving all others unchanged. For instance, a minister of agriculture will be much less concerned about exchange rate policy than about agricultural price policy. This implies that the row in the PAM exhibiting private profits and costs at efficiency prices must be disaggregated to indicate the implications of difference distortions. It could very well be the case that, while a particular price policy *ceteris paribus* has a beneficial effect on private producers, a fiscal, monetary, trade or other non-agricultural policy might reverse the impact. In such a case it is clear that the opposing effects of different policies on the activity should be exhibited, since that might indicate both the relative importance of policies and the necessary reforms.

The third criticism of the PAM concerns the information required for calculation of the different entries in the matrix. The authors suggest that a PAM can be estimated for any activity that can be represented by a budget of costs and returns. It is the recalculation of the costs and returns at different prices, namely efficiency ones, that provides the information necessary to complete the PAM. However, while it is relatively straightforward to compare domestic and international prices at current exchange rates, estimating prices for primary factors of production, as well as foreign exchange, under different policy regimes, is far from trivial. The authors suggest discretion, judgement, and trial and error in order to derive some of these prices. However, this procedure, valuable as it might be under some circumstances, seems quite inappropriate for presenting information of policy relevance. It is quite common, for instance, for difference policies to have opposing effects on the shadow prices of various factors and foreign exchange due to the multi-market general equilibrium nature of the interactions. In fact, seemingly simple issues, such as the response of marketed surplus of staple food to prices, become quite complicated once income effects, to mention only the simplest ones, are included. It can be correctly argued, of course, that what one needs in such a case is a multi-market or general equilibrium model to assess the influence of policies. The authors acknowledge this need, but they dismiss it as too complicated for speedy analysis. However, I suggest that policy reform is not a trivial business, and sound analysis of impacts, no matter how complicated, is a necessary prerequisite to effective recommendations on anything but small policy changes. In fact, foes of policy reforms can all too often latch onto criticisms of simple approaches as a weapon for thwarting meaningful reform efforts.

Given that serious and major policy reforms necessitate debate based on sound empirical analysis, a major part of which is precisely to assess the impact on macro variables such as costs of primary factors and foreign exchange, the information of such exercises will usually be much more than that conveyed by a PAM. I do not think that the major problem of most analysts has been to convey information from sophisticated exercises to policy makers in a parsimonious way. While it is correct to assume, as the authors do, that policy makers are not particularly interested in the methodologies, but rather the results, I believe that the problem with most empirical policy analyses seems to have been their lack of methodologically sound empirical valuation. One might argue that such analyses are time-consuming and might not be available for the critical debates. However, the information on which a PAM is based does not seem any less time-consuming to obtain and analyse. In fact, the micro surveys needed are the most time-consuming exercises in policy analysis.

It must, nevertheless, be emphasized that these criticisms are not meant to suggest that the PAM proposed by Monke and Pearson is a superfluous tool. In fact, its strength lies in exhibiting the 'vertical' pattern of profitabilities of different activities in a relatively simple picture. In particular, I find the idea of separating the pattern of private profitabilities along a vertical market chain rather attractive. This is in keeping with the strength of the approach which is its partial equilibrium nature. However, my belief is that any summation of PAMs across commodities is not as useful, since it is much more likely that it will run up against assumptions that are untenable under general equilibrium. On balance, one should assess the utility of this particular tool in the context of its strengths relative to others and not as a panacea for all types of policy analysis needs.

THOMAS HERTEL, KENT LANCLOS AND MARIE THURSBY*

*General Equilibrium Effects of Trade
Liberalization in the Presence of Imperfect Competition*

INTRODUCTION

The impact of trade liberalization in developing countries is an important issue. Its practical relevance stems both from potential reforms if the GATT negotiations resume and from the heightened interest on the part of individual countries in unilateral policy reforms designed to rationalize their economies. The World Bank has also called for reform, not only because of trade restrictions, but also because of the many market distortions in these economies. Besides distortions related to externalities, the developing economies tend to be characterized by highly concentrated manufacturing sectors (see, for example, Kirkpatrick, Lee and Nixon (1984) and Rodrik (1988)). It is therefore surprising that trade liberalization in developing countries has tended to be evaluated in the context of perfectly competitive models. In surveying applications of computable general equilibrium (CGE) models to developing countries, Robinson (1989) cites only a few efforts which incorporate imperfectly competitive market structures (Condon and deMelo 1986; Devarajan and Rodrik, 1988, 1989).

In contrast to this relative paucity of empirical work, the last decade of research in international trade theory has focused heavily on the impact of trade policy in imperfectly competitive environments. Perhaps the major lesson to come out of this is that market structure assumptions are critical to policy analysis (Helpman and Krugman, 1989). For example, with imperfect competition and scale economies, it is possible for trade liberalization to lower welfare in an economy. This result is counter-intuitive in perfectly competitive models where the only distortions are price wedges created by policy. The excess profits and scale economies associated with imperfect competition imply that welfare effects of trade liberalization will also depend on the scale of production. Naturally, the latter effect depends on the extent to which imperfectly competitive sectors of the economy expand or contract. Because the intellectual foundation of many developing countries' trade policies rests on arguments about potential gains due to scale economies, this extension to the neo-classical analysis of trade liberalization is highly relevant and long overdue.

*Purdue University, USA. The authors wish to thank Shanta Devarajan for helpful comments and for providing the Cameroon CGE model used in the empirical illustrations.

In addition to the normative ambiguity, the presence of imperfect competition and scale economies may also introduce ambiguity in the predicted changes in sectoral output. In this regard, consider the anticipated impact of across-the-board trade liberalization on agricultural output in a developing country with a typical pattern of trade distortions. In a perfectly competitive setting, trade liberalization would be expected to lead to an expansion of agricultural output, since lower income countries typically protect their manufacturing sectors more heavily than the farm sectors. However, when the tariff on manufacturing imports falls, it is possible that the optimal mark-up for imperfectly competitive firms will also fall. This in turn may lead those firms remaining in the industry after the tariff cut to move down their long-run average cost curves. This dampens the degree to which manufacturing contracts, and hence also lessens the expansion of agriculture. Indeed Devarajan and Rodrik (1988, 1989) have argued that, in the case of the Cameroonian economy, this may even cause agriculture to contract, rather than to expand.

The purpose of this paper is to subject this question to rigorous analysis. Is it possible that the conventional neo-classical wisdom regarding the fate of agriculture under trade liberalization could be reversed in the presence of the unexploited scale economies and imperfect competition in the non-agricultural economy? We begin with a theoretical analysis based on a two-sector model. This provides considerable guidance in assessing the empirical work to date. We then turn to a re-examination of the Devarajan and Rodrik results.

A FRAMEWORK FOR ANALYSIS

The question at hand may be addressed in its simplest form with a two-sector model. Sector 1, which may be thought of as agriculture, is an aggregation of all activities which produce a homogeneous product and operate roughly in accordance with the perfectly competitive paradigm. Output price equals marginal cost, and industry average total cost is not affected by changes in the number of firms. Sector 2, non-agriculture, is assumed to produce a product which is differentiated by firm. Furthermore, in order to enter the sector, firms are assumed to incur a fixed (recurrent) entry cost. When coupled with constant marginal costs, this gives rise to increasing returns to scale in the second sector. As profit maximizers, these firms will mark up price over marginal cost according to the inverse of the perceived elasticity of demand for their particular product.

Essential features of this two-sector economy are portrayed in Figure 1. We assume that consumers devote a constant share of their disposable income to each of the two goods. However, composite good two is made up of many individual varieties. The demand for a representative home firm's output D_{2H} is a function of the prices of competing domestic (P_H) and foreign $(P_{2F}(1 + T) = P_{2F}t)$ products. If the domestic market is small, and fully integrated into the world market, then it is reasonable to assume that both the number and the supply prices of foreign firms are unchanged by a perturbation in the local tariff. Thus the proportional change in the power of the tariff (t) equals the proportional change in the price paid by consumers for foreign varieties. The

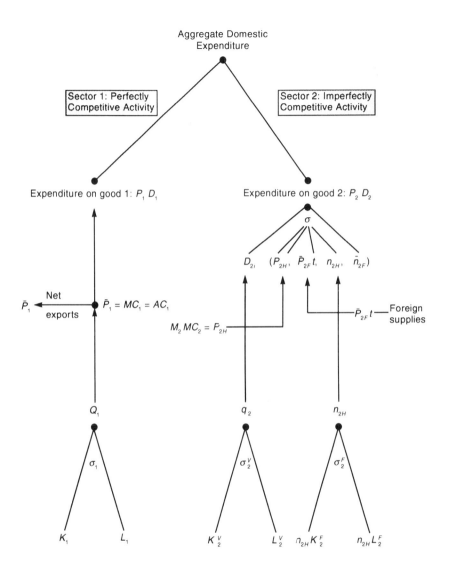

FIGURE 1 *The structure of the two-sector economy*

demand for a representative domestic firm's output also depends on the number of varieties available ($n_{2H} + n_{2F}$). Finally, the elasticity of substitution among varieties of good two (σ) will prove to be an important parameter in this analysis. It must be greater than one if the domestic and imported varieties are to be gross substitutes in consumption.

In order to keep things simple, while capturing the essential features of many developing economies, we assume that sector two does not export any domestic production. In contrast, sector one is a net exporter, facing an exogenous world price for its product (P_1). Both sectors combine labour and capital inputs subject to a constant elasticity of substitution (σ_1 and σ_2). Furthermore, rather than beginning with a tariff-ridden initial equilibrium, and considering the impact of removing these distortions, we consider the mirror image of this experiment, namely the introduction of a tariff into an initially undistorted environment. This vastly simplifies the algebra without altering the intuition. Also, we will consider the case whereby the only border intervention is in section 2. (It is easy to modify the analysis to account for a simultaneous export tax or subsidy on good one). We will then ask how the introduction of imperfect competition alters the prediction that output in sector one will move in the opposite direction of the tariff on imports of good two. Is it possible that a lowering of the tariff on manufactures could cause agricultural output to fall, rather than expand?

NO ENTRY

Theoretical results: As identified by others (for example, Markusen and Venables, 1988), the predictions of this type of model depend on whether or not entry/exit is an option for firms in the imperfectly competitive sector. Accordingly, we consider both possibilities, beginning with the no-entry case. Rodrik (1988, p. 113) argues that this case is particularly relevant to the developing countries where: (a) 'industrial policies have typically been biased toward restricting entry, as investment in many manufacturing sectors are subject to complex licensing and financing arrangements'; (b) 'newcomers to preferred sectors often benefit from special incentive packages, of which latecomers are deprived'; and (c) 'the weakness of capital markets ... means that investment funds are typically internally generated'.

Proceeding with the no-entry case, we manipulate and solve the equations underlying Figure 1, obtaining the following expressions for the general equilibrium (proportional) changes in output per firm ($\hat{q}_2$) and sectoral output ($\hat{Q}_1$) following the introduction of a tariff on the imports of good 2. This yields equation (1):

$$(\hat{q}_2/\hat{t}) = (-\Delta_1/D)\,(\varepsilon_{2F}^* - \beta_{MF}), \text{ and } (\hat{Q}_1/\hat{t}) = (\Delta_2/D)\,(\varepsilon_{2F}^* - \beta_{MF}) \qquad (1)$$

where $\Delta_1, \Delta_2 > 0$ are both intensity-weighted averages of the sectoral elasticities of substitution in production, and $D < 0$ is the determinant of the general equilibrium system. (See Hertel, 1991, for detailed derivations).

The common parenthetic term in (1) is made up of two parameters. The first, ε_{2F}^*, is the cross-price elasticity of demand for a representative domestic product in sector 2, with respect to a change in the foreign price (t in this case), compensated for the income effects of changing tariff revenue and excess profits. It is unambiguously positive for values of $\sigma > 1$. The second term, β_{MF}, is the elasticity of the optimal mark-up with respect to t. It is always positive. Furthermore, it can be shown that when $\sigma > 1$, then $\varepsilon_{2F}^* > \beta_{MF}$. This means that agricultural output will indeed be decreasing in the manufacturing tariff (that is, $\hat{Q}_1 / \hat{t} < 0$).

While it appears that our conventional wisdom regarding the qualitative effect of tariff reform in the presence of imperfect competition will not be reversed in this case, the presence of β_{MF} in (1) warrants some additional discussion. This has been termed the 'pro-competitive' effect of tariff reform; as the tariff comes down, so does the optimal mark-up in sector 2. It can be shown that the absolute value of the quantity changes in (1) are decreasing functions of β_{MF}. This leads to some interesting and useful observations, since β_{MF} depends on the conjectures of individual firms about the reactions of their rivals. The two well-defined cases which economists typically use as benchmarks are the Cournot conjecture, whereby firms assume rivals hold quantities constant and adjust price, and the Bertrand conjecture, whereby it is rivals' prices that are assumed to remain constant. It can be shown that the Cournot case is more 'collusive' in the sense that it generates larger optimal mark-ups. It is also the case that β_{MF} is larger under the Cournot conjecture.

Of course, there is yet another, even more collusive case, which has been explored by Harris (1984). He terms this the case of 'focal point pricing', whereby domestic firms uniformly price their products just below the tariff-inclusive price of competing imports. Although this pricing rule is somewhat *ad hoc*, it provides an interesting benchmark because β_{MF} is equal to one and, for a given number of firms, the optimal mark-up falls in the same proportion as the tariff. At the other extreme is the case where mark-ups do not change at all – as would be the case under perfect competition. In sum, the change in agricultural output following a change in the manufacturing tariff is dampened by the presence of imperfect competition in the latter sector. The more collusive the manufacturing sector, the smaller the subsequent change in the two sectors' outputs.

Empirical illustration

At this point it is instructive to pause for a moment and consider some numerical examples. Table 1 presents a variety of results based on the Cameroon data base utilized by Devarajan and Rodrik. To begin with, we alter the model of imperfect competition to conform with Figure 1. This entails 'recalibrating' each of the five tradable, imperfectly competitive sectors. We choose to assume the same representative firm's share of the market (column one) as Devarajan and Rodrik. The resulting optimal Bertrand mark-up and the implied elasticity of substitution among differentiated products are reported in the next two columns. This is followed by the uncompensated elasticity of demand

TABLE 1 *Results of calibration exercise for imperfectly competitive sectors and effect of tariff on domestic sales under Bertrand conjectures and no entry/exit (implications of Devarajan and Rodrik specification in parentheses)*

	Repr. firm's share: θ_{iH}/n_{iH}	Optimal domestic mark-up P_{iH}/MC_i	Elast. of subst.: $\sigma(\sigma_A)$	Demand elasticity: ε_{iF}	Mark-up elasticity: β_{MF}	Domestic sales per firm $(\hat{q}_i^{PE}/\hat{t})$	General equilibrium effects of *ATB* lib.* $\beta_{MF} \neq 0$	$\beta_{MF} = 0$
							Percentage change	
Food processing	0.295	1.29[a] (1.62)	5.23 (1.25)	1.13[b] (0.07)	0.23[c] (0.01)	0.73[d] (0.06)	n.a. (2.3)	n.a. (2.4)
Consumer goods	0.241	1.39 (1.43)	4.30 (1.25)	0.82 (0.06)	0.22 (0.01)	0.49 (0.05)	n.a. (−1.3)	n.a. (−1.6)
Intermediate goods	0.163	1.29 (1.57)	5.26 (0.50)	1.86 (−0.22)	0.25 (0.05)	1.29 (−0.26)	n.a. (8.4)	n.a. (7.9)
Cement and base metals	0.102	1.30 (1.91)	4.54 (0.50)	2.37 (−0.34)	0.19 (0.03)	1.83 (−0.33)	n.a. (11.9)	n.a. (11.3)
Capital goods	0.012	1.41 (1.49)	3.49 (0.75)	2.38 (−0.24)	0.03 (0.00)	2.28 (−0.24)	n.a. (−4.4)	n.a. (−4.9)

Notes: [a] P_{iH} is the price received by the firm *net of* indirect taxes.

[b] $\varepsilon_{iF} = (1 - \theta_{iH})(\sigma - 1) = \theta_{iF}(\sigma - 1)$. In the Armington case, replace σ with σ_A. ε_{iF}^{*} is estimated using the Cameroon model.

[c] $\beta_{MF} = (1 - \sigma)^2 \, \theta_{iH}\theta_{iF}/d_H^B$, where $d_H^B = \eta_{iH}^B \, (\eta_{iH}^B - 1)$. In the Armington–Cournot case, replace σ with σ_A, and $d_H^A = (\eta_{iH}^A/n_{iH}) \, (\eta_{iH}^A - 1)$.

[d] Results in this column are obtained by applying equation (2).

* *ATB* lib. = across-the-board liberalization.

for a representative domestic firm's product with respect to the price of foreign competitors. It is generated on the premise that composite domestic expenditure on the good in question is held constant. The elasticity of the optimal domestic mark-up with respect to the foreign price is also reported, and it is this pro-competitive effect which distinguishes the perfectly and imperfectly competitive models.

To evaluate the empirical significance of the pro-competitive effect which tariff reform has on mark-ups, β_{MF} must be compared to ε_{iF}. In particular, consider the impact of a tariff on output per firm in partial equilibrium (that is, with marginal cost and composite expenditure on i fixed). This is given by:

$$\hat{q}_i^{PE}/\hat{t} = (\varepsilon_{iF} - \beta_{MF})/1 + \beta_{MF}). \tag{2}$$

Using (2), partial equilibrium results for each of the five sectors are calculated and reported in the next column of Table 1. Since their perfectly competitive analogue is given by $\beta_{MF} = 0$, that is, $\hat{q}_i^{PE}/\hat{t} = \varepsilon_{iF}$, we can see that the pro-competitive effect is indeed significant. For example, in the case of consumer goods, the perfectly competitive model estimates a 0.82 per cent fall in domestic sales/firm for each 1 per cent fall in foreign price. If this industry were, in fact, characterized by product differentiation and Bertrand conjectures, then domestic consumer goods sales/firm would only fall by 0.49 per cent. In other words, the perfectly competitive model would overstate the decline in output per firm by 79 per cent! (That is $((0.82 - 0.49)/(0.49)) \times 100\% = 79\%$.) Note, furthermore, that these discrepancies would be even greater in the case of a more collusive industry (for example, the Cournot case with product differentiation).

Table 1 also reports some information on the Devarajan and Rodrik formulation of imperfect competition in Cameroon. Rather than differentiating all firms products symmetrically, they adopt the 'Armington' specification whereby foreign and domestic products are treated as fundamentally different. They then proceed to assume that domestic products are homogeneous, with domestic mark-ups resulting from Cournot behaviour. Export markets are segmented in their formulation and the associated mark-ups are negligible. Thus, given the same level of profits, initial domestic mark-ups must be higher than in the integrated markets formulation which we have employed. This may be seen by examining the parenthetic entries in the second column of Table 1. For example, the Devarajan and Rodrik domestic mark-up on processed food (P/MC) products is 1.62, as opposed to the integrated markets mark-up of 1.29.

The elasticity of substitution between domestic and foreign goods in the Devarajan and Rodrik model is given in the third column of Table 1. Note that it is quite small, indicating that they are assumed to be poor substitutes. Indeed, the values of $\sigma < 1$ for the last three industries causes domestic and foreign goods to be *gross complements* when domestic expenditure on the Armington composite is held constant. This means that a fall in the tariff on intermediate goods, for example, will increase domestic sales per firm in partial equilibrium! Following equation (2), we see that this is reinforced by the pro-competitive effect. Consequently, $\hat{q}_i^{PE}/\hat{t} < 0$ for these three industries.

It should also be pointed out that the pro-competitive effect is very small for all of the industries in the Devarajan and Rodrik specification. This is because the numerator of β_{MF} involves $(1 - \sigma_A)^2$ and the values of σ_A are relatively close to one in this model. This raises serious concerns about Devarajan and Rodrik's assertion of the importance of this pro-competitive effect in reversing the conventional wisdom regarding the fate of agriculture under trade liberalization. Comparison of the final two columns of Table 1 confirm this suspicion. The first of these sets of general equilibrium results for across-the-board elimination of tariffs in the Devarajan and Rodrik model is taken from their basic model with increasing returns to scale and no entry. As can be seen here, three of the five imperfectly competitive sectors expand.

In their paper, Devarajan and Rodrik compare this model to one in which the imperfectly competitive sectors are *recalibrated* and treated as perfectly competitive sectors. This poses a serious comparability problem, since excess profits are now treated as payments to a fictitious 'fixed factor'. The presence of such a fixed factor now serves to significantly restrain supply response in these manufacturing sectors. A further complication has to do with the asymmetric treatment of export supplies between perfectly and imperfectly competitive sectors. In the former case, domestic and export products are treated as imperfect substitutes, while the imperfectly competitive sectors' outputs may be freely shifted between the two markets. Thus their comparison of trade liberalization results in the presence of perfect and imperfect competition is confounded by the simultaneous use of two different specifications of technology, factor mobility and product differentiation. It turns out that the latter distinctions, not the presence of imperfect competition, are what cause the two sets of results to diverge.

A more straightforward method for isolating the pro-competitive effect of trade liberalization is simply to fix the imperfectly competitive firms' mark-ups, that is set $\beta_{MF} = 0$. The trade liberalization results in this case are presented in the final column of Table 1. As anticipated by the partial equilibrium analysis, they differ little from the results with the pro-competitive effect present. This carries over to all of the other variables in the model, including agricultural output. In sum, as demonstrated in Table 1, the pro-competitive effect of tariff reform (when entry/exit is restricted) is potentially an important empirical phenomenon. However, Devarajan and Rodrik are mistaken to argue that it is this effect which causes a reversal of the 'conventional wisdom' in their model. Indeed, theoretical results suggest that this type of qualitative reversal is unlikely.

ENTRY

Theoretical results

Once entry/exit of domestic firms in response to changing profitability is permitted, an additional constraint is placed on sector 2, namely a zero profit condition. Thus price determination may be characterized as follows:

$$A\hat{V}C_2 + \hat{M}_2 = \hat{P}_{2H} = A\hat{T}C_2 \, (\hat{q}_2 = 0) - \Omega_{F2}\hat{q}_2 \qquad (3)$$

The left-hand side of (3) asserts that the sum of the proportional changes in average variable cost and the optimal mark-up must equal price. It follows from the optimal mark-up condition in Figure 1, along with the assumption of constant returns to scale in variable inputs (that is, $M\hat{C}_2 = A\hat{V}C_2$). The right-hand side of (3) is the zero profit condition, where the change in average total cost is decomposed into two parts: that obtained by holding scale constant, and that attributable to changes in scale ($- \Omega_{F2}\hat{q}_2$ where Ω_{F2} is the cost share of fixed factors in the second sector).

Equation (3) simplifies considerably if we assume, for the model outlined in Figure 1, that variable and fixed costs exhibit the same capital-labour intensities, in which case:

$$A\hat{V}C_2 = A\hat{T}C_2 \, (\hat{q}_2 = 0) \text{ and so } \hat{q}_2 = -\Omega_{F2}^{-1}\hat{M}_2$$

That is, output per firm must always move in the opposite direction of the mark-up. However, the change in the optimal mark-up as a function of the tariff is ambiguous in the presence of entry/exit. The reason for this ambiguity is the fact that the induced change in the number of domestic firms works in the opposite direction of the 'pro-competitive' effect identified in the previous section. While a drop in the tariff lowers the price of competing foreign products and thereby lowers domestic firms' optimal mark-ups, it also has the effect of driving some domestic firms out of the industry. Since the representative firms' optimal mark-up is a decreasing function of the number of competitors, this second effect works to raise mark-ups.

Since the ambiguity in mark-ups translates directly into ambiguity in output per firm, it is hardly surprising that we are unable definitively to sign the change in the price charged for domestically produced, differentiated products. This is yet another manifestation of the maxim that, 'in the presence of imperfect competition, anything can happen'. *But these are not the variables of interest in the analysis at hand.* Rather, we want to sign the change in agricultural output as a function of the non-agricultural import tariff. This, in turn, depends on the change in marginal revenue for a representative domestic firm in sector 2, relative to the exogenous price of sector one's output. It can be shown that, regardless of whether P_{2H} falls, the mark-up falls more rapidly. Thus MR_2/P_1 rises and Q_1 falls with $\hat{t} > 0$. In summary, based on the theoretical model outlined in Figure 1, we are able to conclude the following: when entry/exit is permitted, agricultural output will always move in the opposite direction of the tariff on imports of the differentiated non-agricultural product. It should be noted that this finding is also robust to the form of conjectures postulated.

Empirical findings

Now let us return to the Cameroon model once again. At this point it is relevant to re-examine (3) in the light of the differences between the abstract framework outlined in Figure 1 and the complexity of this empirical model. The most striking difference is in the composition of industry costs. While Devarajan and Rodrik do assume that the relative capital–labour intensity of fixed and variable costs is the same, the presence of *intermediate inputs* in the variable cost component now causes the rates of change in AVC_i and $ATC_i(\hat{q}_i = 0)$ to diverge. In particular, since the first-round effect of tariff elimination is to lower the cost of imported goods, and since these represent a sizable share of variable costs, especially in some of the manufacturing sectors, AVC_i falls much faster than ATC_i ($q_i = 0$). Furthermore, since optimal mark-ups are relatively insensitive to the tariff (Table 1), and since the equilibrating change in firm numbers is small (Devarajan and Rodrik), $M_i \cong 0$. Thus output per firm must increase significantly in order to preserve the two equalities in (3).

Now consider the implications of (3) when manufacturing activity exhibits perfect competition, and hence operates under locally constant returns to scale. In this case $M_i = 1$, $\hat{M}_i = 0$; that is, there is no mark-up over marginal cost. Furthermore, $A\hat{V}C = A\hat{T}C$, and $\Omega_{F2} = 0$. Thus the complement of capital and labour in variable costs is larger, so that when the price of imported intermediate goods falls, the change in the index of average variable costs will be dampened, relative to the case where some of this capital and labour is fixed. Thus the partial equilibrium supply price falls less, and output is lower in the post-liberalization equilibrium. However, the resource-pull effect and hence the implications for agricultural output in these two models will be less dissimilar. This is because the expansion in output per firm under increasing returns to scale requires fewer primary inputs per unit of incremental output.

SUMMARY AND CONCLUSIONS

In summary, we believe that the presence of imperfect competition and unexploited scale economies in the non-agricultural economy can have important implications for the level of agricultural output following across-the-board trade liberalization. The theoretical results in this paper show that a reversal of the direction of change in agricultural output (as compared to predictions based on the perfectly competitive paradigm) is unlikely. However, the degree of adjustment required in both sectors will be less, the more collusive is the non-farm sector. The magnitude of these pro-competitive effects depends importantly on the nature of consumer preferences over domestic and foreign varieties: do consumers distinguish a product by firm, or by country of origin? The divergence in predictions also depends significantly on the source and extent of unexploited scale economies. If these derive from a fixed primary factor requirement, then tariff reform can have a strong stimulative effect on those sectors which also rely heavily on imported intermediate inputs.

REFERENCES

Condon, T and deMelo, J., 1986, *Industrial Organization Implications of QR Trade Regimes: Evidence and Welfare Costs*, World Bank, Washington, DC.

Devarajan, S and Rodrik, D. 1988, 'Trade Liberalization in Developing Countries: Do Imperfect Competition and Scale Economies Matter?', *American Economic Review*, 79, pp.283–7.

Devarajan, S. and Rodrik, D., 1989, *'Pro-competitive Effects of Trade Reform: Results from a CGE Model of Cameroon'*, NBER Working Paper No. 3174.

Harris, R.G., 1984, 'Applied General Equilibrium Analysis of Small Open Economies with Scale Economies and Imperfect Competition', *American Economic Review*, 74, pp.1016–32.

Helpman, E. and Krugman, P., 1989, *Trade Policy and Market Structure*, MIT Press, Cambridge, MA.

Hertel, T.W., 1991, 'Assessing the Effects of Trade Policy in the Presence of Imperfect Competition: Theoretical Insights into Empirical Findings', Impact Project Preliminary Working Paper, University of Melbourne.

Kirkpatrick, C.H., Lee, N., and Nixon, F.I., 1984, *Industrial Structure and Policy in Less Developed Countries*, George Allen & Unwin, London.

Markusen, J.R. and Venables, A.J., 1988, 'Trade Policy with Increasing Returns and Imperfect Competition: Contradictory Results from Competing Assumptions', *Journal of International Economics*, 24, pp.299–316.

Robinson, S., 1989, 'Analyzing Agricultural Trade Liberalization with Single-Country Computable General Equilibrium Models', paper presented at the NBER conference on applied general equilibrium modelling, San Diego, CA.

Rodrik, D., 1988, 'Imperfect Competition, Scale Economies, and Trade Policy in Developing Countries', in R. Baldwin (ed.), *Trade Policy Issues and Empirical Analysis*, University of Chicago Press, Chicago.

DAVID COLMAN*

Section Summary

As a test-bed for the role of international cooperation, the GATT negotiations on agricultural trade liberalization have proved discouraging. In view of its continuing importance this was chosen as the topic for the plenary session. Aart de Zeeuw, in view of his position in the GATT negotiation, possibly had to profess a modestly optimistic view of the outcome of the wrangle, which he did by arguing that the basis for a compromise on tariffication with safeguards and limitation of export subsidies was possible. However, this optimism proved difficult to reconcile with his view that, if compromise cannot be reached by the end of 1991, then the Americas and Europe (west and east) are likely to go their own ways to form separate free-trade areas.

The paper by Sandy Warley and Don McClatchy was opened by Sandy Warley in what he called his swansong to the Association. Swans are often seen as mild, except when provoked, but Sandy was hard-hitting in a presentation, which called, in effect, for the profession to confront the split between the intellectual support it gives to the efficiency of markets and free trade and the entrenched commitment of many of its members to supporting their own country's farmers. This schizophrenia is commonplace among us, but clearly not in Sandy Warley, who is committed to an open trade policy with free transmission of price signals. He called upon us to rise above an agricultural focus and support the higher ideals of free trade and the GATT. This support for the efficiency of free trade contrasts with the earlier call by Norgaard for a higher idealism which stressed the limitations of the neo-classical economic paradigm. Don McClatchy presented the other half of the paper, which discussed some methodological limitations of the models generating quantitative estimates to accompany the trade liberalization agenda. This concentrated on the implications that few modelling exercises have simulated the trade-liberalizing effects of reducing aggregate agricultural support against a fixed external reference price; the hypothesis stated was that, if this were modelled, it might be shown that only modest support reduction (30–50 per cent) would remove most of the distortions in world trade.

Further methodological issues in modelling trade policy liberalization (a euphemism for dismantling much of agricultural policy) were explored in the first associated invited paper session. Tom Hertel, Kent Lanclos and Marie Thursby examined the implications of introducing imperfect competition in sectors downstream of agriculture into our models; the treatment was theoreti-

*University of Manchester, UK.

cal and empirical in relation to the Cameroons. The outcomes reinforced a conclusion of Warley and McClatchey, in that they concluded that allowance for imperfect competition leads to lower estimates of the agricultural supply adjustments of trade liberalization.

The paper by Eric Monke and Scott Pearson also reflected the sub-theme of the domestic impacts of trade liberalization by outlining a matrix system for presenting the impacts of policy change to policy makers. The matrix presents a disaggregated assessment of the impacts at all levels within the system.

The second invited paper session which was to address issues of the possibilities of increasing trade from Eastern Europe and the USSR did not go to plan. Neither paper giver could be (or was) present. The situation was retrieved by Karen Brooks (originally scheduled as a discussion opener) who successfully stepped in to present the Russian view, and Ferenc Fekete and Csaba Forgacs who presented a paper on Hungarian agricultural policy and trade. They retrieved the situation admirably, and a good discussion took place, led by the rapporteur, Tadeusz Huneck.

Chairpersons: Maury Bredahl, Wilhelm Heinrichsmeyer, Jock Anderson.
Rapporteurs: S. Tarditi, J.A.Groenwald, T. Hunek.
Floor discussion: S.Hossimi, K.J.Thomson, H.Mahran, A. Kashuliza, R.Lopez, A.Okorib, M.L.Pereira, A.Sarris, C.Short, O. Soda, D.Sumner, S.Tarditi.

SECTION III

Structural Adjustment Policies and Agricultural Development

ELISEU ALVES, CLOVIS DE FARO AND ELISIO CONTINI*

Government and Agricultural Development

INTRODUCTION

Although we do not intend to deal with the economic role of government in detail some preliminary points are worth mentioning. The tradition stemming from Adam Smith considers intervention as unwise: it causes distortions and is not Pareto enhancing. Furthermore, the government does not always contribute to economic stabilization, improve resource allocation or favour income distribution. Concessions are made in the case of public goods including education, defence, research and law and order, and in the management of macro-economic policy, though intervention should be conducted within clearly established rules limiting the discretionary power of the fiscal and monetary authorities. A similar position is held in respect of strategic policies influencing international trade. Free trade should not be restricted, since that would contribute to loss of global and national welfare. All of these issues, however, are still the subject of intense debate which is far from being settled.

The literature on market failure covers such issues as market imperfection, incomplete markets and information, externalities, public goods and returns to scale. It can be cast in the rigorous tradition of Arrow and Debreu, and include propositions suggesting that policy can be Pareto enhancing. It has also been argued that instances of market failure are the rule rather than the exception (Stiglitz, 1989). So far as public goods are concerned, analysis can be much more complicated. Recent theoretical discussion (Hurwicz and Walter, 1990) has suggested that conditions can exist in which it is impossible to reach Pareto optimality in the provision of public goods. To do so requires that participants in an economic system can be separated into groups across which there are no potentially conflicing interests. In trade analysis there is debate about market imperfections and economies of scale which is often based on game theory (Brander, 1986; Helpman and Krugman, 1986). It has been suggested that strategic policies exist in which the income of a country can be improved and maintained without generating major retaliation.

Given the complexity of debate it is far from easy to be confident that simple rules apply in any situation. It can be suggested, however, that in the developing world there are important externalities which have not been internalized by the private sector. These include information processing, research and development, and education. There are also important issues relating to

*Alves and Contini, Embrapa; de Faro, Vargas Foundation; Brazil.

bias against agriculture, and to trade, in which a government role can be considered. It is stressed that we believe that any intervention should not involve the use of parastatal organizations, which have generally been ineffective.

Furthermore, it must be borne in mind that government action is not costless, and that it can become more expensive as it increases in size and scope. That is one of the implications of the theories of rent seeking through interest group collective action, and of bureaucratic behaviour (Krueger, 1990; Olson, 1971; Tullock, 1965). These stress that potentially Parento-enhancing intervention may have a perverse side, which can be the rule rather than the exception. Government can be more subject than the private sector to problems of imperfect information and incomplete markets; the redistributive and allocative roles of the government give rise to inequities, even when the primary goal is to improve income distribution; corruption and rent-seeking activities are wasteful and weaken people's faith in democracy; and bureaucracy, which sprouts the larger the share of government in the income of the country, contributes to waste of resources. Moreover, there is lack of continuity between current and future governments, and government is slow in adjusting to a changing world; the tendency is to mantain old programmes that are supported by powerful rent-seeking groups. Finally, incentives do not work well within the government, because of lack of competition (Stiglitz, 1989a).

In addition to the above, the services provided by government tend to be provided to most, if not all. Its activities generate rent seeking and, since it is heterogeneous and represents different groups of society, there is great temptation to extend the range of benefits (Krueger 1990). The pressures of insiders and outsiders lead to an increase in expenditures, which is usually financed by inflationary means. (Fishlow, 1990).

The debate on government intervention has focused on polar points: intervention or free competition. But intervention has history. In general, it is a major crisis which brings about conditions favourable to the cause of interest groups. It may be a long period of stagnation or a very skewed income distribution, a major recession or a war, or it may be compensating measures in retaliation to policies of other countries. Such examples are frequently well documented, but there is much less literature dealing with the conditions favouring the elimination or reduction of governmental intervention.

The recent events in East Europe, the results of theoretical and empirical literature, and the postwar experience of policy making have raised strong waves in favour of the free-market model. The developing countries are on the verge of reforming their macro-economic policies, and particularly their agricultural policies. The reforms cannot succeed without symmetric reforms in the policies of the advanced countries that are detrimental to the interests of the developing ones, because of the great and ever-increasing interdependence of the world economy. Without dwelling any further on this theme, we would like to stress that agricultural policies have been used by the developed countries as a weapon in their strategic trade policies. This is a well-known difficulty that has hindered negotiations in favour of free trade (Alston *et al.*, 1990).

Does this mean that there is no role for government intervention, specifically, in our context, in the case of agriculture? We believe that there is a role, although it is important to draw the boundaries with care. The areas most commonly agreed upon are research and development; policies that create a fertile environment for innovation and change of attitudes; sound macro-economic policies; education and health; and infrastructure investments. There is also a place for some strategic policies to foster the development of agriculture (these include policies dealing with prices, exports, and credit), but they should identify the beneficiaries, they should be based upon non-inflationary financing and clearly indicate the cost-return to society, and they should have a well defined time-span. In short, they should pass the test of transparency in every respect.

THE CYCLES OF STRATEGIC POLICIES

To understand the cycles of strategic policies which have affected farming we have to distinguish between modern and traditional agriculture. Traditional agriculture itself supplies most of its inputs, and decision making occurs in the countryside. Modern agriculture buys most of its inputs from industry and the decision-making centres are urban. Traditional agriculture is labour- and land-based. Modern agriculture is science- and industry-based. The two types share a common name – agriculture – but they have markedly different characteristics.

At the time of the repeal of the Corn Laws in Britain, industry was fighting a senile agriculture which was protected from trade competition. The common interests in free trade of both consumers and industry were obvious. They joined forced and the Corn Laws were repealed. Agriculture lost the battle of the Corn Laws because its political rents became much smaller than the industrial ones, and because the many farmers and landowners were less organized. In the postwar period, modern agriculture has flourished in the advanced countries, and more recently in some countries of the developing world. It holds strong links with agribusiness. Protection to agribusiness and to modern agriculture is essentially the same. Modern agriculture, in itself, is an activity with low barriers to entry, because every farmer is free to modernize and the investments required are not especially large. Protection encourages large-scale entry, and consequently the dissipation of the political rents is very intensive. Agribusiness, however, has higher barriers to entry and therefore enjoys the necessary power to maintain trade barriers for a very long period. Its oligopsonic organization facilitates the 'collection' from farmers of the money needed to finance lobby activities or to enhance the political rents, whenever necessary.

In the 1930s, many developed countries were large importers of agricultural products. It was modern agriculture, even though it was less developed than it is now, which was able to organize a strong lobby and secure protection. This also occurred in the United States. It was against that background that strategic policies were put to work to advance modern technology at a very fast pace. The infant-industry argument can also be used to explain why modern agriculture was protected in the early 1950s. The European countries, in particular, were protecting farming from the competition of land-rich countries, at that time mainly the United States, Canada, Australia and Argentina. It is clear that some measures were introduced to compensate for what others were doing.

It is not difficult to explain why traditional agriculture remains so characteristic of the under-developed world. All activities pass through a life cycle of development from the traditional to the modern. In the case of Third World agriculture this is occurring only with difficulty. In the past there was often assistance for specialized export crops, but as the fashion for industrialization spread it was more common to find that agriculture was discriminated against. The aim was to transfer resources to the urban sectors. Modern agriculture could only develop, if at all, against the background of that discrimination. This does not mean that the development intention was absent; there have been numerous programmes aimed at disseminating new techniques. The problem has been that farmers are numerous, and they do not all have the potential to modernize. However, it is particularly difficult to pursue policies which are selective and which exclude backward farmers and backward regions. There can be beneficiaries among larger farmers who are better placed to take advantage of programme allocations. This accounts for some countries having made progress in creating a modern sector. However, it is expensive especially when attempts are made to spread benefits thinly to many potential participants. The effort can fail for lack of finance. Often the emphasis has had to be shifted from modernization to poverty alleviation, which, though obviously commendable in itself, does not foster progress.

In the past, the rural aristocracy established some forms of protection for agriculture of the developing countries, mainly in the export sector. That was the first phase. The second phase came when industrialization policies became more important, and it was then that agriculture was severely discriminated against. Within this phase, the need was felt initially to increase the supply of agricultural products, and the modern sector did receive some encouragement, though it often proved to be short-lived and gave way to less favourable policies. The final phase, whatever it may be, has often yet to be reached. It may be one in which adverse discrimination is eliminated, or it could witness some movement towards protectionism.

With the advent of industrial policies in developing countries it was basically traditional agriculture which was discriminated against, and analysts, by not making a clear distinction between old and modern forms, failed to understand the strategic policies that were introduced to advance up-to-date technology. Such policies would have had to achieve two goals. One was to maintain the discrimination against agriculture in order to transfer resources to the industrial sector; the other was to stimulate modern agriculture.

In an environment characterized by a large number of farmers, only a minority of whom would have the potential to modernize, strategic policies would need to be selective. Furthermore, they would have to compensate for the discrimination against the sector. Obviously the compensation would need to be directed to those able to modernize their farming practices, and would require built-in rules which would not allow the unfit to apply for the benefits of programmes for modern agriculture. In other words, the policies would have to exclude backward farmers (and maybe regions), but to do so in such a way that it would not convey the idea of outright discrimination. As a rule, the medium and larger farmers would become the main beneficiaries. They have a higher level of literacy, better titles to land from the legal point of view, and are normally located in regions with more infrastructure.

The main tools of policy include rural credit, investment in infrastructure in progressive regions, irrigation programmes, research for some groups of crops and regions, promotion of special export policies for processed agricultural products which are mainly cultivated in advanced regions or by advanced farmers, and tax advantages such as income tax exemptions or land tax abatement on account of increased productivity. In some countries policies of this type have succeeded in creating a powerful modern sector within agriculture, while agriculture was simultaneously transferring substantial amounts of resources to industry. Often, however, such a set of policy measures could not be maintained for a long period. Some countries abandoned them half way, before a sufficiently large segment of agriculture was modernized, and before the transformed sector was able to supply most of the needs of agricultural products. In other cases, strategic policies were extended to almost everybody, at great cost, and sometimes with larger farmers being excluded. The basic goals of programmes shifted towards poverty alleviation.

SOME POLICY SUGGESTIONS

The following discussion of the policy suggestions will not be exhaustive and will pinpoint only some aspects. There is no pretension to design a complete programme of policy reform. Emphasis is given to long-run sources of growth. The issue is not whether government should intervene, but what it should undertake and what it should avoid.

The secular decline of agriculture as per capita income grows has been known for a long time. Its share in the national income and employment declines, and the value added outside the farm gate increases continuously because of activities such as transport, processing, storage and modern input production. The decision-making centre shifts gradually to the cities. Either the position of the rich countries, which try to avoid the secular decline of agriculture, is mistaken, or that may apply to the developing countries which are accelerating relative decline (Knudsen *et al.*, 1990).

Education and technology

It is common in democratic societies for the equity principle to lead the state to generate policies that are compatible with the interest of the majority. If a large part of the population is illiterate, technology may appear to need to be simple, and there is a temptation to follow this route, selecting agriculture as a sector in which traditional technology is to be used. It may be an apparently logical choice, since farmers tend to be the group having the weakest levels of education. In our view, however, this is an '*n*th' best solution, where '*n*' is very large. Interest groups which advocate such a stance, although they may have majority support, may nevertheless themselves be backward.

Lack of modernization in agriculture is known to have sad implications for the growth of the economy. The sectoral differential of income grows with the lack of farm modernization, worsening the sector's income situation, and eventually provoking a flood of rural migrants into the cities. Slums and urban violence are the most visible effects.

A more theoretical argument may be mentioned. The literature on human capital has stressed the spillover effect of investment in education, and the appearance of strong positive externalities (Schultz, 1987). Suppose that we have a production function, applicable to industry, with two inputs, labour and capital. Investments in education make the hypothesis of diminishing returns implausible (Lucas, 1988). The marginal product of capital may well be an increasing function within some range of the production function. The implication is that the rate of profit will not decrease in the industrial sector with the increase of the amount of capital, because of investment in education among those living in the cities. While the rural sector is kept backward and discriminated against in human capital investment, the only avenue left open for the convergence of incomes is the labour market, operating through the migration process. Since the illiteracy rate of rural labour is very high, the migrants are cast into the informal sector, or into the sectors that pay lower wages. This is a mechanism which creates slum conditions and generates urban violence. Income distribution in the cities grows worse.

Thus the basic 'solution' worsens income distribution in general, favours discrimination against agriculture (it is much easier to discriminate against a backward sector), and postpones investments in elementary education to a much later date. Initially, the modern sector requires higher-educated people, and so the universities are privileged. Only at a much later period, when the mass of illiterates flowing to the cities becomes a burden to society, is elementary education seen as a priority. Lack of investment in education, as a consequence of the power of backward interest groups over the government, retards the development of democracy, jeopardizes birth control programmes and is a major impediment to a favourable atmosphere for modernization. Programmes such as rural extension, agrarian reform and irrigation become unproductive. It is utopian to believe that illiterate farmers can modernize their enterprises; the scarce factor in modern agriculture is human capital.

Proposing the use of 'intermediate technology' is also an excuse for not investing in education, and a wasteful excuse at that. It is part of the process which generates jobs for city people with a diploma in agrarian sciences. A cursory examination of the size of the extension, irrigation, and rural-credit bureaucracy that is directed at working with small farmers suffices for proof. The term 'intermediate' technology, seems to imply that in an environment where land is scarce relative to labour (the price of land is increasing relative to that of labour), any new production function should have a higher level of marginal product of land than that of labour for every point of the function, in comparison with the older one. Furthermore, the land-saving inputs are divided into two groups. Group 1 refers to simple technologies, while group 2 refers to complex technologies with respect to the cultural background of the population, so that the same relationship between the marginal products of the two production functions must hold globally.

The global properties of the new production function (in comparison with the older one), or more specifically the bias in favour of technologies that are appropriate to the farmers who are illiterate or have a low level of instruction, gives too much scope to bureaucracy and politicians. The priority setting becomes too bureaucratic and dominated by ideology. The chances for individual freedom and creativity shrink. The fundamental need to create new technology is neglected. The market and the intuition of scientists are underrated. Scientists of high calibre tend to be discriminated against if they disagree with the dominant group. It also seems to imply that society is not going to invest in education.

The egalitarian ideology which claims that the sons of the well-to-do must go to the same schools as the sons of the poor, suggestive as it may be, has resulted in the poor staying without schools and public money supporting the education of the rich. The adverse-selection mechanisms of the public school, as regards location and student selection, should be eliminated. The private sector should be stimulated to invest in elementary education too.

To conclude, we consider it wrong to deny modern technology to those working in the rural sector because farmers are illiterate. What should be done is to eliminate illiteracy. This means that education, and most of all elementary education, should be treated as the number-one strategic policy. We also consider it wrong to deprive the rural sector of modern agriculture only because a small proportion can adopt it. It would be better to select a group for modernization and raise a tax on the surplus generated to invest in education. Health policy is also very important, but its effectiveness is low if illiteracy rates are high.

Research

No one questions the fact that research should be a priority in agricultural policy. But lack of (or inadequate) patent laws have kept the private sector out of agricultural research. In an environment without competition, public research has no basis on which to measure its efficiency. Private research saves the public budget for areas that are riskier. Competition improves the efficiency of the public system, and cooperation with private research enhances productivity on both sides.

Public research is now subjected to ideological pressures to give priority to research in intermediate and no-modern-input technologies, and also very pressed by environmental questions. This severely limits the creativity of the researchers. There is nothing wrong with pressures on researchers. The trouble is caused by ideological and political demands that are derived from them. In general they address short-run problems that may not be relevant, but if they are not introduced in the research agenda the odds are higher for a budget cut. It would then be advisable for public research to separate the programme into two areas: one for intermediate technology and environmental research to attend political demands and help the transition to modern agriculture, and the other to support advanced agriculture.

Research is intensive in the use of talented and well-trained manpower. There is a very competitive international market in which it is very difficult for the public institutions of low-income countries to compete. The state is subjected to equity rules which limit wages in the public sector (Stiglitz, 1989). Even if it were possible to establish an exception for research, the odds would be for an increase in the number of politically protected scientists in the ranks of the institutions. There are ways out of this problem. Farmer associations may be allowed to collect tax on some products to do research on them, and the public sector may contract the associations for some projects. Private enterprises, including foreign companies, may also be given special conditions to carry out research. The donor community may be of temporary help, and the international centres have a great contribution to offer.

Infrastructure

Another important area for government intervention is the road network to link farm people to the rest of the economy. This helps to eliminate the barriers separating farm people from city life, brings down the cost of food, and makes farm resources more productive. Also important are investments in means of communication such as radio, telephone and mail delivery. Low-cost transportation and communication increase the chances for a better life for both city and rural people.

Credit

The modernization of agriculture requires investment in areas such as farm machinery, irrigation, infrastructure, soil recuperation and conservation, and pasture. Most of the resources comes through the loan market, be it private or government, formal or informal.

Loans are an exchange of funds by one party with another for a promise of a future return. Loan contracts are heterogeneous, with different probabilities of default. Lending institutions are therefore subjected to restrictions imposed by an environment of incomplete markets and incomplete information. They have to perform the roles of collecting funds, allocating them and monitoring the loan applicants. Thus there are costs of collecting information, screening applicants and monitoring them. The costs tend to be lower for larger loans, higher when the applicants are scattered over an extensive area, and even higher when ignorance about the state of nature is greater. Farmers, especially the small ones, are at a low point on the list. When there is need for credit rationing (and there always is), they are the first to be screened out. In the developing world there are two additional factors: discrimination against agriculture makes it riskier, and the legal weakness of titles to land limits the ability of farmers to offer them as collateral. That is why it is justifiable for government to take action in order to legalize property rights to land.

Banks may prefer to ration credit on a non-price basis rather than increasing the rate of interest (Stiglitz and Weiss, 1981). The selection procedures are based on characteristics associated with relatively low risk. Farmers operate in an environment in which the states of nature are less known or subjected to a larger variation. Thus they may receive less credit from the private sector, and proportionately even less when the aggregate supply of funds declines.

To overcome such problems rural-credit institutions are often established. In the same country one may find public and private institutions lending money according to the rules of the system. One of the means of reducing the risk to the lending institutions is an insurance on outstanding loans, the cost of which is borne by society, at least partially. Sometimes technical assistance is required, which may be paid by the farmers or by government.

Rural credit may be used to provide subsidies to farmers who have great potential to modernize. When a large part of the population is illiterate, located far from bank facilities, and title to land does not exist or is of poor legal status, self-selection or adverse-selection mechanisms tend to appear. Even when the government sets strict rules for both types of banks, they do not always follow them. When they adhere to the small farm segment, it is to benefit farmers who offer less risk. The private bank system offers much more resistance to working with small farmers, because of the cost of searching for information, screening and of monitoring. The laws protect small farmers, and obtaining repayment of a loan can become a complicated legal operation that may create a bad image with public opinion. To induce private banks to finance small farmers is costly to the treasury. The trouble is that the same may happen to public banks. They may induce small farmers into risky operations to be agreeable to them, because they know the treasury will take responsibility for any failure. Generally, however, if it is decided to provide credit to small farmers, there may be no alternative other than the public banks.

If they are likely to lose money on lending to benefit small farmers, one can be sure that adverse selection will be the rule, as is the case of private banks. The wages of bank employees depend on the profitability of the institution, and it is the same with public banks. Why would the institution take the risk of losing money, when it could otherwise gain?

Subsidized credit is a large part of the cost of government programmes to compensate farmers for losses which economic policies impose on agriculture. In this sense, it is a second-best solution. But there is an additional problem, for it induces farmers to cross the line of safe behaviour and take advantage of subsidies since they know that repayment of loans will not be demanded. The experiences of both advanced and developing countries are full of examples in which large numbers of farmers cannot service their debts. The answer to this problem cannot be sought through the credit system; the first-best solution is to remove the underlying distortions which are so characteristic of agricultural policies.

The subsidies incurred are frequently financed by printing money. Rural credit then becomes an important source of inflation. This is a side-effect to add to the distortions of resource allocation and income distribution caused by the subsidization of rural credit. If the monetary authorities push up the interest rate as a means of fighting inflation, then the difference between the interest rate charged to farmers and the market interest rate becomes very large. The treasury authorities would need to cover the difference by non-inflationary means, which is self-defeating: the larger the difference between the two interest rates, the larger the demand of farmers for loans. Credit rationing, which is the compromise solution, is difficult to implement.

Banks monitor their clients to be sure that they follow the contracts agreed upon. But if government becomes liable to cover losses, the monitoring function weakens or may disappear. If there is an insurance on the loans, and government is liable for losses, the same will be true. Whatever method is employed to reduce the costs of rural credit to a comparable basis with other sectors of the economy, the final result is to weaken the monitoring function and to induce farmers to borrow less than wisely.

The scale of farming

It is common to raise the question of whether small farmers are more efficient than larger operators. Both theoretically and empirically, the answer is ambiguous (Binswanger and Elgin, 1989; Stiglitz, 1974). But if the distribution of literacy is bimodal, with a large number of illiterate small farmers, and the medium and larger farmers have a greater degree of literacy, then the latter group is better prepared for modern technology. If that is more profitable, literate farmers will spring ahead on the road to modernization. Income distribution will worsen in rural areas. Some policy makers believe that one should block the advancement of agriculture, or at least that its speed should be reduced, while others argue that the group with potential should be stimulated, if necessary, with subsidies. Without specifying the conditions of the environment, it is difficult to decide on the relative merits of these opposing views. It is, however, very difficult to accept policy measures which may eventually block modernization. There is nevertheless a correct answer: investment in education.

CONCLUSIONS

A summary of the positive roles of government in agrarian reform includes the following points: eliminate the adverse-selection mechanisms within economic policies; improve or reform the law to stimulate sharecropping and land renting; impose a progressive land tax on unproductive land; establish long-term credit to help small farmers to acquire land; and let the farmers' associations manage agrarian reform projects. An extensive discussion of this controversial subject can be found in Binswanger and Elgin (1989).

The view that all government intervention in price mechanisms have negative consequences is well known. For instance, theoretical and empirical literature stresses the high cost for farmers and society of policies such as an over-valued exchange rate, tariffs to protect industry or agriculture, quantitative barriers, voluntary restraints, ceiling prices on wage goods, outright prohibitions on exporting some products or importing requisites, subsidized credit and unsound macro-economic policies. Also well known are the negative effects of the protection which is given by the developed countries to their agriculture, disregarding both their consumers and the farmers of the Third World and distorting world trade (Knudsen *et al.*, 1990).

Nevertheless, governments cannot be denied the role of counterbalancing significant actions of other governments to protect their agriculture or to counterbalance large fluctuations of the world economy. The tit-for-tat type of strategy, however understandable, is a major stumbling-block to free trade and to the reform of the agricultural policies of both developed and developing countries. Every country may not see any advantage in moving towards free trade and a free market economy. These demand good will and cooperation, at least, of the major producers and importers of agricultural products.

The major point in the paper is the need to remove from the economic policies of the Third World the strong bias against investing in rural people, and especially in their education and health. If investments in rural people are not made, either agriculture will not develop or, if it succeeds in advancing, a mass of poor people will be left behind, with sad implications for income distribution, urban unrest and political instability. Investments are highly recommended in research, extension, infrastructure, and in activities that create a general atmosphere which favours innovations. The distortions which act upon agriculture and the adverse-selection mechanisms of economic policies must be eliminated. If for some reason the government intervenes in the price mechanism, the policies must be absolutely transparent, and have a short life span.

REFERENCES

Alston, J.M. *et al.*, 1990, 'Discriminating Trade: The Case of Japanese Beef and Wheat Imports', *Canadian Journal of Agricultural Economics*, vol. 38, August.

Binswanger, H.P. and Elgin, M., 1989, 'What are the Prospects of Land Reform?', in A. Maunder and A. Valdés (eds), *Agriculture and Government in an Interdependent World*, Dartmouth, for IAAE.

Brander, J.A., 1986, 'Rationale for Strategic Trade and Industrial Policy', in P.R. Krugman, (ed.) *Strategic Trade Policy and New International Economics*, MIT Press, Cambridge, MA.

Fishlow, A., 1990, 'The Latin American State', *Journal of Economic Perspectives*, vol. 4, no. 3.

Helpman, E. and Krugman, P.R., 1986, *Market Structure and Foreign Trade*, MIT Press, Cambridge, MA.

Hurwicz, L., and Walter, Mark, 1990, 'On the Generic Nonoptimality of Dominant-Strategy Allocation Mechanisms: A General Theorem that Includes Pure Exchange Economies, *Econometrica*, vol. 58, no. 3.

Knudsen, O. *et al.*, 1990, *Redefining Government Role in Agriculture in the Nineties*, Preworking Series, World Bank, Washington.

Krueger, A.O., 1990, 'Government Failures in Development', *Journal of Economic Perspectives*, vol. 4, no. 3.

Lucas, R.E., 1988, 'On the Mechanics of Economic Development', *Journal of Monetary Economics*, vol.22, no. 3.

Olson, M., 1971, *The Logic of Collective Action*, Harvard University Press, London.

Schultz, T.W., 1987, *The Long View in Economic Policy: the Case of Agriculture and Food*, International Centre for Economic Growth, San Francisco.

Stiglitz, J.E., 1974, 'Incentives and Risk Sharing in Sharecropping', *Review of Economic Studies*, vol. 41.
Stiglitz, J.E. and Weiss, A., 1981, 'Credit Rationing in Markets with Imperfect Information', *American Economic Review*, vol. 71. pp. 393–410.
Stiglitz, J.E., 1989, 'On the Economic Role of the State', in A. Heertje (ed.), *The Economic Role of the State*, Basil Blackwell, Beaufort.
Tullock, G., 1965, *The Politics of Bureaucracy*, University of Michigan Press, Ann Arbor.

DISCUSSION OPENING – JULIO HERNÁNDEZ-ESTRADA*

Alves, de Faro and Contini argue in their paper that it is necessary to remove bias against investment in rural people from Third World economic policies, especially in education and health. If this kind of investment is neglected, agriculture will not develop, and this could worsen intersectoral and intrasectoral income distribution and might cause urban unrest and political instability.

Although the authors' conclusions are valid, it would also be appropriate to urge that economic and agricultural policy reforms in Third World countries be accompanied by similar changes among industrial countries, in order for the poorer countries to be successful in reaching their social, political and economic goals.

The paper pointed out that the secular decline of agriculture as per capita income grows means that agriculture's share in national income and employment also declines, while value added outside the farm gate increases continuously owing to the growing complexity of the processes involved. The richest countries are trying to avoid secular decline of agriculture and the poorest countries are accelerating the process, because they need to produce more manufactured goods. The authors indicated that both policies are misguided. However, international trade in the last two decades has favoured manufactures. Between 1980 and 1987 alone the prices of 33 commodities (basically raw material exports such as copper, iron ore, timber, sugar, cotton and coffee) fell by 40 per cent on average, catching the Third World between the blades of rising debt and falling earnings.

Prospects for the world economy in the 1990s are very much at risk because of the massive balance of payments problems in the major industrial countries, which could result in trade war, and because of uncertainties about changes in the environment. A crisis in either case could sharply reduce the rate of private investment and therefore economic growth. A financial crisis was responsible for the severe depression of the 1930s and for the economic malaise of many developing countries in the 1980s. It is difficult to measure the impact of environmental changes on long-run growth prospects of the world economy since the issues are complex and the nature of the links are not yet understood. Many types of environmental problems cross national frontiers, so their resolution requires international agreements, without which they could be even more costly than a massive financial crisis.

Volatile exchange rates and interest rates are part of any realistic scenario for the 1990s. For example, a decline in the dollar's role as a major exchange

*Colegio de Postgraduados, Mexico.

vehicle could lead to increased financial volatility. The foreign exchange markets have certainly been subject to fluctuation since the late 1970s. In a world of reduced international credit, it has been difficult for the agricultural sectors of the Third World countries to transfer substantial amounts of resources to industry and at the same time create a powerful modern sector within agriculture. Against such an unfavourable background, the authors also stressed that it is utopian to believe that illiterate farmers can modernize their enterprises, especially when modern agriculture depends so much on investment in human capital. To add to that there are signs that land is also becoming scarce.

As we enter the 1990s, the world has little to celebrate on the food front. Between 1950 and 1984, farmers raised world grain output 2.6 fold, an increase that dwarfed the efforts of all previous generations combined. Since then, little progress has been made and the proportion of hungry and malnourished people has increased. Growth in world food output is being slowed by environmental degradation, a world-wide scarcity of cropland and irrigation water, and a diminishing response to use of additional chemical fertilizer. Soil erosion is slowly undermining the productivity of an estimated one-third of the world's cropland. Deforestation is leading to increased rainfall run-off and crop-destroying floods. Damage to crops from air pollution and acid rain can be seen in industrial and developing countries alike. Each year millions of hectares of cropland are lost, either because the land is so severely eroded that it is no longer worth ploughing, or because new homes, factories and highways are built on it. World-wide, the potential for profitably expanding cultivated area is limited. The global decline in grain area per person from 0.16 hectares in 1980 to an estimated 0.14 hectares in 1990 seems certain to continue. The prospect for expanding the world's gross irrigated area is hardly more promising, since it is now lagging behind population growth.

From the middle of the century, increasing use of chemical fertilizer has been the engine powering the growth in world food output. Between 1950 and 1989, world fertilizer use climbed from a meagre 14 million tons to an estimated 143 million tons. If for some reason fertilizer use was abruptly discontinued, world food output would probably plummet some 40 per cent or more. But rapid growth in fertilizer use has depended on the continued spread of high-yielding seeds as well as irrigated area. Once the new fertilizer-responsive varieties are planted on all suitable land, growth in fertilizer use also slows. Many developing countries are now experiencing diminishing returns in fertilizer use and, given its dependence on water availability, the reduced growth in irrigation is almost certain to affect fertilizer use as well.

Alves, de Faro and Contini argue that it is wrong to deny agriculture access to modern technology on the grounds that farmers are illiterate. I agree with that view, but I also contend that education and research must be major policy priorities. There are other contributors to agricultural growth but those two certainly appear to be crucial. A recent study by Frisvold and Lomax (1989) emphasizes the point. To give some examples, the average annual growth rate (1970–80) of agricultural total factor productivity was 0.31 per cent for Mexico, which was the lowest positive value for the selected countries. The highest rates were for Spain and Netherlands, with 4.01 and 3.22 per cent respectively. In contrast, Peru had the poorest performance, with a negative rate of −2.66 per cent, and Pakistan was the second lowest at −1.43 per cent. The total factor productivity levels for 1980 setting USA equal to 100 were as follows: Israel 113.3, Belgium 106.5, Chile 31.9, Mexico 19.2 and India 15.9. In my view, this provides strong evidence that the ability of a country to develop and encourage adoption of new technologies is directly related to public investment in agricultural research and education. It is worth emphasizing that, in the past century, between 60 and 70 per cent of the improvement in living standards in high-income OECD countries can be explained by growth in labour productivity.

The developing countries do face different risks and opportunities from industrial countries. Most of them import new technologies, so their access to technology depends on the availability of the foreign exchange and external financing needed to import appropriate capital goods. Although the comparative advantage of developing countries remains in producing and exporting relatively labour-intensive products, some of them may have to move gradually towards a more capital-intensive strategy. Whatever their position, however, they must invest more heavily in human capital, to take advantage of the opportunities available to them.

In the 1990s, population (and the labour force) in the developing countries is expected to grow substantially faster (roughly 1.9 per cent a year) than in the industrial countries (roughly 0.5 per cent a year). This means that the developing countries on average must grow significantly faster than the industrial countries just to maintain their relative position in terms of real per capita income. To achieve higher growth, the developing countries must improve the level of efficiency of investment and raise the growth rate of labour productivity. It will not be easy, since the external debt problems which many developing countries face show that they are not well prepared, either financially or politically, to cope with major shocks of the kind which occurred in the 1970s and 1980s, including the sharp rise in energy prices and international interest rates. They now have the additional problem of environmental degradation to add to their problems.

REFERENCES

Frisvold, G. and Lomax, E., 1989, *Differences in Agricultural Research and Productivity among 26 Countries*, Economic Research Service, United States Department of Agriculture, Agricultural Economic Report Number 644.

PATRICK GUILLAUMONT*

Adjustment Policy and Agricultural Development

INTRODUCTION

This paper will consider the effects of macro-economic adjustment policy – including stabilization and structural adjustment – on the agricultural sector, and in consequence on economic development. Of course, these effects differ according to countries. Reference is made mainly to African countries, but not only to them.

Most Sub-Saharan African countries have registered balance of payments deficits and have been led to adopt adjustment policies. Some did it at the end of the 1970s or at the beginning of the 1980s, some others more recently. Owing to the large share of the agricultural sector in these countries, the effects of policy reform strongly depend on their agricultural impact. The impact of adjustment on the agricultural sector is not easy to evaluate, because agriculture more than other sectors is affected by exogenous factors, independent of economic policy. Moreover, a long time is often required for agricultural output to react to incentives changes.

Table 1 shows the average trend of constant prices agricultural value added for two nine year periods (1970–9 and 1979–88) and several sub-sets of countries. It seems that, during the 1980s, agricultural growth has remained below that of the other developing countries. The African countries, which for at least three years applied an adjustment policy with the support of the Bretton Woods Institutions, have improved their agricultural growth compared to that of the 1970s, which is not the case for the other African countries. These figures are only (simple) averages, calculated from incomplete and often not very reliable data, and with deviations within the sub-sets.

Although the observed results concerning agricultural growth are not corrected for the influence of exogenous or environmental factors,[1] they cannot allow us to reject the assumption of favourable effects of adjustment policies on agricultural growth. However, the deviations within results lead us to think that adjustment policies, as they have actually been applied, have not always resulted in agricultural development.[2] It may be so because the main instruments of adjustment policies are not acting automatically, as is sometimes assumed.

*Université d'Auvergne, Clermont-Ferrand, France.

TABLE 1   *Agricultural value added rate of growth (1980 constant prices percentages)*

Simple average	1970–79	1979–88	Difference (same sample)
LDC	2.9 (85)	2.8 (92)	–0.2 (82)
adjusting	2.4 (42)	2.4 (41)	–0.1 (41)
without devaluation	1.9 (10)	1.8 (9)	–0.1 (9)
with moderate depreciation	2.3 (15)	2.7 (15)	0.4 (15)
with strong depreciation	2.8 (17)	2.3 (17)	–0.5 (17)
non–adjusting	3.3 (43)	3.1 (51)	–0.2 (41)
SUB–SAHARAN AFRICA	2.2 (35)	2.0 (36)	–0.2 (35)
adjusting	1.9 (20)	2.3 (20)	0.4 (20)
without devaluation	1.7 (7)	1.9 (7)	0.2 (7)
with moderate depreciation	1.5 (6)	2.8 (6)	1.3 (6)
with strong depreciation	2.6 (7)	2.2 (7)	–0.4 (7)
non–adjusting	2.6 (15)	1.6 (16)	–0.8 (15)
LATIN AMERICA	2.7 (24)	1.9 (28)	–1.3 (23)
adjusting	2.7 (12)	1.8 (11)	–0.9 (11)
with moderate depreciation	2.4 (4)	0.2 (3)	–2.2 (3)
with strong depreciation	2.9 (8)	2.5 (8)	–0.4 (8)
non–adjusting	2.8 (12)	1.9 (17)	–1.7 (12)

Notes: Data in brackets correspond to the number of observations in the sample in the period considered. The differences between two periods are calculated on a common sample. This applies to all tables.
         The rate of growth of the agricultural value added at constant prices has been calculated by exponential adjustment.

Sources: *World Tables* (World Bank).

HOW ADJUSTMENT LOGICALLY FAVOURS AGRICULTURAL DEVELOPMENT

Structural adjustment means a lower external deficit for a given overall growth, or more overall growth for a given deficit. Such an outcome needs a change of the structure of production, leading to a larger share for tradable goods (either exports or import substitutes).[3] In a market economy this change results from increased profitability of tradables production, which itself may come from two sources. One is the increase of the relative prices of tradables, the other is an increase of factor productivity.

So defined, structural adjustment is distinct from macro-economic stabilization, which means reducing external deficits through lower domestic demand, and a lower growth. However, these two ways of adjustment are often complementary over time. Structural adjustment involves effective working of markets, avoiding high inflation or acute shortages. It needs the restoration of macro-economic balances, by the way of stabilization. Each of the two sources of structural adjustment (relative prices and productivity) has strong implications for agriculture.

Relative price changes and agricultural growth

An increase of the relative price of tradables (or a decrease of that of the non-tradables, or the so-called decrease of the real exchange rate) is *a priori* favourable to the agricultural sector, in so far as this sector produces tradable goods in a larger proportion than others. Not only export goods but also most food crops are tradables: rice, maize and millet are traded between African countries, often with fewer tariff or administrative obstacles than manufactured goods. Of course, they are also used for farmers' own use.

It is generally admitted that a structural adjustment policy involves an increase of agricultural prices relative to those of other goods and services. Such an increase is often a reversal of the trend registered during the 1970s in numerous developing countries, especially in Africa (the so-called deterioration of the internal terms of trade). As such, it normally leads to lessening 'urban bias'.

If a structural adjustment policy actually leads to an increase of the relative or real price of agricultural products, which is not always the case, as will be seen, agricultural production is expected to rise. How far? This question has been hotly debated. Let us recall some simple conclusions of this debate.[4] The price elasticity of agricultural supply depends on the working of rural markets: when farmers cannot sell their output or when they are unable to find anything to buy with their money, the elasticity is lower, and may even become negative. The long-term price elasticity is higher than in the shorter term, especially for perennial crops and in irrigated agriculture, or more generally when investment is needed to increase output.

In the short term the price elasticity for single products is significantly higher than that of global agricultural supply, since, when the price of a single product increases, output can be increased by replacing other crops. Global supply is limited by the available production factors, which may be fully used. In the long term, an increase of the global agricultural supply involves productivity improvements.

Productivity improvements and agricultural development

For the last ten years structural adjustment has often been considered as 'getting prices right'. However, productivity improvements, the other source of structural adjustment, are needed as well. They are complementary to relative price changes, wherever they occur, in agriculture or in other sectors.

The agricultural production of developing countries, more than that of other sectors, is limited by the availability of primary production factors. Thus productivity improvements are particularly needed to increase production. Agricultural relative price increases provide an incentive to technical progress in agriculture. It is the reason why the long-term price elasticity of supply is probably higher than the short-term one, whatever difficulties there may be to test that proposition.

Indeed, in the long term, factors other than prices influence agricultural productivity, either directly or through the parameters of price reactions: health and education in rural areas, agricultural extension, agronomic research, quality of feeder roads and other communication means, access to credit, and so on.

It is not only the productivity improvements which directly occur in the agricultural sector which are favourable to agricultural development and structural adjustment.[5] Productivity improvements in other sectors can themselves increase the relative price of agricultural goods. This is the case either when productivity improvements lead to a decrease of the price of goods (inputs or consumption goods) bought by farmers, or when they lower processing, transportation or marketing costs of agricultural products, which allows a better price at the farm gate for a given border price.

Briefly stated, structural adjustment, according to the concept itself, is in most developing countries logically favourable to the agricultural sector, even if it is time-consuming. It relies on two complementary means, a rise of the relative price of agricultural goods and productivity improvements. But structural adjustment policy, which aims at these two objectives and to maintain macre-economic balances, requires choices and trade-offs which are sometimes difficult and may or may not favour agriculture.

Diversity of the instruments of structural adjustment policy and of their effects on agricultural development

Structural adjustment policy uses many instruments, covering the various fields of economic policy (monetary and fiscal policy, public-sector management, public investment choice, rate of exchange policy and price and trade policy). Actual adjustment policies combine these different instruments in various ways, so they may focus on the restoration of macro-economic equilibrium, or on the elimination of price distortions, or on productivity improvements. These choices can be more, or less, favourable to the agricultural sector.

The main relevant choices which we consider here are devaluation (strong or moderate) versus maintaining the parity, total price liberalization versus maintaining some regulated or stabilized prices, and public expenditure reduction (more or less strong) versus increase in public receipts in order to reduce budget deficits. Each of these three choices has an impact on agriculture which is never mechanical. Often effects depend on political economy considerations.

RATE OF EXCHANGE POLICY,
REAL PRODUCER PRICES AND PRODUCTIVITY[6]

Most of the countries which have applied an adjustment policy with the support of IMF have been led to devalue several times or to adopt a floating rate. Noticeable exceptions in Africa are the franc zone countries and Liberia, and elsewhere Haiti and Panama.

Expected and observed effects of devaluation on real prices

Devaluation is indeed an essential instrument for structural adjustment, since it normally leads to a change in the ratio of tradable to non-tradable prices, that is, the real exchange rate. It aims at increasing the profitability, and then the volume, of exports or of import substitutes.[7] Currency depreciation actually means a rise of the border price of exported or imported agricultural goods expressed in domestic currency, which *ceteris paribus* increases the profitability of the production of such goods.

However, the expected effect of the devaluation as an agricultural incentive is not an automatic one. The reason is that real producer prices are determined by the real international price of agricultural products, by the real exchange rate, and by the rate of levy which occurs between the border and the farm gate; that is to say, the rate of tax levy, and transportation and marketing costs.[8] So the increase of the nominal border price due to currency depreciation may not lead to an increase of the real producer price, for two reasons. One is the domestic inflation induced by the devaluation, which means that a decrease of the nominal exchange rate results in a lower decrease of the real exchange rate (that is, the devaluation is more or less 'effective'). The other reason is that the increase in the nominal border price is not entirely passed on to the producer, because one part of it is kept either by the state through marketing boards and progressive taxation, or by monopolistic trading networks. So devaluation has a real incentive effect on agriculture only under some conditions.

The figures of Table 2 show that, from 1979–88, these conditions were not always satisfied. Contrary to what could be expected, adjusting countries whose currency has been strongly depreciated (20 per cent a year or more on average during the period) are those where the trend of the real producer prices has been the least favourable, both for food and export crops. Then come the countries with 'moderate' depreciation (less than 20 per cent a year on average). The trend of real producer prices appears to be positive (on average) only in the group of 'without devaluation' adjusting countries. The results hold for the whole sample of 85 (87) developing countries and the African sub-set. In Latin America the few countries with moderate devaluation show the better performance in real producer prices, contrasting with the results of the strongly depreciating countries of these regions.

All these results could indeed come from differences in the trend of the international price of the products exported from the various categories. In

Patrick Guillaumont

TABLE 2   *Trend of real producer prices (1979–88), (linear trend coefficients)*

Simple average	Food crops	Export crops	Trend of the export crop producer price index to the real international price index (ratio)
LDC	−0.8 (85)	−1.2 (87)	7.6 (65)
adjusting	−2.1 (40)	−2.1 (40)	6.5 (33)
without devaluation	0.3 (10)	1.5 (10)	8.4 (8)
moderate depreciation	−1.7 (14)	−1.6 (14)	7.1 (10)
strong depreciation	−4.1 (16)	−4.8 (16)	5.0 (15)
non-adjusting	0.3 (45)	−0.5 (47)	8.8 (32)
SUB-SAHARAN AFRICA	−1.1 (36)	−1.3 (37)	9.8 (26)
adjusting	−2.4 (20)	−2.3 (20)	5.3 (14)
without devaluation	0.9 (7)	2.5 (7)	8.5 (5)
moderate depreciation	−2.8 (6)	−4.9 (6)	1.7 (3)
strong depreciation	−5.3 (7)	−4.7 (7)	4.5 (6)
non-adjusting	0.6 (16)	−0.3 (17)	15.1 (12)
LATIN AMERICA	−0.4 (24)	−0.3 (26)	6.9 (22)
adjusting	−1.9 (10)	−0.9 (11)	8.8 (11)
moderate depreciation	−0.6 (3)	5.8 (4)	16.4 (4)
strong depreciation	−2.4 (7)	−4.8 (7)	4.5 (7)
non-adjusting	0.4 (14)	0.1 (15)	5.2 (11)

Notes: Nominal producer prices for export or food crops have been deflated by the consumer price index (or GDP deflator).
 Real prices for each country have been calculated from an arithmetic average of real producer price indexes, weighted by the relative importance of the main selected products in agricultural production.

Source: FAO annual statistics.

order to eliminate the influence of this factor, we have calculated the trend of the ratio between the real producer prices and the real unit value of exported goods, for the main export crops of each country. This trend, calculated for 65 developing countries, is highest in the non-adjusting ones and in the adjusting countries without devaluation, and lowest in the strongly depreciating countries. Again only in Latin America, does the group of countries with moderate depreciation (only four countries), perform as expected, showing the highest trend.

Economic policy implications

A choice between an adjustment with or without devaluation is not always possible. In countries which registered high inflation, where real exchange rates strongly appreciated, and where simultaneously a strict exchange control associated with foreign exchange shortage resulted in a large parallel market for foreign exchange, with a large gap between the official rate and the parallel one, a strong decrease in the nominal official rate was unavoidable (Ghana, Guinea, Tanzania for example). Its main aim was to unify the foreign exchange market and to integrate the informal or parallel transactions into the official ones. It is difficult to estimate the effect of such depreciations on agricultural activity; indeed a significant increase of agricultural exports occurred in the depreciating countries (Balassa, 1990), but the share of it due to the conversion of unofficial into official transactions is not known.

However, some African countries which did not register a strong appreciation of their real exchange rate could choose to adjust with or without devaluation. Implications of an adjustment without devaluation have frequently been debated.[9] Franc zone African countries which made such a choice succeeded in lowering their real exchange rate. They did so thanks to the depreciation of the French franc (until 1986), but also by adopting a disinflation policy, which lessened domestic price increases, bringing them below the world inflation rate.

Thus one main difference between countries adjusting with or without inflation is the following: the first adjust through inflation and money illusion, the others with disinflation, which makes a difference for agriculture. Inflation may be particularly dangerous in agriculture, for two main reasons. First, small farmers savings may be held principally in cash, which is then reduced by inflation. Second, inflation generates uncertainty about prices which slows down innovation. Another difference between the two categories of countries is a lower real exchange rate depreciation in countries which have not devalued. But maintaining parity gives an incentive to look for other means of adjustment. Thus franc zone countries have been led to lower taxation on agricultural exports (groundnuts, cotton) – a means which is necessarily limited by budget constraints – and to look for productivity improvements in agroindustrial activities. One most interesting case is that of cotton, when facing the fall of world prices in the mid-1980s. Profitability has been maintained without devaluation, mainly thanks to economies achieved in transportation, marketing and processing.

Briefly stated, devaluation is assumed to act on supply mainly through an increase of real agricultural producer prices, but a stronger depreciation does not necessarily lead to higher real prices. It depends on the effectiveness of the devaluation in real terms; that is, on the macro-economic policy applied. The need to reduce budget deficits often leads governments to use devaluation for that purpose, rather than to increase producer prices. As for adjustment policy without devaluation, it is compelled to rely on productivity improvements. These improvements, when achieved outside agriculture have in turn a positive effect on producer prices.

TRADE LIBERALIZATION AND
THE PROBLEM OF PRODUCER PRICE INSTABILITY

In Africa, adjustment programmes have to some degree included a liberalization of domestic trade and prices, in particular of agricultural prices. Liberalization is a natural way to structural adjustment, since market mechanisms and efficient competition are supposed to eradicate price distortions.[10] It also has macroeconomic consequences.

Agricultural prices are necessarily liberalized in countries with floating exchange rates and where currency is fast depreciating (Guinea and Zaire, for example). In these countries administered prices would have to be permanently revised and would have no real meaning. But in countries with a rather stable rate of exchange, the regulation of some prices remains possible, although not necessarily advisable.

Food crops

Liberalization has probably been stronger in the food sector, for several reasons. Boards in charge of stabilization of food prices have often been particularly inefficient (a large share of the transactions remaining outside their operations) and simultaneously costly (for example OPAM in Mali, OPVN in Niger). Moreover, when price fluctuations are linked to supply variations, they stabilize farm income to some degree. These reasons, however, leave room for limited state intervention on markets, for instance through security stocks.

It is difficult to provide a global evaluation of the recent experience of grain trade liberalization, in particular in African countries. Indeed cereals output noticeably increased during recent years in countries where liberalization occurred (Sahel, Madagascar), but a simultaneous improvement in weather conditions has also to be noted.[11]

Export crops

For export crops,[12] much doubt has also appeared about the possibility of an efficient price stabilization scheme, for at least two reasons. One is the risk of a delinking of domestic prices from international prices, then of giving up comparative advantage. Another risk is that stabilization mechanisms were diverted from their initial role and transformed into instruments for excessive agricultural taxation. Actually, the experience of stabilization funds and marketing boards has generally been poor; boards were used to tax the agricultural sector, and the money levied when international prices were high was spent and was not available when prices fell. So adjustment policies often attempt to eliminate these 'parasitic' institutions.

However, the basic principle of producer price stabilization for export crops remains a relevant aim in adjustment policy. It is well known, in particular because of risk aversion, that for a given average price output is higher with a stable price than with an unstable one. Moreover, price instability is probably a brake upon innovation and investment, especially for small farmers. In addition instability does not only affect farmers. Public income instability disturbs macro-economic management, as evidenced by the literature about Dutch disease. Jointly with the objective of producer price stability, an objective of public revenue stability has to be aimed for.

In order to avoid previous errors, the stabilization of producer prices and public income can be efficiently achieved only on two conditions, which are not necessarily unrealizable. One is that price stabilization relies, not on a fixed price, but on the trend of the observed international price, in order to avoid a misallocation of resources. The other is that stabilization funds are absolutely independent of the tax levy and are statutorily deposited outside the reach of the public treasury, in order to avoid both their disappearance and the risk of Dutch disease.[13]

Some countries have maintained some partial stabilization schemes, as is the case in Papua New Guinea for several products, in French-speaking African countries for cotton and for hevea in Côte d'Ivoire. These schemes, the operation of which may, of course, be a matter of debate, aim at establishing some stability of producer prices, without delinking them from international prices, and by depositing the funds beyond the reach of public treasury.

The choice between more or less stability in producer prices is not actually independent of exchange rate policy. It can be noted (Table 3) that during the 1980s the instability of real producer prices (both for food and export crops) was highest in the countries where monetary depreciation was strongest. It was lowest in countries adjusting without devaluation or with moderate devaluation. These results appear in the whole sample and each of the two subsets (Africa and Latin America). In order to check for the influence of the instability of real international prices, we have measured a stabilization coefficient by the difference between this instability and that of real producer prices. The coefficient of stabilization (from 1979 to 1988) was highest in the non-adjusting countries, followed by the countries adjusting without devaluation, then by the countries with moderate devaluation, and was lowest in the countries with strong depreciation.

Briefly stated, liberalization policy involves a mix of the correction of price distortions and the search for an environment favourable to productivity improvement. Monetary stability favours real producer price stability for food crops and is a necessary but not sufficient condition for an efficient stabilization of (real) producer prices for export crops.

DECREASE OF THE BUDGET DEFICIT, AGRICULTURAL TAXATION AND PUBLIC EXPENDITURE

As noted above, structural adjustment involves a preliminary or simultaneous reduction of budget deficits. So it needs a choice between a mix of tax increases and public expenditure decreases. Increasing public revenues appeared particularly difficult in African countries, although the tax–GDP ratio is rather low.

TABLE 3   *Instability of real producer prices (1979-88 percentages)*

Simple average	Food crops	Export crops	Degree of stabilization of producer prices for export crops
LDC	12.4 (75)	13.4 (75)	7.0 (51)
adjusting	13.5 (37)	15.9 (36)	2.8 (26)
without devaluation	7.2 (10)	8.2 (10)	9.6 (7)
moderate depreciation	7.2 (14)	10.5 (14)	4.3 (10)
strong depreciation	25.1 (13)	28.4 (12)	–3.9 (9)
non-adjusting	11.3 (38)	10.9 (39)	11.4 (25)
SUB-SAHARAN AFRICA	13.8 (34)	12.9 (35)	7.2 (25)
adjusting	12.8 (19)	14.9 (19)	1.1 (13)
without devaluation	7.7 (7)	9.4 (7)	9.9 (5)
moderate depreciation	8.4 (6)	11.1 (6)	1.2 (3)
strong depreciation	23.3 (6)	25.2 (6)	–7.8 (5)
non-adjusting	15.1 (15)	10.4 (16)	13.8 (12)
LATIN AMERICA	15.6 (19)	20.1 (19)	3.3 (12)
adjusting	20.7 (8)	24.8 (9)	–0.6 (6)
moderate depreciation	5.8 (3)	12.3 (4)	–0.3 (3)
strong depreciation	29.7 (5)	34.8 (5)	–0.9 (3)
non-adjusting	11.9 (11)	16.0 (10)	7.3 (6)

Notes: Real producer price instability is the average quadratic deviation from linear trends in percentages.
Stabilization degree is the difference between international price instability and real producer price instability for export crops.

This is because taxes are levied mainly on external trade, such taxes being more easily imposed on those transactions than on domestic ones, or on domestic incomes. But structural adjustment logically involves lowering taxes on external trade which are a source of distortion.

However, some countries which devalued, originally to change relative prices, have been tempted to increase levies on agricultural exports. Such a policy made it easier to balance public finances but, as seen above, did not lead to the expected increase of real producer prices (for example, Madagascar from 1982 to 1986). The stabilization objective then prevails over structural adjustment.

In order to reduce the deficit without increasing revenue, adjusting countries are led to cut public expenditure, though the effect on various categories of expenditures have varied. Most often, because it was easier, rural expenditures declined more than urban ones, capital expenditures more than current ones and, among these, working expenditures more than salaries. It seems, therefore, that the expenditures allocated to agriculture, in particular those on which agricultural productivity depend, have been significantly affected:[14]

they are rural, they are for large investment expenditures (infrastructure) or expenditure for services which particularly need working means (health, extension, research, for example).

Of course, the evolution of agricultural public expenditures can vary according to the kind of adjustment policy adopted and especially according to exchange rate policy. Currency depreciation increases somewhat the ratio of public income to GDP, which relies mainly on tradables, and then the public expenditure GDP ratio. But, simultaneously, it tends to increase expenditures allocated to the purchase of tradables, firstly of external debt service, so the relative share of other public expenditures is likely to decrease. This may be the case for expenditures on agriculture, which have a relatively high content in tradables. Finally, the impact of currency depreciation on the agricultural expenditure to GDP ratio is uncertain (as is the comparability of data on such expenditures).

The figures of table 4 show differences among country groups in the evolution of agricultural public expenditures from 1970–9 to 1980–8. The reduction of their share in total public expenditures, as well as in GDP, is higher in adjusting than in non-adjusting countries. Among adjusting countries, the share in GDP decreases on average only in the category of countries with strong depreciation, and it increases only in countries with moderate depreciation.

Briefly stated, the choice or mix between increased taxes and lower expenditures involves some trade-off between the two ways to structural adjustment, correction of price distortions and productivity improvement: choosing less public expenditure may be better for agricultural prices, but is likely to be unfavourable to agricultural productivity. The countries with strongest currency depreciations seem to be those where agricultural expenditures have been the more affected (though they have not succeeded in raising real producer prices).

CONCLUSION

To sum up, structural adjustment policies are theoretically favourable to agricultural development, since they aim at improving price incentives and productivity. Actual policies, in particular in Sub-Saharan Africa, seem on average to have improved the pace of agricultural growth, but they have not always succeeded in doing so. Several reasons can be advanced:

(1) During the application of these policies, the agricultural sector faced several shocks due to climate, or international markets, to a variable degree according to country.

(2) Adjustment policies have been applied for a longer or shorter time and with more or less public involvement; reaction lags in agriculture may be long.

(3) These policies used various combinations of different instruments, the effects of which are never automatic. As a result, illustrated above by exchange rate policy, trade liberalization and budgetary policy, an un-

TABLE 4   *Evolution of public expenditure on agriculture*

SIMPLE AVERAGE	Percentage of GDP			Percentage of total expenditure		
	1970–1979	1980–1988	difference (same sample)	1970–1979	1980–1988	difference (same sample)
LDC	1.7 (84)	2.1 (85)	0.3 (74)	7.4 (87)	7.2 (86)	–0.5 (69)
adjusting	1.6 (37)	1.6 (39)	0.0 (35)	7.5 (37)	6.3 (38)	–1.2 (32)
without devaluation	1.6 (7)	1.6 (9)	0.0 (7)	6.5 (7)	5.6 (9)	–1.8 (6)
moderate depreciation	2.0 (15)	2.5 (14)	0.5 (13)	9.2 (15)	9.2 (13)	–0.4 (13)
strong depreciation	1.2 (15)	0.8 (16)	–0.3 (15)	6.2 (15)	4.4 (16)	–1.7 (13)
non–adjusting	1.7 (47)	2.4 (46)	0.5 (39)	7.3 (50)	7.8 (48)	0.2 (37)
SUB–SAHARAN AFRICA	2.0 (33)	2.3 (30)	0.3 (27)	8.9 (33)	7.6 (30)	–1.0 (25)
adjusting	1.9 (18)	1.8 (19)	–0.1 (17)	8.8 (18)	7.2 (19)	–1.9 (15)
without devaluation	1.5 (5)	1.7 (7)	0.1 (5)	7.1 (5)	6.0 (7)	–2.1 (4)
moderate depreciation	2.7 (6)	3.0 (5)	0.3 (5)	11.3 (6)	10.3 (5)	–1.9 (5)
strong depreciation	1.6 (7)	1.1 (7)	–0.4 (7)	7.9 (7)	6.2 (7)	–1.9 (6)
non–adjusting	2.1 (15)	3.1 (11)	0.8 (10)	8.9 (15)	8.3 (11)	0.3 (10)
LATIN AMERICA	1.4 (25)	1.7 (24)	0.3 (21)	6.2 (24)	5.7 (23)	–1.5 (15)
adjusting	1.1 (9)	0.6 (9)	–0.3 (8)	5.9 (9)	3.7 (9)	–1.7 (8)
moderate depreciation	1.6 (3)	0.8 (2)	–0.3 (2)	6.2 (3)	3.9 (2)	–1.9 (2)
strong depreciation	0.9 (6)	0.6 (7)	–0.2 (6)	5.7 (6)	3.6 (7)	–1.6 (6)
non–adjusting	1.5 (16)	2.2 (15)	0.6 (13)	6.4 (15)	7.0 (14)	–1.2 (7)

Notes: Because of absence of data in some cases averages have been calculated from reduced samples.

Sources: World Bank, Government Finance Statistics Yearbooks and World Tables.

equal weight has been given by the various countries to the elimination of price distortions and to productivity improvements.

(4) Means of structural adjustment are largely complementary and some trade-offs occurred, the result of which has not always been optimal. This means that it remains important to scrutinize how economic policy can promote agricultural development, in Africa and elsewhere.

NOTES

[1] An attempt to eliminate this influence from the observed results is presented by Goi (1988).

[2] We consider here the effect of structural adjustment policies on agricultural output growth, and not the more general problem of their effect on the social situation in rural areas. On this point see Azam, Chambas, P. Guillaumont and S. Guillaumont (1989).

[3] More details on the adjustment concepts can be found for instance in P. Guillaumont and S. Guillaumont (1990b, 1990c, 1991a).

[4] See, for instance, Binswanger (1989), Bond (1983), Bonjean(1990a), Chibber (1988). On the particular problem of the influence of consumer goods markets, see Azam, Berthelemy and Morrisson (1991), Azam and Faucher (1988), Berthelemy and Morrisson (1989), Bevan, Collier and Gunning (1989), Guillaumont and Bonjean (1991a).

[5] The conditions are studied in P. Guillaumont and S. Guillaumont (1990b, 1991a).

[6] The implications of an adjustment with or without devaluation are more thoroughly examined in some of our previous work, especially S. Guillaumont-Jeanneney (1988), P. Guillaumont and S. Guillaumont (1990b), P. Guillaumont, S. Guillaumont and P. Plane (1991).

[7] The growth of some agricultural production, due to adjustment policies may lead to a significant decrease of international prices. It is not discussed in detail.

[8] The relationship between international prices, real exchange rates and real producer prices

is examined in particular by P. Guillaumont and S. Guillaumont (1990c), P. Guillaumont and C. Bonjean (1991b), P. Guillaumont, S. Guillaumont and P. Plane (1991); and, according to a different method, by Krueger, Shiff and Valdés (1988).

[9]This question is debated in P. Guillaumont and S. Guillaumont (1990b), P. Guillaumont, S. Guillaumont and P. Plane (1991).

[10]The problem of the liberalization of external trade, often linked to that of domestic trade, is not considered here and is a complex one. Protectionism raises the price of importable agricultural goods as the objective of structural adjustment advocates, but on the other hand it raises the cost of food and then of labour, which lessens the competitiveness of other production. Moreover, it is often ineffective.

[11]On the experience of Niger, see Bonjean, in P. and S. Guillaumont (1991a) and, on that of Madagascar, Berg (1989).

[12]On this question, see P. and S. Guillaumont (1990a, 1990c, 1991b), Bonjean (1990a, 1990b), O. Knudsen and J. Nash (1990), U. Lele and R.E. Christiansen (1989), J.W. Mellor and A. Raisuddin (1988), M. Schiff and A. Valdés (1990).

[13]See a discussion of these principles, and of costs and advantages of the various schemes of export crops producer price stabilization, in P. and S. Guillaumont (1990a, 1991b).

[14]This has been noted by Mosley (1989) for several African countries.

REFERENCES

Azam, J.P. and Faucher, J.J., 1988, 'Le cas du Mozambique', in *Offre de biens manufacturés et développement agricole,* Centre de Développement de l'OCDE, Paris.

Azam, J.P., Berthelemy, J.C. and Morrisson, C., 1991, 'L'offre de cultures commerciales en économie de pénurie', *Revue Economique,* vol. 42, no. 3.

Azam, J.P., Chambas, G., Guillaumont P. and Guillaumont, S., 1989, *Impact of Macroeconomic Policies on the Rural Poor,* UNDP, New York.

Balassa, B., 1990, 'Incentive Policies and Export Performance in Sub-Saharan Africa', *World Development,* vol. 18, no. 3.

Berg, E. 1989, 'The Liberalization of Rice Marketing in Madagascar', *World Development,* vol. 17, no. 5.

Berthelemy, J.C. and Morrisson, C. 1989, *Développement agricole en Afrique et offre de biens manufacturés,* OCDE, Centre de Développement, Paris.

Bevan, D., Collier, P. and Gunning, J.W., with Bigsten, A. and Hornsell, P., 1989, *Peasants and Governments. An Economic Analysis,* Clarendon Press, Oxford.

Binswanger, H., 1989, 'The Policy Response of Agriculture', Proceedings of the World Bank, Annual Conference on Development Economics, Supplement to *The World Bank Economic Review* and *The World Bank Research Observer.*

Bond, R., 1983, 'Agricultural Response to Prices in Sub-Saharan Africa', *IMF Staff Papers,* vol. 30, no. 4.

Bonjean, C., 1990a, 'Elasticité-prix de l'offre des cultures d'exportation en Afrique: quelques résultats empiriques', in *Revue Canadienne d'Etudes du Développement,* vol. XI, no. 2.

Bonjean, C., 1990b, 'Contribution des facteurs macro-économiques à la variation du prix réel payé au producteur. Exemple du prix du café en Côte d'Ivoire, au Kenya et à Madagascar', *Revue d'Economie Politique,* no.4, July–August 1990.

Bonjean, C., Combes, J.L. and Guillaumont, P., 1991, 'La croissance agricole en Asie et en Afrique. Quels facteurs explicatifs?', Communication au Colloque 'Politique économique et performances agricoles comparées en Afrique et en Asie', Clermont-Ferrand.

Chibber, A., 1988, 'The Aggregate Supply Response in Agriculture : A Survey', in S. Commander (1988).

Cleaver, K., 1985, 'The Impact of Price and Exchange Rate Policies on Agriculture in Sub-Saharan Africa', World Bank Staff Working Papers.

Commander, S. (ed.), 1989, *Structural Adjustment and Agriculture, Theory and Practice in Africa and Latin America,* ODI, with James Curry, London.

Faini, R., 1991, 'Infrastructure and Agricultural Adjustment', communication to the Joint CEPR/ OECD Development Centre Conference on International Dimensions of Structural Adjustment: Implications for developing country agriculture, Paris.

Ferroni, M. and Valdés, A. (eds), 1991, 'Trade and Macroeconomic Linkages and Agricultural Growth in Latin America', special issue of *Food Policy*, vol. 16, no.1.

Ghai, D. and Smith, L. 1987, *Agricultural Prices, Policy and Equity in Sub-Saharan Africa*, Lynne Rienner Publishers, Inc., Boulder.

Goi, I., 1988, 'Performances agricoles', in P. and S. Guillaumont (eds), *Stratégies de développemont comparées,* Economica, Paris.

Griffon, M., 1989, 'Ajustement structurel et politique agricole en Afrique', in *Notes et Documents*, CIRAD.

Guillaumont, P. and Bonjean, C., 1991a, 'Effects on agricultural supply of producer price level and stability with and without scarcity : the case of coffee supply in Madagascar', *Journal of International Development*, vol. 3, no. 2.

Guillaumont, P. and Bonjean, C., 1991b, 'Fonctions de comportement de l'Etat dans la détermination des prix au producteur pour les cultures d'exportation', *Economie et prévisions*, no. 97, September.

Guillaumont, P. and Guillaumont S., 1990a, *Why and How to Stabilize Producer Prices for Export Crops in Developing Countries*, UNDP–World Bank, Trade Expansion Programme, Occasional Paper 6.

Guillaumont, P. and Guillaumont, S., 1990b, 'Exchange Rate Policy and The Social Consequences of Adjustment in Africa', Communication to African Economic Conference, Nairobi, June, in A. Chibber and S. Fisher (eds), *The Analytics of Economic Reform in Sub-Saharan Africa*, 1991.

Guillaumont, P. and Guillaumont, S., 1990c, 'Quels sont les effets des politiques d'ajustement structurel sur le développement agricole?, Seminaire international sur 'l'Avenir de l'agriculture des pays du Sahel – Enseignements et perspectives des recherches économiques', organisé par le Club du Sahel et le CIRAD, Montpellier, 12–14 September.

Guillaumont, P. and Guillaumont S., 1991a, *Ajustement structurel, ajustement informel : le cas du Niger*, L'Harmattan.

Guillaumont, P. and Guillaumont, S., 1991b, 'Politique macroeconomique et stabilisation des prix payés aux producteurs pour les cultures d'exportation", Communication préparée pour le colloque 'Politique économique et performances agricoles comparées dans les pays d'Afrique et les pays d'Asie à faible revenu', 20–22 March, Clermont-Ferrand.

Guillaumont, P., Guillaumont, S. and Plane, P., 1991, 'Comparaison de l'efficacité des politiques d'ajustement en Afrique, Zone franc et hors Zone Franc', *Notes et Etudes*, no. 41, Caisse Centrale de Coopération Economique, April.

Guillaumont, S., 1988, 'Dévaluer en Afrique?, *Observations et diagnostics économiques*, Revue de l'OFCE, no.25, October.

Jeager, W. and Humphreys, C., 1988, 'The Effect of Policy Reforms on Agricultural Incentives in Sub-Saharan Africa', *American Journal of Agricultural Economics*, vol. 70, no.5.

Knudsen, O. and Nash, J., 1990, 'Domestic Price Stabilization Schemes in Developing Countries, *Economic Development and Cultural Change*, vol. 38, no. 3.

Krueger, A.O., Schiff, R. and Valdés, A., 1988, 'Agricultural Incentives in Developing Countries: Measuring the Effect of Sectoral and Economywide Policies', *The World Bank Economic Review*, vol. 2, no.3.

Lele, U., 1988, 'Agricultural Growth, Domestic Policies, the External Environment and Assistance to Africa: Lessons of a Quarter Century', in Colleen Roberts (ed.), *Trade, Aid, and Policy Reform : Proceedings of the Eighth Agricultural Sector Symposium*, World Bank, Washington, DC.

Lele, U., 1989, 'Sources of Growth in East African Agriculture', *The World Bank Economic Review*, vol. 3, no.1.

Lele, U. and Christiansen, R.E., 1989, 'Marchés, offices de commercialisation et cooperatives : problèmes à résoudre dans le cadre d'une politique d'ajustement', *Managing Agricultural Development in Africa*, MADIA, World Bank, June.

Mellor, J.W. and Raisuddin, A. (eds), 1988, *Agricultural Price Policy for Developing Countries*, Johns Hopkins University Press, Baltimore and London, for the International Food Policy Research Institute.

Mosley, P. and Smith, L., 1989, 'Structural Adjustment and Agricultural Performance in Sub-Saharan Africa 1980–87, *Journal of International Development*, vol. 1, no.3.

Schiff, M. and Valdés, A., 1990, 'Synthesis: the Economics of Agricultural Price Intervention in Developing Countries', in World Bank, *A Comparative Study of the Political Economy of Agricultural Pricing Policies*, vol. 4, March.

DISCUSSION OPENING – REBECCA LENT*

The goal of Guillaumont's paper is to evaluate the impact of macro-economic adjustment policy on the agricultural sector. It is certainly a worthwhile objective, since agriculture is an important, if not the most important, sector for most developing countries. He rightly stresses that impact evaluation is not an easy task. Not only is there an insufficient number of observations, there is also a myriad of variables and causal relations to sort out. Five comments will be made.

1. We do need to standardize key words and phrases. For example 'structural adjustment' is first defined as a lower trade deficit at the same growth rate, or a higher growth rate with the same deficit. Later in the text it appears as a global term for free-market oriented monetary, fiscal, institutional and pricing policies. As another example 'macro-economic adjustment policy' is defined as encompassing stabilization and structural adjustment. Further confusion arises in use of the 'real exchange rate' and over the impact of inflation on real rates. The discussion in the paper indicates that an increase in inflation results in an appreciation of the real exchange rate, while textbook definitions indicate an inverse relationship between the two (Lindert, 1986). It appears that definitions may differ in Europe and North America; thus there is all the more need for establishing common ground in discussion.

2. Two results in the paper are particularly interesting. First, countries undergoing structural adjustment are shown to exhibit higher growth rates in value added in their agricultural sectors. This is encouraging and certainly more optimistic than preliminary findings (Jayne and Weber, 1988). Second, producer price increases appear to be greater in those countries which did not devalue their currency. This is more troublesome, not only for policy makers, but also for economists, since it amounts to questioning the very foundation of neo-classical theory.

3. As Guillaumont points out, price response is at the heart of many structural adjustment issues. To a certain extent, arguments for and against currency devaluation also touch on price response issues. There are many estimates of supply responses in the literature which could enrich the discussion, though the author does not report any of them, apparently because of the difficulties associated with estimation. Caution could be in order since over-estimation of supply elasticities led to some disappointing outcomes in the early days of structural adjustment. However, more recent estimates, such as those for rice in Senegal (Martin and Crawford, 1991) are more conservative. I was also struck by the absence of discussion of demand elasticities which can also influence the impact of adjustment policies. In Burkina Faso (Reardon *et al.*, 1988) it has been suggested that cross-price elasticities between imported and

*Laval University, Canada and Economic Research Service, USDA.

local, traditional cereals, are lower than was assumed. It should also be noted that understanding of demand structure is essential if supply is relatively more erratic than demand.

4. Securing the desired results in Third World structural adjustment may be impeded by lack of reciprocity in the developed countries. For example, export subsidies on beef shipped from the European Community to the Côte d'Ivoire simply encourage imports. Furthermore, low-priced supplies are displacing Sahelian meat exports (Burfisher and Missaien, 1987; Josserand, 1990). In addition, record budget deficits in Canada and the United States, along with many barriers to imports throughout the developed world, can only aggravate the situation in the poorer countries.

5. My final point relates to the analysis of adjustment, either in forecasts of possible consequences or in post-evaluation. It is vitally important, since use must be made of the magnitude of critical elasticities, to have a solid understanding of the underlying micro-economic structure of the country under consideration.

REFERENCES

Burfisher, M. and Missaien, M., 1987, *Intraregional Trade in West Africa*, Agricultural Trade Analysis Division, Economic Research Service, USDA, Washington, DC.
Jayne, T.S. and Weber, M.T. 1988, *Market Reform and Food Security in Sub-Saharan Africa: A Review of Recent Experience*, Michigan State University Agricultural Economics Staff Paper No. 88, East Lansing.
Josserand, H., 1990, *Systèmes ouest-africains de production et d'échanges en produits d'élevage: aide-mémoire synthétique et premiers éléments d'analyse régionale*, Document de travail, Club du Sahel, Paris.
Lindert, P.H., 1986, *International Economics*, 8th edn, Irwin Publications in Economics, Illinois.
Martin, F. and Crawford, E., 1991, 'The New Agricultural Policy: Its Feasibility and Implications for the Future', in *The Political Economy of Senegal and Structural Adjustment*, C. Delgado and S. Jammeh, (eds), Praeger, New York.
Reardon, T., Thiombiano, T. and Delgado, C., 1988, *La substitution des céréales traditionnelles pour les céréales importées*, CEDRES (Centre d'études de documentation et de recherches économiques et sociales) Série Rapport de Recherche no. 3, Université de Ouagadougou.

GUILHERME L.S. DIAS*

Inflation and Agriculture:
Ten Years of High Inflation and Government Debt in Brazil

INTRODUCTION: PRICE FLEXIBILITY AND RIGIDITY

The fundamental assumption among structural economists of the early 1980s concerned the existence of a type of market failure. Factor mobility was imperfect when supply was concentrated in a few firms which could sustain a price policy with a fixed mark-up over variable cost, a feature which could arise if the industrial sector enjoyed high levels of protection. Price rigidity would come about in the short run if there was also real wage rigidity, an hypothesis generally accepted when short-run price indexation clauses were widespread in the urban labour market. Adjustments to changes in demand became predominantly changes in the quantity supplied.

In a macro-model formulation the following equations (Ramos, 1986; Olivera, 1967) apply to the rigid price sector of the economy. For urban wage determination:

$$W = f(P_{t-1}) \qquad (1)$$

where:

W	=	nominal wage
$f(\ )\,w$	=	rule for indexation
P	=	general price level

from which we get:

$$dW/W \quad = \quad F\,(dP/P) \qquad (1a)$$

F reflects the capacity of the urban labour sector to keep the real wage constant.

Inflation is a process in which nominal wages temporarily lag behind the desired level, enforcing a solution to the distributive conflict. By the logic of this model, competitive sector prices rise first, followed soon afterwards by wages, with a subsequent adjustment in the 'rigid' price sector. For that

*University of São Paulo, Brazil. Helpful discussions with Joaquim E. C. Toledo are gratefully acknowledged. I have also benefited from comments by Marcos E. da Silva.

reason the agricultural sector is always pointed to as the original source of inflationary pressure (Ramos, 1986). The mark-up price determination equation becomes:

$$Pr = (1+u) \, W/k \tag{2}$$

where:

k = labour productivity
u = mark-up percentage
and

$$dPr/Pr = dW/W - dk/k \tag{3}$$

In the agricultural sector a competitive market prevails with market equilibrium being described as:

$$Q^d c \, (Pc/Pr, y, H) = Q^s c \, (Pc/Pr, Z) \tag{4}$$

where

Pc/Pr = (#) -relative price
y = real income of the urban sector
H = urban population
Z = autonomous expansion of the competitive sector excess supply (technology and/or agricultural frontier expansion)

Logarithmic derivation then gives:

$$-Ep \, \dot{\#} + Ey \, \dot{y} + \dot{H} = Np \, \dot{\#} + \dot{Z},$$

where
E = demand elasticities
N = supply elasticity
and

$$\dot{\#} = \dot{P}c - \dot{P}r = [Ey/(Ep + Np)] \, \dot{y} + [\, \dot{H} - \dot{Z} / (Ep + Np)] \tag{5}$$

It follows that:

$$\dot{\#} > 0 \text{ for } \dot{y} > [\dot{H} - \dot{Z} / Ey]$$

There is a maximum rate of growth for real income in the urban sector, $\dot{y}*$, compatible with constant relative price:

$$\dot{y}* = \dot{Z} - \dot{H} / Ey \tag{6}$$

Finally, a proper definition for the relevant price level P in equations (1) and (2) is needed, given by assuming that:

$$Pt = Pc, t^a Pr, t^{1-a} \tag{7}$$

where a and $1-a$ are respectively the share of rigid and flexible price sector products in the representative basket of consumption goods.

From (7) dynamics of this inflationary process become:

$$\dot{P}t = a [(\dot{P}ct - \dot{P}rt] + \dot{P}rt.$$

From (5), (3) and (2):

$$\dot{P}_t = a [Ey/ (Ep + Np) (\dot{y}_t - \dot{y}^*_t) + (\dot{H} - \dot{Z}) / (Ep + Np)] - \dot{k} + F (\dot{P})_{t-1} \tag{8}$$

Equation (8) can be solved as follows:

If $F=0$, $dP/dt=A$,
where $A=a [Ey/(Ep + Np) (y_t - y^*_t) + (H - Z) / (Ep + Np)] - dk/k$

there will be some inflation because agriculture's relative price will increase if the current rate of growth in real income of the urban sector is greater than the 'warranted' rate of growth – a structural component that reflects the impact of urban population growth and technological rural innovations – as long as these factors are not compensated for by a deflationary effect from urban labour productivity growth. All these factors together can account for a small rate of inflation if there is no indexation (propagation) in the urban wage market.

If $0 < F < 1$ $(dP/P)_t = - A/1 - F [F]^t + A/1 - F$

the inflation rate will converge.

If $F = 1$ $(dp/p)_t = A_t$

the inflation rate will be growing at a constant rate. If $F > 1$ rate of growth of inflation is positive and there will be an explosive path to hyperinflation. This is an acceptable hypothesis if labour is willing to compensate for the loss in the average real wage which occurred in the previous period.

According to the structural economists' consensus, an anti-inflationary policy must have a radical income policy with price and wage control, aiming not only at the invalidation of all indexation clauses on commercial contracts, but also a different and sustainable pattern of income distribution. Recession brought about through fiscal and monetary policy, acting over $[(dy/y) - (dy/y)^*]$, is a very weak element in equation (8). Only if carried on for a long period will it be able to halt an inflationary process of the kind described, and only then at a prohibitive social cost. Political negotiation of such a sensitive income policy is impossible if the unemployment rate is high and long-

lasting, Control over food prices is especially relevant because it would vindi-
cate government intervention in collective wage negotiations. Subsidies on
food imports become common, as do interest rate subsidies on rural credit.
The government primary deficit increases.

The government budget constraint

It will be appreciated that the monetary dimension of the inflationary process
has been placed very much at a secondary level in the system described. The
fundamental reason concerns the role of the central government as the main
actor in the development process. Government has an active role in the process of resource transfer that goes beyond the ordinary institutional framework
of public finance.

Within the structuralist institutional framework, government is directly involved with other groups in the process of transformation of the productive
structure of the national economy. The other main factors are foreign capital,
a fragile urban bourgeoisie, an entrepreneurial class without leadership, and a
disorganized urban labour force. Industrialization is the main target (Hirschman,
1968) and government may become the owner of productive resources, and
bureaucracy the leading source of entrepreneurship, in any strategic sector
(Evans, 1979). National sovereignty is a strong and pervasive ideal legitimat-
ing government power in the appropriation of domestic resources by means
other than ordinary collection of taxes. Money seignorage, inflationary taxation, non-price market rationing devices for scarce foreign currency and over-
pricing of products and services from government agencies are good examples of these heterodox practices (Mendonca de Barros and Graham, 1978).

If this a satisfactory description of government as an aggregate actor in the
structural macro-economic model, the fiscal deficit may be better understood
as an endogenous variable. Deficit financing can be sustained for a long time
through foreign debt: in the Brazilian case from 1968 to 1983. When the
Federal Reserve Bank raised interest rates in 1979, the financial conditions
needed to sustain a process of increasing debt began to deteriorate. In 1982, it
was brought to a sudden halt by the Mexican default. From that time on, in the
absence of a strong fiscal reform, foreign debt service had to be financed
through money creation or public domestic debt.

Interest-paying public debt could be kept increasing by a permanent reduction in the maturity of government paper, and by continuous action by the
central bank in the open market. Money was becoming gradually passive in
the second half of the 1980s (Pastore, 1991). The monetary authorities did not
use money supply as a control variable, but the inflation rate itself became the
'variable' to be monitored by macro-economic policy in order to maximize
seignorage collection. In this framework the government's first priority relates to full employment.

In order to introduce government into the macro-model, the budget constraint needs to be considered. To do so define:

$$(G - T) + i.B + E.i^*.B^* = dBg + E.dB^* \qquad (9)$$

where

$(G-T)$	=	public expenditures less tax collection
$i.B$	=	nominal interest paid on bonds held by private sector
E	=	exchange rate
$i^*.B^*$	=	nominal interest paid on public foreign debt
dBg	=	net increase in total public debt, including the share held by Central Bank
dB^*	=	net increase in public foreign debt (assuming nothing is private)

together with the following identities:

$dBg = dB + dM - EdR$ Central Bank constraint, where dR is accumulation of foreign reserves

$dR = T^* - i^*.B^* + dB^*$ Enforced convertibility, where T^* is current trade surplus; assuming no explicit policy of reserve accumulation

Through substitutions and assuming that bonds are perfectly indexed to current inflation ($dP/P^* = dP/P$), we obtain:

$$[(G-T)/P + E/P.T^*] + r.B/P - dP/P.M/P = d (B/P) + d(M/P) \qquad (10)$$

After many more substitutions we arrive at:

$$db/dt = d(B/py)/dt = \{[(G-T)/P + E/P.T^*] - mdM/M\} + (r - dy/y) b_t \ (11)$$

Equation (11) describes the dynamic behaviour of government interest-paying debt in relation to nominal income. Subjected to an intertemporal budget constraint, meaning that today's debt coefficient must be equal to the present value of future non-financial government budgets coefficients, the solution becomes:

$$b_t - \{m \ dM/M - [(G-T)/P + E/P.T^*]\}/ (r - dy/y) = 0$$

In a steady-state equilibrium, $dM/M = dP/P + dy/y$, and

$$b_t - \{\dot{P} + \dot{y})m - [(G-T)/P + E/P.T^*]\} / (r - \dot{y})=0 \qquad (12)$$

This is an equilibrium condition for b that, when $r > dy/y$, may require an extremely large primary fiscal adjustment. If b is increasing over time, inflation will be expected to increase in order to 'finance' the government deficit. In Figure 1 below there is an upward-sloping curve ($r > dy/y$) that represents the loci of combinations between dp/p and b under the condition that $d(db/dt) = 0$, with the deficit coefficient constant over time. To the right (left) of this curve, at a given level of inflation, seignorage is insufficient (excessive), meaning that b is increasing (decreasing).

The IMF and World Bank have formed the 'Washington Consensus' concerning policy adjustments for countries facing high foreign debt and inflation. Under this agreement stabilization, structural reforms and growth policies are different at subsequent stages in the adjustment process (Williamson, 1990; Selowsky, 1990; Fanelli, Frenkel and Rozenwurcel, 1990). Stabilization comes first with a set of policy actions aiming at a stable path for b without any requirement for an inflationary tax. This usually means an extremely large fiscal surplus in order to service foreign and domestic public debt, together with a strict monetary policy, combining in a set where the interest rate is greater than the real output growth rate for a long period of time.

THE STRUCTURAL MODEL REVISITED:
THE DYNAMICS OF STABILIZATION

In this section the original structural model is modified in order to include government relations with the private sector. In this version (Toledo, 1986) r will become endogenously determined and another equilibrium condition between b and dP/P is required, where a sustainable rate of inflation can be attained. Accordingly the following are used:

(a) wage equation: $W = f_1(P_{t-1})$.
(b) demand for real cash balances from wage earners only, in order to stress the regressiveness of the inflationary tax:

$$M/P = m(WN/P, dP/P, i) \qquad (2')$$

It then follows that

$$\dot{W} = F_1 (\dot{P})_{t-1} + F_2 (P.\dot{M}/P).$$

(c) for the profit equation:

$$Pr = [1 + u(r)] W/k$$
$$dPr/Pr = dW/W - dk/k + u'(r)w$$

Where the mark-up reacts to the real interest paid on government bonds, $u'(r)>0$. Here w = labour share.
(d) the competitive market equilibrium condition is left unchanged.
(e) for the bond market equilibrium

$$r=r(b) \qquad (13)$$

The government increases the interest rate through the open market in order to keep db in the private sector's portfolio, $r'(b)>0$. From these assumptions:

$$dPr/Pr = dW/W - dk/k + u'(r) \ r'(b) \ db \ w \qquad (3')$$

With these new equilibrium conditions we can return to equation (8), obtaining:

$$(dP/P)_t = a\ [Ey/(Ep + Np)\ \{(dy/y)t - (dy/y)'t\} + (dH/H - dZ/Z)/(Ep + Np)] \\ - dk/k + F_1\ (dP/P)_{t-1} + F_2\ (dP/P).M/P) + u'(r)\ r'(b)\ db\ w \qquad (14)$$

It is possible to deduce that this will lead to a downward-sloping locus of combinations of values of b and dP/P in Figure 1, under the condition that $d(dP/P)/dt = 0$. Both partial derivatives of dP/P with respect to b, dp/p are positive, meaning that :

$$d(dp/p)/db{<}0\ d(dp/dt) = 0$$

Any combination to the right (left) side of the curve means that the real rate of interest, given a higher (lower) b, is above (below) the required level to warrant the current mark-up. Prices will be increased at a faster (slower) rate in order to increase (decrease) mark-up ; the inflation rate goes up (down). Figure 1 gives an idea of the instability implicit in equations (11) and (14). It is only at a combination like C that we have a path toward stabilization; at any other combination government debt or inflation may follow an explosive path.

A strict monetary policy would mean less seignorage moving the debt equilibrium curve to the left; a higher interest rate would call for an increase in mark-up moving the inflation equilibrium curve to the right, except that it will also reduce the rate of growth, moving it back partially. The final effect will be a reduction in region C which shows good stabilization properties. If government follows a more flexible monetary policy, reducing the interest rate, and keeping growth not much beyond the warranted rate, it will move the

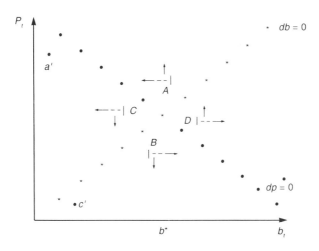

FIGURE 1 *Dynamics of the adjustment process*

debt equilibrium curve to the right without changing the inflation curve significantly. Region C will be enlarged.

Higher indirect taxes will shift the debt equilibrium curve to the right (enlarging the C region) but the equilibrium relative price of food products may well increase (they are price-inelastic in demand). In this case the inflation required to keep wage demands constant would have to be smaller (less inflation tax) as a compensation ; the inflation curve moves to the left (diminishing the C region). Direct taxes or lump sum taxes would be more effective.

Trade liberalization (lower implicit tariffs on agricultural exports and industrial import substitutes, and a higher exchange rate) may increase the trade surplus, worsening the government operational deficit and external debt service position; the debt equilibrium curve shifts to the left (diminishing the C region). Food relative prices will increase, shifting the inflation curve to the left also (further diminishing the C region). This is an unexpected result because trade liberalization is a strongly recommended policy and is a conditionality for IMF and World Bank structural reform programmes.

A perfectly effective income policy would turn our inflation curve into a vertical line. If b is larger than b^* the interest rate is high enough to decrease the rate of growth of income closer to or below the warranted rate, inflation would be reduced. The result is reversed if b is smaller than b^*. Under such an incomes policy there is no role for social conflict in the inflationary process because any movement of prices will be followed by wages in such a way that there will be no income transference. Both curves would be crossing in a saddle point (Messenberg, 1990).

Freezing prices and wages means an instantaneous shift in any combination of dp/p and b to a lower level of dp/p, of a shift from C to c' on Figure 1. Those who were caught below their desired relative price at the moment prices were frozen would be willing, immediately after, to increase prices. Those caught above would try to keep the gains. The inflation equilibrium curve shifts to the left.

Debt repudiation is an instantaneous shift in any combination to a lower level of b, a shift from A to a' on Figure 1. Nobody would be willing to hold government bonds at the same interest rate as before; the debt equilibrium curve shifts to the left. There may be a combination of debt repudiation and price and wage freezing which would reduce the C quadrant to nothing; any debt is too much debt, any primary deficit is too much to be financed by seignorage.

AGRICULTURE AND STABILIZATION POLICY

From 1980 to 1984, the orthodox approach to stabilization predominated; government sustained a high level of investment, first through borrowing, but after 1983 an expenditure switching policy was able to adjust the balance of trade to the external debt crisis (Tables 1 and 2). Relative prices for agricultural products decreased initially but then increased substantially with devaluation and growth recovery. Inflationary acceleration was contained only in 1982, owing to a combination of currency over-valuation and two good harvests in 1980 and 1981. Control over the government deficit meant that for agriculture there was credit rationing and fewer subsidies on the interest rate. After two years of supply adjustment (1983–4) the yield increment was more than

TABLE 1   *Macroeconomic instability in the 1980s*

Year	GDP % growth rate	Deflator (%)	Relative price P_a/P_i	Curr. account Surplus	Real Ex-change rate	PSBR %GDP	Cur. Acc. Gov. Sav. %GDP	Invest-ment %GDP
1980	9.2	90	1.00	−1.4	100.0	n.a.	4.0	22.9
1981	−4.4	107	0.98	−0.4	84.2	6.2	4.4	21.0
1982	0.6	105	0.85	−0.7	78.4	7.3	4.2	19.5
1983	−3.4	140	1.16	2.4	104.6	4.2	4.3	16.9
1984	5.3	213	1.28	5.7	105.9	2.5	4.1	16.3
1985	8.0	232	1.33	5.2	108.9	4.4	3.8	16.4
1986	7.5	146	1.65	2.5	110.1	3.6	5.2	18.7
1987	3.6	204	1.43	3.3	103.0	5.7	1.0	17.9
1988	−0.1	648	1.34	5.2	98.9	4.8	0.3	17.0
1989	3.2	1323	1.17	3.2	77.4	6.9	n.a.	16.7
1990	−4.6	2849	1.56	1.6	67.4	−1.2	n.a.	n.a.

Sources: National accounts, FIBGE; wholesale prices, EGV; current account non-factor payments (expressed as % GDP), FIBGE; exchange rate, national deflator–OECD deflator, public sector borrowing requirement (PSBR), Central Bank.

TABLE 2   *Agriculture's adjustment in the 1980s*

Year	Agric. GDP %	Area crop	Area yield index	NPK cons.	Cred. t-1/ GDPt	Price rec/ paid	Gov. exp. on agric % ag.GDP
1980	9.6	100	100	100	0.58	1.00	0.161
1981	8.2	98	116	102	0.56	1.06	0.110
1982	−0.4	102	108	100	0.60	0.98	0.125
1983	−0.4	90	125	89	0.38	0.98	0.063
1984	3.0	99	123	76	0.25	0.94	0.081
1985	10.1	103	137	89	0.17	1.08	0.158
1986	−7.9	106	117	95	0.21	0.95	0.237
1987	14.0	104	148	116	0.29	0.86	0.339
1988	1.5	109	146	114	0.16	0.84	0.306
1989	2.2	105	158	115	0.18	1.02	0.175
1990	−5.8	97	157	105	0.11	0.98	n.a.

Sources: National accounts, FIBGE; production data, FIBGE; fertilizers, Anda; agricultural credit, Central Bank; prices received/paid, FGV/CFP; government expenditure includes subsidy on credit and on wheat consumption (Villa Verde and Gasques, 1990).

enough to compensate. This provides important evidence concerning the irrelevance of interest rate subsidies as a development policy instrument.

A structural consensus took over in 1985, at first with a more flexible wage policy and a minimum price/subsidy combination for food products. Growth and a high real exchange rate meant a further increase in relative prices. The emphasis of the structural model on relative price shocks meant that food prices had to be controlled. Minimum price support programmes began to look like food subsidy programmes; whenever stocks of basic food items became available they began to be sold in the mid-season at a price below what carrying charges would require. Subsidized interest rates on short-run commercial credit were much in demand by the private sector as an insurance against government sales. Export licensing was required for food and raw materials making up for complex system of implicit taxation of agriculture (Goldin and Rezende, 1990; Brandão and Carvalho, 1989; Dias, 1989; Mauro Lopes, 1986). In the early 1980s, agricultural development policy was concentrated on frontier occupation, though investments in research were kept at a reasonable level (both are important elements making for the autonomous rate of growth in this model). Minimum price determination was common throughout the country, and as a result output from the frontier areas destined for the major consumption centres had to be directly subsidized by government.

The first heterodox anti-inflationary policy, the 'cruzado' plan of 1986, came after a severe drought, which meant relying on a large volume of imported food in order to control prices. A second fundamental element was a price freeze with a legal prohibition on any indexation clause in contracts of less than twelve months' duration. Almost all agricultural transactions lasting less than a year, and even long-term credit contracts, were later exempted from monetary correction, which meant extremely large subsidies. According to the structural consensus a strong emphasis was given to income policies in the urban sector. A 'wage law' was approved by congress, adjusting all wages simultaneously by their average real level in the last six months, and industrial prices were frozen at their level in the last week immediately after the plan was announced.

Public services and government-supplied goods had their prices frozen at a lagged value relative to private sector prices but that was felt to be more than compensated for by increased seignorage and a windfall gain in real tax collection, given that inflation was halted (these were temporary rather than permanent fiscal gains). Government and private sector expenditures increased as a result of the simultaneous monetary correction of all wages, and from saving accounts that were earning an extremely low 'expected' interest rate. Inflationary 'memory' proved to be persistent and the government deficit was another mortal threat to the heterodox anti-inflationary policy.

Failure in the second heterodox trial of 1987 meant a radical shift towards government deficit control. Equation (12) proved itself in the next two years; increasing emphasis on interest-bearing deficit financing in 1988–9 drove the Brazilian economy to hyperinflation because of the explosive behaviour of the financial component of the fiscal deficit. The current strategy is based on partial default in public debt (that is, a lump sum tax on financial assets), the Central Bank does not sustain the short-term liquidity of government debt, there is a substantial cut in real wages in the public sector and price control is in operation. The improvement in the public borrowing requirement is due mostly to partial default and temporary taxes; if it is not a permanent structural adjustment it may be better interpreted as the situation pictured earlier, where region *C* disappears and debt increases (or debt monetization spreads) together with a mild acceleration of inflation. The agricultural relative price is at a high level again due to high real interest rates and the lower level of government expenditure, adding an important component for the acceleration of inflation.

Too much emphasis on the strategy of 'stabilization first-growth later' is a difficult path when government itself is the dominant foreign debtor. Disrupting the traditional role played by the public sector in Latin American development, without the disruption of the growth process requires the activation of a new strategic role for foreign capital and domestic saving. This is a very sensitive political challenge if government is at the same time required to tax heavily the private sector in order to service foreign debt (Fanelli, Frenkel and Rozenwurcel, 1990).

REFERENCES

Brandão, A.S. and Carvalho, J.L., 1990, 'Brasil', in A.O. Krueger, M. Schiff and A. Valdés, (eds.) *Economia Politica de las Intervenciones de Precios Agricolas en America Latina*, Centro Internacional para el Desarrollo Economico, CINDE,World Bank, Washington.

Dias, G., 1989, 'The Role of Agriculture in the Structural Adjustment Process of Brazil', in A.H. Maunder and A. Valdés (eds), *Proceedings*, Dartmouth, Aldershot.

Evans, P., 1979, *Dependent Development–The Alliance of Multinational, State and Local Capital in Brazil*, Princeton University Press, Princeton.

Fanelli, J.M., Frenkel, R. and Rozenwurcel G., 1990, *Growth and Structural Reform in Latin America. Where we stand*, Report for UNCTAD, CEDES, Buenos Aires.

Goldin, I. and Rezende, G.C., 1990, *Agriculture and Economic Crisis: Lessons from Brazil*, OECD-Development Centre Studies, Paris.

Hirschman, A.O., 1968, 'The Political Economy of Import Substituting Industrialization in Latin America', *The Quarterly Journal of Economics*, vol. LXXXII, no. 1.

Lopes, M., 1986, *A Intervencao do Governo nos Mercados Agricolas no Brasil*, CFP-Colecao Analise e Pesquisa, vol.33, Brasilia.

Mendonca de Barros, J.R. and Graham, D.H., 1978, 'The Brazilian Economic Miracle Revisited:Private and Public Sector Initiative in a Market Economy', *Latin American Research Review*, Summer.

Messenberg, R.P., 1990, *'Endividamento, Interno do Setor Publico, Deficit e financiamento Inflacionario'*, unpublished Master Dissertation, Dpto. Economia FEA/Univ. São Paulo, São Paulo.

Olivera, J.H., 1967, 'Aspectos Dinamicos de la Inflation Estructural', *Desarrolo Economico*, vol.7, no. 27.

Pastore, A.C., (1991), *Deficit Publico e Inflacao: Uma Resenha*, Depto.Economia FEA/Univ. São Paulo, mimeo, São Paulo.

Ramos, C.A., 1986, *Agricultura e Inflacao: A Abordagem Estruturalista*, Premio BNDES de Economia-BNDES, Rio de Janeiro.

Selowsky, M., 1990, 'Etapas da Recuperacao do Crescimento Latino-Americano', *Finanças & Desenvolvimento*, vol.10, no.2.

Toledo, J.E.C., 1986, *'Salarios e Ciclos na Economia Brasileira'*. Depto Economia FEA/Univ. São Paulo, mimeo, São Paulo.

Villa Verde, C.M. and Gasques, J.G., 1990, 'Notas sobre o Gasto Publico com Agricultura', in

Proceedings from 20° Encontro da Sociedade Brasileira de Economia e Sociologia Rural, SOBER, Brasilia.
Williamson, J., 1990, 'What Washington Means by Policy Reform, in J. Williamson, *Latin American Adjustment. How much has Happened?*, Institute for International Economics, Washington.

DISCUSSION OPENING – RICHARD L.MEYER*

The highly successful Brazilian economic miracle of the 1960s and 1970s ran into serious trouble in the 1980s. GDP growth rates fluctuated widely between plus 10 and minus 5 per cent. Annual inflation rates rose from two-digit levels at the beginning of the decade to four-digit levels at the end. Investment fell from over 20 per cent of GDP to 16–18 per cent, and the foreign debt grew to over $100 billion. Shock treatments were applied to stimulate growth and slow inflation in 1986 and 1987, but their failure led the new Collor de Mello government to adopt even more stringent policies in 1990. On all three occasions, inflation at first declined but subsequently accelerated. It is obvious that the structural problems contributing to inflation were not resolved.

Professor Dias has undertaken the tough challenge of trying to explain government policy and economic performance in the 1980s. Although the model and supporting arguments are presented in a somewhat frustrating way, I compliment him for the creativity of his work and his courage in tackling a tough assignment. His paper provides us with helpful insights into reasons for the government and the economy behaving as they did, and for growth with stability being so elusive in Brazil during this decade. It also leads me to conclude that Brazilian attempts at 'quick fixes' have about run their course and a fundamental redirection in growth strategy is required.

The author begins his analysis by arguing that Brazilian objectives during the period were to pursue fairly rapid economic growth rates with high levels of employment, and without resorting to increased taxes. They key economic actors are identified as the government, foreigners (or foreign capital), a fragile urban bourgeoisie, an entrepreneurial class 'without leadership' and a 'disorganized' urban labour force. The Brazilian growth strategy for some time has placed heavy emphasis on industrialization, with the government as owner and source of entrepreneurship for strategic economic industries and sub-sectors (such as steel and petroleum). As a result, in the early 1980s, 60 per cent of gross capital formation was handled through state enterprises and government banks. The country has long accepted the government's appropriation of domestic resources by means other than tax collection.

The macro-economic structural model developed by Professor Dias consists of 14 equations, including:

(1)   three equations representing a rigid price structure for (a) urban wages and (b) the mark-up of prices in the non-competitive (non-agricultural) sector;

*Ohio State University, USA.

(2) one market equilibrium equation for the competitive (agricultural) sector;

(3) four equations to explain the dynamics of inflation;

(4) four equations to derive the equilibrium conditions for public sector debt; and

(5) two equations to derive a sustainable rate of inflation.

Figure 1 in the paper is used to show the dynamics of macro-economic adjustment. The upward-sloping line where $db = 0$ represents equation 12 and connects the combinations of price changes and interest rates for which domestic debt is in equilibrium. Points to the right of the curve represent a situation in which interest rates rise faster than inflation to cover fiscal deficits. The downward-sloping line where $dp = 0$ represents equation 14 and connects the combinations of price changes and interest rates for which inflation is in equilibrium. Points to the right of this line represent a situation in which interest rates are higher than the rate of change in prices, so prices are bid up and inflation increases so that real mark-ups are maintained. A leftward shift of either curve decreases area C in the figure, leading to an explosion of government debt, or inflation, or both. Although not explicitly stated in the paper, these shifts occurred in the 1980s, especially at the end of the decade, thereby contributing to the inflation explosion.

Agriculture plays a crucial role in the model. Any shock (such as crop failure or large exports) throws supply and demand out of equilibrium. The rise in prices in the competitive sector is then followed by a rise in wages and prices in the rigid sector. This fact explains government preoccupation with controlling agricultural prices since, by definition, the rigid sector is able to maintain real wages and mark-ups. Therefore, it does not bear the burden of inflation through changes in its relative income. It is surprising, then, that non-farm entrepreneurs are labelled 'without leadership' and the urban labour force 'disorganized' when they are strong enough to maintain their relative incomes during inflation.

Prior to the 1980s, Brazil was able through external finance to grow rapidly with a large government sector and fiscal deficit. This alternative disappeared, however, as a result of the rise in interest rates beginning in 1979 with the change in Federal Reserve policies and the Mexican default in 1982. Thereafter, maintaining rapid growth in a government-led economy required either increased taxation (the alternative not chosen) or increased internal debt financed through high interest rates.

Professor Dias discusses a number of events which occurred in Brazil which reflect the country's struggle to cope with growth and stability, though he does not explicitly relate them to the model. On the one hand, a series of actions were taken to control inflation by keeping agricultural supplies and demand in balance. Frontier expansion was an important source of growth, but investment in research contributed to some yield improvement. Minimum price supports were also used to encourage production. In years of short supply, exports were limited and food imports and subsidy programmes helped hold down food prices. On the other hand, agricultural credit subsidies were at times eliminated, and at other times reinstated. These subsidies have been considered important in Brazil to stimulate production, but they became so large in the 1970s that they are accused of contributing to inflation.

In 1986 and 1987, heterodox anti-inflationary shock treatments were introduced to control inflation and stimulate growth. They involved complicated price freezes, and de-indexation of contracts, but they also included wage increases. Inflation slowed for a time, but later accelerated. Faced with four-digit inflation, the new Collor government took even more drastic action in 1990, including the partial default of public debt through the freezing of private bank accounts. The current resurgence of inflation suggests that the country has not yet found a stable means to finance its desired growth.

I hope that Professor Dias will be encouraged to extend this important work, in four ways. First, a clearer exposition of the model would be useful, along with a more explicit discussion relating the events of the 1980s to the model. Second, it would be helpful to incorporate a more complete discussion of the implications derived from the model of following the standard IMF and World Bank structural reform and growth policies, particularly in a country where expanded agricultural growth and higher prices simultaneously contribute to paying the external debt and to inflation. Third, an attempt to test the model empirically would be interesting in order to quantify its explanatory power and determine the significance of individual variables at specific points of time. Finally, some predictions about future growth strategies could be enlightening. It appears that the simultaneous objectives of rapid government-led growth, with high employment, low rates of interest, inflation and taxes, but without internal or external debt default, are impossible to obtain. Simulation studies might help identify a set of attainable objectives and point to changes which Brazil must implement to return to the 1970s levels of growth and inflation.

EUGENIA MUCHNIK DE RUBINSTEIN*

Impact of Policy Reforms on the Agricultural Sector in Chile

INTRODUCTION

Chile is a country about which much has been said and written over the last decades, partly as a result of the abrupt political changes which it underwent in the early 1970s, but more so because of the successful fundamental economic reforms introduced during the early years of the military regime, which later on became the conventional policy prescriptions of international financial organizations.

During the last six years, the country has been able to achieve economic growth at rates surpassing those of the last two decades, while maintaining moderate rates of inflation. This is a most unusual situation in the area at a time when most Latin American nations have been shaken by recession and widespread inflation. The new democratic government that assumed power in March 1990 has indicated its wish to maintain the overall economic policies inherited from the previous regime, with the additional incorporation of measures which will improve the country's social conditions. No major changes have been introduced so far in terms of agricultural policies, except for the expressed goal to strengthen the support afforded to small farmers.

The purpose of this paper is to describe the most significant policy reforms in Chile since the military coup and the impact these have had on the agricultural sector. Special attention will be afforded to the sequence and timing of the reforms and to the lags which took place in the adjustment process.

BACKGROUND

As in many other Latin American countries, successive governments in Chile intervened increasingly in the economy from the time of the Great Depression of 1930, seeking to isolate the country from external shocks, and to reduce the constant crises in the balance of payments. Instead, except for very short successful periods, the record was extremely poor, with sluggish growth, high rates of inflation, and continued balance of payment problems. The first systematic set of measures to increase farm production and productivity was introduced by the Frei Administration (1965–9), based on improving relative agricultural prices and on agrarian reform. However, higher food prices con-

* Catholic University, Chile

241

flicted with the goal of reducing inflation and with raising real urban wages. Under the Allende regime (1970–3), economic policies were radically changed. The government, in its efforts to achieve socialism, led the country to economic chaos, with rampant inflation, proliferation of black markets, a fiscal deficit of about 24 per cent of GDP in 1973, and with legal and illegal takeover of industries and farms. The outcome was a coup in September 1973, which placed a military junta in power.

The military administration quickly made it known that its main goal was to establish a new economic, political and social order, and inclined itself towards the neo-liberal school of economic thought, under the influence of a group of economists trained in the United States (see Fontaine, 1988). Unlike the typical populist economic policies of other authoritarian systems, the emphasis was placed on trade liberalization, reliance upon market mechanisms, and the adoption of the 'principle of subsidiarity', meaning that government actions were to be restricted to areas where private sector performance was insufficient, or to correct market imperfections and to relieve extreme poverty.

ECONOMY-WIDE POLICY REFORMS SINCE 1974

Amidst a situation of economic turmoil, government took prompt action to lift price controls and subsidies, providing a clear indication of its will to reinvigorate the capitalist economic system. The process of price liberalization was somewhat slower in the case of basic agricultural products. A basic pillar of government policy was to warrant the right to private property, seriously shaken during the preceding years, particularly in the countryside, as a result of a long and profound land reform programme.

The most drastic episode of trade and price liberalization was implemented shortly after the change of government. In order to improve the country's trade position, exchange rates were unified, the peso was devalued, the crawling peg system was re-established and the external debt renegotiated. A progressive reduction of tariffs was implemented according to a planned schedule, culminating with a uniform *ad valorem* tariff of 10 per cent in 1979, with very few exceptions. More important, perhaps, was the elimination of non-tariff barriers to trade and prior deposits on imports, which had been in the past even more restrictive for trade.

In terms of monetary policy, the liberalization of interest rates culminated in May 1975, when it became effective throughout the financial system. Real interest rates became positive, to stimulate savings and capital inflows, unlike the previous decades when negative real rates prevailed. Economic stabilization was achieved by the elimination of the fiscal deficit, reducing public expenditure, privatizing firms and increasing taxes. The austerity of the fiscal policy was such that, beyond its success in eliminating one of the secular sources of inflation, it caused in 1975–6 a serious recession, only to be surpassed in its negative impact on production and employment by the economic crisis of 1982–3.

In terms of labour legislation, until 1978, the government continued to determine minimum wages and salaries on a periodical basis, making functional rather than sectoral distinctions, with rural wages including payments in kind of some significance. Minimum wages continued for some time to be automatically indexed to inflation, which increased the real cost of labour. Subsequently, the policy was modified as a reaction to the observed increasing levels of unemployment, seeking first to keep real wages constant, and later on to reduce them, by de-indexing nominal wages. This led to a systematic decrease of real wages, until 1986. Since then, real wages have increased over time as a consequence of economic growth.

During 1976-80, following the initial recession and later the liberalization process, the Chilean economy experienced high rates of GDP growth (see Table 1), which led to the belief that the development strategy chosen would permit continuous growth, alleviation of poverty and full employment by the mid-1980s. But during 1982–3, Chile went through its worst experience in 50 years. A depression, brought about by the combination of an abrupt ending of foreign financing due to a world recession, coupled with errors in government policies, particularly from pegging the exchange rate while opening the capital account to capital inflows, affected the design of economic policies of the 1980s. There were drastic changes in the exchange rate and monetary policies, including the introduction of important regulation of capital markets. But these modifications left the basic resource allocation mechanisms, implemented by the structural reforms of the late 1970s, untouched (Morande,1990). In spite of the pressures to abandon the neo-liberal model, the government introduced only some marginal counter-reforms, particularly in the agricultural sector, as will be described below. In the case of trade policy, tariffs were transitorily increased after the recession, reaching a maximum of 35 per cent for a brief period, subsequently reduced in several steps, down to 15 per cent since 1987. It is important to indicate that quantitative restrictions were not reintroduced, neither was the size and role of the state changed.

POLICIES TOWARDS AGRICULTURE

During the period 1964–73, the agricultural sector had been the battleground for the social and economic transformation of Chile into a socialist country: 'revolution in freedom' first and 'the Chilean road to socialism', next (Silva, 1982). In 1967, the government initiated a vast programme of agrarian reform, which was intensified during the Allende administration in a way that disrupted production. Over the period 1965 to 1973, about 48 per cent of the country's agricultural land was expropriated. Farm prices and the marketing process were closely controlled. At the same time, the government sought to lower production costs of farmers by fixing output prices and maximum marketing margins and by subsidizing the prices of fertilizers and pesticides.

However, agricultural policy after 1973 and until the economic crisis of 1982–3 had almost no specific elements that would apply to agriculture and not to the other sectors, the only exceptions being the unsuccessful application, for a short period of time, of price bands for a few traditional crops, a forest subsidy programme from 1974, and since 1977 an anti-dumping policy for milk and its derivatives. Agricultural input prices were also liberalized, while marketing activities were progressively handed over to the private sector.

The recession of 1982 did not spare the agricultural sector – the decline in international and therefore domestic prices for agricultural commodities and the appreciation of the exchange rate resulted in an influx of food imports and in low incentives to production. Agricultural GDP fell for two consecutive years and open agricultural unemployment reached levels not known before in the country (see Table 1). The economic authorities responded to the crisis with several agricultural policies of a more interventionist nature, in addition to the trade and macro-economic adjustments mentioned earlier. The most important modification was the introduction of price band mechanisms for three essential commodities:wheat, sugar and oilseed. These price bands were designed to attenuate the adverse impact of short-run fluctuations in international prices on domestic production, while preserving the signals of frontier prices over the medium term for the allocation of resources. Only in the case

TABLE 1	*Evolution of total and agricultural GDP, agricultural labour force and unemployment*

Year	Annual growth rate total GDP (%) (1)	Agricultural sector's GDP participation (%) (2)	Annual growth rate agric. GDP (%)	Agricultural labour force participation (%) (3)	Agricultural sector unemployment rate (%) (4)
1974	1.0	8.2	26.7	n.a.	n.a.
1975	−12.9	9.9	4.8	17.3	4.1
1976	3.5	9.3	−2.9	16.5	4.8
1977	9.9	9.3	10.4	17.2	5.4
1978	8.2	8.2	−4.9	16.4	6.7
1979	8.3	8.0	5.6	15.6	7.3
1980	7.8	7.7	3.6	15.3	5.0
1981	5.5	7.5	2.7	14.7	6.2
1982	−14.1	8.5	−2.1	14.4	9.4
1983	−0.7	8.3	−3.6	14.4	5.8
1984	6.3	8.3	7.1	15.0	5.5
1985	2.4	8.6	5.6	15.3	4.9
1986	5.7	8.8	8.7	19.3	2.5
1987	5.7	8.7	4.5	19.8	3.3
1988	7.4	8.6	5.7	19.4	3.0
1989	10.0	8.1	3.1	19.5	2.3
1990	2.1	8.3	4.8	19.0	2.4

Notes:	(1) original figures at constant prices were expressed in millions of 1977 Chilean pesos.

(2) includes forestry, which over the period accounts for about 4–6 per cent of agricultural GDP.

(3) includes forestry and fisheries; the latter contributes about 2 per cent.

(4) the data for 1986–9 are not comparable to those for 1974–85 owing to differences in methodology.

n.a. = not available.

Source:	Central Bank of Chile and National Institute of Statistics.

of wheat, which is the most important single crop in the country, was the price stabilization scheme supplemented by the operations of a public procurement agency (see Muchnik and Allue, 1991).

In terms of taxation, policy reforms during the 1970s and 1980s did not introduce major changes. Agriculture continued to be taxed according to presumptive income of unimproved land. The amount of taxes paid by farmers evolved as a function of the fiscal values determined by the authorities for agricultural land; these had been increasingly under-valued between 1966 and 1973, but were rapidly adjusted thereafter. Simulation of the tax burden which would have taken place, had a progressive tax on actual income been applied, suggests that the prevailing system implicitly subsidized the highly profitable export activities such as fruit orchards, but not the less profitable traditional annual crops.

Regarding land tenure, expropriations were stopped at the outset of the Pinochet Administration and expropriated land was redistributed in the form of individual parcels. The combined effect of land reform and subsequent land transactions led to a restructuring of land ownership. The most striking changes, comparing 1965 and 1979, were the significant increase in the proportion of irrigated land held by farmers with less than 20 equivalent irrigated hectares, from 22.4 per cent to 52.7 per cent, and a decrease from 55 per cent to 21 per cent for farmers with over 80 equivalent irrigated hectares (Jarvis,1985). Land was distributed to only a fraction of the rural poor. Moreover, the lack of managerial experience of the new landowners in a hostile environment of free competition led many of them to sell their newly acquired properties. Thus the outcome of the land reform process was far from the expectations that had fuelled the previous reform effort. On the other hand, it can be argued that it had a positive unexpected spin-off: an active market for land was created which did not exist before to any significant extent, giving access to land to a new class of medium-sized entrepreneurs willing to introduce modern technology to the countryside. It also gave farmers the possibility of seeking optimum farm sizes in order to realize economies of scale, and thus achieve an efficient allocation of resources.

In terms of agricultural extension and technical assistance, the country had a well-financed extension programme during the land reform period of 1965–73, but technicians concentrated principally on organizing agricultural workers rather than on teaching modern farming techniques (Jarvis, 1985). It is perhaps for this reason that some major changes were introduced during the first years of the policy reform, including a drastic reduction of personnel. Next, in 1977, an innovative programme, based on the incorporation of private firms for the provision of technical assistance to small farmers, was put into effect. About 70 per cent of the cost of the service was covered with public funds,with the expectation that, after a few years, farmers would appreciate the benefits of technical assistance and be prepared to bear the entire cost. This original scheme, seeking to include a private component in the system, did not work out; it was not attactive to farmers or to the providers of assistance, probably because of under-estimation of the true costs involved.

In synthesis, the government played a very minor role, offering a neutral pattern of incentives, letting private agents take the lead. Government expenditures in agriculture decreased, but focused on 'public goods' such as roads, port facilities or agricultural research.

EFFECTS ON AGRICULTURE

Agricultural growth has been affected both by macro-economic policies and by sector specific agricultural policies. The following is a description of the evolution in agricultural output, agricultural trade, technological change, employment and unemployment since the structural reforms of the mid-1970s.

Agricultural growth

A World Bank study by Hurtado, Valdés and Muchnik (1989) of the impact of macro-economic and sectoral policies on incentives to agriculture provides estimates of the effects during the period 1960–3 that these have had on the flow of resources into and out of agriculture. Figure 1 summarizes the estimated flow of resources due to 'direct' price interventions and to total, that is 'direct plus indirect' interventions. This study concluded that both direct price interventions (that is sector-specific) and indirect price interventions (economy-wide policies) prior to 1974, particularly the exchange rate policies and protection to manufactures, had a negative impact on agriculture. On the other hand, policy reforms introduced since 1974 helped to increase agricultural prices relative to non-agricultural prices, particularly as a result of the liberalization programme on the rest of the economy.

The lower level of import protection applied after 1974 decreased the relative prices of non-agricultural import competing activities and also increased the real exchange rate. The latter increased even further, as shown in Table 2, owing to a decline at the time in Chile's terms of trade. This had a positive impact on agricultural prices, considering the relatively high component of tradable commodities within agriculture (see Figure 2).

The substantial improvement in agricultural price incentives,which lasted until 1980, found a poor response in terms of the sector's GDP during that period, as shown in Table 1. Barahona and Quiroz (1989) have suggested three reasons for the relatively poor performance of agriculture during these first years immediately following the economic reforms: land reform had placed an important fraction of the best lands in the hands of inexperienced small farmers, who also lacked capital resources and access to credit; the liberalization of the capital market in the face of scarce foreign savings and the resulting increase in real interest rates negatively affected the traditional crop sector; and finally, the hypothesis that farmers probably considered these incentives to agriculture as transitory.

Another explanation for the lag in output response of agriculture to policy reform can be found in the work of Coeymans and Mundlak (1991). The authors have developed a multisectoral model of the Chilean economy, which provides empirical evidence on the aggregate supply response of Chilean

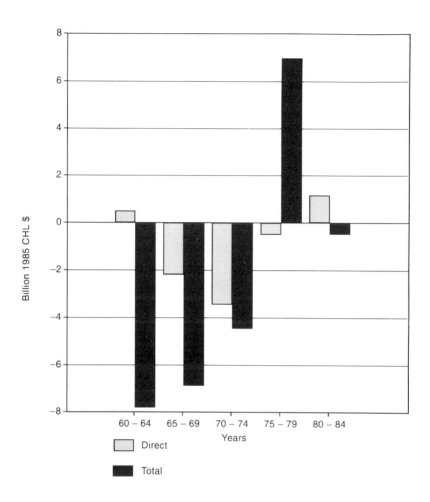

Source: Hurtado, Valdés and Muchnik (1989).

FIGURE 1   *Resource transfer in and out of agriculture*

agriculture to increases in relative agricultural prices. It indicates that the output response is sizable but slow, and this is related to the behaviour of labour and capital, whose sectoral allocation is based on intertemporal considerations and whose adjustment is subject to costs. In their simulation of an increase of 10 per cent in agricultural prices, output increased by only 3.2 per cent in the first three years, but by as much as 10.2 per cent after ten years. A low output response should be expected in the short run due to the impossibility of instantaneous factor mobility across sectors. The long-run response is quite different and emphasizes the importance of persistence in economic policy.

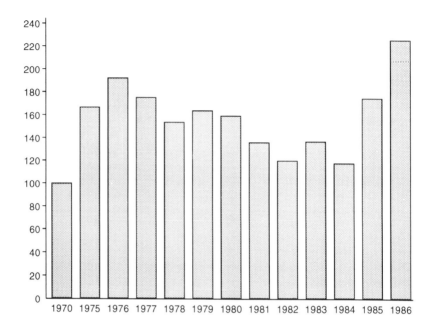

Source: Barahona and Quiroz (1989).

FIGURE 2   *Domestic agricultural/non-agricultural prices, Chile, 1970–86*

The response of the agricultural sector to the substantive policy reforms of the 1970s finally showed itself after 1983, as a result of the serious disruption created in 1981 and 1982 by the economy-wide recession. It was only after the correction of specific policies at the macro level, and the introduction of selected sectoral policies, that agriculture finally experienced the expected sustained rates of high economic growth. Since 1983, agricultural GDP has grown at a record average annual compounded rate of more than 6 per cent (Table 1). Retrospectively, it would have been very difficult to have predicted the changes in output mix which actually took place, including the introduction of exportable crops and fruit trees in regions where they were previously not found.

Agricultural trade

Agricultural exports (including forest products), which had remained stagnant during the previous decade and a half, experienced, from the outset of reforms, a most impressive growth. While all Chilean exports grew 3.9 times between 1974 and 1990 agricultural exports grew 11 times in the same period (see

TABLE 2   *Evolution of trade and real exchange rates (Millions of US$)*

	Total exports	Agricultural exports Value (1)	Participation (%)	Agricultural imports (2)	Sector balance of trade (1) – (2)	Exports of fresh fruits	Real exchange rate index (1978=100)
1974	2 151	189	9	603	–414	18.9	78
1975	1 590	266	17	510	–244	40.0	101
1976	2 116	311	15	428	–117	50.9	91
1977	2 185	410	19	435	–25	63.3	82
1978	2 460	491	20	498	–7	100.1	99
1979	3 835	716	19	579	137	122.8	100
1980	4 705	974	21	788	186	167.7	88
1981	3 836	798	21	766	32	197.1	75
1982	3 706	712	19	565	147	230.4	86
1983	3 831	656	17	511	146	219.2	104
1984	3 651	813	22	463	350	291.7	108
1985	3 804	837	22	250	587	355.7	134
1986	4 199	1 102	26	178	924	477.0	147
1987	5 223	1 360	26	217	1 143	527.4	153
1988	7 052	1 637	23	280	1 357	582.3	162
1989	8 080	1 781	22	269	1 512	552.6	157
1990	8 310	2 123	26	375*	1 748	747.5	153

Notes: (1) also includes paper and paper pulp (except printed paper).
(2) include food and non-food products of agricultural origin.
* corresponds to a new definition of agricultural imports.

Source: Central Bank of Chile and Ministry of Agriculture.

Table 2). This determined that the participation of the sector in total exports increased from 9 per cent to 26 per cent during the same period.

The most dynamic sub-sector in terms of export growth has been fruit orchards, for which the country enjoys natural comparative advantages due to the extended growing periods which are opposite of those of the northern hemisphere. Fresh fruit exports have increased more than 30 times, going from less than 20 million dollars in 1974 to over 747 million in 1990 (Table 2). However, it is important to underscore the dynamism shown by other sub-sectors, such as processed fruits, fresh and processed vegetables, and paper and pulp. At the same time, agricultural imports have steadily declined (except during the economic crisis of 1982) so that, since 1979, there has been an agricultural trade surplus. Perhaps equally important is the diversification which has taken place in terms of number of products exported and number of markets of destination. The former increased from 102 to 374 during the period, although 12 of them still represent 71.6 per cent of the total value of agricultural exports (Ministerio de Agricultura de Chile, 1989). The number of markets of destination increased from 41 in 1973 to 85 in 1988. The above figures illustrate the fact that structural adjustment of the sort implemented in Chile can tremendously improve the capacity of agriculture to generate export surpluses.

Modernization of agriculture

It has been suggested (Hojman, 1990) that the intra-sectoral pattern of development has been very uneven, so that, while those sub-sectors oriented to exportables, such as fresh fruits, show an accelerated growth, high levels of technology and specialization, other sub-sectors mainly oriented to domestic consumption have shown signs of stagnation and have represented sources of poverty and insecurity. A simple comparison of average yields of annual crops for the periods 1974–83 and 1984–8 does not support this hypothesis. In fact, there have been important improvements in farming technology and productivity concerning traditional crops, produced by both small and large farmers. The data in Table 3 show that there has been a remarkable increase of average yields in most annual crops, particularly in the case of corn and wheat. Another example of spectacular increases in productivity and output are poultry and pork production, both of which are mainly oriented to the domestic market (CORFO, 1989).

The observed improvements in yields seem to be due mostly to the adoption of technological innovations, affecting the quality and quantity of agricultural inputs in general. This has also been the case for ancillary inputs such as services, communications and transportation. Fundacion Chile (1989), a leading institution in the introduction of new technologies in agriculture, forestry, fisheries and agro-industry in Chile, suggests that there has been a significant interaction between the liberalization reforms and the above-mentioned transformations. An open economy can rapidly absorb innovations developed elsewhere through the usual market mechanisms because of the permanent need to remain competitive in a rapidly changing and increasingly uncertain economic environment.

Agricultural labour

Agriculture has been able since 1984 to reverse a long-standing trend of a steady decrease in the participation of the agricultural labour force in the total workforce and of the typical rural–urban migration process (see Table 1). This is due to the rapid expansion of activities that produce jobs, particularly the export fruit sub-sector, which is the major employer of agricultural workers. But it took almost a decade after policy reform to reduce unemployment rates within the sector to the levels observed prior to 1974. In fact, since 1975 and until 1982, unemployment rates within agriculture grew from 4.1 per cent to 9.4 per cent, while national unemployment increased from 14.9 per cent to 19.6 per cent. The opposite trend could be observed after the economic crisis, with unemployment in agriculture declining faster than in the remaining sectors, reaching only 2.3 per cent in 1989 (Table 1).

An overall trend in agricultural employment has been a shift from permanent to temporary labour, a trend which in fact began much earlier, as had already become apparent in the agricultural census of 1975–6. The main reasons for this transformation are several and are related to the changes in land tenure patterns, to labour legislation introduced in the late 1960s and to

TABLE 3   *Chile: average yields of annual crops for periods 1965–9, 1974–83 and 1984–8 (metric tons/hectare)*

Crop	Average yield 1965–9	Average yield 1974–83	Average yield 1984–8	Percentage change (84–8/74–83)
Corn	3.3	3.6	6.5	+82
Wheat	1.6	1.6	2.6	+60
Barley	2.1	2.0	2.8	+44
Rice	2.5	3.3	4.1	+24
Beans	1.1	1.0	1.1	+8
Potatoes	8.9	10.3	14.0	+36
Rape-seed	1.2	1.4	1.7	+23
Sunflower	1.3	1.4	1.8	+30
Sugar-beet	39.1	37.1	44.5	+20

Source: National Institute of Statistics.

the induced changes in output composition and in the types of technologies encouraged by the policy reforms. The main concern about temporary workers is that they generally earn lower wages than permanent workers (defined on a yearly basis); they are not covered by the labour laws in the absence of signed contracts, and are exposed to inadequate living conditions. In the more delicate and skilled work of fruit packing for export and other agro-industrial activities, an important fraction of these temporary workers are female, who are generally paid less than men.

CONCLUSIONS

Structural reforms implemented in Chile since the 1970s have had a major influence on the successful performance of agriculture. Exports have provided the engine of growth, led by fruits and vegetables, forest and fish products, while food imports have been reduced to a minimum value. The fear that neutral incentives for agriculture could be regressive in terms of income distribution is not supported by the evidence. Moreover, rural to urban migration has been reversed and agriculture has become a dynamic source of employment and investment opportunities. However, this has not provided a solution for all of Chile's economic and social problems. Small farmers are yet to be incorporated into the modern sector. Transaction costs in markets for inputs and output, lower levels of education, and problems of access to credit and appropriate technology make them less competitive. The solution will require the transfer of appropriate technologies to economically viable farmers, and special programmes to improve the quality of life of the rural poor.

REFERENCES

Barahona, P., and Quiroz, J., 1989, 'Policy Reforms and Agricultural Response: The Case of Chile', in A.H. Maunder and A. Valdés (eds), *Proceedings*, Dartmouth, Aldershot.
Coeymans, J.E. and Mundlak, Y., 1991, 'Aggregate Agricultural Supply Response in Chile,1962–82', in *Food Policy*, vol.16, no.1.
CORFO, 1989, '*Potencial Pecuario:Analisis y Coyuntura de los Sectores Avicola y Carne Porcina*, Santiago
Fontaine, A.,1988, *Los Economistas y el Presidente Pinochet*, Editorial Zig-Zag, Santiago.
Fundacion Chile, 1989, *Economic Policy and Agricultural Development: The Chilean Case: 1974–1988*, Santiago.
Hojman, E. (ed.), 1990, *Neo-Liberal Agriculture in Rural Chile*, Macmillan, London.
Hurtado, H., Valdés, A. and Muchnik, E., 1989, *Trade, Exchange Rates and Agricultural Pricing Policies in Chile*, World Bank Comparative Studies,The World Bank,Washington,DC.
Jarvis, L.,1985, *Chilean Agriculture under Military Rule, From Reform to Reaction, 1973–1980*, Institute of International Studies,University of California, Berkeley.
Ministerio de Agricultura de Chile, 1989, *La Agricultura Chilena Durante el Gobierno de las Fuerzas Armadas y del Orden:Base del Futuro Desarrollo*, Division de Estudios Y Presupuesto, Santiago.
Morande, F., 1990, 'Chile: Recent Past, Prospects and Challenges', Documento de Ensayo No.10, ILADES, Santiago.
Muchnik, E. and Allue, M., 1991, 'The Chilean Experience with Agricultural Price Bands: The Case of Wheat', *Food Policy*, vol.16, No.1.
Silva, P., 1982, 'Estado, Neoliberalismo y Politica Agraria en Chile 1973–1981', *Latin American Studies*, No.38, CEDLA.

DISCUSSION OPENING – PAULO F. CIDADE DE ARAÚJO*

The paper is an overview of the agricultural policy reforms implemented in Chile since the military coup. Special emphasis is given to the sequence and timing of the policy changes, and to the lags which occurred in the adjustment process from 1974 to 1990.

The need for fundamental adjustment in the economic system of many Latin American countries became generally accepted by policy makers, politicians and economic analysts during the 1980s. This need arose out of the external debt crisis and the accelerating inflation which affected countries such as Brazil, Argentina, Peru and Mexico.

The dominant view regarding adjustment favours a move towards (1) reducing the size and increasing efficiency of the public sector through fiscal reform, and (2) implementing trade liberalization through greater reliance on market mechanisms. A major problem in implementing adjustments is that recognition of the need for them only occurs when economic and social conditions have deteriorated to such an extent that demand for public goods and services and government intervention, and therefore for public resources, becomes very strong. In general, the need for unpopular corrective measures becomes very clear only when policy makers have tried many other policy alternatives, which end up driving the economy to deep recession and/or accelerating inflation.

*University of São Paulo, Brazil

The second key difficulty concerns the time-lag involved from the moment changes are implemented until the effective desired results occur, in the form of increased production and employment as well as reduced inflation. The reasons for this apparently unavoidable time-lag are the relatively low degree of short-run factor mobility and the fact that economic agents only react fully to policy changes when they become convinced they are permanent. The time-lag, which may last for a decade or even longer, makes it politically very difficult to sustain adjustment measures in the absence of immediate desirable effects.

These two difficulties suggest that the implementation of necessary, but unpopular, economic measures is a much more complicated matter when it takes place under a democratic regime than under an authoritarian one. I do not mean to suggest that it is impossible to implement democratic reform, but a stronger political willingness and, possibly, a longer period of time will be necessary before a consistent set of corrective measures are put into place. As the Chilean experience teaches us, agricultural production increased only after several years of high agricultural relative prices with de-indexed wages.

Agricultural growth in Chile was made possible, to a great extent, by success in export markets. It is rather impressive that growth took place in the late 1980s, when international markets were dominated by strong protectionism. The reason was that the fresh fruit market was not severely affected. Countries such as Brazil and Argentina, whose exports are heavily dependent on foodgrains and meats, have not been so fortunate. Their chances of securing economic stabilization and preparing for economic growth is more dependent on international cooperation in opening, and stabilizing, external markets. Multilateral effort, rather than internal willingness to implement domestic adjustments, will determine the degree to which stabilization and economic growth are reached. The policy implementation lag is certainly more troublesome in their cases.

As a final consideration, I would like to stress the challenge posed by the conclusions of the paper. In spite of the positive economic aspects of the performance of Chilean agriculture in the second half of the 1980s, direct action has been needed to alleviate major structural and social problems of the rural population. As the author emphasizes, there are at least two ways to secure the welfare of agricultural areas. One is to transfer appropriate technology to economically viable farmers; the other is to increase public expenditure to improve the quality of life of the rural poor. Choice between the two is a difficult one, requiring knowledge of the extent of rural poverty and the degree to which it can be alleviated by the alternative policies available.

KAREN BROOKS* AND AVISHAY BRAVERMAN**

Decollectivization in East and Central Europe

INTRODUCTION

The agricultural transition is approximately a year and a half old, if we date its start from the Polish 'big bang' of January 1990. Like many a recalcitrant toddler, it refuses to behave as expected. A properly behaved agricultural transition is a cornerstone of the framework of stabilization and structural adjustment in East and Central Europe. The agricultural 'supply response' should be an early bright spot in an otherwise bleak picture of slow and costly industrial restructuring and deteriorating real incomes. The supply response is to result from better incentives for producers of food, achieved largely by giving them ownership of land. The redistribution of agricultural land is viewed as simple compared to the complexity of industrial privatization and restructuring, because rural people are close to the land. Once people have possession of their land, they are expected to welcome their unemployed relatives, dismissed from defunct factories.

Agriculture is thus to defy the laws of gravity that pull down production in other sectors. It is to absorb unemployment while contributing to an improved trade balance. These feats are to be accomplished largely on the strength of the land reform and the improved efficiency that new land ownership brings. The foreign community assists this process by encouraging the land reform, lending to the 'emerging private sector', providing newly private farmers appropriate machinery and access to better processing, and offering temporary food aid.

REALITIES OF THE TRANSITION

This is the agricultural transition that many people expect, but it is not the one that we have. Rural people produce less, rather than more, food and have increasing difficulty selling their products. The land reform does not produce many individual private farmers, because few individual farmers can survive the harsh economic realities of the early transition. Consumers would like to have more food, but cannot afford to buy what is available. Donated food aid

*World Bank and University of Minnesota, USA.
**Ben Gurion University of the Negev, Israel.

sits in warehouses unless it is priced significantly lower than international trading prices, raising uncomfortable questions of fair trade practices.

These are not the attributes of a conventionally well behaved transition, but they are fully consistent with the economic logic underlying the process. The supply response needed throughout East and Central European agriculture is a contraction coupled with restructuring to increase efficiency. Both within and outside the country, the need for greater efficiency is recognized, and the resources of the donor community are directed towards this goal. Recognition of the needed contraction has been slow to come, yet its logic is inescapable. Domestic average disposition of food in each of the countries has been close to that of Western Europe, although real incomes are much lower. Price liberalization raises the relative price of food and reduces domestic demand. Intraregional trade in food has collapsed and access to world markets is poor. Traditional collectivized agriculture was enticed into capital intensive production practices by negative real interest rates. The combination of declining domestic demand, poor export prospects, positive real interest rates and discriminatory partial price liberalization overwhelms any positive response that might come from land reform. The contraction is in progress, and in some places it is severe.

In the medium and longer term, domestic demand for food will recover, along with the economy more generally, but economic growth will have to be quite substantial before domestic consumers buy the amount of food they formerly bought at subsidized prices. If the Soviet economy turns around, the USSR can resume its position as a major buyer of East and Central European food. Both the Middle East and Western Europe are potential customers for East and Central European food, depending on economic growth and trade restrictions. With favourable developments in export markets, the traditional supply response, that is more food produced more efficiently, would be good for the sector and the economies as a whole. At present, however, both domestic and export markets are depressed, and will remain so throughout much of the transition.

Depressed demand and falling farm incomes impede the institutional change necessary for the sector's eventual recovery. Before price liberalization, the lack of economic infrastructure supportive of small-scale private farming was enough to keep all but a few producers within the cooperative (Brooks, 1990). Now that the contraction has begun, life as an independent producer is even grimmer. Private producers report that they cannot sell their animals because, with declining demand, processors can get adequate quantities from the cooperatives. As interest rates rise, demand for agricultural credit has fallen. This factual statement inadequately conveys agricultural producers' astonishment and apprehension as they observe the impact of decontrolled interest rates on the capital-intensive farming practices they were encouraged to adopt in the past. The cooperatives have inherited capital assets and a potential to grow their own animal feed, and are thus better able to wait out transitory increases in nominal interest rates. Most private producers do not have that capacity.

COOPERATIVES

In the current contraction, an agricultural sector is emerging that is private in name, but largely collective in fact. Genuine private producers will be squeezed out by the economically stronger cooperatives. households will receive their land rights and sign much of the land in use back to managers of voluntary 'private' producers' cooperatives. These cooperatives will be private in the sense that they will be required to pay dividends to their owners and will operate without automatic state subsidy. They will nonetheless have the conflicts between collective and indidivual incentives that have impeded the competitiveness and long-term economic viability of agricultural producer cooperatives throughout the world (Bardhan, 1989).

These cooperatives, moreover, will not represent a clear enough break with the institutions of the past to bring about new behaviour. The new cooperatives will resemble collective farms of East and Central Europe in the early period after collectivization, when they were relatively small, still paid rent for land and had a greater degree of managerial autonomy and financial independence than they retained later. These may be the necessary institutions of the transition; forced decollectivization should not be pushed on rural people. Surely they are not the foundation of a competitive market-oriented agriculture in the future?

As new producers' cooperatives appear through the land distribution, many observers both within and outside the countries mistake them for the private voluntary marketing cooperatives that have served agriculture well in many economic settings. As long as the new cooperatives have major activities in agricultural production, they should not be grouped with that loose agglomeration of firms called 'the emerging private sector'; they should be sympathetically recognized for what they are, 'the receding collective sector'. Their divestiture of collective production and transformation into marketing and service cooperatives should be assisted.

The contraction is worsened by poor access to European markets, and by the collapse of the Soviet market for East and Central European food. Governments would be well advised to make a strong united regional statement in support of agricultural trade liberalization, and furthermore to demand that food assistance for the region during the transition be purchased from the excess supplies of the region whenever possible, instead of the storehouses of North America and the European Community.

DISTRIBUTION OF AGRICULTURAL LAND

It is in this atmosphere of acute economic uncertainty and declining farm incomes that the distribution of agricultural land is proceeding. Romania leads with swift implementation of a land law passed in February 1991. Many owners expect to take possession of their land after the harvest in the autumn of 1991, although few will farm individually thereafter. The Bulgarian land law was also passed in February 1991 but implementation has been delayed and the approach taken implies a more lengthy process. Land laws in Hungary and Czechoslovakia were passed in April and May, respectively, 1991.

The following paragraphs trace the progress of liberalization of food prices and distribution of agricultural land to date. A detailed exposition of the general framework for the agricultural transition is beyond the scope of this paper, but it is necessary for understanding the origin of the contraction and its impact on the land programmes (Brooks, *et al.*, 1991; Brooks, 1991). The essence of the agricultural transition is the state's withdrawal from its traditional role as residual claimant of (positive and negative) rents to use of agricultural resources. That role will pass in stages to owners of land, where it ordinarily resides in a market economy. A discussion of the new land laws and distribution of land is incomprehensible without attention to conditions that shape the value of land and the income that owners can earn from it.

Czechoslovakia, Hungary, Romania and Bulgaria have each passed laws restoring rights of those who owned land at the time of collectivization. Debate on the legal foundation for re-affirming property rights in land proceeded throughout the region in 1990, and until late in the process it was not obvious that restitution would be the outcome. Parliaments passed land laws in Romania and Bulgaria in February 1991, in April in Hungary, and in May in Czechoslovakia. Since most agricultural land is being returned to people perceived to be rightful owners, recipients do not pay, and the land distribution has little impact on macro-economic balances. In the parts of the Soviet Union in which land was nationalized in 1917 and collectivized between 1929 and 1933, it is difficult to imagine how rights of former landowners could be re-instated. The course of decollectivization is thus likely to be quite different in much of the USSR.

The Romanian land programme embodies the judgement that costs of delay are greater than those of moving ahead before all complications are foreseen and forestalled. Local land commissions in each district were established quickly after passage of the law, and began receiving claims. Households can claim a maximum of ten hectares, and can submit a variety of evidence to support their claims. The period for submission and judgement of claims ended on 20 May, at which date the land commissions were to post their preliminary rulings.

When possible, claimants will be given the land actually owned prior to collectivization. When this is not feasible, a piece of equivalent size and quality will be returned. When the original land was divided into parcels, the process is deliberately duplicated in the returned land. Many households in the Danubian plain area will receive four or five hectares divided into several parcels. Holdings in the hill areas will be larger, and broken into more parcels.

Romanians who receive land through restitution of their rights can sell it immediately if they so choose, or buy more, up to a maximum holding of 100 hectares per household. Family members and neighbours have rights of first refusal on farm land for sale, and this restriction on free sale is intended to address the fragmentation problem. Since in the densely settled areas of intense agriculture almost all land will be distributed through restitution, an active land market could develop rather quickly.

There appears to be little intent in the law or its implementation to create farms of an optimal size, or to look forward to the way in which farming will take place after the land is distributed. This at first appears economically myopic, but may in fact show a much more profound sophistication. The Romanian approach to the land distribution is more like a voucher scheme than a land reform, since it widely disperses claims to the land, but carries little expectation that people will work the land in the units they receive. A small number of people receiving large holdings (for example, eight to ten hectares) plan to manage them as households. Most people plan to keep the land in collective management this season and next. The distribution thus opens a trading period during which households can buy and sell their land, consolidate holdings and prepare to leave the collective when the infrastructure for individual management is more developed. In the meantime the collective will continue to work the land, and land owners will receive a share of returns to land proportionate to their share of the farm's total area.

The IMF/IBRD/OECD/EBRD joint mission to the USSR suggested that the collective serve in a transitional period as a vehicle for the trade and consolidation of members' shares after an initial apportioning (IMF/IBRD/OECD/EBRD, 1991). This role for the collective may be emerging spontaneously from the Romanian land distribution. It is not a consciously assumed role, however, and there is no indication that the new collectives see themselves as transitory organizations. It is thus important that the land distribution be accompanied by new regulations easing procedures by which members can withdraw and take their share of non-land assets with them. Cooperatives should be discouraged from taking on debt that will complicate the future exit of members.

People who worked on cooperative farms in Romania but cannot claim any land through restitution can claim on the basis of their labour input. Since even those with prior claims will receive small allotments, the holdings distributed purely for labour will be quite small. People receiving land in recognition of their contribution of labour cannot sell their land for ten years. This is a curious provision, since young people who chose to remain on collective farms are probably the least likely of their cohort to be the universally despised 'speculators', who might acquire and sell an asset purely to make some money. The quantity of land tied up by this restriction is not significant.

The Romanian approach to land distribution appears to have broken through the confusion about how to start the process. Its progress, and that of the land programmes in Bulgaria, Hungary and Poland, will be monitored in a study jointly undertaken by the World Bank and the member-countries.

The fragmentation of very small holdings implicit in the Romanian approach could plague agriculture in the future. Market-based solutions to fragmentation of farm land in Western Europe after the Second World War were not adequate to consolidate holdings, and administrative consolidation was necessary. The chance for success in market-based consolidation is greater in Romania now, since all rights are distributed simultaneously and many recipients will be trying to adjust their initial claims before removing the land from collective management. Special programmes to promote purchase, sale and trade over the next year could be highly productive in the longer run. Financing for land acquisition may be necessary, and subsidized interest rates for land consolidation would be justifiable. Since the quantities of land traded will be small and the value of land relatively low during the contraction, many buyers will probably choose to pay cash.

The Bulgarian parliament also passed a land law in February 1991, but political stalemate and administrative inertia has delayed its implementation. The National Land Council, the main administrative organ of implementation, was not appointed until 31 May 1991, and appointment of the 248 municipal land commissions was attendant upon the formation of the National Council. As a consequence, people who wanted to claim land in the first half of 1991 had nowhere to take their claims.

The philosophy of land distribution embodied in the Bulgarian law and the implementing regulations is by nature a slow one. Rather than relying on market trades to improve a quick and imperfect distribution of rights, the Bulgarian approach attempts construction of appropriate holdings through administrative assignment. Local land commissions accept and adjudicate claims and, when a substantial number of claims have been verified, turn them over to a team of specialists who draw up a local map of the allocated holdings. This approach is deemed necessary for several reasons. The Bulgarians want to avoid dividing land up into parcels, and doubt the efficacy of market-based consolidation. Market-based solutions are, indeed, unlikely to work, since the law prohibits purchase and sale of land by private individuals for three years, the prime trading period. In many places the amount of land that can be restored is only a proportion of that claimed, since development has changed the contours and use of land, and agricultural area has declined. In these areas all claims will be prorated by the necessary proportionate adjustment. The effort to achieve justice and economic efficiency through administrative meticulousness can be contrasted with the Romanian priority on speed. The costs and benefits of each approach are not yet clear. It is certain, however, that the Bulgarian distribution is much delayed and, three months after passage of the law, not yet ready to move into high gear.

In Hungary, the initial attempt to return agricultural land to prior owners in 1990 was struck down by the constitutional court, with the ruling that restitution of ownership of agricultural land must be considered along with that of other assets. In April 1991, landowners, along with dispossessed owners of other property, were granted vouchers redeemable for agricultural land or other assets. The restitution for those who relinquished title is essentially monetary, and the impact on demand for land depends on economic agents' assessment of the value of land compared to other assets. Landowners who continued to hold title to lands managed by the cooperative are granted the return of their managerial rights unconditionally. This law, too, must pass the constitutional court, and that hurdle remains.

In Czechoslovakia, a law mandating return of agricultural land to prior owners who will cultivate it was passed only in late May 1991, and at the time of passage, little interest in claiming land was reported. Food markets in Czechoslovakia approximately cleared even prior to the price liberalization, and few citizens of the country perceive that they have had or now have a 'food problem'. Thus recognition of the need to change the inherited structure of agricultural production has been late in coming, although a fully open trade regime would demonstrate its high cost relative to world levels. The contraction is just beginning in Czechoslovakia, and difficulties marketing meat and milk are pulling farm incomes down. Pressure for change is increasing, but it is early yet to predict whether the form of change will be protection of the old structure, or the start of decollectivization.

In Poland, the state sector owns only about 20 per cent of agricultural land, since the remainder of land was never collectivized, and remains in fragmented private ownership by smallholders. Although the proportion of marketed output that originated in the state sector was greater than its share of land ownership, the excess supply of food occasioned by the Polish 'big bang' diminished the perceived urgency to reorganize state farms. Those most agitated about the fate of state farms were their employees, who favoured transfer of land and assets to the workforce. The disposition of land in Polish state farms has thus been delayed. In general, property rights on land held by state farms have been considered separately and later than for cooperative land.

In summary, the land distribution programmes in practice are quite diverse, and are not what most people outside the region expected. In surveying the economic options, few outside economists would have chosen physical restitution of rights of prior owners as the preferred solution. The economic difficulties are evident. Moral issues are also relevant: what about the rights of people killed or dispossessed before 1946, or 1948, or the date that serves the interests of those now represented politically? These issues have been raised, but not resolved, in Hungary and have been absent from public debate in other countries.

The restitution approach has an economic advantage to complement its apparent political appeal, and counter some of the economic problems it raises. Had land been distributed without payment to the agricultural workforce with no higher principle than 'land to the tiller', it would have been easy to exclude rural people from further distribution of state-owned assets, on the grounds that they already received their share. Since landowners have instead received back property that was rightly theirs all along, there can be little justification for excluding rural people from a fair share of assets accumulated by the state. Thus, when privatization swings into full force through vouchers or distributed shares, rural people will be integrated into the new capital markets.

PRICE LIBERALIZATION

The speed and apparent success of liberalization of retail food prices is surprising and poorly recognized. Even a year ago the liberalization of retail food prices was considered a political minefield. Governments entered it with great trepidation and varying degrees of caution. All (except Albania and the USSR) are now either in the midst of the process or essentially through it. Curiously, no one has noted that nothing exploded.

The success of the food price liberalization is in part explained because it came first. To that ambiguous honour, plus the fact that the liberalization is in general partial, can be attributed many problems, but the problems must be viewed in the light of the original pessimism that food prices could never be changed without social upheaval. In a world of partial price liberalization with immature markets, many products are sold at essentially world prices, while others are little changed from the days of high Stalinism. A consciously-designed, clean transition would not include price distortions of the kind that are appearing now. Despite the longer-term costs of these price distortions peculiar to the early state of the transition, they explain in part why food price liberalization did not elicit the feared reaction. The retention of controls on other items raised the real value of monetary compensation. These wider distortions increased the relative rise in food prices at the retail level compared to what it will be after a full adjustment.

Since the food processing and retailing industries are not yet privatized and a number of distortions remain throughout, it would be erroneous to argue that retail food prices are free market prices. Prices are free to fluctuate, however, and governments are paying little if anything in direct food subsidy. Few observers would have predicted *ex ante* that this could have been done in a short time without triggering widespread protest.

The success of the price liberalization is all the more remarkable in that it was done in the virtual absence of any safety net to cushion the impact of much higher relative prices for food. Despite widespread discussion of the need for selective food assistance, programmes of direct food relief were not attempted anywhere. In the wealthier northern countries of Poland, Czechoslovakia and Hungary, full liberalization of food prices without directed assistance appears to have been accomplished; generalized compensation was adequate and prices are now largely free. Selective assistance is clearly needed for humanitarian purposes, but in the northern countries it does not appear to have been a political precondition for liberalization.

In Romania and Bulgaria, where consumer incomes are lower but fully free prices will be approximately at world levels, the liberalization which has taken place is incomplete at this writing, in June 1991. Consumers absorbed a large increase when the explicit subsidy was removed, and were compensated through partial adjustment of wages. Governments with shaky political mandates, however, were unable to risk freeing prices to world levels, and retained administrative pressures on wholesale prices. Wholesalers and processors pushed the controls back to the producer level.

In the southern countries, therefore, the contraction is greater and distress at the producer level is more extreme than in the northern countries, where price liberalization has been more complete. Continued sectoral adjustment in Romania and Bulgaria will have to include further freeing of wholesale and producer prices, and introduction of cost-effective selective assistance for needy consumers. The Romanian and Bulgarian experience is likely to be replicated in the USSR unless the harmful impact of partial liberalization on agricultural production can be demonstrated.

The actual course of liberalized food prices is difficult to trace because of the current general weakness in statistics. Both the record and the course of liberalization in Czechoslovakia seem most straightforward. Food prices were raised administratively in July 1990, by 26 per cent on average, to remove the direct budgetary subsidy. Prices were controlled at the new higher levels and consumers received partial compensation. In January 1991, prices were liberalized and jumped quickly by about 30 per cent, before levelling off in March and starting to decline in response to excess supply, particularly of beef.[1]

These price increases, although large by world standards, are modest in the East and Central European current context. Moreover, the larger Czechoslovak incomes, and smaller share of food in family budgets, eased the absorption of the shock. In Romania, in contrast, with partial price liberalization that drives producer prices of grain to approximately half of world levels, the consumer price index for food is reported to have risen in April 1991, to 255 compared to 100 in October 1990.[2]

The announced liberalization at the retail level in Romania and Bulgaria is difficult to reconcile with the continued reports of shortage and declining production. The explanation lies in the considerable degree of control that remains behind the retail level.

CONCLUSION

Return of land to private owners is proceeding in East and Central Europe, and the political commitment behind restitution is strong. The change in property rights, however, will not quickly produce the decollectivization that many observers expect. Under current conditions of acute stress caused by depressed domestic and foreign demand, positive real interest rates and partial liberalization of prices, few producers will leave the cooperatives. Decollectivization will gain momentum as the economic outlook for the sector improves, market infrastructure develops and cooperatives are given incentives to divest their collective production activities. The land programmes are thus important, but are by themselves inadequate to initiate and sustain the dismantling of collectivized agriculture.

NOTES

[1] Interviews during May 1991, Ministry of Finance, Czechoslovakia.
[2] Interviews, May 1991, Romanian National Commission for Statistics.

REFERENCES

Bardhan, Pranab ,1989, *The Economic Theory of Agrarian Institutions*, Oxford University Press, Oxford.
Brooks, Karen, 1990, 'Lease Contracting in Soviet Agriculture in 1989', *Comparative Economic Studies*, vol. XXXII, no. 2.
Brooks, Karen, 1991, 'Decollectivization in East/Central Europe', Draft Working Paper, Agricultural Policies Division, The World Bank, June 1991.
Brooks, Karen, J. Luis Guasch, Avishay Braverman and Csaba Csaki, 1991, 'Agriculture and the Transition to the Market', *Journal of Economic Perspectives*, Fall.
IMF/IBRD/OECD/EBRD, 1991, *A Study of the Soviet Economy*.

DISCUSSION OPENING – RICHARD L. MEYER*

Our speakers have provided an extremely interesting account of the speed and extent to which agricultural reforms are occurring in Eastern and Central Europe. Their central argument is that a deep and well behaved transition in the agricultural sector is the cornerstone for stabilization and structural adjustment in the region, and that a positive supply response would be interpreted as a bright spot in the bleak picture of slow and costly industrial restructuring and deteriorating real incomes. A positive supply response is expected to occur through redistribution of agricultural land, with land ownership providing an associated incentive effect for farmers. Deregulation of product prices, which is also expected to give added incentive to producers, proceeds in parallel.

However, the agricultural transition is not occurring in the expected way. So far agriculture is producing less rather than more food, and is having difficulty in selling what is produced. Individually operated private farms are not emerging, and in several countries collectives appear to be the preferred way to farm for the foreseeable future. This needs explanation.

Brooks and Braverman argue that price liberalization has moved surprisingly quickly and with less social and political upheaval than might have been expected. Furthermore, few social security nets have been provided for those expected to be most seriously affected. Some evidence is provided of the impact of these changes on consumer prices, but nothing is said about producer prices. We are left to assume (as reflected in sketchy media reports) that they have risen as well, particularly for those producers able to sell directly to consumers.

The problem in the agricultural transition, therefore, does not seem to be with price policy but rather, as the authors argue, with the land redistribution process. The land laws passed in 1991 by Romania, Bulgaria, Hungary and Czechoslovakia provide for the restoration of land rights to those who owned land at the time of collectivization. The Romanian programme appears to have progressed most quickly as land commissions were rapidly put in place to receive claims. While the programme runs the risk of creating uneconomic farm sizes, through fragmentation of plots provided to claimants, there is little expectation that people will actually work the land they receive. Most are expected to keep it in collective management in the near future. A market-based process of subsequent land transfers is also expected. The other countries are proceeding more slowly, with systems relying less on market transfer and more on administrative procedures.

Although it is not discussed by the authors, a process of land redistribution with restitution of land rights would seem to introduce great uncertainty about identities of the rightful owners, evaluation of their claims, sorting out competing claims, and determining the interests of recipients, especially if they are not currently farming. The Asian land reform programmes, which are credited with

*Ohio State University, USA.

improved productivity and efficiency, appear to have had a simpler administrative task when they were designed to provide land to the tiller.

The second major problem identified by the authors as impeding decollectivization, in my view, may be the most important. That concerns the lack of economic infrastructure and supporting institutions for small scale private farming. Cooperatives rather than individual farmers may have advantages in gaining access to productive inputs, credit, markets and information. The mere transfer of land ownership rights does not automatically create the supportive environment needed for productive farmers, as any land reform planner in a developing country is painfully aware. It should be no surprise if, during a fairly long transition process, cooperatives and collectives continue to be a preferred method to organize production. Where else are the individual farm owners going to obtain their production inputs and credit until such time as the necessary input systems are being developed? Where else are they to get market information and transport services? Where else are representatives of external markets going to turn to negotiate contracts for purchase of commodities?

While I found this paper an interesting description of the problems being encountered in decollectivizing agriculture, it is not particularly illuminating for anyone who has thought carefully about implementing land reform. It would have been useful if the authors could have given us a more complete understanding of the nature of the economic infrastructure required to support more rapid decollectivization, and their professional assessment of the key bottlenecks faced in each country. What is surprising to me is not that these bottlenecks exist, but that the authors seemed to have expected a smoother or more 'properly behaved' agricultural transition. By definition, an important structural change such as decollectivization, like any profound land reform, is not likely to be a smooth process. Rather, it is likely to be one filled with uncertainty and discontinuities, with short-term disruptions in production and distribution being the price to be paid to achieve long-term growth and efficiency.

JOSEPH NTANGSI*

Agricultural Policy and Structural Adjustment in Cameroon

INTRODUCTION

By comparison with most Sub-Saharan African (SSA) countries Cameroon was a latecomer to structural adjustment.[1] While such countries as Côte-d'Ivoire, Togo, Guinea, Senegal, Ghana, Benin, Nigeria and Zaire, had embarked on a structural adjustment programme during the early and mid-1980s with or without the support of the World Bank and IMF (International Monetary Fund), Cameroon did not sign a stand-by arrangement with the Fund until September 1988 and a structural adjustment loan (SAL) until June 1989.

The reason is that, by SSA standards, Cameroon was considered an island of prosperity until 1989. The country's GDP grew at 4.8 per cent in real terms during the period 1960–78 and at the impressive rate of 8.2 per cent between 1978 and 1986. Agriculture was the major factor in this growth between 1960 and 1978: agriculture's share in GDP and exports averaged 30 per cent and 74 per cent, respectively, during the period. When Cameroon became an oil producer in 1978, agriculture's share declined to 22 per cent of GDP and 51 per cent of exports. Overall, the growth of the agricultural sector is thought to have averaged at least 3.8 per cent since independence, the transformation of agricultural products has constituted the point of departure for industrialization, and Cameroon is one of the few countries in the region to have achieved virtual food self-sufficiency essentially because the government has not intervened in the food sub-sector.

Since 1986, however, Cameroon has been trapped in a deep economic and financial crisis. GDP declined by more than 15 per cent between 1987 and 1990 and, for the first time in the country's history, government operations recorded a large deficit of CFAF 508 billion, equivalent to 12 per cent of GDP. Although the immediate cause of the decline was external stocks, the sudden collapse of the economy also brought to the surface a number of internal structural problems which had been in existence for some time, but which had been concealed and indeed aggravated by the fortuitous oil boom. These structural problems include an overgrowth of public expenditures and a crisis in public finance, a high domestic cost economy, and a stagnant and non-competitive industry in both the international and domestic markets. In agriculture, notwithstanding the overall satisfactory performance, there has been very limited technical change and a virtual stagnation of agricultural exports. These problems have been

*World Bank Resident Mission, Yaounde,Cameroon.

attributed to a hostile macro-economic, institutional and regulatory environment and to a bad conception of the relative roles of the public and private sectors.

The structural adjustment programme (SAP) is a response to the country's structural problems and external shocks. Given the fixed parity of the CFA franc *vis-à-vis* the French franc (CFAF 50 = 1FF), external adjustment through devaluation is not possible without the agreement of the other member-countries. Instead the SAP is based on internal adjustment, with emphasis on increased competition, efficiency and a reduction of costs and prices in the economy.

This study first demonstrates the failure of existing policies and the need for policy reform in Cameroon, with special emphasis on agriculture, and then evaluates the achievements and prospects of reform so far under the SAP. Since the SAP is essentially a macro-economic programme involving an overlapping between macro-economic and sectoral reforms, adjustment in agriculture is analysed within an economy-wide context. Also in the study agriculture is conceived in a broad sense to include agro-industry. The rest of the study is divided into four sections: a review of the existing policy framework and its logic; analysis of the consequences and constraints of existing policies; assessment of the SAP reform; and description of the current state-of-play.

THE EXISTING POLICY FRAMEWORK

The SAP is basically a policy reform programme of the existing policy framework that has been in operation in Cameroon practically since independence and, in some cases, since the colonial period. The existing policy framework has been characterized by two central features: (1) direct, extensive, but inefficient state interventionism in the production and distribution of goods and services, and (2) a system of incentives, controls, regulations and practices that have produced widespread distortions in the country's macro-economic, institutional, and regulatory environment.

Direct state intervention

There has been direct state intervention in the production and distribution of goods and services. Many forms of direct state intervention in Cameroon had their origin during the colonial period. After independence state interventionism was significantly re-enforced and extended, on the conviction that such intervention was necessary for, and in fact synonymous with, rapid development.

The rationale for state intervention

The creation of public enterprises (PEs) in Cameroon has been officially justified on economic and moral grounds and its rationale may be summarized as follows. First, a number of sectors or activities have been considered as strategic because they are seen as being of national or collective interest and therefore should not be left to the private sector. However the term 'strategic' has often been loosely defined to include not only activities of a strictly national or social character (such as public utilities, petroleum and mining), but also activities of a doubtful strategic importance (see below). Second, the Cameroon private sector has been considered to be still at an embryonic stage and therefore incapable of mobilizing the capital, technology and management skills required for investment purposes, at least not in some sectors such as manufacturing. Third, there has been a long-standing prejudice against the private sector, especially by government officials, who consider private traders as unorganized and inefficient (since they operate in numerous small units) or as downright unscrupulous and exploitative of ordinary citizens. Fourth, PEs have been created to sustain the government's import-substitution strategy (see below). Fifth, especially in agriculture, PEs were to act as extension institutions aimed at inducing technical change among small farmers. Sixth, PEs have also been created to enable the state to earn badly needed revenue. Seventh, as we shall see, many PEs in Cameroon, at least until the mid-1970s, were created with the blessing of aid donors who, in many cases, helped finance their operations.

However, in addition to economic motives, state intervention has also been dictated by political imperatives. In Cameroon, as in many SSA countries, the legitimacy and survival of the political leadership and the regime have been crucially dependent upon the support of the elite (which constitutes the group most likely to challenge that legitimacy). The public and parastatal sectors have therefore been used as patronage (in the form of over-staffing, promotions, high wages, fringe benefits and corrupt use of office without accountability and sanctions) to buy that support.

The extent of direct state intervention

An inventory of PEs shows that between 1960 and 1990 a total of exactly 200 PEs were created, of which 120 are still in activity, 61 have been closed and the status of the remaining 19 is unknown (Development Finance Consultants, 1990).[2] In terms of sectoral distribution of the 200 PEs created, 40 were in the primary (agricultural) sector, 57 in the secondary sector, 88 in the tertiary sector (including financial institutions) and 15 were unclassified. Almost 50 per cent of the labour force of PEs is employed in agriculture.[3] The share of PEs in total GDP was 18 per cent in 1985/6 (ibid., p.22).

The legal basis for direct state intervention was provided by the law of 1963 creating the *Société Nationale d'Investissements* (SNI), a public holding with a current portfolio of 63 companies, and by the decree of 1968 on development companies (*sociétés de développement*), and development missions (*missions de développement*), the large majority of which have been agricultural PEs. The agricultural PEs have had widely varying legal forms and have intervened in a vast number of activities. A list of the major agricultural PEs is given in Table 1, showing the year of creation, the main domain of activity, and institutional status.[4] As can be seen from the table, the majority of PEs were created during the decade 1968/78, when the majority of donors apparently still had great faith in the direct role of the state.

Despite their organizational complexity, the agricultural PEs may be classified, on the basis of their activity, into four broad categories. One category covers agro-industries producing such products as sugar, palm oil, rubber, refined cotton, timber, plywood and paper pulp. A second category is rural

TABLE 1   *List of major PEs, by activity, external financing and institutional status*

Public enterprise	Year	Major agric. activity	External funding	Institutional status
WADA	1962	Coffee, rice	FRS	Integrated devt.
SCT [a]	1964	Tobacco	MNC	Agro-industry
COCAM	1966	Timber, plywood	MNH	Agro-industry
SOCAPALM	1968	Palm-oil	IBRD, CCCE	Agro-industry
CREVCAM	1968	Shrimps	MNC	Agro-industry
CENADEC	1969	Cooperative	CARE	Rural devt.
SODENKAM	1970	All crops	FAC	Integrated devt.
SEMRY	1971	Rice	CCCE, FED	Rural devt.
SCS	1971	Textiles	CCCE, FAC	Agro-industry
FONADER	1971	Credit, marketing	—	Financial inst.
ZAPI EST	1972	All crops	IBRD	Integrated devt.
CDC [a]	1973	Rubber palm-oil	IBRD, CCCE	Agro-industry
OCB	1973	Bananas	—	Agro-industry
MIDEVIV	1973	Marketing	USAID	Rural devt.
SODEPA	1974	Cattle	—	Rural devt.
SODECOTON	1974	Cotton	MNC	Agro-industry
SODECAE	1974	Cocoa	IBRD, CCCE, ADB	Rural devt.
HEVECAM	1975	Rubber	UA, CCE	Agro-industry
CAMSUCO	1975	Sugar	MNC	Agro-industry
SODEBLE	1975	Wheat, maize	S. Arabia	Agro-industry
OFFICE C.	1975	Cereals	WFP	Rural devt.
SOFIBEL	1975	Timber, plywood	MNC	Agro-industry
CELLUCAM	1976	Paper pulp	MNC	Agro-industry
ONCPB [a]	1976	Marketing	FED, CCCE	Rural devt.
SODERIM	1977	Rice	IBRD, CCCE	Rural devt.
UNVDA	1978	Rice	FAC	Rural devt.
CENADEFOR	1981	Forestry invest.	CIDA	Public agency
ONDAPB	1981	Poultry	—	Rural devt.
MIDENO	1982	All crops	IFAD, FED	Integrated devt.
ONAREF	1982	Forestry regen.	CIDA	Public agency

Note: [a] Created during colonial period and nationalized or restructured after independence.

Source: Table constructed from individual PE information files.

development-oriented and includes institutions providing extension to farmers growing a specific crop (such as rice, cocoa or coffee); integrated rural development projects; and institutions providing a wide range of other services such as credit, mechanization, and support to farmer cooperatives. Some agro-industries (CDC, SOCAPALM and HEVECAM) also have a rural development component with out-grower's schemes designed to facilitate the transfer of technology to small farmers. State control of cooperatives has reached the point where they are run by government officials in an attempt to keep political power centralized and to use cooperatives as a source of patronage distribution. Furthermore, as we shall see, the PE extension system has largely duplicated and overlapped with the existing traditional extension system of the Ministry of Agriculture.

A third category consists of marketing PEs. The National Produce Marketing Board (ONCPB), which had its origin in the colonial period, has been the main instrument of state intervention in the marketing of cocoa and coffee, the country's traditional export crops.[5] As we shall see below, through marketing, ONCPB has also assumed the functions of stabilization of producer prices, the taxation of farmers and the financing of development. Other state interventions in agricultural marketing have been in the context of inputs and local foodstuffs. Fertilizer, pesticides and small equipment have been subsidized to small farmers since 1973, but, since the demand for subsidized inputs far exceeded supply, a centralized system for input acquisition and distribution was set up to enable the rationing of inputs. The system was managed by a credit institution, the National Fund for Rural Development (FONADER), financed by ONCPB up to 1988, and subsequently from oil revenues. The State has also intervened in foodstuff marketing through the Food Marketing Mission (MIDEVIV). Suspecting that marketing by small traders was inefficient, MIDEVIV sought to reduce marketing costs and therefore urban food prices. A fourth category consists of PEs in the forestry sub-sector. The National Regeneration Office (ONAREF) and the National Centre for Forestry Development (CENADEFOR) have been respectively responsible for forestry regeneration and inventories.

The macro-economic and institutional environment

In addition to direct state intervention, the existing policy framework has also been characterized by a distorted and inappropriate macro-economic, institutional and regulatory environment which has constrained development both within and outside agriculture. The distortions have arisen from a wide range of policies and practices related to trade, investment, pricing, price and wage controls, taxation, legislation, governance, relations between the public and private sector, and so on. In the context of this paper the discussion will be limited to those distortions which are of greater relevance to agriculture.

Cameroon's trade policies have had important consequences for the agricultural and industrial sectors and more than 50 per cent of the industrial sector's value added is in agro-industry. The country's trade policy was dominated by an import-substitution strategy which required high levels of non-tariff and tariff protection of domestic production aginst imports. Non-tariff protection has mainly been in the form of quantitative restrictions (QRs) on imports and of import licences. All products designated as 'sensitive' or 'strategic' were subject to QRs. Regulations affecting Cameroon's international trade are set out in the *Programme Général des Echanges* (PGE) and in the 1988 PGE a total of 156 products were subject to QRs. All imports have been subject to import licences. Additional customs and fiscal advantages have been granted through the investment code and special protection regimes through the context of regional arrangements.[6] Imported finished products are subject to very high tariffs.[7]

Price controls have been applied to most goods and services, exclusive of foodstuffs produced locally by small farmers but inclusive of rice and maize. Domestic prices were either fixed or officially approved on a cost-plus basis with a fixed profit margin of 12 per cent above production costs. Wages are also controlled through the labour code, which fixes minimum wage rates for all categories of workers according to formal training received, regional location and sector of the economy, using civil service salary scales as a reference. Given the high rate of unemployment, labour has been over-priced; wages are rigid and rarely linked to productivity, and the code is highly protective of the worker's employment.[8]

As we have seen, the state has intervened, through ONCPB, in the pricing of cocoa and coffee and this has also been distortionary. State intervention has amounted to direct and indirect taxation of agricultural output with the tax partially offset by input subsidies on inputs (fertilizer, pesticides and small equipment) and other government expenditures in agriculture. Effective rates of taxation have been high overall, despite large variations from year to year due to fluctuations in world prices. Direct taxation has been in the form of export duties and indirect taxation in the form of withdrawals by ONCPB originally intended for stabilization purposes. The over-evaluation of the CFA franc relative to other currencies has also amounted to an indirect taxation of agricultural exports. Table 2 gives the evolution of producer prices for cocoa and robusta and arabica coffee in terms of percentage shares of FOB prices as a general indication of the rate of taxation. During the period 1960/70, the producer's share in FOB prices was 51.2 per cent for cocoa, 69.7 per cent for robusta, and 80.7 per cent for arabica. In 1970/7 they had declined to 39.8 per cent, 41.5 per cent and 51 per cent, respectively, as the government became increasingly in need of funds with which to finance its development programmes. From 1978, the producer's share was increased for cocoa and robusta, reaching a peak in 1988/9 before falling in 1989/90 when producer prices were severely reduced to adjust to the collapse of world commodity prices since 1985/6. The significant increases in the producer's share from 1978 were possible due to the fact that Cameroon became an oil producer that same year and could then afford to lighten the burden on agriculture.

While part of the tax on cocoa and coffee has been returned to agriculture in the form of input subsidies and other government expenditures, a substantial part has been used to cover budgetary expenditures outside agriculture.[9]

TABLE 2   *Cameroon cocoa, coffee robusta, coffee arabica production and prices (production in 000s mt and prices in CFAF/Kg)*

	1960–70	1970–77	1978–88	1988–89	1989–90
Cocoa					
Av. annual production	87.2	107.2	116.9	104.4	115.0
Av. annual producer price	68.7	109.2	349.0	435.0	250.0
Av. annual FOB price	134.2	274.5	770.9	527.1	423.4
Producer price/FOB price (%)	51.2	39.8	45.3	82.5	59.0
Robusta coffee					
Av. annual production	42.8	63.1	84.3	103.3	100.0
Av. annual producer price	107.4	140.0	370.0	455.0	175.0
Av. annual FOB price	154.0	337.3	857.5	501.6	403.9
Producer price/FOB price (%)	69.7	41.5	43.1	90.7	43.3
Arabica coffee					
Av. annual production	17.6	26.6	21.8	21.7	20.0
Av. annual producer price	180.8	206.6	409.0	497.5	250.0
Av. annual FOB price	223.9	405.0	896.0	763.2	610.0
Producer price/FOB price (%)	80.7	51.0	45.6	65.2	41.0

Sources: Ministry of Agriculture/DS; ONCPB.

CONSEQUENCES OF EXISTING POLICIES

The poor performance of public enterprises

Owing essentially to the politicization of PEs and their use for political patronage, the lack of autonomy and accountability in management and the inappropriate macro-economic and institutional environment, the performance of both financial sector and non-financial sector PEs has been generally unsatisfactory. First, the PEs have contributed directly, as well as indirectly, to the crisis in public finance: directly through state operating and investment subsidies to PEs and also through the servicing of external guaranteed debt by the state in the place of PEs in difficulty; and indirectly through default in the payment of taxes by most PEs. Considering only PEs still in activity today, subsidies to PEs stood at CFAF 32 billion in 1983/4, 69 billion in 1984/5, 68 billion in 1985/6 and 98 billion in 1986/7. Following the crisis in public finance, subsidies had to be cut to 13 billion in 1987/8 and 17 billion in 1988/9 (Development Finance Consultants, 1990, p.23). Second, PEs have also been a burden to the rest of the economy through their indebtedness to banks and suppliers. The total indebtedness of PEs in activity stood at CFAF 1030 billion in 1988/9 (ibid., p.7).

Third, state intervention in agricultural marketing has been largely inefficient and inadequate. The ONCPB marketing system has been costly both in terms of the ONCPB structure itself (for example its personnel costs increased from CFAF 1.3 billion in 1978/9 to 7.3 billion in 1986/7, without any increase in the volume of produce handled (AGRER, 1988, p.25)) but also, in terms of the marketing system, licensed private buyers have a monopoly in their zone of purchase and are remunerated on the basis of the scale which represents administratively determined marketing costs and profit margins. Furthermore, the rigidity of the stabilization system has proved to be ill-adapted in times of high international price instability and has recently resulted in the heavy deficits of the cocoa and coffee crop industries. Equally important, the existence of heavy stabilization reserves in the past often resulted in the temptation to misuse the funds. Indeed, when world market prices for cocoa and coffee collapsed in 1985/6, ONCPB could not play its stabilization role because its reserves had been squandered. The inefficiency of state marketing is also evident in the marketing of foodstuffs and inputs; in both cases marketing costs have been higher than for private traders, and in the case of inputs farmers have had to face the additional problem of late delivery. As a result MIDEVIV abandoned food marketing in 1984 and the government also decided to privatize fertilizer marketing in 1987 (even before the SAL was signed).

Fourth, the PE extension system has superimposed itself on the existing, traditional extension system of the Ministry of Agriculture, resulting in two parallel and overlapping systems and in a waste of resources.[10] Fifth, state control of cooperatives has undermined their effectiveness: most became bankrupt and also have not represented the interests of their members – especially not the farmers. Sixth, in the forestry sub-sector not only did the costs of ONAREF and CENADEFOR become unbearable but the rapid pace of forestry exploitation without adequate regeneration threatens the country ecosystem and yet government revenue from the sub-sector has been far below potential. Seventh, the proliferation of PEs in agriculture created a bias against the smallholder sub-sector and in favour of the large-scale plantation sub-sector. For example, during the period 1971/80, 60 per cent of agricultural investment resources were concentrated on PEs (Ministry of Agriculture, 1980, pp.III and 9) which produced only 10 per cent of total agricultural output (p.X).

The overall consequences of macro-economic distortions

The distortions in the macro-economic, institutional and regulatory environment have had severe consequences for Cameroon's industrial and commercial development, both within and outside agriculture. First, the problem with inward-looking policies underscored by an import-substitution strategy is that it has reduced the competiveness of Cameroon's industry in the world market. High rates of protection, coupled with the cost-plus system of price controls, have discouraged competition and resulted in monopolistic and oligopolistic market structures which allow producers to operate with higher costs and higher prices than would be the case without protection and price controls, thus making it difficult for producers to sell in the international market. At any rate producing for export at world market prices is less profitable than producing for a protected domestic market where prices are raised by tariff and non-tariff protection. Second, since the non-tariff protection afforded by the PGE and the incentives provided by the investment code have been granted on a discretionary basis (often with little regard to the contribution of the proposed activity to the domestic economy), the system of protection has tended to distort the pattern of investment and to direct scarce resources into inefficient activities simply because they receive higher protection. The distortions in the pattern of investment have been further aggravated by duty exemptions on imported inputs and low duty on imported capital goods, as well as by the over-pricing of labour. They have encouraged investment in capital-intensive activities and resulted in the poor utilization of local resources and in limited vertical integration of economic activity. Third, the system of non-tariff protection requires extensive bureaucratic procedures, has been too complex to administer, and the monitoring of its impact has been difficult. Moreover, high rates of protection have resulted in incentives for fraud and abuse and the situation is not helped by Cameroon's long open frontiers across which there has been a lot of smuggling. Fourth, the limited size of the domestic market effectively limits the quantity of imports that can be substituted and the potential for achieving economies of scale. Thus a tiny domestic market and limited access to international markets have reduced the resources available to fund further industrialization. Fifth, non-tariff protection has reduced tax revenue to the state and imposed higher costs of living on consumers. Sixth, pricing policies of cocoa and coffee have partly accounted for the general stagnation of cocoa and coffee production since the 1970s (see Table 2). Cocoa production, which averaged 114 metric tons in 1970/5 was 119 in 1985/90; robusta increased from 65 to 95 metric tons, and arabica declined from 29 to 19.6 metric tons during the same period.

THE STRUCTURAL ADJUSTMENT PROGRAMME

The need for structural adjustment

The key macro-economic indicators presented in Table 3 reveal the structural weaknesses of the Cameroon economy and the need for an adjustment programme. The impressive GDP growth of 8.2 per cent in real terms in the period 1980/6 was due to, and coincided with, the oil boom which, indeed, concealed the country's structural problems. Following the external shocks of 1985/6, in the form of the collapse of world prices for Cameroon's major export commodities (oil, cocoa and coffee) which resulted in a 47 per cent deterioration in the terms of trade, the country fell into a deep economic and financial crisis. The oil boom resulted in an over-growth of public expenditures, resulting in an unprecedented deficit in the government budget of CFAF 501 billion in 1986/7, equal to 12 per cent of GDP, compared to a surplus of 2.4 per cent the year before. The deficits were financed essentially through an increase of the external and internal indebtedness. Reductions in domestic demand due to the decline in the purchasing power of the population have aggravated the fiscal deficits by reducing the effective tax base which the government has been unable to compensate for through expenditure reduction measures.

As far as agriculture is concerned, it is difficult to judge performance on the basis of growth in sectoral output, owing to the unreliability of agricultural data in Cameroon (especially for foodcrops). We have already discussed the general stagnation of cocoa and coffee production, which accounts for some 80 per cent of agricultural exports, due to pricing policies and more recently to the decline in world prices and to increasing competition from foodcrops. The PE sub-sector (producing cotton, rubber, palm oil, rice, sugar and so on) despite moderate increases in output, contributes only marginally to exports (less than 10 per cent). The country's greatest achievement in agriculture is undoubtedly its food self-sufficiency; annual growth in food production is thought to have averaged 4 per cent since independence, essentially because the government has not intervened to control food prices (which have been determined by market forces) and also because the country's climatic diversity has permitted the production of a wide variety of foodcrops. Nevertheless, despite past performance, there are serious doubts about Cameroon's ability to sustain its food self-sufficiency. As a result of the elitist educational system and the expansion of the public and parastatal sectors, there has been rural to urban migration of young school leavers, looking for office jobs. This has resulted in massive urban unemployment and in the relative decline and aging of the agricultural population (the average age in agriculture increased from 45 in 1972 to 46.7 in 1984) (Agricultural Censuses of 1976 and 1987). This in

TABLE 3   *Key macro-economic indicators of performance*

	1965–75	1980–86	1986–87	1987–88	1988–89	1989–90
GDP growth (%) (real)	2.3	8.2	−4.7	−13.5	2.2	−2.8
Oil industry	—	23.5	−0.2	−69.0	−53.0	−5.6
Agriculture	4.6	1.9	7.5	2.3	1.6	−5.1
Industry	4.4	15.8	−5.6	−7.1	−11.0	−6.9
Service	0.2	4.6	−4.6	−10.1	−2.0	−7.4
GDP sectoral shares	100	100	100	100	100	100
Oil industry	—	14.7	6.3	6.3	6.2	7.6
Agriculture	32.7	23.6	24.6	25.8	26.9	26.6
Industry	20.3	21.0	22.3	22.3	20.8	20.2
Service	47.0	40.7	46.8	45.6	46.1	45.5
Govt. budget surplus/deficit (% GDP)	—	2.4	−12.0	−6.4	−4.3	−7.2
Current account balance (% GDP)	—	−5.2	−9.2	−8.2	−2.9	−2.3
Public debt to GDP (%)	—	20.9	29.7	36.0	40.2	41.8
Agr. exports (% share of total)	77	35.8	42	48	49	50

Sources: IMF Article IV consultation April, May 1990; World Bank SAL Memorandum to Executive Directors, 16 May 1989, Annex 1.

turn has constrained the expansion of the area under cultivation as well as technical progress in agriculture, which has been limited to cereals (maize, rice, millet and sorghum), while rootcrops (especially cassava and cocoyam) have suffered from declining yields.

The programme and its objectives

The Cameroon SAP is supported by a World Bank structural adjustment loan (SAL) approved in July 1989 in the amount of US 150 million dollars, to be disbursed in three tranches. The SAL was also preceded by an 18-month stand-by arrangement with the IMF, approved in September 1988, equivalent to SDR 69.5 million and a purchase of SDR 46.4 million under the compensatory financing facility. The SAP is also supported by other donors, notably the African Development Bank and the French through the CCCE (Caisse Centrale de Coopération Economique).

The SAP is a comprehensive reform of the existing policy framework and a response to Cameroon's structural problems and external shocks. Given the fixed parity of the CFA franc *vis-à-vis* the French franc, the possibility of external adjustment through currency devaluation is excluded; therefore the Cameroon SAP is based on internal adjustment. More specifically the SAP aims at (1) redefining the role of the state away from direct intervention in the production of goods and services towards greater reliance on markets and the private sector; (2) removing the distortions in the macro-economic, institutional and regulatory environments by introducing simplicity and neutrality in the pattern of relative incentives and reducing the anti-export bias; (3) strengthening the country's economic management; and (4) ultimately restoring the major macro-economic equilibria (including a balanced budget, a positive current account balance, a bearable level of indebtedness and a positive growth in GDP).

STATE-OF-PLAY OF REFORMS UNDER THE PROGRAMME

Limitation of state intervention

Under the programme efforts to reduce direct state intervention have centred on PE and financial sector rehabilitation. In the case of non-financial sector PEs, the strategy has been to maintain in the state portfolio only PEs that are strategic, and to disengage from those that are not, either through privatization or liquidation, although PEs with dubious strategic importance have also been maintained in the state's portfolio. So far, of the 67 PEs on which decisions have been taken 28 are maintained in the state's portfolio to be restructured, 21 have been (or are being) liquidated and 18 are under privatization. Similarly, out of 14 financial institutions (commercial banks and specialized development institutions) 4 have been liquidated, 1 has been transformed and 2 have been fused; furthermore, the state has withdrawn as the majority share-holder from the largest four commercial banks.

In the agricultural sector, the disengagement of the state has been especially significant in the areas of marketing, extension and forestry. The internal and external marketing of cocoa and coffee have been liberalized and ONCPB has been severely reduced from a structure of 3800 employees to one of 500 – the *Office National du Café et Cacao* (ONCC). One of ONCPB's functions, that of financing development activities, had already been discontinued in 1988, when the institution became bankrupt. It is hoped that liberalization will substantially reduce marketing costs and result in a higher residual for the farmer, given that producer prices were slashed by about 50 per cent in 1989, following the collapse of world market prices in 1985/6. Similarly, MIDEVIV marketing of foodstuffs was abandoned in 1984; since 1987, the USAID has been assisting the government in establishing a private marketing system for fertilizer; MIDEVIV is also in the process of privatizing its seed multiplication operations.

Furthermore, there has been a general disengagement of the state from extension and rural development activities. There has been a move away from the intensive and expensive PE extension system to the less expensive system of the Ministry of Agriculture which has involved, or will involve, the liquidation of many extension PEs such as SODECAO, SEMRY, UNVDA and CENADEC, as well as 'out-grower' schemes such as those attached to CDC and SOCAPALM. Consequently, farmers will have to assume the full costs of inputs as subsidies are being phased out which have been provided by the state, carry out certain operations (such as spraying pesticides against cocoa and coffee diseases) and be expected to survive with little or no extension. Equally important, a new cooperative law is under elaboration which will guarantee the autonomy of cooperatives, and farmers will be allowed to manage their own affairs. In short, there is a general movement away from the existing paternalistic approach to development.

Finally, in the forestry sub-sector, CENADEFOR and ONAREF have been liquidated and the establishment of inventories and regeneration activities will be handled by the private sector on the basis of large-scale concessions. Only the general conception and regulatory roles are retained by the Forestry Department.

Rationalization of the macro-economic and institutional environment

There has been substantial progress with the rationalization of the macro-economic and institutional environment. In the domain of external trade, QRs were removed for 68 products in June 1989, for a second group of 46 products in February 1990, and for a third group 26 of products in January 1991. Only 16 strategic, mostly agricultural products (for example rice, sugar, vegetable oils and maize) remain subject to QRs. The objective is to replace QRs with tariff protection. Concomitantly, import licences were abolished for all products not subject to QRs. Similarly, price controls were relaxed in January 1989 for all except 35 products and 10 categories of services; during the second phase in June 1989 only 26 products and 7 categories of services were subject to price controls; then in January 1991 the products were reduced to 16. The reforms also provide greater incentives for industrialization and export. A new investment code was adopted in November 1990 with, as major innovations, the elimination of duty exemptions in favour of fiscal advantages, greater competition, utilization of local resources and increased incentives for export. The incorporation of the one-step window (*quichet unique*) in the code ensures the simplification of procedures for creating new businesses and for access to the benefits of the code. Furthermore, export taxes have been eliminated for all exports except timber.

Other reforms under way are aimed at modernizing the legislative environment for business activity. The new labour code aimed at deregulating wages, reducing the cost of labour and facilitating its mobility and productivity is being finalized with bank assistance. Similarly, the general statute for companies (*Statut Général des Entreprises*) and company law (*Code des Sociétés*) are under elaboration and are expected to be adopted by December 1991.

Problems and prospects

The SAP has been beset by a number of problems. First, while the liberalization of external and internal trade is central to the reform programme, the will to change has not always been matched by the ability to effect the change. The capacity of Cameroon producers to adjust to lower levels of protection requires a more detailed analysis to determine the tariff rates necessary to induce progressive adjustment and the effects of over-supply (which give rise to dumping) on the world market. So far there has been insufficient statistical data permitting the monitoring of the effects of liberalization. Similarly, in the liberalization of internal marketing for cocoa and coffee, the main problem is how to go from the existing to the new system. It remains to be seen whether the preconditions for the private sector to operate efficiently exist – including free entry, information, factor mobility, existence of financial markets and access to credit. It is not so much the principle of liberalization as its sequencing, phasing and pace. This point was underscored recently by the fact that the Minister of Industrial and Commercial Development had to re-introduce QRs for some products that had already been liberalized to save the producers from going out of business. This means that if liberalization is not closely monitored it could have disastrous consequences.

Second, adjustment to liberalization is not helped by the fact that it is taking place in the context of an acute financial crisis both for the state and the private sector. The state is not in a position to pay its arrears to the private sector, nor does the liquidity problem of the banking sector permit it to grant loans to the private sector, particularly long-term credit for manufacturing. Furthermore, the state cannot afford the funds with which to rehabilitate the PEs and financial sectors and to finance agricultural research and extension. Third, there has for long been a lack of strong political commitment to the adjustment programme, a lack of policy dialogue between the public and private sectors, and a general lack of confidence in the latter from the former.

'Many in the private sector do not know where the economy is going, as they are unsure about the nature of the rules of the game' (FIAS Report, 1990, p.3). The situation is compounded by the fact that the private sector is generally opposed to the thrust of the liberalization programme (which breaks their monopolistic and oligopolistic power). Fourth, there has been a basic contradiction between the short-term goal of stabilization of macro-economic aggregates generally pursued by the IMF and the medium-term objective growth sought by the World Bank. For example, cuts in the public investment programme have improved the state's financial situation in the short run but compromise growth in the medium run. Fifth, the problem of over-valuation of the CFA franc remains a major counteracting factor to the adjustment effort and demonstrates the limits of internal adjustment. For example, despite the good management of HEVECAM and CDC (which produce rubber) the two PEs remain uncompetitive in the world market. Finally, the advent of multi-party politics and democratization in Cameroon may make it more difficult for political leaders to take tough adjustment measures.

On the positive side, recent experience with the programme seems to show that the crisis may turn out to be a blessing in disguise. In the public sector, the crisis in public finance, the bankruptcy of many PEs and the illiquidity of the banking sector have forced the state to disengage from many activities in favour of the private sector, which would almost never have happened without the crisis. With subsidies cut off, PEs have to survive on their own or die. In the face of massive lay-offs of workers in the private and parastatal sectors, and the existence of widespread unemployment, new recruits (for example, young university graduates) are willing to accept even meagre wages (even before the new liberal labour code is voted in) and public and parastatal staff are willing to accept important cuts in fringe benefits rather than risk the prospect of lay-offs. Political leaders are suddenly realizing that cutting public sector wages is politically less dangerous than not paying wages at all. Thus many components of the reform programme considered only a year ago as revolutionary are now being adopted wholesale. Ironically, therefore, the severity of the crisis may actually result in a faster pace of adjustment. Furthermore, democratization will almost certainly result in a depoliticization of management, and therefore in greater accountability, transparence, efficiency and punishment of poor or corrupt political leaders and managers. Finally, of vital importance for agriculture is the fact that mass unemployment and low wages in the urban sector are bound to force young people to return to agriculture, and there are signs that this is already happening.

NOTES

[1]Cameroon covers an area of 475 000 sq. km and has a population of almost 12 million, with widely varying densities. The country has been referred to as the microcosm of Africa because of its extreme geoclimatic diversity, its complex ethnic compostion (about 200 ethnic groups, speaking 24 major dialects have been identified), making Cameroon the 'racial crossroads of Africa' and because it is the only African country with a triple colonial experience. Cameroon was a German protectorate and subsequently divided between France and Britain.

French and British Cameroons became independent in 1960 and 1961, respectively, and reunified to form modern Cameroon.

[2]This inventory was carried out under the auspices of the Technical Commission for the Rehabilitation of PEs. The 19 are likely to be PEs which were created but have never functioned, or PEs which have ceased activities but are not officially closed.

[3]Agriculture's high share in total PE labour force is to be attributed largely to the Cameroon Development Corporation (CDC) which alone employs over 14 000 workers.

[4]The minor PEs as well as PEs in which the state is a minority shareholder have been omitted owing to space limitations. It should be noted that, even when the state has been a minority shareholder, it has often behaved as though it were a majority shareholder, influencing key decisions.

[5]The Marketing Board was created in British Cameroons in 1955 and the *Caisses de Stabilisation* in French Cameroons in 1957. At independence and reunification of the two Cameroons in 1961 to form a Federal State, the two institutions were maintained. However, in 1976, following the change from a Federal to a Unitary State (in 1972), the Marketing Board and *Caises* were fused to form ONCPB, which was assigned a substantially enhanced role.

[6]For example, in the 1984 investment code, incentives are provided through the different regimes derived from the UNDEAC (*Union Douanière des Etats de l'Afrique Centrale*) general guideliness and the single regional tax (*taxe unique*) and internal turnover tax (*taxe interieure à la production*) regimes which grant import duty exemptions on inputs and a maximum of 5 per cent import duty on capital goods.

[7]Nearly one-third of the products had tariff rates in excess of 70 per cent, while tariff rates on strategic products, which are meant to be banned, vary from 53 per cent to 202 per cent (Maxwell Stamp Associates, 1989, p.8).

[8]These policies had their origin in the colonial period. Following the rocketing inflation and disruptions of the Second World War, the French issued in 1945 an ordinance introducing price controls in all its overseas territories. Similarly, the French also introduced a labour code (*Code de travail Outre mer*) in 1952 which sought to protect the worker from what was seen as the excessive power and abuses of employers.

[9]Agriculture has contributed substantially to the state's budgetary expenditure outside agriculture. For example, in 1978/9 the total contribution of agriculture to the state budget was CFAF 72.6 billion (42.6 billion in the form of export taxes and 30 billion in the form of ONCPB withdrawals), whereas total spending in agriculture by the state was only CFAF 25 billion, leaving a net contribution by agriculture of CFAF 47.6 billion.

[10]The overlapping of two extension systems has resulted in a waste of resources, in power conflicts, and above all in the confusion of farmers. The Ministry of Agriculture organized a seminar in December 1985 to examine this problem, but no lasting solution was found. Now, under the SAP, the solution has been to liquidate the extension PEs and to re-enforce the Ministry of Agriculture's extension system (through the introduction of the Training and Visit System).

BIBLIOGRAPHY

AGRER, 1988, *Secteur Agricole: Rapport Final,* Mission de Rehabilitation des Entreprises du Secteur Public et Para-Public, MCM/PS, Presidency of the Republic.

AGRER, 1989, *Etude Diagnostique de l'ONCPB: Rapport Définitif,* MREP, MCM/PS, Presidency of the Republic.

Development Finance Consultants, 1990, *Impact du Programme de Réforme des Entreprises Publiques au Cameroon,* MREP, MCM/PS, Presidency of the Republic.

Foreign Investment Advisory Service (FIAS), 1990, *Cameroon Agricultural Sector Report,* Washington, DC.

Maxwell Stamp Associates, 1989, *Etude du Système de Protection Tarifaire au Cameroun: Rapport Final,* MINDIC.

Maxwell Stamp Associates, 1990, *Etude Complémentaire sur le Coût de la Protection et l'Impact de la Libéralisation du Commence au Cameroon,* MINDIC.

Ministry of Agriculture, 1980, *Bilan Diagnostique du Secteur Agricole (1960–1980)*, Direction Etudes et Projets.
Ministry of Agriculture, 1990, *Politique Agricole du Cameroun*, Division des Projets Agricoles.
World Bank, 1989, *Cameroon Agricultural Sector Report*, Washington, DC.

DISCUSSION OPENING – AJA OKORIE*

The value of this paper lies in its comprehensive review of agricultural policies in Cameroon between 1960 and 1990. Description of policy and implementation changes over the period, however, raises issues but does not provide answers. That is evident in the 'problems and prospects' sub-section. The author appears to be presenting World Bank views on structural adjustment in the Cameroon. These are sometimes over-stated. For example, the assertion about over-priced labour and the emphasis placed on increases in nominal personnel costs in public enterprise are both made without consideration of general inflation in the economy. Real wages are ignored. In addition, the increase in unemployment, with its attendant evils, does not seem to be seen as a problem accompanying structural adjustment implementation. The paper also fails to examine the successes or failures of policy reforms in Cameroon, difficult though that might be at such an early stage.

There are some contradictions in the views expressed by the author which need discussion. First, the economy of Cameroon (including the agricultural sector) is said to have done well in the decade 1968–78, which coincides with the period in which most of the agricultural public enterprises (PEs) were created and funded by donor agencies. Subsequent decline could be due to withdrawal of aid and support by donor agencies, and not necessarily to the discovery of oil in 1978 alone, given that it could be argued that high dependency on external funding made the economy of Cameroon highly susceptible to external shocks (see Table 1). These appeared in 1985/6 (resulting from the collapse of the world commodity prices and deterioration of the terms of trade) and, more than any other factor, seem to be primarily responsible for the collapse of the Cameroonian economy.

Related to this is a second point. The author dwells on such issues as 'a distorted and inappropriate macro-economic, institutional and regulatory environment' as a source of difficulty. The crucial point is that he fails to show how distortions and inappropriate policies could possibly have hindered the 'good performance' of the economy between 1960 and 1978. My view is that the problems were basically external rather than internal.

Finally, I have to take issue with the author about his complaint that the pricing of cocoa and coffee, which implicitly involves the taxation of agriculture, has been unwise. This is basically consistent with a once popular theory of economic development in which agriculture is regarded as a source of funding in the critical early stages of growth. Without some reliance on such mechanisms it is difficult to see how another policy failure, namely lack of diversification and industrialization, is avoidable in the 1990s.

*University of Agriculture, Makurdi, Nigeria.

MOKHTAR BOUANANI AND WALLACE E. TYNER*

Structural Adjustment Policies and Agricultural Development in Morocco

INTRODUCTION

Structural adjustment is generally defined as a combination of macro-economic stabilization and sectoral policy reform (World Bank, 1988, p.32). Stabilization involves reducing balance of payments and government budget deficits and inflation to levels that can support sustained growth. Sectoral policy reform is aimed at providing incentives for the economy to function more efficiently and to earn more foreign exchange (Nelson, 1990, pp. 3–4). This paper reviews the recent history of structural adjustment in Morocco, with particular reference to the agricultural sector.

Structural adjustment is usually deemed necessary when a country finds itself facing serious economic difficulties. Luther Tweeten (1989, p.1102) has delineated five symptoms indicating that structural adjustment is needed:

(1)   persistent government budget deficits,
(2)   national debt becoming a large proportion of national income,
(3)   high inflation rates,
(4)   over-valued national currency and
(5)   shortage of foreign exchange.

ECONOMIC CONDITIONS PRIOR
TO THE ADJUSTMENT PROGRAMME

All of these conditions (except perhaps high inflation) existed in Morocco in the early 1980s as Morocco began its structural adjustment programme. The Moroccan economy had seen moderate and stable growth during the 1960s and 1970s. Economic growth had attained an average of 4 per cent during the 1970s, reaching a high of 7.5 per cent during the 1973–7 period, only to fall rapidly to a negative 3 per cent growth rate during the 1981–5 period. The phase of rapid expansion had been permitted in part by rising world prices of raw materials in the early 1970s. Between 1973 and 1974, the price of phosphate more than tripled. This rapid price rise had contributed to the launching of public works programmes of physical infrastructure: construction of dams and roads, school and university buildings, industrial facilities, and others. At

*Ministry of Agriculture and Agrarian Reform, Morocco and Purdue University, USA.

the same time, prices of basic food products were maintained at a relatively low level, thanks to a system of subsidies on food consumption. The increase in world prices of phosphates proved to be ephemeral, since prices began to drop significantly in 1975–6. That drop was translated into a major increase in the government budget deficit because expenses did not drop as quickly as budget receipts. At the same time, easy access to external finance permitted the government to continue investment programmes and to sustain public sector support of the social system. These policies stimulated domestic demand that grew more than 11 per cent per year in real terms during the period 1973–7. At the same time, there was a reduction in competitiveness of Moroccan exports (because of the increase in energy prices and the prices of intermediate goods) and an increase in the import bill.

In the second half of the 1970s, international finance became more difficult and the rate of interest increased. The weight of the debt became heavier and the investments realized during the 1970s could not finance reimbursements of the debt (Seddon, 1988, p. 1). The increase in world prices of raw materials imported by Morocco worsened the balance of trade deficit (oil, grains, sugar and vegetable oil) and brought about public sector deficits because of the subsidy policies.

Conditions in the agricultural sector

In the agricultural sector, as for the rest of the economy, the solid growth of the 1960s and much of the 1970s could not be sustained in the 1980s. Food production did not follow the rapid growth in food demand, which was stimulated by rapid population growth (2.6 per cent), rapid urbanization and a fall in real prices of principal food products (flour, sugar and vegetable oils). At the same time, agricultural exports stagnated during this period, and the balance of agricultural trade went into deficit ranging between $200 and $250 million per year.

The analysis in global terms of the agricultural sector does not take into account the differences in performance in the irrigated and rain-fed agricultural sub-sectors. Irrigated agriculture, which produced nearly 40 per cent of agricultural production, 80 per cent of sugar and more than 50 per cent of milk, had seen during the 1960s and 1970s a rapid expansion, which was followed by a slowing down in the beginning of the 1980s. This performance of the irrigated sector was the consequence of investments in irrigation equipment, increased use of factors of production (fertilizer, improved seeds and so on) and use of modern techniques of production. Moreover, changes in consumption habits in urban areas led to increased revenues for irrigated agriculture due to increased demand for more remunerative crops such as tomatoes.

On the other hand, the rain-fed sector that produces the major part of basic food commodities (cereal grains, vegetable oils and legumes) did not experience a growth in production comparable to the irrigated sector. Public investment had favoured irrigated areas, and the impact of drought made production gains highly variable in rain-fed areas. But the structural constraints and climate (one drought every two years on average during the last two decades) could not by themselves explain the problems of Moroccan agriculture. The performance of the agricultural sector had been negatively affected by both macro-economic and sectoral policies in agriculture. The macro-economic policies followed during the 1970s and the beginning of the 1980s had disadvantaged Moroccan agricultural exports because of the over-valued exchange rate and regulations on foreign trade. The controls on prices of basic food products also had negative impacts on agricultural production. To compensate for the negative impacts of public sector macro-economic policies on the agricultural sector, policy makers attempted to stimulate directly the production and income of farmers. State interventions were either indirect, such as investments in infrastructure and improved technology, or direct, such as subsidies on inputs and elevated price supports for production. The principal instruments of intervention by the state have been as follows:

(1)   price supports for major agricultural commodities,
(2)   restrictions on domestic and foreign trade and a proliferation of controls and administrative regulations in the formation of prices and exchange rates throughout the whole agro-industrial chain,
(3)   subsidies for most inputs, such as fertilizers, improved seeds, irrigation water, and livestock feedstuffs, as well as farm investments, and
(4)   subsidies on basic food products such as flour, sugar and vegetable oil.

These interventions have had repercussions on public finance as well as causing distortions in production and exchange.[1]

STRUCTURAL ADJUSTMENT IN MOROCCO

Morocco actually began structural adjustment on its own in the late 1970s and early 1980s with the plans of 1978–80 and 1981–5. The factors mentioned above, plus the effects of the drought of the 1980s, accentuated the weight of the external debt and caused the country to undertake, in collaboration with external financial agencies, structural adjustment reforms of the economy. The principal objectives of these reforms were the reduction of the balance of payments and budget deficits. A whole range of measures at the macro and sector levels were undertaken to achieve these objectives. Macro measures included reducing the growth in the money supply, reducing the rate of growth of government expenditures and increasing government revenues, and adopting a flexible exchange rate to devalue the dirham.

The first adjustment programme with the World Bank was called the Industrial and Trade Policy Adjustment (ITPA). It was designed to reduce domestic price controls and distortions and to reduce tariff protection to a maximum of 45 per cent. ITPA also included a number of financial sector reform measures. The structural reform measures in ITPA were coordinated with the general economy measures discussed above.[2]

Conditions in agriculture leading to adjustment

The measures adopted in the agricultural sector are discussed in detail below. The agricultural sector measures were designed to supplement and complement the general economic measures and those taken in the industrial sector. A study of prices and incentives in the agricultural sector helped to identify the existing distortions and opportunities for policy change (Ministère de l'Agriculture et de la Réforme Agraire, Royaume du Maroc, 1986, pp. 177–86). The study revealed that the domestic resource costs for dryland cereals, oilseeds, legumes and export crops were all less than one, meaning that Morocco had a comparative advantage in production of these crops. Most sugar production had domestic resource costs greater than one, and milk was generally equal to or slightly over one.

Clearly, one of the principal effects of state intervention in the agricultural sector has been the reinforcement of the dual structure of the sector (modern and traditional). Before 1980, more than 65 per cent of agricultural investments were devoted to irrigated zones that represent only 10 per cent of cultivated land and 15 per cent of the agricultural population. It was only beginning with the 1981–5 five-year plan that investments were re-oriented in favour of rain-fed zones. The dualism encouraged by state intervention appeared also in the application of new techniques. Programmes have existed for encouraging utilization of modern techniques and factors of production in irrigated zones, but rain-fed zones remain largely traditional and little touched by systems of subsidies for farm investments. Moreover, products cultivated in rain-fed zones, notably the cereals, have been penalized relative to irrigated crops because the majority of crops cultivated in irrigated zones receive fixed prices and a sure market outlet provided by the state.[3] The large rain-fed crops (mainly cereals) have a price support that, because of food subsidies and the lack of an adequate collection network, are known to be inefficient. In a year of good harvest, market prices for certain cereal grains have often been considerably below official prices.

The policies of subsidizing inputs have benefited irrigated zones more than rain-fed zones. About 40 per cent of total input subsidies have been absorbed by irrigation water and 35 to 40 per cent of fertilizer subsidies have been consumed in irrigated zones.

MAJOR COMPONENTS OF AGRICULTURAL ADJUSTMENT

The agricultural sector structural adjustment programme in Morocco was supported by several loans from international lending agencies and by grants from USAID and other donors. The World Bank provided $100 million for the first structural adjustment load in agriculture in 1985. In 1987, a second World Bank loan was signed for $225 million. The African Development Bank loaned $150 million in 1985. The World Bank provided two other loans for improving irrigation, and the European Investment Bank and EC provided other credits for agricultural development (The Economist Intelligence Unit, 1990, p.27).

The major components of the agricultural structural adjustment programme are as follows (World Bank, 1987, pp. 9–18):

(1)   liberalize agricultural and food product pricing and marketing,
(2)   restructure priorities for the public investment programme,
(3)   streamlining and/or privatize government support services,
(4)   improve the management of natural resources,[4] and
(5)   strengthen the Ministry of Agriculture's policy analysis capability.

Price and marketing policies

The largest group of conditions of the external loans fell into this category. In general, the aim of the conditions was to de-control prices and de-regulate domestic and international marketing of agricultural products. The agreements called for changing the approach to establishing domestic support prices for cereals to a linkage with world market prices instead of domestic cost of production.[5] Most agricultural products would move from a restricted trade status to free trade with a tariff. Much of the structure of control and regulation of internal markets for cereals and other agricultural products would be removed or simplified.

The cost of inputs and services that benefit essentially irrigated zones would be adjusted in such a manner as to approach their real economic value. Subsidies would be removed from fertilizer, irrigation water, services provided by the irrigated perimeters and the animal feed by-products, bran and sugar beet pulp. Perhaps the most controversial components called for the progressive removal of subsidies for flour and vegetable oil. There were also numerous conditions concerning changes in the regulations of flour mills and flour products.

Restructuring public investment programmes

Public investments were to be restructured in a manner to allocate priority according to the following criteria:

(1)   fundamental contribution to the sectoral development objectives,
(2)   little or no impact on the operating budget,
(3)   maximum profitability of the investments,
(4)   positive effect on the balance of payments, and
(5)   creation of employment in the short and long term.

In addition, private sector incentives for larger participation in agricultural investments would be provided. The state would cease to furnish support services that are commercially viable. Also, in order better to direct state actions, agricultural subsidies would be transmitted by a special fund called the Fund for Agricultural Development.

Restructuring and privatizing government support services

The structural adjustment programme sought to rationalize and render more efficient public sector interventions to recover costs and to encourage the state to cease to provide services that could be transferred to the private sector. Improvements were planned in agricultural research and extension.

Services that could be efficiently provided by the private sector were to be wholly or partially privatized. Veterinary services were to become essentially private. Input marketing for fertilizer was to be reformed, so that state companies and the private sector would be on an equal footing. Production and distribution of seeds was to be gradually privatized.

Strengthening agricultural policy analysis capability

The structural adjustment programme sought to reinforce the means of analysis available to the Ministry of Agriculture to permit the following:

(1) analysis of price, trade, and agricultural marketing policies;
(2) examination of the level of investment programmes and expenses as a function of social and economic criteria;
(3) undertaking longer-term analyses; and
(4) monitoring the evaluation of projects and investment programmes in the agricultural sector.

ASSESSMENT OF THE
AGRICULTURAL ADJUSTMENT PROGRAMME

The agricultural adjustment programme effectively began in 1985 and continues to the present. Hence, at this point, we cannot evaluate total impacts of the programme. Rather, we can describe the measures that have been undertaken and assess the results to date.

Price and marketing policies

Substantial progress was made on the reduction of input subsidies. The programme to eliminate subsidies on fertilizer has proceeded essentially as planned, causing the price of fertilizer to rise substantially. The fertilizer price is now free. Even with the price increase, use of fertilizer increased substantially, except in 1987, which was a poor rainfall year. Studies have shown that the marginal value product of fertilizer is still considerably higher than its price, so usage is not likely to be affected in the future by the higher prices. Also, about 30 retail outlets for fertilizer have been closed to shift more of the distribution to the private sector. Subsidies on seed production were frozen at the 1984 level in real terms. Steps have been taken to privatize the production of seeds.

Prices of wheat bran and sugar beet pulp were increased by 60 and 17 per cent, respectively. These price increases were determined by using a feed ration model to determine shadow values of nutrients, given the prices and nutritional characteristics of other livestock feed ingredients. Today there is an official price of bran that is used only to calculate flour subsidies – the bran market is free.

During 1985–8, the official prices of agricultural products were frozen in nominal terms. The first official price increases occurred for the 1989 crop. The approach of linking domestic price to world price for bread wheat was adopted, with the caveat that the domestic price would not fall below the 1986 level in real terms. The marketing and prices for durum wheat, maize and barley were completely freed. For milk, the price differential between high lactation and low lactation was increased by more than 20 per cent to provide a greater incentive to increase production during low lactation.

A farm budget analysis was carried out to calculate the impacts of these changes on producer income. It appears that major changes occur only for irrigated farms that practise a crop mix largely composed of wheat, and for producers of barley in rain-fed zones. Irrigated wheat producers are better off because the price of wheat went up in 1989 more than the increases in cost of production.[6] For rain-fed zones the impact of structural adjustment on farm income is closely linked to variation in yield that is characteristic of these zones. This variability is amplified by the effect of size of farm and by the portion of the farm that goes into wheat or into barley. Barley producers have seen a drop in their income by 10 per cent in real terms, despite increases in yield. This drop is explained by the fall in the price of barley that has not been supported as it was before the programme. In conclusion, one can say that it is probable that farm income has been affected more by measures outside the structural programme (devaluation, food subsidy policy and so on) than by the application of structural measures. The major beneficiaries have been the producers of wheat.

Changes were also made in subsidies and regulations on production and distribution of food products. The price of 'deluxe' flour was de-controlled. Quotas were established by mill for the production of the lower-quality 'national', flour, which is still subsidized. The extraction rate for 'national' flour was increased from 78 to 80 per cent.

In 1987, the government of Morocco, in cooperation with USAID and the World Bank, launched a Compensatory Food Programme designed to help lower-income groups who might be adversely affected by the price increases and other measures in the structural adjustment programme. About $80 million has been allocated to the food component of this programme.[7]

Public investment programmes

In the early years of the adjustment programme, public investment fell in agriculture both in absolute terms and as a percentage of total investment. Agricultural investment fell from Dh 1 387 million in 1981 to Dh 582 million in 1986. The percentage of total investment fell from 13.5 in 1981 to 7.0 in

1986. However, investments in agriculture increased substantially in later years bringing the nominal level in 1988 nearly back up to the 1981 level. There have been changes in the mix of investments within agriculture as well. Research and extension rose from 4 per cent of the total in 1985 to 8 per cent in 1987, while large-scale irrigation dropped from 50 per cent to 42 per cent over the same period.

Government support services

Major changes have been undertaken in irrigation policy. A key objective is financial autonomy for the irrigation offices. Recovery of water charges has increased from 47 per cent in 1984 to 67 per cent. Three irrigation offices out of nine have moved from deficit into surplus. Marketing of inputs is being transferred to the private sector.

Veterinary services have been privatized in certain areas. The public sector still pays costs for any large-scale epidemics and for artificial insemination services. In agricultural research, legislation was adopted to strengthen the autonomy of the National Agricultural Research Institute. Experiment stations that were mainly engaged in seed production were closed, so that activity could be transferred to the private sector.

For extension, the most important realization concerns the elaboration of the extension plan of which the principal objective is the re-enforcement of training of farmers by the establishment of interdisciplinary teams that operate on the ground and that will be in contact with other more specialized teams at the level of the province. The specialized teams will be in permanent contact with the national research agency at the central level. This structure is going to permit contact between the research institute and farmers and will facilitate the results of agro-economic research and equally the feedback from farmers to researchers concerning problems encountered on the ground that could not be solved by the multi-disciplinary teams.

Agricultural policy analysis capability

Substantial progress was made in increasing the Ministry of Agriculture's policy analysis and monitoring capability. The Ministry economic policy staff have conducted several studies that have been used in debating and establishing agricultural policy. The staff are all computer-literate and most have received training in micro-economics, quantitative methods and policy analysis. Several have received advanced degree training. Some of the staff participated in a joint policy study with the World Bank on market interactions in the agricultural sector.

CONCLUSIONS AND RECOMMENDATIONS

The assessment drawn from the execution of the structural adjustment programme to date suggests the following principal conclusions:

(1) Many of the changes have been quite positive and have laid the groundwork for a more vigorous agricultural sector in the future.

(2) The timing and sequencing of implementation could have been improved considerably. In the early years, the agricultural sector bore many of the costs of adjustment (subsidy removal, reduced protection and so on) without receiving the benefits. This situation has brought about an increase in cost of production that is not compensated by the level of prices of products.

The lessons drawn from this experience permit us to formulate the following recommendations:

(1) In the identification of a programme of adjustment, it is vital to associate all of the concerned parties with the objectives and process to help create shared ownership and responsibility and to ensure complementarity and good integration of all of the components of the programme.

(2) Before putting the programme of adjustment into operation, it is important to begin with an initial evaluation of the impact of the measures to be undertaken, not only at the aggregate level but also at the level of the producer and the consumer. Successful evaluation requires the availability of strong tools of analysis adapted to the social economic context of the country. Also it is necessary to develop these tools to proceed with the evaluation, and undertake simulations throughout the application of adjustment measures to take into account new situations that can appear.

(3) Structural adjustment does not necessarily need to be synonymous only with solving problems of budget resources. It seems, in effect, that the tendency in the application of adjustment programmes is to favour especially the execution of measures that have positive incidence on the public budget. It is important to consider all of the public objectives, to create a plan of structural adjustment measures and to follow the totality of that plan in order to achieve the combination of objectives. Partial execution can result in undesired outcomes.

NOTES

[1] Parts of this section and other sections in this paper are taken from, or draw heavily upon, a previous paper by Mokhtar Naanani (Bouanani, 1989).

[2] See the contribution on Morocco by B. Hamclouch in S.El-Naggar (1987).

[3] A. de Janvry *et al.* (1990) conclude that the current agricultural sector adjustment programme should thus be seen as a highly attractive context to mount a serious effort at rural development to assist small farmers to take maximum advantage of the opportunities offered by reducing the historical bias against dryland agriculture.

[4] Improvement of natural resource management included soil conservation, reforestation,

reducing desertification, and other programmes. Because of space limitations, it will not be discussed in this paper.

[5]There was a safety clause whereby the domestic support price would not fall below the 1986 real support price in the event of a fall in world market prices.

[6]The prices and incentives work has shown that Morocco has a comparative advantage in production of dryland wheat, but not irrigated wheat, in general.

[7]An evaluation of this programme is currently under way (personal communication with Assistant Professor Joanne Csete, Department of Nutritional Sciences, University of Wisconsin).

REFERENCES

de Janvry, A., Fafchamps, M., Raki, M. and Sadoulet, E., 1990, 'Structural Adjustment and the Peasantry in Morocco', Pre-publication draft paper, University of California at Berkeley, Stanford University, and Institut Agronomique et Vétérinaire Hassan II (Rabat).
Economist Intelligence Unit, 1990, *Morocco Country Profile 1990–91*, London.
El-Naggar, Said (ed.), 1987 *Adjustment Policies and Development Strategies in the Arab World*, IMF, Washington, DC.
Ministère de l'Agriculture et de la Réforme Agraire, Royaume du Maroc, 1986, *Politique de Prix et d'Incitations dans le Secteur Agricole*, in association with Associates for International Resources and Development, Inc., Somerville, MA.
Naanani (Bouanani), Mokhtar, 1989, *Principaux Problèmes Relatifs A l'Analyse des Politiques Agricoles au Maroc*, Rabat.
Nelson, Joan M. (ed.), 1990, *Economic Crisis and Policy Choice: The Politics of Adjustment in the Third World*, Princeton University Press, Princeton, New Jersey.
Seddon , David, 1988, *Structural Adjustment and Agriculture: Morocco in the 1980s*, Discussion Paper No.206, School of Development Studies, University of East Anglia, Norwich, Great Britain.
Tweeten, Luther, 1989, 'The Economic Degradation Process', *American Journal of Agricultural Economics*, vol. 71, no.5.
World Bank, 1987, *Kingdom of Morocco: Second Agricultural Sector Adjustment Loan*, Washington, DC.
World Bank, 1988, *World Development Report 1988*, Washington, DC.

DISCUSSION OPENING – RASHID M. HASSAN*

I have read the paper by Dr Bouanani and Professor Tyner in conjunction with that by Alberto Valdés, on which I will also open the discussion (see below). They are, to a large extent, complementary, though it will be realized that Valdés provides a somewhat more general review of policy reform, whereas Bouanani and Tyner have been asked to concentrate on the experience of a single country. They have certainly produced a particularly interesting analysis of the economic conditions and the policy environment in Morocco before and after reforms were introduced. Time may not be opportune for a comprehensive assessment of the impact of policy reforms owing to their being so recent as well as there having been only partial application of adjustment measures. Country experiences, however, provide the only laboratory for economic policy simulation experiments.

For that reason I value the contribution of Bouanani and Tyner since they provide a valuable review of the symptoms which indicate that policy reform

*International Maize and Wheat Improvement Centre, Kenya.

is necessary and clearly describe the package of rather typical measures which have been applied, with a focus on agriculture. Part of my interest in their work, which will be reflected in my later comment on Dr Valdés's paper, relates to the matter of timing and sequencing of policy implementation. This appears to be one case in which agriculture was faced with re-adjustment at a very early stage, and in which it bore burdens without receiving speedy benefit. The other feature of the paper is the emphasis which it gives to the importance of policy analysis, which I will also refer to later.

My criticism of the paper is that it does tend, understandably in a very short presentation of a case study, to be somewhat weak in analysis of the overall background. A small table summarizing data on key indicators of protection and economic incentives, such as the real exchange rate for the important sectors and commodities in Morocco before and after the policy change, would have made this paper very illuminating. Such information is necessary for evaluating the relative importance of the various components of the re-form programme, especially if combined with data on resulting changes in the volume and composition of commodity trade and factor allocation. Useful lessons about the relative importance and timing of the different measures of economic adjustment can be learned from specific case studies. This does require a very full analysis, since the linkages involved can be complex. It is still important, however, to understand the descriptive background and it is in this that the authors have made a useful contribution.

ALBERTO VALDÉS*

Agricultural Trade and Pricing Policies in
Developing Countries: Implications for Policy Reform

INTRODUCTION

In many developing countries there is a new attitude towards economic policies and the role of the public sector. The changing economic and political realities in so many developing countries call for a new development strategy. This seems to be an unprecedented opportunity for their agriculture. A large number of countries are embarking on a revision of trade and pricing policies, moving towards a more open economy and recognizing the importance of maintaining a realistic exchange rate for achieving broad-based, sustainable, economic growth.

The first section of this paper sets up a basic concept of an efficient structure of incentives from an economy-wide perspective against which actual policies can be evaluated. The second and third sections present evidence on the effect of price interventions in the past, to support the arguments for policy reform. For most of the past three decades, government policies regarding agriculture have adversely affected prices, production and farm income. The second section presents a synthesis of findings on the patterns of agricultural protection and taxation in 18 LDCs, based on studies by Krueger, Schiff and Valdés (1988) and Schiff and Valdés (1991). The third section examines how this pattern of incentives affected agricultural growth and farm income. The last section on policy reform identifies critical elements for a new agricultural trade strategy.

One of the complex questions for policy makers is how broad an effective reform process must be and whether the reform measures should be introduced in a specific sequence. The analysis indicates that there could be a strong interaction between the macro-economic process and the response to reforms by the agricultural sector. Furthermore, delineating the boundaries for the micro-economic aspects of liberalization raises complex issues which remain to be fully analysed. What should be the role of state agencies in agricultural marketing during and after trade reform? How to deal with a variety of markets currently subjected to extensive regulations, such as financial, labour and land markets, is a critical issue which deserves more analysis. Bottlenecks in related transport and communication sectors could also inhibit

*The World Bank, Washington, USA. The author wishes to thank Maurice Schiff for his valuable comments.

the output response and the credibility and sustainability of agricultural trade and price reforms.

ON THE CONCEPT AND
MEASURE OF AGRICULTURAL INCENTIVES

In respect of agricultural incentives, a feature of the approach of Krueger, Schiff and Valdés (1988) was the distinction between direct (or sectoral) and indirect (or economy-wide) price interventions. Agricultural incentives were defined as the domestic price of agricultural goods relative to the price of non-agricultural goods ($P_a/_{na}$). Price interventions were then measured as the percentage departure from the relative price of agricultural goods that would have prevailed in the absence of sectoral price intervention, as well as in the absence of trade intervention in the non-farm sector, and corrected for any exchange rate misalignment. For a given farm product, a negative price intervention occurred whenever the price of that good relative to the non-agricultural sector appeared below its counterpart price under a non-intervention scenario. Broadly, the non-intervention price is efficient under the assumption that: (1) the country is a price taker in the product in question, and (2) there are no externalities or economies of scale in production.

Direct (sectoral) price interventions were measured by the direct nominal and effective rate of protection at the official exchange rate, after adjusting for quality differences and transport and storage costs, in order to estimate the domestic price which would have prevailed under free trade. Policies underlying direct price interventions include tariffs and quotas, prior import licences, direct price controls, taxes and subsidies on products and inputs often operating through the activities of parastatals involved in the marketing of these products.

Indirect price interventions were defined as those arising from policies operating in the rest of the economy, such as trade restrictions in the non-farm sector and macro-economic policies resulting in exchange rate misalignment. Let P_a denote the price of agricultural tradables, P_{nat} the price of tradables outside agriculture, and P_{nah} the price of home goods outside agriculture. There are two economy-wide effects. As a result of higher industrial protection, P_a/P_{nat} and P_a/P_{nah} fall (the latter falls because of the appreciation of the real exchange rate caused by industrial protection, but it falls less than the former) and expansionary macro-economic policies lead to a further appreciation of the real exchange rate and to a further fall in P_a/P_{nah}. Defining $P_a/P_{na} = P_a/[\alpha P_{nat} + (1 - \alpha)P_{nah}]$ as agriculture's terms of trade (where α is the share of tradables in the non-agricultural sector) then the indirect effect of economy-wide policies was measured as the weighted average of the effects on P_a/P_{nah} and on P_a/P_{nat}.

The sum of the direct (*NPRd*) and indirect (*NPRi*) nominal protection rates was defined as the total nominal protection rate (*NPRt*) which measures the joint impact of sectoral and economy-wide policies. A synthesis of the findings on direct, indirect and total nominal protection rates for 18 developing countries during 1960–84 is presented in Table 1.

TABLE 1   Direct, indirect and total nominal protection rates (average, per cent)

Region	Country	Period	Indirect	Importable	Direct Exportable	Direct all	Total all
Sub-Sahara Africa	Côte d'Ivoire	60–82	–23.3	26.2	–28.7	–25.7	–49.0
	Ghana	58–76	–32.6	42.9	–29.8	–26.9	–59.5
	Zambia	66–84	–29.9	–16.4	–3.1	–16.3	–46.2
			–28.6	17.6	–20.5	–23.0	–51.6
North Africa	Egypt	64–84	–19.6	–5.1	–32.8	–24.8	–44.4
	Morocco	63–84	–17.4	–8.2	–18.5	–15.0	–32.4
			–18.5	–6.7	–25.7	–19.9	–38.4
AFRICA			–24.6	7.9	–22.6	–21.8	–46.3
East Asia	Korea	60–84	–25.8	39.0	n.a.	39.0	13.2
South Asia	Pakistan	60–86	–33.1	–6.9	–5.6	–6.4	–39.5
	Sri Lanka	60–85	–31.1	39.0	–18.4	–9.0	–40.1
			–32.1	16.1	–12.0	–7.7	–39.8
South-East Asia	Malaysia	60–83	–8.2	23.6	–12.7	–9.4	–17.6
	Philippines	60–86	–23.3	17.4	–11.2	–4.1	–27.4
	Thailand	62–84	–15.0	n.a.	–25.1	–25.1	–40.1
			–15.5	20.5	–16.3	–12.9	–28.4
ASIA			–22.8	22.4	–14.6	–2.5	–25.3
Latin America	Argentina	60–84	–21.3	n.a.	–17.8	–17.8	–39.1
	Brazil	69–83	–18.4	20.2	5.4	10.1	–8.3
	Chile	60–83	–20.4	–1.2	13.5	–1.2	–21.6
	Colombia	60–83	–25.2	14.5	–8.5	–4.8	–30.0
	Dominican Republic	66–85	–21.3	19.0	–24.8	–18.6	–39.9
LATIN AMERICA			–21.3	13.1	–6.4	–6.5	–27.8
TOTAL AVERAGE			–22.5	14.4	–12.6	–7.9	–30.3

294

PATTERNS OF AGRICULTURAL PROTECTION

A most striking result in Table 1 is the high level of price intervention in agriculture. On average for all products and for all years, the total (*NPRt*) averaged approximately –30 per cent. In other words, in the absence of intervention, the relative price of agricultural goods would have been 42 per cent higher (30/70). Important differences among countries emerged. Côte d'Ivoire, Ghana and Zambia exhibit the highest degree of negative price intervention (*NPRt* equal to –51.6 per cent on average). The degree of price intervention for Malaysia and Brazil was substantially lower (*NPRt* between –17 per cent and –8 per cent) although still negative, and in Korea agriculture was protected (*NPRt* of 13 per cent).

A second important result concerns the source of price effects. On average, indirect price intervention accounted for approximately three-quarters of the total disadvantage affecting agriculture. This high negative indirect effect arose mainly from the high prevailing levels of industrial protection and, to a lesser although still influential extent, from exchange rate misalignment resulting from both macro-economic imbalances and industrial protection.

An important finding concerning direct price intervention is the systematic difference observed in the treatment of importables *vis-à-vis* exportables, resulting in a strong anti-trade bias. While direct price interventions to agricultural import-competing activities were in most cases positive (*NPRd* between 7.9 per cent and 22.4 per cent for the country groups in Table 1), direct protection to agricultural exportables was in most cases negative (between –6.4 per cent and –32.8 per cent). On average for all countries, direct intervention resulted in a protection rate of about 15 per cent for importables and in a tax rate close to 13 per cent for exportables.

This pattern of direct protection is attributed to the desire to achieve a certain minimum level of self-sufficiency in food production, in the case of importables, and collecting government revenues, in the case of exportables. For example, it was estimated that for the sample countries the latter contributed to approximately 20 per cent of total public expenditure during 1960–9, 11 per cent during 1970–9, with a lower figure of 5.8 per cent in 1980–3 (Schiff and Valdés, 1991). Supporters of agricultural protection in Japan, Sweden and the EC have made persuasive use of the food security objective to make their case (Honma and Hayami, 1986). This is often justified by the gloomy picture of world demand and supply for cereals and the risk of food shortages. Our findings for this sample of LDCs suggest that a relatively high weight was also given to the food security objective, regardless or whether or not the relevant world food market in question was very thin (as for white maize) or fairly well developed with central transaction points (as for wheat).

These findings suggest that there was substantial resource misallocation between importables and exportables. The optimal export tax argument cannot be used as a defence of taxation of exportables, except in a few cases. Preliminary findings of a recent study by Panagariya and Schiff (1990) indicate that for 1986 the optimal export tax in Côte d'Ivoire was 25 per cent, and in Ghana about 20 per cent; hence the level of their export taxes was not too far from the optimum tax (however, rice was highly protected, so P_m/P_x was still distorted). For Egyptian cotton the actual direct export tax (32 per cent) was below the optimum (53 per cent). Coffee in Brazil, Colombian rice and Thai rubber are other relevant cases. Similarly, Zambia has market power in white maize. Hence the marginal import cost is higher than the border price and an import tariff should have been imposed; however, our findings indicate an average direct *tax* on importables of 16 per cent. Thus, for non-price takers, allowing for optimal trade intervention does reduce the degree of the actual distortion (from direct price intervention) to exportables in a few cases. However, the case of non-price takers is not common, and when it does apply the actual tax levels often fail to coincide with the optimum levels. Relative prices within agriculture are distorted even when adjusted for the optimal trade tax.

POTENTIAL EFFECTS OF A TRADE AND PRICE POLICY REFORM: IMPLICATIONS FOR AGRICULTURAL GROWTH AND FARMERS' INCOME

Price intervention can affect agricultural growth, consumption and trade flows. In addition it may have other, broader, economic implications through its influence on the government budget, and on the real income of urban and rural households. As with most policy intervention, there are winners and losers, an issue on which more empirical research is needed if we are to understand the motivation underlying price intervention and the political and economic constraints impeding policy reform.

In this section we focus on two features: the consequences for agricultural growth and farm income. The background material providing the estimation of these effects is found in Schiff and Valdés (1991). An understanding of the effect of incentives on agricultural growth requires an economy-wide view of returns. This is because agricultural growth is influenced by intersectoral resource flows. The partial equilibrium methodology so common in the literature is inadequate for the task. Aggregate output responsiveness will depend on how depressed prices are, the credibility of reforms, the time-frame involved, and responsiveness to a given price adjustment.

We find that the difference in the rates of protection across sectors in LDCs is dramatic. While agriculture in the sample of African countries had an average direct protection rate of *minus* 20 per cent, the importable non-agricultural sector enjoyed a protection rate of 36 per cent.

Incentives and agricultural growth

An assumption of much of the development strategy followed in the past was that aggregate agricultural supply was relatively unresponsive to incentives. If that were the case, taxing agriculture would have redistributive effects (from agriculture to the rest) but no major impact on overall output. It is true that most empirical studies of aggregate response in LDCs show a weak effect of price changes; though it is also true that the empirical foundation of most of this analysis is still quite fragile. Most of the literature is dominated by studies using a single-equation time-series approach which fails to capture the under-lying migration and investment processes in a dynamic framework (see Binswanger, 1989, and discussion by Valdés in the same issue). More recent econometric work, by Mundlak, Cavallo and Domenech (1991) on Argentina, and Coeymans and Mundlak (1991) on Chile, explicitly include intersectoral resource re-allocation over time through migration and investment responding to prices. These studies obtain a larger supply response. Unfortunately, there are still unsolved questions such as the extent to which parameters change in response to policy changes (the Lucas critique), how to control for exogenous changes in infrastructure provision, and the consequences of the prevailing uncertainty about the future course of the domestic terms of trade.

Based on the sample of 18 LDCs during 1960–84 (373 observations), two tests were performed to examine the relationship between price intervention and agricultural GDP, one parametric and the other non-parametric. In the non-parametric test, the average growth rate of agricultural GDP was com-pared with the average protection across two groups, those with high and low levels of protection (Schiff and Valdés, 1991). In the case of total nominal protection, the difference in the average annual growth rates of the low and high protection cases is large (2.5 percentage points or a 93 per cent difference) and significant. The evidence of such an association for direct price interven-tion is weaker. Consequently, this test suggests a strong association between high total taxation and low growth rates of agricultural GDP.

For the parametric test, a model of growth of agricultural GDP consisting of a long-term growth equation supplemented by an error-correction equation was estimated. It was found that removal of total taxation would have in-creased the annual rate of agricultural growth from 2.5 per cent to 3.1 per cent, or about 22 per cent. In fact, this provides a lower bound of the actual effect because agricultural growth was found to depend on overall growth, which we assumed to be exogenous. Hence removal of total intervention could have had an additional positive impact on agricultural growth through the positive effect on overall growth, considering that total interventions were also found to be negatively associated with overall growth.

Effect on farm income

A second important consequence of interventions is that they can generate substantial resource transfers within and between sectors. In fact, one expects that the prime motivaton of the policies is to have precisely that effect. Transfers may take a variety of forms. In some export products, it may appear as higher government revenue from export taxes. If the exportables are also food products, it lowers food prices to urban consumers. Or it may take the form of input subsidies (on fertilizers and credit) or appear as parastatal monopoly of agricultural trade, capturing revenues from trade or bearing losses from selling at lower domestic prices. Transfers from indirect price intervention can result from exchange rate misalignment, and from the effect of industrial protection on the prices paid by farmers for inputs and consumer goods.

Price-related transfers were defined as the change in real income of agriculture resulting from direct and indirect price interventions affecting output and input prices and those of consumer goods purchased by farm households (Schiff and Valdés, 1991). Specifically, these transfers were measured as the change in value added, resulting from both price interventions measured at the actual level of production and adjusted for the change in the rural consumer price index. Non-price transfers were defined to include public investments that can be considered public goods, such as irrigation, roads, research and extension. Marketing-related expenditures by state agencies, such as on storage, were excluded on the grounds that (1) their effects are reflected in domestic prices paid or received by farmers and thus appear under price intervention, and (2) they do not clearly constitute a transfer to agriculture since these activities could be, and in many countries are, undertaken by private traders.

As an illustration of the magnitudes of income transfers, Table 2 presents estimated effects of total price interventions for three Sub-Saharan African countries during the period 1960–84, all expressed as a percentage of agricultural GDP. Similar computations are available for direct price transfers and for the other 14 LDCs (Schiff and Valdés, 1991). The results show that total price interventions on outputs reduced agricultural GDP by about 28 per cent, and transfers into agriculture through input subsidies raised agricultural income by approximately 8 per cent. Expanding the output coverage to the rest of agriculture (given that input subsidies apply to most of the sector) and assuming that there are no direct price interventions for the rest of agriculture (that is nominal rates of protection equal to zero) raises the average total net transfer out of agriculture to a staggering 103.3 per cent. Finally, the non-price-related transfers into agriculture amounted to approximately 8 per cent, resulting in a net overall transfer out of agriculture of approximately 96.3 per cent of agricultural GDP.

While input subsidies and public investment do to some extent compensate for the negative transfer through output prices, albeit in an inefficient form (Table 2), this compensation is equivalent to only a fraction of the income loss, particularly when the indirect effects are taken into account. The magnitude of the net transfer out of agriculture is so large that its cumulative effect must have had a profoundly harmful effect on farm investment and income in the long term.

SOME GUIDELINES FOR A
NEW AGRICULTURAL TRADE STRATEGY

Recognition is growing that governments are burdened with economic functions which they are incapable of performing efficiently. Simultaneously, many government roles which cannot be performed by the private sector, such as primary education in rural areas, management of land titles, construction of roads, and agricultural research, are neglected. Broadly speaking, there

TABLE 2 *Net income transfers to (+) and from (−) agriculture as a result of total (direct and indirect) price and non-price interventions, 1960–84 (percentage of agricultural GDP)*

Country	Period	Total price transfers				Sum of total price transfers		Non-price transfers	Sum of total price and non-price transfers		
		Output of selected products	All inputs	Output of other agricultural products		Assum. 1	Assum. 2		Assum. 1	Assum. 2	Avg. of assum. 1 & 2 (half of (8) + (9))
				Assum. 1	Assum. 2	(1) + (2) + (3)	(1) + (2) + (4)		(5) + (7)	(6) + (7)	
		(1)	(2)	(3)	(4)	(5)	(6)	(7)	(8)	(9)	(10)
Côte d'Ivoire	1960–69	−13	1	−10	−55	−22	−67	6	−16	−61	−38.5
	1970–79	−32	3	−42	−126	−71	−155	18	−53	−137	−95
	1980–82	−15	2	−27	−78	−40	−91	20	−20	−71	−45.5
	1960–82	−21	2	−26	−89	−45	−108	13	−32	−95	−63.5
Ghana	1962–69	−28	1	−65	−154	−92	−181	3	−89	−178	−133.5
	1970–76	−25	4	−60	−218	−81	−239	3	−78	−236	−157
	1976–84	n.a.	n.a.	n.a.	n.a.	n.a.		n.a.	n.a.	n.a.	n.a.
	1962–76	−26	2	−63	−184	−87	−208	3	−84	−205	−144.5
Zambia	1960–70	n.a.	n.a.	n.a.	n.a.	n.a.	n.a.	n.a.	n.a.	n.a.	n.a.
	1971–79	−19	9	−80	−144	−90	−154	5	−85	−149	−117
	1980–84	−71	36	−304	−411	−339	−446	4	−335	−442	−388.5
	1971–84	−37	19	−160	−239	−178	−257	5	−173	−252	−212.5
Group average[a]	1960s	−20.5	1.0	−37.5	−104.5	−57.0	−124.0	4.5	−52.5	−119.5	−86.0
	1970s	−25.3	5.3	−60.9	−162.7	−80.7	−182.7	8.7	−72.0	−174.0	−123.0
	1980s	−43.0	19.0	−165.5	−244.5	−189.5	−268.5	12.0	−177.5	−256.5	−217.0
	1960–84	−28.0	7.7	−83.0	−170.7	−103.3	−191.0	7.0	−96.3	−184.0	−140.2

Notes: n.a. indicates that data were not available.

[a] Simple, unweighted group average.

(1) The change in the gross value of output of selected agricultural products as a result of direct price interventions (relative to the counterfactual simulation without intervention).

(2) Transfers resulting from the price interventions on inputs, including credit subsidies (and replanting subsidies for rubber), on all agricultural products.

(3), (4) The additional effect of price-related transfers on the gross value of output for the rest of agriculture. Under assumption 1, the rate of nominal protection for the rest of agriculture is assumed to be zero (not shown); under assumption 2, it is assumed to equal nominal protection (or taxation) for the selected products.

(7) Nonprice transfers include public investment in irrigation, agricultural research and extension, and land improvements.

are four economic policy reform issues which affect agriculture in LDCs: (1) policy reforms to improve the economic environment for agriculture; (2) strengthening the public sector to support technology development and transfer, education in rural areas and infrastructure projects supportive of agriculture; (3) encouraging opportunities for increased economic participation of the historically disadvantaged (that is, small farmers and landless workers); and (4) natural resources management.

In this paper we address the first of these four categories. The section is structured around three issues: trade and macroeconomic factors; guidelines for successful agricultural price and trade reform; and the need for simultaneous reforms in selected sectors which impinge on the success of the agricultural trade policy reform process.

Trade and macroeconomic factors

These are perhaps the most important influences on the success or failure of agricultural price reform. A reduction in industrial protection alone would produce a major improvement in agricultural incentives, as indicated in an earlier section. For the sample of 18 countries analysed, reduction in industrial protection to, say, a uniform tariff of 15 per cent would induce an increase in relative prices for agricultural tradables by approximately 22 per cent relative to industry, and by about 15 per cent relative to the entire non-agricultural sector.

Moreover, there has been a strong interaction in the past between the macro-economic circumstances and the prevalence of government controls of individual agricultural markets. For example, the majority of price controls has been imposed in an effort to reduce inflation, and price controls and quantitative restrictions have been intensified because of inflationary pressures and/or balance of trade difficulties. Thus the persistence of macro-economic disequilibrium will create strong pressures against the removal of price controls on farm products, particularly on food.

If there is one clear lesson from the experience of the bold trade liberalization programmes in the Southern Cone countries in South America during the late 1970s (Corbo, Goldstein and Khan, 1987), and in New Zealand after 1984 (Sandrey and Reynolds, 1990) it is the considerable risk for agriculture which can arise from the macro-economic management of the economy. At the time, the financial strategy of the governments in these countries resulted in a very high real interest rate which attracted a considerable inflow of funds from abroad. As a result, high interest rates adversely affected agriculture, both because of the high cost of capital and because of the impact of higher capital inflows on the real exchange rate, adversely affecting agricultural investment and its international competitiveness, delaying the agricultural output response to the trade reform.

Thus reducing the indirect effects of fiscal deficits and real exchange rate appreciation, and avoiding sharp fluctuations in real interest rates and exchange rates, are necessary and fundamental elements of a policy reform package as it affects agriculture.

Guidelines for successful agricultural price and trade reform

Four significant results have come out of recent studies on the agricultural trade regime in LDCs. First, there is a marked contrast between the direct policies adopted towards traditional export crops and those directed to import-competing food products; governments heavily tax the production of exportables and protect the production of food. Second, quantitative restrictions (QRs) on agricultural trade (such as quotas, licences and state trading) are widespread in most LDCs. Third, a characteristic of the trade regime in farm products has been its discretionary and selective nature, its lack of transparency and the implicit discrimination against sub-sectors of agriculture. Fourth, in some countries, revenues from trade taxes represent a significant share of government revenues, and thus a removal of trade distortions without increasing revenues from other sources might not be possible in those countries.

It is submitted here that dismantling of QRs, even if some degree of protection is maintained, is a condition of the liberalization package. Replacing QRs with tariffs has several advantages, most importantly that the role of the price mechanism is enhanced. QRs are more selective and less visible than tariffs – they mask the level of protection and insulate the domestic markets from world market changes. In addition, dismantling of QRs could greatly reduce the role of state agencies in trade. An additional advantage of replacing QRs with tariffs is that the latter generate government revenue, removing one of the obstacles to trade reform in some countries. Bold steps to eliminate QRs on output and inputs, and to dismantle the administrative machinery, were an important element in the successful trade liberalization programme in Chile, and are an explicit component of the ongoing trade reform programmes in Mexico, Bolivia, Colombia, Venezuela, Peru and several other countries.

Furthermore, an important goal of trade reform is to achieve more neutrality in the trade regime; that is, narrowing the range of nominal and effective rates of protection. As documented in the earlier section, agricultural price interventions were found to have a strong anti-trade bias, with a wide dispersion in tariff equivalences within importables as well as in the export tax equivalent on exportables. There are strong arguments against selectivity in the pattern of trade restrictions. One is a strictly economic argument. Many farm products are intermediate inputs to processing industry and, depending on their share in the cost structure, even small differences in nominal tariffs across the economy can result in wide variations in effective rates of protection to processors, unrelated to the initial goal of the reform.

Another argument derives from political economy considerations. The experience with trade interventions suggests that, through time, a selective approach to interventions tends to be captured mostly by the more powerful pressure groups, deviating considerably from the initial motives for interventions. This has been the case, for example, with credit and input subsidies which tend to be captured mostly by larger farmers, and with protection to the products of particular regions or by certain classes of farmers. Thus liberalizing agricultural trade means not only lowering the average levels of protection and removing export taxation and restrictions, but also narrowing the range of nominal and effective rates of protection.

Even though economic analysis identifies several economic motives for trade and price intervention in agricultural markets, the case for deviations from the uniform tariff rule are very few (Valdés and Siamwalla, 1988). These include the optimum tariff case and the fiscal revenue motive (both mentioned earlier in the text), interventions to deal with world price instability, and food subsidies for the most vulnerable households. To avoid a capricious and distorting pattern of trade intervention, we submit that the goal should be in the direction of equality of nominal rates of protection on inputs and final products throughout the economy, including agriculture. Special cases should be only those – very few exceptions – where the burden of proof is to demonstrate the merits of the special case.

Need for simultaneous reforms in related sectors

A number of markets are subject to controls of varying degrees of severity, including financial and labour markets, transport and communications, the importance of which will vary from country to country. As in the case today of Eastern European agriculture, delineating the exact boundaries for a successful micro-economic reform package is obviously a very complex issue of which we have a very limited experience.

It is widely recognized that the cost of adjustment (in terms of unemployment and financial pressure for farmers) precedes the benefits of liberalization and trade reform. There is, however, a real risk that the benefits in terms of agricultural output, employment and farm income could take many years, reducing political support for reform. Owing to the biological nature of agriculture, some adjustment lags are inevitable. However, the challenge is to identify the possible bottlenecks that could slow down the output response. In most countries, these related markets are still subject to extensive regulation, and this could inhibit the agricultural output response to trade and price liberalization, reducing the credibility of the reforms.

We highlight the following: (1) security of property rights and deregulation in land markets with respect to rentals, very important at least in Latin America; (2) developing medium- and long-term credit lines with competitive interest rates and methods for dealing with accumulated farm debt; (3) development of competitive services on transport and communications, particularly important for the growth of nontraditional exports; and (4) public sector reforms to improve productivity in the use of public sector resources and privatization of state agencies whose continued holding by the public sector is not justified on policy grounds. These are believed to be particularly significant in respect of the sequence of reforms.

Our experience in the analysis of agricultural policy reforms in LDCs is still very limited. In the future we should be able to offer more precise guidelines about the order in which reforms must be undertaken. For example, should agricultural trade and price policy reforms follow stabilization, and not be attempted before macro-economic equilibrium and stability are firmly established? Similarly, should changes in agricultural trade and price policy reform, which could occur rapidly, be delayed because of others which would take longer (like improving physical infrastructure, providing security of property rights and developing an efficient service sector)? It will take several years to have these elements in place. It is important to move ahead to initiate the necessary reforms. My intuition tells me that the only case for delaying agricultural trade and price policy reform is in countries suffering unsustainable macro-economic policies accompanied by high and variable rates of inflation, and by variability in the key macro-economic variables, namely the real exchange rate and real interest rate.

REFERENCES

Binswanger, Hans, 1989, 'The Policy Response of Agriculture', *Proceedings of The World Bank Annual Conference on Development Economics*, Washington, DC.

Coeymans, Juan Edo and Mundlak, Yair, 1991, 'Aggregate Agricultural Supply Response in Chile: 1962–82', in *Food Policy*, vol. 16, no. 1, February.

Corbo, Vittorio, Goldstein, M. and Khan, M., 1987, *Growth-Oriented Adjustment Programs*, IMF World Bank, Washington, DC.

Honma, Magayoshi and Hayami, Yugiro, 1986, ch. 4, in K. Anderson and Y. Hayami (eds), *The Political Economy of Agricultural Protection*, Allen & Unwin.

Krueger, Anne O., Schiff, M. and Valdés, A., 1988, 'Agricultural Incentives in Developing Countries: Measuring the Effect of Sectoral and Economywide Policies', *World Bank Economic Review*, September, vol. 2, no. 7.

Mundlak, Yair, Cavallo, D. and Domenech R., 1991, 'Agriculture and Growth in Argentina', *Food Policy*, vol. 16, no. 1, February.

Panagariya, Arvind and Schiff, Maurice, 1990, 'Commodity Exports and Real Income in Africa', World Bank PRE Working Paper WPS 537, November.

Sandrey, Ron and Russell Reynolds, (eds), 1990, *Farming Without Subsidies: New Zealand's Recent Experience*, New Zealand Ministry of Agriculture and Fisheries.

Schiff, Maurice and Alberto Valdés, 1991, *A Synthesis of the Economics of Price Interventions in Developing Countries*, Johns Hopkins University Press, Baltimore, Md.

Valdés, Alberto, 1989, 'Comment' on H. Binswanger, *Proceedings of The World Bank Annual Conference on Development Economics*, Washington, DC.

Valdés, Alberto and Siamwalla, Ammar, 1988, 'Foreign Trade Regime, Exchange Rate Policy, and the Structure of Incentives', ch. 7, in John W. Mellor and Raisuddin Ahmed (eds), *Agricultural Price Policies for Developing Countries,* Johns Hopkins University Press, Baltimore, Md.

DISCUSSION OPENING – RASHID M. HASSAN*

Economists, policy makers and even politicians all subscribe today to the urgent need for rigorous economic reforms in developing countries to remedy the serious imbalances in their domestic and foreign sectors. There is no consensus, however, on the components and order of reform programmes. The paper by Valdés provides a realistic framework for analysing the influences of trade and pricing policies and macro-economic adjustments on output, income and growth in agriculture. His paper summarizes the strong evidence for the negative impacts of government intervention in the agricultural and non-agricultural tradables and home-goods markets on the performance of the

*International Maize and Wheat Improvement Centre, Kenya.

farm sector. Measures of protection in 18 country case studies have shown the bias built into the structure of incentives against agricultural tradables and in favour of the urban-based manufacturing and home-goods sectors. The case for a new agricultural development strategy is well established, and important lessons learned from the past are synthesized in a set of useful guidelines for broad-based sectoral policy and macro-economic reform. I want to emphasize some important issues raised in the paper and add comment on what I see as grey areas.

A comprehensive list of measures, ranging from lower indirect taxation to exchange rate alignment and financial liberalization, is proposed by the World Bank and the IMF for economic recovery in developing countries. As pointed out earlier, the current challenge to economic research and policy making concerns choice of the appropriate combination of prescribed adjustment measures, and the sequence in which to apply them. I find myself in full agreement with Valdés's guidelines suggesting movement towards the use of neutral tax and exchange rate regimes as the basis for promoting efficient intersectoral allocation of resources. Accordingly, policy reform programmes should begin by removing relative price distortions caused by the use of differential tariff, tax and exchange rates. Replacing quantitative restrictions with tariffs on trade is also an important step towards a more effective role for the price mechanism in reacting to market signals.

While unification is crucial and relatively easy to adopt, determination of optimal tax levels and a realistic rate of exchange depend on key structural features of the economy in question. Elasticities of demand for exports and imports, as well as domestic supply responsiveness to economic incentives, are important factors to consider. The small country assumption, for example, is critical for an effective devaluation of over-valued currencies. An elastic supply structure of agricultural exportables is also essential for exchange rate adjustments to have positive impacts on the trade balance.

Most studies, including this one, focus on measuring the effect of intervention policies on the structure of incentives within the economy. Unfortunately, little information is available about the degree of responsiveness of supply and demand for agricultural tradables to changes in relative prices. Proper estimation of elasticity parameters is therefore needed in order to verify key assumptions underlying strategy proposals. The trade-off between dynamic specifications and the Lucas critique on parameter stability, discussed by Valdés, is important.

Another issue to consider in determining optimal taxation relates to impact on macro-economic equilibrium. While structural adjustment calls for reduced indirect taxation, taxes on trade are the main source of government income in developing countries. Lower tax revenues may therefore worsen the budget deficit and induce monetary expansion in the absence of alternative financing sources, placing unfavourable inflationary pressures on the real exchange and interest rates. One way out of this apparent conflict between structural adjustment and stabilization policies is through increased reliance on direct (income) taxes and open market operations (borrowing from the private sector) to finance the budget deficit. While these are effective measures of monetary control in developed economies, they are rarely used in developing countries, where capital markets are non-existent and tax collection institutions are inefficient. Liberalization of highly controlled domestic credit markets and associated interest rates, together with substantial improvements in the institutional efficiency of tax collection, are therefore necessary for balanced economic growth.

Most studies analysing the influence of intervention policies and economic reform concentrate on commodity markets. While this takes care of demand for factor services, the implicatons of changing economic incentives for suppliers of primary factor resources have not been equally stressed. Factor immobility reduces output response to changes in relative commodity prices, and limits the efficiency of intersectoral re-allocation of resources. Like food prices, however, factor markets are difficult to liberalize. Nevertheless, inflexible land, labour and capital markets reduce the capacity of the economy to adjust to changing economic conditions. More research is needed to evaluate the relative importance of removing factor market distortions. That is important for several reasons. The order of liberalization and the distributional impacts of reforms are two important issues, about which our knowledge is rather deficient, as Valdés points out.

Much more also remains to be done in assessing the distribution of costs and benefits from structural adjustment and stabilization programmes. We need to provide answers to questions concerning effects on the functional and regional distribution of income and poverty, and to consider the identity of gainers or losers among smallholder farmers, landless farm workers and the urban poor.

My last problem relates to sequence of reform. From the results in Table 1 of the paper, the indirect effects of non-agricultural policies on the incentive bias against agriculture are much higher than the influences of direct trade and price policies. This indicates the key importance of macro-economic reforms. However, whether macro-economic adjustment and liberalization of factor markets should precede, follow, or go hand-in-hand with sectoral commodity trade and price reforms remains an unresolved research question challenging economic policy analysis.

ANTONIO BRANDÃO*

Section Summary

Within the last decade structural adjustment programmes have been applied in most parts of the developing world. Successful and unsuccessful experiences exist and the analysis of the many examples has provided a deeper understanding of their role and their potential. With the deepening of the discussion, new issues emerge and the design of new projects becomes more complex. Moreover, with the transformation of the political systems in former communist countries yet more new requirements are added to the list. The new and overwhelming question concerns the route for the transformation of the centrally planned into market-oriented economies. This session provided an assessment of structural adjustment programmes focusing both on basic conceptual issues and on a wide range of country experiences.

In the discussion opening on Patrick Guillaumont's plenary paper Rebecca Lent spoke of lack of clarity in the definition of structural adjustment. For Bouanani and Tyner it is a combination of macro-economic stabilization and sectoral policy reform: 'Stabilization involves reducing balance of payments and government budget deficits and inflation to levels that can support sustained growth. Sectoral policy reform is aimed at providing incentives for the economy to function more efficiently and to earn more foreign exchange'.

While a general definition can be provided, one should not overlook the fact that the exact nature of reform is a function of specific characteristics of each country. A comparison among the various experiences discussed in this section of the programme soon reveals this. The issues in Africa are distinct from those in Latin America. Not only does Eastern Europe pose new issues, but the institutional conditions are very different from those in the other two regions. Since the ultimate goal of adjustment is to increase long-run growth rates, reforms must be able to increase investment rates. But this requires consistency and clear messages indicating that the process will not be reversed. Otherwise the private sector will not respond with higher investment. While the government can be instrumental in providing infrastructure, education and public goods in general (Alves, de Faro and Contini made this point in the first plenary paper) reforms must create conditions for an increased role of the private sector in the economy. Government investment, by itself, cannot be expected to sustain growth.

Ntangsi, analysing the recent experience of Cameroon with structural adjustment, identifies the lack of confidence of the private sector as one of the

*World Bank, Washington, USA.

306

problems for the success of the programme. In his words: 'there has for long been a lack of strong political commitment to the adjustment programme, a lack of policy dialogue between the public and private sectors, and a general lack of confidence in the latter from the former. ... The situation is compounded by the fact that the private sector is generally opposed to the thrust of the liberalization programme which breaks their monopolistic and oligopolistic power.'

A combination of sectoral and macroeconomic reforms is the usual prescription arising from World Bank and IMF-sponsored structural reforms. The correction of the macro-economic imbalances is usually the initial step. Otherwise, sectoral reforms will face severe problems. Dias's paper may be seen as an attempt to model this in the case of Brazil. Wage rigidity with high government deficits may transform the required changes in relative prices that follow the sectoral reforms into further inflationary pressures and higher deficits. The difficulties faced by the Brazilian economy are, to a large extent, caused by the inability to make the fundamental macro-economic changes required. Valdés raises the question of whether the trade and price reforms can wait until macro-economic stabilization is achieved. He indicates that we may not have enough experience with reforms in LDCs to support either position. However, he indicates 'that the only case for delaying the agricultural trade and price policy reform is in countries suffering unsustainable macro-economic policies accompanied by high and variable rates of inflation, and by variability in the key macro-economic variables, namely the real exchange rate and the real interest rate'.

The impact of policy reforms on the poor was brought into the discussion by Eugenia Muchnik. She argued that, in the case of Chile, a country which undertook a successful reform process, a large number of small farmers have not benefited and, in consequence, specific actions should be taken. This is indeed an area in which the government should have an active role, not only to guarantee that the benefits of improved growth will be widely distributed (more on this later), but also to facilitate the adjustment of the groups more vulnerable during the transition period. In particular, higher relative food prices require specific actions to alleviate the impacts on low-income groups.

One of the urgent questions associated with these policy reforms is the role of the government in the process. It seems that one can usefully distinguish what can be done in terms of coping with the problems associated with the transition and what can be done to foster growth. There was very little said with respect to the former role. On the second issue, Alves, Faro and Contini analyse a range of areas in which intervention is indeed required. They emphasize investments in human capital, especially education and health: 'If investments in rural people are not made, either agriculture will not develop or, if it succeeds in advancing, a mass of poor people will be left behind, with sad implications for income distribution, urban unrest, and political instability.' According to the authors, other areas for government action include infrastructure, land titling (an issue also stressed by Valdés), insurance for outstanding loans to reduce the risk of lending institutions and, in their words, 'The governments cannot be denied the role of counterbalancing significant actions of other governments to protect their agriculture or to counterbalance large fluctuations of the world economy'. Even though highly controversial, the last two statements seem to have been ignored in the discussions that followed the presentations.

Brooks and Braverman argue that, in the case of Eastern European countries, 'The essence of the agricultural transition is the state's withdrawal from its traditional role as a residual claimant of (positive and negative) rents to use of agricultural resources. That role will pass in stages to owners of land, where it ordinarily resides in a market economy.' The legal and economic aspects of the distribution of land in Romania, Bulgaria, Hungary, Czechoslovakia and Poland occupy a central part of this paper. Nonetheless, it is argued, 'The change in property rights, however, will not quickly produce the decollectivization that many observers expect. Under current conditions of acute stress caused by depressed domestic and foreign demand, positive real interest rates and partial liberalization of prices, few producers will leave the cooperatives. Decollectivization will gain momentum as the economic outlook for the sector improves, market infrastructure develops and cooperatives are given incentives to divest their collective production activities.' The authors also call attention to the fact that the liberalization of food prices at the retail level, which many feared, has been completed in some countries and is fairly advanced in others. So far the worst fears of many policy makers and governments have not materialized.

The debate on structural adjustment is a lively one. This was clear in this meeting from many of the discussion openers statements and in the large number of questions coming from the floor in the Plenary and in the Invited Paper sessions. Our purpose was to bring this debate once more to the IAAE meeting so that we could exchange views, increasing our knowledge and, perhaps, contributing to the elaboration of better programmes in the future.

Chairpersons: Antonio Brandão, Harold Riley, Theodor Dams. Rapporteurs: H.S.Dillon, D.Roldan, D.Metzger.
Floor discussion: W.Ampousah, T.Akiyama, D.Belshaw, J.Kydd, J.Ntsangi, M.Petit, J.van Rooyen, A.Sarris, C.Short, I.J.Singh, L.D.Smith, E.Tollens, J.Torres-Zorrilla, E.S.de Obschatko.

SECTION IV

The Potential of Biotechnology for Agriculture and the Food Industry

W. JAMES PEACOCK*

Key Elements of Modern Biotechnology of Relevance to Agriculture

INTRODUCTION

In the present day it would be unthinkable to mount a research programme into a topic of functional biology without it having a component of recombinant DNA technology. The advances of molecular biology have led to remarkable increases in our knowledge, and in our powers to acquire new knowledge, at the cellular and molecular level of understanding of biological processes.

The integration of recombinant DNA technologies into modern biology has reached a stage where they are now influencing our agriculture. I believe it is not an overstatement to say that, as from now, it would also be unthinkable to mount a programme in agricultural research without considering the opportunity for recombinant DNA technologies to be used. This is especially true in the production of improved cultivars. Genetic engineering, one aspect of recombinant DNA technology, is poised to make significant contributions in the modification of the genetic instruction that underlies the performance of our agricultural production plants.

We now are able to consider either the addition or subtraction of particular genes in the genetic make-up of a plant. We are also able to modulate the level of expression of genes, thus influencing the amount of a particular gene product being made. In the not too distant future we will be able to consider gene replacement, so that we can upgrade the existing genetic software by replacing one version of a gene with a newer, improved version for a more desirable product. These manipulations will close the gap between yield and potential yield. They will provide plants with more robust resistance to pests and disease. They will enable the production of high-performance hybrid seed in many crop species and will allow us to construct plants whose seasonal requirements are complementary to each other in agricultural production systems.

As well as providing a more flexible and valuable entry of plant products to the food production sector, genetic engineering will enable agricultural production to have a greater impact in the pharmaceutical and industrial business sectors.

*CSIRO Division of Plant Industry, Canberra, Australia.

WHAT HAS MADE BIOTECHNOLOGY POSSIBLE?

Genetic engineering is the most important way in which we will use advances in biotechnology in agriculture. Basically genetic engineering is the precise manipulation of the genetic make-up of a plant, where the manipulation involves the addition of a known gene construct. The key advance that has made this possible is the development of a series of technologies which allow us to isolate and mass-produce particular segments of DNA. We have an increasing ability now to isolate a particular gene sequence and to experiment with it. This has led to huge increases in our knowledge and understanding of gene structure and function.

Another important step has been the development of gene transfer systems. We now have methods which enable us to place new genetic material into recipient plants. Invariably this means introducing a new gene into a single cell and then recovering a whole, transgenic, plant from that single cell. Regeneration of plants from single cells, often from somatic cells, the cells from vegetative parts of plants, is an essential part of genetic engineering.

A third major factor has been the realization that a gene has two major components; a product-coding region which dictates the specific amino-acid sequence of the gene product, and a control region which determines the expression (use) of the coding segment. The control component of a gene, the promoter region, provides, to a large degree, the specificity of gene expression, controlling where the gene product will be produced in a plant, when it will be produced and in what yield it will be produced.

These three key elements of genetic engineering have given us the ability to make gene constructs, working genes with an appropriate control or promoter segment and the desired coding segment, both of which are necessary for a needed adjustment to a plant. This has opened up the doors of genetic variation to researchers involved in plant improvement. We are able to take parts of genes from many different organisms and combine them into working gene constructs. We are able to give properties to plants that are quite new; properties that previously, for example, might have been found only in bacteria or in animals.

These same advances that have allowed for the addition of novel characteristics have also provided us with the ability to modify the existing genetic material of a plant to provide a slightly changed product or to alter the expression of existing genes so as to direct a gene product to be formed in a part of a plant in which it was not previously present. The new knowledge base has also meant that we are now beginning to be able to modify the physiological responses of plants to environmental stresses and to modify the architecture of plants so they better fit the conditions of our agricultural production systems.

In general concept many of these kinds of change have been carried out by plant breeders using the technologies that have been available to them up to the present time. But breeders have been limited in what they have been able to achieve by their control of genetic variation being restricted to the sexual confines of any one species. Breeders have also been limited in synthetically producing variation by having to use random procedures to change the basic genetic code (irradiation or chemical mutagenesis), this leading to great inefficiencies in the process of selecting required variation. The plant breeder will no longer be limited to the gene products normally found in the particular plant. If the breeder can define the need or the problem there will be a good chance that genetic engineering will provide a solution, often using gene components from other species.

WHERE ARE WE NOW?

Just what is the reality at present – will we see genetically engineered plants in our horticulture and agriculture in the near future? The answer is more dependent on legislation and societal acceptance than on limitations of our science.

Gene transfer

We are not yet able to introduce genes into every plant important in our agriculture. We are still limited, for example, in the cereal plants by not being able to introduce genes into wheat or barley. Recently, methods have been developed for rice and maize, although these methods have not yet reached a stage of general availability. With the broad-leafed plants there has been much more success. Mostly this is due to the availability of a natural gene transfer system. The soil bacterium, *Agrobacterium*, has the unique ability to introduce specific gene segments into the chromosomal DNA of its host plants. This system has been adapted for use in genetic engineering to provide a powerful gene transfer system. *Agrobacterium* acts as a gene taxi, taking our laboratory-made gene constructs into the chromosomal DNA of the target plant.

In the cereals and other monocotyledonous plants we mostly rely on physical methods for introducing specific gene segments into the host cells. Most spectacularly, gene segments can be carried into cells by high-velocity particles, often propelled by an explosive charge.

In all of these gene transfer systems the host plant provides many of the enzymes that are needed for the actual insertion of a DNA segment into chromosomal DNA. This reliance on the specificities and powers of enzymes associated with nucleic acid replication and repair has been an underlying strength in the development of recombinant DNA technologies – a genetic engineer uses natural processes to great advantage. We can confidently expect gene transfer systems for all species of plants that are important in agriculture, forestry or horticulture.

What genes can we transfer?

Coding sequences that are obviously important in plant improvement are those which dictate the production of insecticides and those which give resistance to disease organisms and to herbicides. Frequently, these are coding sequences which can be taken from species other than the agricultural plant. They represent novel genetic properties. Increasingly, the new coding sequences may be synthetic, based upon naturally existing sequences or constructed by the genetic engineer on the basis of fundamental biological knowledge that has become available, largely as a result of molecular biology-based research.

An important point in regard to disease and pest resistance is that our increased understanding of interactions between organisms is enabling us to construct resistance sequences which will be longer lasting than many of the naturally occurring genes. The resistance genes can be constructed so as to present a far more difficult barrier to the pathogen or pest to overcome with its natural genetic processes. Other coding sequences that are already being used are sequences that provide for new proteins in the seed, tuber or leaf or for different fatty acids in seed oils. These changes lead to improvements in food quality, for either animals or humans. Specific modification of the nutritional properties of plant material is one aspect of genetic engineering which will be of great significance in meeting the pressures of growing world population. A genetic software system providing for a more balanced diet, with delivery through a plant seed, is likely to be one of the most user-friendly solutions to the food problem, especially in developing countries using low-technology production systems.

We will be modifying other proteins, for example the enzymes which are critical in various metabolic pathways of a plant. Again, we will be relying on increasing understanding at a cellular and molecular level of what is actually taking place in plant cell metabolism. Perhaps we will provide additional enzymes to create a branch biosynthesis pathway yielding a new product, or we may provide a modified code for a gene which will yield an enzyme with different performance characteristics from the one already existing. We can also plan to produce proteins or other molecules in parts of plants where previously they may not have been present. An example is the production of tannins in leaves of fodder plants where previously those same tannins might have been restricted to the seeds or other parts of the plant. If the addition of one or two genes could provide for this activity, it could have important consequences in the suitability of the leaf material for animal feed, avoiding problems such as bloat for the animals.

Thus far I have described the addition of coding sequences of a plant genome. It is likely that there will be as much need to subtract coding sequences. In many of our plants there are some metabolic activities that we would prefer to be there. For example, removal of an anti-nutritional factor from legume seeds (soybean) could improve the value of the seed meal in animal feeds. The technology to remove a gene product, rather than add one, has been developing rapidly in the last two or three years. We are not able in plants to do what can be done in yeast; that is, to remove a specific gene or replace a gene with a non-operative copy. But we can interrupt the information flow by causing the mRNA from the gene to be nullified in terms of informational content for protein making, or to be destroyed. If we add a gene construct which produces a mRNA containing the anti-sense strand of a gene-coding segment, the simultaneous production of the sense and the anti-sense mRNAs leads to a molecular pairing of the strands, so that the sense mRNA is not available to direct the production of a protein product. Anti-sense technology is still in its infancy but already it has been used in modifying the genetic make-up of tomato to improve its storage and ripening qualities. The genetically engineered tomato may well be the first transgenic plant foodstuff to reach the market-place.

We also have a developing technology where a gene can be constructed to produce a mRNA containing a ribozyme which seeks and destroys a target mRNA in the nucleus. We can build a ribozyme-making gene specifically directed against an unwanted gene product. Similarly, a ribozyme-making gene can be directed against an invading virus, providing an entirely new way of combating some of the most serious diseases of agricultural and horticultural crops. In Canberra we are developing a powerful, general-purpose ribozyme technology which we call Gene Shears. It has remarkable specificity and wide applicability and is likely to be an important tool in the armoury of plant breeders of the future.

The other coding sequences that are of great importance and which will become a more common substrate in our genetic engineering activities in the future are those genes determining complex agronomic characteristics such as performance and yield. Also we will pay increasing attention to those genes controlling key stages in plant development. Most of the genes involved in yield and other multi-genically determined characters are unknown, but they are no longer beyond our reach. There are two major developments in technology which have made it possible to track genes contributing to a complex phenotype. One is the use of DNA sequence markers. Variations in nucleotide sequences are sufficiently frequent in most plants for any given DNA sequences to be likely to have a range of nucleotide substitutions surrounding it. If these substitutions create or destroy restriction enzyme recognition sites then the tag DNA sequence can be expected to be found in different length DNA segments in different genetic stocks. These DNA segment lengths are called restriction fragment length polymorphisms, RFLPs, and they can be linked to other genes by standard genetic linkage analysis.

This and related techniques of sequence tagging provide us with a powerful method to track DNA segments of major value in contributing to a complex phenotype. These segments can be tracked at a DNA level without worrying about phenotypic expression throughout a plant breeding programme. Segments with an additive or complementary effect can be brought together at a critical late state in the production of an elite genotype. These sequence markers are beginning to increase the efficiency of breeding programmes, saving generations of biological testing for many characters and permitting precise selection of wanted individuals in pedigrees. In most breeding programmes this will bring about a saving in time and resources.

Arabidopsis: a powerful new force in plant biology and agriculture

Analysis of mutants has been a valuable technique in plant breeding and genetics, but it has now become vastly more powerful, with the adoption of *Arabidopsis*, a cruciferous weed, as an experimental model plant. *Arabidopsis* has a small genome and a rapidly expanding array of molecular aids to assist in the analysis of its genome. It is also a self-pollinating plant (important in the production of mutations), capable of being cultured readily in growth cabinets under laboratory conditions, the life cycle taking only a few weeks.

These attributes have persuaded plant molecular biologists to focus on this plant and develop it as a tool for analysis and manipulation of plant genomes. It is possible to use *Arabidopsis* to obtain mutants for almost any phenotype under investigation. But beyond a genetic analysis, the traditional follow-up of mutagenesis, the other features of *Arabidopsis* make it possible for us to be confident that we can map the mutant in the genome and proceed to isolate it physically, enabling detailed analysis of the gene to be carried out.

The advent of *Arabidopsis* is leading to an acceleration of the understanding of many aspects of plant biology. It is also leading to the use of gene sequences isolated from *Arabidopsis* as probes to acquire comparable sequences from the genomes of agriculturally important plants, ones not usually so amenable as *Arabidopsis* in molecular analysis. Weed genes are leading us to crop genes!

The control segments of gene constructs

An introduced coding sequence in a transgenic plant is only as good as its system of expression control. Analyses of the control regions of genes have been of enormous importance in tackling some of the most basic concepts in plant biology. For example, we are beginning to see that cell differentiation and the processes of development in a plant's life cycle depend in large part upon differential gene expression – cell tissue-specific expression patterns.

Knowledge that control sequences immediately upstream of the product-coding region of a gene are sufficient to provide correct expression has meant that we have been able to approach agricultural problems and design solutions for them at the gene construct level. This has probably been the most important bridge enabling a meeting of these different disciplines. We already have a lot of knowledge on the types of controls that are coded into the promoter regions of many genes. We know of controls that provide for a developmental or cell-specific pattern of expression, controls that provide for expression in response to particular environmental cue or stress, and controls that modulate the level of gene expression. We know that similar control sequences in front of different genes provide for coordinate expression of the genes, despite their being distributed widely throughout the genome.

Maybe most important of all is that the majority of gene controls in plants are conserved across the plant kingdom. This is of critical importance because it has meant that gene constructs can be put together with components from a wide variety of species with a high degree of confidence as to the maintenance of correct expression in the heterologous recipient plant species.

Our knowledge of promoter (control) elements has progresed to a stage where we are now able to join them in different combinations to provide entirely new controls influencing the patterns and levels of expression of genes. We can tailor our production plants to fit new requirements.

WHAT WILL WE BE ABLE TO ACHIEVE IN THE FUTURE?

Our skills in genetic engineering will enable us to make small, precise and significant changes to the genetic make-up of our production plants, changes which will improve the capability of plants to cope with the prevailing climatic and soil conditions. We will change the architecture of plants, their proportional structuring, to make them fit better as complementary partners in production systems, so that intensive double-cropping may be achieved with a number of alternative crop species.

Apart from tailoring plants to our agricultural environments, we will make precise changes to the products we derive from these plants. We will change the chemical and biochemical make-up of the products so that they fit market requirements. This will be of great importance to the food-processing industry. Not only will we make changes to existing plant products, we will introduce entirely new products to plants. We can expect to use plants as factories for biochemicals of value in pharmaceutical uses and for enzymes for industrial purposes.

The marriage of biotechnology to agriculture is a powerful and critical development. The plant-based food and commodity production industry will never be the same again.

DISCUSSION OPENING: A COMMENT FROM THE PERSPECTIVE OF THE LESS DEVELOPED COUNTRIES – C.H. HANUMANTHA RAO*

The relevance of biotechnology to the agriculture of less developed countries is a subject about which hard information usable by economists is yet to be generated. Yet the promise and potential of biotechnology are so glaring that economists cannot shy away from analysis, however speculative and conjectural it might be, if they are to contribute to the setting up of correct priorities in research and the framing of socio-economic policies for deriving full benefit from, and adjusting to, these new developments.

Dr Peacock highlights the enormous potential of modern biotechnology for the future of agriculture and provides insights sufficient to stimulate speculative thinking on its possible socio-economic consequences. The experience of

*Institute of Economic Growth, University of Delhi, India.

green revolution technology, based on hard data, provides the necessary building-blocks for such exercises to become constructive and useful.

The main techniques of genetic engineering enable improvements to be achieved in the nutritive value of existing products, or the introduction of entirely new products of higher nutritive value. The suitability of legume seeds or of leaf material for animal feeds can also be substantially enhanced. Apart from quality improvement there is also scope for increasing physical yields. Although the achievements so far are limited to cereals such as wheat and barley, methods have recently been developed which show promise in the case of rice and maize, which are particularly important in poorer parts of the world.

Genetic manipulation of plants will close the gap between actual and potential yields by providing more robust resistance to pests and diseases. In addition, changes in genetic characteristics will result in plants able to cope with abiotic stress from adverse soil and climate conditions. It is thus possible to modify the physiological responses of plants to environmental stresses and to modify their architecture so as to promote their use as complementary partners in production systems involving intensive double-cropping. The biochemical constitution of products can also be changed to fit the requirements of storage, marketing and processing. For example, anti-sense technology has already been used in modifying tomato genetics to improve storage and ripening qualities.

In view of the significant potential of biotechnology the introduction of genetically engineered plants in agriculture now depends more on legislation and social and economic acceptability than on limitations in science itself. Perhaps the most important feature of biotechnology which distinguishes it from the 'seed fertilizer' green revolution package is the possible saving which it might bring in chemical inputs, including pesticides. Dr Peacock did not deal with the potential of bio-fertilizers and biological fixation of nitrogen, though my own view is that genetic engineering could be important in replacing the 'chemicalization' associated with the green revolution. 'Seed' becomes central to biotechnology.

Through new plants having better capability to withstand biotic and abiotic stress, biotechnology will impart stability to crop yields, close the gap between actual and potential yields by reducing crop losses, and enable wider geographical coverage to occur when compared to green revolution technology which is often suited only to agro-climatically favoured areas. Because there is likely to be time saving in the engineering of new seeds when compared with the evolution of strains through traditional plant-breeding methods, the rate of technological advance is likely to be quite fast. Usually when technical change is rapid the rate of obsolescence is also likely to be high in quite a few cases. Hence biotechnology will be knowledge- and skill-intensive, necessitating greater investments in research and a need to improve the capabilities of farmers. There are a number of potentially important socio-economic consequences stemming from the new developments.

Food security

Ceilings on yields have been reached in most of the favourable areas affected by the green revolution, and prospects for further growth have become bleak except where heavy investment has been provided in infrastructure in new areas. Given high rates of growth in population in many less developed countries, the threat of their becoming heavy importers of foodgrains in the near future looms large. The new tools of biotechnology offer significant possibilities for breaking these yield barriers and overcoming food and population imbalance. A significant decline in the relative price of foodgrains, given greater food security, would be a most important outcome. However, there is one difficulty. It may not be possible to ensure access to food through the generation of employment and purchasing power for a growing agricultural labour force. Some recent evidence suggests that, although the absolute income of labour may rise, its relative share may decline steeply on account of much higher increases in returns to fixed resources such as land (Ahmed, 1989). The experience of the green revolution also cautions us against expecting higher labour absorption within agriculture in the wake of the bio-revolution. Therefore maintaining access to food through general employment generation has to become a major concern of overall development strategy.

Stability

Owing to its need for controlled irrigation and its vulnerability to pests and diseases, seed-fertilizer technology resulted in higher yield variability, particularly in the case of crops grown in less favourable agro-climatic conditions. By imparting stability to crop yields biotechnologies can be expected to raise the investment incentives for small farmers who are deficient in resources and risk-averse. Farm labour would experience more stable incomes because of the reduction in annual variability in farm employment. More stable yields would also reduce costs of storage and distribution and thus strengthen national food management systems.

Productivity

There will be significant savings of conventional resource inputs of land, labour and capital. However, these gains will depend crucially on access to knowledge and skills. Therefore investment in basic and applied research, and in farm extension work and education, will be required.

Equity

Biotechnology could have a pro-poor bias arising from scale neutrality at the farm level, saving on chemical inputs, stability in yields, and improved prospects for crops grown in less favoured areas. However, the realization of pro-poor potential depends on research priorities. Biotechnology presents a wide range of options which could be slanted to suit entrenched interests rather than favouring the poor. The classic example is the choice between pest-resistant versus pesticide-resistant types of seed. The former can be pro-poor as well as environmentally sound because of savings in pesticide costs, whereas the latter can promote the interests of multinationals supplying pesticides (Ruivenkamp, 1988). The advanced research capacities of the developed countries could also be harnessed to developing import substitutes, displacing the products of less developed countries (Panchamukhi and Kumar, 1988).

Sustainability

In many of the less developed countries, at their present stage of development, damage to the environment arises not so much out of the extent to which chemicals are used in farming as from the extension of cultivation to marginal lands, and the prevalence of widespread poverty which causes undue pressure for the poor to clear forests to augment their incomes (Rao, 1990). Biotechnology could contribute immensely to sustainability through the protection and regeneration of the environment, partly from reduction in chemical inputs, but more significantly by raising yields and thus releasing marginal lands from cultivation. Large-scale afforestation could also be facilitated through using tissue culture techniques.

Realization of gains depends on research and policy orientation. Here there are three points to consider.

Research

The high flexibility of biotechnology enhances the importance of choice in research strategy. The role of governments of developing countries in framing priorities appears to be even more essential than in the case of green revolution technology. Since so much of the area under crops is located in unfavourable environments, it may be desirable, from the point of view of growth as well as of equity, to invest in techniques suited to them. There is likely to be a high return to such investment since the gap between actual and potential yields is quite high in unfavourable areas (Widawsky and O'Toole, 1990).

Policy

The distribution of gains will depend on access to new inputs and new knowledge, particularly among smaller farmers and in lagging regions. Public intervention to strengthen the capabilities of disadvantaged groups and to provide critical inputs and services will be essential for the equitable sharing of benefits. It will also be necessary to safeguard the interest of private sector entrepreneurs, among other things by formulating norms regarding access, sharing and utilization of germ-plasms for crop improvement by participants.

The role of economists

Because of the wide range of options opened up by biotechnologies there is great room for economists and social scientists to contribute to the evolution of appropriate techniques. Economists have so far been engaged mainly in *ex post* analysis relating to the consequences of adoption of seed-fertilizer technology, and their contributions towards the evolution of appropriate methods, in collaboration with agricultural scientists, at the policy-making level have been rather limited. It is therefore necessary to initiate collaborative ventures to bring agricultural economists and agricultural scientists together in shaping technology policy.

REFERENCES

Ahmed, Iftikhar, 1989, 'Advanced Agricultural Biotechnologies: Some Empirical Findings on their Social Impact', *International Labour Review*, 128 (5).
Panchamuki, V.R. and Kumar, Nagesh, 1988, 'Impact on Commodity Exports', in Research and Information Systems for the Non-Aligned and Other Developing Countries, *Biotechnology Revolution and the Third World:Challenges and Policy Options*, RIS, New Delhi.
Rao, C.H.Hanumantha, 1990, 'Some Inter-Relationships Between Agricultural Technology, Livestock Economy, Rural Poverty and Environment: An Inter-State Analysis for India' in Golden Jubilee Volume, *Agricultural Development Policy: Adjustments and Reorientation*, Indian Society of Agricultural Economists, Bombay.
Ruivenkamp, Guido, 1988, 'Emerging Patterns in the Global Food Chain', in RIS *op. cit.*
Widawsky, David A. and O'Toole, John C., 1990, *Prioritizing the Rice Biotechnology: Research Agenda for Eastern India*, The Rockefeller Foundation, New York.

DISCUSSION OPENING : A COMMENT FROM THE PERSPECTIVE OF DEVELOPED COUNTRIES – CORNELIS L.J.VAN DER MEER*

Introduction

Modern biotechnology makes headlines.[1] It occupies a significant share of the columns of journals and magazines. It receives huge research budgets from governments and private business. It is regularly on the agenda of politicians, research administrators, interest groups and professional organizations. It arouses heated debates about its possible benefits, its risks, its impact on economy and society, and its ethical aspects. Some people see it as a likely bonanza, others as an alchemist's dream and some fear it is Pandora's box. Some have great expectations of the spread of benefits of biotechnology,

*Ministry of Agriculture, The Netherlands.

others fear that it will enhance dictatorship by technocrats and monopolistic power of multinational corporations. Biotechnology may provide great opportunities for agriculture, though sceptics expect it to result in agriculture becoming a dependent part of big agro-chemical, pharmaceutical and food-processing companies. Obviously such issues make headlines. This comment examines the potential impact of biotechnology on the structure of agriculture and related sectors in the next 10 to 20 years.

Assessing the impact of biotechnology on structural change

Agricultural sectors in developed economies have experienced tremendous structural change over the past century, in particular in the period since the Second World War. Many factors have contributed to that change and there is no reason to expect that trends will alter drastically in the next 10 to 20 years, even without biotechnology. The question of impact concerns the extent to which it will affect the process of change. This requires an assessment of alternative scenarios.

The concept of the 'structure of agriculture' refers to such issues as the relation of factors of production within farms, the size distribution of farms, their characteristics of production and productivity, and their relations with supplying and processing sectors. Alterations in structure are the outcome of many factors such as technical change, social and institutional development and the effects of shifts in the price structure. Productivity is also closely related to changes in structure. Trends in agriculture over the past decades can be characterized as follows.[2] Since the Second World War, labour productivity in the agricultural sectors of developed countries, with the notable exception of Japan, has increased more rapidly than in industry and services (Van der Meer and Yamada, 1990). In general this has also been true of net factor productivity. Since demand for agricultural products grew only slowly the agricultural sector exhibited a persistent tendency towards excess capacity and excess production, and faced pressure on prices and income. In many cases, price support given by governments to mitigate the depressing effect on incomes resulted in over-production. The major factor contributing to total productivity growth was the rapid decline of labour input. Although agricultural research is mainly devoted to achieving yield increases and biological efficiency, the effect of these improvements on total productivity is usually much less than that of the decline in volume of labour. Despite its rapid development, the agricultural sector still largely consists of small independent farms, operated by a farm family working alone or at most employing only a few paid labourers. Farm sizes have increased, but only the biggest 10–20 per cent of farms achieve economies of scale. Although farms are small, and are often engaged in some form of contract farming, they still have a great deal of freedom in their choices.

An assessment of the impact of biotechnology on these characteristics and trends is difficult to give. There are often conflicting expectations and speculations about technical possibilities, and assessment of the likely economic and structural impact is even more complex. It is, after all, the producer and the consumer who will decide. Economists are better at explaining the past than at predicting the future. So what can economists contribute to all the claims about biotechnology? One useful contribution can be to discuss biotechnology in the light of experiences with technological change, growth of output and productivity and structural change in agriculture, and to make conditional statements about the possible impact on present trends. Attention should also be paid not only to direct effects on the structure of agriculture by applications of biotechnology in the sector itself, but also to indirect effects which can result from applications within research and development programmes and within input-supplying and product-processing industries.

Impact of biotechnology in research and development

Biotechnology is a generic term, which means that it has a large number of possible applications in many fields. When genetic codes were deciphered and techniques for modifying genetic properties developed it was very useful to bring this kind of research together in special programmes, but gradually applications have become integrated within other research activities. This implies, for example, that knowledge of genetic codes accelerates and enhances effectiveness of traditional breeding programmes. Similarly, it may help all kinds of research by developing better detection methods. In other words, a new set of tools has been added to those already available to researchers. On the other hand, by its very nature, research and development in the field of biotechnology requires a fairly advanced general research infrastructure. If such a broad base is not available, research in biotechnology is probably rather ineffective and inefficient.

All this has two implications. First, it means that biotechnology is not an appropriate technology for countries that have not yet built up a good research system. Second, since it can only properly function jointly with other research, the estimation of the return to investment in biotechnology as such is difficult to isolate and in most cases therefore over-stated by its proponents.

Impact of applications within agriculture

How should the techniques that together form biotechnology be characterized from the point of view of agriculture? In the literature of agricultural development a distinction is usually made between land- and labour-saving technologies. The former consist largely of biological and chemical techniques and the latter mainly of mechanical techniques. Although in practice a clear-cut distinction is not always possible, since some techniques exhibit characteristics of both, the distinction is important for conceptual reasons. Green revolution technologies, involving use of high-yielding varieties, fertilizers and water control, are typically land-saving. Tractorization is an example of labour-saving innovation. Comparative research shows that land-saving technologies are most important in situations of land scarcity and at a lower level of economic development, whereas labour-saving technologies are important in land-abundant and labour-scarce situations and in advanced economies (Hayami and Ruttan,1985). In general, land-saving technologies are scale-neutral, whereas labour-saving technologies are characterized by economies of scale. Both types of techniques are to some extent embodied in purchased inputs, but they usually require farmers' knowledge for successful application. This knowledge can be obtained from other farmers, extension workers or from education.

Biotechnology that can sooner or later be applied in agriculture seems to be a typical example of biological and chemical techniques; modified properties of products, resistance against diseases and better technical input–output relations. They are not likely to generate significant economies of scale. From this perspective, therefore, no change in the pattern of agricultural development is likely. However, the possible impact of biotechnology does not only depend on the characteristics of the technologies but also on the pace and the intensity with which they are becoming available.

Although biotechnology applications are likely to become more important in the next 20 years it seems unlikely that they will exert a strong effect on the pace of technical change. There are several reasons for this. First, the commercially viable bio-techniques are emerging slowly because of technical and financial obstacles. For most products it may take quite some time before genetically modified and commercially attractive varieties become available. This is a general experience with generic technologies, from the development of electricity to the beginnings of information technology (OECD, 1989, ch. III). The diffusion of genetically modified varieties in agriculture is likely to take quite some time as well. Genetic modification of micro-organisms is technically easiest and therefore likely to result in significant applications first, though these will be made mainly in industry, not in agriculture. Genetically modified plants will have more impact on agriculture but developments in this field are slower because of technical difficulties. Applications of biotechnology on animals are still more complicated than on plants.

Second, in several cases biotechnology applications may be technically possible but still less cost-effective than traditional breeding techniques. Third, if the present GATT negotiations result in liberalization of markets and decoupled income support, then in most of the developed countries prices will decline. This will make yield-increasing technologies less attractive and probably slow down the pace of land-saving technological change in countries that have at present relatively high price levels. Fourth, there are risks and uncertainties about safety of applications for health and environment, which can probably be dealt with, but which will initially increase costs and result in lengthy and sometimes complicated procedures for admission.

Fifth, opposition to biotechnology seems firmly rooted in different groups. There are ethical questions about its applications, in particular for animals. Among some fundamental Christian groups, the ethical belief is widespread that genetic manipulation is perhaps not within the range of acceptable activities involving nature and life. Among ecologists and environmental groups, many see biotechnology as a dangerous and undesirable set of tools which should not be used. Among political activist groups on the left, there is also opposition, which can perhaps best be understood as a continuation of an age-old movement against capitalist development and the role of technology in a capitalist world.[3] Since the industrial revolution there have been continuous objections against new techniques. In most cases in recent history, however, ethical and political objections gradually disappeared or were overruled. This may also turn out to be the case with biotechnology, but it is also possible that strong opposition will remain. In this respect there are likely to be significant differences between countries, such as is already the case between countries in south and north-west Europe.

Sixth, consumer acceptance, which is partly related to the two previous points, is still far from certain. The attitude of consumers towards food has changed significantly during the past decades. If, for example, products have to be labelled, some of them may receive discounted prices, which would partly offset potential productivity benefits for producers.

Many uncertainties are evident and may result in significant setbacks in the rate of adoption of biotechnology. Even if everything is going smoothly the rate of application of biotechnology within agriculture may still be slow in the next 10 to 20 years. The net benefit of applications is the difference between value added in the with and without cases. In practice benefits seem often to be much over-estimated. Claims by biotechnology lobbies are sometimes exaggerated in the sense that they suggest high market shares for biotechnological products and incorrectly equate the net benefits to the share in value of production. Moreover, as argued already, increases in total productivity are more dependent on decreases in labour input than on biological efficiency. So, from an economic point of view, it seems realistic to have only moderate expectations about the net economic benefits of biotechnology in agriculture in the next two decades.

Impact of applications in input-supplying industries

Agriculture obtains considerable amounts of input from supplying industries. In developed countries this often amounts to more than 50 per cent of total value of production. The quality and price of inputs are crucial factors for international competitiveness of the agricultural sector. However, biotechnology is mainly applied in pharmaceutics, plant breeding and propagation and animal breeding,[4] and it follows that the relevant inputs account for only a modest share of the total input in farming. Nevertheless, there is much concern that agriculture will become increasingly dependent on a few multinational corporations in this field, because of the increasing role of concentration, patents and plant breeders' rights. It is true that, since the middle of the 1970s, there have been many mergers among seed companies. This development was related to the increasing economies of scale in this activity as well as to the fact that the oil crises in the 1970s stimulated interest in utilizing renewable resources, not least among oil companies. However, the prospects for producing bio-energy and developing non-food applications are now much reduced and returns have been below expectations. In one recent case in the Netherlands, an oil company sold a seed company to an agricultural cooperative and informed sources claim that this is not an isolated case.

Still there is a persistent strong concern among farmers, and in particular among Third World activists and radical groups on the left, that farmers and Third World countries are becoming dependent on breeders' rights and patents and that they may be exploited by multinational companies. These groups have little confidence in the role of competition or in countervailing power. There is certainly over-sensitivity with respect to seed companies. The argument is not advanced by pointing to the fact that the world market for chicken supply for production of layers and broilers is served by scarcely a dozen companies, that four-wheel tractors are supplied by even fewer, and that there are also few pharmaceutical companies left. The present sentiments about dependence on seed companies seem to be a continuation of those voiced by similar groups with respect to the green revolution. Not rarely in debates they still refer to 'the failure of the green revolution' when talking about the possible adverse effects for farmers of applications of biotechnology by seed companies.

Impact of applications in processing industries

In the processing industry, biotechnology is likely to be applied on a significant scale, both in food production and in non-food applications. In processing of agricultural products two developments that are already taking place could be accelerated by biotechnology. First, there is a trend for farmers to be encouraged to produce certain products under carefully specified conditions. This has resulted in various forms of sub-contracting. Diversification in consumer markets partly results in diversification of demand for raw materials. Some people, and in particular those critical groups mentioned above, believe that such developments will make farmers more and more dependent on big companies and that the application of biotechnology will strengthen this trend. Second, industries are continuously looking for possibilities to substitute cheap for expensive raw materials and they have been successful in doing so. It is assumed that biotechnology will enhance this process. This is often marked down as a negative impact of biotechnology, since it forces agriculture to compete with synthetics and also introduces competition among groups of farmers who previously produced for separate markets. It is believed that, as a conseqence, total value added will decrease. Artificial sweeteners and substitutes of vegetable origin for dairy products are the most common examples.

Although changes in the processing industries induced by applications of biotechnology can have adverse effects for particular groups, there are offsetting positive effects, hence the view which stresses negative effects only is rather superficial and biased. It ignores the consumers' interests, it fails to see the relation between substitution and protection in the sugar and dairy markets and it narrowly focuses on some selected effects of some processes without considering the wider impact of processes of technological change and economic development. One particular future contribution of biotechnology to the competitiveness of agriculture could be that plants and animals become new, or more attractive, sources of special chemicals, or that their products are better processed. Such developments could enhance the competitiveness of some branches of a more differentiated agriculture.

The dependence of farmers cannot be properly understood unless the dependence of processing industries is taken into consideration as well. Once processing industries have invested in specific products they require a reliable supply of raw materials of good quality. So there is usually a mutual dependence of farmers and processors, which is likely to become of increasing importance for the competitiveness of agriculture. Indeed, it is very probable that traditional bulk-producing farmers are more dependent on powerful outsiders than well-educated and properly organized groups of modern farmers.

NOTES

[1]Some definitions of biotechnology are very broad and include all traditional uses of biological processes. In this paper a more narrow definition is applied. Here 'biotechnology' refers to the collection of techniques which use knowledge of genetic codes and genetic modification, in particular by recombinant DNA techniques and cell fusion.

[2]For a detailed discussion of growth and development in agriculture see Van der Meer and Yamada (1990) and Van der Meer (1983 and 1989).

[3]From an economic point of view an interesting review is found in Kitching (1982). Van der Pot (1985) has given a broad overview of schools of thought from a philosophical perspective.

[4]To some extent it is applied in the feed industry as well. In the Netherlands, additives are used to reduce the phosphate content in compound feed in order to reduce environmental problems in areas with intensive livestock raising.

REFERENCES

Hayami, Y. and Ruttan, V.W., 1985, *Agricultural Development, An International Perspective*, Johns Hopkins University Press, London.

Kitching, G., 1982, *Development and Underdevelopment in Historical Perspective*, Methuen, London.

OECD, 1989, *Biotechnology: Economic and Wider Impacts*, OECD, Paris.

Van der Meer, C.L.J., 1983, 'Growth and Equity in Developed Countries', in A.H. Maunder and K. Ohkawa (eds), *Growth and Equity in Agricultural Development*, Proceedings of the Eighteenth International Conference of Agricultural Economists, Gower, Aldershot.

Van der Meer, C.L.J., 1989, 'Agricultural Growth in the EC and the Effect of the CAP, in A. Maunder and A. Valdés (eds), *Agricultural and Governments in an Interdependent World*, Proceedings of the Twentieth International Conference of Agricultural Economists, Dartmouth, Aldershot.

Van der Meer, C.L.J. and Yamada, S., 1990, *Japanese Agriculture, A Comparative Economic Analysis*, Routledge, London and New York.

Van der Pot, J.H.J., 1985, *Die Bewertung des Technischen Fortschritts*, Van Gorcum, Assen.

DISCUSSION OPENING : A COMMENT FROM THE PERSPECTIVE
OF INTERNATIONAL AGRICULTURAL RESEARCH CENTRES –
RANDOLPH BARKER*

I have been asked to comment on the paper by Dr Peacock from the perspec
tive of the International Agricultural Research Centres (IARCs). As a current
member of the Board of Trustees of the International Institute of Tropical
Agriculture (IITA) in Nigeria and a former Head of Economics at the International Rice Research Institute (IRRI) in the Philippines, I have had occasion
to be concerned about priorities in research, particularly between biotechnology
and other activities.

The organization of the Conference programme has allowed us to consider
some of the invited paper material before our plenary session. Yesterday we
heard an invited paper by Collinson and Wright on 'Biotechnology and the
International Agricultural Research Centres of the CGIAR' and the very interesting discussion which followed by Dr Evenson and others. My comments
will attempt to provide continuity between today's presentation and yesterday's discussion.

The key issue raised by Dr Evenson is why, given the potentially high pay-
off of investment in biotechnology described in Dr Peacock's paper, the
Consultative Group on International Agricultural Research (CGIAR) had
committed so few resources to biotechnology research. As Collinson and
Wright indicated, research priorities at the IARCs are demand-driven. But I
believe that research priorities (or demand) are primarily determined by the
donors to the CGIAR. Collinson and Wright suggest that the CGIAR wants to
strengthen research capacity in resource management and environment. However, the budgets for the IARCs have been steady or declining in the past
several years. Thus the new priorities of the CGIAR can only be achieved by a
reduction of capacity in traditional agricultural sciences. This has a direct
bearing on the capacity to transfer biotechnology to developing countries.
Without strong programmes in traditional agricultural sciences such as plant
breeding, this capacity will be greatly reduced. Meanwhile, at the national
level, the donors to the developing countries are busy transferring external
resources for research from ministries of agriculture to ministries of environment or natural resources, again reflecting the priorities of the developed, not
the developing, countries.

At IITA we are beginning to develop an applied biotechnology capacity. We
have a link with Purdue University for biotechnology on cowpeas and with
George Washington University (and indirectly Monsanto Chemical Company)
for research on the cassava mosaic virus. We are also receiving financial and
technical support from the Italian government for our biotechnology programme. We have one of the handful of laboratories in Africa capable of
training scientists in applied biotechnology research, not the capacity to do
recombinant DNA that Dr Peacock suggests, but simpler techniques such as
tissue culture, embryo rescue and use of gene markers. Work on DNA transfer
must be done in the advanced laboratories in the developed countries. Mean-

*Board of Directors, International Institute of Tropical Agriculture.

while, with these simple techniques, progress has been made at IITA in identifying varieties resistant to one of the most serious fungus diseases in plantain, black sigatoga.

The IARCs current involvement with biotechnology will expand in the future. A recent evaluation of international biotechnology efforts recommended that the IARCs consider undertaking the following initiatives (Plucknett *et al.*, 1989; Barker and Plucknett, 1991):

(1) Identify and transfer high-priority technologies to developing countries. This would involve the centres in helping to determine the needs and abilities of their national programme partners and to ensure the transfer of high-priority technologies.

(2) Explore opportunities for establishing commercial relations with private industry. This could operate in a way similar to that used by universities for the receipts of royalty payments. It could generate an additional source of core funding to the centres. It should be emphasized, however, that the first responsibility of the public funded IARCs is to ensure widespread access to the technology. Formal links with the private sector that establish priority rights to the technology may be in conflict with these goals.

(3) Establish institutional biosafety committees (IBCs) to coordinate the safe use and development of technologies in international research. Each IBC could coordinate testing with host-country approved mechanisms, as well as with current regulatory standards adopted by the developed countries.

(4) Establish a standing group of experts to deal with the role of biotechnology in world agriculture. While each centre is capable of determining its own course, active dialogue with the group of experts should help each institution make better decisions. The IARCs do not now have the capacity to undertake all of the above activities. At IITA, for example, there are only two scientists dedicated full-time to biotechnology research. The pace of growth in research capacity in biotechnology at IARCs will be determined in large measure by the price of the major foodgrains and by the demonstrated capacity of biotechnology to raise agricultural productivity. With foodgrain prices at low levels, the donors to the CGIAR will continue to give priority to research on the environment and management of natural resources, as opposed to research designed to raise agricultural productivity. Whether or not this proves to be short-sighted only time will tell.

REFERENCES

Plucknett, Donald, Cohen, Joel I. and Horne, Mary E., 1989, 'Future Role of IARCs in the Application of Biotechnology in the Developing Countries', in *Agricultural Biotechnology Opportunities for International Development*, CAB International, Wallingford.

Barker, Randolph and Plucknett, Don, 1991, 'Agricultural Biotechnology: A Global Perspective', in Bill R. Baumgardt and Marshall A. Martin (eds), *Agricultural Biotechnology: Issues and Choices*, Purdue University Agricultural Experiment Station, West Lafayette, Indiana.

CARL E. PRAY*

*Plant Breeders' Rights Legislation,
Enforcement and R&D: Lessons for Developing Countries*

INTRODUCTION

Declining government budgets, pressure from donors and agribusiness firms and the failure of some government seed corporations are encouraging policy makers in a number of less developed countries (LDCs) to privatize plant breeding, seed production and seed distribution. Developed countries are pressing for stronger intellectual property rights such as plant breeders' rights (PBR) and utility patents as a means of encouraging private firms to transfer technology and conduct more research and development (R&D). A number of countries are considering PBR legislation.[1]

This paper examines the development of PBRs and their impact on private research and technology transfer in the United States, France, the United Kingdom, Argentina and Chile. From their experiences some policy implications for developing countries are derived.

INTELLECTUAL PROPERTY RIGHTS
LEGISLATION RELATED TO PLANTS

At present 22 countries have PBRs. Most Western European countries passed their legislation in the 1960s and 1970s. The USA adopted its Plant Variety Protection Act in 1970, while Kenya, Argentina and Chile followed within the decade. Australia and Canada have recently passed legislation. France and the United Kingdom were selected for this study as examples of European PBR legislation with different historical backgrounds. The USA was chosen because it has what is generally considered to be a weaker version of PBRs. Argentina and Chile appear because they are the only developing countries which have PBR rules which they are attempting to enforce.

Protecting varieties before plant breeders' rights

The legal structures for protecting intellectual property in plants in France, the United Kingdom, the United States, Argentina and Chile are shown in Table 1.

*Rutgers University, USA.

France, the USA and Argentina had laws which helped companies capture the gains from research before PBR laws were passed.

In France, a trademark law was passed in 1927 to protect consumers against fraud. Firms could register a trademark on a plant variety and prevent other firms from using that trademark without their permission. Compulsory registration of new plant varieties was established in 1932. Before a new variety could be registered and marketed it had to be tested for three years to establish its superiority over currently used varieties. The combination of these laws provided plant breeders with the equivalent of property rights. Trademarks prohibited other firms from selling a variety under the breeding firms' name and compulsory registration made it almost impossible for a rival firm successfully to register and sell the variety under another name.

The ability of firms to appropriate the benefits from research was weakened in the 1950s, when a French court decided that plant varieties could not be protected with trademarks. Plant breeders continued to collect some royalties from firms selling their varieties, but their property rights had been seriously weakened.

Argentina had a set of regulations which gave breeders some protection from 1935, when the government established the National Commission for

TABLE 1   *Comparison of intellectual property rights on plants*

Years			Countries		
	France	USA	Chile	Argentina	UK
1929	Trademark Law (1927)	Trade secrets (common law)			Trade secret (common law)
1930	Compulsory registration (1932)	Plant Patent Act (1930)		Compulsory Registration and 3-year monopoly (1935)	
1940					
1950	Trademarks on varieties disallowed (mid-1950s)		Voluntary certification system (1958)	Closed pedigree registration for hybrids (mid-1950s)	
1960					Seed Law-included PBR and compulsory registration (1964)
1970	Plant breeders' rights (1971)	Plant Variety Protection Act (1970)	Law of Seeds (PBR) (1977)	PBR Law (1973) Implementing regulations (1978)	
1980		Utility patents applicable to plants (1985)			

Grain and Elevators to regulate commerce in grains and promote improved varieties. The law required all varieties of the main crops (wheat, oats, barley, rye, corn, sunflower and flax) to be registered. It needed several years of trials to prove that the variety was (1) high-quality, (2) disease-resistant, and (3) higher-yielding. The seed of registered varieties could be sold either in certified or identified form. The former had to meet germination and purity standards and producers of certified seed had to go back to the breeders for new breeders' seed every three years. Identified seed could be multiplied indefinitely without purchasing new seed from the original breeders.

An additional provision law was that for the first three years a variety could only be sold by one company. This was supposed to prevent any one variety from spreading too fast, but it also had the effect of giving a breeder property rights over his variety for a period of time. In the 1950s, hybrids were allowed to be registered without revealing the parental inbred lines.

The first legislation explicitly providing intellectual property rights on plants in the USA was the 1930 Plant Patent Act (PPA) which allowed the patenting of asexually propagated plants, except potatoes. This did not cover any major field crop in the USA because they are propagated by seeds.

The development of hybrid corn by university and government scientists in the USA in the 1920s and 1930s allowed breeders of new hybrid varieties a monopoly on their hybrid as long as they could keep the inbred lines secret. It was possible to take another firm's inbred lines and then reproduce its hybrids. For example, employees could move to another firm or start their own company, and bring inbreds from their former firm along with them. In the United States, trade secret law, which is based on English Common Law, protects a firm's exclusive rights to use an inbred as long as it takes all reasonable precautions not to disclose it. This did not become very important until the 1950s, when firms developed some very good private inbreds which replaced the public inbreds that they had been using earlier.

Plant breeders' rights legislation

The United Kingdom passed a new seed law in 1964. It included PBRs and compulsory registration of varieties. The French made PBRs law in 1971. Both countries require that a new variety first be registered before the breeder can be granted ownership. To be registered, a variety must be field-tested for several years to ensure that it is distinct, uniform and genetically stable. In France it also has to be superior to other varieties over a range of key characteristics. European farmers may grow their own seed, but they do not have the right to sell seeds to other farmers. The owners of a variety cannot prevent researchers from other companies from using it to produce a new variety.

The US Plant Variety Protection Act (PVPA) of 1970 provides protection for 18 years to novel, sexually propagated varieties and inbred lines of hybrids. The Department of Agriculture (USDA) checks applications against the descriptions of varieties in its data bank. If the variety is different, a certificate is issued. PVPA has several key exclusions from protection. Farmers can reproduce seed for themselves and sell seed as long as sales are less than 50 per cent of the total product of their farm.

Argentina passed PBR legislation as part of a new seed law in 1973, though the regulations for implementation were not developed until 1978. This law, like those of Europe, requires that varieties be distinct, uniform and genetically stable. A government agency conducts appropriate field tests. The breeder may then request proprietary rights which are granted for 15 years for annual crops. First-generation hybrids are excluded from protection, but inbred lines can be protected.

Chile passed a new seed law that included PBRs in 1977, with enforcement procedures being drawn up and approved in 1979. The law provides protection for 15 years. A new variety is given a temporary permit immediately upon application for breeders' rights and then has to be grown for two years to ensure that it meets the description in the application. If government tests confirm the breeder's claims, the breeder is given the final permit.

PBR legislation in all five countries reviewed above provides the breeder or his company with 15 to 18 years of ownership. They all have the research exemption which allows a protected variety to be used in breeding other varieties without payment to the owner. In some other important characteristics the US law differs from PBRs in the other four countries. The US PVPA requires neither major differences from current varieties nor field testing, while the others require field testing to prove that new varieties are distinct and superior in some way to old varieties. In addition only the USA has an explicit exemption which allows farmers to sell seeds.

Evolution of PBR Enforcement

Legislating plant breeders' rights does not actually give breeders any rights unless someone identifies violators, and courts stop the violation and impose penalties. In most countries enforcement has grown as new institutions are developed to enforce breeders' rights and court precedent is established.

French and British breeders' rights are enforced by a plant royalty bureau which establishes and collects the amount of royalties and distributes them to the owners of the varieties. The bureau also inspects seeds in the market for violation of breeders' rights. If it finds violations, it offers the violator the opportunity to sign a contract and start paying royalties. If the violator refuses, then the company whose rights have been violated can take court action to obtain an injunction and damages.

To enforce the US PVPA, firms must identify violators and bring them to court to seek injunctions against further infringement, royalties and punitive fines. Adherence to the law has varied over time. When it was passed, firms mounted a publicity campaign to inform farmers and other firms about its provisions. In most cases when companies discovered violations, they just needed to write to farmers or cooperatives informing them about the legal provisions and the violations stopped. Adherence to the law was weakened by the court case *Asgrow* v. *Kunkle* 1987, which found that even very large sales by farmers were legal as long as the farmer sold less than 50 per cent of his crop as seed for reproductive purposes.

Adherence to the Argentine PBR law has varied over time. The new regulations of 1978 were not enforced immediately by government agencies because they were among the last acts passed by an unpopular government before it was replaced. In the first year after the law was passed, cooperatives and firms which were selling private wheat varieties voluntarily paid royalties to the firms which bred them, though by the second year only about half of the sellers paid royalties. In the fourth year no one bothered to pay any royalties at all (Gutierrez, 1990). Firms did not take to court cooperatives owned by big farmers who violated the law because to do so would have meant taking action against some of their most important customers.

A major step in reducing the production of 'black' seed took place when three of the largest cooperatives and the national Farmers' Federation started producing and selling the national agricultural research institute's (INTA) proprietary varieties of soybeans, wheat and maize in joint ventures with INTA. Since the cooperatives were now selling proprietary varieties or trying to collect royalties on them, they stopped violating other firms' proprietary rights.

The Argentine Seed Association, INTA and the Ministry of Agriculture are establishing a private enforcement agency modelled on the French association for collecting royalties and licensing protected varieties. This is first being implemented for wheat. In 1990, Buck, Klein and INTA hired a lawyer to enforce their property rights on wheat varieties. They first arranged for government agents to inspect the seed in several key markets, when it was found that some companies were selling proprietary varieties without paying royalties. These companies were then asked to pay, and about 60 per cent did so. Legal proceedings against the remainder have begun. This is the first occasion on which that has occurred in Argentina (Gutierrez, 1990).

In Chile the government takes a more active role in enforcement than in Argentina. A company believing rights to have been infringed can bring a claim to the Ministry of Agriculture. If the claim is found valid, the Ministry can immediately stop further sales. Violators almost always obey without the need for court action. There are three or four complaints each year, of which on average one is found to be legitimate.

IMPACT OF PBRS ON
RESEARCH AND TECHNOLOGY TRANSFER

Research

If PBR legislation has the expected incentive effect, private R&D expenditure on self-pollinated crops like wheat, barley and soybeans should increase as enforcement grows stronger. Five to ten years after research starts, new private varieties will appear on the market. The experience of the United Kingdom provides the strongest empirical support for the effectiveness of PBR legislation. The French, Argentinean and American cases also provide some support for PBR effectiveness, but it is not as dramatic as in the British case.

In the United Kingdom in 1960 'commercial plant breeding in the major agricultural crops was virtually non-existent' (Murphy, 1981, p.30). Three years after the 1964 seed law ten firms were engaged in research and by 1981 the number had risen to 23. In 1981, 90 per cent of spring barley, which is the most extensively grown crop in the United Kingdom, was comprised of private varieties. Winter wheat was the only major crop in which public varieties covered a greater area than private strains at that time (Mastenbroek, 1982). Most public varieties came from the Plant Breeding Institute, Cambridge, which the government sold to Unilever in 1987.

French plant breeding started with farmers who were also breeders. During the 1930s, some of these farmers evolved into wheat-breeding firms. By the beginning of the Second World War, 15 to 20 medium-sized firms conducted most of the private wheat breeding (Joly, 1990). In 1981, private varieties covered 97 per cent of the winter wheat area and all of the spring wheat (Mastenbroek, 1982).

Private firms started breeding wheat in Argentina around 1925. After the grain control law in 1935, nine companies conducted wheat-breeding programmes. By 1939, at least 25 per cent of the wheat area was planted with private varieties (Figure 1). These traditional large farmer/breeding firms declined to two major companies by 1970. In the 1960s, the local firms were joined by several USA based multinationals which started hybrid wheat-breeding programmes. Despite the decline in the number of firms, the share of area sown to private varieties grew to 90 per cent by 1965. The decline in private varieties during the 1970s was due, in part, to the government research system's success in developing semi-dwarf varieties of wheat based on CIMMYT lines. In 1989, officials interviewed from Pioneer, DeKalb, Northrup-King, Cargill, Morgan and several smaller Argentine companies reported that there were no increases in private research on self-pollinated crops or on hybrids, owing to the 1973 PBR legislation (Pray, 1989).

Private plant breeding began in the USA at the beginning of the century. Most was done by farmers, with USDA and agricultural universities entering only after the First World War. Seed companies did not start breeding programmes until about 1920, concentrating on maize. Major companies did not commence wheat and soybean research until just before PBR was passed in 1970. Real private R&D on wheat, maize and soybeans in the USA increased rapidly after 1965 (Figure 2a). The growth in R&D as a percentage of the value of cereals and soybeans after 1965 (Figure 2b) is the strongest evidence available at present that PVPA fostered more private R&D. The high levels of hybrid corn and sorghum R&D in Figure 2b indicate the importance of hybrids in combination with trade secrets in inducing private research.

The increase in private R&D did not, however, lead to a steady increase in the area under private varieties in all crops. The share of soybean area under private varieties was negligible in 1960, increased to 3 per cent in 1969 and then to 21 per cent in 1984. The share of wheat area under private varieties declined from 25 per cent in 1939 to 16 per cent in 1969 and 11.5 per cent in 1984 (Huffman and Evenson, 1987). Maize shifted from 100 per cent public varieties to 100 per cent private hybrids between 1940 and 1960 (Kloppenberg, 1988) and has remained 100 per cent private since them. Thus soybeans is the only crop in which the share of private varieties increased after PVPA was passed.

Most Chilean private wheat and hybrid maize research started in the 1950s. Private plant-breeding expenditure remains very limited – less than US $150 000 in 1984 (Venezian, 1987). Figure 3 shows R&D expenditure on wheat by the two principal private plant-breeding organizations – SNA, which is a farmers' cooperative, and Semillas Baer, which is a private company. Baer's R&D declined after the PBR law was passed in 1977, while SNA's R&D increased. The share of wheat acreage covered by private varieties declined from 44 per cent in 1971 to 36 per cent in 1984, after PBR was passed (Venezian, 1987). The maize area was primarily planted with private hybrids by 1977. The one change that has taken place is an increase in the area under imported US hybrids, which rose to 40 per cent in the late 1980s (Nodine, 1990).

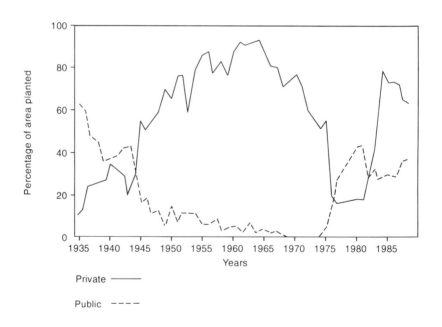

Source: Gutierrez (1990).

FIGURE 1   *Public and private wheat varieties in Argentina*

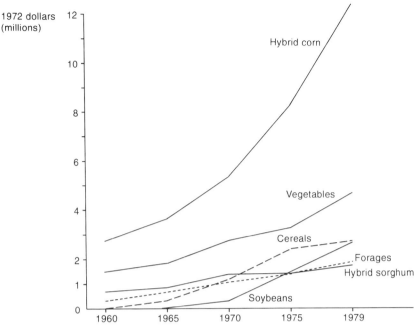

Source: Perrin *et al.* (1983).

FIGURE 2a   *Crop-breeding research expenditures by 59 firms for various crops, in constant (1972) dollars (adjusted by implicit GNP Price Deflator)*

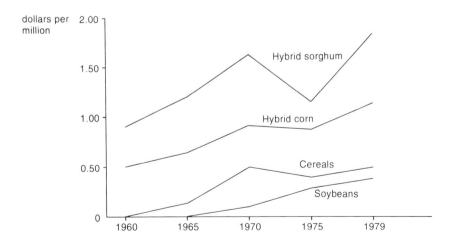

FIGURE 2b   *Crop-breeding research expenditures (59 firms) per million dollars of annual US crop value in the preceding five years*

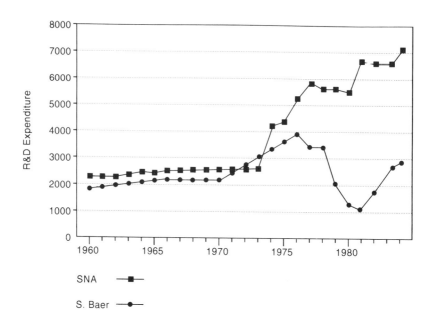

Source: Venezian (1987).

FIGURE 3   *Private wheat R&D in Chile (thousands of 1984 pesos)*

Technology transfer

PBRs have stimulated the transfer of varieties of self-pollinated crops in some
countries. In the United Kingdom it provided foreign firms with an incentive
to transfer technology:

> the initial improvement following from the introduction of our [PBR] legislation
> arose from the introduction into the United Kingdom of varieties bred abroad. The
> Act did in fact encourage very significantly the establishment in the UK of com-
> panies which did very little else initially than evaluate varieties bred abroad...It is
> interesting that many of these companies which started off by introducing foreign
> varieties developed their own specialized breeding programmes and are now pro-
> ducing varieties bred in the UK (Murphy, 1981, p.32).

Chile appears to be going through the initial stages of a similar process.
With the combination of trade liberalization, export promotion, a more positive business climate and the passing of PBR legislation, many maize, fruit,
and vegetable varieties have been imported from the USA.[2] Maize seed imports are associated with increased yields (Nodine, 1990) and imported varieties of fruits and vegetables have been very productive.

In contrast, in Argentina, where property rights are less secure, foreign companies are not using the single-cross hybrids which are the highest-yielding varieties in the USA and Europe. Instead they continue to transfer and breed double-cross hybrids which give firms better control of proprietary lines but may give farmers lower yields.

CONCLUSION AND POLICY IMPLICATIONS

PBRs or similar legislation appear to have increased research in all five countries, except Chile. The large impact of PBR legislation in the UK,[3] relative to the USA, is due to loopholes in the US legislation, compulsory registration in the UK and perhaps to the presence of the UK Plant Royalty Bureau. In Argentina, the 1935 law induced research but the absence of an enforcement mechanism for PBRs, and the unwillingness of seed companies to sue cooperatives and seed companies, rendered the 1973 law ineffective. The absence of any impact on private R&D on wheat in Chile is due to the small size of the Chilean market and the decline in demand for wheat when trade policies changed.

These case studies suggest a number of lessons for policy makers in LDCs. First, some type of property rights, either legal or technical (hybrids), does seem to be a necessary, but not sufficient, condition for private research on self-pollinated crops. Second, without enforcement mechanisms such as plant royalty bureaux, an efficient court system and companies which are willing to take violators of PBRs to court, PBRs provide only a limited amount of protection to breeders. Third, even if PBR legislation does effectively protect breeders' rights, markets have to be sufficiently large to justify private investments in research. Fourth, PBRs may stimulate the transfer of technology.

What are the implications of these findings for countries that are trying to stimulate more biotechnology research and technology transfer? Private firms will transfer technology and conduct biotechnology research in those countries and crops which have the best property rights and largest markets. The most effective property rights in most LDCs are provided by hybrids because only Argentina and Chile have PBRs and even those are not very effective. The largest markets are for hybrid seeds in LDCs and large countries such as Argentina, China, Brazil, India and Mexico. Thus companies will concentrate their biotechnology research on hybrid corn, sunflower, sorghum and a few other crops in large LDCs. If hybrid rice is successful in the tropics it will also attract a number of companies. Stronger intellectual property rights laws may also be able to stimulate more private biotechnology research on self-pollinated crops in large countries such as Brazil and India and may encourage more transfer of biotechnology in smaller countries.

NOTES

[1]The governments of Bangladesh, the Philippines and India are debating the merits of PBR legislation.

[2]Interviews in 1989 with the President of ANUSAC, the largest local company, and Dr Eduardo Venezian, Dean of the Catholic University.

[3]Also probably in France, though less evidence is available to support PBR impact.

REFERENCES

Gutierrez, Marta, B. 1990, INTA, Buenos Aires, Argentina, Personal communication, October.

Huffman, W.E. and Evenson, R., 1987, 'Crop Varietal Improvements', in *The Development of U.S. Agricultural Research and Education: an Economic Perspective. Part III Iowa State University Staff Paper No.170*, Ames, Iowa.

Joly, Pierre-Benoit, 1990, INRA, University of Grenoble, personal communication, August.

Kloppenberg, J.R., 1988, *First the Seed: The Political Economy of Plant Biotechnology 1492–2000*, Cambridge University Press, Cambridge

Mastenbroek, C., 1982, 'The Significance of Plant Breeding by the Private Sector', in Records of a Symposium at the 15th ordinary session of the Council of the International Union for the Protection of the International Union for the Protection of New Varieties of Plants (UPOV), Geneva.

Murphy, P.W., 1981, 'Plant Breeders' Rights and Improvements of Plant Varieties', in Records of a Symposium at the 14th ordinary session of the Council of the International Union for the Protection of New Varieties of Plants (UPOV), Geneva.

Nodine, L., 1990, 'The Impact of Seed Policies on the Research and Productivity of Wheat and Maize in Chile', MSc thesis, Department of Agricultural Economics, Rutgers University.

Perrin, R.K., Kunnings, K.A. and Ihnen, L.A., 1983, *Some Effects of the US Plant Variety Protection Act of 1970*, Department of Economics and Business Economics Research Report No. 46, North Carolina State University, Raleigh, August.

Pray, C. E., 1989, 'The Seed Industry and Plant Breeders' Rights', Report of 4–13 October, 1989, Trip to Argentina and Chile, mimeo, Department of Agricultural Economics, Rutgers University.

Venezian, E., 1987, *Chile and the CGIAR Centers: A Study of Their Collaboration in Agricultural Research*, World Bank, Washington, DC.

DISCUSSION OPENING – DOUGLAS D. HEDLEY AND W.T. BRADNOCK*

Carl Pray's paper is intended to draw the attention of agricultural economists to an important issue in agricultural development. Taken in its broadest context the issue of plant breeders' rights (PBRs) falls under the third dictum of development. 'Getting prices right' and 'getting the human capital right', as the first two tenets, are supported by a considerable body of literature which continues to grow. The third element, 'getting institutions right', is equally important, but has a far smaller research base on which to provide guidance in development.

The issue being considered is an example of the complementarity between institutional development and scientific advance in the form of the green revolution HYV package of inputs. Part of the institutional change has also involved the host of pricing and trade policies that encouraged the use of new technology. The paper makes an important contribution to our understanding

*Agriculture Canada, Ottawa, Canada.

of one particular facet of institutional change which has potential influence in augmenting the impact of the scientific effort which is occurring around the world.

Dr Pray examines two aspects of PBRs. The first covers the historical experience of the development of rights in five countries; the second involves discussion of their impact on research and technology transfer. In both cases, partly no doubt because of the space limitations on a single paper, he has taken a relatively narrow view of PBRs. Nonetheless, wider issues closely associated with PBRs and the 'bio-revolution' need consideration.

It is necessary to distinguish between the requirements under national seed laws and plant breeders' rights legislation. In many countries, the national seed law restricts varieties which can be sold to those which have been tested and found to have merit for farmers and users. This testing can be unrelated to the granting of PBRs. The seed laws can override the latter, since a variety may be eligible for rights but may not be allowed to be sold for lack of merit. This is noted by Dr Pray, but he does not clearly bring out the issue in attempting to assess the impact of PBRs on the level of research or its funding origin, whether public or private. This would seem to be of critical relevance in the analysis of whether PBRs can effectively complement HYV technology.

There are several other closely related institutional arrangements which can affect performance in research under PBRs. These include:

(1) the level and nature of PBR enforcement;
(2) collection and use of royalties from PBRs;
(3) access for further research, both domestically and internationally;
(4) the relationship beween patent rights for genetically altered life forms (DNA and RNA recombinant research among others) and PBRs; for the farmer, the patents appear to apply to the 'upstream' bio-technology research, while the PBRs relate to the finished product of research ready for the farmer and the consumer; and
(5) the international trade rules and the legitimacy of rights across national boundaries; the GATT Uruguay Round includes the question of intellectual property rights, although this part of the negotiations appears to be driven by the more traditional patent and copyright issues rather than the emerging life form patenting and PBRs.

The conclusion we draw is that all of these aspects need to be much more carefully defined to enable assessment of the impact of PBRs alone.

The paper also implies that the objective of PBRs is to encourage only private sector plant breeding, seed production and seed distribution. Our view is that this objective may be too narrowly based to capture the range of institutional design that may be necessary for combined public and private sector varietal development and distribution.

There are two concerns which are not raised in the paper. The first relates to the moral and ethical issues surrounding the patenting of life forms and varieties. The argument involves a number of issues:

(1)   the 'free' collection of germ-plasm from around the world, the genetic alteration of this material and the subsequent collection of royalties in selling the material back to the countries of origin;
(2)   allowing the large transnational companies to extract quasi-rents on varieties when the greatest share of the early investment in germ-plasm collection and cultivar development is public investment; and
(3)   food resources and the ability to develop them should remain open and available to all the world, and not be subject to private appropriation of benefits.

These are powerfully emotive points which need careful attention in organizing future research. In the Canadian experience of establishing PBRs as one component in research, development and technology transfer they were the biggest stumbling-blocks, for which no well documented answers existed. For over a dozen years the major political parties attempted to pass PBR legislation in the face of these moral and ethical issues.

The second point relates indirectly to the objective for PBRs cited by Dr Pray, namely the decline of public funding to agricultural research and the need to marshall private funds to maintain or expand research, development and technology transfer. For the past 20–30 years, many of the productivity advances in Third World farming can be traced back to the work of the international research centres. It has been noted, however, that, while the green revolution was primarily fuelled by public funding, the bio-revolution is dominated by the private sector (Buttel *et al.*, 1985). This appears likely to continue. The implications of drawing private funding into research through PBRs, and the shift in the basis for productivity gain from varietal selection to bio-technology, raises major questions about long-term strategies for international and national research centres around the world. Dr Pray's paper is not designed to shed light on this strategic issue, though it is a major step along the way.

REFERENCE

Buttel F., Kenney M. and Kloppenburg J., 1985, 'From Green Revolution to Bio-Revolution: Some Observations on the Changing Technological Bases of Economic Transformation in the Third World', *Economic Development and Cultural Change*, 34, (1), October.

DAVID B. SCHWEIKHARDT AND JAMES T. BONNEN*

*Financing Agricultural Research in the Presence of International
Benefit Spillovers: The Need for Institutional Coordination and Innovation*

INTRODUCTION

The financing of agricultural research presents a unique policy problem. While research is often funded by national or subnational governments, the benefits of research often spill across the boundaries of the government financing the research. Many observers of international agricultural research have noted that the spillover problem exists and that policy changes are needed if an internationally optimal level of investment in research is to be provided (Judd, *et al.*, 1987; Idachaba, 1981; Ruttan, 1987b). Some empirical studies have attempted to measure international benefit spillovers (Evenson, 1977; Davis, *et al.,* 1987; Edwards and Freebairn 1984), and others have attempted to establish criteria for allocating research resources (Idachaba, 1989, p. 6; Paz, 1981; Norton and Pardey, 1987; Fishel, 1971; de Castro and Schuh, 1977; Carter, 1985). None of these studies has addressed two central policy questions: what policy tools are available for achieving an optimal level of investment in research and what institutional innovations are required to make these policy tools operable? This paper uses public finance theory to address these questions. First, the sources of international benefit spillovers will be examined. Second, public finance theory will be used to examine the policy tools available for achieving institutional coordination in financing research. Third, empirical estimates of international benefit spillovers will be reviewed. Finally, the policy implications of this research will be examined, with special emphasis on the institutional innovations needed to finance an optimal level of agricultural research.

SOURCES OF INTERNATIONAL
SPILLOVERS IN AGRICULTURAL RESEARCH

Since agricultural technology must be adapted to the ecological conditions in which it is used, agricultural research is often a location-specific enterprise that must be conducted in the same ecological conditions in which production takes place (Ruttan, 1982). At the same time, research that yields significant increases in the supply of farm products will ultimately provide a large portion of its benefits to consumers as the supply of farm products shifts along an inelastic demand for such products (Ruttan, 1982). As a result of this combi-

*Mississippi State University and Michigan State University, USA.

nation of location specificity and diffused benefits, agricultural research can provide three types of international benefit spillovers:

(1) benefits can accrue to consumers outside the investing country when research increases production and causes prices to decline in world markets;

(2) benefits can accrue to producers and consumers outside the investing country when research produces new technology that can be adopted by producers outside the investing country (either with or without further adaptive research); and

(3) benefits can accrue to many individuals outside the investing country when research produces scientific knowledge that enhances research in other (sometimes unrelated) areas of science (Davis et al., 1987).

THE ROLE OF INSTITUTIONAL
COORDINATION IN FINANCING AGRICULTURAL RESEARCH

The question raised by A.C.Pigou (1946) remains a central problem in public policy: under what circumstances should government intervene to align private and social costs (or private and social benefits)? Modern public finance theory recognizes that this question is not simply a matter of whether government should provide public goods, but also of which unit of government should do so. This question arises because some goods cannot be classified into the polar cases of pure private or public goods. Instead, there are a number of 'non-private' goods whose benefits are available to individuals in unequal amounts (as opposed to a pure public good which is available to all individuals in an equal amount). In addition, the problem of under-investment in non-private goods is further complicated when the problem of 'imperfect mapping' arises (Breton, 1965).

If a good is perfectly mapped, the benefits of that good will accrue only to those within the boundaries of the government financing the good. Assuming that the government of that jurisdiction can determine the preferences of its citizens, it will provide the optimal quantity of the perfectly mapped good. On the other hand, if the benefits of the good are imperfectly mapped, or spill across the jurisdictional boundaries of the financing government to those outside the jurisdiction, the investing government, like the investing individual in Pigou's analysis, will under-invest in the good. A higher level of government can encourage the provision of a socially optimal quantity of the spillover-generating good by providing a subsidy to the lower level unit of government.

An analysis of two types of subsidies is shown in Figure 1 (Scott, 1952; Boadway and Wildasin, 1984). A jurisdiction of government (a national government in the case of agricultural research) is assumed to allocate its budget (represented by the budget line A_1A_2) between the consumption of a public good that generates benefit spillovers in other countries (such as agricultural research) and all other goods (either private goods consumed by citizens or public goods that create no benefits outside the jurisdiction). The country's indifference curve I_1 indicates that its welfare is maximized at point E_1, and the quantities purchased will be Y_1 and X_1. Another unit of government (such as an international grantor[1]) may wish to provide a subsidy to compensate the country for the benefits that spill across its boundaries.

The subsidy could take the form of an unconditional, lump-sum grant. Such a grant has no restrictions on its use and may be allocated by the recipient for any purpose. Thus some of the grant will be allocated to the spillover-generating public good and some will be allocated to non-spillover public goods or to private goods (via a reduction in taxes in the recipient community). Such a grant is shown in Figure 1 as a shift in the recipient's budget line from A_1A_2 to B_1B_2. The recipient's new allocation will be Y_2 of the spillover-generating good and X_3 of other goods.

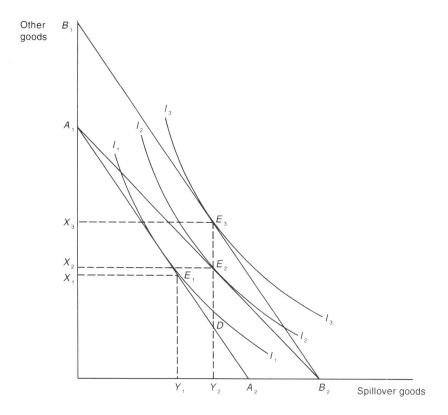

Source: Boadway and Wildasin (1984).

FIGURE 1   *Comparison of an unconditional lump-sum grant and a conditional matching grant*

As an alternative to a lump sum, the grant could take the form of a conditional matching payment. In this case, the grant will only be received on the condition that (1) the recipient use the grant for consumption of the spillover-generating good, and (2) that the recipient match the grant at a specified rate with its own funds. Assuming the original slope of the budget line is β and that the matching rate (defined as the number of international dollars granted for each national dollar invested in the spillover-generating good) is established such that δ is the share of the cost of the good Y paid by the grantor, the new budget line will have a slope of $\beta(1 - \delta)$ and will rotate from A_1A_2 to A_1B_2. The recipient's new allocation will be X_2 and Y_2. Thus, if Y_2 is the socially optimal level of the public good, the grantor can achieve this level of output at least cost by use of a conditional matching grant (that is, the grantor's cost of achieving output Y_2 is DE_3 if a lump-sum grant is used, but only DE_2 if a matching grant is used). This result arises because the lump-sum grant produces only an income effect, while the matching grant reduces the recipient's price of the spillover-generating good, thereby creating a price effect and providing a more powerful incentive for the recipient to increase its spending on the spillover-generating good. Several studies of grant programmes in the USA confirm that lower-level jurisdictions of government do respond to such price effects and, as a result, recipient spending is stimulated more by a matching grant than by a lump-sum grant of equal size (Gramlich, 1977).[2]

To find the optimal matching rate for financing agricultural research, a model of optimal public investment will be employed (Harford, 1977). For each country, the model is defined by two equations:

$$NIB_i = B(R_i) - C(R_i) \tag{1}$$

$$NNB_i = \alpha_i B(R_i) - (1 - \delta_i)C(R_i), \text{ where:} \tag{2}$$

NIB_i = the net international benefit from research conducted in country i;
NNB_i = the net national benefit from research conducted in country i;
$B(R_i)$ = the benefit function of agricultural research conducted in country i;
$C(R_i)$ = the cost function of agricultural research conducted in country i;
α_i = the share of research benefits that accrue to country i as a result of agricultural research conducted in i;
δ_i = the share of the cost of agricultural research conducted in nation i paid by the international grantor;
R_i = funds spent on agricultural research in country i.

The optimal share of the cost of research in country i that should be paid by the international grantor can be determined by maximizing the national and international net benefit equations. Differentiating equations (1) and (2) provides the international and national conditions for optimal research spending:

$$\frac{dB}{dR_i} - \frac{dC}{dR_i} = 0 \tag{3}$$

$$\alpha_i \frac{dB}{dR_i} - (1 - \delta_i) \frac{dC}{dR_i} = 0. \tag{4}$$

Solving (3) and (4) simultaneously and rearranging provides the international grantor's optimal share of the cost of research conducted in country i:

$$\delta_i = 1 - \alpha_i \tag{5}$$

Thus the share of the cost of research paid by the international grantor varies directly with the proportion of marginal research benefits that spill out of country i. Equation (5) can also be expressed as the optimal matching rate for a conditional matching grant:

$$m_i = \delta_i/(1 - \delta_i) \tag{6}$$

Thus, to achieve the internationally optimal level of investment in agricultural research, the international grantor should provide country i with m_i dollars for each dollar of agricultural research provided by country i. When this condition is met, the marginal benefit of research that accrues to country i is equal to the marginal cost of research paid by its taxpayers, and the marginal benefit of research that accrues outside the country is equal to the marginal cost paid by the international grantor.

EMPIRICAL EVIDENCE OF RESEARCH BENEFIT SPILLOVERS

Few studies have attempted to measure the international spillovers produced by national investments in agricultural research. Edwards and Freebairn (1984) used economic surplus models to estimate the international welfare effects of wheat and wool research conducted in Australia. Their results indicated that anywhere from 2 per cent to 98 per cent of the total benefits of such research accrued to consumers or producers outside Australia. These estimates were extremely sensitive to (a) the relative elasticities of supply and demand and (b) the applicability of such research to production in other countries (that is, when producers in other countries can adopt the technology developed by research, the spillover benefits are much larger than when no outside adoption is possible).

Davis *et al.* (1987) also used economic surplus models to estimate the distribution of benefits that would result from research that yielded a 5 per cent reduction in the cost of producing 12 major commodities. Their results indicated that between 64 per cent and 82 per cent of the total benefits of such research would accrue to producers and consumers outside the country financing the research.[3] As with the Edwards and Freebairn study, differences in the spillover patterns among commodities were the result of differences in either the relative price elasticities of supply and demand or the size of the geographic area over which the research was assumed to be applicable.

Evenson (1977) used a production function model to measure the international spillovers that result from cereal grains research.[4] His results indicated that between 55 per cent and 69 per cent of the marginal benefits of cereal grains research conducted in developed countries accrue outside the country financing the research. Between 47 per cent and 82 per cent of the marginal benefits of cereal grains research conducted in developing countries accrue outside the country financing the research.

These results suggest that research benefit spillovers can be significant and that individual countries are unlikely to provide an internationally optimal level of investment in research. In such an environment of pervasive spillovers, institutional innovations are required if these positive externalities are to be internalized by national policy makers.

THE NEED FOR INSTITUTIONAL INNOVATIONS: SOME GUIDELINES FOR POLICY MAKERS

If an internationally optimal level of research investment is to be achieved, cost-sharing arrangements must be developed to coordinate research investments and compensate countries for the benefit spillovers they create. Public finance theory, combined with past empirical studies of research spillovers, provides several guidelines for designing these institutional innovations.

The share of the marginal cost of research paid by the international grantor should reflect the share of the marginal benefits of research that accrue outside the country in which the research is conducted

As indicated by equation (5), the international grantor should compensate the investing country according to the share of the research benefits that accrue outside its borders. This suggests that three factors would influence the share of the cost of research paid by the international grantor. First, the cost share paid by the international grantor will be greater when a greater share of the investing country's production is exported. Since a large share of research benefits accrues to consumers, research spillovers to foreign consumers will be greater when a larger share of the product is exported. Second, the cost share paid by the international grantor will be greater when the research is applicable by agricultural producers over a wider geographic area. When producers outside the investing country adopt the technology developed by research, benefits are created for both producers and consumers in the adopting countries (and perhaps for consumers in many countries if the commodity is exported by the adopting countries). Once again, benefits are spread beyond the borders of the investing country and a larger international investment is required. Third, basic research that is likely to provide widely dispersed benefits when it is adapted for use in a wide range of production regions, or that may enhance research in other areas of science, will also require the international grantor to bear a larger share of the cost of such research. While some observers have suggested that higher levels of government should finance basic research and lower levels of government should finance applied research (Idachaba, 1981, p. 106), it must be reiterated that public finance theory suggests that both applied and basic research must be funded through a cost-sharing system if an optimal level of investment is to be achieved, since both exhibit significant spillovers. From a public finance perspective, the problem of financing an optimal level of investment is the same; the only difference in these two types of research is the share of the cost paid by the international grantor.

When research is partially funded by a sub-national level of government, a three-tier cost-sharing system should be used to finance research

Since research is often funded in part by sub-national governments, especially in geographically large nations, any institutional innovations could also require a three-tier cost-sharing system. Evenson and Kislev (1975) estimated that 29 per cent of the benefits of research conducted by state governments in India accrue to states other than the one in which the research was conducted. Similarly, empirical studies indicate that 33 per cent to 68 per cent of the benefits of research conducted by the state agricultural experiment stations in the United States accrue outside the state in which the research was conducted (Ruttan, 1982). To solve this spillover problem, matching grants could be provided to the states by both the national government and the international grantor. In a three-tier cost-sharing system, each level of government (state, national and international) would pay the share of the marginal cost of research that equals the share of the marginal benefits of research that accrue within its jurisdictional boundaries (Harford, 1977).[5]

Matching grants could be used as a means of building political support for research and strengthening research institutions in developing countries

Although the spillover problem exists for both developed and developing countries, matching grants can play a dual role in strengthening the national agricultural research systems of many developing countries. The first role is that of compensating the developing countries for the research spillovers they create. The spillover problem is especially burdensome in developing countries that face severe resource constraints and that cannot afford to finance research that provides significant benefits outside their borders (Ruttan, 1987b,). A matching grant system would be the most efficient means of internalizing the research spillovers created by these countries.

Some observers contend that the major problem faced by the agricultural research systems of many developing countries is the lack of a political constituency to provide long-term political support for agricultural research (Eicher, 1989, p. 37; Ruttan, 1987b, p.92). Thus the second role for matching grants would be that of building a national constituency in support of stronger national research systems in developing countries. The experience of developed countries (for example, Japan and the USA) indicates that a cost-sharing system can provide the needed incentive for a local constituency to develop and support agricultural research. This local constituency can then play a major role in defining the research agenda and obtaining resources to support the research system (Ruttan, 1989, p.200; Pray 1988). Such a cost-sharing system would also provide greater funding stability than is currently available in many project-funding systems used by international donors (Ruttan, 1987a, 1987b, 1989; Trigo, 1987).

Institutional innovations are also needed to finance the international grantor's budget for research

Although this paper has focused on the policy options for allocating research resources, it has not considered the means of providing the resources to be used by the international grantor. Musgrave (1986) contends that a combination of direct consumer taxes and an *ad valorem* tax on cost payments would be the most equitable means of financing a public good that creates international spillovers and reduces all private costs equally. Such a tax would allocate the cost of the international granting agency according to the benefit principle (that is consumers and producers would pay for agricultural research according to the share of benefits accruing to each group). Given the lack of an international institutional structure for levying such taxes, the establishment of a resource base for the international grantor would be the most challenging problem faced by policy makers.

IS INTERNATIONAL COORDINATION
OF RESEARCH INSTITUTIONS POSSIBLE?

Although the problem of financing agricultural research in the presence of research spillovers is widely recognized and the economic prescriptions for financing research are available, the task of establishing the necessary coordinating institutions is formidable. No institutions of the sort outlined above are available at the international level. The international research centres have had some success in spawning regional research programmes directed at the common problems of developing countries (Ruttan, 1987b), but these efforts do not provide an adequate level of research funding or a *coordinated system* for internalizing research externalities or financing research in the presence of benefit externalities. Similarly, research funding provided by foundations or national donors is likely to be episodic rather than a continuing, systematic means of internalizing research externalities and developing national and local support for research.

The first major problem in designing such institutions is the need to develop effective power-sharing arrangements that preserve national (or national and local) autonomy in establishing a research agenda and still maintain effective financial coordination of research. National policy makers, quite rightly, can be expected to protect their autonomy in establishing the research agenda. Because of the location-specific nature of agricultural production and the many unique ecosystems in agriculture, local autonomy in establishing the research agenda is necessary if users' problems are to be articulated to research managers and political decision makers. At the same time, an international granting agency would demand accountability for any funds granted to national research institutions and must be able to articulate international research priorities that require attention at the national level. Some nations with multi-level systems of government have been able (with mixed success) to reconcile these tensions within their own research systems, but the prospects for political jealousy and opportunistic behaviour are escalated by an order of magnitude when the problem of international coordination is considered.

Second, we must recognize that no system of coordinating institutions can ever achieve more than a rough measure of equity in financing research. The problems of establishing reliable measures of research benefit spillovers are substantial and, under the best of circumstances, such estimates are likely to be highly sensitive to the assumptions underlying such analysis.[6] Nevertheless, agricultural economists must continue to pursue this line of work if we are to make a meaningful contribution to the research policy debate and to the design and development of appropriate supporting institutions.

Finally, larger political obstacles must be overcome before the problem of international research will be placed on the policy agenda in many countries. The need to build local political support for research institutions is not yet appreciated by many policy makers, and domestic agricultural research – let alone international research coordination – is rarely viewed as a high priority. In addition, agricultural research must ultimately compete with many other public investment opportunities for funding. In some cases, these alternative investments may also require international coordination of benefit externalities, and efforts to coordinate these investments may have a higher priority than agricultural research investments.

CONCLUSION

The theoretical and empirical results presented in this paper indicate that the achievement of an optimal level of investment in agricultural research will not be possible unless significant investments are made in institutional innovations that internalize the international benefit spillovers created by research. These innovations must establish coordinated cost-sharing arrangements that provide appropriate incentives for financing research, while still maintaining national autonomy in establishing the research agenda based on the problems articulated by diverse local constituencies. Such a system may be the only means of promoting an efficient level of research investment, adapting research to local conditions and establishing the political base necessary for the long-run survival of research institutions.

The problems of international research coordination are likely to increase in importance for both developed and developing countries. Any reduction of trade barriers – through the GATT process, bilateral agreements or free trade areas – will increase the magnitude of research spillovers and heighten the need for coordination in financing research. At the same time, developing countries must establish stronger local constituencies in support of research and the means to overcome political boundaries that are inconsistent with the geographic boundaries of research spillovers. Economists cannot by themselves cause the creation of international institutions to internalize spillovers, but they can help to avoid institutional misdirection and failure in the original design when and if such institutions are established. Equally importantly, economists can define and give visibility to the problem that politicians and others must understand before nations are likely to act on the problem of spillovers.

NOTES

[1]This paper will refer to an 'international grantor' as the institution responsible for providing grants to the national government. No such institution exists at the present time. The final section of this paper will examine the form such an institution might take.

[2]The Hatch Act research funding system is a matching grant programme designed to compensate states in the USA for the benefit spillovers they create. The national government matches each state dollar spent on agricultural research with one national dollar for research (Knoblauch, *et al.*, 1962).

[3]The spillover shares (per cent) for the commodities were: rice (65), sugar (68), coconuts (72), groundnuts (73), maize (75), bananas and plantains (75), sweet potatoes (77), sorghum (77), pulses (79), sheep and goats (79), wheat (80), and potatoes (82).

[4]The production function approach uses regression methods to estimate the impact of changes in conventional inputs and public inputs (including research) on output. When specified to include research in other countries, estimates of research benefit spillovers can then be derived from the regression results.

[5]The Smith-Lever system is an example of a three tier cost-sharing system for financing agricultural extension services in the United States (that is, the cost is shared by national, state and county governments through a matching grant system).

[6]Consider, for example, Edwards and Freebairn's estimate (1984) that anywhere from 2 per cent to 98 per cent of the benefits of Australian wheat and wool research can accrue outside the country. This wide range of estimates arises from different assumptions about the elasticities of supply and demand and the applicability of such research to production in other countries.

REFERENCES

Boadway, R.W. and Wildasin, D.E., 1984, *Public Sector Economics*, 2nd rev ed, Little, Brown and Company, Boston.

Breton, A., 1965, 'A Theory of Government Grants', *Canadian Journal of Economics and Political Science*, 31, pp. 175–87.

Carter, C.A., 1985, 'Agricultural Research and International Trade', in K.K.Klein and W.H.Furtan (eds), *Economics of Agricultural Research in Canada*, University of Calgary Press, Calgary.

Davis, J.G., Oram, P.A. and Ryan, J.G., 1987, *Assessment of Agricultural Research Priorities: An International Perspective*, Australian Centre for International Agricultural Research, Canberra.

de Castro, J.P.R. and Schuh, G.E., 1973, 'An Empirical Test of an Economic Model for Establishing Research Priorities: A Brazil Case Study', in T.M. Arndt, D.G. Dalrymple, and V.W. Ruttan (eds), *Resource Allocation and Productivity in National and International Agricultural Research*, University of Minnesota Press, Minneapolis.

Edwards, G.W. and Freebairn, J.W., 1984, 'The Gains from Research into Tradable Commodities', *American Journal of Agricultural Economics*, 66, 41–9.

Eicher, C.K., 1989, *Sustainable Institutions for African Agricultural Development*, International Service for National Agricultural Research, Working Paper number 19.

Evenson, R.E., 1977, 'Comparative Evidence on Returns to Investment in National and International Research Institutions', in T.M. Arndt, D.G. Dalrymple, and V.W. Ruttan (eds),

Resource Allocation and Productivity in National and International Agricultural Research, University of Minnesota Press, Minneapolis.

Evenson, R.E. and Kislev, Y., 1975, *Agricultural Research and Productivity*, Yale University Press, New Haven.

Fishel, W.L., 1971, 'The Minnesota Agricultural Research Resource Allocation Information System and Experiment', in W.L. Fishel (ed.), *Resource Allocation in Agricultural Research*, University of Minnesota Press, Minneapolis.

Gramlich, E.M., 1977, 'Intergovernmental Grants: A Review of the Empirical Literature,' in W.E. Oates (ed), *The Political Economy of Fiscal Federalism*, Lexington Books, Lexington, MA.

Harford, J., 1977, 'Optimizing Intergovernmental Grants With Three Levels of Government', *Public Finance Quarterly*, 5, pp. 99–115.

Idachaba, F.S., 1981, 'Agricultural Research Resource Allocation Priorities: The Nigerian Experience', in D. Daniels and B. Nestel (eds), *Resource Allocation to Agricultural Research*, International Development Research Centre, Ottawa.

Idachaba, F.S., 1989, *State-Federal Relations in Nigerian Agriculture*, World Bank MADIA Discussion Paper number 8.

Judd, M.A., Boyce, J.K. and Evenson, R.E., 1987, 'Investment in Agricultural Research and Extension', in V.W.Ruttan and C.E.Pray, (eds), *Policy for Agricultural Research*, Westview Press, Boulder, Co.

Knoblauch, H.C., Law, E.M. and Meyer, W.P., 1962, *State Agricultural Experiment Stations: A History of Research Policy and Procedure*, USDA, Miscellaneous Publication Number 904.

Musgrave, R.A., 1986, 'Inter-nation Equity', in *Public Finance in a Democratic Society: Fiscal Doctrine, Growth, and Institutions, Volume II of the Collected Papers of Richard A. Musgrave*, New York University Press, New York.

Norton, G.W. and Pardey, P.G., 1987, *Priority-Setting Mechanisms for National Agricultural Research Systems: Present Experience and Future Needs*, International Service for National Agricultural Research, Working Paper number 7.

Paz, L.J. 1981, 'A Methodology for Establishing Priorities for Research on Agricultural Products' in D. Daniels and B. Nestel (eds), *Resource Allocation to Agricultural Research*.

Pigou, A.C., 1946, *The Economics of Welfare*. 4th rev ed, Macmillan, London.

Pray, C.E., 1988, 'Mobilizing Support for Agricultural Research at the University of Minnesota', *Agricultural Administration and Extension, 28,* pp. 165–80.

Ruttan, V.W., 1982, *Agricultural Research Policy*, University of Minnesota Press, Minneapolis.

Ruttan, V.W., 1987a, *Agricultural Research Policy and Development*, Research and Technology Paper 2, United Nations Food and Agriculture Organization, Rome.

Ruttan, V.W., 1987b, 'Toward A Global Agricultural Research System," in V.W.Ruttan and C.E.Pray (eds) *Policy for Agricultural Research*.

Ruttan, V.W., 1989, 'The International Agricultural Research System'. in J.L. Crompton, (ed.), *The Transformation of International Agricultural Research and Development*, Lynne Rienner Publishers, Boulder, Co.

Scott, A.D., 1952, 'The Evaluation of Federal Grants', *Economica*, 19, pp. 377–94.

Trigo, E.J., 1987, 'Agricultural Research Organization in the Developing World: Diversity and Evolution', in V.W.Ruttan and C.E.Pray, (eds), *Policy for Agricultural Research*.

DISCUSSION OPENING – HSI HUANG CHEN*

Discussion of the problems of financing agricultural research in the presence of international benefit spillovers leads the authors to conclude that research and innovation must be based on coordinated cost-sharing arrangements which provide appropriate financial incentives. Although I want to qualify that conclusion I have no reason to argue with the basic logic of the paper. However,

*National Taiwan University, Taiwan.

at a global level, my view is that some of the major themes could have an alternative interpretation.

Agriculture has become increasingly integrated into the national and world economy, and this redefines the context in which agricultural research is funded by national governments and international donors. We should distinguish between economic and financial profitability of research so that the investing country can weigh the real value of pursuing research. All too often, however, investing countries base their decisions on financial rather than economic analysis. We know that resource use in research responds to the benefits to be received and that national governments should be more concerned about research returns. In this context, the authors provide some very useful guidelines for designing institutional settings for research. These are of particular interest, given renewed emphasis on cost-sharing arrangements to coordinate activity and compensate countries for benefit spillovers.

However, the cost-sharing proposal is theoretically sound but practically weak. It is difficult to estimate benefit spillovers prior to research, or to give relative weights to the various parties who might be affected when the results are finally adopted. Yet it is precisely that flow of benefits which needs to be taken into account in assessing the extent to which cost sharing needs to occur. It is also important to emphasize that analysis needs a long-run focus, since there may be many feedback effects at work influencing the distributive impact of research. My view is that we know too little about techniques of measuring the returns to research, in the long run and at the global level, to make significant progress.

The other interesting issue raised in the paper relates to the key question of raising revenue to fund research, though it was covered only briefly in the section discussing Musgrave's proposal. This is an issue concerning the distinction between efficiency in research itself and its ultimate equity effect. We do not know how much would be raised by a combination of a tax on consumers and a charge on cost payments to impinge on producers, though my own guess is that a levy of only one per cent could yield a large revenue base. I have already remarked on the problems of estimating the magnitude of research benefits in an appropriate way. What I am emphasizing now is the equally important issue of creating an international fund to provide the means of fostering the supply of research effort.

M. P. COLLINSON AND K. WRIGHT PLATAIS*

Biotechnology and the International
Agriculture Research Centres of the CGIAR

INTRODUCTION

The euphoria of the 'green revolution', which was at once the first-born and the father of the Consultative Group on International Agricultural Research (CGIAR), is on the wane. Herdt and others have drawn attention to the fact that existing yield potential in rice, the basic food staple for the majority of the world's poor, is already being achieved by farmers in areas where population pressures are greatest. There is no more slack (Herdt, 1988; De Datta *et al.*, 1988). W. David Hopper, recently retired Chairman of the CGIAR, has commented:

> Unless there is a significant advance in productivity, far greater than I see evident from the present data from the CGIAR centres and others engaged in tropical agricultural research, I do not see us being able to beat the Malthusian proposition by 2000 or 2030....the next significant advances must come from genetic engineering. (Hopper, 1990)

The CGIAR system will be one set of institutions helping to disseminate the benefits of biotechnology.

THE CGIAR IN PERSPECTIVE

The CGIAR perceives itself as an international component in a global agricultural research system. One role is to link scientific capacity in the developed and developing worlds for the benefit of small Third World farmers and poor consumers. Historically, the reputation of the CGIAR system has rested on the development and transfer of improved germ-plasm, particularly the short-strawed rices and wheats which contributed to the 'green revolution'. Improved germ-plasm of other crops from the newer International Agricultural Research Centres (IARCs) is increasingly moving into farmers' fields. This germ-plasm improvement is paralleled by world-wide efforts to conserve crop genetic resources and to improve the management and productivity of both crops and animals in small farming systems.

*Consultative Group on International Agricultural Research, Washington, USA.

Global and regional research remains ivory-towered, unless effectively mobilized by Third World national agricultural research and development systems. A vital goal for the CGIAR is the build-up of national agricultural research institutions to meet the needs of their small farmer clients with new technologies. Some 20 per cent of the annual CGIAR budget is spent on training, institution building and food policy analysis towards this goal.

After the current expansion from the 13 to 18 IARCs is complete, the expected budget of approximately US$320 m. a year for the CGIAR will still represent significantly less than 10 per cent of total annual investment in developing country agricultural research and less than 0.7 per cent of official development assistance. The CGIAR annual budget is just over 50 per cent of the US$612 m. R&D budget of Monsanto, a major agricultural chemical company ranked twenty-fourth for corporate R&D spending in the USA. As another comparator, the number of scientists employed in agricultural research in developing countries, excluding the Peoples Republic of China, was estimated to be over 45 000 for the period 1980–5 (Pardey and Roseboom, 1988). The CGIAR employs some 1 700 scientists with MSc qualification or higher.

In the USA, a dual pattern has emerged in the private sector. On the one hand, large agricultural chemical and seed companies such as Monsanto have moved into biotechnology to exploit complementarities in their product range. Monsanto has 240 senior scientists working in animal and plant biotechnology with an annual expenditure of about US$55m., approximately 9 per cent of the company's total R&D budget (Fraley, personal communication, March 1991). On the other hand, small venture capital companies have also emerged. One such company is Calgene of Davis, California. In 1989, Calgene had an annual R&D budget of US$10.5m. supporting 50 research scientists, equal to one-half of the value of the company's product sales for the year (Calgene, spokesperson, personal communication, March 1991).

The 1990 CGIAR budget included approximately US$14.5m. for biotechnology, about 4.5 per cent of the budget for the system as a whole. Half of this was for plant biotechnology, distributed across nine crop-improvement centres working on some 15 crops. The other half went to animal biotechnology, primarily at ILRAD, a centre working on animal diseases and focused exclusively on state-of-the-art molecular biology. While ILRAD expenditures have grown slowly over the years, the budget for plant biotechnology in the CGIAR centres has increased to US$7.5m from US$4.3m. over the last two years. Currently, across the CGIAR system, some 65 senior scientists, including post-doctoral fellows, work on biotechnology – half of these in the plant sciences.

BIOTECHNOLOGY IN THE CGIAR

The diverse range of biotechnology research activities found in the IARCs is indicative of the wide scope of the available techniques. Current research includes plant tissue and cell culture (anther culture, somaclonal variation, meristem culture, rapid clonal propagation), *in vitro* germ-plasm conservation, molecular diagnostics (nucleic acid probes, monoclonal antibodies, enzyme-linked immunosorbant assay (ELISA)), embryo rescue and genetic engineering (Plucknett, Cohen and Horne, 1990).

The CGIAR crop improvement centres have been cautious, some would say over-cautious, in embracing biotechnology. The CGIAR first addressed issues in biotechnology at the Inter-Centre Seminar on 'The IARCs and Biotechnology', hosted by IRRI in 1984. The following year a paper was presented at a joint CGIAR/Japanese forum on biotechnology (Plucknett, 1985). The Technical Advisory Committee (TAC) – a group of eminent scientists which guide the scientific programme of the CGIAR – also approached biotechnology cautiously. In 1988, the Committee issued the first policy statement on biotechnology in the CGIAR. Most recently, a joint statement by the CGIAR Centre Directors at the Mid-Term CGIAR Meeting, in May 1990 noted:

> It is for each centre to determine the role that biotechnology can play in solving the problems that it confronts in fulfilling its mandate, to assess its fitness to apply biotechnological solutions, and its comparative advantage as a player within the CGIAR system and the scientific research community as a whole. (CGIAR, 1990)

CGIAR biotechnology strategy

Genetic engineering of crops requires a sophisticated research effort. Private companies are particularly adroit at organizing the complex and expensive teams needed. For example, the successful engineering of *Bacillus thuringensis* endotoxin genes in several companies appears to have involved 5–10 full-time scientists working directly on each effort for several years to develop resistant plant materials worth taking to the field. At $100 000 per scientist per year, this represents an investment of the order of $1.5–3m. to get one gene transferred into one crop – a significant investment (Mueesen, 1990). Such modern biotechnological research in plant science provides intermediate products, rarely useful in isolation and dependent on the traditional routines of plant breeding to bring the improved product to farmers.

'Understanding of the biology of plants at the molecular level needs to be dovetailed with understanding of the plants at the tissue, whole plant and population levels, and finally reconciled with the socio-economic circumstances of producers' (Javier, 1990).

The strategy of the IARCs embraces this perspective of biotechnology as the tip of the R&D iceberg in agriculture. A balance between biotechnology and conventional research is crucial to the CGIAR role in the sustainable improvement of agricultural productivity in developing countries. Thus the strategy of the CGIAR crop improvement centres is to use relevant, tested, biotechnology techniques for more efficient resolution of their existing research agendas.

The well established linkage strategy of the IARCs, coopting and passing on useful research techniques to National Agricultural Research Systems (NARS), will also be a way forward in biotechnology, particularly for smaller

NARS. One route may be to link the IARCs with LDC universities as future centres of strategic research. As skills mature, the universities can then exploit IARC connections to begin their own interactions with advanced institutes and companies in the industrial economies.

Organization

The 'in-house' management of biotechnology in the IARCs is rooted in this strategy and is still in a formative stage. Centres each have one, two or sometimes three scientists who act as coordinators. They listen to programme needs in the centre and monitor 'the biotechnology world'. The coordinators bring demand and supply together with the help of senior managers who have increasing exposure to contractual, legal and safety concerns. In some centres individuals have been the catalysts in institutionalizing biotechnology with equipment and laboratory space allocated to new techniques. One or two of the larger centres have invested in laboratories and equipment as separate units for their biotechnology initiatives.

Networks

The use of networks to exchange research personnel, techniques and information as a means of acquiring new technologies is extensive in the CGIAR and has found ready application in the field of biotechnology. Research tools are found in both private and public advanced institutions interested in similar problems. Monitoring these collaborative opportunities requires informal networks of professionals who respect each other's views. It requires knowledge of the market ambitions of the private and public sector, and the ability to negotiate with biotechnology proprietors in a creative fashion (Plucknett, Cohen and Horne, 1990). Informal collaborations with scientists in industrial countries also enable the IARCs to explore new and more relevant research techniques. Formal and informal networks, involving both advanced institutes and developing country NARS enable the transfer of both IARC products and new research techniques.

The first biotechnology network is in rice and was established in 1985 with major assistance from the Rockefeller Foundation. It links advanced laboratories in Europe, the United States and elsewhere with the International Rice Research Institute (IRRI) and CIAT. Molecular maps are being developed to allow breeders to determine whether an individual rice plant contains known genes. Many potentially useful genes have been cloned from rice, including the gene 'oryzacystatin', an inhibitor of the digestive enzymes of insect pests (Toenniessen, personal communication, March 1991).

The formation of the Cassava Biotechnology Network in 1988 at CIAT, involving scientists in Latin America, Europe and the United States, brought new techniques to difficult research problems in cassava. Most of the network's research is taking place in developed country university laboratories. Activities include studying the biochemistry and genetics of cyanogenesis.

Social scientists will assist with farming system studies in areas where cyanogenic cassava is grown (CIAT, 1990). Findings will help define research objectives: whether to eliminate cyanide throughout the plant, or to increase specific enzyme activity to reduce the presence of cyanogenic compounds. A further goal is the production of true seed as a joint venture between the International Institute of Tropical Agriculture (IITA) and CIAT. This is sought through the incorporation of apomictic genes to obtain unfertilized embryos. The research will be complemented by a socio-economic study to examine the acceptability of seed as an entirely new planting method for cassava.

The Centro Internacional de la Papa (CIP) and CIMMYT are also effectively using collaborative networks. CIP has established an extensive network of collaborating institutions in the United States, China, Israel and Europe for activities in genetic engineering and RFLP analysis. Efforts such as this have afforded CIP access to scientific expertise in a cost-effective manner. As of 1990, CIP had received over 30 gene constructs at a cost of only US$60 000 (Dodds and Tejada, 1990). CIMMYT has established the CIMMYT–North/Latin American RFLP Network which will map the genome of maize (Plucknett, Cohen and Horne, 1990). Currently seven US universities, EMBRAPA (Brazil) and CIVESTAN (Mexico) are participating. Network objectives are to examine RFLP's ability to mark quantitative traits, as well as to serve as a marker-assisted selection aid in breeding.

Technology transfer and training

ILRAD is an exception among the CGIAR centres as 'an advanced institute' using biotechnology in the same manner as university and private sector institutions in the industrialized countries. ILRAD's 15 years of research experience incorporating both the traditional and the newer biotechnology techniques have enabled it to demonstrate a degree of bridging with developing country laboratories.

The main goals of the centre are the development of vaccines against theileriosis and trypanosomiasis. These involve recombinant DNA research relevant to analyses of (1) organization and expression of specific genes in both parasite and mammalian hosts; (2) production of antigens to serve as vaccine candidates; and (3) expression of identified genes in bacterial and viral systems. Embryo transfer and embryo splitting is also used for analyses of immune response to parasite infections and for trait/gene mapping to determine the genetic base of resistance (Doyle, personal communication, 1991). Regular training courses and graduate research programmes pass on information and techniques to national programmes and ILRAD scientists provide technical support and expertise at the request of national governments. In 1989, ILRAD participated in national disease-control planning seminars in Ethiopia, Kenya, Tanzania, Uganda and Zimbabwe (ILRAD, 1990).

Across the CGIAR the diffusion of research techniques is actively fostered by the training programmes and the collaborative and consultative networks supported by the IARCs. The 13 IARCs in the CGIAR gave training, mainly short courses in research techniques, to an estimated 25 000 developing country scientists over the five-year period 1985–9. Research fellowships in biotechnology at the higher degree level for up to two years have been offered at several centres. Most common, however, are short courses. Researchers are trained in aspects of biotechnology before returning to their home institution to apply what they have learned. These same people then become valuable collaborators of the IARCs in future activities with the NARs.

Centres have a responsibility to be aware of the needs of national programme scientists in biotechnology research. CIAT recently surveyed Latin American and Caribbean institutions on their needs for advanced training in modern biotechnology and its application to agricultural problems. Of the 60 national universities and research institutions surveyed, 51 responded. The results from this survey will help CIAT plan for advanced training courses to be offered over a subsequent three-year period (Roca, personal communication, April 1991). Short-term personnel exchanges with developed country institutions, including private laboratories, also help the centres establish the basis for effective collaborations.

Institution building

The Netherlands government commissioned ISNAR to study existing agricultural biotechnology research in Colombia, Indonesia, Kenya and Zimbabwe. These studies are part of a series of studies commissioned in conjunction with the World Bank/ISNAR/Australian government agricultural biotechnology study. Study teams will work with scientists and policy makers to identify opportunities for building biotechnology into national systems. The studies are focused on priorities in research, public- and private-sector investments, regulatory procedures, management of intellectual property and technology transfer in the international arena (ISNAR, 1990). These studies will provide collaborative learning opportunities and guidelines for similar work in other countries.

As a group, the IARCs are hesitant to promote biotechnology training in countries without significant plant-breeding programmes and without appropriate laboratory facilities. Experience shows that researchers returning to a national system with no means of mobilizing the skills learned are frustrated and often become brain-drain candidates. Such outcomes are also a waste of IARC resources. Approaches which lend themselves more directly to commercial investment offer an alternative. In 1986, a biotechnology programme was established by the Andean Development Bank (CAF – Corporacion Andina de Fomento). The IARCs in the Latin American region were enlisted to assist in some of CAF's activities on a 'backstopping' basis. A biotechnology unit was established by CAF in Caracas, Venezuela, to work on problems relevant to the Andean region. CIP assists in CAF-supported projects in five Andean countries; Peru, Bolivia, Ecuador, Colombia and Venezuela. These range from the use of tissue culture for improved seed production and virus irradiation to *in vitro* tuber production for seed to facilitate transport of planting materials to remote areas (Dodds and Tejada, 1990).

Examples of direct help for NARs in building biotechnology facilities are few to date. Sigatoka, a devastating fungal disease of plantains, entered Nigeria in 1986 and IITA has since helped establish two plantain and banana tissue culture laboratories there. This is the first large-scale tissue culture operation for production and distribution of planting material in Nigeria (IITA, 1989).

ISSUES IN THE CGIAR'S ROLE IN BIOTECHNOLOGY

Biotechnology has been likened to the microchip – a generic technology with spin-offs to wide areas in both industry and agriculture. Expectations have drawn many developing countries into significant investments in biotechnology R&D. Biotechnologies are perceived to have particular advantages for LDC agriculture: lower cost to research budgets and to farmers, enhanced stability, an 'environmentally friendly' image resulting from less dependence on purchased pesticides and less reliance on sophisticated infrastructure by reducing the volume of inputs to be distributed (Mueesen, 1990).

The CGIAR strategy will help bring these advantages to developing country farmers for the commodities in which it is mandated. Nevertheless, the CGIAR strategy, limited as it is, faces formidable challenges. These demand active and innovative management from individual centres and the system as a whole. At the heart of these challenges is the changing research environment particularly in developed countries.

> The changes are driven by dynamic interactions between emerging opportunities in biotechnological research, increasing interest by public research institutions in developed countries in controlling and commercializing research outputs, restructuring of, and closer actions between private and public agricultural research industries, constraints on public research budgets and by an extending domain of intellectual property rights... Inappropriate response to the changes by the IARCs could adversely affect their working relationships with national agricultural research systems, and reduce their attractiveness to research collaborators. (ICRISAT, 1991).

IARCs: future access to biotechnologies

The IARCs experience and expertise in crop improvement is a strong attraction to the private sector. This can be attributed in part to the extensive germplasm collections. Each IARC holds a major collection of the germ-plasm for its mandated crops. IBPGR promotes gene banks world-wide, including banks in NARS. It collaborates closely with the CGIAR crop-improvement centres in maintaining their collections. It is widely believed that the value of these collections will escalate as biotechnology expands and the genes they contain gain importance. They represent a major asset for the IARCs in dealings with the private sector. The real strength of the crop-improvement IARCs is in conventional plant breeding. The centres organize global and regional networks for international testing sites and have long-standing collaborative arrangements for testing and for distribution of materials with national systems in developing countries.

The universities and multinationals in the industrial countries respect these strengths. Yet the IARCs in the CGIAR have traditionally operated an open-door policy. All bona fide clients, whether public or private, in the developed or developing countries, have access to IARC germ-plasm. The IARCs have monitored its use and have to date successfully deterred commercial companies from patenting their materials. The ongoing shift to private sector research in the industrial economies juxtaposed to the advent of modern biotechnology, creates a major dilemma for the IARCs.

It should be borne in mind that current discussions remain hypothetical. Currently no 'economic genes' have been engineered into CGIAR crops. Even when such engineering is achieved, intellectual property rights (IPR) negotiations (and eventual contracts) will be crop- and site-specific. Genetically engineered maize for the US Mid-West will in all probability be unsuitable for Africa. Much of the ongoing discussion is aimed at shaping a CGIAR position.

Orphan crops

It is generally agreed that the IARCs and NARS will have access to private sector biotechnologies for commodities in which there are no extensive markets at stake. The fact that venture capital companies and university initiatives in biotechnology are discrete operations relying on client contracts creates opportunities for the centres as clients and as partners. While access looks feasible, the stock of research experience and the breadth of biotechnology in the so-called 'orphan crops' activity are likely to remain limited. Within this 'orphan crops' group, cassava, sorghum, millet and plantains are of particular importance. The CGIAR may provide the only means of cassava improvement, yet the cost of leadership can already be seen in the investment levels for conventional research. The IARCs currently provide one quarter of the investment in cassava research, with NARS in cassava-producing countries the other three-quarters. That is approximately 400 per cent of the comparative ratio for the IARCs/NARS as a whole (Bertram, 1990). Orphan crops such as cassava have little free-rider potential and the CGIAR costs for similar products in the orphan group of crops are expected to be high.

The private sector is particularly interested in the fact that the IARCs dominate strategic research in the 'orphan crops' group which is already producing sources of genes widely useful in industrial country commercial crops. Such generically useful genes will also be a focus for inter-IARC collaboration. An issue here is how aggressive the IARCs should be in capturing the benefits from multiple-use genes derived from the orphan crops. It seems likely that the advanced institutes see such genes as the major reason for collaboration. Contract prices are likely to reflect the expectations of such benefits and, where IARC germ-plasm is one factor in negotiations, arrangements may be wholly collaborative.

Commercial crops

In terms of access to biotechnologies, the most difficult category of crops for the CGIAR is those important in both industrial and developing economies. Maize, rice, potato and wheat are the major ones. The three cereals listed account for 80 per cent of starch staple food production in the developing world.

The real arena for the IARCs and the private sector is commodities of commercial interest in both industrial and developing economies. The policy debate on reducing subsidies to the agricultural sector of the industrial world threatens the domestic markets of the large agricultural multinationals and increases their interest in new markets overseas. Their biotechnology capacity may become a major bargaining chip for market access in developing countries. In 1980, the Chinese signed an agreement with Ringaround Seed Products (Occidental Petroleum) and Cargill, giving them the exclusive rights to the production and marketing of hybrid rice in certain parts of the world. Since that time the private companies have pressured the Chinese not to share the information and germ-plasm involved in their hybrids with IRRI. In another case, rather than accept the restrictions imposed by a private company on the use of its research in wheat gene mapping, CIMMYT has sought public sector funding to support wheat gene mapping research in the USA (Barker and Plucknett, 1991).

Intellectual property rights

To date few centres have been exposed to intellectual property issues. In a response to a query from the CGIAR Chairman in November 1989, five centres indicated that they had experience with managing intellectual property. Two centres had patented farm equipment or fertilizer preparations; one shared a patent with a collaborating institution in a developed country; the other two had filed for patent protection, one for a variety, the other for a biotechnologically developed vaccine. None has yet filed for plant variety protection for varieties developed at centres (Barton and Siebeck, pre-print version).

In the context of these discussions, two recent reports (Persley, 1990a; DGIS, Netherlands, 1991) have recommended a break with the open-door tradition. They emphasize that effective bridging of the benefits of biotechnology to the developing countries requires the IARCs to work with the private sector, itself heavily dependent on product protection to earn a return on capital invested in R&D.

If the IARC's adopt a protective strategy and patent their products they can license NARS to use the products without cost. Yet commentators perceive the private sector as nervous about the IARCs' ability to maintain the security of IPR agreements. Can the IARCs ensure that protected products from the private sector will not go astray? Even more difficult is the question of whether the IARCs can condition their own clients, particularly public sector institutions, not to 'leak' products to companies competing in the established markets of the industrial economies. The private sector badly needs assurance of the security it can expect from the IARCs. Current uncertainty is a major inhibitor to collaboration (Mueesen, 1990). Some see the legal problems of access increasing in intensity as the public universities and private sectors grow closer. Others cite the helpful initiatives by USAID in gaining access through US companies, in part by providing funding for collaborative work (Dodds, personal communication, March 1991). Countries in their roles as donors may provide leverage for the developing countries where damaging restrictions can be anticipated.

A recent report (Barton and Siebeck, pre-print version) commissioned in consultation with the IARC Centre Directors, draws the following conclusions for the IARCs access to biotechnologies and IPR:

(1) Much of the activity of the centres can continue without intellectual property protection and without breaking the centres' traditions of open scientific exchange. This is appropriate for centres working in countries that have not extended intellectual property protection to plants and biotechnology and for centres whose mandate crops have little commercial interest for industrial country markets.
(2) As it will be imperative for the centres to continually gain access to new proprietary technologies of potential use to developing countries they will require understanding of patents and licensing. This will also entail acceptance of restrictions on the free inter-system exchange of materials.
(3) Each centre should carefully review the institutional linkages as to how its innovations reach the farmer, and for protection when marketing the innovations to developing country farmers.

There will be many management challenges to the IARCs as new patterns develop. One particularly difficult area will be in retaining the trust of the NARS in countries with a long-standing concern about the motives of the multinational corporations. It may be that some corporations value the IARCs as a channel for access to developing country markets. The IARCs will have to remain very clear, in a difficult balancing act, that their allegiance remains with developing country farmers. There are concerns, given the highly political nature of the government/multinational interaction in many countries, that the IARCs will occasionally fall foul of the process. Commentators have made a case for centralized support on legal questions accessible by all IARCs (Persley, 1990b; Barton and Siebeck, pre-print version; Beachy, personal communication, 1991).

IARCS and NARS organizational issues

There are organizational issues at each end of the 'biotechnology bridge'. The type of staff the IARCs will need for effective monitoring of biotechnology, given its rate of expansion and increasingly sophisticated applications, may be difficult to attract. Some centres already feel inadequate to evaluate the potential of emerging techniques. Doubts have been expressed about their ability to hire and retain state-of-the-art molecular biologists. Attracting them away from the 'cutting edge' laboratories, asking them to monitor and implement techniques developed elsewhere, and wondering whether they can keep up with mainstream research while remaining somewhat isolated are serious aspects for consideration. A system of retainers or joint appointments between IARCs and university institutes may be a more viable alternative (Dodds, personal communication, 1991).

Even larger LDCs which have set up full biotechnology laboratories have faced stumbling-blocks. Two barriers seem to be: (1) a failure by governments to complement the new institution with friendly policies to minimize the logistical and licensing problems of importing raw materials and equipment; and (2) the failure of central units to link with the institutions implementing the main R&D process in agriculture. Central units may be effective, if they are vertically integrated with a demand-driven agenda, as well as a channel reaching the rural areas. Many commentators advocating biotechnology investment in LDCs under-estimate the institution building required throughout the R&D process to bring appropriate technologies to farmers. In some cases, where fears of renewed dependency can be allayed, the private sector will offer the less torturous alternative.

The logistics and licensing of imported materials also affect the ability of some IARCs to exploit biotechnology at the molecular level. In IARC host countries new developments in legislation on both bio-safety and IPR will be important. Most centres have internal bio-safety committees and there are adequate guidelines (from the National Institutes of Health, the Animal and Plant Health Inspection Service (APHIS) in the USA and from the EC) on the laboratory conditions for containment in genetic engineering and the field testing of engineered plants. One question still at issue is the release of genetically engineered plants in areas of natural diversity. It will be important to those IARCs deliberately located near centres of origin of their mandate crops. In one or two cases the IARCs have been able to help host countries draw up their own bio-safety codes. It is clearly important that the IARCs operate within the laws of their host countries. As legislation is passed, current memoranda of agreement between host country and centre may be affected.

A LOOK AHEAD

The IARCs are well positioned to contribute to disseminating appropriate biotechnologies to LDC agricultural research. Their strengths are the relevance of their agenda and their growing links with biotechnology in the industrial countries. The IARCs are building these strengths on both sides. On the research agenda side several have recently brought regional groupings of NARS into their priority-setting processes. The result is an increasingly demand-driven CGIAR, using its global perspectives on productivity, equity and sustainability to allocate resources across regionally identified priorities.

Currently there are strong donor pressures on the CGIAR on both funding and programmes. The long-term nature of strategic and applied research brings such pressures when donor domestic budgets are tight. This time it is reinforced by other funding needs: the environment, Eastern Europe and the needs of the NARSs themselves. These have drawn attention away from the CGIAR. Yet donors see the CGIAR as successful. They want stronger programmes in three areas — biotechnology, resource management and the environment. They also want greater support to strengthen NARS. These demands pull the CGIAR in opposite directions, upstream to more strategic research and downstream to build adaptive research capacity in the NARS. This contradiction also pervades biotechnology for which the NARs require strong plant-breeding programmes in which the IARCs continue to invest heavily in training.

The CGIAR is planning to restructure itself around two types of research and support mechanisms. Global commodity centres will engage primarily in strategic germ-plasm research, including biotechnology, and limit their training to specialized techniques. Such centres would divest themselves of general training in agronomy and plant breeding and would have direct interactions only with strong NARS. Eco-regional mechanisms will take two forms: first, networks among strong NARS, in which research and training are contracted out to competent institutions in both developing and developed countries; second eco-regional centres for regions dominated by weak NARS. Both mechanisms will assume responsibilty for eco-regionally defined resource management research. Centres will have the added responsibility of coordinating CGIAR interactions with the NARS, including institution building and training.

Increasing the scale of CGIAR activities depends on funding, which is part of the larger question on the level of development aid. The rapidly emerging concerns about the environment and global warming demonstrate the overwhelming interdependencies and highlight the global nature of the problem. This may eventually generate the political will to look beyond national boundaries in drawing up the social and environmental cost/benefit evaluations. When this effects the level of development aid the CGIAR should benefit.

REFERENCES

Barker, R. and Plucknett, D., 1991, 'Agricultural Biotechnology: A Global Perspective', ch. 8 in Bill R. Baumgardt and M.A. Martin (eds), *Agricultural Biotechnology Issues and Choices. Information for Decision Makers*, Purdue University A.E. Station.

Barton, J.H. and Siebeck, W.E., 1991, 'Intellectual Property Issues for the International Agricultural Research Centres – What Are the Options?' (pre-print version), Draft for the Consultative Group on International Agricultural Research.

Beachy, R.N., 1991, Director, Centre for Plant Science and Biotechnology, Washington University, personal communication, April.

Bertram, R.B., 1990, 'Cassava', in G.J. Persley, (ed.), *Agricultural Biotechnology: Opportunities for International Development*, CAB International, Wallingford, UK.

Calgene Inc., 1991, Spokesperson, personal communication, March.

CGIAR, 1990, 'Biotechnology in the International Agricultural Research Centres of the Consultative Group on International Agricultural Research – A Statement by Centre Directors', for the CGIAR Mid-Term Meeting, the Hague, the Netherlands, May.

CIAT, 1990, Amaya, S. (ed.), *CIAT Report/ Informe CIAT*, Centro Internacional de Agricultura Tropical, Cali, Colombia, p.xvi.

DeDatta, S.K., Gomez, K.A. and Descalsota, J.P., 1988, 'Changes in Yield Response to Major Nutrients and Soil Fertility Under Intensive Rice Cropping', *Soil Science*, vol. 146, no. 5.

DGIS, 1991, 'The Impact of Intellectual Property Protection in Biotechnology and Plant Breeding on Developing Countries', a study commissioned by the Stimulation Programme, 'Biotechnology and Development Cooperation', of the Directorate General International Cooperation (DGIS), The Hague, the Netherlands.

Dodds, J.H., 1991, Head, Genetic Resources, Centro Internacional de la Papa (CIP), personal communication, March.

Dodds, J.H. and Tejada, M., 1990, 'Potato', in G.J. Persley (ed.), *Agricultural Biotechnology: Opportunities for International Development*, CAB International, Wallingford, UK.

Doyle, J.J., 1991, Director of Research, International Laboratory for Research on Animal Diseases, personal communication, March.

Fraley, R.T., 1991, Director Plant Science and Technology, Monsanto Agricultural Company, personal communication, March.

Herdt, R.W., 1988, 'Increased Crop Yields in Developing Countries: Sense and Non-Sense', for the American Agricultural Economics Association, August.

Hopper, W.D., 1990, 'Preface' in G.J. Persley, *Beyond Mendel's Garden: Biotechnology in the Service of World Agriculture*, CAB International, Wallingford, UK.

ICRISAT, 1991, 'Managing Intellectual Property at the International Agricultural Research Centres', summary report: The Workshop on Consequences of Intellectual Property Rights for the International Agricultural Research Centres held at the International Crops Research Institute for the Semi-Arid Tropics, Andhra Pradesh, India, November 1990.

IITA, 1989, *International Institute of Tropical Agriculture Research Briefs*, vol. 9, no. 2, June.

ILRAD, 1990, *International Laboratory for Research on Animal Diseases Annual Report 1989*, Nairobi, Kenya.

ISNAR, 1990, *International Service for the National Agricultural Research Annual Report 1989*, the Hague, the Netherlands.

Javier, E., 1990, 'Issues for National Agricultural Research Systems', in G.J. Persley, *Agricultural Biotechnology: Opportunities for International Development*, CAB International, Wallingford, UK.

Mueesen, R.L., 1990, 'Insect and Disease Resistance', in G.J. Persley (ed.), *Agricultural Biotechnology: Opportunities for International Development*, CAB International, Wallingford, UK.

Pardey, P.G. and Roseboom, J., 1988, 'A Global Evaluation of National Research Investments, 1960–1985', in E. Javier and U. Renborg (eds), *The Changing Dynamics of Global Agriculture – Report of a Seminar on Research Policy Implications for National Agricultural Research Systems*, ISNAR/DSE/ZEL, September.

Persley, G.J., 1990, *Beyond Mendel's Garden: Biotechnology in the Service of World Agriculture*, CAB International, Wallingford, UK.

Persley, G.J. (ed.), 1990b, *Agricultural Biotechnology: Opportunities for International Development*, CAB International, Wallingford, UK.

Plucknett, D.L., 1985, 'An Overview of Biotechnology Research in the CGIAR', presented at a joint CGIAR/Japanese Forum on Biotechnology, Tokyo, June.

Plucknett, D.L., Cohen, J.I. and Horne, M.E., 1990, 'Role of the International Agricultural Research Centres', in G.J. Persley (ed.) *Agricultural Biotechnology: Opportunities for International Development*, CAB International, Wallingford, UK.

Roca, W.M., 1991, Head, Biotechnology Research Unit (BRU), Centro Internacional de Agricultura Tropical (CIAT), personal communication, April.

Toenniessen, G.H., 1991, Associate Director for Agricultural Sciences, the Rockefeller Foundation, personal communication, March.

DISCUSSION OPENING – R.E. EVENSON*

In their informative paper Collinson and Wright Platais note that advances in the biological sciences have created important opportunities for agricultural research organizations. Private firms have invested heavily in agricultural biotechnology research. A single chemical company, Monsanto, has 240 senior scientists working on plant and animal biotechnology with expenditures of some $55 million, and numerous other private firms are engaged in agricultural biotechnology research. Public agricultural research systems in developed countries have been more cautious, but have also invested heavily. It is therefore notable that the authors, in considering the CGIAR, report that about 4.5 per cent of the budget was allocated to plant and animal biotechnology research in 1990. For plant biotechnology, they indicate spending of $7.5 million (up from $4.3 million in 1988). This is less than 3 per cent of the CGIAR budget for plant research. Most of this spending is actually 'special project' funding and it is unlikely that core spending on biotechnology is even one per cent of plant research spending in the CGIAR.

One could perhaps understand why many research programmes in developing countries would find it difficult to mount substantive biotechnology research programmes, given staffing and funding constraints. One could also understand the rationale for a 'wait and see' strategy on their part, given the apparently high degree of uncertainty and risk in biotechnology projects. But why should we accept the limited and timid response to biotechnological research opportunities by CGIAR institutions? These institutions cannot argue that funding and staffing constraints have prevented them from making a more vigorous response. Furthermore, they have accepted a responsibility to be the conduits through which developed country scientific advances are made accessible to developing country research programmes and to provide leadership in scientific matters.

Collinson and Wright Platais mention various 'network' activities by IARCs (for example, rice biotechnology at IRRI, cassava at CIAT and IITA, and genetic projects at CIP and CIMMYT). These networks are in sharp contrast to the normal networks by which IARCs convey scientific findings to developing country NARCs. In these networks the IARCs are the recipients, not the initiators.

Collinson and Wright Platais state that the centres have a responsibility to be aware of the needs of national programme scientists and that they are engaged in training and institution building. They note cases of limited capacity to absorb technical assistance by developing countries. They also discuss some of the conflicts with private sector research and the possibilities of limitations of germ-plasm exchange imposed by stronger intellectual property rights.

On the whole, the authors conclude that there are a considerable number of cases of IARC responses to the opportunities and problems inherent in the field. They note a broad degree of awareness and indications of planning and

*Yale University, USA.

prospects for future work, yet, even though they are not highly critical of the IARC response, they do not paint a picture of aggressiveness of leadership.

Plausible explanations for the CGIAR's failure to provide aggressive leadership in this field are:

(1) that the expectations underlying investments made in developing countries have been unrealistically high;
(2) that expected biotechnology products are inherently better suited to the market conditions in developed countries and that they will have limited markets in less developed areas;
(3) that NARCs have a limited 'absorption' capacity for advanced skills and would not be able to benefit from more IARC effort; and
(4) that the IARCs find a leadership and conduit role in biotechnology beyond their inherent capacity and their own view of their role.

In this reviewer's judgment, the arguments regarding unrealistic expectations and developing country location specificity of biotechnology products have validity and justify caution and a wait-and-see attitude on the part of CGIAR institutions. One would not expect CGIAR institutions to respond to biotechnology opportunities in the same way as leading US universities (and certainly not as Monsanto Chemicals has). However, the timid response actually observed is surely not consistent with the acceptance of the science conduit role by the IARCs.

This inability or unwillingness on the part of CGIAR institutions to respond more aggressively appears to be heavily ingrained in the system and part of a broader set of institutional problems with applied research centres. Over the past 15 years or so, the CGIAR system has responded to exhortations to move downstream to more applied activities such as farming systems research and on-farm research. This move away from science has been seen as a way to exhibit more relevance to the donor community.

The history of agricultural research systems has generally been one of continuous conflict and tensions between the demand-side interests of farm groups pressing for more applied and more relevant research and the supply-side interests of scientists. Productive research systems have managed a complex resolution of most of these conflicts through the development of 'pre-technology' science fields to complement the applied agricultural sciences (Huffman and Evenson, 1991).

The CGIAR system is subjected and has responded to pressures on the demand side for more applied research. It appears to have relatively weak pressures, as yet, to respond to the supply-side opportunities afforded by the sciences. Unfortunately, historical experience with agricultural research institutions, including the US agricultural experiment stations, is that they do not respond to scientific opportunities on the supply side until their applied research programmes reach states of obvious exhaustion. In other words, they do not 'reach upstream' to the sciences until it is clear that their present research efforts are unproductive. Only then do they attempt to convince their 'donors' that they must become more scientific. If this historical experience holds for the CGIAR institutions, we have some time to wait before we can expect more response to scientific opportunities such as those afforded in biotechnology.

REFERENCE

Huffman, W. and Evenson, R.E., 1991, *Science for Agriculture*, Iowa State University Press, Ames.

R.K. LINDNER*

The Role of the Private and Public Sectors in the
Development and Diffusion of Biotechnology in Agriculture[1]

The dividing lines between public and private knowledge in the evolution of a technology is a topic to which economists have given little attention. (Nelson, 1982)

INTRODUCTION

Much ado has been made about the revolutionary nature of the so-called new biotechnologies. For instance, the OECD (1988) argued that 'in the agricultural sector, biotechnology clearly represents a means for pivotal change'. Similarly, Lacy and Busch (1989) state: 'In the past ten years dramatic new developments in the ability to select and manipulate genetic material have generated a new basic science frontier in the public research sector and ignited unprecedented interest in the industrial use of living organisms.'

In this paper I want to suggest that the forthcoming biotechnology revolution is not just the product of a scientific watershed flowing from the discovery of the double helix, but also the culmination of a more gradual evolutionary process over the past century involving revision of intellectual property rights to biological research. While some of the key issues from a scientific perspective hinge on improved knowledge of basic life processes and the potential of the consequential new biotechnologies to overwhelm technical constraints on increased production, the question of intellectual property rights is much more significant for the topic of this paper. Not only is the delineation of these property rights an important determinant of the respective roles of the public and private sectors in agricultural biotechnology research, but arguably it is the single most important policy instrument available to governments to influence the extent to which one sector substitutes for the other.

However, this is not the only significant policy issue, and the following questions are indicative of some of the issues of relevance to the respective roles of public and private sector R&D in the development and diffusion of agricultural biotechnologies. First, was the recent extension of intellectual property rights to allow new knowledge embodied in life forms to be patented in the public interest? If so, is there still a role for public sector research to complement that taken over by the private sector? Finally, are there remaining

*The University of Western Australia.

forms of market failure not addressed either by extended intellectual property rights or by complementary public sector research, and do any such distortions justify government regulation of private sector R&D? Owing to space limitations, I plan to touch only on the second and third issues, in order to focus on the first.

Because of the perspective taken in this paper, no attempt will be made to use the term 'biotechnology' in a precise manner. In the literature, it is possible to find different definitions of biotechnology, ranging from the quite specific to the very general. According to Persley (1990b), biotechnology includes both 'traditional biotechnology' and 'modern biotechnology'. The latter encompasses technologies based on use of recombinant DNA technology, monoclonal antibodies (MCA) and new cell and tissue culture techniques. Of these modern forms of biotechnology, it is recombinant DNA (rDNA), or genetic engineering as it is popularly known, that has aroused the greatest public interest because of the prospects of transgenic organisms, as well as the greatest controversy because of the alleged risk of environmental problems. For these reasons, and because genetic engineering is the form of biotechnology research which brings the issues into sharpest focus, this paper has been written mainly with this particular form of modern biotechnology in mind. However, the essential theme is that the role of public and private sector research in biotechnology development and diffusion merely epitomizes the issues for a broader class of patentable self-reproducing innovations.

To provide background for these issues, the next section presents a fairly brief overview of the findings of several recent studies of the relative roles of public and private sector agricultural biotechnology R&D, and of how they have evolved over the past decade or two. These findings are then placed in historical perspective relative to the rather longer time-span of the history of agricultural R&D in general. In the third section, normative considerations are introduced in a review of the efficiency losses generally associated with publicly funded research on the one hand and patent protected invention on the other. This is followed by a discussion of the special characteristics of biotechnologies likely to influence the magnitude of these efficiency losses, or otherwise to provide grounds for a role for the government sector to substitute for private sector R&D. Other considerations pertinent to a role for the public sector to complement or guide private sector R&D are briefly discussed in the final section.

THE EVOLUTION OF AGRICULTURAL BIOTECHNOLOGY R&D

The eventual impact of biotechnology R&D on the roles of the public and private sectors in agricultural R&D is yet to be determined, but trends to date indicate that the combined effect of the development of rDNA technology in combination with recent legal decisions establishing property rights in biological organisms will further shift the division of labour in agricultural R&D away from the public sector institutions and toward private industry. Comprehensive evidence on the extent of involvement of private industry in agricultural rDNA technology development comes from the United States, where private industry has also been investing heavily in public sector agricultural research. For instance, Lacy and Busch (1989) found that agribusiness contributed an estimated $40 million to university bio-engineering research in 1983, and that the amount spent by biotechnology companies in grants and contracts to universities trebled in the following year: 'This funding represented approximately 16–24% of the public sector universities' total funds expended for biotechnology in 1983 and 1984 compared to an average of 3–5% that industry provides for all research funds expended in institutions of higher education.'

At least in the USA, such statistics under-state the importance of private sector research because they do not include in-house research. For instance, new biotechnology firms which were started almost exclusively to commercialize innovations in biotechnology, and which spend a considerable proportion of their funds on development, if not pure research, were estimated to have raised about $450 million in a five-month period in 1983. Lacy and Busch (1989) note: 'A 1985 survey of the agriculture biotechnology firms indicated that they employed more than a thousand molecular biologists and invested over $200 million in agricultural biotechnology research and development in 1984.' There is now some evidence emerging that the relatively small but entrepreneurial start-up biotechnology firms which were so prominent during the early stages of the industry will be swallowed up, or at least supplanted in importance by in-house research conducted by established companies in the agribusiness industry.

Persley (1990b) presents more recently available global information on the relative importance of public *vis à vis* private sector funding of biotechnology research which is summarized in Table 1.

TABLE 1   *World-wide R&D expenditure on biotechnology (1985 estimates, in US$m)*

Type of biotechnology	Seeds	Agricultural Microbiology	All	Other	Total
Private	350	200	550	2 150	2 700
Public	250	100	350	950	1 300
TOTAL	600	300	900	3 100	4 000

A persuasive case can be made that the history of agricultural R&D has involved a gradual transfer from public sector institutions to private industry, and that the recent outburst of private sector investment in biotechnology merely represents the continuation of a trend that has been going on for at least a century, if not longer. Lacy and Busch (1989) suggest that this trend towards industrialization started in the food-processing sector with the development in 1785 of the first fully automated continuous process flour mill.

This was followed in 1789 by the introduction of bread-kneading machines in Genoa. Hog slaughtering was partially automated in 1830, and pasteurization of dairy products in the late nineteenth century marked a further step forward before full industrialization of poultry production in the twentieth century. The rise of the farm equipment industry followed closely behind the development of processing, with the introduction of stationary steam engines on US farms in the early 1830s, and of tractors for cultivation by the turn of the century. By the first part of this century, the role of private firms in agricultural R&D was 'confined largely to the manufacture of farm machinery and the processing of farm products', but, after the Second World War, the private sector also became heavily involved in chemical R&D.

The transition from public sector to private sector R&D started rather later for plant-breeding research. Lacy and Busch (1989) state:

> Until 1923, USDA regularly distributed free seed, collected from around the world, to farmers who wanted to test the seeds on US soils. As the techniques and effectiveness of plant breeding improved, USDA discontinued the distribution of free seeds. Instead, the state agricultural experiment stations (SAES) began to disseminate new varieties developed by their breeders to seed companies who multiplied and sold the new varieties to farmers. The development of hybrid corn shifted the division of labour again. Experiment stations gradually shifted to the development of parent material and significantly reduced the production of finished products for hybrid crops. With the growth of the seed development (as opposed to multiplication) industry, came increased presssure on the public sector to cease producing finished varieties.

This tendency for the private sector to move into a research area once intellectual property rights have been established, and then to crowd out public sector R&D, may well continue in the field of biotechnology. Lacy and Busch claim that the recent development of private biotechnology research capacity has resulted in 'pressure on public sector scientists from industry to abandon varietal breeding, pressure from administrators for greater productivity'. In addition, the lure of large amounts of private money for biotechnology research has led to a change in disciplines in the SAES, although the authors also quote survey results indicating that the experiment stations have been hiring large numbers of molecular biologists, and that many of the positions were obtained by reducing the scope of conventional breeding programmes.

PUBLIC VERSUS PRIVATE SECTOR R&D – A CONCEPTUAL FRAMEWORK

What considerations need to be taken into account in evaluating whether a particular area of inventive activity should be conducted by private or public sector R&D organizations when it is feasible for the state to establish intellectual property rights? One way of approaching this issue is to treat it as synonymous with the question of whether patent legislation should or should not cover the particular class of inventions under consideration.[2]

It is commonly accepted that the potential for market failure inherent in the public good characteristics of knowledge produced by research provides both the justification and guidance for governmental intervention in the market for inventive activity. Of the two most widely employed forms of government intervention, one has been directly to fund and organize R&D activity in public sector institutions, and the other has been to confront the appropriability problem by establishing legal protection in the form of patents for the intellectual property rights of the inventor. In much of this paper, these two alternatives will be treated as polar stereotypes of public and private sector R&D, respectively.

In a world of perfect knowledge, allocative losses associated with public sector R&D would be limited to the excess burden associated with raising the revenue to fund the research. In a world of imperfect information, public sector R&D is likely to incur additional welfare losses due to misallocation of research resources in aggregate and/or between alternative areas for investigation. The wealth of evidence on high rates of return to public sector agricultural R&D is often cited as *prima facie* evidence of under-investment by governments in research.[3] Conversely, in the private sector, the race to discover and patent new technology is the more likely to result in over-investment in research. This is the so-called common pool problem, in which individual inventors ignore the marginal effect of their research activity on aggregate research productivity. Unless the conditions of the patent are optimized in some way, such behaviour will lead to over-investment in research and associated allocative losses because the expected average benefit to a competitive R&D firm from successfully developing and patenting an invention will exceed the expected marginal social product. In addition, the resulting technology will be under-utilized if patent holders attempt to appropriate the benefits. This widely recognized deadweight loss associated with the patent system arises because potential adopters are charged for knowledge which is non-rival in use. Less widely recognized is the likelihood that the patent system will retard the rate of innovation diffusion because appropriation of research benefits by inventors will reduce innovation profitability to potential adopters.

Wright (1983) has shown that any potential advantage of patents over publicly funded contract research lies in informational asymmetries between public funding agencies and private profit-maximizing firms with regard to market opportunities, the cost of research and its probability of success. This conclusion is supported by the findings from an analytical model in which the choice of the superior alternative took account of the following three allocative difficulties: lack of appropriability of knowledge, deadweight loss of the patent, and the common pool problem. As Wright acknowledges, there are other considerations besides those discussed above that are relevant to the choice between public and private sector R&D. Of the other allocative difficulties, the most important stems from the joint product nature of research output. As Nelson (1987) puts it, there are two different aspects of technology, one being operative knowledge,[4] which can be embodied in patentable inventions or 'techniques', and the other being the so-called 'logy' or body of generic knowledge about the way technologies work. The public good dimension of basic research[5] has been part of conventional wisdom for many years, but Nelson (1982, 1987) has gone further and argued that even applied R&D adds to the stock of knowledge about where and how to search for new technologies. Empirical support for such an externality is provided by Jaffe (1986) who found that R&D productivity is increased by the R&D of 'technological neighbours,' and that firms adjust the technological composition of their R&D in response to technological opportunity.[6]

Because this 'logy' has strong public good properties, including being non-rival in use as well as reducing the costs and/or returns of subsequent research, it is neither possible nor desirable for the private sector to capture all of the benefits from this component of research output, even if 'ideal' patents could be designed. However, while such knowledge has the potential to enhance future research productivity, this potential will only be realized to the extent that it is freely available as an input into further research by all parts of the R&D industry. In particular, if the establishment of intellectual property rights to biotechnologies adversely affects this positive information externality, then the short-run stimulus to private investment in R&D provided by patents may be offset in the long-run by declining research productivity. This aspect may well emerge as the key issue in any evaluation of the respective roles of the public and private sectors in biotechnology R&D.

Proponents of the patent system argue that the requirement to disclose the scientific basis of the innovation as part of the patenting process makes sure that this information externality is not lost. However, there are some counter-arguments. In particular, there is no incentive under the patenting system for commercial R&D firms to disclose any of the results of research projects which do not generate a patentable innovation/invention. Thus the pool of knowledge put into the public domain as a result of the patenting requirement for disclosure is only a small fraction of the information generated by commercial R&D firms. The other point to note is that, even in the case of patentable inventions, the disclosure requirement only achieves partial release of relevant information. Furthermore, even if no attempt is made to keep the 'logy' secret, the much higher search costs faced by other firms seeking to discover such private information (compared to the more traditional scientific approach of publication in journals or in public conferences) will seriously reduce the extent of the informational externality. The finding by Scotchmer and Green (1990) that the 'stringency of the novelty requirement in patent law affects the pace of innovation because it affects the amount of technical information that is disclosed among firms' tends to support this argument.

Unfortunately, virtually nothing is known about the size of the potential allocative losses identified above as relevant to the choice between public and private sector R&D for any area of agricultural research, let alone one as novel as biotechnology. Clearly, institutional arrangements specific to the country involved will be critical determinants of these losses. However, even for a simple choice between the two highly over-simplified and stylized systems outlined above, there is an almost total dearth of empirical information about the nature or magnitude of the trade-offs between the various potential losses. In the next section, some of the characteristics peculiar to biotechnologies which might influence the size of these allocative losses or otherwise be relevant to the role for public sector R&D are discussed in qualitative terms.

CHARACTERISTICS OF THE EMERGING BIOTECHNOLOGIES

Biotechnology research is distinguished from other more conventional forms of agricultural research by a number of characteristics, only some of which are relevant to the optimal mix of public versus private sector participation in the development and diffusion process. The most commonly commented on is the revolutionary nature of the scientific basis for molecular biology. In terms of Evenson and Kislev's (1976) stochastic model of applied research, the significance of this property is that it implies that the expected rate of return to applied rDNA research is likely to be much higher in the foreseeable future than from more conventional types of agricultural research which, arguably, are suffering from *technological exhaustion*.[7] Given that biotechnology research is likely to follow a similar productivity cycle to other areas,[8] and will eventually become less profitable than is apparently the case at present, determination of the optimal level of investment in R&D is an optimal control problem which is too difficult to solve, given current knowledge. However, if, as seems likely, this optimal level of aggregate investment varies over time, then the evident greater flexibility of the private sector to adjust its investment in research is an advantage.

It has already been suggested above that biotechnologies belong to a subset of biogenetic inventions which are both embodied and patentable, and that, within this class, their only distinguishing characteristic is novelty. Conversely, as Stallman and Schmid (1987) point out, what distinguishes most biotechnologies, and rDNA technology in particular, from other embodied and patentable technologies is that they are embodied in a living organism with the intrinsic capacity for self-replication.[9] This unique property has crucial implications, both for the ability of the private sector to appropriate the benefits from its research, and for the utility of the disclosure provisions in patenting provisions as a device to offset the information externality described above.

Owing to this innate capacity for self-reproduction, the cost of imitation of biotechnologies is negligible, and less likely to be subject to economies of size than for mechanical, electrical or chemical inventions. Equally important, potential imitators have no need for access to the 'information' component of the invention[10] in order to reproduce it, so lack of knowledge is no impediment to the ability to free-ride. Consequently, the detection of imitations and enforcement of any property rights conferred by the legal system are much more difficult and costly. By the same token, this same characteristic reduces the cost of innovation adoption, and thereby speeds up the rate of diffusion of the technology and reduces any deadweight loss from under-utilization of knowledge.

Schmid (1985) notes that 'Because of high information costs, patent laws for plants and micro-organisms cannot provide exclusivity of use without eliminating original and non-copied substitutes.' The impotence of computer companies to counter software piracy provides a graphic example of the possible dimensions of this problem. After reviewing experience on the operation of the 1930 Plant Patent Act, which permits asexually propagated plants to be patented, Stallman and Schmid (1987) found that 'plant patents in fruits and variety protection in field crops have not alone given the protection and exclusivity necessary to allow market returns to cover research costs.' They further concluded that, owing to differences between biology and chemistry, patent protection for biogenetic inventions is unlikely to allow biological researchers successfully to appropriate benefits in the manner achieved for chemical inventions. If this prediction proves to be accurate, there may still be a case for public sector funding of applied biotechnology research simply to ensure a high enough level of aggregate investment.

This reproductive capacity of rDNA technology is also relevant to the debate about the impact of patents *vis-à-vis* public funding on the possible under-utilization of the 'logy' component of research output. Since the disclosure provision of patents is less effective for such innovations than for other classes of inventions where the cost of imitation is a function, *inter alia*, of knowledge about the way to reproduce the technology, private R&D firms may try to keep as much as possble of this type of knowledge secret, thereby exacerbating the information externality. There is much anecdotal evidence relating to biotechnology at the current stage of development which suggests that there are considerable and very significant restrictions on the dissemination of information between scientists in competing firms, and that the contractual relationships which these firms form with public sector scientists extend these constraints to public sector R&D as well. On the other hand, Nelson (1987) argues that the private sector has a vested interest in keeping the 'logy' of its science in the public domain, and in other areas has developed a variety of mechanisms (which he documents at some length) to ensure efficient utilization of such knowledge.

The discussion above presumes a traditionally organized public research sector, which is funded mainly or totally from government revenues, and which publishes research results without attempting to exploit commercially any new technologies developed as part of the research programme. In some countries at least there has been a trend away from this 'pure' model during the past couple of decades. For instance, in Australia, government funding is gradually being displaced by industry funding (financed partly by production levies) to the point where the wool industry research funds are now providing the majority of funding (some 58%) in the Division of Wool Technology and over one-third of the funding (36%) in the 'Division of Animal Production' in CSIRO, which is the major public scientific research organisation in Australia. Public research organizations, including universities, are also attempting to market commercially valuable technologies developed from their research programmes. In part this is simply a response to cut-backs in government funding, in part it reflects altered opportunities brought about by changes to the law regarding intellectual property rights, and in part it is a response to pressure from industry funding bodies to appropriate the returns from technologies generated by 'their' research programmes.

Clearly this trend towards commercialization of public sector research diminishes the differences between public and private sector research. James and Persley (1990, p.372) suggest that industry will have a comparative advantage in these circumstances because of superior access to capital markets, more diverse organizational form, better ability to consolidate a critical mass of scientific resources within a core research group to capitalize on complementarities between agricultural and medical biotechnology, greater marketing expertise, and access to global markets and associated economies of scale. As a result, they predict that the private sector will emerge as the predominant provider of agricultural biotechnologies. Such a trend would be of concern if it increased the likelihood of the public sector being crowded out by the private sector, and especially so if there are unfavourable externalities or distributional consequences from the use of biotechnologies.

Another concern if the possibility that industry will distort the direction of public sector research rather than crowding it out. One of the features of the biotechnology industry has been the forging of such extensive public/private sector links that it has been described as the new university–industrial complex. It is outside the scope of this paper to discuss possible advantages as well as problems which could flow from this association, but it is already the subject of an extensive literature.[11] In terms of the topic of this paper, it is worth noting that many of the researchers involved in the emerging alliance between public sector basic science and private sector technology development do not belong to traditional agricultural research establishments. A key question is whether they will prove to be a complement to , or a substitute for, existing publicly supported agricultural research, or merely a vehicle for cross-subsidization or private R&D.

OTHER POLICY ISSUES FOR
PUBLIC SECTOR BIOTECHNOLOGY RESEARCH

A number of authors have argued that the greatest benefits from genetic engineering are likely to be found in Third World agriculture, while the greatest propensity to pay for its technologies are to be found in the developed world. The contribution that biotechnology could make to agricultural productivity and increased food production has been documented in Persley (1990a). Barker (1990) has suggested that the potential to continue increasing food production without biotechnology is limited, since yield plateaux have been reached for several major food crops, and because the opportunities to expand the area under cultivation are almost exhausted. If the biotechnology revolution were to by-pass developing countries, the distributional consequences would be disastrous, but to date private sector R&D has been concentrated almost exclusively in the industrialized countries.

There are two other potential characteristics of biotechnologies which do not seem to have been widely recognized and which, if they materialize, could have important implications for government policy towards rDNA research. The first relates to the capacity of rDNA technologies to break free of, or at least diminish the importance of, environmental constraints on the production of food and other agricultural products. For instance, the OECD (1988, p.27) predicts that the 'effects will be felt in an increasing convergence of agriculture and industrial practice.' If this hypothesis proves to be correct, then biotechnology-based supply curves will be more elastic than the equivalent supply from conventional agriculture, which could result in major distributional effects. For instance, if this led to a marked reduction in the real price of food, then, as Pinstrup-Andersen *et al.* (1976) have pointed out, the distributive implications would be strongly progressive, as well as helping to alleviate malnourishment. A related but separate consideration is the capacity of biotechnology to stabilize production variability and, at least in that limited sense, to make agricultural production more comparable to industrial production.

These possibilities suggest that the potential exists for public sector research to play a complementary rather than competitive role to private sector research. However, there are concerns that international differences in protection afforded by the legal system to intellectual property rights, as well as in the cost of imitation, will adversely influence the respective roles of private and public R&D on biotechnology for agriculture on the world scene. Evenson and Putman (1990) clearly demonstrate that, while products of biotechnology R&D are afforded significant patent protection in the USA, and to a similar extent in a number of other industrialized countries, the situation in the developing world is very different, with the legal systems of most countries affording very little protection to all forms of agricultural inventions, but in particular to technology embodied in living organisms. In the context of the international transfer of technology, the significant consideration is the capacity of the national agricultural research systems of the countries to imitate or adapt technology generated overseas. In this regard, there are important differences between countries in the developing world, with some scientifically advanced countries such as India and Brazil having a relatively strong capacity to copy and/or adapt biotechnologies, while a range of other countries have very little capacity (Evenson and Putman, 1990). As a result, the issue of intellectual property rights is a contentious matter in the Uruguay round of GATT negotiations. Failure to resolve this issue could result in advanced developing countries being shunned by private sector R&D, and having to rely largely if not entirely on public sector research devoted to imitating and adapting technologies developed in the industrialized countries. Alternatively, less scientifically advanced developing countries have an incentive to pass relatively strong intellectual property right legislation, so multinational companies are likely to establish adaptive R&D programmes in such countries.

There is at least one important caveat that needs to be noted with regard to the above projections, and that concerns the role which the international agricultural research centre system (IARC) will play in the development of new biotechnologies for the developing world. Several authors have argued that the main role ought to be to complement private sector R&D by carrying out work on so-called orphan commodities. For instance, certain plants, such as cassava, coconuts and coffee, are likely to be ignored by the private sector R&D system because of lack of ability of Third World producers to pay for advanced technologies, or because imitation costs are too low and exclusion costs too high, or because the size of the market is regarded as being too small in relation to the cost of R&D. Two considerations will be crucial to success in fulfilling this role. One is the outlook for the IARC system not only to maintain its existing funding base, but also to expand it sufficiently to be a significant player in the world biotechnology R&D system. Another is the ability of the system to collaborate effectively with the private sector biotechnology R&D companies in the industrialized countries, or to mount an effective R&D programme in the area of biotechnology if it is not able to secure such collaboration.

Finally, biotechnology in general, and rDNA technology in particular, differs from more conventional technologies because of the widespread public perception that there is a much greater risk of environmentally catastrophic outcomes associated with this form of technology. While the concern with the environmental aspect might be more or less contemporary, a sense of history suggests that public apprehension about the consequences of new technologies during the early stages of their development and diffusion is not unique to rDNA research. Indeed, at least to date, there has been nothing comparable to the widespread social disruption caused by the Luddites during the industrial revolution. Moreover, the net effect on the environment is a highly contentious issue, and many experts believe that genetic engineering will produce technologies which are at worst environmentally benign, and at best advantageous because they significantly reduce the need to control pests and diseases by chemical means. Apart from this possible externality, the potential problems associated with environmental release of genetically engineered organisms would seem to imply a regulatory role for government, but that is a topic for another paper.

NOTES

[1] I am grateful to Rob Fraser for helpful comments on an earlier draft of this paper, and to the Australian Agricultural Economics Society for financial support to attend this conference.

[2] On the basis of available evidence such as that presented above, it seems reasonable to presume that, when private sector R&D firms can appropriate a significant part of research benefits, then sooner or later they will 'crowd out' public sector applied research in the same area. Similarly, Nelson (1987, p. 117) argues that 'The advantages of giving firms incentives, and hands-on capability for innovation can be seen most sharply by considering the poor innovation performances of socialist countries where neither of these conditions exist.'

[3] For instance, see Ruttan (1982, pp. 242–8).

[4] That is, knowledge about specific ways of doing things, or about current operating methods in the industry.

[5] In this paper, the term 'basic research' will be used to describe research where the intended output is simply more information, while applied research will be used to describe the search for new technologies, and which may or may not produce more information as a by-product.

[6] Jaffe (1989) also found that an information externality exists between academic research in universities and corporate patent activity, and that university research appears to have an indirect effect on local innovation by inducing industrial R&D spending.

[7] In terms of their model, basic research on rDNA has opened up a whole new series of distributions of new technologies to search with applied research, or, as they describe it, experimentation.

[8] See Evenson (1974, 1976) for selected case studies.

[9] It is likely that some biotechnologies will be developed which deliberately do not incorporate the characteristic of self-reproduction so as to emulate the precedent of hybrid corn, which does not rely on patents or plant variety rights to protect the intellectual property rights of inventors.

R.K. Lindner

[10]See Evenson and Putman (1990) for a detailed discussion of the distinction between the information discovered by the inventor and that embodied in the device to which patent law provides property right protection against unauthorized reproduction, sale or use.

[11]See Kenney (1986) for an extensive treatment of this topic.

BIBLIOGRAPHY

Arrow, K.J., 1962, 'Economic Welfare and the Allocation of Resources for Invention', in *The Rate and Direction of Inventive Activity: Economic and Social Factors*, Arno Press, New York.

Barker, R., 1990, 'Socio-economic Impact', in G.J. Persley, (ed.), *Agricultural Biotechnology: Opportunities for International Development*, CAB International, Wallingford, UK.

Centner, T.J. and White, F.C., 1987, 'Protecting Inventors' Intellectual Property Rights in Biotechnology', *American Journal of Agricultural Economics*, 69, (5).

Evenson, R.E., 1974, 'International Diffusion of Agrarian Technology', *Journal of Economic History*, 34, (1).

Evenson, R.E., 1976, 'International Transmission of Technology in the Production of Sugar Cane', *Journal of Development Studies*, 12, (2).

Evenson, R.E. and Kislev, Y., 1976, 'A Stochastic Model of Applied Research', *Journal of Political Economy*, 84, (2).

Evenson, R.E. and Putman, J.D., 1987, 'Institutional Change in Intellectual Property Rights', *American Journal of Agricultural Economics*, 69, (2).

Evenson, R.E. and Putman, J.D., 1990, 'Intellectual Property Management', in G. Persley (ed.), *Agricultural Biotechnology: Opportunities for International Development*, CAB International, Wallingford, UK.

Hueth, D.L. and Just, R.E., 1987, 'Policy Implications of Agricultural Biotechnology', *American Journal of Agricultural Economics*, 69, (2).

Jaffe, A.B., 1986, 'Technological Opportunity and Spillovers of R&D: Evidence from Firms' Patents, Profits, and Market Value', *American Economic Review*, 76, (5).

Jaffe, A.B., 1989, 'Real Affects of Academic Research', *American Economic Review*, 79, (5).

James, C. and Persley, G.J., 1990, 'Role of the Private Sector', in G.J. Persley, (ed.), *Agricultural Biotechnology: Opportunities for International Development*, CAB International, Wallingford, UK.

Kenney, M., 1986, *Biotechnology: The University–Industrial Complex*, Yale University Press, London.

Lacy, W.B. and Busch, L., 1989, 'The Changing Division of Labour Between the University and Industry: The Case of Agricultural Biotechnology', in J.J. Molnar, H. Kinnucan, (eds), *Biotechnology and the New Agricultural Revolution*, Westview Press, Boulder, Co, pp. 21–50.

Longworth, J.W., 1987, 'Biotechnology: Scientific Potential and Socio-economic Implications for Agriculture', *Review of Marketing and Agricultural Economics*, 55, (3).

Nelson, R.R., 1982, 'The Role of Knowledge in R&D Efficiency', *Quarterly Journal of Economics*, 97.

Nelson, R.R., 1987, *Understanding Technical Change as an Evolutionary Process*, (Professor Dr. F. de Vries Lectures in Economics, vol.8) Elsevier Science Publishers, Amsterdam.

Organization for Economic Co-operation and Development, 1988, *Biotechnology and the Changing Role of Government*, Report by OECD, Publications Service, Paris.

Organization for Economic Co-operation and Development, 1989, *Biotechnology: Economic and Wider Impacts*, Report by OECD, Publications Service, Paris.

Perrin, R.K., 1990, 'Economic Analysis of Biotechnology Research', *Southern Journal of Agricultural Economics*, 22(1).

Persley, G. (ed.), 1990a, *Agricultural Biotechnology: Opportunities for International Development*, CAB International, Wallingford, UK.

Persley, G., 1990b, *Beyond Mendel's Garden: Biotechnology in the Service of World Agriculture*, CAB International, Wallingford, UK.

Pinstrup-Andersen, P., Ruiz de Londono, N. and Hoover, E., 1976, 'The Impact of Increasing Food Supply on Human Nutrition: Implications for Commodity Priorities in Agricultural Research and Policy', *American Journal of Agricultural Economics*, 58, (2).

Ruttan, V.W., 1982, *Agricultural Research Policy*, University of Minnesota Press, Minneapolis.
Scherer, F.M., 1972, 'Nordhaus' Theory of Optimal Patent Life: A Geometric Reinterpretation', *American Economic Review*, 62, (3).
Schmid, A.A., 1985, 'Intellectual Property Rights in Bio-Technology and Computer Technology', *Zeitschrift für die gesamte Staatswissenchaft*, 141,(1).
Scotchmer, S. and Green, J., 1990, 'Novelty and Disclosure in Patent Law', *Rand Journal of Economics*, 21(1).
Stallman, J.I. and Schmid, A.A., 1987, 'Property Rights in Plants: Implications for Biotechnology Research and Extension', *American Journal of Agricultural Economics*, 69(2).
Wright, B., 1983, 'The Economics of Invention Incentives: Patents, Prizes, and Research Contracts', *American Economic Review*, 73,(4).

DISCUSSION OPENING – R.K. PERRIN AND J. BEGHIN*

Professor Lindner notes that the increasing strength of intellectual property rights (IPRs) in biotechnology should have a strong impact on private R&D and that this in turn has implications for the role of public R&D. The message of the paper seems to be that public sector research should *not* be abandoned or distorted by public/private linkages, but there is little positive guidance on what the public sector *should* do. Neither is there any advice about IPRs as a policy instrument, whether they should be strengthened further in the Third World, or weakened in the developed world.

The paper does offer us a synoptic view of the IPR issue as it relates to biotechnology. The issue is whether to have them and, if so, in what form and strength. The accepted economic viewpoint can be summarized as follows. All forms of knowledge have public goods characteristics (knowledge is non-rival in consumption, has low costs of access by marginal users, and tends to have high costs of exclusion), hence there is a theoretical presumption of market failure in the form of under-allocation of resources to R&D. One solution to this market failure is to establish IPRs. Another solution is public sector R&D. Still other alternatives are legal protection for trade secrets, public prizes for new knowledge, and public–private syndicates.

None of these solutions offers a panacea. They are difficult to analyse and compare theoretically because of the myriad possible dimensions of statutes and enforcement mechanisms, and they are difficult to analyse empirically because of the paucity of useful social experiments and the difficulty of obtaining data in any case. In theory, IPRs should increase R&D but they also wastefully limit the diffusion of the new knowledge so created (the appropriability–diffusion dilemma). IPRs might in theory create excessive R&D because of 'patent races' (in the case of negative externalities in the production of knowledge). The most obvious theoretical problem associated with public R&D, on the other hand, is the distortion of incentives that is inherent in bureaucracies.

Lindner's paper elaborates on this summary of received economic wisdom, and describes the evolution of IPRs as they affect biotechnology. In the process of these discussions, he emphasizes three key points with which we would like to take issue. First, he asserts that a key issue for IPRs is whether

*North Carolina State University and Iowa State University, USA.

the concomitant restrictions on knowledge transfer will result in a long-run decline in research productivity. Decline compared with what? Presumably, compared to that which would hold without IPRs and only public R&D. His argument is that public scientists exchange information freely, while private firms would disclose only the minimum necessary to meet the 'enabling disclosure' requirements of the IPR system. We think this is an exaggerated concern. But current IPR systems require an enabling disclosure; that is, sufficient information for those competent in the art to be able to 'practise' the invention. In the case of biotechnologies, this frequently includes the deposit of living material to which others can gain access. This precludes a lot of secrecy. Furthermore, it has been suggested that, under the US patent system, the median time-lag for detailed information on new products and processes to fall into the hands of at least some competitors is less than 18 months (Mansfield, 1984). It seems to us that the weight of this theoretical and empirical evidence suggests that the problem of secrecy is at best a short-run problem, with no possibility of offsetting the long-run productivity effects of IPRs, as is suggested by Lindner. The author also seems to minimize the possibility of government failures that could be induced by (public) information asymmetries, institutional design and incentives faced by public researchers. The disclosure of failed innovation attempts is not likely to be reported by public researchers (for instance, 'negative' regression results are not reported in economic journals).

Second, we dismiss his assertion that the unique self-reproduction property of rDNA technologies has crucial implications for (a) the ability to appropriate returns and (b) the utility of disclosure provisions for offsetting the information externality. In the first place, the self-reproduction property is not unique – it is shared by other forms of knowledge which can be 'reproduced' in new applications at very low or zero marginal cost. There are no implications for rDNA because of this self-reproduction property that do not also hold for other forms of intellectual property. The crucial implications for disclosure escape us, unless the author is referring to the potential need for an enabling disclosure to include an accessible deposit of living material, which is indeed important but is already an accepted component of most current IPR systems that affect biotechnology. Professor Lindner could also have related the disclosure issue to patent design (Ordover, 1991) where some patent rules more than others foster early disclosure and diffusion of information (for example, first to file versus first to invent). It would have been interesting to see what design features would be unique or specific to biotechnology.

Third, the author suggests that the public research sector should not be crowded out 'if there are unfavourable externalities or distributional consequences from the use of biotechnologies'. A public reasearch sector could only solve the problem of such 'bads' if public scientists were to invent things that not only avoid the 'bads' but would also be adopted in preference to the 'bad' inventions offered by the private sector. This does not seem very likely. A better use of public funds would be to study the potential bad effects and the potential for government interventions that might avoid them (in other words, hire economists rather than scientists!).

We also have two remarks to make on issues not addressed in the paper. First, the paper could have provided more guidance on the role of the public sector. Some guidance can be found in Scotchmer (1991) for the case of cumulative research. Public basic research is appropriate for the case where the first technology in a cumulative process has low expected profit, while it is valuable for further innovations and where the innovator cannot appropriate the social value of the innovation and its positive externalities. More could have been said also on the state as a facilitator of cooperation and collusive behaviour among firms to integrate/cooperate on technological developments.

Finally, we would like to stress the importance of the role of IPRs in developing country agriculture, as opposed to developed country agriculture. LDCs have the opportunity, by eschewing IPRs altogether, to free-ride on (to 'pirate') technology that is invented in developed countries. Enforcement of IPRs would provide an incentive for invention of custom-developed technology, but it would also subject farmers to payment of royalties on both borrowed and custom technology. Would not a small country (with therefore limited scale incentives for customer R&D), or one with an environmental niche similar to a developed economy (with therefore limited pay-off for customized over pirated technology) be better off as a biotechnology pirate? Just when is it in the interests of developing economies to adopt the IPR systems of developed economies?

REFERENCES

Mansfield, E., 1984, 'R & D and Innovation: Some Empirical Findings', in Z. Griliches, (ed.), *R & D, Patents and Productivity*, University of Chicago Press for the NBER, Chicago.

Ordover, J.A., 1991, 'A Patent System for both Diffusion and Exclusion', *Journal of Economic Perspectives*, 5.

Scotchmer, S., 1991, 'Standing on the Shoulders of Giants: Cumulative Research and the Patent Law', *Journal of Economic Perspectives*, 5.

HERBERT STOEVENER*

Section Summary

Major developments in agricultural research, and the protection of intellectual property rights in its results, have important implications at the national and international levels. Within our general conference theme it can be said that it was the role of international cooperation which was of major concern in the whole section. The sessions also provided an opportunity for us to learn more about the underlying scientific techniques in biotechnology, a feature which spills over to some extent into Section VII.

The lead-off paper on this topic, by James Peacock, described the biological concepts and techniques of biotechnology and commented on the potential for application in agricultural production. It cited several examples of current developments in genetic engineering in the plant sciences which are likely to lead to practical applications in the field within three to four years. The speaker was optimistic about a rapid rate of technical progress in this field. He emphasized the 'user-friendly' nature of this technology as it comes to the farmer in the form of improved seed, an input with which he is familiar. Bio-engineering also makes possible the expansion of the range of uses to which plant materials can be put. For example, plants may find greater uses in the pharmaceutical industry because of these developments in the future.

Subsequent discussion by the formal programme participants in the plenary session (Rao, Van der Meer and Barker were asked to make extensive comments from their different perspectives) as well as from the floor, reinforced many of the points made by the main speaker, but also pointed to some potential problems in the development and application of this new technology. On the positive side, and especially from the viewpoint of the developing countries, the possibilities were seen for improved food security, cheaper food through lower production costs, greater stability in crop yields and farm incomes, and the possibility of natural resource conservation by reducing use of chemical inputs and withdrawing highly erodible lands from production. The very specific ways in which these techniques can be used to affect plant characteristics were also seen as a way to make plant science research more immediately responsive to a socially determined research agenda.

Several speakers were concerned about public reaction to the real or perceived risks associated with the new technology and about consumer acceptance of products resulting from it. The need was pointed out to have regula-

*Virginia Polytechnic Institute and State University, USA.

tory mechanisms developed by a well-informed public. These educational and regulatory processes may lag behind the potential technical developments taking place in this field.

Invited paper sessions covered plant breeders' rights (Pray), problems of international spillovers (Schweikhardt and Bonnen), the international research centres (Collinson and Wright Platais) and the role of private and public sectors in development and diffusion of research (Lindner). During the discussions emphasis was given to the economic and institutional environment in which research resources are allocated to biotechnology. During the current era of flat price trends for agricultural products and high priorities for using public research funds for other purposes, public interest is low in research that expands agricultural output. This was seen as an important brake on bio-engineering research in general. Specifically, this was suggested as the explanation for the apparently very cautious movement into this research area by the international agricultural research centres. Some thought that this situation might improve as bio-engineering research becomes less expensive as the start-up costs are behind us, and as this research replaces some of the former field work and becomes a standard part of laboratory procedures and academic curricula in biology.

The appropriate mix of public and private sectors in financial support and conduct of bio-technology research was also given considerable attention. On the one hand, there are circumstances under which the nature of this research allows the developer of the technology to capture a larger share of the benefits derived from it. This provides powerful incentives for the private sector to conduct the work. On the other hand, there are also possibilities for large spillovers of benefits, for example, when consumers in general enjoy lower prices of the genetically engineered products or when there is scope to duplicate application of the research results by other producers than those for whom the research was conducted. Broad gains in new knowledge which have public goods characteristics may also result from the research. These spillovers and public goods would argue for public conduct of the research.

There was agreement that one should expect society to do a considerable amount of experimentation with the design of institutions as it struggles with the economic and ethical issues associated with biotechnology research. In this process it may be possible to draw lessons from past experience with protecting the rights of plant breeders. It may also be necessary to develop entirely new modes of cooperation among the various participants in the private sector, the international agricultural research centres, educational institutions and public agencies that support and conduct research at national and sub-national levels. Agricultural economists will have many opportunities to contribute their analytical insights as these new institutions evolve.

Chairpersons: W.F.Musgrave, Michelle Veeman, Alex Dubgaard.
Rapporteurs: Per Halvo Vale, Toshio Kuroyanagi, Hans Jansen.
Floor discussion: D.Belshaw, M.Lopez-Pereira, R.Dumsday, J.Strasma, K.J.Thomson, T.Horbulk, J.Peacock.

SECTION V

Environmental Enhancement Strategies and Policy Linkages

SANDRA S. BATIE*

Sustainable Development: Concepts and Strategies

INTRODUCTION

There is increasing recognition that economic growth will not necessarily or automatically lead to an improvement in human welfare, in obtaining justice, or in the protection of the environment. Indeed, many critics have indicated economic growth as being exploitative of both people and the environment. However, there also appears to be considerable validity in the expression 'Wealthier is healthier' (Wildavsky, 1988). That is, throughout the world, rising standards of living have accompanied economic growth. Furthermore, the results of economic growth have been spectacular: freedom from many diseases, protection from many natural disasters, the elimination of famine for much of the world, and the freeing of people from drudgery by the substitution of machines for human labour. Thus economic growth can simultaneously be viewed as both the problem and the solution, which in itself poses a dilemma for the designing of informed economic and environmental policies.

These contradictions become apparent when exploring the concept of sustainable development. It has been defined by the Bruntland Commission as development that meets the needs of the present generation without compromising the ability of future generations to meet their own needs (World Commission, 1987). It has evolved to mean the selection of development paths that protect ecosystem functioning as well as protecting traditional cultures.

There are many different schools of thought concerning interpretation of the concept (Batie, 1991; Colby, 1989). While differing substantially, they are for the most part united in the belief that sustainable development does not mean the *status quo*. Also, the neo-classical economics measure of gross national production as a monetized proxy for human welfare is rejected. Furthermore, if sustainable development is to have meaning, it must include consideration of the environment and of human needs and aspirations. Thus sustainable development incorporates the idea of transformations of relationships between people and between people and nature – both now and through time. There remains, however, considerable tension between those schools of sustainable development thought which draw their strength mainly from the ecological science paradigm, and those based on the economic science paradigm. Furthermore, the desires to protect the environment, to develop econo-

*Virginia Polytechnic Institute and State University, USA.

mies and people, to enhance personal liberties, and to maintain a stable and just government can be, and frequently are, in conflict. Recognition of paradigm differences and the nature of sustainable development issues has implications for policy reform and institutional design. Specifically, the recognition of our ignorance and the inherent uncertainty of important future events requires the design of sustainable development strategies that prepare us for the unpredicted and unpredictable.

SUSTAINABLE DEVELOPMENT: ECOLOGY AND ECONOMICS

The two main paradigms underlying concepts of sustainable development, those of economics and ecology, incorporate substantially different assumptions about ecological and economic relationships. I explore these assumptions in depth in another paper (Batie, 1990), but I will highlight a few of the differences here. First, economics and ecological paradigms differ in their assumptions about relative scarcity. Economics incorporates a belief in the almost unlimited possibility of the substitution of man-made and human capital for natural resource capital. As one economist, Ed Schuh (1987), succinctly describes this belief:

> economic development, rather than creating economic scarcity, in its general force tends to create economic abundance. The reason is obvious ... . The engine of economic growth does not lie in physical and natural resources ... but in science and technology – [that is] knowledge. (p.373)

That is, as resources become more scarce, market prices (or shadow prices) will rise, which will induce a search for more abundant substitutes as well as for the technology appropriate to their exploitation. Because of this belief, the traditional economic model does not incorporate limits to growth, at least not in a meaningful time-frame. While recognizing that the earth is a finite collection of living and non-living systems, most economists believe that the inventiveness of the human mind and the responsiveness of institutions has, and will, avoid absolute constraints to growth for centuries.

Ecologists tend to incorporate the idea of absolute scarcity and, therefore, real limits to economic growth as a key assumption in their paradigm. The biosphere is conceived of as posing absolute limits on economic growth (but not on economic development); of particular concern is the limit of the assimilative capacity of the environment with respect to waste residuals from human activities. The foundation of this belief in absolute limits stem from the laws of thermodynamics:

> These laws guide the interaction of energy-matter on the planet and are immutable. It is ultimately the laws of nature, not of man, which determine the biospheric constraints imposed on the level of economic activity. If, for example, increasing entropy is a reality, then knowledge cannot infinitely expand the domain of human material progress at the expense of natural environment. (Underwood and King, 1989, p. 324).

The first law of thermodynamics – the Conservation of Matter Law – states that energy-matter can neither be created nor destroyed. When this law is considered with the second law of thermodynamics, or the Entropy Law, the usual conclusion is that all consumption and production ultimately increases entropy and irrevocably diminishes our future ability to use resources (Underwood and King, 1989; Daly, 1991).

Both the economists' belief in relative scarcity and the ecologists' belief in absolute scarcity can be thought of as untested. The second law of thermodynamics, on which rests the absolute limits hypothesis, applies to closed, non-living systems; its applicability to a solar energy-receiving earth which is pocketed with complex interacting, living systems remains highly debated (Zhu, Batie and Taylor, 1991). Whether the imperative for ever-increasing entropy can be applied to such a large system is not yet known[1]; there is no scientific 'high ground' for those who advocate absolute scarcity or for those who advocate relative scarcity.

The second major difference in assumptions between the traditional economics paradigm and the ecological paradigm stems from their perspectives of the economic and natural system. Neo-classical economics, like many sciences, emulates Newtonian physics (Mirowski, 1988) – a situation that more than one author has referred to as 'physics envy'. Mechanistic systems predominate in neo-classical economics. 'The neo-classical model is mechanistic in the assumption that the economic system can operate in equilibrium at any position along a continuum and move back and forth between positions.... Atomistic–mechanistic models are characterized by a range of stable equilibria and the reversibility of system changes' (Norgaard, 1985, p. 383).

The mechanistic view of the world results in most (but not all) neo-classical economists searching for optimal solutions, equilibrium positions and reversible actions (and not incidentally, developing a fetish for formal mathematical rigour and quantification). Mechanistic systems are particularly suitable for the analysis of stable, predictable systems. However, the ecologist tends to see the world as one of irreversibility, unstable systems, unpredictable system changes, disequilibria and non-incremental events; therefore an ecologist is much more likely to draw lessons from the Darwinian revolution – not the Newtonian one. The paradigm of evolution places great emphasis on the interconnectiveness of ecosystems that have coevolved as well as on ecosystems and traditional human cultures that have coevolved. As a result of these and other fundamental differences in assumptions, the meaning of the concept of sustainable development can differ substantially between the two disciplines.

'Progressives' and 'Environmentalists'

However, sustainable development advocates, no matter what their science, tend to agree that past and current development strategies are not desirable and are not sustainable in the long run. However, the foundation of their thinking – economics or ecology – tends to influence their perspective on the corrective action that should be pursued.

Norgaard (1991) draws the distinction between 'progressives' and 'environmentalists.' 'Progressives' tend to draw many lessons from economics: 'They argue that sustainability will require a significant expansion in agriculture, forestry, and other research to implement more environmental compatible technologies, significantly more environmental monitoring and assessment, and design new institutions to internalize external costs. They envision sustainability as a matter of fully optimizing people's interaction with nature' (p.11). This 'progressive' perspective on sustainable development has been termed that of 'resource management' (Colby, 1989) and leads towards discovering the 'right' incentives to produce solution-oriented technologies, implementing the 'right prices' to internalize the externalities as well as using natural resource accounting (Ahmad, Serafy and Lutz 1989; Repetto 1986). It can also involve advocacy for collective action to reduce the use of 'throughputs' in the system (Batie, 1989). Resource management policies are also those of improving imperfect capital markets, investing in education and infrastructure, health and nutrition, as well as in productivity-enhancing, pollution-controlling technologies, (Schuh, 1987; 1988; Mellor, 1988). For example, in the developed nations, there are numerous resource management strategies that can be pursued to achieve energy efficiency, to reduce agrichemical use, and to reduce pollutants (Repetto, 1990). These strategies encompass those of relating vehicle taxes to fuel consumption, eliminating 'below-cost' forest timber sales, using accurate marginal cost pricing for all energy production and reforming farm policy to eliminate incentives for surplus monoculture production (Repetto, 1990).

In contrast, 'environmentalists', who tend to draw their lessons from ecology, believe the corrective course of action should be the reduction of the overall level of economic activity. Whereas the 'progressives' goal may be to lift the poor closer to the rich through the adoption of non-polluting, efficiency-enhancing technology, the 'environmentalist' is more likely to advocate the pulling of the rich towards the poor through land tenure reform, redistribution of income and adoption of appropriate small-scale technology. While the 'progressive' is more likely to use economic incentives to achieve desired goals, an 'environmentalist's' first instincts are to regulate and to use command control institutions. This tendency stems, in part, from less knowledge of (or faith in) economic incentives. That is, the 'environmentalist' is likely to feel that the problems of protecting the environment are so complex, with numerous scientific uncertainties and with such potential for catastrophic outcomes, that no amount of tinkering with markets will suffice. However, another reason 'environmentalists' usually do not use economic incentives to achieve goals is in part their desire to stigmatize undesired behaviour as ethically, morally and legally wrong. This major school of sustainable development thought has been termed 'eco-protection' (Colby, 1989) and is preservationist in nature; it has as an objective the 'maintenance of the resource base'. That is, it is a minimization, steady-state concept that implies minimizing the use of the natural environment.[2]

In this eco-protection perspective, not only is a highly managed nature not desirable, it is not feasible. 'Environmentalists' claim that pursuit of managed environments may lead to ecosystem collapse and perhaps even the eventual extinction of the human species. Thus eco-protection goals imply limiting human management of resources as well as use of non-solar, non-renewable energy inputs.

There is also a high level of risk aversion among the eco-protecting 'environmentalists'. There is a willingness to bear high costs in terms of bureaucracies, implementation costs, foregone income, foregone personal freedoms or other opportunity costs, in order to reduce the risk of ecological damage. When in doubt – about whether there is a risk of ecological damage or whether certain behaviours or technologies are contributing to an environmental risk – the 'environmentalist' is likely to advocate that society adopt the more risk-averse court of action. In the language of statistics, the 'environmentalist' is willing to pay high costs if necessary to avoid a Type II error. A Type II error in this case would be wrongly assuming that there be no long-lasting environmental harm from a chosen action. 'Environmentalists' are also likely to estimate unknown risks to the environment as more costly than do resource managers; that is, 'environmentalists' are more pessimistic than 'progressive' resource managers. Such differing attitudes can be seen, for example, in estimates of, and willingness to bear the costs of, carbon dioxides and other pollutants to reduce the possibility of global warming.

Norgaard (1991) further distinguishes between 'technocratic environmentalists', 'who think the new environmental scientists have reasonable answers' and 'populist environmentalists', who strive to change the world into what they desire. This attitude is in contrast to the pragmatism of the 'progressive' resource managers who work with the world and its values as they find it. Thus the 'populist environmentalists' emphasize changing peoples' values, limiting population growth, redistribution of society's income and wealth, and protecting traditional cultures.

THREE PILLARS OF A SUSTAINABLE SOCIETY

Regardless of which perspective of sustainable development is adopted, however, all have as goals a sustainable, humane and just society. There appear to be three pillars to such a society: economic stability, political stability and ecological stability. All are related; all are dependent on one another; all can be in confict. Many of the dependencies are well known; however, economists as 'progressives' tend to perceive mainly the social dependencies, while ecologists as 'environmentalists' tend to perceive mainly the ecological and physical dependencies.

For example, economists tend to recognize the social complexities encompassed in the relationships between economic stability and ecological stability with respect to rates of human population growth. They perceive the problem to be one where fertility rates outstrip death rates, owing to a variety of understandable social and economic factors. Poor families must ensure their survival. When children enhance a family's economic security through their labour and when they enhance their mother's limited social status by their existence, then families tend to be large. Large poor families, on the other hand, frequently have no choice (particularly when faced with a variety of economic influences such as limited capital markets and no off-farm employment opportunities) other than to exploit and degrade their environment. The evidence is overwhelming that rising standards of living and enhanced status for women can reduce the birth rate. Therefore economists tend to look for ways to enhance income and/or women's status as one way to reduce environmental degradation in developing countries. Furthermore, because economic development causes population rates to decline as parents have fewer children (but provide each child with more education), ultimately there can be a decline in the demand for natural resources. Thus economic and ecological stability are perceived as being strongly linked with economic development.

In constrast, ecologists tend to see the ecological complexities in the relationship between economic and ecological stability. They recognize that there are complex linkages that result in unintended ecological damage by well-intended economic actions. For example, they were the first to call attention to the complex ecological relationships that are damaged when wetlands or forests are converted to agricultural uses. Ecologists understand that ecosystem health is important to all species including the human species, and they recognize that protecting species' diversity and habitat also protects future human choices.

While interdependencies between economic, political and ecological stability are real and common, so are conflicts. For example, in order for natural resources to be conserved and used wisely, they must not be priced too cheaply. Thus, whether one refers to the American West or the Soviet Union Aral Sea, one finds that inexpensive irrigation water, for example, can lead to serious ecological disfunctioning. In these examples, economic stability (for example, pricing resources appropriate to their scarcity) and ecological stability (such as protecting ecosystem functioning) appear to be compatible, but political stability may imply pricing food and clothes inappropriately low relative to the true opportunity costs of their production, or may mean maintaining populations on fragile lands. Thus political stability may be counter to economic ecological goals. Furthermore, there are many examples where economic and political stability are seemingly in conflict with ecological stability. The closing of old growth forests to the logging industry on which local cultures depend is such an example.

Furthermore, with environmental issues, the potential for surprise and system breaks – that is, sudden and drastic changes in the parameters of the system – is high and potentially catastrophic: 'Examples include the greenhouse effect, the hole in the ozone layer, algal bloom from fertilizer runoff, acid rain, … all of these are relatively new phenomena, and they were all unexpected and unwelcome' (Faber, Manstetten and Proops, 1990, p. 14). As a result, there are some profoundly different issues surrounding the problems of natural resource depletion and those of ecological stability. The economic policies that apply to resource depletion may well be inappropriate for protection of ecosystem stability.

Informed sustainable development policies should incorporate appreciation of social, political and ecological interrelationships. As the past-president of the World Bank stated:

A purely technical approach to the environmental challenge, insensitive to social, cultural, and public health considerations, results in a wide array of social problems. Profligate industrial policies assail the world's climate. The basic requirement of food for ceaselessly growing populations is met at the expense of degraded soils, making future agricultural efforts more costly. Development resting only on exploitation of nonrenewable resources leaves us poorer in the long run. All these issues and others are intertwined and must be addressed. (Conable, 1989, pp. 2–3)

Dynamic concepts

In addition, it is important to recognize that these three pillars – economic, political and ecological stability – are dynamic concepts whose definitions change with changes in culture, time and scale. The interpretation of these terms embodies value-judgements not resolvable by science. The ecological concept of carrying capacity, for example, must be defined for a certain scale of an economy, the size of its population and the available level of technology, as well as its use of finite resources. A 'slash and burn' nomadic culture may be sustainable when there are few people using a large amount of land; such practices only become unsustainable when populations grow out of proportion to the land base. Such large populations may be sustainable, however, if they adopt an alternative technology to 'slash and burn' as their agricultural production practice.

Furthermore, embodied in the concept of carrying capacity are value-judgements concerning appropriate standards of living, levels of biological diversity or limits to a managed nature. While scientists may be able to assist in identifying the ultimate limits of carrying capacity – that is, the circumstances which would be present when the entire ecosystem or social system ceases to function – the selection of the optimal carrying capacity short of these ultimate limits is not a scientific decision. Furthermore, in many cases, even the ultimate limits are subject to fierce debate, not just between social and physical scientists, but between physical scientists themselves. The current debates swirling around the reality and significance of global warming or acid rain are testament to this point. There are great uncertainties associated with many of the economic–ecological linkages: 'The sobering prospect is that most of the major public decisions about resource use and environmental management will be made in the face of large uncertainty deriving from ignorance of physical and biological systems and from evolving techniques and social values' (White, 1980, p. 183). In many cases, there are also significant political impediments to achieving ecological and/or economic stability goals.

IMPLICATIONS FOR SUSTAINABLE DEVELOPMENT POLICIES

What are the implications of these insights for economists interested in sustainable development policies? The first is simply that there will be differences in the way well-informed, well-intentioned people view the concept. Ultimately one's views depend on such factors as one's:

- faith in technological progress and institutional capacity to react to change,
- belief in the appropriate scale and relationship between humans and nature,
- estimate of the time-frame in which any physical limits to growth apply,
- concern over and belief in the potentiality of complete ecosystem collapse, and
- definition of intergenerational and intragenerational justice.

It is no wonder that debates are so common, so intense, and so unresolved.

One's view on these factors influences the answers to fundamental questions: how do we determine who and what we want to sustain and how do we organize society to achieve sustainability? Much of the literature on sustainable development presumes we know what would comprise a sustainable world and that our sustainable development goals are predetermined. Some authors directly or implicitly argue for command control policy instruments (for example, emission controls, regulation of equipment, production processes, inputs and outputs) and/or highly planned centralized decision making to achieve these goals. Others proclaim that we must rely on decentralized decision making, preferably market-based incentives (such as tradable permits, taxes, subsidies). In practice, neither of these approaches, historically, have proved totally satisfying for obtaining environmental quality or sustainability. Markets will never 'get the prices right' on unpriced resources, particularly when the non-market value of the natural functioning of the resource is unknown. Markets involve trade at the margin; any physical limits, ecosystem collapse, or future generations' needs are beyond the traders' vision. Markets thrive on disequilibrium, instabilities and uncertainties; on trial and error, not single-minded pursuit of stability goals or steady-state economics. Furthermore, the market does not encourage a sense of a global community of interests.

On the other hand, centralized decision making has proved to be exceptionally poor at providing even market goods, let alone environmental quality and ecological stability. Centralized planning does not reduce rent-seeking behaviour or uncertainty, and it tends to stifle entrepreneurship for innovative solution-providing technologies and institutional design. Similarly, command control instruments require deterministic centralized planning relating to what the environmental standards should be and how they are to be achieved. The implicit and generally wrong assumption is that we know the who, what and how of sustainability.

Sustainable development strategies need to reflect recognition of knowledge uncertainties and the potential for both market and government failures. Furthermore, there must be more knowledge sharing between the physical and social sciences as to the socio-economic, political and ecological reasons underlying both environmental degradation and poverty. Such knowledge sharing is essential if we are to adequately analyze supply and demand for unsustainable actions (Idachaba, 1987). For example, the world is replete with examples of environmental degradation resulting, not from population pressures or economic growth, but rather from public policies. Sugar policy in the USA, for example, by artificially raising the US domestic price of sugar, encourages sugar production, encourages corn production (since corn fructose is a substitute for sugar), injures non-US producers of sugar and (without intending to) results in the pollution of southern United States wetlands as well as soil erosion in the mid-West United States. US transportation, land use and energy policies encourage the use of cars and airplanes as well as the growth of suburbs but discourages railway development and (without intending to) results in air pollution and habitat destruction.

There are a variety of socio-economic, political and ecological factors underlying these examples. Only careful examination of these factors will provide the knowledge necessary to redesign the institutions to implement sustainable development policies. Such institutional design can have as much or more social pay-off as can new production technologies (Ruttan, 1987), particularly if it addresses uncertainties and provides for non-deterministic ways of addressing trade-offs between economic, ecological and political stability.

Sustainable development requires new strategies that differ from those that emanate from strategic, rational, deterministic planning. These strategies should allow for technological progress to solve environmental problems, but not depend solely on such progress. Furthermore, any strategy should be open to revision when an unexpected 'novelty' or currently unforeseen possibility occurs.

Adaptive management

This broad, flexible strategy has been termed in environmental planning: 'adaptive management' (Walters, 1986). Adaptive management involves trial and error, monitoring and feedback in the development of alternatives, as well as the exploration of values. Rather than develop a fixed goal and an inflexible plan to achieve that goal, adaptive management recognizes the imperfect knowledge about the interdependencies that exist within, and among, natural and social systems. Furthermore, this imperfect knowledge is seen as requiring plan modifications with improved technical knowledge and changing social preferences. In effect, adaptive management is a learn-by-doing approach to decision making.

Adaptive management is capable of incorporating either market incentives or command control instruments but includes feedback and redesign as fundamental components of the policy implementation. Thus, if adaptive management is to be sustainable, it requires knowledge of ecological as well as social systems. Feedback mechanisms must include those that feedback, in a timely effect manner, information about ecological dysfunctioning (Dryzek, 1987) as well as economic and political cost. Flexibility and responsiveness of institutions to changing environmental or economic conditions is necessary as well.

Adaptive management, sustainable development strategies establish objectives through an opportunity cost decision-making approach. For example, within the opportunity cost framework, the answer to the question 'how much environmental protection is enough?', emerges from legitimate social choice processes within governments which determine the degree of environmental protection that is desirable. Confronting the decision process with economic, ecological and political cost information elicits 'values' from that process. Continually focusing the question on whether an action is 'worth' its economic, ecological and political opportunity cost is seen as the most practical way to answer the question 'how much is enough?'

The development of institutions capable of adaptive planning is a tremendous challenge. In the United States, for example, most environmental strategies do not incorporate monitoring, feedback and re-design. For example, the 1985 Farm Bill authorized the United States Department of Agriculture (USDA) to pay farmers not to farm highly eroding lands for ten years. The bill did not include any evaluation and re-design as the programme was implemented. Subsequently some monitoring has taken place, conducted, not by the USDA, but rather by environmental groups who were suspicious that environmental goals were not being met. Re-design had to wait until congressional re-authorization of the farm bill in 1990, an awkward, slow, inflexible institutional approach. Further re-design must wait until 1994.

There are a few examples of adaptive management. One is found in the United States Chesapeake Bay programme's nutrient management strategy. The four state governments in the region initially established a goal to reduce non-point nutrient loads to the Bay by 40 per cent and agreed to a variety of approaches to achieve the goal. They also committed themselves to continuous study of the goal itself as well as the cost and effectiveness of the means to attain the goal. The 'value' of goal attainment would be discovered in relation to the cost of goal attainment. As a result, both goals and approaches are subject to revision over time. However, adaptive planning such as is incorporated in the Chesapeake Bay programme is rare, still quite imperfect, and in need of refinement.

CONCLUSION

The design of adaptive management institutions represents a significant change from the status quo and hence is a tremendous challenge. However, it appears to provide one of the more promising answers to the fundamental questions of what we want to sustain and how we organize to achieve such sustainability. Such design requires contributions from many disciplines and a willingness to experiment with alternative sustainable development strategies. Sustainable development will not become reality if the three pillars of economic, political and ecological stability are leaning in opposite directions as a result of unco-ordinated development and environmental policies. Sustainable development, as implemented in adaptive planning strategies, is an imperfect and dynamic process that attempts to reflect the social–political value context necessary for systems' stability.

NOTES

[1]However, the entropy law clearly applies to the conversion of raw material ultimately to 'waste' material that is returned to an environment characterized by a finite capacity to absorb wastes. That is, there is a limit to ecosystems' ability to absorb pollutants and still continue to provide natural services such as clean air.

[2]This latter eco-protection view can include those of deep ecology 'ecocentrists' whose beliefs are dominated by concerns with rights for non-human species and systems.

REFERENCES

Ahmad, Y. J., Serafy, S.E. and Lutz, E. (eds), 1989, *Environmental Accounting for Sustainable Development*, World Bank, Washington, DC.

Batie, S. S., 1989, 'Sustainable Development: Challenges to the Profession of Agricultural Economics', *American Journal of Agricultural Economics*, 71,(5), pp. 1083–1101.

Batie, S. S., 1990, 'Economics and Sustainable Development: Compatible?', paper presented at Duke University, Durham, NC, 14 September.

Batie, S. S., 1991, 'Three Perspectives of Sustainable Development', paper presented at Princeton University, Princeton, NJ, 16 April.

Colby, M. E., 1989, 'The Evolution of Paradigms of Environmental Management in Development', Strategic Planning and Review Discussion Paper 1, October, World Bank, Washington, DC.

Conable, B. B., 1989, 'Development and the Environment: A Global Balance', *Finance and Development*, 26, (4), pp. 2–4.

Daly, H. E., 1991, *Steady-State Economics*, Island Press, Washington, DC.

Dryzek, J., 1987, *Rational Ecology: Environment and Political Ecology*, Basil Blackwell, Oxford.

Faber, M., Manstetten, R. and Proops, J., 1990, 'Towards an Open Future – Ignorance, Novelty and Evolution', Discussion Paper 149, Department of Economics, University of Heidelberg, FRG.

Idachaba, F., 1987, 'Policy Issues for Sustainability', in T. J. Davis and I. A. Schirmer (eds), *Sustainability Issues in Agricultural Development*, World Bank, Washington, DC.

Mellor, John W., 1988, 'The Intertwining of Environmental Problems and Poverty', *Environment*, 30(9), pp. 8–30.

Mirowski, P., 1988, 'Shall I compare thee to a Minkowski–Ricardo–Leontief–Metzler Matrix of the Mosak–Hicks type? Or, rhetoric, mathematics and the nature of neoclassical economic theory', pp. 117–45, in A. Klamer, D. W. McCloskey and R. M. Solow, (eds),*The Consequences of Economic Rhetoric*, Cambridge University Press, Cambridge.

Norgaard, R. B., 1985, 'Environmental Economics: An Evolutionary Critique and a Plea for Pluralism', *Journal of Environmental Economics and Management*, 12, pp. 382–94.

Norgaard, R. B., 1991, 'Sustainability as Intergenerational Equity: A Challenge to Economic Thought and Practice', Asia Regional Studies Report IDP97 (June), World Bank, Washington, DC.

Repetto, R., 1986, 'National Resource Accounting For Countries with Natural Resource-Based Economics', paper presented at World Resources Institute, Washington, DC, 3 October.

Repetto, R., 1990, 'Promoting Environmentally Sound Economic Progress: What The North Can Do', The World Resources Institute, Washington, DC.

Ruttan, V. W., 1987, 'Institutional Requirements for Sustained Agricultural Development', in T. J. Davis and I. A. Schirmer (eds), *Sustainability Issues in Agricultural Development*, World Bank, Washington, DC.

Schuh, G. E., 1987, 'Some Thoughts on Economic Development, Sustainability, and the Environment', in T.J. David and I.A. Schirmer (eds), *Sustainability Issues in Agricultural Development: Proceedings of the Seventh Agriculture Sector Symposium*, World Bank, Washington, DC.

Schuh, G.E., 1988, 'Sustainability, Marginal Areas, and Agricultural Research,' paper presented at International Fund for Agricultural Development, 22 September.

402   *Sandra S. Batie*

Underwood, D. A. and King, P. G., 1989, 'On the Ideological Foundations of Environmental Policy', *Ecological Economics*, 1, pp. 315–34.

Walters, C.J., 1986, 'Adaptive Management of Renewable Resources', MacMillan, New York.

White, G., 1980, 'Environment', *Science*, 209, pp. 183–90.

Wildavsky, A. B., 1988, *Searching for Safety*, Transaction Books, New Brunswick, N. J.

World Commission on Environment and Development, 1987, *Our Common Future*, Oxford University Press, Oxford.

Zhu, M., Batie, S. and Taylor, D., 1991, '*Thermodynamic Laws and Sustainable Development Concepts*', Department of Agricultural Economics, Virginia Polytechnic Institute and State University, unpublished.

DISCUSSION OPENING – RICHARD R. BARICHELLO*

Let me open by commending Sandra Batie for the fine job she has done in pulling together the several schools of thought and lines of argument which surround the concept of sustainable development. The topic remains only vaguely defined because it is applied to a wide range of problems and has attracted an exceedingly diverse set of authors, disciplines and prescriptions. Confronted with this array of material, Professor Batie has organized it well and has extracted the key issues clearly.

The paper focuses on the two paradigms of ecology and economics, and begins with the beliefs, assumptions and implicit models which each uses to analyse the sustainability issue. Two major differences are highlighted: (a) the interpretation of relative scarcity, and (b) the question of whether the economic and natural systems are characterized by equilibria and smooth marginal changes, or by disequilibria, instability, irreversibility and non-incremental changes.

The paper goes into greater detail by discussing prescriptions for dealing with sustainability problems through the eyes of 'economic progressives' and 'environmentalists'. These characterizations show some differences in objectives, but primarily a difference in the underlying model of the way the world works.

To find common ground and policy responses to these problems, the paper emphasizes three factors or pillars underlying a sustainable society which must collectively guide us in the search for policies to help resolve the sustainability problem: economic, political and ecological stability. The paper emphasizes the interrelationships among these three factors, changes in their relations over time, recognition of where our knowledge base is weak, knowledge sharing between physical and social sciences, and a re-design of our institutions to implement more sustainable policies. An adaptive management strategy incorporating economic and ecological information, and which features monitoring and feedback, is recommended.

Let me now summarize what I think we have learned from this ongoing debate and the speaker's review, with a focus on appropriate resolution of sustainability problems.

*University of British Columbia, Canada.

(1) A number of externalities, particularly relating to sustainability issues, have been much better identified and defined. This information has been largely provided by ecologists and scientists, and a continued flow of data documenting such externalities is a necessary part of an effective policy process. Dr Yadav's paper, within this section of the programme, provides a nice illustration of bringing important local data to bear on practices.

(2) The environmentalist goal of preserving certain resources or resource stocks, with the interests of future generations in mind, has been identified and emphasized anew. This may be legitimate, but requires some measurement and weighing of costs and benefits.

(3) Let me restate the importance of combining economics, ecology and politics in developing appropriate policies and strategies, although with a somewhat different emphasis than that found in Batie's paper. There are three headings:

— from scientists and ecologists, we need more information to overcome 'knowledge uncertainty' about the technical details of ecological problems;
— economic analysis is required, not only in measuring costs and benefits, but also in identifying the incentives faced by farmers or others whose actions are causing problems;
— understanding of the political economy of government action or inaction would help us to understand the interests involved, and how policies might be changed.

Above all, it should be emphasized that there is in practice a very great distance between current policies and institutions and those which would be recommended by economic analysis to provide a solution to sustainability problems. We do not live in a 'first-best' or even a 'second-best' world. The policies and institutions which are in place often do little to help solve environmental externalities; indeed, many actually create or exacerbate them. This suggests that there is a wide range of problems which are amenable to resolution with more appropriate economic instruments. For example, soil degradation problems are significantly resolved by reducing the incentive to expand production to marginal lands, and over-cutting of forest lands is reduced by improved land tenure arrangements (see below). This may sound like a repetition of familiar arguments, but when we examine a selection of particular problem areas, such as irrigation water control or sugar policy, we are so far from sensible economic policies in many areas that it is worth restating the priority of using economic tools carefully and more persuasively.

The challenge facing economists is to design incentive systems which lessen or solve our environmental problems, and to construct local institutions which can provide those incentives. Farmers' responses to the incentives they actually face, given their information base and other constraints, are not in doubt, although determining farmers' true incentives is not a trivial problem.

It is also my view that many of the problems we face are the result of local circumstances, institutions and ecological conditions. Their solution similarly requires the use of problem-specific data rather than general prescriptions.

Without this focus on specific data, one could easily be drawn into a quasi-religious approach to solving sustainability difficulties.

One example can be provided to illustrate my points; it concerns tropical deforestation, where there are key issues relating to institutional conditions. Are timber concessions awarded for 20 years when the optimal felling cycle is 35 years, giving concession-holders no financial interest in the next crop? How do taxation and royalty policies encourage logging practices? If taxation is on the basis of timber removals rather than marketable stems, one should expect high grading, increased site damage and wasted wood, lower productivity in the next crop, and lower government royalty and tax revenues. What kind of replanting policy is in place and what are the resulting incentives to replant? Are there other restrictions, such as selective felling, which may increase environmental damage and lower forest productivity? If enforcement of concession rights against localized small farmer logging or slash and burn agriculture is difficult, what do concession holders do to their forest stocks to deal with illegal harvesting and squatting? The sensible response of concession holders may be to cut the forest stock faster.

The point is that many externalities are generated by chosen government policies and existing local institutions. Attention to these details is often critical to (a) understanding why unsustainable practices are being followed, and (b) designing appropriate solutions. Last, and not least important, is the question of why the policies were designed in such a way in the first place. What are the political interests which may prevent effective policy and institutional reform? If the responsible ministry is reluctant or unable to change forest policies because of the influence of powerful vested interests, what are the reasons for opposition from those interests? Further, can other policies or regulations be put in place which would enlist interest group support?

In closing, I would hazard the guess that a large proportion of the environmental problems which have spurred the sustainability debate are not caused by irreversibilities or by an unwillingness to consider future generations: for example, the smallholders who are damaging forest stocks are often saving prodigiously to enable their children to own their land and to better their education at family expense. Rather, the problems are the result of government policies and local institutions which are serving some political interest, or which have inadvertent side-effects injurious to sustainability. Dealing with these sustainable development problems requires a mix of careful economic analysis, adequate local environmental and policy data, and an attempt to understand the political economy considerations which have led politicians to choose the policies in place.

R.P. YADAV*

Sustainable Development Strategies in
Less Favoured and Marginal Production Areas*

INTRODUCTION

'Sustainable development' has become an integral part of the terminology of development jargon. It is still an elusive concept. The most profound definition of sustainable development is found in the Brundtland Report, 'Our Common Future', where it is defined as 'development which meets the needs of the present without compromising the ability of future generations to meet their own needs'. It assumes three things: a long-term perspective, equity between generations, and dynamic phenomena. Sustainability with respect to intergenerational equity cannot be achieved as long as a glaring inequity in the distribution of resource endowments exists. At the global level, 'only one-fifth of the world's people live in the rich countries but are using up about four-fifths of the world's annual resource output' (Trainer, 1989). Similarly, at the national level, a small proportion of the population uses most of the resources, while the larger proportion of the population subsists on meagre resources. This widely-quoted conceptual definition of sustainability can be realized only if major steps are taken towards redistribution of resources both at global and at national levels. The current socio-political structure remains a major obstacle in the way of such a redistribution. Consequently, people in poor countries must often sacrifice long-term environmental stability for their short-term subsistence needs (Mellor, 1988).

On the more operational level, in the context of resource use, 'Sustainability is the ability of a system (e.g., the fragile resource – agriculture) to maintain a certain well-defined level of performance (output) over time, and, if required, to enhance the same, including through linkages with other systems, without damaging the long-term potential of the system' (Jodha, 1990). In other words, sustainability implies an increase in use intensity and higher productivity of land resources without permanently damaging such resources.

This paper is principally focused on issues related to sustainability in less favoured and marginal production areas, in general, in mountain regions in particular. The prospects of sustainability for agriculture in marginal areas are severely constrained by the specific features of their natural resource endowments. Sustainability was maintained in the past because of low pressure on

*Mountain Farming Systems Division of the International Centre for Integrated Mountain Development (ICIMOD), Nepal.

these resources through the application of traditional, land-extensive practices. Indigenous systems were oriented towards resource use with conservation without raising the use intensity and the productivity of the land. In most cases a sustainable balance was achieved between the needs of the mountain people for food, shelter, fuel and water, and the capacity of the natural environment to provide these. Unfortunately, continued population growth and market-induced demands have increased the pressure on land and other natural resources, and this has prompted greater resource use intensity, consequently leading to negative changes known as 'indicators of unsustainability' (Jodha, 1990).

Less favoured and marginal production areas generally have a fragile resource base. Areas which have a low potential for crop farming (mountain regions with steep slopes, deserts, rain-fed arid and semi-arid tropical areas with low and irregular rainfall, and coastal areas prone to salinity and waterlogging) fall into this category. Such areas cannot tolerate the higher intensity of use associated with specialized usage and have higher input–output ratios than better land resources (Jodha, 1990).

The conditions of biophysical resources bases and their proper use, through technological and institutional measures, constitute the supply aspect of sustainability, whereas the management of population and of market-induced pressures on mountain resources constitute the demand aspect of sustainability. The efforts, on both the supply and the demand side, to reverse the unsustainability situation, or to avert unsustainability prospects, should constitute the essence of development strategies.

THE SUPPLY ASPECTS OF SUSTAINABILITY

The biophysical resource base

The long-term productivity of the natural resource base is determined by the inherent capacity of the resource and the pattern and method of its usage. Given its inherent characteristics, the natural resource base of marginal areas is limited in use. More intensive uses can be productively maintained for a short period with a high degree of artificial support (for example, subsidies in chemical, biological and physical forms) or by causing serious damage to the inherent capacities of the resource base itself. In either case, intensive resource use is a definite step towards long-term unsustainability. This problem is more pronounced in regions with fragile land resources, for example the mountains and dry tropical areas. Generally, in this situation, the range of options ensuring a proper match between resource characteristics and resource use is extremely narrow. However, as a result of human adaptation mechanisms and their ingenuity over the generations, the range of options has been widened. Features of traditional, integrated farming systems in these regions corroborate this (Whiteman, 1988; Moock, 1986; Altieri, 1987; Jodha, 1990). However, these options, having evolved in the context of low demand on fragile resources, are becoming increasingly unfeasible or ineffective in the context of the pressures generated by population growth, market-induced forces and public interventions (Liddle, 1975; Reiger, 1981; Jodha, 1990). Consequently, to compensate for falling incomes or to meet basic needs, cultivation has been extended to sub-marginal areas, monoculture has been induced by promotion of HYV crops (the RR21 variety of wheat is grown on 80 per cent of the total wheat area in Nepal); grazing land has been over-stocked and deforestation has become excessive. Thus there is a mismatch between resource potential and usage, leading to the emergence of indicators of unsustainability. In such situations, the re-establishment of a 'match', between resource characteristics and their use patterns, is an important step in enhancing the sustainability of fragile resources (Jodha, 1990).

Any conceptual or operational framework for mountain area development must take into consideration the factors of slope and altitude which separate the mountains from other areas. Compared with the two-dimensional spatiality of the plains, mountain habitats are characterized by three-dimensional spatiality. This additional dimension, verticality, obstructs the applicability of development of other experiences of plains to mountains, since it disconnects the highlands physically, economically and culturally from the plains because of the barrier of physical inaccessibility. Because of slope and altitude and associated conditions, mountains have six distinct characteristics: inaccessibility, fragility, marginality, diversity, 'niche' or comparative advantages of mountain areas, and human adaptation mechanisms. These are also called mountain 'specificities'. The explicit consideration of these mountain characteristics in designing policies, programmes and projects for the hills and mountains is referred to as the 'mountain perspective'. The principal lacuna in development strategies in the past has been the lack of this mountain perspective. The characteristics are not only interrelated but also demonstrate considerable variability within mountain areas. For instance, all locations throughout mountain areas are not equally inaccessible, fragile or marginal, nor do human adaptation mechanisms exist uniformly throughout mountain regions. These mountain specificities could serve as a useful filter to screen public interventions in order to understand their sustainability implications and the changes required to enhance the sustainability of mountain agriculture (Jodha, 1990).

Mountain characteristics

Six mountain characteristics are briefly described here:

Inaccessibility The vertical dimension of mountain areas encompassing altitude, slope and undulating topography contributes significantly to the inaccessibility characteristics of the mountains. The direct manifestations of inaccessibility are isolation, poor communications, high transport costs and a slow pace of transformation. Measures to mitigate these constraints are transport subsidies, processing of value-added products, and construction of roads and other transport infrastructure such as short take-off and landing (STOL) airports, mule tracks and ropeways.

Fragility   Fragility refers to the vulnerability of mountain resources to rapid degradation through even a small degree of disturbance (DESFIL, 1988). This poses a serious constraint in the greater use of resources that are prone to ecological degradation. The management of fragile land is usually carried out through conservation by 'non-use', afforestation and improved watershed management practices with a strong focus on resource-centred technologies.

Marginality   Marginality is an attribute of identity (be it in the context of physical and biological resources or social groups) that reflects a relatively lower status or priority in the context of the 'mainstream' or dominant features of society. The basic factors contributing to marginality are remoteness and physical isolation, fragile and low-productivity resources, and a number of man-made handicaps that prevent participation in the mainstream of activities (Chambers, 1987; Lipton, 1983). In terms of its biophysical aspects, marginality shares the attributes and consequences of fragility, and its socio-economic dimensions are poverty, isolation and neglected minority/tribal communities. The solutions for marginal areas are similar to those for fragile areas and a 'target-group' approach to address specifically poor, disadvantaged or ethnic groups.

Diversity or heterogeneity   Because of the immense variations in biophysical and socio-cultural factors among and within ecozones, even over short distances, diversity or heterogeneity is a major characteristic of mountain areas. This extreme degree of heterogeneity is a function of the interaction of various factors, ranging from climatic to geologic and edaphic conditions (Troll, 1988). In terms of its operational implications, the diversity of the mountains offers both a range of opportunities and a set of constraints.

Mountain 'niche'   Mountains have specific ecological and physical features that offer several unique opportunities (products and activities) for which mountains have comparative advantages over the plains. Examples include special medicinal plants, fruits and flowers, and hydropower production sites. While harnessing these opportunities, exchange with other regions on equitable terms can contribute towards the sustainable development of mountain regions but reckless exploitation causes resource degradation and unsustainability. The sustainable development of comparative advantages provides a basis for trade and exchange between different altitudes and ecozones (Calkins, 1976).

Human adaptation mechanisms   Over many generations, mountain communities have evolved, through trial and error, their own adaptation mechanisms to handle the constraints indicated by several of these mountain characteristics. Accordingly, either the resource characteristics were modified (for example, through terracing, ridging and irrigation) or activities were designed to adjust to the requirements of resource conditions or mixed (integrated crop–livestock–forestry) farming. Thus farmers have traditionally made sustainable use of fragile resources. The proper understanding of indigenous knowledge could, therefore, help in designing better options (technological and institutional) for sustainable agriculture in fragile resource areas.

Mountain perspectives: country experiences

China Before 1978, the Chinese government placed an emphasis on foodgrain production throughout China, notwithstanding the unsuitability of some areas for this activity. Production brigades and production teams were required to meet certain targets determined by high echelons of government. This policy led to the production of foodgrains even on unsuitable marginal lands (which were otherwise fit for growing grasses, forests and other perennial crops). Consequently, this led to negative changes and environmental degradation. Thus, prior to 1978, there was a general insensitivity to the needs and limitations of mountain people and mountain environments (Wenpu and Qinfa, 1988). The collectivization, restriction on subsidiary activities, and emphasis on self-sufficiency in food reduced the individual incentive as well as the scope for harnessing the unique and diversified opportunities offered by mountain areas. Thus the government policy completely disregarded the mountain perspective and consequently led to increased environmental degradation. It was only in the post-1978 period that, through the introduction of the Household Responsibility System and agri-ecological regionalization, farmers began to be involved in diversified production activities to suit the different mountain 'niche'. Permission to retain and privately re-invest surplus encouraged them to choose a mix of the most appropriate options. A substantial increase in agricultural production took place after the reform. However, in some areas the focus on private incentives has also encouraged people to pursue private gains at the cost of society and this has led to the degradation of natural resources in mountain areas. The immediate cases are negligence in the collective management of common property such as local irrigation systems or common pastures or farm roads, and the high use intensity of fragile lands to increase incomes without conservation measures (Dafu, 1988).

Himachal Pradesh (India) The experiences in Himachal Pradesh provide a successful example of the consideration of comparative advantages. The main emphasis in agriculture was the shift from self-sufficiency in foodgrains to maximization of farm income through cash crops (fruits and vegetables) that are highly remunerative and for which the state has a comparative advantage because of climatic and other factors. The necessary road access and market facilities, support prices, strong research and extension systems for generation and dissemination of appropriate technologies, and requisite support services were provided. The state, as well as the central government, made substantial investments in road construction throughout the country. This in a very real sense provided a road to development in Himachal Pradesh. In addition, the state's investment in the development of hydropower led to the generation of government revenue by selling electricity to the neighbouring states. In

Himachal Pradesh almost all villages, including those in remote and difficult areas, have access to roads and electricity. This has facilitated the transformation of the economy.

Over the period 1951–81, there has been a noticeable shift in the pattern of labour use and the composition of aggregate output. During this period, the share of agriculture in the total employment figures fell from 91 per cent to 70.8 per cent and correspondingly the share of the non-agricultural sector rose from a mere 9 per cent to 29.2 per cent. In three decades (1950/1 to 1982/3) the aggregate output of the agricultural sector fell to 43.4 per cent from 69.4 per cent, while the share of the non-agricultural sectors increased to 56.6 per cent from 17.3 per cent (Sharma, 1987).

In 1954, when the state was established, hardly 1 000 hectares of land were used for fruit growing. By 1988/9, the area and production reached 150 000 ha and 3.1 million tons, respectively (Verma and Partap, 1990). Assessments made in Himachal Pradesh (and also in Nepal) reveal that off-season vegetables generate more employment opportunities for small and marginal farmers and agricultural labourers. Summer vegetable production in Himachal Pradesh rose from less than 30 000 tons in 1966 to 312 000 tons by 1985/8. About 200 000 small and marginal farmers were engaged in vegetable growing (Tiwari, 1990).

Marketing and other support services were developed to support both fruit and vegetable production. The Himachal Pradesh Horticultural Produce Marketing and Processing Corporation Ltd (HPMC) was established as a public sector company to undertake marketing and bring together all other post-harvest services for the promotion of horticultural development (Rana, 1990).

Pakistan   The government in Pakistan seems to have disregarded the mountain perspective in its policies and programmes which are uniformally applied throughout the country without making any distinction between the mountains and the plains. Thus most development interventions have failed because of their plains biases. Inaccessibility has been a major constraint in the economic development of people in the mountain areas. Isolation, poor communications and the lack of a marketing infrastructure generally compel farming communities to disregard the opportunities offered by diversity and comparative advantages in favour of food security considerations.

Nevertheless, in the last decade, some innovative programmes have been initiated in Northern Pakistan, where mountain specificities are duly considered, and these have brought considerable developments to the area. First, the inaccessibility barrier has been broken through the building of the Karakoram Highway (KKH) connecting Pakistan with China in the north. Link roads have been built to connect villages with the KKH. The Aga Khan Rural Support Programme (AKRSP), commencing in 1982, introduced an innovative institutional mechanism in the project area with the objective of increasing the capacity of local people to identify their needs and problems and to use opportunities to resolve them. The institutional model emphasized the formation of village organisations (VOs), where the majority of village residents expressed an interest in working together, for village development. Other important principles of this model have been to promote the solidarity of the

VOs through collective management of common property regimes with a specific focus on the participatory approach, upgrading and creation of appropriate skills by providing training, and capital formation through group savings. VOs undertook many productive physical infrastructure, irrigation channel, link road and storage construction activities.

These developments have facilitated the gradual change from a subsistence crop-dominated production system to a cash crop-dominated production system taking agro-ecological characteristics into consideration. Input supplies increased and agricultural marketing expanded. It has also become cheaper to transport subsidized flour from the plains to Gilgit. Similarly, the opportunities for employment in tourism, construction, common public sector agencies and other non-agricultural activities have multiplied.

Nepal   In Nepal, because of the problem of inaccessibility in many hill areas, emphasis on foodgrain production has been the policy of the government. This has adversely affected the environmental conditions of the mountains since increasing areas of marginal land on steeper slopes are brought under grain cultivation on highly erodible and fragile soil.

The inaccessibility of substantial areas of the mountains in Nepal has made the import and distribution of improved agricultural inputs and foodgrains costly, and it has also constrained the development of high-value cash crops because of the lack of marketing infrastructure. Thus farmers are compelled to concentrate on foodgrain production, and this has been further necessitated by the rapid increase in the mountain population. Although previous Five Year Plans have recognized the need for regional specialization, based on comparative advantages, this recognition was not put into operation because of the lack of adequate transport, poor marketing, and inadequate research and support services in the mountains.

Despite the various efforts made to increase the production of cereals, the productivity of major foodgrains has declined overtime (Yadav, 1987). Improved varieties of cereals that have been recommended for the hills and mountains require high inputs of fertilizer and assured irrigation, whereas most hill agriculture is carried out under rain-fed conditions. Both unreliable supplies and the high cost of fertilizer inhibit development. Nepal has not yet shown any definite signs of transformation in hill agriculture, from subsistence to commercial, except for a few products around its urban centres.

Innovative programmes such as Small Farmers' Development and Production Credit for Rural Women, which are oriented towards the specific target groups of small farmers and poor rural women, have addressed the marginality problems in the mountains. However, programme coverage is still extremely limited.

THE DEMAND ASPECT OF SUSTAINABILITY

It is the continuing additional pressures on existing resources that create unsustainability in fragile resource areas in general, and in the mountains in particular. Four factors, among others, seem to place additional pressures on mountain resources.

(1) *Human population* In mountain areas, there are too many people in relation to available resources, and projections for the future indicate that the population is likely to increase. The population in the hills of Nepal increased from 6.34 million to 8.46 million between 1961 and 1981, in spite of the fact that 1.19 million people migrated from the hills to the plains during the same period (Banskota, 1985). Seasonal migration and trading with other regions are common features of mountain communities. These activities have, to some extent, reduced the pressure on existing resources and, at the same time, supplemented family income through earnings outside the region. The development of roads in the mountain areas has facilitated further migration and trading activities.

(2) *Livestock population* Along with the increase in human population the livestock population has increased. This also brings additional pressure to bear on limited resources. Since livestock are an integral part of the hill farming system, they have to be maintained and over-stocking leads to the degradation of forest and grazing land.

(3) *Market-induced demand* Mineral resources, hydropower, forests and other cash crops, for which the mountains have comparative advantages over the plains, are extracted by private and government agencies for profit and revenue generation. These market-led interventions have been instrumental in pushing self-sufficient economies towards increased commercialization, with negative consequences (Mehta, 1990). This process has led to the substantial extraction of resources from the mountains, to the increasing benefit of the plains and at the cost of the marginalization of mountain communities. These agencies are insensitive to local needs and to the negative effects of the over-extraction of resources. There is ample evidence regarding deforestation for commercial use and the environmental insensitivity of mining activities, and hydropower and irrigation schemes and so on from various areas in the Hindu-Kush-Himalayas region (Banskota and Jodha, 1990). Similarly, in terms of interregional trade, based on timber exports and the harnessing of irrigation and hydropower, North Pakistan's mountain areas continue to be net exporters of resources to the plains (Khan, 1989).

(4) *Tourism* Mountains have special advantages for the promotion of tourism. In some mountainous countries, tourism has emerged as a major source of foreign exchange earnings and a potential source of off-farm employment. For example, in Nepal, 20 per cent of the total foreign exchange earnings came from tourism in 1987/8. At the same time, tourism places additional stress on the environment through excessive use of fuel wood, by creating a demand for food in areas where food is already deficient and by polluting the environment with rubbish. Firewood consumption by trekkers (porters) has had the most critical impact on the environment. Thus mountain tourism has the potential for bringing about both positive and negative changes in mountain habitats. Therefore attempts should be made to minimize the pressures on the limited re-

sources by looking for different options such as alternative energy sources, promotion of afforestation and strict monitoring of litter, and by building appropriate site for garbage disposal, by imposing restrictions on the number of tourists allowed in an area within a given time period, and by opening up more trekking routes (Sharma, 1989).

DEVELOPMENT STRATEGIES

Broadly, development strategies for mountain areas should include the following considerations:

(1) *Sensitivity to mountain characteristics* The review of country experience in the HKH Region indicates that there is a strong association between the success of development initiatives and their sensitivity to mountain characteristics (Jodha, 1990). Therefore there should be explicit consideration of specific mountain characteristics in designing policies, investment programmes and projects for the hills and mountains.

(2) *Population control* During the past three decades, the accelerated population growth in most of the HKH region has resulted in a considerable imbalance between population and resources. Significant efforts need to be made to control the population. Since population growth, poverty and resource degradation are mutually reinforcing, a joint assault needs to be launched on all fronts simultaneously to reverse the downward trend.

(3) *Massive investment in the development of transport, hydropower, and marketing networks* The transformation of stagnant (or even declining) hill economies into rapidly growing economies requires massive investments in transport, hydropower and marketing networks. Himachal Pradesh provides a good example of this type of investment. The more infrastructure a region has, the more development potential it acquires. A precondition of specialization is obviously the development of transport to reduce the cost of bringing in inputs and services and exporting products. Because of the number of rivers and streams criss-crossing the mountain terrains, there is tremendous potential for the installation of mini- and micro-hydroplants. The installation of large hydropower plants has all too often benefited the people in the plains and not the mountain communities, whereas the installation of small hydro-plants benefits mountain communities. They are easy to construct and operate, and even local manufacturers are capable of supplying the necessary equipment to install small plants.

(4) *Highland–lowland interaction* The ecological and development problems of the mountains are inextricably linked with the lowlands (plains) through watersheds and river systems, migration and markets and these often have transnational and national political boundaries. International cooperative action is needed to deal effectively with the destructive forces of nature, the short-sighted interventions of man, and the movement of people and products. The sustainability of mountain regions is greatly

dependent upon the lowlands that provide a market for the products and resources for investment. Similarly, if watersheds are managed properly in the highlands, the siltation of river basins and productive land is reduced. On the other hand, the continued deterioration of highland resources will ultimately affect the lowland areas as well.

Sustainability can be maintained through complementarities in production between the hills and the plains. Surplus food production in the plains could benefit the hill people in two ways. Firstly, the constraint on wage goods that is essential for labour-intensive programmes could be released. Low-income people spend the bulk of their additional income on food, and a high employment policy increases the income of the poor, with demand for food subsequently increasing (Mellor, 1976). This increase in demand for food could be met from the marketable surplus generated in the plains. Secondly, the surplus would have an effect on the domestic price of foodgrains, depending upon the quantities exported. It is innovations in the production of foodgrains, for which the demand is inelastic, that are passed on to consumers, including hill people.

Conversely, the hills will produce fruits and vegetables, livestock products and handicrafts, the demand for which is elastic. In the production of such commodities, the benefits of technological innovations are passed on to the producers. Thus technological innovations in foodgrain production in the plains will not only benefit plains producers, through an increase in production, but will also benefit the hill people who are potential consumers. Hill people will further benefit from technological innovations in horticulture, livestock and cottage industries (Yadav, 1987). Thus investment policies should be designed so that upland/lowland linkages are strengthened to benefit mutually the people in both regions on a sustainable basis.

(5) *Integration of crop–livestock–forestry*   Farming systems, consisting of the integrated activities of crops, livestock and forestry, have been practised for generations by hill people. It is important to have a clear understanding of their interrelationships and investment programmes to emphasize the strengthening of their linkages and to maintain a proper balance. This integrated system must remain central in designing research and extension activities. A technology package, taking into account the farming system perspective rather than emphasizing the development of a single component technology, is necessary. This would require more time and resources to understand local constraints, opportunities and interrelationships among the various farming activities. Also, because of the wide variation in the conditions of the numerous micro-regions in the mountains, location specificity and diversity should be the central focuses of technological developments for agriculture.

(6) *Leasehold forestry*   In the mountains, large areas of forest are denuded. For example, in Nepal, the heavily denuded forest and shrub area, with an average crown density of 25 per cent is about 2.12 million hectares (HMG/LRMP, 1986). This constitutes about 37 per cent of the total forest and shrub area in the mountain region. This should be leased on a long-term basis to poor households and put to better use by integrating

forestry, livestock and horticulture. This arrangement will give the poor users the right to plant tree crops and grasses and to generate income to meet their basic needs and to bring about ecological rehabilitation at the same time, thus providing an innovative mechanism to combat both poverty and environmental degradation.

(7) *Farmer-managed irrigation systems*   These should be the primary mode of irrigation development in the mountains. This would require an emphasis on three aspects: (1) farmers' participation in need assessment to design, to operate and to manage; (2) consideration of low-cost structures; and (3) users' group management systems for irrigation, giving them a sense of ownership (Pradhan, 1989).

(8) *Institutional innovation*   It is commonly seen that many development projects, in spite of adequate financial and technical assistance, have failed because of the lack of proper institutional arrangements to carry them out. Even if they succeed in achieving certain targets, the continuity of a programme ceases after the completion of the project. The main reason is the lack of a clear mechanism for the equitable participation of the people in design, implementation and benefit sharing. Therefore institutional innovation is the key to increased productivity, environmental sustainability and greater equity. Experiences from successful projects in the HKH region indicate that the main basis of such institutional innovation is collective management in a participatory manner, investment in human capital through skill upgrading and capital formation through group savings (Bajracharya, 1990).

(9) *Off-farm employment generation*   Agriculture alone is unlikely to meet the needs of mountain communities. In the absence of sufficient employment and income rural communities are continually faced with increasing poverty, resulting in migration and rapid environmental degradation. It is therefore essential to identify viable off-farm alternatives and to promote practical approaches to employment generation, subsequently enhancing income in the hills. This requires an emphasis on four interelated factors: technical innovation, improved extension and support services, integration with the wider market and enhancement of organization and management (Bajracharya, 1990).

In designing specific activities, the hills and the mountains should be partitioned conceptually into four categories, based on accessibility.

Area with road access   Here the object is to shift gradually from self-sufficiency in foodgrains to maximization of farm income through cash crops (fruits and vegetable) and improved animal husbandry, in order to take advantages of the hill and mountain 'niche'. Both horticulture and livestock are of special significance to the poor and to small farmers in terms of employment and income, because they are more labour-intensive, both on and off the farm, than foodgrains (IFAD, 1979). Farmers could undertake these activities on a commercial scale only if they have a guaranteed market for their products. Because of the factor of bulk and perishability, the rapid movement of these commodities is essential. Roads linking the production area with the consumer (market) centres facilitate rapid movement. Horticulture and livestock activities are supportive of each other in the sense that livestock provides manure (compost) for horticultural crops and orchards provide fodder for livestock.

Traditionally, the foodgrain sector has received emphasis in the hills and mountains and consequently greater priority for research and extension has also been given to foodgrains. This gradual shift from food self-sufficiency to horticulture and livestock would necessitate a higher intensity of research and extension for horticulture and livestock. Also it would require the strengthening of other agricultural support services such as credit, input supply and promotion of small entrepreneurs in trading and marketing.

Accessible areas provide better opportunities for promoting the development of mountain-specific, natural resources such as hydropower, minerals, tourism and small-scale industries. Here, of course, in promoting these activities, environmental impact analysis must be taken into consideration so that the mountain 'niche' is maintained without serious negative consequences. Small market towns and service centres should be developed at different places along the road to provide services to the rural hinterland.

Areas where road access is economically justified and which will become accessible within the next ten to fifteen years   Initially, the emphasis should be on the subsistence, mixed farming system but preparatory activities should be undertaken for the transformation of subsistence agriculture into commercial agriculture. Thus most of the activities would be similar to those in areas that already have access to roads, with some time-lag.

Areas where there is sufficient potential to justify limited access improvements (such as suspension bridges, trails, mule-tracks, airports); these could be made accessible by road in the next 20 to 50 years   Here the objective is to meet basic needs (self-sufficiency in foodgrains) through improvement, in income and productivity at subsistence or sub-subsistence levels. Agricultural research is oriented towards subsistence crops, mixed farming systems and low-cost technology. Emphasis is given to the production of compost and not to the use of inorganic fertilizers. Extension and other support services should be oriented towards meeting the needs of the mixed enterprise farming system, in which grain crops, livestock and horticulture are given proper emphasis. In terms of cash crops, emphasis is on low-weight and high-value crops such as medicinal herbs.

On and off-farm employment opportunities should be promoted through the intensification of agriculture in low valley areas, expansion of selected cottage and small-scale industries (processing), and public works – essential for an increase in the income of small farmers. Agro-forestry practices should be encouraged on sloping lands to meet food, fodder and fuel needs. In steep upland areas, the Sloping Agricultural Land Technology (SALT), widely practised in the hilly areas of Mindanao in the Philippines, should be adopted to reduce soil erosion and to stabilize and enrich the soil. This technology can increase a farmer's annual income almost threefold in a period of five years (Tacio, 1988).

Food-for-work and rural, public employment schemes should be carried out to reach the large number of poor and unemployed. Also in food-deficit areas, food distribution with transport subsidies should be provided. However, caution is needed to make sure that the low price of imported food does not depress the production of grain crops in the area.

Areas with no hope of access – expenditure on a road not justified in the long run   These areas should be depopulated and the people from these areas resettled, in other mountain areas in the three categories discussed above or in the plains. These vacated areas should be declared natural reserves in order to promote biological diversity with natural vegetation.

CONCLUSIONS

Attainment of constant or high productivity levels on a continuous basis without damaging the long-term potential of resources is known as the sustainable use of resources. The prospects of sustainability in fragile mountain areas are severely constrained by the specific features of their natural resource endowment. In the past, sustainability could be maintained because of low pressure on these resources. However, continued population growth and market-induced demands have put greater pressures on land and other natural resources, and brought about higher resource use-intensity, consequently leading to negative changes, known as indicators of unsustainability.

The biophysical conditions of the resource base and their use constitute the supply side of sustainability, while the factors that place pressure on resources, such as population growth and market-induced demands, constitute the demand side. It is the management of both supply and demand aspects which constitutes the essence of development strategies for sustainable development.

The mountains have a fragile resource base and therefore fall into the category of less-favoured and marginal production areas. Mountainous areas distinguish themselves from the plains because of their characteristics of verticality which obstruct the applicability of development frameworks conceived for the plains to the mountains. Slope and altitude are manifestly in six important characteristics of the mountains: inaccessibility, fragility, marginality, diversity, 'niche' and human adaptation mechanisms. Of these characteristics, inaccessibility is the central (pivotal) factor influencing resource use and determining the adjustment of other characteristics. The road developments in Himachal Pradesh and the northern hill regions of Pakistan paved the way for bringing other components of development, consequently orienting agriculture away from subsistence practices towards commercialisation. This transformation process was accelerated by the provision of strong backward linkages, in terms of input supplies, and forward linkages in terms of processing and marketing. In addition, technological development, price incentives and subsidies further helped the process. The key factor for the success of these endeavours was the explicit consideration of specific mountain characteristics in designing policies and investment programmes.

On the demand aspects of sustainability, the control of population growth and market-induced over-extraction of resources are the most important factors for reducing pressure. Similarly, emphasis on skill formation and support facilities for agro-based processing and marketing activities and promotion of public works to increase off-farm employment are essential to reduce the pressure on land.

REFERENCES

Altieri, M.A., 1987, *Agroecology: The Scientific Basis of Alternative Agriculture*, Westview Press, Boulder, Co.

Bajracharya, D., 1990, 'Local Institutional Innovation for Rural Transformation and Poverty Alleviation: Perspectives from Mountain Areas', in Jodha *et al.* (eds), *Sustainable Development of Mountain Agriculture*, Oxford & IBH Publishing Co. Pvt. Ltd., New Delhi.

Banskota, M., 1985, 'Hill Farmers as Watershed Managers', paper presented at the International Workshop on Watershed Management Experiences in the Hindu Kush-Himalayan Region, 14–19 October 1985.

Banskota, M. and Jodha, N.S., 1990, 'Mountain Agricultural Development Strategies: A Comparative Perspective from the countries of the Hindu Kush-Himalayas', in Jodha *et al.* (eds) *Sustainable Development of Mountain Agriculture*, Oxford & IBH Publishing Co. Pvt Ltd., New Delhi.

Calkins P.H., 1976, '*Shiva's Trident*: The Impact on Income Employment and Nutrition of Developing Horticulture in the Trisuli,Watershed, Nepal, (PhD thesis) Cornell University, Ithaca, NY.

Chambers, R., 1987, *Sustainable Rural Livelihood: A Strategy for People, Environment and Development*, IDS Discussion Paper 240, Institute of Development Studies, Sussex, England.

Dafu, Y., 1988, 'Mountain Agriculture under the Responsibility System: Experiences from the Mountain Areas of West Sichuan, China', paper presented at the Workshop on Agricultural Development Experiences in West Sichuan and Xizang, China, held at Chengdu, 6–10 October, 1988.

Development Strategies for Fragile Lands (DESFIL), 1988, *Development of Fragile Lands: Theory and Practice*, Washington DC.

HMSO, 1986, *Land Resource Mapping Project Report*, HMSO, London.

International Fund for Agricultural Development, 1979, *Report of the Special Programming Mission to Nepal*, IFAD, Rome.

Jodha, N.S., 1990, *Sustainable Agriculture in Fragile Resource Zones. Technological Imperatives,* International Centre for Integrated Mountain Development, Mountain Farming Discussion Paper Series No. 3. ICIMOD, Kathmandu.

Khan, A.R., 1989, 'Soil Conservation and Watershed Management in the Mountain Areas of Pakistan', paper presented at the Workshop on Agricultural Development Experiences of Pakistan, held in Swat, 15–18 February 1989.

Liddle, M.J., 1975, 'A Selective Review of Ecological Effects of Human Trampling on Natural Ecosystems', *Biological Conservation*, vol. 7,(1).

Lipton, M., 1983, 'The Poor and the Ultra-poor: Characteristics, Explanation and Policies', *Development Research Digest*, 10, Institute of Development Studies, Sussex, England.

Mehta, M., 1990, 'Cash Crops and the Changing Context of Women's Work and Status: A Case Study from Theri Garhwal, India', *Mountain Population and Employment, Discussion Paper Series No.2*, ICIMOD, Kathmandu.

Mellor, J.W., 1976, *The New Economics of Growth: A Strategy for India and the Developing World*, a Twentieth Century Fund Study, Cornell University Press, Ithaca and London.

Mellor, J.W., 1988, 'The Intertwining of Environment Problem and Poverty', *Environment*, vol. 30, no. 9.

Moock, J., 1986, *Understanding Africa's Rural Households and Farming Systems*, Westview Press, Boulder, Co.

Pradhan, P., 1989, *Increasing Agricultural Production in Nepal: Role of Low Cost Irrigation Development Through Farmer Participation*, Country Paper No. 2, IIMI, Nepal.

Rana, R.S., 1990, 'Role of the Himachal Pradesh Horticultural Produce Marketing and Processing Corporation in the Development of Horticulture', *MFS Discussion Paper No. 9*, ICIMOD, Kathmandu.

Reiger, H.C., 1981, 'Man Versus Mountain: The Destruction of the Himalayan Ecosystem', in J.S.Lall and A.D.Moodie (eds), *The Himalaya: Aspects of Change*, Oxford University Press, Delhi.

Sharma, L.R., 1987, *The Economy of Himachal Pradesh*, Mittal Publications, Delhi.

Sharma, P., 1989, 'Assessment of Critical Issues and Options in Mountain Tourism in Nepal', unpublished internal document of ICIMOD, Kathmandu.

Tacio, H.D., 1988, 'SALT: Sloping Agricultural Land Technology', *ILEIA, Newsletter*, March, vol.4, no. 1.

Tiwari, S.C., 1990, 'Role of Off-season Vegetables in the Development of Hill Agriculture in Himachal Pradesh, India', *Mountain Farming Systems Discussion Paper Series No. 8*, ICIMOD, Kathmandu.

Trainer, Ted, 1989, *Developed to Death: Rethinking Third World Development*, The Merlin Press Ltd., London.

Troll, C., 1988, 'Comparative Geography of High Mountains in View of Landscape Ecology: A Development of Three and a Half Decades of Research and Organisation', in J.A. Allan *et al.* (eds), *Human Impacts on Mountains*, Rowman & Littlefield, New Jersey.

Verma L.R. and Partap, T., 1990, 'The Experiences of an Area-based Development Strategy in Himachal Pradesh, India', Paper presented at the International Symposium on Strategies for Sustainable Mountain Agriculture, held at Kathmandu, 10–14 September 1990.

Wenpu, L. and Qinfa, L., 1988, *Agricultural Development Strategies in the Chinese Himalayas*, ICIMOD commissioned paper presented at the Workshop on Agricultural Development Experiences in Chengdu, 6–10 October, 1988.

Whiteman, P.T.S., 1988, 'Mountain Agronomy in Ethiopia, Nepal and Pakistan', in J.A. Allan *et al.* (eds), *Human Impacts on Mountains*, Rowman & Littlefield, New Jersey.

Yadav, R.P., 1987, *Agricultural Research in Nepal: Resource Allocation, Structure and Incentives*, Research Report No. 62, International Food Policy Research Institute, Washington DC.

DISCUSSION OPENING – TONGROJ ONCHAN*

Dr Yadav presented an excellent paper on the problem of sustainable resource management and development in marginal areas, particularly mountains. It is generally known that, owing to rapid population growth and increasing demand for food and other agricultural products, marginal and sub-marginal lands, including mountains, have been extensively used in many countries of Asia. This has resulted in serious degradation of natural resources and environmental quality. If left unchecked, it will certainly have an adverse effect on the sustainability of agricultural development.

In discussing strategies for mountain areas, Dr Yadav correctly emphasized the importance of both the demand and the supply sides. On the supply side it is very useful to understand mountain characteristics, while on the demand side population growth is very important. Another interesting demand factor is tourism. The negative environmental effect of tourism in protected areas such as national parks and wildlife sanctuaries has been evident in many countries, including Thailand. To help alleviate this undesirable effect, so-called 'ecotourism' may be promoted. This simply means tourism in which

*Kasetsart University, Thailand.

eco-protection is incorporated in order to preserve nature. This concept, I believe, is receiving increasing attention in many countries.

Of the nine strategies for sustainable development in mountain areas proposed by Dr Yadav, I would like to comment on three: crop–livestock–forestry, leasehold forestry, and off-farm employment generation. First, crop–livestock–forestry strategy is already widely practised by upland and highland people, and should promote increasing economic and ecological stability. The important issue is that a technology package which will provide a proper balance of the three activities is required, though it is not yet clear that one has been developed.

The strategy on leasehold forestry deserves attention. Tree planting will help restore the forests. I agree that it is important to provide long-term land leases to rural households for use in forestry, livestock and horticulture production. However, issues of distribution and equity also have to be considered. We have an interesting experience in Thailand, where a very ambitious programme of massive reforestation has been launched. The government has been trying to lease out large areas of denuded forest to the private sector and the rural people, though up to now only big companies have been able to obtain leases. This has provoked a great deal of protest among the rural poor who for years have illegally occupied the land which, in fact, they believe they own. The issue of receipt of benefits from this programme is therefore important. Furthermore, there are also other factors which must be carefully considered in implementing the programme, such as the area to be leased out to a family, the types of tree to be planted, and the farming systems and technology to be recommended.

Off-farm employment generation is particularly important in reducing pressure on land and raising the income of the poor and the landless. Dr Yadav is right in pointing out that, although a great deal of data are now available on characteristics, problems and potentials of non-farm enterprises and off-farm employment, research has tended to be neglected and much remains to be done.

In the final part of my comment, allow me to mention a few issues which are relevant when considering development strategies in mountain areas. This discussion is mainly in the context of Southeast Asia.

(1) Upland and highland people are usually indigenous tribal and ethnic groups but, increasingly, there are large numbers of newly arriving immigrants, many from lowland areas. To design a programme for a hill tribe, one must understand the people, their culture and their way of life. In some cases, we still do not know the exact population of the highlands. We suspect that it is increasing over time, but it is, of course, difficult to conduct a survey or a census in highland areas.

(2) Land tenure in mountain areas usually involves public land use, though much of the land has been set aside as conservation reserves. Hence land is technically illegally occupied, although hill tribe people may have lived there for generations. Many people also believe that they are the rightful owners of the land they cultivate. Land tenure insecurity is therefore a serious problem. Some uplanders in the Philippines and

Thailand have been trying to obtain private property rights or legal land documents, but without success. In fact, in Thailand, we try to resettle them in lowland areas, though this has proved to be difficult, for various reasons. The issue of land rights is important as it has profound economic, social and environmental implications.

(3) Technology and research efforts in the highlands are limited and technology being used is mainly indigenous. If production is to be increased, new technologies must become available. Further, one must realize that, in highland areas, low production and high instability are caused by the fragility of the environment, where steep slopes, high rainfall and unstable soils contribute to soil erosion. Hence major concerns of highland development are increasing, stabilizing and sustaining the productivity of highland farms and the rehabilitation of marginal highland areas. Soil erosion is perhaps the most critical factor. Engineering devices such as terraces, dams, dikes and channels can change landslope characteristics to reduce their effects. Reforestation is an important element of rehabilitation. Therefore soil erosion, fallow periods and reforestation, as well as farming system technologies, are important research topics. In recent years, government agencies, research institutions and private organizations have been paying increasing attention to the problem of highlands, though research is still too limited to have an impact on development.

(4) Relationships between highlands and lowlands are such that migration occurs in both directions. In Thailand, an increasing number of lowlanders migrate to highlands and many of them settle there permanently. It is apparent that economic change and the well-being of the lowland people will have an effect on resource use in the highlands. Hence developing lowland areas is as important and necessary as developing the highlands.

It must be emphasized that it is important to pay attention to the welfare of highland people and help them to carry out economic activities which will simultaneously conserve fragile natural resources, with the ultimate objective of achieving sustainable development. Until now it is very unfortunate that public interest in the uplands has arisen less from a concern to improve the livelihood of local people than from a realization that lowland environmental and agricultural problems are directly related to upland deforestation and ecosystem degradation. This attitude and policy direction must change if sustainable development strategies in mountain areas are to be properly designed and effectively implemented.

TROND BJØRNDAL*

Management of Fisheries as a Common Property Resource[1]

BACKGROUND: WORLD FISHERIES

Introduction

Man has always looked to the ocean as an important food source, and expectations as to what resources the oceans can bring have often been high: 'It is said that the last frontier of inner space lies in the oceans of the world, and that man, by thrusting back this frontier, may gain almost limitless resources to feed future generations' (Christy and Scott, 1965). However, as with any other resource, the ability to obtain maximum benefits from its exploitation rests on the ability to utilize it efficiently.

Traditionally most fisheries were common property resources characterized by free entry or open access. This meant that the resource was open to anybody, and no one had the right to preclude others from fishing. However, free access to any scarce resource inevitably leads to inefficient exploitation. For fisheries, this involves over-exploitation of the resource and the application of excessive amounts of productive resources such as capital and labour to the production process.

World catches of fish

World catches of fish, crustaceans and molluscs in marine areas for the period 1970–88 are given in Table 1. From 62 million tonnes in 1970, and a slight decline in the early 1970s, catches were fairly stable for the rest of the 1970s, but then increased gradually in the 1980s to almost 85 million tonnes in 1988. This increase of about 36 per cent in a 19-year period is substantial, and occurred after the introduction of 200-mile exclusive economic zones (EEZs) in the second half of the 1970s.

It has been estimated that, with proper management, the total yield from currently exploited fish resources can be increased to at least 100 million tonnes. Although this hardly involves the ability to obtain 'almost limitless resources to feed future generations' there is still potential for increased output. There is also further potential from the exploitation of resources that are currently

*Norwegian School of Economics and Business Administration, Norway.

TABLE 1 *World catches of fish, crustaceans, molluscs, etc. in marine areas, 1970–88*

Year	Quantity (tonnes)	Year	Quantity (tonnes)
1970	61 982 400	1980	64 467 700
1971	59 678 800	1981	66 628 000
1972	55 466 700	1982	68 350 700
1973	55 915 500	1983	68 318 100
1974	59 957 100	1984	73 810 600
1975	59 171 200	1985	75 403 300
1976	62 634 700	1986	80 961 100
1977	61 544 600	1987	80 501 200
1978	63 335 200	1988	84 560 700
1979	63 797 900		

Source: FAO *Yearbook Fishery Statistics, Catches and Landings*, vols. 46, 50, 64 and 66.

not being utilized, from ocean ranching and from aquaculture. In addition, there is scope for substantial savings from reducing inefficiences in fishing.

World production (including inland areas) of fish, molluscs and crustaceans from aquaculture increased from 7.6 million tonnes in 1985 to 10.8 million tonnes in 1988, a remarkable 42 per cent increase in a three-year period, and is dominated by carp, barbel and other cyprinids, representing 43 per cent of total output in 1988. This group of fish, along with shrimps and prawns, salmon and scallops, showed the largest relative increases in the 1985–8 period. For aquaculture, the potential for further increases in production is greater than for capture fisheries.

Catches of the ten most important fishing nations for 1970 and 1988 are shown in Table 2. The pattern has remained fairly stable over the past two decades, although the ranking of the different nations has changed. In 1970, Peru was the leading fishing nation in terms of quantity harvested, with a total catch of 12.5 million tonnes, representing 20.2 per cent of the world catch. The collapse of the anchoveta fishery in the early 1970s, however, dramatically reduced the Peruvian catch. The bottom level was reached in 1983, with a catch quantity of 1.6 million tonnes. Subsequently, Peruvian catches have increased, reaching a level of 6.6 million tonnes in 1988, placing Peru as the fourth largest producer of fish in the world.

After the collapse of the anchoveta fishery, Japan replaced Peru as the most important fishing nation. The Japanese catch of 11.9 million tonnes in 1988 represented 14.1 per cent of the world catch. It is remarkable that Japan has not only maintained its catch quantity in the period under consideration but even increased it, despite having lost access to many of its traditional high-sea fishing grounds in the late 1970s and early 1980s as the result of the imposition of 200-mile EEZs. This increase is due to greater catches in offshore areas, partly as a result of substantial stock-enhancement programmes.

TABLE 2   *Catches by Country of Fish, Molluscs, Crustaceans, etc. in Marine Area (Tonnes)*

Country	1970	1988
Japan	8 658 400	11 896 935
USSR	6 386 500	11 332 101
China	2 192 500	10 358 678
Peru	12 532 900	6 637 106
USA	2 729 300	5 965 598
Chile	1 200 300	5 210 201
India	1 085 600	3 145 650
Korea Rep.	725 500	2 727 059
Indonesia	804 000	2 703 260
Thailand	1 343 400	2 350 000

As noted above, the world catch of fish has been increasing. It is particularly noteworthy that China, which had an annual catch of about 3 million tonnes throughout the 1970s, reached a catch quantity of 10.4 million tonnes in 1988. With the exception of Peru, all other nations included in Table 2 increased their catch.

Fish consumption and trade

Reduction into fish meal and oil is the major usage of fish. In 1988 27.1 million tonnes of fish, corresponding to 32 per cent of the world catch, were used for this purpose. The species generally used for reduction purposes are clupeids such as herrings and sardines, for which there has been an upward trend in catches for the past 15 years.

Frozen usage has shown a substantial upward trend in this period, and fresh usage, while remaining stable for quite a long period, has shown substantial increases in the last few years. Combined, fresh and frozen usage represented 44.0 million tonnes in 1988, or 52 per cent of the world catch. On the other hand, curing and canning exhibited lower upward trends. As prices for fresh and frozen products in general are higher than for most other usages, this development indicates that increases have been most important for the higher-valued usages.

Major fish markets are found in Japan, North America and Europe. Japan, the USA and the European Community (EC) import substantial quantities of fish, although the EC and the USA have considerable fish exports. In terms of total fish trade, the EC is the most important market. However, Japan is most important in terms of net imports, with a value of $9.6 billion in 1988, followed by the EC and the USA.

In Japan, the most important fish market in the world, consumption has increased substantially in the 1980s to a level of 72.5 kg per capita in 1988.

Compared to Japan, US per capita fish consumption is low, but has also been increasing. Owing to the size of its population, the USA is a very important fish market. The trend towards increased fish consumption is also seen in European markets. Increased fish consumption may be due to changes in consumer preferences towards a healthier diet. This effect is also likely to be increasingly important in the future. For some fish products, increased consumption may be a result of increases in real income. On a world-wide basis, population increases have also led to shifts in the demand for fish.

Management issues

While property rights in the course of history have been extended to most natural resources, this has proved difficult in the case of fish. Common property is still associated with many fisheries, largely because the fish do not respect man-made boundaries.

In the postwar period, substantial attempts have been made to reduce the undesirable effects due to common property exploitation, culminating in the extended fisheries jurisdiction, with introduction of 200-mile exclusive economic zones (EEZs) in 1977. As more than 90 per cent of all catches are taken within the EEZs of all coastal nations (Eckert, 1979, p. 116), this new institutional arrangement may appear to have the potential for solving important management problems by assigning property rights to both national and international fish stocks. However, common property is still associated with the utilization of fish stocks in the EEZs of most countries. Moreover, the fact that many fish resources migrate or are spread across national boundaries causes substantial management problems.

Even if the problem of optimal exploitation of fish resources can be solved, for example through cooperation between countries sharing a resource about the fixing of quotas, the problem remains as to the efficient management of a given quota. This involves keeping the fishing effort at an efficient level through regulatory arrangements. This issue has in practice proved to be very difficult. However, extended fisheries jurisdiction has facilitated the adoption in some countries of new institutional arrangements for regulating fisheries. In particular, individual transferable quotas (ITQs) appear to be promising in terms of improved fisheries management.

BIO-ECONOMIC MODELLING

A fish stock is a renewable resource. Accordingly, through proper management, it can provide harvests indefinitely. Furthermore, stock size may be increased by reducing harvests below natural growth for some time period: that is, by investing in the resource. In an analogous manner, one can disinvest by increasing harvest quantities. In other words, the resource may be considered a capital stock. What distinguishes fisheries managements from the exploitation of most other renewable resources, such as forests, is that fish resources traditionally constitute common property. This institutional arrangement has profound consequences for fisheries management.

Economic analyses of fisheries are based on a biological model. The investment problem referred to above can be considered as follows:

$$\text{Net Growth of the Stock} = \text{Natural Growth} - \text{Harvest}$$

Natural growth is sometimes referred to as nature's production function, being zero for non-renewable resources. Harvesting represents the 'regular' production function.

Optimal management

The objective of this paper is to study the management of fisheries as common property resources. However, first, optimal management will be considered. This will be a 'yardstick' against which common property, that is non-optimal exploitation, can be compared. It is assumed that the fishery is managed by a social planner or sole owner whose objective is to maximize the present value of net revenues from the fishery by determining optimal time-paths of effort for boats in the fishery.

The social planner aims to maximize the sum of net present values over all boats subject to stock dynamics and initial conditions. For each boat effort should be determined so that the marginal product of effort, evaluated at present value market price minus shadow price of the resource, should be equal to present value marginal cost. The optimal level of effort depends on output and input prices as well as stock size. The solution to the optimization problem will give an optimal time-path for the stock which will converge on the target stock level.

Although efficient management is a prerequisite for maximizing the economic returns from fishing, most fisheries have traditionally been common property and characterized by over-exploitation.

Common property exploitation

Traditionally, most fisheries were common property characterized by free entry or open access. The consequences of free entry have been analysed by many authors. The institutional arrangement can be considered the polar extreme of the sole owner case considered above.

Bionomic equilibrium is defined as the individual firm, the industry and the stock being in equilibrium simultaneously. The fleet will be in equilibrium when there is no incentive for further entry to or exit from the fishery. For the marginal firm, the present value of profits will be zero. If fishing effort is homogeneous, this condition applies to all boats. When stock size in bionomic equilibrium is compared to the sole owner case, it may be noted that the shadow price of the resource is set to zero in the free entry fishery. Thus, in bionomic equilibrium, the stock will be over-exploited compared to the social optimum and all resource rent will be dissipated. However, if fishing effort is heterogeneous, some fishing firms may be earning intra-marginal rents.

Biologists commonly recommend that a fishery should be regulated to achieve a stock size yielding maximum sustainable yield (X_{msy}). When economic considerations are included in the analysis, the optimal stock level is equal to X_{msy} only under very special conditions. Whether the optimal stock level is greater than, equal to or less than X_{msy} is found to be an empirical question.

ECONOMIC INEFFICIENCIES IN COMMON PROPERTY EXPLOITATION

In the above analysis, optimal management has been compared to common property exploitation. Open access to a fish resource was seen to involve over-exploitation of the stock with too much fishing effort (capital and labour) being applied to the fishery. However, in a purely 'technical' sense, bionomic equilibrium does not involve over-capacity. This is because, for the given stock level, the associated fishing effort (and cost) is the minimum required to catch the open-access harvest quantity. The inefficiency of bionomic equilibrium is due to too much effort being devoted to the fishery from society's point of view and the concomitant over-exploitation of the stock. This is caused by what is denoted the stock externality; that is, the failure of individual fishermen to take into consideration the effect of their own harvest on all other fishermen. This externality is internalized by a sole owner.

In principle, a quota system can be used to achieve the optimal (or 'desirable') stock level. The problem to be addressed here is that the introduction of catch quotas, without regulating effort, will lead to over-capacity in the fleet. This is another way in which the resource rent can be dissipated and gives rise to what Munro and Scott (1984) denote the Class II common property fishery.[2]

The Class II common property fishery can be illustrated with reference to a simple seasonal model for a fishery due to Clark (1976). In the model, it is assumed that the fishery is regulated by a total allowable catch (TAC) quota, and that stock size is stabilized so that surplus growth is equal to the quota. The stock size in question could, for example, be the sole owner optimum stock level. Maximum season length is assumed given by nature. The fishery remains open until the TAC is harvested and is then closed by the fisheries authorities. With maximum season length given, it is a straightforward matter to find the minimum fleet size that would maximize resource rent from the fishery.

If actual fleet size initially is equal to minimum fleet size, fishermen will be making pure profits (provided the fishery is not marginal). If effort is not regulated, this will attract new entry to the fishery. In order to prevent over-fishing, regulation of season length will have to be introduced. Season length will have to be reduced, and the boats will be idle for a greater and greater part of the season. The fleet will expand up to the point where total costs equal revenue. The resource rent is dissipated through over capacity in the processing industry. As the same quota of fish will be landed in a shorter period of time, larger storage and processing facilities will be required. As with fishing effort, this capacity will be idle for part of the season.

Although simple, this model is fairly realistic as a positive description of real-world fisheries management. Many fisheries are regulated by TACs. When the quotas are strictly enforced, this instrument can be used to maintain stock size at a level higher than the one associated with bionomic equilibrium. However, unless effort is regulated, over-capacity will develop in the fleet and cause rent dissipation.

In fisheries models, fishing effort is commonly assumed to be homogeneous, with a fixed relationship between all inputs. In other words, a Leontief production function is assumed, with no substitution possibilities between inputs. Many fisheries are regulated not only by a TAC, but also by some measure of effort such as the number of boats, possibly in combination with, for example, boat size. If again it is assumed that stock size is stabilized at some level greater than bionomic equilibrium and fleet size (initially) is at the minimum level, boats will again be making pure profits. As the number of boats is assumed to be fixed, and if there were no substitution possibilities, there would be no further entry to the fishery and boats would be enjoying pure profits *ad infinitum*.[3]

This scenario, however, hardly fits the real world. In reality, substitution possibilities exist and, when pure profits are made, boats will substitute unregulated inputs for the regulated one, for example substituting unregulated engine size for regulated cargo capacity. Boats will be faced with an optimization problem where the objective is to maximize its share of the resource rent. Additional investments will be undertaken as long as expected returns exceed the cost. As all boats face the same problem, this process will continue until the resource rent is dissipated in added input costs.

This adjustment process may in part explain the proliferation of fisheries regulations. Initially, only one or a few elements of effort may be regulated. With pure profits being earned, there will be investments in some other element of effort. To prevent total effort from expanding, fisheries authorities must introduce additional regulations. This process may well continue as long as substitution possibilities exist.

In addition to dissipating the resource rent, this kind of incentive structure also has other undesirable effects. First, boats will be unable to choose the least-cost combination of inputs and will thus be unable to operate at the minimum point on their long-run average cost curves. Second, research and development may be directed towards circumventing regulations rather than towards 'pure' technological research.

MANAGEMENT ISSUES

The exploitation of international fish resources, which traditionally were considered common property, has been characterized by over-exploitation and excessive use of effort. While biologists and conservationists have been concerned about over-exploitation, economists have been concerned about the associated inefficiencies. Attempts to regulate fisheries were primarily aimed at stock conservation and, in some cases, also at reducing conflicts between fishermen competing for the dwindling resources (resulting for example from externalities such as gear collisions and over-crowding).

The objective of fisheries management, both nationally and internationally, was generally to achieve the stock level that would yield maximum sustainable yield (MSY). A number of international bodies were set up to regulate fisheries, for example, the North East Atlantic Fisheries Commission (NEAFC) and the International Commission for the Northwest Atlantic Fisheries (ICNAF). Although membership of these organizations included all major countries participating in the fisheries they monitored, they had no power to implement their recommendations.

The international bodies were not successful in terms of achieving the objectives for which they were set up, such as desirable stock levels and controlling fishing effort. This could be considered the Prisoner's Dilemma applied to fisheries (Dasgupta and Heal, 1979). These international organizations were primarily concerned with stock management, with little attention given to economic considerations. However, as problems of over-exploitation became more evident, the same was true for economic inefficiencies. The need for improved management was therefore evident both to biologists and to economists.

Extended fisheries jurisdiction (EFJ)

As international fisheries resources are common property resources, exclusive property rights could not be assigned to them; only by capture could property rights be established in fish. Thus the attempts to regulate international fisheries were largely unsuccessful. The results were felt to be particularly unsatisfactory by coastal states, with closest access to the fishing grounds.

In the post Second World War period, several conflicts arose between coastal and distant-water fishing nations, as the former attempted to assert greater control over resources close to their shores by extending territorial waters or declaring extended fishing zones. The 'cod wars' between the United Kingdom and Iceland are just one example.

Extended Fisheries Jurisdiction (EFJ) is based on the 1982 United Nations Convention on the Law of the Sea. Although to date this has not been ratified by the required number of countries, it has acquired the status of customary international law. According to this Convention, the coastal state has 'sovereign' rights for exploiting and managing natural resources in its 200-mile EEZ. In practice, this means that the coastal state has the sole right to determine TACs for fish stocks in its zone and whether or not it has sufficient harvesting capacity to catch these quotas. Moreover, if other countries are allowed to fish in the EEZ of a country, the coastal state determines under what conditions this may take place. The coastal state may, for example, impose fees on catches by foreign vessels. McRae and Munro (1989) trace the development towards EFJ in the past decades and the nature of the rights of a coastal state over the resources in its 200-mile EEZ.

Trans-boundary and migratory resources

As most fish catches are made within the EEZs of coastal states, EFJ may
superficially appear to have the potential of 'solving' all important management problems. However, fish do not respect man-made boundaries, and for
this reason considerable management problems remain, particularly with respect
to trans-boundary and migratory resources.

Trans-boundary resources are those where a fish stock is divided between
the EEZs of two or more coastal states. In the case of migratory resources, the
stock migrates between the EEZs of two or more states, for example, according
to a seasonal pattern or over the life cycle of the fish. Thus, in one period, a
stock may be under sole control of one state, while in another it is fully in the
EEZ of another country. Although this distinction between trans-boundary
and migratory resources may not be clear-cut in real life, it is convenient for
analytical purposes.

Analysis of the management of trans-boundary resources is closely associated with the name of Munro (1979, 1990, 1991a; see also McRae and Munro,
1989). Munro considers the case where a resource is shared between two
countries, assuming each country to have full control over its fishing effort.
Stock size in bionomic equilibrium and under optimal management depends
on economic and biological parameters. Thus, with differences in economic
parameters, the two countries will have different perceptions about optimal
stock size and bionomic equilibrium. For the case of competing exploitation
of a resource, Munro's analysis is based on non-cooperative game theory. The
author demonstrates how the outcome with respect to stock size and rent
dissipation depends on economic configurations in the two countries involved.
Under all conditions, competitive exploitation is found to involve stock depletion and rent dissipation. Thus a *prima facie* case exists for cooperative exploitation of the resource. For the case of cooperative management, Munro's
analysis is based on cooperative game theory. If economic parameters are the
same in both countries and they are willing to enter into a binding agreement,
the two owners would maximize total profits and negotiate the sharing of
returns.

The more interesting (and realistic) situation arises when economic parameters differ, owing to differences in prices, discount rates, costs of effort or
fishing technology. The outcome of the cooperative game depends on whether
harvest shares are constant over time and whether side-payments are allowed.
When economic parameters differ, this means that the two countries put a
different value on the resource. If the price of fish is higher in country 1 than
in country 2 (and there is no trade in fish), or the unit harvesting cost or the
discount rate is lower in country 1, then country 1 will place a greater value
on the resource than country 2. For the case of cooperative management with
a binding agreement, this would involve country 1 being sole manager of the
resource, buying out country 2. This is feasible, as the resource is worth more
to country 1 than to country 2. In the words of Munro, this would represent
the *optimum optimorum*.

Commonly, side-payments are not feasible, and harvest shares are usually
constant over time, based on historical catch records or the dispersion of the

stock between the EEZs of the two countries, for example. Munro considers in detail the case where the discount rates in the two countries differ (assuming prices and costs of effort to be equal). For cooperative management with a binding agreement, the (compromise) optimal stock level is found to be a function of time. Greater weight is placed on the preferences of the high-discount country in the near future and less in the more distant future, as a consequence of its time preferences. The converse is true for the low-discount country. In the long run, the compromise stock level approaches the optimal stock level of the low discount country asymptotically. When side-payments are not allowed and harvest shares are to be fixed over time, these can be considered constraints to the optimization problem. Therefore the outcome in this case will be less than in the *optimum optimorum* described above.

Under more complicated game-theoretic conditions involving non-binding agreements, the possibility of a stable equilibrium is somewhat more doubtful (see Kaitala, 1985, on cooperative management and the consequences of non-binding agreements). However, even in those cases, an equilibrium bargaining solution may exist. Nonetheless, a fundamental conclusion of the economic theory of trans-boundary resources is that the parties involved are always better off in a bargaining solution than under open access.

Although trans-boundary issues are very important, only a few empirical applications have been undertaken. Munro (1991b) describes the problems involved in the exploitation of the highly migratory and very valuable tropical tuna and Pacific salmon stocks. Both of these cases provide prime examples of the problems of cooperative management of fish stocks. This is especially true for the tropical tuna, which is found in the EEZs of several diverse Pacific island states, with distant-water fishing nations harvesting the resource. Munro describes the setting, that is, the problems facing both the owners and the harvesters, the subsequent negotiations and their ultimate resolution. Having described how the Pacific island states greatly benefited from extended fisheries jurisdiction, the paper concludes on a rather optimistic note. It appears from these examples that mutually beneficial cooperative fisheries management is possible even when the parties involved are apparently facing formidable bargaining obstacles.

The analysis of Armstrong and Flaaten (1991) is set in the framework of a trans-boundary species exploited by Norway and the Soviet Union, which share the Arcto-Norwegian cod stock as a result of migratory behaviour over the life cycle of the species. While the nursery and adolescent grounds are primarily within Soviet fisheries jurisdiction, the spawning grounds are almost exclusively in Norwegian waters. The migratory behaviour is not explicitly modelled. Rather, exploitation is defined in trans-boundary terms, and a model of the bargaining situation due to Munro is employed. Based on economic parameters for the two countries, the authors find that the actual agreements reached by Norway and the Soviet Union concerning the utilization of this stock are reasonably efficient and conform broadly with the predictions of Munro's theory. Although the management of trans-boundary resources has received much attention in the literature (see also Levhari and Mirman, 1980), this is less the case for migratory resources. The consequences of migration may require refinements or modifications to the transboundary model.

Arnason (1991) analyses the exploitation of migratory species. The migratory behaviour means that harvesting conditions change over time even when the size of the biomass is constant. This non-autonomous nature of the situation significantly increases the complexity of the optimal harvesting problem. Among other things, Arnason shows that, with continuous migrations, the optimal stock level generally does not converge to a constant equilibrium level. Moreover, a general characterization of the nature of the optimal biomass and harvesting paths does not appear to be easily available. The situation becomes even less tractable in the case where the migratory stock periodically moves between EEZs. In those cases, the nations in question are periodically faced with intervals when their access to the resource is blocked, while their competitors have temporarily exclusive access to the resource. Obviously, this makes for a very complicated game.

It appears that further research in this field will be directed towards analysis of the management of migratory resources and empirical research into the exploitation of both trans-boundary and migratory resources.

Individual transferable quotas (ITQs)

Extended Fisheries Jurisdiction (EFJ) and the establishment of 200-mile EEZs have meant that most fish stocks are now under the control of coastal states. EFJ provides a new basis for the management of fisheries, with potential for substantial improvements, although to date in many instances these remain to be realized.

For stocks under the sole control of a coastal state, the latter has full control over its management. Essentially, the coastal state is a sole owner and may optimize resource use. For the case of trans-boundary resources, the states involved are the combined sole owner and may cooperate over resource management. This may take the form of setting a joint TAC for the stock, which in turn is divided into national quotas. The coastal state will then have full control over its share of the TAC. This arrangement is predominant, for example for the management of shared resources in the North-east Atlantic. In other words, one may envisage the management of fisheries as being undertaken in two steps. First, the TAC is determined through an optimization procedure. Second, the quota must be allocated among fishermen in a way that ensures harvesting efficiency in order to avoid the Class II common property problem. Harvesting efficiency is a requirement for capturing the rent that the resources can yield.

An individual transferable quota (ITQ) is a legal right to catch a quantity of fish over a certain period of time. The ITQ is divisible and transferable in a quota market. Thus it is to be considered a private property right. It is considered a right to a certain fraction of the *surplus production* from a given stock, but not to the stock itself. Over time, most natural resources that originally were characterized by open access have come under private ownership. The reason this did not happen to fish earlier was that it appeared both unnecessary and impossible to assign property rights to fish in the sea. This was because fisheries resources were believed to be inexhaustible, and because property rights could not be enforced (Grotius, 1608; quoted in Christy and Scott, 1965, p.155). With modern technology, many fish stocks are exhaustible, but at the same time enforcement is more practicable than in earlier times, although enforcement remains a very serious problem in the management of many fisheries.

The changes that have taken place in the last decades have thus been in two stages. In the first stage, through the Law of the Sea, property rights to fish stocks were assigned to coastal states. In the second stage, a number of countries, including New Zealand, Australia, Canada and Iceland, have 'privatized' some of their resources through the establishment of ITQs in an attempt to ensure harvesting efficiency. This represents a fundamental institutional change and will make the management of fisheries more 'similar' to that of other resources, such as forestry.

From a management perspective, important objectives may be achieved through a combined TAC–ITQ system.[4] First, by setting the TAC, the country can achieve a sustainable harvest level that maximizes the net economic benefits from the fishery. Second, by introducing a harvest quota for each individual fisherman, the incentive problem facing the fisherman is changed from competitive behaviour to that of cost minimization. Even in the short run, with no other institutional changes, this could improve harvesting efficiency. However, with transferability of quotas, each boat will, over time, find the effort level that will minimize its long-run harvesting costs, provided the quotas are of sufficient duration to permit long-run investments in capital equipment. Thus quotas will improve harvesting efficiency in both the short run and the long run, although important benefits can only be achieved in the long run.

A further objective is to achieve management goals at minimum cost. In a sense, with ITQs, a greater part of the enforcement of fisheries may now be left to the industry itself. From the fishermen's perspective, ITQs permit flexibiity in terms of which species to fish. Moreover, they permit easy entry and exit to the fishery, which is another precondition for harvesting efficiency.

NOTES

[1]This paper is dedicated to Yukiko Kageyama, without whose assistance this paper could not have been written. I also thank G.R. Munro and K.J. Thomson for helpful comments and suggestions. An extended version of this paper has been published by the Institute of Fisheries Economics, Norwegian School of Economics and Business Administration.

[2]In the terminology of Munro and Scott (1984), the Class I common property fishery is the one analysed in Section 2.

[3]The rents from the fishery would, of course, become capitalized in the value of the boats.

[4]Essentially, economic optimization is here assumed to take place through a two-stage procedure. Arnason (1989) analyses the way in which ITQs can simultaneously achieve both these objectives.

434   *Trond Bjørndal*

REFERENCES

Armstong, C. and Flaaten, O., 1991, 'The Optimal Management of a Transboundary Fish Resource: The Arcto-Norwegian Cod Stock', in R. Arnason and T. Bjørndal (eds), *Essays on the Economics of Migratory Fish Stocks*, pp. 137–52.

Arnason, R., 1989, 'Minimum Information Management with the Help of Catch Quotas' in P.A. Neher, R. Arnason and N. Mollett (eds), *Rights Based Fishing*, pp. 215–41.

Arnason R., 1991, 'On Optimal Harvesting of Migratory Species', in R. Arnason and T. Bjørndal (eds), *Essays on the Economics of Migratory Fish Stocks*, pp. 21–40.

Arnason, R. and Bjørndal, T., 1991, *Essays on the Economics of Migratory Fish Stocks*, Springer-Verlag, Berlin.

Christy, F.T. and Scott, A.D., 1965, *The Common Wealth in Ocean Fisheries*, Johns Hopkins Press, Baltimore.

Clark, C.E., 1976, *Mathematical Bioeconomics*, Wiley, New York.

Dasgupta, P.S. and Heal, G.M., 1979, *Economic Theory and Exhaustible Resources*, Cambridge University Press, Cambridge.

Eckert, R.D., 1979, *The Enclosure of Ocean Resources: Economics and the Law of the Sea*, Hoover Institution Press, Stanford.

Kaitala, V.T., 1985, 'Game Theory Models of Dynamic Bargaining and Contracting in Fisheries Management', Institute of Mathematics, Helsinki University of Technology, Helsinki.

Levhari, D. and Mirman, L.J., 1980, 'The Great Fish War: An Example Using a Dynamic Cournot-Nash Solution', *Bell Journal of Economics*, 11, pp. 322–34.

McRae, D. and Munro, G.R., 1989, 'Coastal State "Rights" Within the 200-Mile Exclusive Economic Zone', in P.A. Neher, R. Arnason and N. Mollett (eds), *Rights-Based Fishing*, pp. 97–111.

Munro, G.R., 1979, 'The Optimal Management of Transboundary Renewable Resources', *Canadian Journal of Economics*, 12, pp. 355–76.

Munro, G.R., 1990, 'The Optimal Management of Transboundary Fisheries: Game Theoretic Considerations', *Natural Resource Modeling*, 4, pp. 403–26.

Munro, G.R., 1991a, 'The Management of Transboundary Fishery Resources: A Theoretical Overview', in R. Arnason and T. Bjørndal (eds), *Essays on the Economics of Migratory Fish Stocks*, pp. 7–20.

Munro, G.R., 1991b, 'The Management of Migratory Fish Resources in the Pacific', in R. Arnason and T. Bjørndal (eds), *Essays on the Economics of Migratory Fish Stocks*, pp. 85–106.

Munro, G.R. and Scott, A.D. 1984, 'The Economics of Fisheries Management', Vancouver: Department of Economics, University of British Columbia, Discussion paper 84–09.

Neher, P.A., Arnason, R. and Mollett, N., 1989, *Rights-Based Fishing*, Kluwer Academic Publishers, Dordrecht.

RANDALL A. KRAMER AND VISHWA BALLABH*

Management of Common-pool Forest Resources

INTRODUCTION

The use and management of the world's forests has become one of the more hotly debated issues in natural resource policy. In developing countries, the rapid depletion of forest stocks has led many to question the sustainability of economic development efforts. Pressures on forest resources for fuel wood, fodder, timber and new agricultural land have led to rapid rates of deforestation of both moist and dry forests. Deforestation has caused environmental degradation, increased poverty and reduced agricultural productivity. In some developed countries such as the USA, there have been controversies over forest management on public lands, because of disagreements about the appropriate balance of timber versus recreation, wildlife and other service flows. At the global level, there have been concerns raised about maintaining the global environmental services provided by forests, such as biodiversity protection, carbon sequestration and climate regulation.

A common thread linking these forestry issues is the common-pool nature of many forest resources. Certain goods and services flowing from forests are characterized by individual use but not individual possession. Examples include the flow of fuel wood and medicinal products from open-access forests, the wildlife and fish which inhabit forest ecosystems, and the genetic pool inherent in a natural, tropical forest. Each of these examples poses a considerable challenge: how to coordinate use by individuals to attain desired levels of consumption or production for the larger community (Oakerson, 1986). This paper focuses on a broad array of products and services which flow from forest land and its associated resources. Several alternative management approaches at the local, national and global level are examined. As illustrations of these management approaches, we will draw on research we have conducted with colleagues on (1) Van Panchayats, communal forest management systems in northern India, (2) nature reserves and parks in Madagascar, designed to reduce deforestation and depletion of biodiversity, and (3) a proposed Global Nature Fund for addressing 'global commons' issues related to forest resources.

*Duke University, USA and Institute of Rural Management, Anand, India. The authors appreciate comments by Mimi Becker.

A FRAMEWORK FOR
ANALYSING COMMON-POOL MANAGEMENT

Over the past several decades, writers have often used the term 'common property' in a variety of ways to refer to natural resources not subject to private or state ownership. A point of confusion has been that the term suggests resources held in common, when in fact writers often referred to resources unmanaged by any individual or group. This has created the mistaken impression that group-owned or managed resources were subject to inevitable degradation (Bromley and Cernea, 1989). In this paper, we will use the term 'common-pool' to avoid confusion associated with the term 'common property', since we will describe forest resources that are held under a variety of ownership (property) arrangements.

We define common-pool forest resources as forest ecosystems that produce one or more flows of outputs where exclusion from resource use is costly or impossible and use by one individual reduces the welfare of others. Thus there are two key characteristics of the common-pool forest resource: non-exclusion and separability of consumption. This implies that common-pool forest resources share one characteristic of pure public goods (high exclusion costs) and one characteristic of private goods (rival consumption). (See Olstrom, 1986).[1] The multiple outputs (such as fuel wood, minor forest products and biodiversity protection) and multiple communities (for example, local, regional or global) of users that characterize common-pool forest resources make them quite different from the typical common-pool resource described in the literature (Price, 1990).

Exclusion difficulties arise for common-pool forest resources because of the physical nature of the forest resources. It may be impossible or extremely costly to restrict access to the flow of goods and services from a forest. For example, it has been difficult in many countries to keep people from collecting fuel wood or practising slash-and-burn agriculture in forested areas. Controlling access is also a problem for fugitive resources, such as fish and wildlife, that are a part of forest ecosystems. Furthermore, to the extent that forests provide watershed protection services or maintain gene pools, it is virtually impossible to exclude users or to charge them a price.

The other distinguishing characteristic of common-pool resources, rival consumption, also poses difficulties from the standpoint of efficient management. Since consumption from the pool by one individual reduces the welfare of others, what is economically rational for the individual may not be rational from the collective viewpoint. Hence joint use can lead to over-exploitation. For example, the harvesting of medicinal plants in a natural forest by individual collectors may increase the gathering costs of other collectors. This is analogous to the common-pool ground-water problem; as more wells are drilled into an aquifer, existing users find their pumping costs increasing as water is extracted at rates in excess of recharge capacity.

In analysing the use and management of common-pool forest resources, it is necessary to specify the property rights regimes governing resource access. These property rights can be categorized as: (1) private property, (2) state property, (3) common or communal property, and (4) open-access. Large amounts of forest resources are found in each ownership category. In fact, a given forest may produce flows that are subject to different property regimes. For example, a forest may be located on land owned by the state, contain trees that are managed and harvested by private firms with concession rights, and provide habitat for wildlife that is available to hunters on an open access basis. A further complexity with management implications is that some environmental services (such as biodiversity protection) can only be provided if the forest is maintained as an intact ecosystem.

Private property is the regime existent when individuals or corporations control access to a resource. Privatization is often promoted as a solution to common-pool resource problems, but there is a danger that privatization (or nationalization) of large forest areas will deprive large groups of individuals of their livelihoods without fulfilling the expectation of better resource management (Bromley and Cernea, 1989). However, particular service flows (such as recreation) may be privatized without privatizing forest land.[2] State property rights regimes give governments (local, provincial or national) complete control over use of natural resources. There appears to be a move towards increasing public management of forest resources in developing countries. For example, the government of Costa Rica has set aside 9 per cent of its forests as parks and reserves and requires permits for all trees harvested, whether from public or private land (World Resources Institute, 1990). Common or communal property occurs in situations where an identifiable group, using formal or informal rules, is able to exclude non-group users and to regulate use of the resource. Various institutional arrangements have been adopted by communal groups in many parts of the world, with varying degrees of success in managing resource use (Berkes and Feeny, 1990). Open access is the absence of specified ownership or rules governing resource use. Because common-pool resources subject to open access are available on a first-come basis, there is no incentive to manage wisely. Valuable forest resources are available on an open access basis either because they have never been brought into a social regulation system or because traditional communal property regimes have been undermined (Bromley and Cernea, 1989).

Each of these property rights regimes implies different institutional arrangements for managing resource use. The confluence of forest resource phenomena with forest institutional arrangements is depicted in Figure 1. The solid, downward-pointing, lines depict the different types of goods resulting from the physical attributes of different forest phenomena. The solid, upward-pointing, lines show the management approaches resulting from different institutional arrangements. The dotted lines depict matches of management approaches and forest output types. With private property rights, there is free entry to the resource, transactions are *quid pro quo*, and markets can work well to allocate those forest resource flows that are excludable and rival in consumption. On the other end of the spectrum, public management is probably the 'best match' for non-excludable and non-rival public goods such as scenic landscapes. However, for the common-pool resources depicted in the middle of Figure 1, the policy debate is centred on what institutional arrangements will work best to manage resource use (Olstrom, 1986). While open-access arrangements are widely deplored, each of the other property rights regimes

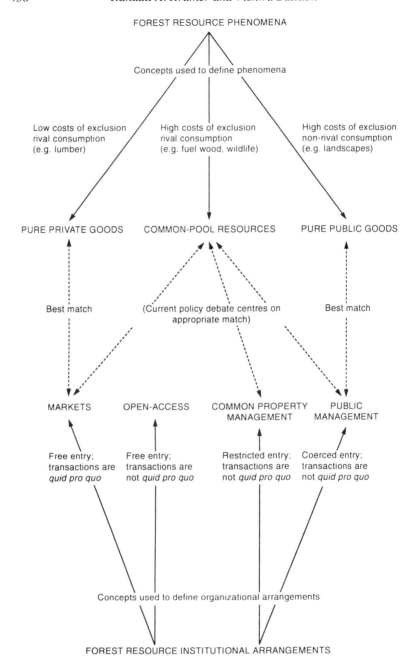

FIGURE 1   *The confluence of forest resource institutions and forest resource phenomena*

Note: Adapted from Olstrom (1986)

has advocates. Some analysts focus on the separability of consumption and argue for privatization (Smith, 1988). Others focus on exclusion problems and recommend public management (Myers, 1989). A third camp argues for common property management, citing examples of successful community institutional arrangements (Berkes *et al.*, 1989). In the next part of the paper we will present one case of common property management (with a long history) and one case of public management (still evolving), and will conclude with a potential management approach involving international cooperation.

MANAGEMENT OF VAN PANCHAYATS

Van Panchayats in India's Uttar Pradesh hills were born out of the conflicts and compromises that followed the settlements and reservations of the forests in the hills at the turn of the last century (Ballabh and Singh, 1988a). In pre-British India, the cultivated land produced a great variety of crops, and the non-cultivated land produced a variety of plant and animal products, largely for fulfilling the subsistence needs of the local populations (Gadgil, 1984). In the Uttar Pradesh hills, people had unrestricted rights in the use of forest resources, except when some forest products were to be exported.

Industrialization in England enlarged the demand for and enhanced the value of forest products. Around the middle of the nineteenth century, the Indian government thought to regulate the utilization of forest products by enacting the Indian Forest Act of 1865 (which was modified in 1878 and 1927) and created the Forest Department to manage India's forests (Ballabh and Singh, 1988b). The passing of the Forest Act encountered resistance in various parts of India, including the Kumaon region of the Uttar Pradesh hills (Guha, 1985). After several periods of social unrest, a grievance committee recommended that forests should be reclassified, and that in areas where local demand was heavy Van Panchayats should be formed to manage the forests.

Following the enactment of Van Panchayat (VP) rules in 1931, over 4000 Van Panchayats were formed by the villagers covering about 15 per cent of the total forest land in the Uttar Pradesh hills. Several observers have claimed that the vegetative coverage of Van Panchayat forests is better managed than in the surrounding reserved forests in their vicinity (Ballabh and Singh, 1988a; Gadgil, 1984; Guha, 1983).

At the village level, the Van Panchayat committee (five to nine members, depending upon the size of the village and forest area) headed by the Sarpanch (President) is the sole arbitrator for management of the Van Panchayat forest. The members of Van Panchayat are elected (informally and not by secret ballot) by the village people every five years. Generally, castes are represented in the committee in proportion to the number of households of that caste in the village. Two important factors help ensure that informal elections are not manipulated by those who are socially and economically strong: (1) the homogeneity of households in terms of caste; and (2) a relatively less skewed distribution of landholdings in the Uttar Pradesh hills compared to the plains. The system provides room for participation across the groups and moderate members of the opposition group are generally included.

The VP forests provide grazing space, fodder, dried and fallen leaves which are used as litter and for making compost, grasses, fuel wood, poles and timber for house construction. The availability of these products is not uniform and depends largely on the type and size of the forest. The land, trees and other resources contained within the VPs are clearly common-pool resources, since consumption of any of these outputs by one villager reduces the amount available for others, and exclusion mechanisms must be employed to prevent people from other villages extracting these valued products. Three methods are used to detect and guard against forest access by non-group members: (1) keeping a paid guard; (2) villagers reporting encroachment to the VP committee (since every villager has a stake in the forest); and (3) the Sarpanch and Panchas visiting the forest occasionally to ensure rule enforcement. The most important of these is the employment of a paid guard.

Ballabh and Singh (1988a) have examined the institutional arrangements in four Van Panchayats. They found that not only does the entitlement of property rights vary, but methods of utilization also vary from one VP to another VP. Even within VPs, both entitlement and methods of utilization vary over time (Table 1). Some of the important characteristics of these methods are worth mentioning. First, the method of utilization varies across the VPs and the rights of the people appear to diminish as resource availability becomes less (for example, see the rights in Parwara *vis-à-vis* Naikada and Devikhal). Second, the VPs have adopted mechanisms to distribute the produce fairly and equally, as in Naikada and in Jeharikhal, and weighing and measuring it accurately.

In spite of successful protection and use of VP forests, recent rule revisions have eroded the capability of Van Panchayats to manage the forest. A major weakness of the present structure of the VP is the generally weak support given to them by the Revenue and the Forest Departments. Although these agencies are responsible for providing technical, personnel and financial assistance to the VPs, these services are rarely provided. Nearly all aspects of the 1976 revisions increased the power of the state government and decreased the power of the VP committee. Now, the Van Panchayats cannot provide any product except grasses and leaves without cumbersome, prior approval from the Forest Department.

Building appropriate institutions is one of the most difficult aspects of any development programme, particularly those programmes concerned with environmental management. Given the great potential for misuse of common-pool forest resources, there is a strong need to adopt strategies including regulatory rules and norms which reduce transaction costs associated with collective management and enable people's participation. Several factors have helped people's participation in Van Panchayats forest management: (1) the high stake of local people in the forest resources, (2) open and informal elections for management committee members, (3) homogeneity in terms of caste and class within a village, and (4) assured and fair distribution of the product. Even factionalism within a village has not hindered proper management of the Van Panchayat forests (Ballabh and Singh, 1988b). The model of Van Panchayats could be effectively utilized in other areas of India (and

TABLE 1 *Rights, restrictions and methods of forest product utilization in selected Van Panchayats in Uttar Pradesh, India*

Particular	Parwara	Devikhal	Naikada	Jeharikhal
1. Area of Van Panchayat forest (hectares)	248.8	20	42.5	22.6
2. No. of compartments	5	3	7	2
3. Rights to outsiders	For dry and fallen leaves to a neighbouring village	nil	nil	nil
4. Grazing	Unrestricted	Prohibited	Prohibited in 3 compartments, open in 4	Prohibited
5. Lopping for fodder	Restricted for 20–25 days in Jan.–Feb. in 1–2 compartments	n.a.	n.a.	A few oak trees are lopped in some years in January
6. Years of rotation for lopping	3–4	—	—	4
7. Grass cutting	Unrestricted	1. Collective harvesting or 2. Parcelling and selling to households for 15 days in Oct.–Nov.	Parcelling and allotment to each household for 15–20 days in October – November	Collectively harvested and weighed or measured by ropes, 15–20 days in October–November
8. Collection of dry and fallen leaves	Unrestricted	—	Unrestricted	Nobody collects, but unrestricted
9. Fallen twigs	Unrestricted	Unrestricted	Unrestricted	Unrestricted
10. Fallen branches and dried trees	Auctioned or given to needy at nominal charges	Auctioned	Auctioned	Auctioned
11. Timber	One tree, for house construction	Not available	One, for house construction, if dead and dry trees are available/ household/year	One, if someone constructs house per household/year
12. Any other	1. 8–10 poles if someone is constructing a house per household/year 2. Wood for agricultural implements	One tree for funeral without any charges	Branches are lopped in some parts for fuel and given equally to each member	Every third year branches are lopped for fuelwood. Equal distribution to each household by weight.

Source: Ballabh and Singh (1988b)

perhaps other parts of the world) to rehabilitate degraded and denuded forest lands, provided conditions are created for local participation.

MANAGEMENT OF RESERVES AND PARKS IN MADAGASCAR

Situation off the east coast of Africa, Madagascar is a 1000 mile-long island inhabited by over 11 million people. It has been singled out by the international environmental community as one of the ecologically richest countries in the world and one whose biological diversity is at great risk. As much as 80 per cent of the island's plants, all of its mammals, and half of its birds occur nowhere else in the world. At the same time, Madagascar is one of the world's economically poorest countries, with a per capita income of $250 (World Bank, 1989).

With international assistance, the Madagascar Department of Water and Forests is establishing a system of parks and reserves throughout the country. To assess the effectiveness of this approach to managing common-pool resources, two key questions must be addressed. First, is the institutional structure in the country capable of taking on the responsibilities of managing a large system of parks and reserves and creating the right incentives to reduce encroachment pressure on the protected forests? Second, given that the new public management system is replacing a combination of open access and communally controlled access, will there be positive net benefits from these reforms?

The initial answer to the question of institutional capability appears to be negative. The Department of Water and Forests is under-staffed, has traditionally focused on timber outputs, and has been unresponsive to new funding initiatives for managing protected areas. Recently, the government has been willing to establish a new entity to take over management of the protected areas. The success of the new entity remains to be seen.

To address partially the second question about net benefits from public forest protection efforts, there is a joint Duke University and World Bank research effort to quantify the benefits and costs of a new park being established in the eastern rain forest (Kramer, Mercer and Shyamsundar, 1991). To assess the economic impact of the park, it is necessary to determine how it will affect the total value of the forest. The total value of a moist forest in Madagascar is comprised of both use and non-use values. The use values include what households derive from the flow of fuel wood, building poles, crayfish, fruits, medicine and other products from the forest, as well as nutrients obtained from the forest by means of slash-and-burn agriculture. Use values also include the value to foreign and domestic tourists who engage in recreational activities in the forest, locals who use the forest for religious and cultural purposes, and scientists who use the forest for biodiversity investigations. The non-use values are primarily existence values. Because of the rich habitat provided by forests in Madagascar and the unusually high number of endemic species, many people may value forest protection even though they never plan to visit the forest.

To measure the change in total forest value engendered by the park, there are three empirical analyses under way: (1) use of the travel cost method to measure the value to international and domestic tourists of the new national park; (2) use of contingent valuation and opportunity cost analysis to measure the benefits and costs of the park to local people; and (3) use of productivity analysis to estimate health benefits resulting from decreased deforestation.

The results will provide insights into the potential net benefits of public management of common-pool forest resources in the humid tropics.

A GLOBAL NATURE FUND FOR
FOREST MANAGEMENT AND PROTECTION

On a global scale, forests cover about one-third of the total land area. They provide a number of economic and environmental services. At present there is a sharp contrast between forest management in the developed and developing countries. In the industrialized countries, forests are managed with clearly defined private and public property rights. Large forested areas have been cleared in the past, but forests in Europe and North America have now been stabilized. In the developing countries, significant proportions of forests are officially owned by governments but large forest areas are treated as open access. The forested area of developing countries has declined by almost half this century. Especially in the tropics, forests are undergoing rapid degradation or conversion to other land use (World Bank, 1990).

Many developing countries have taken steps to slow deforestation and increase reforestation. However, with limited financial resources and other pressing development needs, deforestation has continued at a rapid rate. At the same time, developing countries have been reluctant to invest in conservation measures (such as biodiversity protection) which would largely benefit the world community.

It is increasingly recognized that the world's forests represent a global common. Forests preserve numerous species and ecosystems which may have present or future commercial value (Kramer, Healy and Mendelsohn, 1990). Of the millons of forest species, over half are found in tropical forests. It is estimated that 4 000 to 6 000 species are lost each year to deforestation (World Bank, 1990). The use of forests is also related to global climate change. Deforestation is second only to fossil fuel use as a human activity increasing carbon dioxide in the atmosphere (World Resource Institute, 1990). Other global concerns related to deforestation are changes in global biogeochemical cycles which affect climate and regional hydrological cycles and contributions to acid precipitation (World Bank, 1990). These are clearly common-pool problems, since the resource use decisions of individuals are affecting the welfare of the larger global community. The benefits of protecting biodiversity, wildlife habitats, and critical forest ecosystems accrue to present and future generations of people within a country, as well as to the world community at large.

One way to address these global environmental concerns, overcome financial constraints in developing countries and enable beneficiaries world-wide to contribute to forest protection measures would be to establish a Global Nature Fund to support forest conservation and sustainable management (Sharma and Kramer, 1990). The fund could be established to supplement the international funding for forestry activities available through the Tropical Forestry Action Plan and other current sources. The fund could be supported by several alternative funding mechanisms. One would be a contribution of one-tenth of one per cent of GNP from countries with at least US$6 000 of per capita GNP. This would generate about US$11 billion and could be replenished every three years. Another mechanism would be to place an import tax in the industrial countries on tropical timber products, including logs. A third mechanism would be to increase International Development Association funding. The fund could be used for investments supporting (1) forest conservation (reserves, parks and so on), (2) sustainable management of natural forests, (3) reforestation and afforestation activites, and (4) institution building and human resource development.

Countries eligible for funding would be all developing countries and Eastern European countries now eligible for funding from multilateral agencies. Countries applying for funds would have to demonstrate strong commitment to sustainable management of critical forest ecosystems. The fund could be operated through the World Bank and the regional development banks (Asian Development Bank, African Development Bank and Inter-American Bank). Through international cooperation, the fund would combine financial commitments from the world community with resolve from sovereign nations to use and manage forests more efficiently for present and future generations.

CONCLUSIONS

Many of the current debates over forest policy and management have their roots in the common-pool nature of forest resources. Forests are characterized by a large number of outputs with both use and non-use values. A single forest may serve different user groups, depending on the particular good or service. For example, local people may consume fuel wood and food products, tourists may consume recreational services, regional residents may benefit from watershed protection services, and the global community may benefit from protection of the genetic stock represented in the forest ecosystem. A variety of institutional arrangements are available for managing the forest resource common pools. In contrast to the single output resource serving an easily identifiable user community, forest resource systems are much more complex from both a social system and natural system point of view. We have presented an overview of three contrasting approaches at three different levels of organization: the communal, the national and the international. We urge social scientists to devote greater attention to the management of these complex, valuable, natural and social systems. The cases we have examined offer some possible directions for this attention.

NOTES

[1]Non-exclusiveness is the inability to restrict access and results from an attenuation of property rights. Unrestricted grazing in a forest is an example of non-exclusiveness. Non-rivalness exists when consumption by one individual does not diminish the amount available for other consumers. One person enjoying the scenic beauty provided by a forest does not reduce the beauty available to another viewer. Non-rivalness is a physical characteristic of the resource, not an institutional choice, as a non-exclusiveness may be (Randall, 1987).

[2]Privatization has also been used to reduce pressure on endangered wildlife species and create new economic opportunities. Through the development of ranching systems, opportunities have arisen for individuals to produce marketable products from captive animals (see Smith, 1988, for examples).

REFERENCES

Ballabh, V. and Singh, Katar, 1988a, 'Managing Forests Through People's Institutions: A Case Study of Van Panchayats in Uttar Pradesh Hills, *Indian Journal of Agricultural Economics*, vol. 43, no. 3.

Ballabh, V. and Singh, Katar, 1988b, '*Van (Forests) Panchayats in Uttar Pradesh Hills: A Critical Analysis*', Research Report No. 2, Institute of Rural Management, Anand, Gujarat (Limited distribution).

Berkes, F. *et al*, 1989, 'The Benefits of the Commons', *Nature*, vol. 340, July.

Berkes, Fikret and Feeny, David, 1990, 'Paradigms Lost: Changing Views on the Use of Common Property Resources', *Alternatives*, vol 17, no.2.

Bromley, Daniel W. and Cernea, Michael M., 1989, 'The Management of Common Property Natural Resources: Some Conceptual and Operational Fallacies', World Bank Discussion Paper 57, Washington, DC.

Gadgil, M., 1984, '*Towards an Indian Conservation Strategy*', paper presented at Workshop on a New Forest Policy, Indian Social Institute, Delhi, 12–14 April.

Guha, Ramchandra, 1983, 'Forestry in British and Post British India – a Historical Analysis', *Economic and Political Weekly*, vol. 18, p.45–6.

Guha, Ramchandra, 1985, 'Forestry and Social Protests in Kumaon', in Ranjit Guha, (ed.), *Subaltern Studies*, vol. IV, Oxford University Press, New Delhi.

Kramer, Randall A., Healy, Robert and Mendelsohn, Robert, 1990, 'Forest Valuation', paper prepared for World Bank forest policy study, Washington, DC.

Kramer, Randall, A., Mercer, Evan, and Shyamsundar Priya, 1991, 'Economic Valuation of the Environmental Impacts of Development Projects: Study Design, Concepts, and Progress', Centre for Resource and Environmental Policy Research, Duke University, Durham, NC.

Myers, Norman, 1989, *The Primary Source: Tropical Forests and Our Future*, W.W.Norton, New York.

Oakerson, Ronald J., 1986, 'A Model for the Analysis of Common Property Problems', *Proceedings of the Conference on Common Property Resource Management,* National Research Council, Washington, DC.

Olstrom, Elinor, 1986, 'Issues of Definition and Theory: Some Conclusions and Hypotheses', *Proceedings of the Conference on Common Property Resource Management*, National Research Council, Washington, DC.

Price, Martin F., 1990, 'Temperate Mountain Forests: Common-Pool Resources with Changing, Multiple Outputs for Changing Communities', *Natural Resources Journal*, vol. 30, Summer.

Randall, Alan, 1987, *Resource Economics: An Economic Approach to Natural Resource and Environmental Policy*, John Wiley, New York.

Sharma, Narendra P., and Kramer Randall A., 1990, 'A Global Nature Fund for the Stabilization of the World's Forests', Centre for Resource and Environmental Policy Research, Duke University, Durham, NC.

Smith, Robert J., 1988, 'Private Solutions to Conservation Problems', in T. Cowen, (ed.), *The Theory of Market Failure: A Critical Evaluation*, George Mason University Press, Fairfax, Virginia.
World Bank, 1989, 'Environmental Action Plan for Madagascar', Washington, DC.
World Bank, 1990, 'Forest Policy Brief', Washington, DC.
World Resources Institute, 1990, *World Resources*, Washington, DC.

DISCUSSION OPENING – BEATRICE KNERR*

Beyond any doubt the retreat of the world's forest cover is among the most important challenges to mankind today. Numerous international conferences and an increasing number of government hearings have turned to this problem over recent years, without yet finding any practical solution. Countless governmental and non-governmental organizations are currently concentrating their efforts on projects which might contribute to halting the deforestation process and on reafforestation activities, without achieving more than 'a drop of cold water on a hot stone'. Forest destruction has reached its most dramatic level in Third World countries where forests are often treated as a safety-value for increasing population pressure, unequal land distribution, economic decline and social unheaval. Each year an estimated 0.6 per cent of the remaining tropical forests disappear (with the highest losses in West Africa and Central America) and the rate of destruction is still accelerating (FAO, 1988).

The paper by Kramer and Ballabh is concerned with the assessment of different methods of managing common-pool resources in order to contribute to the preservation of forests in developing countries. In its first part it establishes the definition of the concept of 'common-pool forests' as being characterized by non-exclusion and separability of consumption. This is followed by the presentation of three systems of management of such resources: the communal management of Van Panchayats in India, the state management of parks and reserves in Madagascar, and a potential approach involving international cooperation through a Global Nature Fund.

Three questions arise: (a) do the definitions refer to a significant proportion of the forests which are threatened today; that is, do they provide an appropriate framework for finding solutions? (b) are their models transferable to other regions? (c) are their models viable and stable in the longer run, and under rapidly changing conditions? I believe that the concept of 'common-pool forest resources' provides a helpful basis for analysing the problem of forest destruction and for elaborating measures which could be taken to help solve it. However, I question the authors' statement that deforestation 'has caused … increased poverty'. Starting a search for ways to stop deforestation with this assumption in mind is misleading and obstructs the search for the most important causes of present-day forest destruction. In the long run, the retreat of global forest cover will contribute to the depression of the world's productive potential, but one of the most fundamental conflicts affecting forest destruction is that between the short-term increase in economic well-being

*Universität Hohenheim, Germany.

which is achieved by deforestation, and the long-term deterioration of a country's economic base. This applies at the private as well as at the governmental level. Both try to improve the present economic situation by exploiting the seemingly free forest resources. Most of the external effects which are produced by present-day deforestation are shifted to future generations, notwithstanding the fact that a lot of negative consequences are already felt today.

The Indian example of village Van Panchayats appears to be an ideal solution at first glance. India, however, is a country where the forests are already greatly denuded. Deforestation has slowed down over the last decades only because almost no forest is left, except in some mountainous regions. The last patches of dense forest are highly endangered for economic reasons, as in Kerala, where expansion of plantations into the forested mountain sites leads to increasing fluctuations of water supply in other parts of the state through the loss of the forest protection on the slopes. Also, in the Himalaya region of Jammu and Kashmir, the local people, in spite of a legal ban, find many sophisticated ways to cut and transport big trees under almost uncontrollable circumstances. The major problem with regard to India's forest resources is not that of stopping deforestation but of reafforestation. It would be a valuable contribution to save the remaining patches of forest, but it would be a small one.

In my opinion, the Van Panchayat system has very limited applicability. The special village structure presented in the paper does not seem to apply to most of India, where the villages are dominated by social groups who control the largest part of the economic resources. Moreover, experience of Indian land reform demonstrates that even the people who belong to the poorest and most under-privileged groups tend to elect persons from the upper social strata for the Panchayats (Raj and Tharakan, 1983). This might be an advantage for forest conservation because the economically more well-to-do are less dependent on exploiting forest resources for their survival. But it would be disturbing for the establishment of Panchayats for forest control as the poorer strata of the village population might not be interested in establishing such a system.

As regards other parts of the world, in particular Africa, the opinion is growing that traditional systems of communal control might, in principle, be the best ones for conserving forests (Niamir, 1990). However, traditional social systems are increasingly dissolving, younger people are migrating from their traditional regions, and central governments are gaining more and more influence over traditional power structures. At the same time, inherited values which formerly contributed to forest preservation (such as ideas of 'holy forests') as well as adherence to traditional group pressures, are declining.

The management of reserves and parks in Madagascar is another example cited by Kramer and Ballabh. Unfortunately they say nothing about its transferability to other regions, and not much about its contribution to general forest preservation. I fully agree with the authors in their pessimism about the limited institutional capacities for forest control in many developing countries. The state is often not interested in protecting the forests but rather feels urged to exploit them. National parks are accepted only as long as they attract foreign tourists or some private or public institution is paying for them. In many countries it would not even be possible to establish and maintain natural reserves and parks out of public funds, simply because not enough money is available. The possibilities for establishing parks for tourism are also restricted to a few areas, as the global tourist sector is highly competitive and there is limited demand for specialisms such as the observation of wildlife. Madagascar might be a special case (and therefore an unrepresentative example) since its unique flora and fauna make it of particular interest and therefore put it high on the priority list of funding.

The authors' proposition of establishing a 'global nature fund for forest management' starts from the realistic assumption that a large number of developing countries are not in a position to protect their forests, even if their governments are willing to do so. If the richer nations, in their own interests, want to preserve the forests of the Third World, they will have to pay in one form or another. However, while I agree with the idea I do not believe that it is very likely to be implemented at the moment. In any case, the problem of control over forests still remains. Empirical evidence has shown that state control of forests is not effective, and governments themselves are often the main culprits in deforestation (Repetto, 1988).

So, what can realistically be done to save at least a part of the tropical and sub-tropical forests? In my opinion, a solution can only come from the countries concerned and from support by richer nations. Under the circumstances which prevail in many countries, there is often a very strong incentive to clear forests for growing crops, gathering fuel wood, feeding animals, and other things which are necessary for survival. Governments in Third World countries, faced with the reality of those incentives, often pay much lip-service to forest conservation, but allow the remaining forests to be exploited as a buffer against economic and social deterioration and to secure their own political survival. Given that background, if the industrialized countries wish to preserve endangered forests in the Third World, in their own interest, then there is no other way than to pay for their conservation. Maybe, for the time being, it would be most helpful if international organizations modified policies which contribute to forest destruction and if industralized nations provided freer access to the exports of developing countries which are not land-dependent. An example of the former is World Bank structural adjustment programmes which explicitly urge many developing countries to expand their production of export crops and to reduce the subsidization of fertilizer, both of which lead to an enlargement of deforested area in many indebted countries. The latter is a matter for GATT negotiatons. I must add that I am not very optimistic on either count and can only regret the implications for the situation of forests in two or three decades.

REFERENCES

Food and Agriculture Organization of the United Nations (FAO), 1988, *An Interim Report on the State of Forest Resources in the Developing Countries*, FAO, Rome.

Niamir, M., 1990, 'Traditional woodland management techniques of African pastoralists', *Unasylva*, 41,(1).

Raj, K.N. and Tharakan, M., 1983, 'Agrarian reform in Kerala and its impact on the rural economy – a preliminary assessment;, in A.K. Ghose (ed.), *Agarian Reform in Contemporary Developing Countries*, World Resource Institute, London, Canberra, New York.

Repetto, R., 1988, *The forest for the trees? Government policies and the misuse of forest resources*, World Resource Institute, Washington, DC.

RICHARD M. ADAMS*

Climate Change, Agriculture and the Environment: Some Economic Issues

INTRODUCTION

Potential global climate change arising from increases in atmospheric CO_2 and other trace gas concentrations is an environmental issue of growing concern. While there is controversy about the timing and magnitude of climate change, forecasts from general circulation models (GCMs) of the earth's atmosphere suggest that the earth's climate will warm substantially ($1.0°$ to $4.5°C$) over the next century (Hansen, 1986; Houghton, Jenkins and Epraums, 1990). These increases in global mean temperature are expected to be accompanied by substantial changes in regional and seasonal temperature and precipitation (National Research Council, 1982; Manabe and Weatherald, 1987).

Climate change can affect virtually all ecosystems and ecosystem service flows (Wigley, Ingram and Farmer, 1981; Lough, Wigley and Palutikof, 1983). One ecosystem that is sensitive to climatic variability and of particular importance to human welfare is agriculture. The human consequences of crop failure arising from short-term climatic variation are apparent, as exemplified by recurring famine in the Sahel region of Africa caused by prolonged drought. The flora and fauna of natural or unmanaged ecosystems, including forests, estuaries, lakes and streams, are also likely to be affected by climate changes of the magnitude forecast by some of the GCMs. Changes in forest productivity, a rise in sea levels (due to partial melting of the polar ice caps) and alterations in hydrological balances induced by climate change could affect human welfare.

Implications for human welfare of alterations in ecosystem flows have been drawn largely by natural/physical scientists and engineers. Thus the role of human adaptations, either to slow the rate of CO_2 build-up or mitigate any resultant effects, has not been a central feature of most estimates of the social consequences of climate change. As the 'limits to growth' controversy of the early 1970s demonstrates, failure to account for the role of price mechanisms and other social incentives in altering human behaviour can often result in misleading predictions about paths of resource use (Meadows *et al.*, 1972).

Policy makers are focusing on strategies both to abate the build-up of greenhouse gases and to plan for climatic change (Lashof and Tirpak, 1989). Strategies for abatement include long-term reductions in fossil fuel consumption, reduced use of inputs and production practices that contribute to other

*Oregon State University, USA.

trace gases, and establishment of forest plantations as 'carbon sinks'. Each strategy implies some costs. Several questions then seem relevant. For example, among the problems faced by mankind, does climate change warrant this current attention (and research expenditure)? Are the costs of slowing a CO_2 build-up justified by the benefits (of avoided damage)? What is the role for economic analysis in view of the long time horizons and the associated uncertainties in critical natural and physical science data?

The objective of this paper is to provide an economic perspective on these questions, including (1) an overview of contemporary economic thinking and research on potential benefits and costs of climate change abatement, (2) a discussion of problems and challenges in performing economic analysis of climate change, and (3) an assessment of research directions. The discussion draws heavily upon a sizable and growing economic literature on global climate change, including several recent articles by the author.

BENEFITS VERSUS COSTS OF CLIMATE CHANGE ABATEMENT

As agencies and political bodies consider major investments in research or regulations pertaining to climate change, it is instructive to examine existing conceptual and empirical analyses of similar problems. While most economists would question the ability of the standard benefit – cost analysis (BCA) framework to assess a problem of this temporal complexity, economic input is critical to the design and structure of research of the topic (Sonka and Lamb, 1987; Adams, 1989). For example, despite their limitations, comparative static economic analyses can help to identify the important from the unimportant consequences of climate change and hence guide research and regulatory priorities. Some preliminary economic analyses of climate change effects and abatement strategies, dealing primarily with agricultural resource issues, are reviewed briefly in this section.

Climate change can affect crop yields through a number of mechanisms, including elevated atmospheric CO_2 concentrations and temperatures, as well as changes in precipitation (Decker, Jones and Achutuni, 1986; Rosenzweig, 1988). These changes can lead to increased pest and pathogen populations and reduced irrigation water supplies. Secondary effects of the greenhouse phenomenon, such as increased tropospheric ozone and surface incidence of UV-B radiation may also affect yields. While forecasting the magnitude and geographical distribution of these yield changes is difficult, some preliminary yield effects are available for the USA, Canada and selected areas of the remainder of the world in some recent US EPA-funded studies (Rosenzweig, 1988; Peart *et al.*, 1988). Several recent exploratory economics studies translate such yield forecasts to economic consequences (Adams *et al.*, 1991; Kane *et al.*, 1991; Fischer *et al.*, 1991; Kaiser *et al.*, 1991).

Most of the existing economic studies focus on consequences for US agriculture. For example, Adams *et al.* (1990) consider climate effects on US agriculture, including some preliminary analysis of changes in crop water use and irrigation supply. A similar study by Kane *et al.* (1991) investigates the consequences of climate changes for US trade in agricultural commodities.

Economic analyses that focus on state or multistate regions of the US include studies by Dudek (1988), Kaiser *et al.* (1991) and Easterling, Parry and Crosson (1989). A common finding of the studies which measure aggregate welfare effects is that climate change does not appear to be a food security issue, at least for North America. While locational shifts may occur in land and water use, the net economic effect could actually be positive under some climate change forecasts, owing primarily to the yield-enhancing effect of increases in atmospheric CO_2. Even under more pessimistic GCM forecasts, the annualized loss is relatively modest (for example, for the USA. Adams *et al.*, 1990, estimate an annual loss of approximately $11 billion or less than 7 per cent of the farm-gate value of production). Preliminary results from sensitivity experiments on global food production suggest a similar pattern to that observed for the USA; that is, likely regional gainers and losers but with relatively modest total changes in food production and economic welfare (Fischer, *et al.*, 1991).

While these studies suggest only modest losses (or gains) to the agricultural sector from climate change, predicted shifts in crop production and expansion in irrigated acreage imply demands or pressure on environmental and natural resources, including water quantity and quality, wetlands, soil, fish and wildlife and other resources. For example, a northward shift in corn and soybean production in North America (into Southern Canada) forecast by Adams. *et al.* (1990) would exacerbate the loss of prairie wetlands by making drainage and conversion to crop production more profitable. Similar northward shifts in European agriculture are noted in Easterling, Parry and Crosson (1989). Obtaining water to facilitate any increases in irrigated acreage also implies more and larger reservoirs, which in turn implies greater pressure for water development. Other water resource effects can be inferred from the regional GCM temperature and precipitation forecasts. Specifically, increased spatial and temporal variability in rainfall and snow packs will also increase pressure to build dams for flood control. Increased competition for remaining streamflows seems likely.

Other resource effects are anticipated under climate change. Forest ecosystems will change, with some coniferous forest species, such as Ponderosa pine and Douglas fir, being displaced. Climate change, in connection with acid rain and photochemical oxidant air pollution, could be particularly stressful to forests in Europe and the eastern USA, where the forests die back to cover a smaller area. Wildlife populations may be reduced following habitat losses and more extreme weather variability. The implication for global wildlife resources, including endangered species, must be recognized (Batie and Shugart, 1989). A rise in sea level has implications for the productivity of coastal estuaries as well as agricultural production in low-lying areas of Bangladesh and Indonesia.

Climate change will impose some indirect effects on agriculture, particularly in terms of input costs. Perhaps the most important of these will be energy costs of control measures imposed on fossil fuel combustion. Current technology for CO_2 emissions reductions on fossil fuel combustion is costly. Construction of nuclear power plants is also costly. As a result, real per unit costs of electricity will rise. The feasibility of expanded or even current irrigation will then be dependent on increases in real commodity prices sufficient to offset energy cost increases.

Agriculture itself generates various greenhouse gases. Methane from flooded rice fields, livestock and deforestation, nitrous oxide from fertilizer use, carbon dioxide from fossil fuel use and deforestation, and CFCs used in food processing are all greenhouse gases. The global increase in livestock numbers, irrigated rice acreage and nitrogenous fertilizer use is occurring as the result of population increases. Much of the deforestation in tropical areas, which contributes about a third of the total annual CO_2 flux, is for agricultural purposes. Control measures for such greenhouse gases could generate additional financial stress for farms and rural communities and exaggerate spatial adjustments. Indeed, as shown recently by Adams *et al.* (1991), the cost to US agriculture (consumers and producers) of controlling some trace gas emissions is substantial, ranging from $550 per ton of methane from rice production to $4 000 per ton of nitrous oxide. Additionally, controls on inputs such as fertilizer in developing countries would occur exactly when productivity increasing inputs would be in high demand.

Finally, the agricultural land base in some countries may be under pressure as a means of mitigating CO_2 increases. An increasingly popular proposal is to expand forested regions (through tree plantations) to remove carbon dioxide from the atmosphere (Woodwell, 1987). The preliminary analysis by Adams *et al.* (1991) indicates that a forest plantation programme on agricultural lands in the USA sufficient to sequester 20 per cent of annual US CO_2 production would require 50 million acres of land and cost between $21 and $38 per ton of carbon. A programme to sequester half of US CO_2 production would require over 270 million acres, at costs in excess of $100 per ton of carbon. It is not clear what would be done with the timber produced on these plantations. The increase in timber supply from 50 million acres of additional forest area is estimated to reduce timber prices by up to 50 per cent, discouraging private timber production. This suggests the need for major subsidies or other incentives to encourage such a forest plantation programme.

In summary, the limited economic analyses on benefits and costs of climate change in North America, while crude, suggest that, in the aggregate, the effects may not be serious for agriculture. Indeed, gains in social welfare may be possible if the CO_2 fertilizer effect holds at elevated levels. However, regional or country-level effects are likely, with both gainers and losers. Costs of reducing trace gas emissions for agriculture seem large, suggesting that agriculture is not a cost-effective means of achieving such reductions. The lack of scientific information on a range of other resource effects under climate change precludes meaningful economic analysis at the present time. Unfortunately, these unaccounted consequences may be among the most important long-term effects of climate change. Technological change, and indeed economic analysis itself, may have little to contribute to the resolution of ethical issues involving intergenerational transfers of a degraded natural environment.

PROBLEMS IN PERFORMING ECONOMIC ASSESSMENTS

There are several dimensions of climate change assessments that pose special analytical challenges. One important feature is the dominant role played by intertemporal considerations. Climate change is likely to alter the stream of returns that assets produce, whether the assets are buildings, public infrastructure, forests or agricultural products. Unfortunately, the benefit–cost calculation process is likely to be severely stressed when dealing with the time horizons associated with climate change.

The intertemporal features of climate change introduce the effects of risk on producer and consumer behaviour. However, most climate change damage assessments are *ex post* in nature. Because an *ex post* representation establishes a number of contingent states and proceeds to treat each of them as if it were certain, it is incapable of accounting for the agent's attitude towards risk; that is, it disregards the expenditures that the agent makes in preparing for states that go unrealized. The alternative, *ex ante* representation of the economic agent, addresses the consequences when the magnitude of the climate change effects is not yet known. The implications of these two analytical challenges are discussed in detail in Adams and Crocker (1989).

There are at least two conceptual dimensions of climate change assessments that need special consideration. Because climate change is a global phenomenon, assessing the consequences for other regions may require a different assessment perspective than is found in the US studies reviewed earlier. The first challenge relates to the different orderings of economic processes, markets and institutions across countries. This diversity requires an assessment framework that exploits the common relationships between economic orderings while adequately representing the relevant individual characteristics of each country or region. The second challenge relates to a fundamental lack of biological data from which to predict crop and other ecosystem responses across matrices of pollutant, site and time combinations. Innovative procedures are required to determine the 'transferability' of response parameters within this matrix.

A crucial starting-point in economic analysis is the correct representation of the choice problem facing decision makers. In market-oriented economies the neo-classical paradigm of producer and consumer behaviour provides an appropriate basis for such a representation. However, in many settings, individual supply and demand behaviours are essentially subsumed into planned, aggregate supply and demand decisions based on perceived social objectives. In addition to such planned, socialistic economies, a sizable percentage of world food production and consumption occurs in developing countries, where the interaction of subsistence, village-oriented agriculture and aggregate food demand/production goals creates a third possible configuration of market structures.

The challenge in measuring the international consequences of climate change on the service flows provided by agriculture, forestry and other ecosystems is to fold these diverse economic structures into a framework which provides some common measure of social welfare. The market-oriented and planned economies have a somewhat common goal, namely, to produce efficiently a socially optimal level of agricultural output. Market-oriented economies work towards this goal through the market signals created by the interaction of supply and demand forces. Planned economies rely on indirect approaches based on planning goals (that is, perceived consumption needs) and associated allocations of resources to meet those goals. Interactive, iterative decision processes are frequently used to measure the contribution of alternative re-source allocations to the attainment of the specified goals. However, a funda-mentally different decision problem confronts the subsistence grower, who is both a producer and consumer of his output. Numerous behavioural models have been advanced to capture this dual role, but in general it is believed that such growers are risk-averse and thus produce first for family or local con-sumption, using historically proven techniques. Production moves into domestic or international markets only after the primary goal of subsistence has been achieved. In this producer/consumer setting, climate change may affect the individual in ways distinct from the cases treated heretofore. Specifically, while the grower's quantity (yield) and quality of commodities may be affected by climate change, as in any other economic setting, the effects on the grower may be more severe in terms of individual welfare. First, any productivity loss reduces marketable surplus and possibly even the subsistence component of production. Second, changes in quantity and quality of production may effect labour productivity if the grower's health is influenced, given that the grower (and immediate family) provide the major input, labour, into the agricultural production process. Third, climate-induced yield reductions may intensify risk-averse behaviour, thereby reducing the likelihood of the grower adapting new, yield-increasing technologies.

These unique dimensions of subsistence agricultural economies suggest that measuring the economic effects of pollution on developing countries will require economists to incorporate some of these producer/consumer/labour supplier linkages into the assessment framework. Available literature on eco-nomic development may provide some guidance. In addition, the concept of the 'household production function' proposed by Tinbergen (1956) and ex-panded by Lancaster (1966) offers a possible framework within which to explore the effects of pollution on the decision problem facing the producer/consumer.

The second general challenge in assessing climate effects is the scarcity of natural and physical science data. If more exact natural science information is to be of substantial value, this will probably be because of its contribution to better understanding of the manner in which the consequences of climate change impacts are distributed across groups and regions. Evidence provided by Adams and Crocker (1989) indicates that production and consumption patterns, and thus distributional consequences, are often sensitive to rather small changes in ambient environmental quality. Currently, it is not known whether estimates of the human behavioural responses which are the root sources of this sensitivity would be greatly altered by more exact natural science information.

The policy relevance of improved information concerning the sensitivities of production and consumption patterns to climate change is demonstrated by the consistent finding in many air pollution studies that farmers substitute other inputs, especially land, for poor environmental quality; that is, reduced environmental quality causes more land and inputs such as fertilizer and pesticides to be used. The magnitude of substitution and the conditions that cause this magnitude to vary are at present little understood, although the finding reported by Adams and Crocker that producer losses often increase exponentially implies that the worth of available substitution possibilities declines. To the economist, good natural science information is synonymous with a precise mapping of the substitution possibilities. The importance of such a mapping becomes apparent when one recognizes that land, fertilizer and pesticide substitutions could be responsible for a good deal of the pollution and trace gases that originate from agricultural practices. Policy makers might then confront the paradox that the environmental change from other sources (such as CO_2-induced climate change) which affects agriculture encourages agricultural practices which themselves contribute to climate change; for example, more nitrogen fertilization, more deforestation. Conversely, a reduction in exogenous environmental change from other sources could reduce some trace gases and other forms of pollution that originate in agricultural practices. A thorough understanding of whether this quite plausible story is, in fact, true will require an as yet unavailable detailed natural science and economic understanding of influential substitution possibilities and behavioural responses.

CONCLUSIONS

Most economic inquiries concerning agriculture and related resource use suggest that, for moderate climate change, agricultural production appears capable of meeting projected demands. While these analyses are based on uncertain data and assumptions, the findings do suggest that economic adjustments can mitigate some adverse climate effects, but with attendant regional gainers and losers. Planning may therefore be needed to soften the negative impacts on specific regions, countries and resources.

The evidence to date on the possible consequences of climate change is largely drawn from the North America and European experience. While tropical deforestation and rice production attract attention as major sources of trace gases, there is little research on the relationship between the behaviour of farmers, including input use patterns, and both the causes and effects of climate or other environmental change in developing areas of the world. If BCA is to be used in debating the role of agriculture in global decisions pertaining to the regulation of trace gases, then economists need to influence the scientific research agenda to eliminate current gaps in both the natural and social sciences, particularly in terms of our understanding of effects in the LDCs.

In summary, a couple of points seem important. First, economic processes have the potential to mitigate the direct effects of climate change on agricultural production and consumption. In fact, such potential adjustments are likely to be understated in the economic studies recorded here, given the long time horizons involved. While the adequacy of food and fibre production may not be an issue, it seems probable that the conflict between those adjustments that fill the stomach and those which nurture the human psyche will intensify. Thus a more rapid loss of environmental and other non-market assets and consumption opportunities seems likely under the climate projections. This implies that future generations will be the real losers. Second, most of the implications drawn here do not open vast new areas for economic research. Rather, they suggest a continuation (and perhaps renewal) of much of the resources-oriented research performed by agricultural economists over the last several decades, but with greater attention to the processes underlying agricultural development.

REFERENCES

Adams, R. M., 1989, 'Global Climate Change and Agriculture: An Economic Perspective', *American Journal of Agricultural Economics*, 71, p.5.

Adams, R. M. and Crocker, T.D., 1989, 'The Agricultural Economics of Environmental Change: Some Lessons from Air Pollution', *Journal of Economic Management*, 28, pp. 295–307.

Adams, R. M. *et al*, 1990, 'Global Climate Change and U.S. Agriculture', *Nature*, 345, p.6272.

Adams, R. M., McCarl, B.A. and Callaway, J.M., 1991, 'The Role of Agriculture in Climate Change: A Preliminary Evaluation of Emission Control Strategies', in J. Reilly and M. Anderson (eds), *The Economics of Global Climate Change: Implications for Agriculture, Forestry, and Natural Resources*, Western Press, Boulder, Co.

Batie, S.S. and Shugart, H.H., 1989, 'The Biological Consequences of Climate Change: An Ecological and Economic Assessment', in N.J. Rosenberg, W.E. Easterling, P.R. Crosson and J. Darmstadter (eds), *Greenhouse Warming: Abatement and Adaptation*, Resources for the Future, Washington, DC.

Decker, W.L., Jones, V.K. and Achutuni, R., 1986, 'The Impact of Climate Change from Increased Atmospheric Carbon Dioxide on American Agriculture', U.S. Department of Energy, DOE/NBB-0077, Washington, DC.

Dudek, D.J., 1988, 'Climate Impacts Upon Agriculture and Resources: The Case of California', in J.B. Smith and D.A. Tirpak (eds), *The Potential Effects of Global Climate Change in the U.S.*, Environmental Protection Agency, Washington, DC.

Easterling, W. III., Parry, M.L. and Crosson, P.R., 1989, 'Adapting Future Agriculture to Changes in Climate', in *Greenhouse Warming: Abatement and Adaption*, Resources for the Future, Washington, DC.

Fischer, G., Frohberg, K., Kayzer, M. and Parikh, K., 1991, *Linked National Models: A Tool for International Food Policy Analysis*, International Institute for Applied Systems Analysis, Kluwer Academic Publishers, Amsterdam.

Hansen, J., 1986, 'The Greenhouse Effect: Projections of Global Climate Change', statement presented to US Senate Subcommittee on Environmental Pollution of the Committee on Environment and Public Works, 10 June.

Houghton J. T., Jenkins, G. J. and Epraums, J.J., 1990, *Climate Change: The IPCC Scientific Assessment*, Cambridge University Press, Cambridge.

Kaiser, H. S. Riha, Rossiter, D. and Wilks, D.S., 1991, 'Agronomic and Economic Imports of Gradual Global Warming: A Preliminary Analysis of Midwestern Crop Farming', in J. Reilly, and M. Anderson, (eds), *The Economics of Global Climate Change: Implications for Agriculture, Forestry and Natural Resources*, Westview Press, Boulder, Co.

Kane, S., Reilly, J. and Tobey, J., 1991, 'Effects of Climate Change and World Trade Flows of Agricultural Commodities', in J. Reilly and M. Anderson (eds), *The Economics of Global Climate Change: Implications for Agriculture, Forestry and Natural Resources*, Westview Press, Boulder, Co.

Lancaster, K. J., 1966, 'A New Approach to Consumer Theory', *Journal of Political Economy*, 74, pp.132–57.

Lashof, D.A. and Tirpak, D.A., 1989, *Policy Options for Stabilizing Global Climate Change*, draft report to Congress, US Environmental Protection Agency, Washington, DC.

Lough, J.M., Wigley, T.M.L. and Palutikof, J.P., 1983, 'Climate and Climate Impact Scenarios for Europe in a Warmer World', *Journal of Climate and Applied Meteorology*, 22, pp. 1673–84.

Manabe, S. and Weatherald, R.J., 1987, 'Large-Scale Changes in Soil Wetness Induced by An Increase in Carbon Dioxide', *Journal of Atmospheric Science*, 44, pp. 1211–35.

McQuigg, J.D., 1981, 'Climate Variability and Crop Yield in High and Low Temperature Regions', in W. Bac, J. Pankrath and S.H. Schneider (eds), *Food-Climate Interactions*, R. Reidel Publishers, Dordrecht, The Netherlands.

Meadows, D.H., Meadows, D.L., Randers, J. and Behren, W.W. III, 1972, *The Limits to Growth*, Universe Books, New York.

National Research Council, 1982, *Carbon Dioxide and Climate: A Social Assessment*, National Academy Press, Washington, DC.

Peart, R.M., Jones, J.W., Curry, R.B., Boote, K. and Allen, L.H., 1988, 'Import of Climate Change in Crop Yields in the Southeastern US.', *Report to Congress on the Effects of Global Climate Change*, US Environmental Protection Agency, Washington, DC.

Rosenzweig, C., 1988, 'Potential Effects of Climate Change on Agricultural Production in the Great Plains: A Simulation Study', *Report to Congress on the Effects of Global Climate Change*, US Environmental Protection Agency, Washington, DC.

Sonka, S.T. and Lamb, P.J., 1987, 'On Climate Change and Economic Analysis', *Climatic Change*, 11, pp. 291–311.

Tinbergen, J., 1956, 'On the Theory of Income Distribution', *Weltwetschaftliches Archiv*, 77, pp. 155–74.

Wigley, T.M.L., Ingram, M.J., and Farmer G. (eds), 1981, *Climate and History: Studies in Past Climates and Their Impact on Man*, Cambridge University Press, Cambridge, Mass.

Woodwell, G.M., 1987, 'Biotic Implications of Climatic Change: Look to Forests and Soils', in J.C. Topping, (ed.) *Proceedings of First North American Conference on Preparing for Climate Change: A Cooperative Approach*, Climate Institute, Washington, DC.

DISCUSSION OPENING – J. ARNE HALLAM*

The analysis of significant environmental events provides important challenges to economists in both theory and measurement. In order to understand their impact, physical as well as economic information is needed. Adams discusses recent attempts to measure the impact of climate change and introduces some of the many theoretical issues that must be confronted. The author has done an excellent job in surveying the recent literature, presenting us with a number of important points that should be considered in all future attempts to measure the impact of climate change. While specific criticisms might be levelled at each of the papers reviewed by the author, which some discussants might choose to dispute, these comments will attempt to provide a framework that might be used to model climate change. This will provide a basis on which to discuss the problems presented by the author. In evaluating the impact of exogenous factors on the economy, three systems must be considered. The interactions of the physical (including biological), economic and social systems jointly determine outcomes.

*Iowa State University, USA.

Physical models of climate change

One simple type of model used by scientists is based on time series, extrapolating past trends into the future. More complicated models attempt to predict the future by simulating the physical system and then using forecasts of inputs to the system to predict outputs. For example, a scientist might predict global temperature based on total carbon dioxide emissions, determine emissions using the physical properties of current production practices, and then extrapolate the use of such practices in the future to forecast future climate conditions. The accuracy of such forecasts depends on the ability of the model to simulate dynamically the physical process, and on the precision of the exogenous extrapolation.

Models of climate change depend on data collected and synthesized over long time-periods. Much of the data used is also subject to significant errors of measurement. Atmospheric models, in particular, often depend on the solution of partial differential equations which are very sensitive to starting values and to parameter variation. When possible, all point estimates of change should be supplemented with precision estimates, so that induced models can both report standard errors and perform reasonable sensitivity analyses.

Most physical models also assume that input–output relationships are constant, or at least change according to some fixed and known rule. Given that production processes are operated by economic agents, there is inherent potential for both exogenous and induced technical change to occur. To the extent that physical models do not properly account for changes in technology, they may miss the essence of changes in the system. For example, if crop breeders were able (unlikely though it may be) continuously to produce new varieties having a constant rate of yield growth across all possible environments, then global climate change would have no impact.

A simple way to forecast system inputs is to extrapolate past trends. While this may be useful for short-term analysis, the essence of economics is that agents will respond to price signals. To the extent that system outputs change the economic environment, decision makers are induced to change the production process, and thus change the results of the system model. This is the major hypothesis of the next section.

Integrating economic and physical process models

The physical and economic environment influence the choices of individual agents. In an equilibrium and dynamic context, the decisions of agents can influence both technology and the economic environment. These interactions are important in measuring and predicting the effects of changes in the physical system.

Consider a model of the decision process for an individual. There are three components of this process. The first is technology. This is the production function or input–output relationship, which can be considered fixed in the short run. The second component is the physical and economic environment. Elements of this component include prices, government policy variables, physical parameters such as sunlight and rainfall, and elements of the legal and social system. The third component is the objective function reflecting the preferences of the decision maker. This includes individual attitudes towards income, wealth and risk. Given these three components, the decision maker or the economist can determine the optimal decision, given information about technology and the environment. The solution of this decision problem leads to economic relationships such that decision variables are expressed in terms of exogenous variables. An example is a supply function which depends on price movements and fixed input levels. If environment variables remain explicit in the reduced form economic model, then their impact can be evaluated directly. Alternatively, the effects can be evaluated by changing the levels of the climate parameter and resolving the optimization problem directly.

While some economic parameters, such as prices, are fixed for the individual, they are endogenous in the economic system. Thus individual models must be aggregated and solved to obtain market equilibrium effects. An important decision in modelling is the level of aggregation that accurately reflects the economic system. The conditions for aggregating micro-level models are stringent and often not reasonable, while models formed in the aggregate often inadequately reflect technological and site-specific environmental factors. To the extent that increased yields in one region have a price impact, economic conditions in all regions will be affected. More critically, over time, the outcomes of the individual decision process, when integrated into the system, may affect the underlying technologies available, and thus the underlying decision problem.

Some other modelling issues

As pointed out by the author, individual preferences as regards risk and return are difficult to measure. Global climate change is a long-run phenomenon, and so intertemporal preferences are important. The appropriate discount rate for the individual is not obvious. Whether intertemporal preferences are time-separable, and thus amenable to the consequent simplifications in modelling, is an unresolved issue.

The current literature in macro-economics makes clear the need to model appropriately the way in which generations are linked together. Models with an infinite time-horizon and a single decision maker may have very different solutions from those with overlapping generations and different ways to transfer or trade wealth. Thus classical extraction models may lead to different conclusions, depending on whose wealth is maximized. While altruism may be a rational economic action, selfishness and maximization of one's own wealth are also possibilities.

Few detailed models of the economic system solve for a full intertemporal rational expectations equilibrium. A key aspect of modelling is how complete a model to build and how to link periods over time. Does the modelled system reach full equilibrium in each period or does the system muddle through a series of temporary equilibria, always shooting for a more complete solution?

The appropriate measure of welfare in an intertemporal context depends on whose welfare is to be measured. Simple sums of consumer and producer surplus seldom accurately measure the implications of regime change. Since work to date has shown more distributional than aggregate consequences for climate change, the appropriate way to measure such effects seems critical.

At a more macro level, shocks to the economic system may induce changes in the very fabric of the underlying political and social system. The induced innovation hypothesis is but one example of the way in which changes in technology may lead to other changes, with perhaps significant impacts on the social system. Global climate change could have an impact in such a way as to reduce the viability of the current world economic and social system. For example, suppose that climate change leads to a gradual northward movement in crop production, with resultant further impoverishment of much of the Third World. The currently unequal distribution of wealth would become worse and past unheeded calls for a new international economic order might become more poignant and more powerful. Such movements, if successful, could lead to the overthrow of the current system, with significant impacts on future technology, trade flows and wealth accumulation. Alternatively, the beginnings of global climate change may sensitize individuals in developed countries so that they call for significant changes in the production system in order to protect the environment. In either case, the underlying economic predictions from a model based on the current system lose credence.

Conclusions

These comments have been very cautionary about the use of physical and economic models to evaluate global climate change. Nevertheless, the author firmly believes that such forward-looking analysis is necessary. Only by attempting to understand the future can correct decisions about the present be made. Those who undertake such analysis are to be commended for bravery and concern for society, but they should make the assumptions of their work transparent and attempt to address in one way or another the issues raised by these comments. Sensitivity analysis, complete with standard errors, would be a most welcome addition to much current work. Finally, all results from forward-looking work should be viewed with caution, since all crystal ball gazing is only as good as the current technology in crystal balls.

STEN NILSSON*

Air Pollutants and Options for Their Control:
Experiences from the European Scene

BACKGROUND

During the first half of the twentieth century, air pollution had a predominantly local character in Europe. But since the late 1960s, Europe has been
aware of the threat to the environment from the effects of pollutants from
large emission sources transported over long distances.

At the end of the 1970s, the air pollution problem (mainly acid rain) was
recognized as one of the most severe threats to the environment in Europe. As
an effect of the alarming reports on the acidification problem the Convention
on Long-Range Transboundary Air Pollution was adopted in 1979 within the
framework of the United Nations Economic Commission for Europe (ECE). A
first result of the implementation of this convention was an improvement of
the Co-operative Programme for the Monitoring and Evaluation of the Long-
Range Transmission of Air Pollutants in Europe (EMEP). A second breakthrough for the implementation of the Convention was the *Protocol on the
Reduction of Sulphur Emissions or Their Transboundary Fluxes* by at least 30
per cent adopted in Helsinki in July 1985. A third important step in implementing the convention was a protocol conceiving control of emissions of
nitrogen oxides, which was signed in Sofia in November 1988. The protocol,
signed by 25 countries in Europe and North America, calls for a stabilization
of the emissions of nitrogen oxides. In Sofia, 12 countries signed a declaration
on a 30 per cent reduction of nitrogen oxide emissions. Thus the abatement of
air pollution effects has a high ranking on the political agenda in many
European countries.

EUROPEAN EMISSIONS

The emissions of SO_2 are presented for sub-regions of Europe in Table 1. The
individual countries are grouped according to the ECE classification:

Nordic: Finland, Norway, Sweden

*International Institute for Applied Systems Analysis, Laxenburg, Austria.

462

EEC-9:   Belgium, Denmark, France, former Federal Republic of Germany, Ireland, Italy, Luxembourg, Netherlands, United Kingdom
Central: Austria, Switzerland
Southern: Greece, Portugal, Spain, Turkey, Yugoslavia
Eastern: Albania, Bulgaria, Czechoslovakia, former German Democratic Republic, Hungary, Poland, Romania
USSR

TABLE 1   *European SO₂ emissions (in thousand tonnes SO₂)*

Region	(1) 1980	(2) 1989	(3) CRP 1995	(4) Unabated 2000	(5) Maximum feasible reductions
Nordic	1 200	554	475	1 048	211
EEC-9	17 264	9 414	9 064	13 297	1 564
Central	455	168	138	412	100
Southern	5 935	6 876	8 748	10 656	3 024
Eastern	16 381	15 870	13 726	15 690	2 714
USSR	12 800	9 300	8 820	11 960	1 370
Total Europe	54 035	42 182	40 971	53 063	8 983

Source: Amann and Sörensen (1991).

Columns 1 and 2 describe the emissions as computed by the RAINS-model (Alcamo *et al.*, 1990), column 3 illustrates emission levels as expected in the year 1995 after implementation of the current emission reductions agreement. Column 4 presents the estimated emissions with the projected fuel consumption without any emission control measures. Column 5 illustrates the emission levels by applying all currently available emission control technologies.

There has been a strong reduction in the SO_2 emissions between 1980 and 1989 for the whole of Europe. At a sub-regional level, this is not the case for the Eastern and Southern regions. Further reductions, according to current plans for 1995, will improve the conditions slightly. It is also questionable whether the Eastern, Southern and USSR regions will in reality achieve any improvements by the year 1995; from the last column in Table 1 it can be seen that there are big possibilities for reductions by implementing available emission control technologies.

The emissions of NO_x are presented in Table 2. In this case, total planned reduction by 1995 in comparison with the emissions in 1980 is only about 5 per cent. There are also uncertainties concerning NO_x emissions; it is far from clear that the Southern, Eastern and USSR regions will be able to achieve the reduction goals. As in the case of SO_2, there is potential for reductions of NO_x if available emission control technologies are implemented.

TABLE 2   *NOₓ emissions (in thousand tonnes NO₂)*

Region	1980	CRP 1995	Maximum feasible reductions
Nordic	736	514	350
EEC-9	10 003	7 751	3 691
Central	425	297	164
Southern	2 088	3 254	1 456
Eastern	4 594	4 581	1 676
USSR	9 454	9 454	3 678
Total	27 317	25 854	11 027

Source: Amann (1989).

For ammonia emissions, no protocols on reduction have yet been signed. The development of emissions between 1980 and 1987 is presented in Table 3. About 95 per cent of the totals stem from the agricultural sector. During the period studied, there have been no reductions of total ammonia emissions in Europe.

In addition to the specific emissions mentioned above, it can be suggested that ozone concentrations are much above established critical limits in all of Europe. There are also emissions of various trace elements of at least 375 000 tons per year (Nilsson, 1991a).

TABLE 3   *Ammonia emissions (in thousand tonnes NH₃)*

Region	1980	1987
Nordic	169	166
EEC-9	2 769	2 034
Central	141	136
Southern	1 187	1 206
Eastern	1 635	1 661
USSR	1 146	1 255
Total	7 045	7 205

Source: Klaassen (1990a).

IMPACT OF AIR POLLUTANTS

The different air pollutants have harmful effects – individually or in combination – on several sectors of the European economy. Leipert and Simonis (1990) have identified the following problems or problem areas:

- health hazards,
- material damage,
- vegetation degradation,
- agriculture production,
- forest production,
- water pollution.

It is not possible to discuss all of the cases in the space available, hence the presentation will concentrate on the impacts of air pollutants on forestry as an example of an approach to analysis. This is based on more detailed work by IIASA (Nilsson *et al.*, 1991a).

Modelling forest decline attributed to air pollutants

The first step in quantifying the effects of air pollutants is to classify the distribution of the forests of each country into several sensitivity classes with respect to sulphur and nitrogen depositions. The sensitivity classes are based on capabilities of forest soils to act as buffers against acidification from deposition of sulphur and nitrogen compounds. Highly sensitive sites have low buffer capacity, the opposite applying to low-sensitivity sites.

Specific *critical loads* and *target levels* of sulphur and nitrogen deposition have been assigned to the individual sensitivity classes. Critical loads are quantitative estimates of an exposure to one or more pollutants, below which significant harmful effects to specific sensitive elements of the environment do not appear to occur. Target loads are less restrictive with respect to deposition loads, in that they incorporate consideration of other pollution-control factors, such as the economic.

Critical loads for sulphur and nitrogen have been defined by the United Nations Economic Commission for Europe (UN-ECE, 1988). Target loads for sulphur have been proposed at levels somewhat higher than the ECE critical loads by the Beijer Institute Centre for Resource Assessment and Management, University of York, U.K. (Chadwick and Kuylenstierna, 1988). The target loads used for nitrogen are the same as the critical loads set by UN-ECE (1988). The analyses carried out by IIASA are based on the target loads presented in Table 4.

By combining the IIASA Forest Study data base on forest resources in Europe (Nilsson *et al.*, 1991b) and the IIASA RAINS model (Alcamo *et al.*, 1990) it was possible to estimate the extent of forest area and standing volume, with depositions exceeding target loads today and in the future. The deposition estimates generated by the RAINS model for the year 2000 are used in the analyses. These are based on the current plans announced by

TABLE 4 *Target loads for sulphur and nitrogen deposition used in the forest-decline scenarios (grams of substance per m² per year)*

Substance Sensitivity	Conifers			Deciduous		
	Low	Medium	High	Low	Medium	High
Sulphur[a]	2.0	1.0	0.5	4.0	2.0	1.0
Nitrogen[b]	1.5	1.0	0.3	2.0	1.2	0.5

Notes: [a]Target loads set by the Beijer Institute (Chadwick and Kuylenstierna, 1988) based on critical loads set by ECE (UN–ECE, 1988).
[b]Target loads for nitrogen are the same as critical loads set by ECE (UN–ECE, 1988).

individual governments to reduce emissions of SO_2 and NO_x (see Tables 1 and 2). Basic calculations were carried out at the country level; the aggregated results are presented in Table 5.

Table 5 shows that planned pollution-abatement strategies will not be effective in reducing risk to forests from pollutants. In the year 2000, most of the

TABLE 5 *Exposure of European forests to significant amounts of air pollutants*

Pollutant Forest type/period	Region					
	Nordic	EEC-9	Central	Southern	Eastern	European USSR
Sulphur						
Coniferous						
1985	59	88	98	62	98	27
2000	48	76	93	84	98	21
Deciduous						
1985	19	34	50	18	84	4
2000	7	24	46	40	76	3
Nitrogen						
Coniferous						
1985–2000	75	83	100	34	76	53
Deciduous						
1985–2000	52	55	86	21	47	36
Ozone						
1985–2000	1–2xCL	1.5–2.0xCL	2.0–2.5xCL	n.a.	1.5–2.5xCL	?

Notes: Data for sulphur and nitrogen are percentages of the total forest area where target loads for the pollutants are exceeded. Data for ozone are based on diurnal concentration distributions, April–September 1986.
n.a. = not available owing to insufficient data.
CL = critical load
For a quantification of the damage cycle and growth effects, see Nilsson *et al.* (1991b)
Source: Nilsson and Posch (1989); Nilsson (1991b).

European forests will still have depositions and concentrations exceeding the target loads for SO_2, NO_x, NH_3, and O_3.

Researchers in Berlin have developed an elaborate tool known as PEMU (Bellman *et al.*, 1991) for analysis of cause–effect relations between air pollution and forest conditions. Input data to the PEMU system are based on field observations made since the early 1960s at a set of test sites along deposition gradients. This analytical tool has been employed in estimating the decline effects on forests if depositions exceed target loads. Results from the PEMU system are expressed in terms of a damage cycle and growth losses. The international criterion for monitoring forest decline attributed to air pollutants is loss of foliage, different degrees of which define different decline classes. The damage cycle describes how many years a forest stand of a particular sensitivity class stays in different defoliation classes at a specific rate of pollutant deposition. Using the German system, it has been possible to generate quantitative estimates of the damage cycle showing that the decline process is more rapid for more sensitive sites and with increasing depositions. Growth effects are linked to the loss of foliage and have consequently also been estimated by the German system. From these results it can be suggested that:

- the nitrogen depositions compensate the individual decline effect of sulphur depositions up to the critical load limit for nitrogen (during the enrichment phase);
- during the disintegration phase (depositions above critical load limits) the combined effects of sulphur and nitrogen deposition lead to increased decline in comparison with individual effects of sulphur and nitrogen.

To mitigate the negative effects of the decline process in forests, some silvicultural measures can be taken. The objectives are to increase stand vitality, delay the decline process and save commercial wood. Examples of the measures used are intensified thinning, shortened rotation periods and changed species composition. The quantified mitigative measures are presented in Nilsson *et al.* (1991b).

THE TIMBER ASSESSMENT MODEL

A matrix-type simulation model has been built to generate scenarios of the development of forest resources. A detailed country-by-country data base for Europe has been assembled by the IIASA Forest Study to link with the simulation model. The model generates scenarios of growing stock and timber-harvest volumes over time by country (and sub-regions of a country) and by species group and age, making it possible to undertake a general timber supply assessment. Two major forest-model concepts were used within the matrix approach: (1) the unit area and its characteristics, and (2) the tree and its characteristics. The first concept is developed by Sallnäs (1990) and the second by Houllier (1989). The decline component described in the former section has been incorporated in these model concepts.

IMPACTS OF AIR POLLUTANTS
ON FUTURE EUROPEAN WOOD SUPPLY

Work using the Timber Assessment Model has been carried out for two types of conditions relating to air pollution: the 'with' (decline) and 'without' (no decline) effects. Simulations cover a period of 100 years (1985–2085). The decline simulations are based on the effects of sulphur and nitrogen without consideration of the effects of other pollutants. The emissions are only quantified and taken into account up to the year 2000. The impact of the air pollutants on future wood supply is presented in Table 6.

Using the calculations, it can be seen that the loss of potential harvest in Europe caused by emissions of air pollutants according to current reduction plans is about 112 million m^3 per year over 100 years. However, it should be stressed that the results for European USSR are preliminary.

TABLE 6   *Aggregated results concerning potential harvests*

	Potential harvest in million m^3 o.b./year for all species and average for 100 years		
	Without air pollution effects	With air pollution effects	Removals in 1987
Nordic	155	144	121
EEC-9	150	126	109
Central	25	19	22
Southern	78	71	72
Eastern	126	92	100
European USSR	272	242	264
Total Europe	806	694	688

Source: Nilsson *et al.* (1991a) and Sallnäs and Hugosson (1991).

ECONOMIC IMPACT

It is hard to visualize a political debate concerning concrete air pollution abatement strategies being based only on the volume effects, as described above. To obtain a more complete picture there is a need to express the decline in monetary terms. However, the economic calculation should not be the only foundation for the formulation of required abatement strategies.

The emission of air pollutants today will cause harmful effects, not only for the existing European generation, but also for those of the future. The long-term effects will influence both welfare and well-being. These aspects lead us to conclude that economic valuations must be conducted in a long-term, political – economic perspective. Thus the evaluation should be centred at the level of the national (political) economy. The basic approach to show the aggregate impact of different policies and effects on the national economy is to study changes in the income accounts of the GNP (OECD, 1990). Therefore the simple concept of adjusting the national accounts for the losses caused by air pollution has been followed. The starting-point for the valuation has been based on multiple-use forestry and the primary forest products industry. Multiple-use forestry, as a concept, involves taking into account the social welfare aspects of forestry. The value-added component for industrial production only takes into account the incremental losses caused by air pollutants and their effects on primary industries. The economic impact, according to this calculation, is presented in Table 7.

The economic impacts deal only with forestry. As illustrated earlier, other sectors can also be influenced. Based on work by Leipert and Simonis (1990) a rough estimate can be made, distinguishing between economic impacts on forestry and those on other environmental sectors. According to this estimate, the forest decline is between 11 and 18 per cent of the total economic impact of air pollutants. By employing this estimate, the total costs of air pollutants in Europe can be calculated to be in the range of 160–260 billion dollars per year. Of course, there are large uncertainties in this calculation.

TABLE 7 *Economic impacts on multiple-use forestry/primary forest products caused by air pollutants*

Region	Billion dollars/year
Nordic	2.9
EEC-9	7.4
Central	1.6
Southern	1.8
Eastern	8.5
European USSR	6.6
Total Europe	28.8

Source: Nilsson (1991c).

CONTROL EXPENDITURES

Shaw and Nilsson (1991) have analyzed the control expenditures needed to place all European forest resources within the target loads (see Table 4). With existing control technologies they found that to be impossible. The best result (most of the forests within the target loads) was achieved by implementing the best available control technology, for which expenditures are presented in

Table 8. Implementation of best control technology means a reduction by 70–80 per cent of SO_2 emissions and by about 60 per cent of NO_x emissions.

The costs for ammonia reduction are not calculated on a European scale, though Klaassen (1990b) has presented some calculations for the Netherlands. A 70 per cent reduction (which is the plan for the country) would cost 1.4 bilion dollars per year and a 30 per cent reduction 78 million dollars per year. Thus, to save the European forests from negative impacts, control expenditures of at least 90 billion dollars per year are required. However, if this could be achieved, most harmful effects in the other environmental sectors would also disappear (total economic impact: 160–260 billion dollars per year).

TABLE 8   *Control expenditures in billion dollars/year*

	SO_2 emissions	NO_x emissions
Best available control technology	50.2	39.7
Current reduction costs (plan to1995)	8.0	~8.0

Source: Shaw and Nilsson (1991).

INTERNATIONAL IMPLICATIONS

The major policy-related international implications are the following:

(1) *Doing what is planned is not enough.* It can be concluded from the analysis that the current reduction plans will be insufficient to halt negative impacts of air pollutants. Alcamo *et al.* (1991) stress that the air pollution problem may first appear in sectors other than forestry after the year 2000. Therefore it is crucial to take long-term environmental consequences into account when assessing the effectiveness of emission reduction plans.

(2) *The best will be expensive.* As illustrated above, the control expenditures will be very high. Many countries cannot afford the required measures. Therefore strong international cooperation is required to generate the necessary funds. It is obvious that implementing the best available technology will be worthwhile and have a high rate of return for European society.

(3) *International cooperation will save money.* Alcamo *et al.* (1991) have shown that international cooperation in implementing the new technologies is required and will save money.

REFERENCES

Alcamo, J., Shaw, R. and Hordijk, L. (eds), 1990, *The RAINS Model of Acidification. Science and Strategies in Europe*, Kluwer Academic Publishers, Dortrecht, Netherlands.

Alcamo, J., Shaw, R. and Hordijk, L., 1991, 'The RAINS Model of Acidification. Science and Strategies in Europe', *Executive Report 18*, International Institute for Applied Systems Analysis, Laxenburg, Austria, January.

Amann, M., 1989, *Potential and Costs for Control of NO$_x$ Emissions in Europe*, International Institute for Applied Systems Analysis, Laxenburg, Austria, November.

Amann, M. and Sörensen, L., 1991, *The Rains Energy and Sulphur Emission Database, Status 1991*, International Institute for Applied Systems Analysis, Laxenburg, Austria, May.

Bellman, K., Lasch, P., Schulz, H., Suckow, F., Anders, S., Hofman, G., Heinsdorf, P. and Kalweit, B., 1991, 'The PEMU Forest Decline Model. Cumulated Dose-Response Approach to Evaluate Needle Losses in Pine Stands of the North-East German Lowlands under S- and N-depositions', in S. Nilsson, O. Sallnäs and P. Duinker (eds)., *Potential Forest Resources in Eastern and Western Europe*, Springer Verlag, Berlin.

Chadwick, M.J. and Kuylenstierna, J., 1988, 'The Relative Sensitivity of Ecosystems in Europe to the Indirect Effects of Acidic Depositions', unpublished manuscript, The Beijer Institute, University of York, UK.

Houllier, F., 1989, 'Data and Models Used for France in the Context of the Forest Study', unpublished manuscript, French National Forest Survey Service and International Institute for Applied Systems Analysis, Laxenburg, Austria.

Klaassen, G., 1990a, 'Emissions of Ammonia in Europe', *Working Paper WP-90-68*, International Institute for Applied Systems Analysis, Laxenburg, Austria.

Klaassen, G., 1990b, 'Cost Functions for Controlling Ammonia Emissions in Europe', *Working paper WP-90-71*, International Institute for Applied Systems Analysis, Laxenburg, Austria.

Leipert, Ch. and Simonis, U.E., 1990, 'Environmental Damage – Environmental Expenditures: Statistical Evidence on the Federal Republic of Germany', *The Environmentalist*, vol. 10, no. 4.

Nilsson, S. (ed.), 1991a, *European Forest Decline: The Effects of Air Pollutants and Suggested Remedial Policies*, IIASA, InterAction Council and the Royal Swedish Academy of Agriculture and Forestry, Stockholm.

Nilsson, S., 1991b, 'Air Pollution in the European USSR' in S. Nilsson (ed.), *European Forest Decline: The Effects of Air Pollutants and Suggested Remedial Policies*, IIASA, InterAction Council and the Royal Swedish Academy of Agriculture and Forestry, Stockholm.

Nilsson, S., 1991c, 'Economic Impacts in S. Nilsson (ed.), *European Forest Decline: The Effects of Air Pollutants and Suggested Remedial Policies*, IIASA, InterAction Council and the Royal Swedish Academy of Agriculture and Forestry, Stockholm.

Nilsson, S. and Posch, M., 1989, 'Pollutant Emissions and Forest Decline in Europe', *Draft working paper*, International Institute for Applied Systems Analysis, Laxenburg, Austria.

Nilsson, S., Sallnäs, O. and Duinker, P., 1991a, 'Forest Potentials and Policy Implications: A Summary of a Study of Eastern and Western European Forests by the International Institute for Applied Systems Analysis', *Executive Report 17*, International Institute for Applied Systems Analysis, Laxenburg, Austria, Februrary.

Nilsson, S., Sallnäs , O. and Duinker, P., 1991b, *Potential Forest Resources in Western Europe*, Springer Verlag, Berlin.

OECD, 1990, *The Economics of Sustainable Development: A Progress Report*, OECD, Paris.

Sallnäs, O., 1990, 'A Matrix Growth Model of the Swedish Forest', *Studia Forestalia Swecica*, No. 183, Uppsala, Sweden.

Sallnäs, O. and Hugosson, M., 1991, 'The State and Future Development of Forests in the European USSR, in S. Nilsson (ed.), *European Forest Decline: The Effects of Air Pollutants and Suggested Remedial Policies*, IIASA, InterAction Council and the Royal Swedish Academy of Agriculture and Forestry, Stockholm.

Shaw, R. and Nilsson, S., 1991, 'Commercial Wood at Risk from Sulphur and Nitrogen Deposition', in S. Nilsson (ed.), *European Forest Decline: The Effects of Air Pollutants and Suggested Remedial Policies*, IIASA, InterAction Council and the Royal Swedish Academy of Agriculture and Forestry, Stockholm.

UN-ECE, 1988, 'Critical Levels Workshop Report', United Nations Economic Commission for
Europe, Geneva, Switzerland.

DISCUSSION OPENING – LAURI KETTUNEN*

Air pollution is a very suitable topic for an international conference because
no state can avoid its harmful effects. Pollutants spread with the wind far
away from national boundaries, so both polluters and neighbouring states
suffer the consequences. Most countries are polluters and, subsequently, must
address the problems of harmful side-effects, both at the national and international level. Although this conference is not making decisions, it is distributing information which will affect decision makers all over the world.

Dr Nilsson's paper deals with three issues: (a) air pollutants in Europe; (b)
their effect on the growth of forests; and (c) their estimated economic effects.
It is rooted in the system analytical work conducted by IIASA, which is
oriented towards the future. Analysis is based on well established knowledge,
and there can be no doubt about the importance of his message. I confess that
I found the material rather hard to understand and I would have welcomed the
use of maps to help get to grips with the rather abstract material. Nevertheless,
Dr Nilsson has provided a good overall view of the problem and of the
importance of reducing it, which, as an agricultural economist, I welcome.

The analysis indicates that SO_2 pollutants have decreased considerably in
Europe in 1980s. This is an indication that the protection of the environment
has produced results. During the 1990s, the situation will continue to improve, although uncertainties in forecasting exist for some areas. The results
of the IIASA study demonstrate that attaining pollution target levels is not
enough. Looking towards the year 2000, air pollutants might slow forestry
growth by up to 27 per cent, even if target levels are reached. This being the
case, future growth of forests will correspond to present felling levels.

The report states that damage caused by air pollutants costs 160–260 billion
dollars annually. It is unclear how this figure is estimated. Dr Nilsson points
out that European states might avoid these expenses by utilizing modern
technology. This approach would cost approximately 80 billion dollars a year.
How well decision makers understand these figures, and to what extent they
are willing to act on the recommendations, is difficult to determine. Investments are being made to improve the situation. Industry and heating and
electricity plants are continuously improving their technology, so we can
expect further improvements in reduction of pollutants. Whether these actions
are sufficient is another issue. It is also worth noting that Dr Nilsson only
provides the final results, and some explanation of the types of models behind
his calculations would have been helpful. For example, what are the confidence limits of the forecasts? I believe they are quite sizable. Is there any
sense then, in making 100-year forecasts?

As an example of some of the complexities involved in considering costs
and benefits, as they occur over time, experiences from my home town in

*Agricultural Economics Research Institute, Finland.

northern Finland, located close to the Soviet border, may be useful. My home town is situated in one of the poorest provinces, known as Hunger Land, where the unemployment rate is consistently the highest in Finland. Consequently, people have moved to southern Finland or abroad in search of better opportunities. In the 1970s, the construction of a large industrial complex, nearby in the Soviet Union, helped to improve the economic situation. The industrial complex planned to exploit large iron ore deposits in the region. Since this area in the Soviet Union was unpopulated, Finnish workers were recruited to build the whole processing plant and the housing complex. This provided employment opportunities for many men and women of my home town for over a decade, the quality of life improved, and the number leaving in search of other opportunities diminished. Today, however, the industrial complex is a big air polluter. The forests in my home town are suffering from the acid rain caused by this complex; even the trees in my small forest plot are suffering. For those people who originally built the factories, the blessing of the 1970s has become a burden in the 1990s. Finland has pushed the Soviet Union to clean up the whole industrial complex and other polluting industries located close to the Finnish border. We are providing economic assistance to help the Soviet Union in this task. Finland recognizes the sacredness of the environment and the consequences of industrial development, and more importantly, is willing to act.

Pollution can also have important effects on agriculture, both directly and indirectly. In Finland and neighbouring Scandinavian countries, forests play a significant role in the agricultural economy. In Finland, for example, forest income makes up about 11 per cent of farmers' total income, and any long-run decline in the growth of forests will have an indirect impact on agriculture. Furthermore, it is also important to consider the way in which air pollutants directly affect farming, and to note that the effects can be complex. Agriculture mainly suffers from the increase of ozone in the atmosphere. Some studies show that, of all air pollutants, ozone is the most harmful, causing about 80 per cent of the reduction in crop production. There are also problems with heavy metals and with polluted rain which increases the acidity of the land and requires more lime to be added to the soil. According to research done in Finland, the cost increase of the latter is small. It totals about 8 per cent of the annual liming costs, or about 0.6 per cent of fertilizer costs. The important point, however, is that the full balance sheet of effects is still imperfectly understood and requires more research. Part of the complication is that effects can be reciprocal. Agriculture can also pollute the environment. For example, ammonium deposits, produced only by agriculture, are harmful to forests. These pollutants account for about 20 per cent of the acid rain in Finland. The reduction in the number of dairy cows and of animals for fur raising has decreased the amounts of ammonium deposits. Further progress in reduction is possible through better feeding and housing of animals. Improved handling, storage and spreading of manure would also reduce pollution.

HERBERT STOEVENER*

Section Summary

Given the theme of this Conference, 'Sustainable Agricultural Development: The Role of International Cooperation', it is appropriate that the part of the programme dealing with environmental issues be focused, first of all, on the subject of sustainability. This topic has been given increasing attention during the past five years by professionals and the public in general. The plenary session was opened with a broad conceptual paper presented by Dr Sandra Batie, who reviewed the various schools of thought concerning sustainability. She contrasted the views of ecologists and economists, concluding that the policy positions associated with these various viewpoints differ greatly from one another. However, there appears to be concurrence among all of them to hold the values of a sustainable, humane, and just society as fundamental. Such a society has three pillars: economic stability, political stability and ecological stability. These goals are interdependent and can be in conflict with each other.

Dr Batie drew on the concept of 'adaptive management', a pragmatic, trial-and-error approach to environmental management. This approach focuses on the three pillars and provides continuous feedback on the economic, political and ecological costs associated with a given action. Instead of seeking a deterministic answer to the question of how much environmental protection is warranted, feed-back on the interaction of these goals calls for a continuous re-evaluation and respecification of the environmental protection strategy. Dr Batie does not find wide application of this approach at the current time. However, she sees it as holding promise for answering important questions of what to sustain and how to go about it.

The discussion opening by Richard Barichello expanded on many of the issues raised in the paper, mentioning in particular the need to combine scientific information, positive economic analysis and political economy in approaching sustainability issues. It was subsequently pointed out in floor discussion that environmental problems are so pervasive that they need to be considered in the formulation of what may appear to be 'higher-level' policies. For example, trade liberalization in grains and grain substitutes could lead to there being greater incentives in Third World countries to clear forests and strain water resources. Further, adaptive management requires more than a 'forward look' at new policy proposals, whatever they are; it needs to contain evaluation of the implementation of what has gone before. Analytic

*Virginia Polytechnic Institute and State University, USA.

techniques for implementation studies are already well tested in the development literature and could be utilized within Dr Batie's framework.

The second plenary session paper by Dr Ram Yadav was directed at the question of agricultural sustainability in primarily mountainous areas. In the management of these resources, Dr Yadav suggested the adoption of a 'mountain perspective' which includes considerations of the inaccessibility, fragility, marginality and heterogeneity of these areas, as well as their special cooperative advantages and human adaptive mechanisms, the results of which are embedded in a region's physical resources or social institutions. Dr Yadav evaluated development experiences in China, India, Pakistan and Nepal from the standpoint of this mountain perspective. Dr Yadav was also concerned with increasing resource demands faced by such regions. These stem primarily from population growth, increases in tourism and exogenous demand changes for minerals, hydropower and commercial timber.

The focus of the first invited paper session was on the management of common property resources. The management of fisheries has long been analyzed using this framework and was the subject of the first paper presented by Dr Trond Bjørndal. He introduced the subject with a description of current developments in the production and consumption of international fisheries. Both have increased substantially during the past two decades.

He then turned to the development of the common property framework, pointing out how the optimal levels of stock size recommended by biologists and economists diverge from one another under all but the most trivial economic circumstances. Without regulation there will generally be over-exploitation of the resource. The paper concluded with an application of this framework to the current institutional setting in international fisheries. Recently, individual countries have assumed jurisdiction over their coastal waters within 200 miles of their borders, where it is possible to regulate the total catch. Some issued individually transferable quotas to their fishermen. These can lead in the long run to an efficient and sustainable level of fishing effort.

In the next paper, common property resource concepts were applied to forest management, although Dr Kramer and Dr Ballabh prefer the term 'common pool', to avoid confusion with the actual ownership arrangements under which forested properties are held. They applied the common pool concept in three settings.

First, there are the commonly used forest resources managed through the Van Panchayats in northern India. This is potentially a flexible system with local user participation in decision making and maximum information feedback (to use Batie's concepts). They see possibilities for applying this system in other areas in India or perhaps the world. Unfortunately, recent initiatives of the state government have reduced the effectiveness of the local institutions.

They then commented on current moves to establish a system of parks and reserves in Madagascar to manage common-pool resources. They are concerned about the administrative capacity of the government to manage this scheme. The authors are now trying to measure the value of one park to tourists and to local residents, and to estimate the health benefits from public management of this common-pool resource.

Finally, the authors discussed a proposal to establish a Global Nature Fund for Forest Management and Protection. This proposal follows from the recognition that tropical forests represent a global resource and are under-managed mainly because of funding constraints in the developing countries. Financial support for the fund would come from the rich countries and would be received by those now eligible for aid from the multilateral agencies. This proposal would be most productively applied in those circumstances where the public goods nature of the services from forest environments is most pronounced. As pointed out by the discussion opener, there may be formidable political difficulties with this approach.

More generally, there was considerable discussion of the assumptions and wider applicability of the particular illustrations of resource management schemes within the papers. It was also emphasized that distributional issues should not be ignored. Allocation of property rights, auctioning of quotas, and framing appropriate tax and compensation schemes were all mentioned as key issues for examination. The papers also provoked discussion of appraisal methods and of the value of the 'new institutional economics'.

The final invited paper session consisted of a review of information relating to climate change and its agricultural consequences, prepared by Richard Adams and presented by Timothy Mount, and another on air pollution, the most pervasive internationally linked environmental problem, by Sten Nilsson, in which the focus was on forestry in Europe. It was clear from both papers that we have only very incomplete information about the physical relationships between air pollution and climate changes and production in agriculture and forestry. If we could list standard errors with our economic estimates, they would be very large. Among the tentative conclusions one might list from these two papers are the following:

(1) climate change is not likely to be a food security issue,
(2) under some climate change forecasts, the net economic effect might actually be positive.
(3) there are likely to be regional shifts in agricultural production patterns,
(4) costs of mitigating pollution problems in agriculture are likely to be high,
(5) there has been considerable progress in controlling air pollution in Europe,
(6) impacts of air pollution on forestry production are considerable, and
(7) much more air pollution control is possible, but using the best technology is expensive.

At this point, and in conclusion, it may be good to remind ourselves of Sandra Batie's proposal for an adaptive management approach. When so little is known about the physical–biological relationships with which we are concerned, when consequences of our actions may extend over many generations, and when we may have opportunities to adapt our technologies and social institutions in response to changing circumstances, we may be well advised to re-evaluate our environmental strategies on a continuing basis for their opportunity costs in terms of economic stability, political stability and ecological stability.

Chairpersons: Herbert Stoevener, Kenneth Farrell, Giuseppe Barbero.
Rapporteurs: Francisco Amador, Kenneth Thomson, Ewa Rabinowicz.
Floor discussion: P. Calkins, A. Dubgaard, D. Belshaw, A.Brun, B.J. Revell,
L.D. Smith, M. Merlo, R.Tiffin, D.R. Shah, T. Engelhardt, G.T. Jones, F.G.
Mack, H. Jensen, H. Alfons, M. Petit, L. Tweeten, L. Venzi, V. Zachariasse.

SECTION VI

Farm Households as the Dominant Institutional Units in Agriculture

ANTHONY IKPI*

Household Time Allocation – The Ultimate Determinant of Improved Agricultural Technology Adoption in Nigeria: An Empirical Activity Interphase Impact Model

INTRODUCTION

Nigeria's small-scale and resource-poor rural farmers are very conscious of the one production resource they possess and have complete control over: family labour (time). They know that food production in the continent has not been keeping pace with population growth, and that crop technological advances abound that could appreciably improve the situation. However, they have been equally aware of the time demands that accompany the adoption of such new improved agricultural technologies. They have, therefore, been rather careful in adopting available technologies. They carefully determine their household activities and attitudes towards improved technology adoption according to family time allocation considerations.

Because of their seeming reluctance to adopt available improved crop technologies that could dramatically increase food production in the continent, many observers have described them differently. For instance, whereas some classify average Nigerian farmers as irrational resource allocators who are conservative, ignorant and superstitious and so cannot operate a viable farming system (Aribisala, 1983), others still have enough confidence in them to use their experience and capacity to manage their meagre resources and produce food for the growing population most economically (Swaminathan, 1983; Hartmans, 1984). Many others attribute the poor performance of these small-scale, resource-poor farmers to unnecessary preponderant intervention in agricultural production by the continent's public sector (Olayide, 1976), increasing pressure of population growth, poor extension services and contact with farmers (Okigbo, 1983), increasing environmental degradation and adoption of non-sustainable agricultural practices (Eicher, 1985; Brown and Wolf, 1985) and insufficient investment in agricultural research and technology (Stifel, 1986).

Thus, although so many know something about the factors that cause failure in the agricultural systems of Nigeria, very few appear to understand and appreciate the real reasons for the average Nigerian farmer's reluctance to adopt available improved agricultural technologies. The problem really centres around farming systems research and intra-household dynamics in the

*University of Ibadan, Nigeria.

Nigerian rural farm family – especially with respect to the farmers' ability to manage time *vis-à-vis* the labour demands of the agricultural technology packages being offered by research centres.

RESEARCH METHOD

This study was designed to generate information and data that would facilitate an understanding of the Nigerian rural farmer and household so that agronomic and biological research designs of national and international agricultural research centres could focus more on relevant beneficiary-perceived farming needs. Emphasis was on the determination of the existing priority in rural household activities and family time allocation in order to support the estimation of an empirical time budget and the establishment of a realistic basis for predicting the impact of new technological packages on Nigeria's agriculture.

Consequently, the study was a participatory observation survey in which 420 representative households (with 1978 respondents in them having and operating farm plots) were selected from seven states of Nigeria and studied for nine months. The selection of these households was based on a four-stage random sampling procedure in which seven states with notable agro-ecological differences were selected. From these, a total of 18 representative local government areas (LGAs) were covered. In each selected LGA, five representative villages or village communities were selected from a complete village listing obtained from each state's Agricultural Development Programme (ADP) and/or the affected Local Government Secretariat. The final stage of random sampling involved the selection of the 420 households from a comprehensive listing of households in each included village or village community. Thus, the basic sampling frame for this study was a comprehensive village/household listing – the former obtained from ADPs or LGA secretariats, and the latter by enumerators themselves with the cooperation of the village head or chief prior to the survey proper.

In all, 24 male and 25 female enumerators were employed to administer the 38-page survey instrument (structured questionnaire) in the seven states. They were recruited in such a way that they came from their States/ LGAs of study, so that language of communication with respondents was never a barrier. They were trained for seven days and paired during interviews to facilitate quick rapport with the respondents. Arrangements were made for them to reside throughout the survey period with the selected households in the villages.

EMPIRICAL ANALYSIS AND FINDINGS

Rural household time allocation and activity clocks

Rural household activities are dictated by family time allocation which shows the order of importance of these activities. Although intercultural and interstate differences exist in the relative importance of, and time allocation to these activities, a detailed family time use analysis shows that there are basically three principal activity sectors within the rural household. These are the farming activity sector, the non-farming commercial activity sector, and the non-monetized home production activity sector. Whereas the first two ativity sectors are monetized, the third is not. Time for leisure in rural households is treated as a residual of the overall time available to household members after netting out the sum of the three activity sectors identified above. Because it is a residual, where individual activities are dependent upon whether its value is positive or zero, it is not treated or recognized as an activity sector within the household.

The farming activity sector encompasses all activities related to agricultural production starting from land preparation for crop and livestock production to the marketing of the agricultural output. Table 1 lists the activities in this sector and summarizes the average time in hours that is spent by family members and hired labour. The table shows that, for an average farm size of 6.10 hectares, a total of 6368 work hours is spent by family and non-family labour on all aspects of farm production in one cropping season. The gender disaggregation of the time input shows that male labour supplied 49.94 per cent of all the work hours (that is, 3180 hours), while female labour accounted for the remaining 50.06 per cent (or 3188 hours) of all the work hours. On an age basis, adults (male and female) accounted for 70.57 per cent (or 4494 hours) of the total, while children who were up to working age (male and female) accounted for the remaining 29.43 per cent (or 1874 hours) of the total work hours. At a time that the educational policy of the Nigerian government is calling for universal primary and secondary education, 29.43 per cent is too high a labour time demand on children for agricultural activities. It implies that, if children's time is withdrawn, most of the technological and financial impacts that are envisaged for most new improved agricultural technologies will be unrealized, or children will continue to be held back from going to school in order to provide child labour, for most of which they are grossly underpaid, if paid at all. Finally, Table 1 shows that family-supplied labour time accounted for 63.30 per cent, while hired labour accounted for 36.70 per cent, of all the work hours put into farming activities in the surveyed states.

The non-farm activity sector includes those endeavours of household members for the sole purpose of making money. They range from outright labour provision for a fee to direct trading in things not even related to agricultural production. Sometimes family members may engage in the trading of agricultural produce that is not from their own farms; such trading is not classified under agricultural produce marketing of the farming activity sector but comes under the non-farming commercial activity sector. Table 2 summarizes the six main activities identified under the non-farm commercial sector and indicates the average number of hours spent each year by members of the family.

Rural farmers spend on the average a total of 2212 work hours on non-farm commercial activities in a year. Thus the time spent by these same people on farming activities is approximately thrice that spent on the non-farm commercial sector. Among the six major non-farm commercial activities, trading takes the highest proportion (26.40 per cent) of the farmers' non-farming

TABLE 1 *List of farming activities and the combined average time (hours) spent on them per season per average total family farm size of 6.10 ha in the states studied, 1987/8*

			Family			Hired labour		
					Children			
Farming activities		Husband	Wife	Male	Female	Male	Female	Total
(i)	Land preparation							
	Land clearing	46	24	9	11	52	2	144 (2.26)
		(0.1)	(0.3)	(0.2)	(0.4)	(0.1)	(0.1)	
	Tree felling	21	12	10	13	25	14	95 (1.49)
		(0.2)	(0.6)	(0.4)	(0.9)	(0.2)	(1.1)	
	Farm burning	13	16	13	18	20	22	102 (1.60)
		(0.2)	(0.5)	(0.3)	(0.8)	(0.3)	(0.9)	
	Stumping and raking	18	54	19	24	24	45	184 (2.89)
		(0.3)	(0.9)	(0.7)	(2.3)	(0.3)	(1.6)	
	Land tillage	23	54	21	25	54	29	205 (3.22)
		(0.1)	(0.3)	(0.3)	(0.5)	(0.1)	(0.3)	
(ii)	Planting							
	Seed dressing	18	34	17	24	21	36	150 (2.36)
		(0.2)	(0.9)	(0.8)	(0.9)	(0.6)	(3.5)	
	Crop/seed planting	21	49	21	24	24	55	194 (3.05)
		(0.1)	(0.2)	(0.2)	(0.4)	(0.2)	(0.3)	
	Fertilizer application	19	11	10	11	22	12	85 (1.33)
		(0.1)	(0.3)	(0.3)	(0.5)	(0.3)	(0.5)	
(iii)	Weeding							
	Hand weeding	12	106	31	50	13	107	319 (5.01)
		(0.4)	(0.2)	(0.3)	(0.3)	(0.2)	(0.2)	
	Herbicide application	19	20	8	11	20	22	100 (1.57)
		(0.2)	(0.6)	(0.5)	(0.5)	(0.5)	(1.4)	
(iv)	Harvesting							
	Harvesting	182	186	77	43	189	202	879 (13.80)
		(0.1)	(0.2)	(0.2)	(0.6)	(0.2)	(0.3)	
	Collection/	44	82	50	74	44	74	368 (5.78)
	transport	(0.2)	(0.2)	(0.7)	(0.3)	(0.2)	(0.5)	
(v)	Processing							
	Direct processing	23	72	17	19	18	86	235 (3.69)
	(cassava)	(0.3)	(0.4)	(0.6)	(0.6)	(0.6)	(0.6)	
	Threshing	21	32	19	19	24	32	147 (2.31)
	(other crops)	(0.2)	(0.2)	(0.2)	(0.3)	(0.5)	(0.6)	
	Cleaning	24	23	26	22	—	26	121 (1.90)
	(other crops)	(0.2)	(0.2)	(0.4)	(0.5)	(—)	(0.4)	
	Crop grading	25	23	9	9	35	35	136 (2.14)
	(other crops)	(0.2)	(0.4)	(0.6)	(0.8)	(0.4)	(0.8)	
(vi)	Compound gardening	165	255	135	185	108	66	914 (14.35)
		(0.3)	(0.4)	(0.3)	(0.6)	(0.9)	(0.9)	
(vii)	Livestock husbandry:							
	Within home	160	150	440	220	560	130	1 660 (26.07)
		(0.1)	(0.2)	(0.4)	(0.9)	(1.1)	(0.3)	
	Slaughtering work	60	40	46	94	36	54	330 (5.18)
Total		914	1 243	978	896	1 288	1 049	6 368 (100.0)
Percentage of total		14.35	19.52	15.36	14.07	20.23	16.47	100.00

Note: Figures in parentheses below the hours spent on each activity are the calculated standard errors of the means, while those in parentheses beside the entries in the last (total) column are the calculated percentages.

Source: *Nigerian Rural Household Economics Field Survey*, 1988.

time, closely followed by the gathering of wild edible fruits (with 22.20 per cent). This is one activity in which time could be saved and transferred to the home production sector if both the yield and naira returns from the farming sector were sufficiently high to make wild food gathering unnecessary and/or uneconomical. Labour provision for a fee, interestingly, came fourth (with 17.09 per cent) indicating that the rural farmer would rather spend time on home production activities if income generated from the farming sector was sufficient for the daily needs. It is kept from the last position by fishing (10.67 per cent) and hunting wildlife (5.65 per cent) because children do not participate in those activities.

TABLE 2   *List of non-farm commercial activities and the combined average time (hours) spent on them per annum in the states surveyed, 1987/8*

	Time allocation (hours) Family				
			Children		
Non-farming activities	Husband	Wife	Male	Female	Total
Trading	143	154	120	167	584(26.40)
	(0.3)	(0.3)	(1.4)	(5.9)	
Handcrafts making	182	144	72	—	398.4 (17.99)
	(0.8)	(—)	(0.3)	(—)	
Hunting wildlife	125	—	—	—	125 (5.65)
	(0.5)	(—)	(—)	(—)	
Fishing	82	154	—	—	236 (10.67)
	(0.5)	(0.6)	(—)	(—)	
Gathering wild edible fruits	124	144	103	120	491 (22.20)
	(0.6)	(0.6)	(0.8)	(1.3)	
Labour provision	63	140	85	90	378 (17.09)
	(0.2)	(0.3)	(0.9)	(0.9)	
Total	719	736	380	377.4	2 212.4 (100.0)
Percentage of total	32.51	33.27	17.18	17.04	100.00

Note: Figures in parentheses below the hours spent on each activity are the calculated standard errors of the means, while those in parentheses beside the entries in the last (total) column are the calculated percentages.

Source: *Nigerian Rural Household Economics Field Survey*, 1988.

The non-monetized home production activity sector covers those activities that do not fall into the other two sectors but which relate to home care and maintenance, food preparation and childcare. Detailed activities of this sector are summarized in Table 3, along with the gender and age disaggregated time spent on them by an average household. According to the table, wives alone contribute 42.23 per cent (or 750 hours) of the total 1776 work hours per

TABLE 3   *List of home production/consumption activities and the combined average time (hours) spent on them per annum by an average family of 4 working members in the states surveyed, 1987/8*

Home productivity consumption activities	Husband	Wife	Children Male	Children Female	Total
Food preparation					
Peeling	20	55	25	30	130 (7.32)
	(7.0)	(0.1)	(.05)	(0.1)	
Grating	10	40	18	30	98 (5.52)
Grinding (pepper, etc.)	5	10	5	7	27 (1.52)
	(0.2)	(0.3)	(0.1)	(—)	
Dehusking/milling	15	20	15	20	70 (3.94)
	(1.2)	(0.1)	(0.3)	(0.1)	
Pounding	15	35	15	20	85 (4.79)
	(0.1)	(—)	(0.1)	(0.1)	
Cooking					
Breakfast	10	40	15	20	85 (4.79)
	(0.1)	(0.9)	(—)	(0.1)	
Lunch	10	35	10	20	75 (4.22)
	(0.2)	(0.1)	(0.1)	(0.1)	
Dinner	12	45	15	20	92 (5.18)
	(0.1)	(0.6)	(—)	(0.1)	
Dish and pot washing	5	20	10	15	50 (2.82)
	(0.7)	(0.6)	(—)	(0.1)	
Firewood gathering	35	35	15	20	105 (5.91)
	(0.1)	(0.1)	(0.1)	(0.1)	
Water fetching	10	35	30	35	110 (6.19)
	(0.1)	(0.1)	(0.1)	(0.1)	
Childcare:					
Washing and dressing child	12	75	15	30	132 (7.43)
	(0.3)	(0.1)	(0.5)	(0.4)	
Feeding	8	35	15	20	78 (4.39)
	(0.2)	(0.1)	(0.1)	(0.1)	
Child petting, mothering	15	80	20	35	150 (8.45)
	(0.4)	(0.3)	(0.4)	(0.3)	
Tending the sick	7	25	5	10	47 (2.65)
	(0.5)	(0.2)	(—)	(0.1)	
Clothes washing and ironing	12	20	10	15	57 (3.21)
	(0.1)	(0.1)	(0.1)	(0.1)	
Home Maintenance					
House cleaning	40	90	40	10	180 (10.13)
	(0.3)	(0.3)	(0.1)	(0.1)	
Fence repair	25	20	10	5	60 (2.38)
	(1.1)	(0.3)	(0.4)	(0.7)	
House repairs/construction	35	20	15	10	80 (4.50)
	(0.2)	(0.4)	(0.7)	(1.3)	
Digging/repairing latrines	10	5	5	5	25 (1.41)
	(0.4)	(0.5)	(0.9)	(1.9)	
Other maintenance	15	10	8	7	40 (2.25)
	(0.3)	(0.3)	(0.3)	(0.3)	
Total	326	750	316	384	1776 (100.0)
Percentage of total	18.36	42.23	17.79	21.62	100.0

Note: Figures in parentheses below the hours spent on each activity are the calculated standard errors of the means, while those in parentheses beside the entries in the last (total) column are the calculated percentages.

Source: *Nigerian Rural Household Economics Field Survey*, 1988.

annum, while husbands contribute 18.36 per cent (or 326 hours) of the total home production work hours. The children put in a total of 39.41 percent (or 700 hours) of all home production time. Taken on a gender basis, male members of an average household contribute 36.15 per cent (or 642 hours), while the female members contribute 63.85 per cent (or 1134 hours) of the work hours needed for home production. Thus the wife of such a household alone contributes more to home production than do all the male members put together. On an activity basis, Table 3 shows that the three principal activity groups in the home production sector are food preparation (which takes up 40.10 per cent of the family time), followed by home maintenance (which consumes 21.67 per cent of the time), and childcare (which accounts for 20.27 per cent of the time spent on home production/consumption).

Household activity clock

The usual assumption when attempting to assess rural farmers' output is that they spend 12 hours in their farms doing nothing else but farming. Practical experience, however, shows that the household activity clock of a rural farmer is made up of many activity segments whose dimension depends on the day of the week, the gender of the farmer and his or her religious affiliation. There are also some discernible inter-state differences when a detailed data-pool is analysed.

Figures 1 to 3 show the generalized time use pattern or activity clock of households Monday to Saturday and on Sunday using the aggregated data from the seven states surveyed.

With minor inter-state differences, the generalized 24-hour activity clocks for Nigerian rural farmers show that, on the average:

(1)   Nigerian rural farmers (male and female and Christians and Muslims) stay awake for 16 hours each day from Monday to Saturday. On Sundays, however, the Christian farmers sleep longer, with the men staying awake for 14 hours and the women for 15 hours. Both men and women wake up on Mondays to Saturdays around 5.30 am and stay up till 9.30 pm. On Sundays, the women wake up by 6.00 am, while the men generally get up around 7.00 am.

(2)   The general order of activities, Monday to Saturday, for *male Christians* is:

5.30 am to 7.00 am: morning prayers, greeting of neighbours and break-
                             fast;
7.00 am to 7.30 am: checking of traps;
7.30 am to 1.00 pm: morning farmwork;
1.00 pm to 4.00 pm: lunch, rest and relaxation in the farm;
4.00 pm to 6.00 pm: evening farmwork;
6.00 pm to 7.00 pm: setting of traps and fetching of firewood;
7.00 pm to 9.00 pm: dinner, visiting of friends;

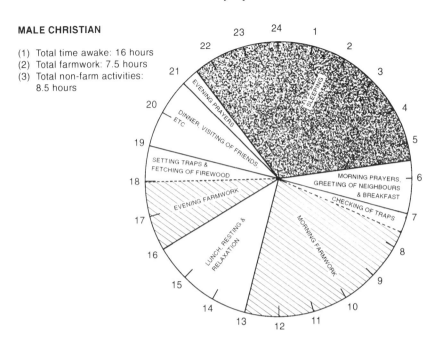

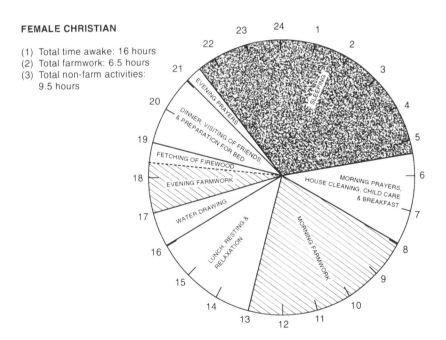

FIGURE 1   *Household activity clocks in selected states of Nigeria,*
Monday to Saturday

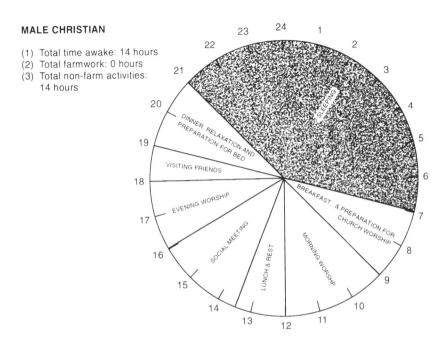

MALE CHRISTIAN

(1) Total time awake: 14 hours
(2) Total farmwork: 0 hours
(3) Total non-farm activities: 14 hours

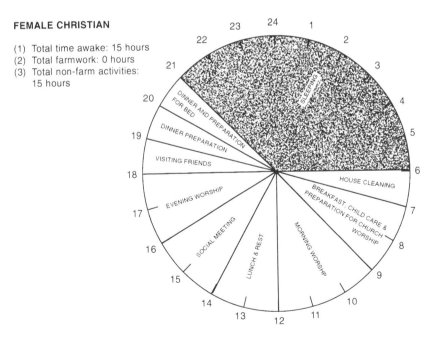

FEMALE CHRISTIAN

(1) Total time awake: 15 hours
(2) Total farmwork: 0 hours
(3) Total non-farm activities: 15 hours

FIGURE 2 *Household activity clocks in selected states of Nigeria, Sundays*

490 *Anthony Ikpi*

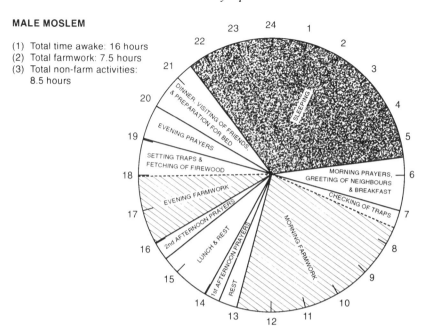

MALE MOSLEM

(1) Total time awake: 16 hours
(2) Total farmwork: 7.5 hours
(3) Total non-farm activities: 8.5 hours

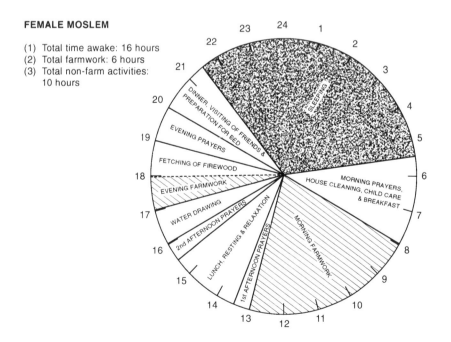

FEMALE MOSLEM

(1) Total time awake: 16 hours
(2) Total farmwork: 6 hours
(3) Total non-farm activities: 10 hours

FIGURE 3 *Household activity clocks in selected states of Nigeria, Monday to Saturday*

9.00 pm to 9.30 pm: evening prayers and preparation for bed; and
9.30 pm to 5.30 am: sleeping.

For the *female Christians*, the general pattern of activities, Monday to Saturday, is very similar to that of the male, but with a few differences:

5.30 am to 8.00 am: morning prayers, house cleaning, childcare and breakfast;
8.00 am to 1.00 pm: morning farmwork;
1.00 pm to 4.00 pm: lunch, rest and relaxation in the farm;
4.00 pm to 5.00 pm: drawing of water from stream or well;
5.00 pm to 6.30 pm: evening farmwork;
6.30 pm to 7.00 pm: fetching of firewood;
7.00 pm to 9.00 pm: dinner, visiting friends;
9.00 pm to 9.30 pm: evening prayers and preparation for bed; and
9.30 pm to 5.30 am: sleeping.

For the *male Moslems*, the Monday to Saturday (excluding Friday) daily routine is:

5.30 am to 7.00 am: morning prayers, greeting of neighbours and breakfast;
7.00 am to 7.30 am: checking of traps;
7.30 am to 1.00 pm: morning farmwork;
1.00 pm to 1.30 pm: rest;
1.30 pm to 2.00 pm: first afternoon prayers;
2.00 pm to 3.30 pm: lunch and rest;
3.30 pm to 4.00 pm: second afternoon prayers;
4.00 pm to 6.00 pm: evening farmwork;
6.00 pm to 7.00 pm: setting of traps and fetching of firewood;
7.00 pm to 8.00 pm: evening prayers;
8.00 pm to 9.30 pm: dinner, visiting friends and preparation for bed; and
9.30 pm to 5.30 am: sleeping.

For the *female Moslems*, the Monday to Saturday (excluding Friday) time use pattern is:

5.30 am to 8.00 am: morning prayers, house cleaning, child care and breakfast;
8.00 am to 1.00 pm: morning farmwork.
1.00 pm to 1.30 pm: first afternoon prayers;
1.30 pm to 3.30 pm: lunch, rest and relaxation in the farm;
3.30 pm to 4.00 pm: second afternoon prayers;
4.00 pm to 5.00 pm: drawing of water from stream or well;
5.00 pm to 6.00 pm: evening farmwork;
6.00 pm to 7.00 pm: fetching of firewood;
7.00 pm to 8.00 pm: evening prayers;
8.00 pm to 9.30 pm: dinner, visiting friends and preparation for bed; and

9.30 pm to 5.30 am: sleeping.

On Fridays, the afternoon prayer times take one hour each instead of the usual 30 minutes, so that there is no evening farmwork in most cases. Like their male counterparts, Moslem women devote two hours to afternoon prayers on Fridays and usually do not perform any evening farmwork on Fridays.

For Christian farmers (male and female), Sundays are primarily for worship and socializing throughout the 14 to 15 hours the farmers stay awake. No farmwork is undertaken at all on Sundays.

(3)   Total farmwork hours range from 6 hours for Moslem women, through 6.5 hours for Christian women to 7.5 hours for both Moslem and Christian males.

(4)   Non-farm activities (including both commercial and home production ones) take between 8.5 hours for Christian and Moslem men and 10 hours for Moslem women; Christian women average about 9.5 hours on non-farm activities.

Although, these activity clocks suggest some element of rigid routine, there is a lot of flexibility in the time scheduling of rural farmers, except when it comes to farmwork related to certain specific practices like land preparation, planting and weeding, when they do not like other activities to disturb them. The relative amount of time spent on each of these identified activities actually depends on the pressure of work and the interest of the farmer in the job, but, where 'emergencies' do not arise, one could literally tell the time of day in rural areas by examining what a farmer is doing.

Rural household activity interphase impact model

Data generated from the household activity/time allocation study provide a basis for impact modelling. For example, from the evidence above, it becomes obvious that, although it is normally considered to be primarily agricultural, a rural household in Nigeria is basically an economic entity comprising three principal components; a farm firm component, a non-farm entrepreneurial firm component and a home production and consumption firm component.

The farm firm component usually carries out farming activities from which the entire household not only feeds itself but also derives income from the sale of some of its produce to meet other family needs. On the other hand, the non-farm enterprise component normally conducts non-farming commercial activities that principally generate funds with which farm incomes are supplemented to meet household expenses. The home production/consumption component basically carries out non-monetized home management activities that are necessary for the existence of the household as a unit. The first two components can thus be seen to be economic activity sectors, while the third may be viewed more as a social activity sector, but, because they are all so closely related, competing for time and decision making from the same individual(s) (especially the household head) in the family, the behavioural principle of the rural household as an entity is one of the maximization of the following utility function:

$$U = U(h,y)$$

where h is the amount of time in hours per year that is at the disposal of all family members up to working age; and y is the total household money income obtained during the same year by working members of the household.

The basic assumptions of this utility function are that: (1) the marginal utility of labour, U_h, is negative (that is $U_h < 0$) since labour in use causes physical and/or mental pain – a direct disutility; and (2) the marginal utility of the money income earned, U_y is positive (that is $U_y > 0$) since money in use results in pleasure (a direct utility) (see Nakajima, 1986). These two simple assumptions therefore make it imperative that time available to a rural household be shared out among the component sectors in such a way that maximum benefits are derived whenever an externality is introduced into the household that tends to destabilize its equilibrium time allocation among its activities.

However, the individual maximization behaviour of each of the three component sectors of the rural household differs from this general utility maximization principle. For instance, the farm firm component is basically an economic production unit whose central interest is to maximize the total farm output of the household. Similarly, the non-farm enterprise component is an economic unit whose basic objective is to maximize the income earned from those non-farm commercial activities including the provision and 'sale' of household members' time as direct labour to outsiders. Finally, the home production and consumption component of the household acts more as a social buffer than an economic coordinating unit whose objective function is to maximize the utility of family time involved in home production, although it is not paid for in cash.

In other words, using the Marshallian concept of 'economic surplus' (that is, economic benefits over costs obtained by economic entities or units from their economic activities), the first household component (the farm firm) continuously tries to maximize its *producer's surplus* resulting from its farming activities, while the second household component (the non-farm commercial enterprise) attempts to maximize its *labourer's surplus* resulting from its commercial labour supply activities. The third component attempts to maximize its *'prosumer's surplus'* which is obtained from the family's home production and consumption activities. In order to explain how all three surpluses can be maximized, it is necessary to describe how the three component firms are brought together in a juxtaposition determined by the commonality of household decision making. The result of such a juxtaposition is what constitutes the empirical rural household activity interphase impact model shown below in a diagrammatic form.

To get the exact relationship between the three component sectors, each activity sector is represented by a circle whose relative area is determined by the proportional distribution of work hours spent by the household on that sector. Since all three activity sectors are controlled by the same decision maker(s) and their basic factor relationship is defined by time (hours), the relative positions of the centres of the circles representing these sectors have to be equidistant from one another and so are delineated by the vertices of an equilateral triangle. With each vertex of the triangle acting as the centre of one of the circles, the three sectors are drawn in according to their proportionally determined areas. Under such geometrical construction, the three activity sectors interlink and create interphases at three different overlapping areas. Each interphase defines a predetermined impact area whose size gives, in a pictorial format, the relative magnitude of that impact area.

Figures 4 to 6 show the three typologies of activity interphase impact models derived respectively for Ondo and Oyo States (Figure 4), Imo and Katsina States (Figure 5) and Bauchi, Cross River and Kwara States (Figure 6). Categorized information received from farmers shows that there are four principal impacts that an exogenous factor like improved crop technology can create if introduced into a rural household and is imbibed or used to such an extent that it affects the household's existing time allocation equilibrium. These are the *technological, social, financial* and *economic* impacts (Gittinger, 1984). Each principal impact interphase is a composite of several sub-impact indices.

For instance, the technological impact index comprises sub-indices such as: the rate and level of adoption, the level and mode of land development resulting from the adoption of the new technology, the yield of (or resulting from the introduction of) the improved variety of species, the resulting resource allocation demands such as labour flows, and the level of secondary and tertiary post-harvest infrastructure for food handling and processing. The social impact index is a composite of the volume of employment created by the newly introduced factor, the income-generation and distribution capacity caused by the factor among gender and class, the quality of life, the effect on the nutritional and health statuses of the affected people, the effect on the environment, the creation of markets for the resulting new products, and the effect on household, regional and national food security. The financial impact index comprises the change in output prices emanating from the new intro-duction, the effect of the introduction on input cost subsidization, and the level of government price support on output, if any. The economic impact index consists of the value and level of product prices, and the contribution to the gross domestic product (GDP) (of the community or state) resulting from the new introduction.

The interphase created by the interlink of the farming and non-farming commercial activity sectors constitutes the technological impacts, while that created by the farming and home production activity sectors is the social impacts. The interphase between the non-farming commercial and home pro-duction activity sectors constitutes the financial impacts and the common grounds (the union) of these first three impact interphases are what constitute the economic impacts.

Agricultural technology adoption primarily affects the farming activity sec-tor (or the farm firm component) of the household. Simultaneously, however, because of the dynamic relationship of this sector with the rest of the household

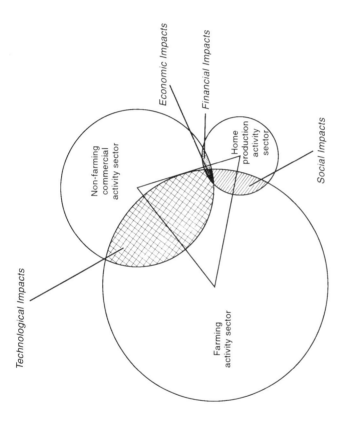

Technological Impacts

Non-farming commercial activity sector

Economic Impacts

Financial Impacts

Home production activity sector

Farming activity sector

Social Impacts

FIGURE 4   *Rural household activity interphase impact model, Ondo and Oyo States*

495

sectors, once a newly introduced improved agricultural technology is adopted by that household, it constitutes an exogenous factor which immediately disorganizes the existing household time allocation equilibrium. The net effect is a shift in time use patterns within the household but between the three activity sectors, resulting in the creation of new sizes of all the major impacts – the technological, social, financial and economic. The magnitude of shock created by the introduced externality determines the new impact equilibrium and the size of the resultant impact indices that are now established.

Thus, in Figure 4, we see that, where the introduction and adoption of some new agricultural technologies had caused the household time to be distributed among the farming, non-farming and home production sectors in the approximate ratio 3:2:1 (as is the case in Ondo and Oyo States), there is created a disproportionate impacting on the household. In other words, the time demands of the new technology concentrated relatively too many hours in the farming sector and caused what superficially might have appeared an attractive technological impact (such as yields increasing dramatically, or more land being brought under cultivation). But then the accompanying social, financial and economic impacts are so relatively small that desired economic welfare effects are not achieved in those states. If this continues, the farmers will in future years subsequently cut down on the adoption of the new technology.

There are two immediate possible explanations for this outcome. One is that the new technology was not time-saving, with the result that in order for it to be adopted, the household had to withdraw time from the home production sector. Secondly, the new technology may by itself have been rather expensive to acquire and/or involved the complementary and simultaneous adoption of other expensive 'attachments', so that all financial and economic benefits were drastically reduced, leaving behind a poor producer's surplus.

Figures 5 and 6 present other actual typologies of impact that were observed in the other states surveyed. In each case, this household activity interphase impact model is sending out a strong research policy implication message or signal: that each new agricultural technology is intended ultimately to benefit the adopter by greatly enhancing general economic and social welfare. This can only be done by increasing the level and magnitude of financial, economic and social impacts without necessarily losing much on the technological impact indices. This will require a major shift in the number of hours spent on the farming actitivy sector to, especially, the home production activity sector. In other words, the crop technologies, for instance, that have been found to be high-yielding and resistant to diseases should be sufficiently time-saving that their adoption will preferably release the farmer's time from the farm firm sector to the home production one, or at best not upset any previously established time allocation equilibrium that gave a better economic and financial reward to the adopting family or household.

To arrive at some optimum time allocation between the three principal activity sectors of the rural household, there has to be a systematic and measured transfer of hours from the farming to the non-farming, and/or home production sectors. Theoretically, such optimum time allocation between the household sectors will require the determination of a time contract or conflict curve which will be the locus of all points of tangency between one sector's

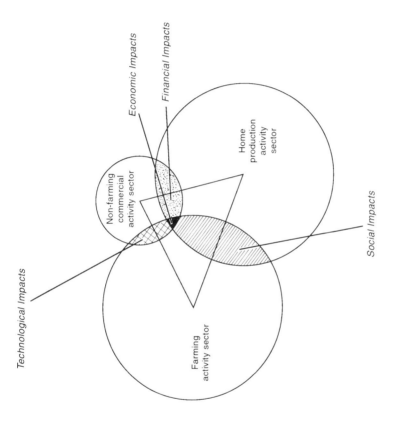

FIGURE 5   *Rural household activity interphase impact model, Imo and Katsina States*

497

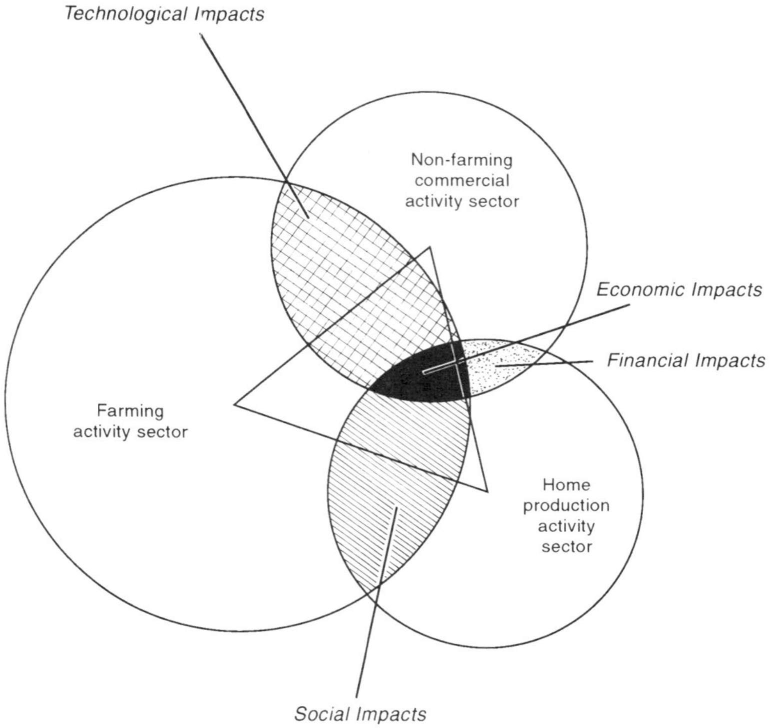

FIGURE 6 *Rural household activity interphase impact model, Bauchi, Cross River and Kwara States*

time indifference curves and those of another sector, given a particular technology being adopted. In practical terms, however, the closest approximation to this optimum condition will be the point where the marginal cost of transfer (or hour substitution) of farming to (or for) non-farming hours is zero.

The average income-generating capacity of rural households under existing conditions and present component activity analysis shows that, for all the surveyed states combined, the average farm income generated by a respondent household amounted to ₦ 5395.20 (US$830.00) per annum, while the average non-farm income generated by the same household from commercial activities was ₦ 1384.51 (US$213.00) per annum. With the hours put into farming activities by the average household totalling 6368 hours and those spent on non-farming commercial activites amounting to 2212 hours, an hour of farming activites brought in ₦ 0.85 (US$0.13) to the household, while an hour of non-farm commercial activities grossed ₦ 0.63 (US$0.10) to the same household. This immediately suggests that 'extra' hours from the farming sector should not be transferred to the non-farming commercial sector, since the marginal rate of hour transfer (or substitution) from farming to non-farming enterprise will be negative.

The most rational thing to do will then be to implement the earlier suggested use of 'extra' time (hours), namely to move such time from the farm production sector into the home production sector or even add it to the leisure time of the household. With such hour transfer, the relative adjustment factor between the various impact indices shows that the technological impacts will reduce a little while the social, financial and economic impacts will increase. The exact amount of increase of the latter indices will depend upon the criticality of those aspects of the farming activity time savings that are actually transferred to the home production sector. For instance, time savings from those practices such as weeding and fertilizer/pesticide application that usually demand a peak labour profile at the same time as some non-farm commercial and home production activities will be more critical and desirable than those from, say, land preparation.

CONCLUSION

Such savings in time can only come from the derived advantages of using new improved agricultural technology packages from research centres like the International Institute of Tropical Agriculture (IITA) in Ibadan, Nigeria. If such time savings cannot be achieved by crop breeder scientists, the alternative will be that agricultural technology packages from centres like the IITA and other international and national agricultural research centres will have to increase the relative net farm income of rural households to at least one and a half times its present level in order to induce the farmers to adopt such technologies. Given the fact that prices of agricultural produce are usually low and fixed outside the rural households' control, this means that existing yield levels of improved crop varieties will have to be increased more than three-fold in the farmers' fields when cultivated within the traditionally established cropping systems of rural Nigeria.

REFERENCES

Aribisala, T.S.B., 1983, Nigeria's Green Revolution: Achievements, Problems and Prospects',
     Distinguished Lecture No. 1, Nigerian Institute of Social and Economic Research (NISER),
     Ibadan, Nigeria.
Brown, L.R. and Wolf, E.C., 1985, 'Reversing Africa's Decline', *Worldwatch Paper 65*,
     Worldwatch Institute, Washington, DC.
Eicher, C.K., 1985, 'Famine Prevention in Africa: The long view', *Food for the Future: Pro-
     ceedings of the Bicentennial forum*, Philadelphia Society for Promoting Agriculture, Phila-
     delphia, USA.
Gittinger, J.P., 1984, *Economic Analysis of Agricultural Projects*, 2nd edn. Economic Develop-
     ment Institute of the World Bank, Washington, DC.
Hartmans, E.H., 1984, 'Prospects for Nigerian Agriculture, *Distinguished Lecture No. 2*, NISER,
     Ibadan, Nigeria.
Moock, J.K.L., 1986, *Understanding Africa's Rural Households and Farming Systems*, Westview
     Press, Boulder, Co. and London.
Nakajima, C., 1986, *Subjective Equilibrium Theory of the Farm Household*, Elsevier, Amster-
     dam, Oxford, New York and Tokyo.
Okigbo, P., 1983, 'Planning the Nigerian Economy for less Dependence on Oil', *Distinguished
     Lecture No.3*, NISER, Ibadan, Nigeria.
Olayide, S.O., 1976, 'The Food Problem: Tractable or the Mere Chase of the Mirage?' Inaugu-
     ral Lecture, 1975/76, University of Ibadan, Nigeria.
Stifel, L.D., 1986, 'Director-General's Report, in *IITA's 1985 Annual Report and Research
     Highlights*.
Swaminathan, M.S., 1983, 'Agricultural Progress – Key to Third World Prosperity', *Third World
     Quarterly, 5(3) pp.55–66*, (Third World Lecture).

DISCUSSION OPENING – WILLIS OLNOCH-KOSURA*

Professor Ikpi deserves congratulations for making a bold attempt to revisit
the important, but somewhat neglected, issue of time allocation in households,
considered to be both production and consumption units. To my knowledge,
since Becker's theory of time allocation was published in 1965, very few
empirical studies have appeared in anything like the detail provided for us. I
think part of the reason for that neglect is the complexity of the field work
required. It requires careful organization and much patience for good results
to be obtained. This is evident from the research method used in Ikpi's study
in Nigeria, where we find a participatory observation survey of 420 households (1978 respondents), a 38-page structured questionnaire, and the need
for enumerators to reside with the selected village households. No wonder
support for such studies has been limited.

Nevertheless, it is clear that relevant work has great potential for providing
valuable results, as Ikpi's study shows. For instance, until now, there have
been those who think that there is surplus labour in contemporary developing
countries and that agricultural technologies which are developed should be
labour-intensive (or labour-using). To the extent that most agricultural operations in rain-fed agriculture are 'time-bound', if they are to be optimal, the
reason for lack of effectiveness of the green revolution in some parts of Africa
may lie in lack of understanding of the mechanics of household time allocation.

*University of Nairobi, Kenya.

When one travels casually through some villages in Africa, one may get the impression that a great deal of labour is available but that it is under-utilized. However, transforming apparently idle hands into effective farm labour requires the individuals concerned to calculate their own opportunity cost.

Ikpi has divided time allocations into three groups: farm activity, non-farm activity and non-monetized home production activity. I would suggest that these activities be collapsed into two in order to distinguish only monetized and non-monetized activities. In effect his category of non-monetized home production activity actually either facilitates commercial production on the farm or elsewhere, or home consumption. Thus the opportunity costs of the under-utilized labour seen in villages may be high or low, depending on the season, in rain-fed agricultural systems. To the casual observer many individuals who are doing little more than loitering, both during slack agricultural seasons and at other times, may represent surplus labour which could be exploited for agricultural work. Yet these individuals may genuinely be enjoying well deserved leisure after working extremely hard during the peak season. In any case, working in the field for more than five hours a day in the tropics is almost impossible. Improved technology should therefore be aimed to fit within the framework of known activity clocks among individuals or communities. If the activity clocks are not recognized, the rate of technology adoption will be relatively low, as Ikpi has demonstrated in his paper. This will of course reinforce the other reasons for lack of adoption, such as lack of awareness, high cost, or complexity of technology.

The other potential contribution of this type of study is in the area of differences in productivity by age and gender. For a long time, farm management specialists have been arbitrarily allocating productivity weights in such a way that women and children are weighted at providing half the amount of work provided by men. Such biases can be removed by fitting production functions, using the type of data collected by Ikpi, to determine the marginal productivity of each category of individual. This is not done by Ikpi, but it could be a worthwhile extension. In any case, it may add weight to the call for more support for time allocation studies, since there is current emphasis on the way to improve the welfare of women who are actually major participants, as producers and managers, in smallholder farms.

Ikpi's finding that children devote 30 per cent of their time to farmwork has serious implications for education by diverting them from schooling. It also has an important bearing on the whole question of the 'demand' for children by rural households, and the consequent high rate of population growth. If available technologies are time-using during critical periods of farm production, there may be pressure on households to have more children to support farmwork. This also explains why children are not 'inferior goods' in the new household economics framework.

I would like to conclude with a plea to researchers and research-funding agencies to take greater interest in time allocation studies if agricultural technology adoption is to be increased in farm households.

GÜNTHER H. SCHMITT*

*The Theory of Resource Allocation by Farm Households: The Role
of Off-farm Employment, Household Production and Transaction Costs*

INTRODUCTION

'The process of economic development is in part a shift from household production to market production', as Locay (1990) has explained. This 'partial shift', in fact, has three different, although very often mutually interrelated, dimensions. First, numerous economic activities originally carried out within households have been transferred to emerging firms, resulting in an institutional division of labour between households supplying resources such as labour and capital, and firms demanding resources. Supply and demand of resources are coordinated by factor markets, which therefore have a role to play in affecting the economic activities of households and firms. Basically, the division of labour is affected by technological innovations which are reflected in lower costs of production due to economies of scale, requiring inputs of labour exceeding the amounts which could be supplied by a single household.

Second, in contrast to firms employing more workers than can be provided by a single family, there may be other 'firms', even in advanced economies, whose optimal labour input does not exceed, or even falls short of, the labour capacity of single households. This can occur when economies of scale of size in production of specific goods and services are rather restricted, and/or when household labour capacity has been enlarged by labour-saving technological innovations. Instead of large-scale production resulting in an institutional division of labour between households and separate firms, in cases of small-scale production the household and the 'firm' remain as an integrated organizational unit, conveniently designated as a 'family firm', or 'family managed firm'. In agriculture the concept is that of a family farm.[1] In fact, official statistics reveal that the family farm – typically worked jointly by a married couple and their children, or, in many countries by members of an extended family who live together in a single household – is the dominant form of agricultural organization in the United States and in most non-socialist developed and developing countries (Pollak, 1985). Family firms or farms are, therefore, not subject to an institutional division of labour, but only to a functional division of labour with respect to the production of various goods and services.

*University of Göttingen, Germany.

Third, all households whether they are only supplying resources to relevant markets, or supplying goods and services to commodity markets, combine family labour with goods and services bought either on markets or produced by the family firm or farm in order to obtain goods and services which the household ultimately desires (household production). Since Becker's (1965) 'new home economics' household production is seen as the economic foundation of households behaviour. Household production plays an important role even in developed countries (Eisner, 1988) and is almost by definition a concomitant of family farming.

Family farming, therefore, implies that farm households have to be seen as the decisive institution organizing farm and household production. This implies, furthermore, that farm family resources are allocated to farm as well as household production. In fact, however, use within the family will compete with employment outside the farm and farm household and thus part-time farming comes to the fore. By extending Becker's non-farm household model to farm households, it can easily be demonstrated that equilibrium in the allocation of household members' time is achieved if the marginal labour product in each of all three competing forms of employment is equal and equal to the marginal utility of leisure time.

In this paper I will present only the fundamentals of the economic theory of farm households, because it is already available (Schmitt, 1989b, 1990a and 1990b). The empirical relevance of that theory, rather than a theory of 'farms as firms', will be demonstrated by analysing the problem of optimal farm size. The implications of principal–agent relations, which appear in all form of economic organization, are considered next. Finally, the role and implications of household production by farm households will be discussed, with some further, final conclusions.

THE THEORY OF FARM HOUSEHOLDS

As already explained, farm households differ from non-farm households only in so far as they allocate resources to farm production in addition to household production and, very often, to off-farm employment. Most non-farm households allocate resources only to household and off-household production of non-farm goods and services. Whether, and to what extent, farm family labour as the major resource is allocated to farm and/or off-farm production alongside household production is determined by the respective marginal value of labour product, which has to be equal if family income is to be maximized.

In Figure 1, that optimal allocation of farm household members' time T is demonstrated, neglecting transaction costs at the start. In the upper part of Figure 1, Y^A reflects the farm's income possibility curve,[2] E, and Y^{A+H} reflects the farm plus household production and (imputed) income possibility curve, whereas Y^{A+H+0} reflects the farm household's aggregated income possibility curve, including off-farm income (the linear segment of that curve). Whereas both the farm and the household's production income curve are subject to the law of diminishing returns, the non-farm income possibility curve is not. This reflects the fact that off-farm wages are not affected by the size of a single

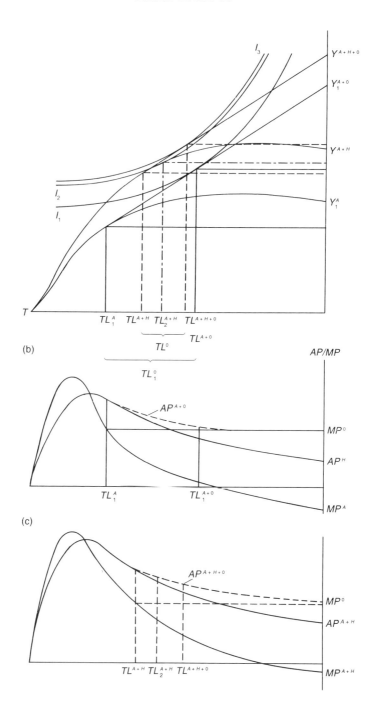

FIGURE 1   *The fundamentals*

household's labour supply (that is, there is perfect competition in labour markets). Assuming perfect markets, the off-farm wage rate represents the opportunity costs of family labour allocated to farm and household production.

Optimal allocation of the farm family's time is achieved if marginal value labour product (*MP*) is equal in all three activities; that is, MP^{A+H} is equal to the off-farm wage rate (MP^0), as the lower part of Figure 1 reveals. In this equilibrium, optimal labour input to farm and household production is TL^{A+H} and to off-farm employment TL^0, so that total labour time is TL^{A+H+0} and leisure time $T-TL^{A+H+0}$ (Figure 1c). As Figure 1 reveals, at the optimal allocation of time, the highest indifference curve I^3 is achieved, and any time allocation different from this equilibrium will be sub-optimal.[3]

As Figure 1 also demonstrates, optimal farm labour input and, hence, the optimal farm size in terms of labour input, are determined by, first, the economies of farm size (reflected in Y^A) and, second, by the opportunity costs of farm labour (MP^0). The higher (lower) opportunity costs are, the smaller (larger) is the optimal farm size *ceteris paribus*. Furthermore, provided that, within the range of household labour capacity, marginal farm labour product achieved is below opportunity cost, off-farm employment is optimal and, hence, part-time farming has to be seen as an efficient resource allocation by farm families. This is contrary to the conventional wisdom, according to which part-time farming is seen as inefficient and not competitive relative to full-time farms.

With respect to Y^A, the second factor affecting optimal farm size in addition to the opportunity costs of farm labour is determined, of course, by the relevant production technique currently prevailing. This incorporates physical production conditions such as the quality of farm inputs and prevailing economic conditions such as sectoral terms of trade. Other things being equal, more efficient production techniques, as well as more favourable physical and economic conditions, favour larger farms. These could employ hired as well as family labour and would, therefore, be organized as family-managed farms. However, it has also to be remembered that household production is important with respect to optimal farm sizes, because household and farm production have to be seen as competitive with respect to the allocation of farm household's time. Figure 1 may also be interpreted in this respect by assuming that household production is more efficient than farm production. This can be demonstrated by simply interpreting Y^A as the imputed income possibility curve of household rather than farm production. In this case, optimal farm labour input (farm size) would be $TL^{A+H} - TL^A_1$ and, thus, it would be smaller than TL^A_1 (the original optimal farm size).

Before discussing the problem of optimal farm sizes, as such, with respect to the form of organization in agriculture, the empirical relevance of off-farm wages to the opportunity cost of farm labour has to be discussed. Of course, markets for (farm) labour are subject to various forms of imperfections and, furthermore, the suitability of farm labour for off-farm activities is restricted by discrepancies in qualifications due to age, sex, education, training and experience (Huffman and Lange, 1989; Gunter and McNamara, 1990). All of these factors may be unfavourable to farm household labour. In this context,

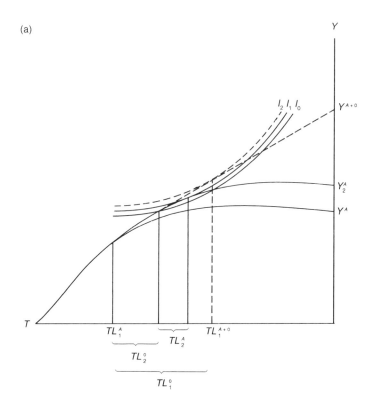

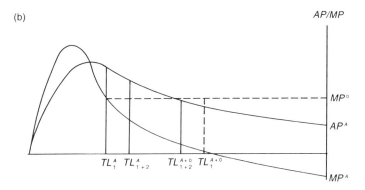

FIGURE 2   *Working time restrictions*

however, a specific, but rather common, imperfection of (formal) markets for labour has to be stressed, namely restrictions concerning minimum and maximum (daily, weekly and lifelong) working time as fixed by relevant regulations.

In Figure 2 the implications of such working time restrictions for optimal time allocation of (part-time) farm households are demonstrated. Without such restrictions, optimal farm labour input would be TL_1^A and off-farm labour input TL_1^0, so that total labour time would be TL_1^{A+0} (neglecting household production) and, thus, indifference curve I^2 would be achieved. By contrast, if maximum off-farm labour time is restricted, say to TL_2^0, optimal labour time offered by the farm household to off-farm employment TL_1^0 exceeds that maximum labour time. Without any adjustment to the relevant regulations, the farm household could only achieve I^0. However, owing to the fact that opportunity costs of labour time exceeding $TL_1^A + TL_2^0$ are below the market wage rate MP^0, it is efficient to enhance farm labour input to TL_{1+2}^A so that marginal farm labour product is below MP^0 but still contributes positively to the household's income. Thus the relevant income possibility curve is Y_2^{A+0} (instead of Y^{A+0} under perfect market conditions) and indifference curve I^1, above I^0, can be achieved. Therefore the optimal farm size is larger than under perfect labour market conditions (Figure 1).

THE PROBLEM OF OPTIMAL
FARM SIZES: FAMILY VERSUS HIRED LABOUR FARMS

Whether farms are organized by farm families or other forms of organization, such as hired labour farms or producer cooperatives, depends *inter alia* on economies of size in farm production, as already stated. In fact, agricultural economists rely almost exclusively on economies of scale, size and scope as determinants of the lowest level of average cost of production and, hence, maximum profits become the decisive criterion of optimal farm size. Furthermore, technological innovations in agriculture are favouring large-scale production, so that smaller than optimal family farms are expected to be replaced by large hired labour farms. This was the view expressed by Karl Marx in 1848.

In analysing the relationship between average production costs and farm sizes, the Office of Technology Assessment (1986, p.113) came to 'two major conclusions: first, most economies of size are apparently captured by moderate-size farms. Second, while the lowest average cost of production may be attainable on a moderate-size farm, average costs tend to remain relatively constant over a wide range of farm sizes. Thus farmers have a strong incentive to expand the sizes of their farms in order to increase total profits.' However, US statistics reveal that the share of moderate-size farms, which are almost exclusively family farms, is rather small and has not increased very much over several decades.[4] The share of those farms which are above moderate size is even smaller, so that the majority of farms are organized by farm families and consist of smaller than moderate-size farms, often seen as less efficient. A large share of these farms are, however, part-time farms.

Therefore three closely interconnected questions have to be raised: first, why are most farms smaller than optimal sized farms? Second, why are these farms organized as family farms? Third, are these smaller than optimal sized family farms really inefficient? Seven explanations are listed, as follows, although only three of them will be discussed more fully later on.

(1) Cost-reducing economies of scale, size and scope in agriculture favour-ing large hired labour farms are rather restricted as compared with many (but not all) forms of production of non-farm goods and services. Such an assessment is supported by OTA, at least implicitly.
(2) Most technological innovations in agriculture are biased towards in-creases in labour productivity. Statistics reveal that labour productivity in agriculture has increased two to four times as much as in the non-farm economy. Increases in labour productivity imply corresponding expansion of labour capacities of farm families, thus enabling them to capture economies of size of enlarged farms.
(3) Opportunity costs of family labour are to a large extent below market wage rates of hired labour in industry, and also in agriculture. Low opportunity costs are due not only to specific adverse characteristics of family labour, but are also due to labour time restrictions. The latter may result in part-time farming being an efficient use of labour which is not fully employed outside agriculture. This has already been demonstrated (Figure 2).
(4) Farm households react to changing economic conditions within agricul-ture by adjusting not only the size and structure of farms but also the size and structure of families. So the migration of farm labour to off-farm employment takes two forms: it may involve remaining in the family's farm household, or leaving that household and founding a new one.
(5) The (farm) family as a small 'team' (Radner, 1987) implies advantages in lower transaction costs relative to other forms of organization, such as hired labour farms and producer cooperatives. That proposition will be discussed more fully later.
(6) The more efficient solution of the 'principal–agent problem' within and by farm families, resulting in lower transaction costs, also enables fami-lies to allocate family labour efficiently and in the most flexible way, to farm, off-farm and household production, according to prevailing and rapidly changing economic conditions. This has also been demonstrated theoretically (Figure 1).
(7) Household production provided by farm households is subject to similar production costs (low opportunity costs of family labour) and transaction cost advantages as farm production itself. Therefore household production increases the welfare (real income) of farm families (Fast and Munro, 1991).

THE PRINCIPAL–AGENT PROBLEM IN AGRICULTURE

Whereas the first four propositions mentioned above need no further comment, the role and significance of transaction costs linked to the principal–agent problem, and to that of household production, has to be commented upon.

Since the famous article of Coase (1937) on 'The Nature of the Firm', institutional economists have been well aware of the fact that factor allocation is not automatically achieved by the 'invisible hand' of markets as conventional economics assumes. It involves organizations acting within markets or, more precisely, executives managing the organizations. However, because of imperfections of markets, especially imperfect information resulting in uncertainties and risks, managers' decisions on factor allocation are burdened by organizational or transaction costs. According to Matthews (1986), 'the fundamental idea of transaction costs is that they consist of the costs of arranging a contract *ex-ante*, and monitoring and enforcing it *ex-post*, as opposed to production costs, which are the costs of executing a contract'. Institutional economics also teaches that transaction costs differ between various forms of organization. Thus in a competitive world that organizational form will succeed which has the lowest transaction costs, provided that transaction cost advantages are not neutralized by higher costs of production (Williamson, 1975, 1985). Basically, transaction costs are costs to the organizations which are founded in order to reduce uncertainties and risks of market transactions by better information.

A major problem of allocating resources efficiently, by and within organizations, relates to the coordination of the activities of their employees. The problem is examined within the 'principal–agent' literature which is 'concerned with how one individual, the principal (say an employer), can design a compensation system (a contract) which motivates another individual (his agent, say the employee), to act in the principal's interests' (Stiglitz, 1987). The problem arises because the principal's interests are different from those of his agents. Whereas the principal may be concerned with maximizing profits, the agents pursue utility-maximizing objectives. However, any compensation system which is to work efficiently has to be operated, and that requires agents' actions being monitored and supervised. It is an activity burdened with transaction costs.

The fact that the majority of farms in developing and developed countries are organized by farm families has to be explained mainly by lower transaction costs compensating, or even over-compensating, lower production costs of larger non-family units. Thus Pollak (1985) comes to the conclusion that 'the family farm can be regarded as an organizational solution to the difficulty of monitoring and supervising workers, who, for technological reasons, cannot be gathered in a single location'. Although transaction costs are difficult to measure, as Pollak and others maintain, there is some empirical evidence supporting the conclusion. For instance, Riebe (1961) in analysing the advantages and disadvantages of family and hired labour farms in West Germany, has suggested that 'hired labour farming requires large supervising and administrative activities. According to our analysis, in family farms only 10 working hours per hectare are needed for management contrary to 20 to 25

hours per hectare by hired labour farms. Such differences of labour input are
of special importance because administrative personnel are paid relatively
high wages. Furthermore, the differences in administrative labour input has to
be multiplied by the total area of farms ...'. It has to be added that labour
costs of monitoring and supervising hired labour are increasing rather quickly
owing to the fact that organizational innovations increasing productivity of
administration are relatively scarce. Higher transaction costs of hired compared
with family labour implies that hired labour has to be seen, therefore, as an
imperfect substitute (Pollak, 1985).

Before the factors affecting those transaction cost differentials are discussed, their implications with respect to optimal farm sizes are demonstrated
in Figure 3. This is almost identical to Figure 1(a) except that (opportunity)
costs of transactions TC are taken into account. Assuming that these transaction

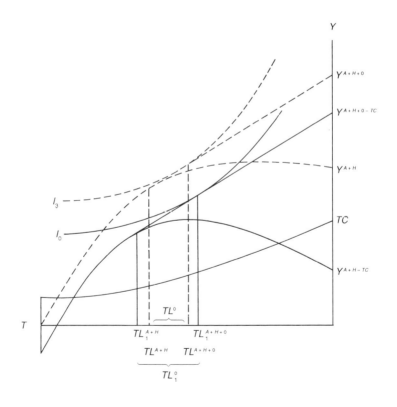

FIGURE 3   *Transactions costs and optimal farm size*

costs have a fixed cost element (Niehans, 1987), average costs are declining (or are constant) as long as family labour is expanded, but are rising if, in addition to family labour, hired labour input is used and further enlarged. As Figure 3 reveals, the relevant farm income possibility curve is Y^{A+H-TC} and the relevant total income possibility curve $Y^{A+H+0-TC}$. Therefore, optimal farm labour input is TL_1^{A+H} (instead of TL^{A+H}) so that the optimal farm size is smaller as compared with Figure 1. Although optimal off-farm labour time (TL_1^0) is larger than without considering TC and, therefore, total labour time supplied by the farm family (TL_1^{A+H+0}) is larger, only indifference curve I^0, instead of I^1, can be achieved.

Why are costs bound to an efficient solution of the principle–agent problem just described higher for hired as compared to family labour, as Figure 3 implies? Pollak (1985) gives the following answer:

> When agricultural tasks can be monitored easily in terms of inputs or outputs, family farms are often overshadowed by other forms of agricultural organization. For some crops and tasks, hired labour can be concentrated into work gangs and supervised directly, so plantation agriculture is possible. For other crops and tasks (e.g. harvesting) output can be measured directly and workers paid on a piece-rate basis ... Nevertheless, since most farm tasks are not susceptible to either of these forms of supervision or monitoring, the family farm is the dominant form of agricultural organization.

Owing to the fact that a farm family as a small team provides better information concerning the actions, comparative advantages and behaviour of family members, as well as the local conditions of the farm, transaction costs are low as compared to large hired labour farms. Pollak's conclusion that the family farm can be regarded as an efficient solution to the principal–agent problem had already received support from Aereboe's (1928) much neglected comparison of the working conditions in agriculture and industry, as well as by his (1923) comparison of family and hired labour farms.

However, there is still an open question concerning the compensation system applied in farm families as a necessary prerequisite of the efficient solution of the principal–agent problem which also faces them. Whereas, in hired labour farms, workers have to be paid whatever their individual performance, family members are most often paid much less. Therefore questions arise about the nature of incentives stimulating the family member's interest in the farm's economic success. That question can only be answered by referring to the role of home production by farm households. Of course, altruism and solidarity may be important in this respect. However, those factors may not be sufficient to explain the monitoring advantages of family farming.

THE ROLE OF HOUSEHOLDS AND
THE ECONOMIES OF (FARM) HOUSEHOLD PRODUCTION

According to an American definition a household can include all persons who occupy a housing unit, including related family members and all unrelated

persons, who share the housing unit. For behavioural analysis use is made of
the economics of household production (Becker, 1965). The specific economies of household production are due to low opportunity costs of family
labour and low transaction costs (Pollak, 1985). These are similar to the
economies of farm production by farm families, but are also due to some
economies of scale. According to Nelson (1988), the latter are achieved because
'some goods are public within households ... increasing returns in household
production of goods and services [and] advantages of bulk discount in purchasing' inputs to that production.

Eisner (1988) has systematically grouped various home production activities,
such as 'housework', 'obtaining goods and services', 'care of the family and
others including child care', 'helping and teaching children', and 'medical care
provided to children and others in the household'. Such an enumeration implies
that household production has two important characteristics. First, household
production involves a large amount of provision of social security to children,
plus provision for family members who may have lost their gainful employment
because of age, illness, accident, unemployment, separation or divorce. Of
course, private and collective insurance schemes as well as obligatory training
and education systems have eroded the role of families as an 'insurance company'
(Becker, 1981). However, that role has not yet vanished, even in economically
advanced countries, especially with respect to farm households.

Second, household production implies income redistribution between gainfully active and inactive members. That redistribution is affected by the fact
that inputs supplied by markets for use in household production have to be
financed by income achieved by active members, and also by the fact that the
output of home production is distributed both to inactive and active members
according to their specific needs. In fact, income redistribution, as effected on
the input side of household production, is reflected in the point, mentioned
earlier, that pecuniary remuneration of family members engaged in household
and farm production is smaller than opportunity costs (market wages). In
addition, family members engaged in off-farm employment, and remunerated
by market wages, are expected to transfer parts of that income to the household as a compensation for household goods and services consumed.

With respect to income redistribution within (farm) families it might be
assumed that incentives stimulating members' efforts and activities would be
extremely weak. However, in the longer run, active family members will
become inactive because of age, while children currently inactive will become
active. Furthermore, the activity of present active members is subject to
uncertainties and risks due to unexpected unemployment, illness or accident.
The provision of social security by households reduces those uncertainties
and risks. Therefore, in the shorter run, there are incentives provided to
household provision of social security to guard against unemployment, illness
and accident. In the longer run, social security provided by households is
valuable in old age, and that again strengthens the efficiency of the incentive
systems which operate in farm households. Thus the compensation system
provides an efficient solution of the principal–agent problem facing family
farm households. They are able to exploit the monitoring and supervising
advantages which that form of organization offers to the small team.

SOME MORE GENERAL CONCLUSIONS

Five more general conclusions can be added:

(1) The economics of agriculture can only be explained and analysed adequately by applying the theory of farm households instead of the theory of farms as firms. This will apply as long as agriculture is mainly organized by family farms.

(2) Resources in agriculture and elsewhere are not allocated by the 'invisible hand' of markets but by the visible hand of organizations such as firms, farms, households and government agencies.

(3) Institutional economics offers a broader base for the explanation of the organization of agriculture and its economic operation than does traditional (neo-classical) economic theory. Nevertheless, institutional economics cannot, and should not, replace neo-classical economic theory. The latter must be used and applied as an analytical tool within institutional economics.

(4) Institutional economics can therefore help in explaining why agriculture is mainly organized by farm families and why that organization is an efficient one.

(5) The organization of agriculture by family farms implies that household production, transaction costs and, very often, off-farm employment, do play an economically important role which affects resource allocation and enumeration in a way quite different from that which traditional agricultural economics presumes.

NOTES

[1] There is no strict definition of family firms (farms) with respect to the number of family and non-family members who are gainfully employed. However, in general, family farms may be defined as such if the number of non-family workers does not permanently exceed the number of active family members. Pollak (1985) distinguishes family firms from 'family-managed firms' and from 'firms that are merely family owned in which a single family member participates in management'. According to that definition, most farms in many western countries are predominantly family farms, and to a much smaller extent family-managed farms.

[2] The farm income possibility curve reflects farm families' net farm income, as defined by USDA, which includes earnings of unpaid family members. It differs because capital gains are excluded.

[3] Figures 1(b) and (c) also reveal that, by neglecting the family's allocation of labour time to household production (and off-farm employment), the average (AP) and marginal 'farm' labour product (MP) (and hence family 'income') are under-estimated. Labour productivity in agriculture is most often estimated by relating labour statistically attached to agriculture to that sector's value added (Kuznets, 1971), which is therefore misleading. See Schmitt (1989b).

[4] According to OTA (1986), in 1969 and 1982, 3.1 per cent and 8.1 per cent, respectively, of all farms in the USA were defined as moderate-size farms. In 1987, more than 80 per cent of that category were defined as family farms.

514   *Günther H. Schmitt*

REFERENCES

Aereboe, F., 1923, *Allgemeine landwirtschaftliche Bertriebslehre*, Parey, Berlin.
Aereboe, F., 1928, *Agrarpolitik*, Parey, Berlin.
Arayma, Y., 1989, 'Outmigration from Agriculture and Time Allocation within Farm Households' in *Agriculture and Government in an Interdependent World*, Dartmouth, Aldershot, pp. 63–77.
Becker, G., 1965, 'A Theory of the Allocation of Time', *Economic Journal*, 75, pp. 493–517.
Becker, G., 1981, *A Treatise on the Family*, Cambridge University Press, Cambridge.
Coase, R., 1937, 'The Nature of the Firm', *Economica*, 4, pp. 386–405.
Eggertson, T., 1990, *Economic Behaviour and Institutions*, Cambridge Surveys of Economic Literature, Cambridge.
Eisner, R., 1988, 'Extended Accounts for National Income and Product', *Journal of Economic Literature*, 26, pp. 1611–83.
Fast, I. and Munro, B., 1991, 'Value of Household and Farm Work: Evidence from Alberta Farm Family Data', *Canadian Journal of Agricultural Economics*, 39, pp. 137–50.
Gunter, L. and NcNamara, K.T., 1990, 'The Impact of Local Market Conditions on the Off-farm Earnings of Farm Operators', *Southern Journal of Agricultural Economics*, 22, pp. 155–65.
Huffman, W. and Lange, M., 1989, 'Off-Farm Work Decisions of Husbands and Wives: Joint Decision Making', *Review of Economics and Statistics*, 71, pp. 471–80.
Kuznets, S., 1971, *Economic Growth of Nations. Total Output and Production Structure*, Harvard University Press, Cambridge, MA.
Locay, L., 1990, 'Economic Development and the Division of Production between Households and Markets', *Journal of Political Economy*, 98, pp. 965–82.
Matthews, R., 1986, 'The Economics of Institutions and the Sources of Growth', *Economic Journal*, 96, pp. 903–10.
Nelson, J., 1988, 'Household Economies of Scale in Consumption: Theory and Evidence', *Econometrica*, 56, pp. 1301–14.
Niehans, J., 1987, 'Transaction costs', *The New Palgrave. A Dictionary of Economics*, vol. 4, Macmillan, London, pp. 676–9.
Office of Technology Assessment (OTA), 1986, *Technology, Public Policy, and the Changing Structure of American Agriculture*, Washington, DC.
Pollak, R., 1985, 'A Transaction Cost Approach to Families and Households', *Journal of Economic Literature*, 23, pp. 581–608.
Radner, R., 1987, 'Teams', *The New Palgrave. A Dictionary of Economics*, vol. 4, Macmillan, London, pp. 613–16.
Riebe, K., 1961, 'Arbeit und Betriebsgröße', *Das landwirtschaftliche Betriebsgrößenproblem im Westen und Osten*, Sonderheft 13 der Agrarwirtschaft, Hannover, pp. 83–91.
Schmitt, G., 1988, 'Wie optimal ist eigentlich die "optimale" Betriebsgröße in der Landwirtschaft?', *Agrarwirtschaft*, 37, pp. 234–45.
Schmitt, G., 1989a, 'Warum ist Landwirtschaft eigentlich überwiegend bäuerliche Familienwirtschaft?', *Berichte über Landwirtschaft*, 67, pp. 161–218.
Schmitt, G., 1989b, 'Simon Kuznet's Sectoral Shares in Labor Force: A Different Explanation of His (I + S)/A Ratio', *American Economic Review*, 79, pp. 1262–76.
Schmitt, G., 1989c, 'Farms, Farm Households and Productivity of Resource Use in Agriculture', *European Review of Agricultural Economics*, 16, pp. 257–84.
Schmitt, G., 1990a, 'Is Structural Change Really a Source of Economic Growth? The Case of Agriculture', *Journal of Institutional and Theoretical Economics*, 146, pp. 470–9.
Schmitt, G., 1990b, 'Die ökonomische Logik der Einheit von Haushalt und Betrieb in der Landwirtschaft', *Agrarwirtschaft*, 39, pp. 209–20.
Stiglitz, J., 1987, 'Principal and Agent (II)'.*The New Palgrave. A Dictionary of Economics*, vol. 3, Macmillan, London, pp. 966–72.
Williamson, O., 1975, *Markets and Hierarchies: Analysis and Antitrust Implications*, Free Press, New York.
Williamson, O., 1985, *The Economic Institutions of Capitalism*, Free Press, New York and London.

DISCUSSION OPENING – MARY C. AHEARN*

In his paper Günther Schmitt seeks to establish two major points: first, that the appropriate model for understanding the behaviour of farms is the household production model pioneered by Becker (1965) and extended by many, for example, Pollak and Wachter (1975); second, that hired labour is an imperfect substitute for family labour in agriculture, owing largely to the high level of transaction costs associated with non-family labour, and that these costs are an important force in maintaining family farming and moderating the forces of economies of size on agricultural structure.

The importance of the 'new household economics' for addressing many issues affecting farm behaviour cannot be refuted. The relevant optimizing unit overall is a farm household, not a farm firm. I agree with the author's view that traditionally the firm is taken as the unit of analysis. Historical data series are often a mirror of the ruling concepts of the times. In the USA, although the number of days worked off the farm by the operator were collected by the Census of Agriculture from 1929, estimates of off-farm income were not collected until 1959, and even then only for the farm spouse. In the same vein, until 1988, the USDA has assumed that all of the sector's farm income accrued to the farm operator household – in other words it has assumed that all farms were sole proprietor, single household operations. In general, data on farm firms have taken a back seat to data on farm households.

In the new approach the production, consumption and labour supply decisions of the farm household are addressed simultaneously. Households are assumed to maximize utility subject to constraints of time, income, farm production technology and local labour market opportunities. Utility is a function of leisure and consumption goods. Some of the consumption goods may be produced in the household with inputs from either the market or the farm business.

The dominance of off-farm sources of income in the farm operator household's total income is well established and that dominance will obviously affect resource allocation decisions. In the USA in 1988, some 85 per cent of the cash income of farm operator households came from off-farm sources, two-thirds were more dependent on off-farm income than farm income, and about 90 per cent of farm operator households received some income from off-farm sources (Ahearn, 1990). Depending on how one wishes to define part-time farming, anywhere from one-third to two-thirds of US farms are part-time farms (Ahearn and Lee, forthcoming).

In fairness to the contributions of many agricultural economists, off-farm income and the trade-offs in labour allocation within the farm household have been studied for some time (for example, Lee, 1965). At the recent 1991 American Agricultural Economics Association Meetings, at least seven presentations directly focused on the topic of off-farm income. Many of the farm applications of the household production model have been with respect to developing countries where consumption and production decisions are fully intertwined (Singh, Squire and Strauss, 1986). With reference to the USA,

*Economic Research Service, US Department of Agriculture.

Wallace Huffman has applied the model of the new household economics extensively to the labour allocation issue of farm households (Huffman, 1980).

Having stated my basic agreement with one of Schmitt's major points, namely that a more comprehensive model of the farm decision process is appropriate, I would also add that there are times when this model is less appropriate. In empirical applications, the new household model has extensive data requirements and, because of its comprehensiveness, is mathematically complicated. As with many conceptual models, these characteristics impose very real constraints on its applicability. In response to those constraints, economists very often assume separability within a system. In the case of farm firm decisions, that may be a valid assumption for many types of research questions, especially those decisions viewed in the short run. So I do not share Schmitt's apparent view that the most appropriate model for all allocation issues is a household-based model, rather than a firm-based model. Furthermore, a disadvantage of the new household model for some purposes is that it relies on the existence of a functioning rural labour market. In many agricultural areas of the USA, and I assume other places in the world, there is no rural labour market outside agriculture.

Schmitt's second major point is that significant transaction costs exist in hiring non-family labour on the farm. These added costs are the primary reason why family labour is superior, and farm size is seen as constrained by its availability. For example, Schmitt states: 'The fact that the majority of farms in developing and developed countries are organized by farm families has to be explained by lower transaction costs compensating, or even over-compensating, lower production costs of larger non-family units.' In developing his argument, Schmitt draws on the transaction cost literature of the 'new institutional economics' which focuses on the *ex ante* costs of obtaining information about contract selection and the *ex post* costs of the transaction, such as monitoring the contract. He also draws our attention to the relevance of the principal–agent literature and contrasts the motives of family labour versus hired non-family labour with respect to the performance of the farm business.

There are two criticisms. First, I believe he over-states the importance of transaction costs when he argues that the reason that farming is largely a family business is the transaction costs of hiring non-family labour. Many factors shape the agricultural structure, which Schmitt acknowledged earlier and then seems to overlook. The traditional factors viewed as affecting structure, such as technological development and market conditions, cannot be dismissed. This is not to say that transaction costs are unimportant, and may become even more important if the market for hired agricultural workers becomes thinner as a result of regulations and immigration reform. However, ignoring all other factors but transaction costs can lead to faulty results. For example, government policies must have enormous impacts on agricultural structure world-wide, given the massive subsidies and taxes imposed on the sector (World Bank, 1986). As an example of the danger, I refer to a recent analysis of Tunisian agriculture where transaction costs have recently been found to be an important explanation for a move towards sharecropping and away from cash renting by absentee landlords (Matoussi and Nugent, 1989). I have no knowledge from which to question that specification for Tunisian agriculture and thought the research was interesting and well done, but a specification of a model to explain share renting of major crops in the USA would be meaningless without inclusion of government programmes. Take rice, for example, which is one of the most highly subsidized commodities in US agriculture. About 63 per cent of all rice land in production is share rented, 16 per cent is owned and the rest is cash rented. The major reasons for the high proportion of share rented land can be ascribed to the nature of the programme: (1) share rent landlords are eligible for payments, but not cash rent landlords, and (2) there is a payment limit to individuals, thereby discouraging concentration of ownership in the hands of a few individuals. Transaction costs probably play a minor role in this common tenure arrangement for rice production.

My second criticism of Schmitt's treatment of the role of transaction costs in explaining farm structure is a general one. His explanation for the role of farm households in minimizing transaction costs is tautological: he maintains that the efficient form of farm organization will predominate and, since the family farm predominates, it is the most efficient. This type of a functionalist assumption is a common criticism of the new institutional economics (Datta and Nugent, 1989). Schmitt would have been on much firmer ground had he proposed the importance of transaction costs as a hypothesis to explain the predominance of family farms, and then proposed a means for empirical testing.

The types of transaction costs associated with hiring non-family labour in the past have often been grouped with general management expenses. For example, such expenses serve as the explanation for a theoretical cost curve having a U shape, with costs rising for the largest firm sizes. However, I strongly agree with the author that labour is a unique input and the transaction costs associated with labour merit individual study.

Much can be said about the issue of agricultural labour. We know relatively little about it. In the USA, there is even a controversy about how many annual labour hours we have in agriculture, as well as how to value paid and unpaid family hours (Huffman, Sumner, forthcoming). In addition, valuation of unpaid labour is often critical in the conclusions one can draw about economies of size in agriculture (Vlastuin, Lawrence and Quiggin, 1982). Hired non-family labour is not a perfect substitute for family labour, but it is a partial substitute. It is useful to think of labour as a composite input. Its component parts have implications for the firm and, in the case of family labour, directly for the household. With respect to the firm, labour can offer a physical and a management component. What family labour generally offers relative to hired non-family labour is a large management component. One advantage to family labour is that often the management function (of assimilating relevant production and marketing information) occurs simultaneously with the physical labour. Timmer has stated that one of agriculture's unique features is that labour and management cannot be separated without a loss of efficiency (Timmer, 1988). This is a rather strong statement which probably does not hold for all types of agriculture, for example for commodity production requiring large amounts of hand harvesting or irrigating. However, the message in his statement is consistent with the issue of simultaneity of physical labour and management that occurs with family, and to a much lesser degree non-family, workers. Schmitt footnotes USDA's practice of valuing unpaid labour at the hired farm wage rate as 'a more or less irrelevant assumption'. The practice is certainly controversial, but one justification for the practice is that the physical labour component of the contribution of the unpaid worker is what is being valued in this treatment. USDA then includes return to management in the residual.

Two important quality differences are that family labour may generally be more flexible in terms of scheduling of work and since less 'shirking' occurs, requires less supervision. The importance of flexibility in agriculture cannot be over-stated. Crop farmers, especially, must wait patiently for the right mix of conditions, most of which are not under their control, to perform many of their field operations. For an operation dependent on hired labour, bottlenecks can occur.

From a household perspective, there are additional distinctions between family and hired non-family workers. First, there is the issue of the intergenerational transfer of wealth, which Schmitt mentions, and other considerations of preservation of the family heritage. Finally, there are benefits which accrue from providing employment and a rural life style to family members. Because utilities are interdependent in a family, both the individual enabling others to have employment and a rural life style and those in the family seeking those arrangements are benefited.

In closing, I would like to commend the author for integrating some very useful literature in the examination of the issues of farm resource allocation.

REFERENCES

Ahearn, Mary, 1990, 'The Role of the Farm Household in the Agricultural Economy', *Agricultural Income and Finance Situation and Outlook*, AFO-37, May.
Ahearn, Mary and Lee, John E. Jr., forthcoming, 'Multiple Job Holding Among Farm Operator Households in the U.S.: Historical Perspective and Future Prospect', in J. Findeis, M. Hallberg and D. Lass (eds), *Multiple Job Holding Among Farm Families in North America*, Iowa State University Press, Ames.
Becker, G., 1965, 'A Theory of the Allocation of Time', *Economic Journal*, 75, September.
Datta, S. and Nugent J., 1989, Ch. 2 in M. Nabli and J.B. Nugent (eds), *The New Institutional Economics and Development: Theory and Application to Tunisia*, North Holland, Amsterdam.
Huffman, Wallace, 1980, 'Farm and Off-Farm Work Decisions: The Role of Human Capital', *The Review of Economics and Statistics*, 62, (1).
Huffman, Wallace, forthcoming, 'A Perspective on Estimating Costs of Human Capital Services and More', Ch. 18 in M. Ahearn and U. Vasavada (eds) *Costs and Returns for Agricultural Commodities: Advances in Concepts and Measurement*, Westview Press, Boulder, Co.
Lee, John E., Jr., 1965, 'Allocating Farm Resources Between Farm and Nonfarm Uses', *Journal of Farm Economics*, 47, (1).
Matoussi, M.S. and Nugent, J.B., 1989, ch. 4 in M. Nabli and J.B.Nugent (eds) *New Institutional Economics*.
Pollak, R. A. and Wachter, M.L. 1975, 'The Relevance of the Household Production Function and Its Implications for the Allocation of Time', *Journal of Political Economy*, 83, (2).
Singh, I., Squire L and Strauss J. (eds), 1986, *Agricultural Household Models: Extensions, Applications, and Policy*, Johns Hopkins University Press, Baltimore, Md.

Sumner, D., forthcoming, 'Human Capital Issues in Measuring Costs of Production', ch. 17, in M. Ahearn and U. Vasavada, (eds), *Costs and Returns*.

Timmer, C. Peter, 1988, 'The Agricultural Transformation', in H.B. Chenery and T.N. Srinivasan (eds) *Handbook of Development Economics*, vol. 1 of *Handbooks in Economics*, 9, North Holland, Amsterdam.

Vlastuin, C, Lawrence D. and Quiggin J., 1982, 'Size Economies in Australian Agriculture', *Review of Marketing and Agricultural Economics*, 50, (1).

World Bank, 1986, *World Development Report 1986*, Oxford University Press, New York.

YOAV KISLEV*

Family Farms, Cooperatives and Collectives[1]

THE FAMILY FARM

What is more natural than to expect modern agriculture to be organized in large-scale food factories? Such expectations notwithstanding, a large part of agricultural production – though not all of it – is still done on family farms. Evidently, economies of scale in production, to the extent that they exist, are outweighed by countervailing forces.

Transaction costs and control

Family organization of production minimizes transaction costs through 'specializaton by identity' and the harmony and trust that comes with it (Ben Porath 1980).[2] In agriculture, this advantage is augmented by the specific nature of control as discussed already by Brewster (1950) who observed that (a) the biological production process in agriculture is time-dependent: food and fibres cannot be produced in an establishment in which different stages of the product are manufactured simultaneously; and (b) the farm product cannot be moved along a production line: rather the worker has to go to the field to perform the necessary tasks. Both these characteristics make control of labour in agriculture difficult relative to manufacturing and increase the comparative advantage of the family farm versus large units relying mostly on hired hands.

Labour according to Brewster, is a fixed factor on the family farm. The farmer strives, therefore, for a balanced product mix with stable labour requirements throughout the year. As a result, the advantage of the family farm is visible in areas suitable for diversified farming, while large farms dominate the regions of monocultural agriculture. Brewster's analysis can be augmented with modern theoretical insight and historical experience. Where land ownership is highly unequal, the problem of control can be solved by sharecropping (Otsuka and Hayami, 1988). That this is only a partial solution can be seen from the fact that even livestock production is still mostly a family enterprise. The staged production nature of the biological processes, with waiting periods in between seasonal tasks which prevents factory-like organization of production in agriculture, enables farm operators and other family members to seek part-time, off-farm employment. Monoculture is no more an obstacle; it is an

*Hebrew University, Rehovot, Israel.

advantage. Professional career, farm size, product mix and employment are now determined simultaneously – affected by considerations of income and risk.

The above arguments are analytical and hypothetical, but would not size-associated efficiency gains on large farms compensate for the loss of the aforementioned advantages? To answer this question, we have to examine the empirical evidence.

Economies of scale and farm growth

In an often quoted paper, Griliches (1963) found the sum of the coefficients in a Cobb-Douglas production function for the American farm sector to be 1.28; others report a similar value. Consequently, growth of farm size has been attributed to economies of scale. Hence we now face two questions: should these high-scale estimates be taken at face value and, if not, how can farm growth be explained?

I wish to argue that the evidence does not support unequivocally the existence of economies of scale in agriculture. Because of data shortcomings, every measurement and method of estimation can and should be questioned. When this is done, the case for economies of scale is weakened significantly. The argument is detailed in Kislev and Peterson (forthcoming). Only the main points are repeated here. Conventional production function estimates, Griliches' included, do not allow for differences in ability and local conditions on farms. Indeed, virtually all reported estimates of covariance analysis – taking care of the unobserved specific factors (Mundlak, 1968) – fail to find economies of scale in agriculture. The alternative method of synthetic firm analysis, which has also produced large estimates of economies of scale, assumes away the crucial issue of control.

For further examination, consider expected consequences. If scale economies exist in a competitive industry, they must be a disequilibrium phenomenon and growth should bring farms down the average cost curve towards its minimum. Yet re-estimates of the Cobb-Douglas production function for American agriculture with data spanning the period 1949–87 repeatedly came up with sums of coefficients of 1.3. No convergence to constant cost is discerned in the data with this method of analysis.

Between 1929 and 1987, output per farm in constant dollars grew in the USA by some 6.4 times. If economies of scale are not important, what can explain such fast growth? Changing prices appear to be the answer (Kislev and Peterson, 1982): between 1929 and 1969, machine rental declined relative to alternative labour cost in agriculture by 3 per cent per year. Farmers reacted by increasing the machine–labour ratio. An operator with more machines cultivated a larger area, and farms grew in size. The trend changed when the cost of operating machines increased during the 1970s and, indeed, farm growth was then halted. Farm size seems to have resumed its upward trend in the 1980s, when increases in real wages again overtook growth in machine rentals.

Hence structural characteristics of production and control make the family farm the dominant form of organization (though not the only one, as should be emphasized) and price ratios affect optimal size of operation. The amounts of land, capital and labour on the farm are determined simultaneously with lines of production and non-farm employment.[3]

Decision making

The economic decisions of the farm household in a market economy are guided by prices and optimization can often be done recursively. In the first stage, income is maximized by deciding on allocation in the production 'department' of the family farm; in the second stage, income is distributed between consumption items and saving (Singh, Squire and Strauss, 1986).

However, the recursive nature of decisions is not always maintained. Examples to the contrary arise when household members draw different utility from working on or off the farm (Lopez, 1986) and when uncertainty affects the price of a product that is both produced and consumed on the farm (Finkelshtain and Chalfant, 1991). Though recursiveness and relying on market signals simplify decision making, with experience and necessity even traditionally bounded peasants allocate optimally the resources at their command (Schultz, 1964).

COOPERATIVES AND COLLECTIVES

The advantage of the family organization is usually presented in comparison with commercial farms relying on hired labour. Cooperation facilitates the realization of scale economies in services and trade, overcoming local monopolies (Sexton, 1990), risk sharing and credit enhancement. The problem of control can be much reduced in cooperatives and collectives (voluntary, not forced cooperation of course), but it cannot be eliminated completely, as evidenced by the Israeli experience which culminated in a severe financial crisis in the mid-1980s. I draw on that experience and base my discussion to a large extent on Zusman (1988). For an analysis of the economy of the kibbutz, see Barkai (1977).

Four types

The major types of cooperatives in Israel are moshavim, kibbutzim, regional cooperative enterprises, and supply cooperatives. They are now undergoing revolutionary changes and their description is somewhat outdated. The recent changes will be discussed towards the end of the paper.

A moshav (moshavim is plural) is a farming community in which all farms are family-operated, and all farmers are members in the multi-purpose, democratically run, village cooperative. In principle (practice varies) the cooperative association in the moshav purchases all farm supplies for its members and markets their farm products. A kibbutz is a commune. Members work together and receive from the kibbutz all their needs. Again in principle, a member in the kibbutz owns his or her personal belongings but no other property. Moshavim and kibbutzim are members in two types of second-order cooperatives: regional service enterprises (such as feed mills, slaughter houses and transport services) and supply cooperatives set up to purchase farm requisites for their members the moshavim and the kibbutzim. Owing to space limitations, the supply cooperatives will be considered only to the extent necessary to explain developments in the first–order cooperatives – moshavim and kibbutzim – and the regional enterprises will not be dealt with at all.

Starting with the transfer of suppliers' credit to their members, the village associations of the moshavim and the supply cooperatives expanded into full-scale financial intermediation and the domineering position that financial activities occupied among their functions greatly affected both their well-being and their structure. It is useful to commence the discussion with a review of balance sheets.

Structure and accounting

Differences in their organizational forms are reflected in the accounting practices and balance sheet composition of the cooperatives (Jensen, 1983). Three randomly chosen cases are presented in Table 1.

TABLE 1   *Balance sheet composition (percentages, September 1984)*

	Moshav	Kibbutz	Supply coop
Fixed and financial assets	2.8	62.4	8.3
Loans to members	75.7	0.1	77.8
Other current assets	21.5	37.5	13.9
Total assets	100.0	100.0	100.0
Equity	2.9	33.7	0.4
Outside debt (including suppliers)	51.7	54.7	84.1
Loans from members	5.4	0.1	15.5
Loans from supply coops	40.0	11.5	—
Total liabilities	100.0	100.0	100.0

Source: Lerman (1989)

The accounting framework of the moshav and its balance sheet are for the cooperative association of the village; the economic activities of the family farms are not covered. In this way, the privacy of the members is respected, but the practice also limits the moshav's monitoring ability, representing a weakness which contributed to the financial crisis. The importance of credit intermediation can be seen from the share of loans to members in its balance sheet, estimated at 75.7 per cent of the assets. The moshav raises credit from outside sources, almost a half of it from the supply cooperative, and distributes it to its members. To a lesser extent, the moshav also functions as a clearing house, receiving deposits from members with surpluses for others in need (5.4 per cent of the liabilities). Like the association in the moshav, the supply cooperative is also mainly a provider of services, including financial intermediation. The functional resemblance is reflected in the similarity of the balance sheet composition.

The kibbutz conducts its accounting like a family farm, combining its business and household books. The assets in the balance sheet are machines, buildings, orchards, livestock, plus the members' dwellings, pension funds and other savings. Consistently consumption, not labour, is considered as part of the cost of running the economic enterprise. Such mixed accounting practices mask the distinction between business and household, and between ownership and management, and are obstacles in the control of the economic affairs of the kibbutz.

Kibbutzim are profit-maximizing entities, engaged mostly in production and aiming to accumulate equity capital–a third of the liabilities in the kibbutz in Table 1. As zero-profit cooperatives, concentrating on financial intermediation, the moshav and the supply cooperative naturally accumulated smaller share of equity.

STRUCTURAL CONSTRAINTS
AND ORGANIZATIONAL BEHAVIOUR

The moshav, the kibbutz and the regional cooperative are contractual institutions. Members contract, explicitly or implicitly, with the cooperative association and with each other to perform together certain economic and social activities. In principle, members are bound by the rules and regulations of the cooperative or the collective they joined. Practice is dictated by the democratic governance of the kibbutz and the moshav and by the cooperative ethics of their members.

Conflicts of interest – the moshav

By tradition, cost (of marketing, for example) is allocated in cooperatives according to 'patronage'. This results in average pricing, which may differ from optimal pricing if, in the quantity of services provided, marginal costs differ from their average. In an attempt to improve upon this rule, the moshav may choose a two-part cost allocation rule: each member is charged a given amount α (perhaps to cover part of the fixed cost) and an additional sum β per unit of product marketed through the cooperative. Now there is room for conflicts of interest. Members with a large volume of production will try to allocate most of the cost to the fixed element α, small producers will favour charging mostly on a per unit basis. If this issue comes to voting and the distribution of members by size of production is not symmetrical, the value of β chosen will reflect the interest of the majority. The minority may find itself shouldering a larger than proportional share in the cost of the marketing service.

Consider now the construction of a feed-processing plant by the moshav for the service of its members. When the investment is financed by the general fund of the moshav, the risk of the venture will be shared by all members. If only livestock farmers participate in the investment, others will avoid the risk, but the moshav will not make use of one of the great advantages of cooperation. The possibility of any degree of risk sharing raises new issues of moral hazard which are nowhere more pronounced than in credit and will be discussed below.

Conflicts of interest – the kibbutz

A major source of structural conflict in the kibbutz is its set of operational constraints: equality, own labour (no exploitation of hired employees) and standard of living in parity with the standard of reference groups outside the kibbutz. These constraints are not always consistent. Own labour implies that unskilled work in agriculture and manufacturing is also done by members. These tasks seldom yield the income needed to support the expected standard of living. During the 1970s and early 1980s, ample supply of credit assisted in maintaining the desired private and social consumption levels, but also created the background for the financial crisis to follow.

By sharing income equally among its members, the kibbutz avoids much of the kind of conflict of interest that plagues the moshav. The outcome of an economic undertaking affects similarly all members and it is generally not in the immediate interest of any one group to tilt decisions in its direction. And if conflicting interests arise, since personal consumption is not affected, the intensity of the conflicts and the social antagonism they may generate are more often than not much weaker than in similar situations in the moshav.

The identity of the society and the community with the economic activities is the source of another kind of conflict. The majority of the members cannot comprehend fully the economic situation of the kibbutz – particularly with accounting practices that do not separate business from the community – but everyone understands social problems. Consequently, unlike the rational traditional peasants whom Schultz (1964) praised, kibbutzim are often subject to the logic of collective action (Olsen, 1965). They tend to have bloated services (particularly in children's care), readily purchase new and convenient machines, continue with failing activities to avoid painful labour re-allocation, and invest in dwellings and community services even if the necessary capital is too costly. They also tend to permit their young members long leave periods to experience outside life. The result is an unequal age distribution of the labour force, which is manifested particularly as the kibbutzim get older.

Recapitulation

As we have just seen, the structure of the kibbutz and the moshav breeds conflicts which may hamper optimal operation. Other difficulties can be regarded as free-riding and moral hazards. They are manifested strongly in the financial activities of the moshavim and the kibbutzim to which we now turn (for details, see Kislev, Lerman and Zusman, forthcoming).

THE COOPERATIVE AS A FINANCIAL INTERMEDIARY

Moshavim and kibbutzim cultivate national land leased to the farmers on a long-term basis. Farms cannot be used as collateral against loans. The alternative is cooperation in credit. Both the associations in the moshavim and the supply coops of the moshavim and the kibbutzim function as credit intermediaries. To augment monitoring and to facilitate convenient collection, members are required to market farm product through the cooperative.

Advantages

With financial cooperation, members in the moshavim, and moshavim and kibbutzim in the supply cooperatives, enjoy economies of scale in loan processing, professional financial management, and stronger bargaining position in the credit market. However, the greatest advantage of cooperative credit, both in the moshav and in the supply cooperative, lies in risk pooling, the implementation of which is founded on mutual liability and guarantee. Members in the moshav sign mutual guarantee agreements for the moshav association and representatives of moshavim and kibbutzim pledge similarly for loans raised by the supply cooperative. The social pressure to comply with cooperative norms is strengthened under mutual liability arrangements. The probability of default is reduced. Banks evidently recognize the advantage inherent in this arrangement, as credit is often made conditional on renewal of mutual liabilities.

Weaknesses

Several kinds of structural difficulties afflict the moshav and the supply cooperative, particularly (a) moral hazard – members tend to invest on their farms in risky projects knowing that with mutual liability they will be bailed out should the returns on the investment be disappointing; (b) free riders – a member in the moshav, or a moshav and a kibbutz in a supply cooperative, may choose to market farm products privately, thus weakening the association's standing in the credit market; and (c) agency cost – banks and other lenders view the cooperative associations as their agents and expect them to protect their interest (for example, by limiting credit to failing members) but the associations are guided by interests which are not always those of the lenders. Similarly, officers in the associations may be tempted to expand operations and to assume risks which prudent members would avoid.

Enforcement of the cooperative's norms and rules – in practice, mainly enforcement of the inter-linkage arrangements of product marketing through the moshav and through the supply cooperative – is critical to its continued functioning as a credit cooperative. However, compliance with the behavioural code requires high standards of cooperative ethics and will to enforce. But enforcement is difficult in the internal political environment of the cooperatives.

PUBLIC POLICY

Cooperation in agriculture has always been supported by the government in many ways, but the most profound public involvement was in credit. The government raised capital on the markets in Israel for its budgetary needs, thus crowding out private sources of investment capital. To remedy the shortage of its own creation, the government distributed credit and subsidized it. Farm cooperatives were among the beneficiaries of this policy. The dependency on the government and the expectation that it would bail out moshavim and kibbutzim in trouble created moral hazard problems. Cooperatives at all levels were willing to rely on large amounts of debt, and banks were willing to lend, all trusting the government to save them in case of difficulty. This problem of moral hazard was recognized by the government, but the will to maintain a strict policy could not withstand the flood of credit in the late 1970s, when Israel participated in the global credit expansion. The situation was aggravated as inflation accelerated (to an annual rate of 440 per cent in 1984) while interest rates lagged and real rates were negative on many kinds of loans for most of the 1970s and early 1980s. The combination of ample supply of credit with the weaknesses of cooperative financial intermediation resulted in over-expansion and excessive reliance on debt in moshavim, kibbutzim and regional cooperative enterprises. In July 1985, inflation was abruptly halted by severe measures, including tight monetary policy. Real interest rates rocketed.

Crisis

The financial crisis in agriculture erupted at the end of 1985, when creditors realized that agriculture, particularly cooperative agriculture, could not continue to service its debt in view of the exceedingly high real rates of interest and that the government – bound by a stringent fiscal regime – could not bail the sector out any more. Most regional cooperatives and many of the associations in the moshavim collapsed. Farm production has continued, often with private credit arrangements and the farmers' own resources. But this cannot be a complete solution to the crisis, and banks and other creditors are still demanding repayment of their loans. For most members in the cooperatives, the heavy burden is not their own debt but their share in the mutual liabilities

– their share in covering the debt of several heavy borrowers in the moshav and the debt of the regional enterprises.

Agriculture could not repay or service its debt in full. Once this was realized, the government moved in, offering support in an effort to reach a debt settlement between the banks, on the one hand, and the moshavim and the kibbutzim on the other. Agreements have been formulated but their implementation has been slow, as many in the sector still hope that they can gather political support for a more favourable settlement.

A major victim of the crisis has been cooperation. Many of the village associations in the moshavim ceased to function as cooperatives and most of the supply cooperatives had to give up financial intermediation. It is practically impossible to get credit guarantees, and banks became suspicious of borrowers. These changes, coming in the wake of the crisis, are affecting different sectors of agriculture in various ways. Wealthier farmers in moshavim can offer collateral in the form of private property and saving. Poorer members have only their farms to offer, but these are not acceptable and such operators are often driven to expand off-farm work.

The crisis also accentuated the differences between moshavim and kibbutzim. A moshav can function as a village even if the farmers desert the cooperative and each fends for himself. Kibbutzim are made up of their membership and, should the members leave, the kibbutz will disappear. Some young members are already leaving (not all for economic reasons, to be sure). The crisis is therefore much more dangerous for the kibbutz. And indeed, many of the kibbutzim reacted by adopting revolutionary structural changes including the division of the kibbutz into several semi-independent economic units, each with its own board of directors and reporting obligations. The kibbutzim are showing here – not for the first time – both their commitment to the collective idea and their practical flexibility. It seems that, economically, most of them may survive the crisis and emerge from it strengthened, provided the younger generations stay and continue in their parents' tradition.

CONCLUSIONS

Cooperation was in the forefront of agricultural development in Israel. Much of the sector's institutional building and technological advancement was achieved through cooperative efforts. For many years, members in moshavim and kibbutzim reached satisfying income levels and maintained stable cooperatives. The late 1970s were particularly favourable for cooperative agriculture: with its access to credit, it succeeded in accumulating large amounts of equity capital, much of it due to inflationary gains resulting from negative real interest rates. With prudent housekeeping, moshavim and kibbutzim could have emerged from the inflationary experience stronger than ever. Instead, driven by weaknesses of cooperative action, combined with irresponsible government policy, they have sunk deeper into debt and prepared the ground for the devastating crisis.

It may well be that, with time and with changing government attitudes and public atmosphere in Israel, cooperation would have lost in the evolutionary struggle to private modes of organization in agriculture. However, in the intensely unstable economic environment of the last 20 years, the institutional competition has been unfair and cooperation may have retreated too much. Time will tell whether cooperation will return to its pivotal position in Israeli agriculture.

NOTES

[1]Writing was supported by the Maurice Falk Institute for Economic Research in Israel. The paper reflects ideas I received from Zvi Lerman, Willis Peterson, Gadi Rosenthal and Pinhas Zusman. The responsibility is solely mine.

[2]Schmitt (1990) also attributes particular importance to transaction costs in explaining the survival of the family farm.

[3]Here I differ with Schmitt (1990) who assumes, perhaps implicitly, that land and capital are given exogenously and farm size is determined (actually, defined) solely by labour allocation.

REFERENCES

Barkai, Haim, 1977, *Growth Patterns of the Kibbutz Economy*, North-Holland, Amsterdam.

Ben Porath, Yoram, 1980, 'The F-connection: Families, Friends, and Firms, and the Organization of Exchange', *Population and Development Review*, 6, pp. 1–30.

Brewster, John M., 1950, 'The Machine Process in Agriculture and Industry', *Journal of Farm Economics*, 32, pp. 69–81.

Finkelshtain, Israel and Chalfant, James A., 1991, 'Market Surplus Under Risk: Do Peasants Agree with Sandmo?', *American Journal of Agricultural Economics*, 73, pp. 557–67.

Griliches, Zvi, 1963, 'The Source of Measured Productivity Growth: United State Agriculture, 1940–60', *Journal of Political Economy*, 71, pp. 331–46.

Jensen, Michael C., 1983, 'Organization Theory and Methodology', *Accounting Review*, 58, pp. 319–39.

Kislev, Yoav and Peterson, Willis, 1982, 'Prices, Technology and Farm Size', *Journal of Political Economy*, 90, pp. 578–95.

Kislev, Yoav and Peterson, Willis, 1991, 'Economies of Scale in Agriculture: A Reexamination of the Evidence', in John M. Antle and Daniel A. Sumner (eds) *Essays on Agricultural Policy in Honour of D. Gale Johnson.*

Kislev, Yoav, Lerman, Zvi and Zusman, Pinhas, forthcoming, 'Recent Experience with Cooperative Farm Credit in Israel', *Economic Development and Cultural Change*, 39, pp. 773–89.

Lerman, Zvi, 1989, 'Capital Structure of Agricultural Co-operatives in Israel', *Yearbook of Co-operative Enterprise*, Plunkett Foundation, Oxford, UK.

Lopez, Ramon E., 1986, 'Structural Models of the Farm Household That Allow for Interdependent Utility and Profit-Maximizing Decisions', in Singh, Squire and Strauss (1986).

Mundlak, Yair, 1968, 'Empirical Production Functions Free of Management Bias', *Journal of Farm Economics*, 43, pp. 44–56.

Olsen, Mancur, 1965, *The Logic of Collective Action*, Harvard University Press, Cambridge, MA.

Otsuka, Keijiro and Hayami, Yujiro, 1988, 'Theories of Share Tenancy: A Critical Survey', *Economic Development and Cultural Change*, 36, pp. 30–68.

Schmitt, Günther, 1990, 'Why Collectivization of Agriculture in Socialistic Countries Really Has Failed: A Transaction Cost Approach', Discussion Paper 9002, Göttingen University.

Schultz, T.W., 1964, *Transforming Traditional Agriculture*, Yale University Press, New Haven, Conn.

Sexton, Richard J., 1990, 'Imperfect Competition in Agricultural Markets and the Role of Cooperatives: A Spatial Analysis', *American Journal of Agricultural Economics*, 72, pp. 709–20.

Singh, Inderjit, Squire, Lyn and Strauss, J., (eds), 1986, *Agricultural Household Models, Extensions, Applications, and Policy*, Johns Hopkins Press, Baltimore, MD.

Zusman, Pinhas, (1988) *Individual Behaviour and Social Choice in a Cooperative Settlement, The Theory and Practice of the Israeli Moshav,* Magnes Press, Jerusalem.

DISCUSSION OPENING – LAURENT MARTENS*

It is a common occurrence to accept the role of discussion opener on the basis of nothing but a title. As long as the discussant does not have any other information, he may have rather conflicting expectations concerning the content of the paper on which he will eventually have to comment. On the one hand, he can hope to find a paper with which he substantially agrees, since we all like our biases to be reinforced by others. In such a case the discussant can explain how much he enjoyed reading an excellent paper and he can complement the paper from his personal experience. On the other hand, he may look forward to a paper leaving plenty of opportunities for disagreement with the author, enabling the discussion opener to point out major theoretical or empirical shortcomings and leaving all participants with the impression that the topic still holds scope for further research.

In this particular case it appears that there is ample scope for both approaches. The title alone suggests that the theme is an evergreen one and an old classic as well. It is evergreen because the issues covered remain exceptionally topical in the present debate on structural transformations in agricultural production world-wide, not just in what used to be centrally planned economies in Central Europe, but also in less developed economies and in the European Community. It is an old classic because throughout the 70-year history of IAAE one can hardly find a single conference during which the institutional organization of agricultural production was not on the agenda. Yet the overall scope of the paper has been scaled down considerably from the initial title to the final content. The initial title sounded like 'Competitive Institutional Arrangements in Farming: Theory and Evidence', from which one could expect a paper developing a universal theory, based upon empirical evidence, which would cover all aspects of alternative institutional structures. Participants who have chosen to attend this particular session on the basis of such high expectations might feel somewhat misguided or even disappointed by the much narrower scope of the paper which has been presented. Indeed the concept of family farms has been scaled down to that of developed market economies, and even mainly to United States experience, while agricultural cooperatives and collectives are even further reduced to the very specific institutional arrangements of moshavim and kibbutzim in Israel.

The section on family farming focuses almost exclusively on economies of scale and is introduced by the question 'What is more natural than to expect modern agriculture to be organized in large-scale food factories?' I would rather have expected this question to be phrased in the opposite sense: 'Is there anything natural about expecting large-scale food factories?' Indeed, it is soon made clear that the evidence does not support unequivocally the existence of economies of scale in agriculture. Kislev is certainly in good

*University of Ghent, Belgium.

company in suggesting that fast growth in farm size and the maintenance of the dominant position of the family farm can be reconciled. In his stimulating paper presented at the 19th IAAE conference, Boussard (1985) concluded that farm size heterogeneity is a consequence of absence of economies of scale, and that farm structure heterogeneity is a result of the interactions between a dynamic process of adjustment towards optimal price-dependent structures and of market constraints which perturb this adjustment. At the same conference, Newby (1985) stated that the tendency towards increased concentration of production has not been accompanied by the disappearance of the family farm or peasantry, and that the emergence of a dual farming economy can be witnessed in many countries.

At the 5th Congress of the European Association of Agricultural Economists, Nielsen (1985) concluded that its ability to adjust to changing external conditions is the main reason for the family farm having been such a persistent institution. Schmitt (1989, and at this Conference), focuses more specifically on the argument that transaction costs related to farming are smaller in family farms, so that economies of scale are insufficient to compensate higher transaction costs in commercial farms. All this points us towards household production theory as an integration of the neo-classical theory of producer and consumer behaviour. However, here too there is hardly anything new. Almost four decades ago Heady (1953) wrote: 'Motivational forces behind the farm producing unit are consumption-inspired as well as profit-inspired ... The optimum use of resources in production or the optimum allocation of income in consumption cannot be defined unless the two basic sets of economic relations are related'.

This approach also clarifies the similarity between the kibbutz and the family farm. Indeed, Kislev points out that in both cases business and household accounts are combined and that the labour of the extended kibbutz family is not recorded as an expense. Some similarities could also appear with farming systems based on extended families throughout the world. Here, too, the unity of the firm and the household can be a source of conflict, either because it can result in the exploitation of family labour or because it leaves ample scope for free-riding.

The institutional framework of the moshavim is clearly that of a cooperative characterized as a voluntary association of people, in which capital subscription does not form the basis of voting power and in which the reward is primarily seen as the patronage rebate or discount based on the value of business done with the society. Kislev points out some conflicts of interest in the management of the moshavim, such as cost allocation according to patronage. Cobia (1989) formulates this problem in more general terms, in relation to the heterogeneity in size and structure of farms mentioned earlier, stating: 'The size disparity among farmers challenges existing cooperatives to serve patrons with very different needs'. Olson (1965) is even more specific: 'Unless the number of individuals in a group is quite small, or unless there is coercion or some other special device to make individuals act in their common interest, rational self-interested individuals will not act to achieve their common or group interests'. This phenomenon of conflicting individual and group interests is also experienced in some cases of group farming in Western

Europe (Martens, 1973) and certainly helps us to understand the financial crisis of the kibbutzim to which Kislev is referring.

Somewhat to my surprise, the author states that 'institutional competition has been unfair to cooperatives'. I would like to know how the author arrives at such a conclusion, especially since in the same paper we also read that 'cooperation in agriculture has always been supported by the government in many ways', and that the greatest involvement was in credit, with negative real interest rates. Moreover, kibbutzim cultivate national land leased on a long-term basis. My personal, very subjective, impression is that in Israel institutional competition has been unfair to family farms and to non-family corporate farms. If fair competition prevailed, and considering the unbalanced age distribution of the labour force as well as the fact that some of the initial objectives behind kibbutzim and moshavim are somewhat outdated, I would rather expect that family farms or non-family corporate farms would take over. Of course, the answer could be a matter of definition and it is my impression that many arguments concerning institutional arrangements in farming have to do with a lack of clearly defined concepts.

REFERENCES

Boussard, J.M., 1985, 'Changing environment and structural heterogeneity in agriculture', in *Proceedings*, 19th IAAE Conference.

Cobia, D., 1989, *Cooperatives in Agriculture*, Prentice-Hall, New York.

Heady, E.O., 1953, *Economics of Agricultural Production and Resource Use*, Prentice-Hall, New York.

Martens, L., 1973, 'Nouvelles formes de collaboration dans le domaine de la production agricole'; in *Commission Communautés, European, Information internes sur l'Agriculture*, 110.

Newby, H., 1985, 'The changing structure of agriculture and the future of rural society', in *Proceedings*, 19th IAAE Conference.

Nielsen, A.H., 1985, 'The family farm in a changing technological and economic environment', *European Review of Agricultural Economics*, 14, (1).

Olson, M., 1965, *The Logic of Collective Action*, Harvard University Press, Boston.

Schmitt, G., 1989, 'Farms, farm households and productivity of resource use in agriculture', *European Review of Agricultural Economics*, 18, (3/4).

HELEN H. JENSEN*

*Contributions of Women and
Household Members to the Economy in Rural Areas*

INTRODUCTION

The economies of rural areas in the United States have been in economic transition since settlement. Today, changes in agricultural production have fostered an economic environment in rural areas which is moving away from that traditional reliance on agriculture and towards greater economic diversity. The expended use of capital-intensive agricultural technologies and the increase in farm size have led to changes in the farm sector, and spurred an outward migration of labour from agriculture. Many formerly prosperous regions and residents face problems of economic adjustment and, in many cases, changed standards of living.

The ability to adapt to changes in the economic environment is a valuable human resource (Schultz, 1975). Households faced with changes in economic conditions may follow alternative adjustment strategies to stabilize or improve incomes. These adaptive strategies include moving labour resources off the farm to non-farm labour markets. The ability to adjust to these economic changes has important implications for well-being and incomes in rural areas.

In the United States, as well as in many other economies, agriculture may not be able to provide an economic base sufficient to sustain population and income levels consistent with those in the non-agricultural sectors of the economy. This situation underscores the importance of the non-farm sector to rural incomes. Although there is no single well-developed paradigm to describe the development process in rural areas of the United States, the strategies that households have used to meet these changing conditions provide insight into ways in which households adapt to the new economic environment of rural areas. Successful approaches to improving incomes often involve enhancing the ability of farm households to gain access to the economic base within the non-farm economy.

This paper is based on evidence from the United States and develops three major themes. First, the agricultural base of traditional rural areas alone is unlikely to support households in non-metropolitan areas[1] through the economic adjustments without significantly changing their standard of living relative to residents of metropolitan areas. Development policies that will successfully enhance economic growth in rural areas must entail increased

*Iowa State University, USA.

off-farm income generation. Second, farm households have successfully used the non-farm economy of rural areas to enhance household income. This is an adaptive strategy and has occurred primarily through off-farm employment and most often through the activities of women and household members other than the farm operator. Third, both for farm and non-farm households, a major limitation to higher incomes is lack of access to higher paying jobs in rural areas. These conditions imply that more effective development strategies will benefit households in rural areas. The objective here is to provide a better understanding of the contributions of household members in rural areas to generate non-farm income in the United States, and the economic environment in which this occurs.

LINKS BETWEEN AGRICULTURE AND THE RURAL ECONOMY

Returns to agricultural production have historically been the backbone of rural economies. However, agriculture has become more limited in its ability to support the economies of rural America. Evidence suggests that there will be relatively slower growth both for agriculture and the associated rural economy. Furthermore, dependence on the non-farm rural economy will increase.

There are several reasons for the more limited economic outlook for rural areas. First, the long-term projections of real prices of major agricultural commodities, especially those important in the Mid-west, such as soybeans, corn and wheat, are projected to be relatively constant (Food and Agricultural Policy Research Institute, 1991). These projections incorporate consensus macro-economic conditions and current agricultural policies, which include relatively large subsidies to the agricultural sector. Should policies change, such as through ongoing GATT trade negotiations, the outlook on prices may improve. However, without major policy changes, the projected farm income base of the rural economy is likely to continue its downward secular decline.

Second, the growth in non-farm sectors in primarily agricultural states has been relatively slow. Evidence for the mid-western states in the United States, an area that has generated a relatively large share of US farm income, shows that non-farm employment in the Mid-west has grown more slowly than that in the United States as a whole during the 1970s and 1980s (Johnson *et al.*, 1989). Employment in agriculturally related industries has grown more slowly than employment in non-agricultural manufacturing. Off-farm income as a share of total cash income of farm operator households have averaged nearly 50 per cent in the 1980s (US Department of Agriculture, 1990).

Another indicator of the changes in the economic environment in rural areas is the presence of relatively high levels of poverty, as shown in Table 1. In the past, rural poverty in the United States has been characterized by pockets of poverty. Today, rural poverty has become more widely dispersed, with rates higher than those in urban areas among selected groups of the population: the elderly, children and female-headed households.

TABLE 1   *Poverty in non-metropolitan and metropolitan areas, 1985*

	Poverty rate (%)	
	Rural	Urban
Total population	18	12
Elderly	18	11
Children	24	19
White	16	10
Non-White	24	22
Persons in female-headed households	44	36

Source: US Department of Commerce, Bureau of the Census (1985).

ADAPTIVE RESPONSE OF RURAL HOUSEHOLDS

Households which face such structural re-alignment in non-metropolitan areas make decisions about allocating physical and human resources in order to achieve efficiently their desired standard of well-being. With declines in the agricultural economic base, we would expect to see labour resources adjusted from farm work into non-farm work.

The farm household labour allocation model (see Singh, Squire and Strauss, 1986; Huffman and Lange, 1989) provides a theoretical structure for the decision-making process within which household members allocate resources to gain income and other non-wage benefits. Household labour is available from the operator, spouse and other members of the household. The allocation of labour and other physical resources occurs as household members allocate time among farm and non-farm productive activity and leisure time, choosing to allocate marginal units where the marginal return is greater than the marginal value of other uses of their time resource. Improvements in the returns from off-farm work increase participation in work off the farm. In general, the findings of empirical studies show that labour supply has a positive supply elasticity and human capital has a positive effect on off-farm work participation and supply.

Enhanced ability of households to respond in such an economic environment is determined by their ability to adapt efficiently to changes (Schultz, 1975; Huffman, 1985). Differences in adaptive ability lead to different decisions made with respect to change. We expect households with superior adaptive ability to perceive the change (or disequilibrium) better, to evaluate the situation better and to make more efficient decisions on allocating resources with respect to the changing conditions. Thus human capital in the form of adaptive abilities leads to differences in behaviours and differences in observed objective measures of well-being such as money income.

Both farm and non-farm rural residents encounter the same off-farm labour opportunities. When returns in the agricultural sector fall relative to the non-

agricultural sector, adjustments can be expected, particularly through movement of labour resources out of farming and into non-farm employment. Such
a strategy may not only raise the level of income for the farm household, but
also is likely to reduce the risks associated with depending solely on farm
income and the greater variability attached to farm income.

Evidence of the importance of off-farm employment to the income of farm
households in the United States is widely available. Off-farm work by members of the farm household has become a well-established strategy for using
farm-based resources. Does non-farm income enhance the economic well-
being of farm households? Recent evidence comes from the USDA 1988
Farm Costs and Returns Survey (El-Osta and Ahearn, 1991) which categorized
farm operator households into income quartiles and by asset levels. As shown
in Table 2, non-farm income, which includes non-farm business, wages and
salary, represents a significant share of income for all farm operator households.
The off-farm incomes are relatively lower for the lowest quartile, but high for
farm households in the upper quartile. Non-farm income of high-asset farms
averaged more than $95 000 initially. The largest share of non-farm income
for the upper quartile came from non-farm business, wages and salary. Why
are the farm households in the highest income quartile successful? Among
major reasons, as Table 2 shows, are that more operators and their spouses
worked more hours off the farm; more considered occupations other than
farming as their major occupation.

In another study, Tokle and Huffman (1991) analysed the joint work participation decisions of farm and non-farm households in rural areas where both
husband and wife were present. They found that wage work participation
decisions by males and females in households are joint decisions. This suggests that the observed off-farm labour market activity results from an intra-
household allocation process; and, consistent with earlier studies, schooling
had a positive effect on the probability of wage work for the married farm and
non-farm males and females. As expected, children aged under 18 had a
negative effect on the probability of work by married females. When leisure
was assumed to be a normal good, higher farm output prices reduced the
probability of both husband and wife participating in wage work for farm
households.

In addition to increasing the overall level of income, the diversification of
income sources may also help to stabilize household income through market
wages and, potentially, through additional compensation in the form of fringe
benefits.

THE NON-FARM ECONOMY

Despite evidence linking the economic choices of farm households to the non-
farm economy, the interdependencies between the two sectors are not well
understood. Labour markets in rural areas become a critical link in the ability
of farm, and non-farm, residents to gain access to the off-farm economic base.

The empirical evidence drawn from the work by Tokle and Huffman indicates that both rural farm and non-farm households respond to economic

TABLE 2   Characteristics of US farm operator households and farm operator households based on income levels of farm operator households and farm asset values, 1988[a]

Item	Lower quartile		Middle quartile		Upper quartile	
	<$500 000 in farm assets	>=$500 000 in farm assets	<$500 000 in farm assets	>=$500 000 in farm assets	<$500 000 in farm assets	>=$500 000 in farm assets
Number of farms	365 015	71 990	798 576	75 935	307 499	129 730
Share of all farms (%)	21	4	46	4	18	7
Farm operator household characteristics						
Income by source ($)						
Net cash farm income	−8 768	−48 963	1 150	6 620	18 734	59 922
Nonfarm business, wages, and salaries	3 053	5 251	14 894	8 802	48 012	75 100
Interest and dividends	583	2 660	1 445	3 832	4 352	12 423
Income from all other nonfarm sources	2 350	2 023	4 065	4 137	4 636	8 214
Average total income	−2 782	−39 029	21 554	23 391	75 734	155 659
Operator characteristics						
On-farm average hours of work (per week)	32	51	27	48	28	48
Off-farm average hours of work (per week)	10	4	19	6	25	11
Major occupation (%)[b]						
Farming	66	89	48	86	40	77
Other	34	11	52	14	60	23
Spouse characteristics						
% of farm operators with spouses	75	77	89	85	92	93
On-farm average hours of work (per week)	8	15	7	14	8	12
Off-farm average hours of work (per week)	9	7	13	10	21	11
Major occupation (%)[b]						
Farming	23	31	18	32	17	22
Home-making	51	49	48	48	29	47
Other	26	20	34	20	55	32

Notes: [a]The cut-off incomes for lower quartile, middle quartile and upper quartile are $8401 or less; greater than $8401, or $38 240 or less; and greater than $38 240; respectively.
[b]Numbers may not add up to 100 per cent, owing to rounding.

Source: El-Osta and Ahearn (1991). Data from the *Farm Costs and Returns Survey*, 1988. Note that farms that are organized as non-family corporations, or managed by an operator who does not share in the net income of the business are excluded.

conditions in local labour markets when making labour supply decisions. Increased wages and expectations of improved labour market offers (specifically those viewed as permanent changes) led to higher levels of participation in off-farm labour markets. The effects of anticipated economic changes through employment growth and unemployment rates appear to be stronger for males in the labour force than for women (Tokle and Huffman, 1991).

If households rely on employment in rural areas, how 'good' are the jobs? Are there differences in the returns from employment for those working in non-metropolitan areas compared to metropolitan areas? These could be due to different types of jobs, different costs of living, or relatively slow adjustment of labour and lack of mobility caused by the fixity of capital (for farm households) and an unwillingness to migrate.

There is some evidence that compensation offered by non-metropolitan employers is not as high as that offered by those in metropolitan areas for the same occupations (Jensen, 1982; Jensen and Salant, 1985). A national survey of employer compensation practices indicated that neither average wages nor non-wage compensation was as high in non-metropolitan areas. Both the structure of industry and the location contributed to this outcome.

A 1980 national sample of women who head families provides evidence on rural–urban differences in women's earnings. Among women who worked full-time, those living in rural areas had earnings comparable to those living in small towns; the earnings of the rural women were significantly lower than the earnings of those in central cities or suburban areas. The differences were found to be due more to differences in pay scales than to occupational differences (Cautley and Slesinger, 1989). That is, lower pay does not seem to come from having different occupations, although aggregated occupational classification may mask some differences in jobs.

Other evidence from recent surveys of farm households shows that women who worked off the farm earned less than males. This was the case both when compared at different levels of education (Saupe, 1990) and by industry of employment (Salant, 1983), as shown in Tables 3 and 4. These differences indicate that rural women face different off-farm job prospects than men, and that the differences are not fully explained by education.

CONCLUSIONS

Changes in agricultural technologies and in the size of farms are altering the economic base of rural areas. Households in rural areas are in the process of adapting to these economic changes. The major shift has been to move labour resources off the farm, away from farm-based home activities and into non-farm jobs. The dependence of farm households on agriculture as a source of income has diminished.

Women and other household members have increased off-farm work. The success of this adaptive strategy to meet the declines in agricultural prospects is apparent. When women (farm operator spouses) earn off-farm income, the economic prospects for the farm family household are improved.

TABLE 3   *Average off-farm wage rates of individuals in south-western Wisconsin farm households, 1986 (dollars per hour)*

| | | | Other farm household members | | | |
| | | | < Age 25 | | ≥ Age 25 | |
Category	Male farm operator	Wife	Male	Female	Male	Female
Years of schooling completed						
< 12 years	6.95	3.68	3.47	2.50	6.00	3.68
12 years (high school grad)	8.29	5.83	4.90	5.00	7.73	6.61
> 12 years	13.06	8.15				
Average	10.09	6.71				

Source: Saupe (1990).

TABLE 4   *Mean hourly wage rates, by sex of worker, selected occupations and industries, 29-county Mississippi – Tennessee area, 1980 (dollars)*

Job classification[a]	Female	Male
Industrial		
Manufacturing		
Durable	4.52	6.36
Non-durable	3.88	5.14
Trade (wholesale & retail)	4.28	5.26
Services		
Education	5.88	6.71
Other services	4.69	4.99
Occupational		
Administrative, professional & technical	6.36	7.66
Marketing, sales & clerical	4 71	6.45
Service	3.41	3.88
Production work	4.03	6.10

Note: [a] Standard Occupational Classifications and Standard Industrial Classification.

Source: Salant (1983).

It is unlikely that agriculture will lead the economic development of rural areas in the United States. Other industries, including those that can overcome geographic separation through new communications technologies, are likely to be relatively more important in rural areas. Public policies and programmes that enhance the ability of rural household members to have access to such jobs will improve employment prospects and incomes. This would include information services to reduce labour market transaction costs. For women, especially, this would include childcare services, and, although this is less easy to identify, we need to understand better the basis for lack of equality in pay levels. This becomes increasingly important to incomes of farm households as women move to work in off-farm jobs.

NOTES

[1]The terms 'non-metropolitan' and 'rural' are used interchangeably throughout this paper.

REFERENCES

Cautley, Eleanor and Slesinger, Doris P., 1989, 'Labor Force Participation and Poverty Status among Rural and Urban Women Who Head Families', in Harrell R. Rodgers, Jr. and Gregory Weiher (eds), *'Rural Poverty: Special Causes and Policy Reforms*, Greenwood Press, New York.
El-Osta H. and Ahearn, M., 1991, 'Income Distribution of U.S. Farmers', (USDA/ERS, unpublished manuscript).
Food and Agricultural Policy Research Institute, 1991, *FAPRI 1991 World Agricultural Outlook*, Staff Report 2–91. Iowa State University, Ames, and the University of Missouri-Columbia.
Huffman, W.E., 1985, 'Human Capital, Adaptive Ability, and the Distributional Implications of Agricultural Policy', *American Journal of Agricultural Economics*, 67, pp. 429–34.
Huffman, W.E. and Lange, Mark, 1989, 'Off-Farm Work Decisions of Husbands and Wives', *Review of Economics and Statistics*, 71, pp. 471–80.
Jensen, Helen, H., 1982, 'Analysis of Fringe Benefits for Nonmetropolitan versus Metropolitan Employee Compensation', *American Journal of Agricultural Economics*, 64(1), pp. 124–8.
Jensen, Helen H. and Salant, Priscilla, 1985, 'The Role of Fringe Benefits in Operator Off-Farm Labor Supply', *American Journal of Agricultural Economics*, 67, (5), pp. 1095–9.
Johnson, Stanley R., Otto, Daniel, Jensen, Helen and Martin, Sheila A., 1989, 'Rural Economic Development Policies for the Midwestern States', in Harrell R. Rodgers, Jr. and Gregory Weiher (eds), *Rural Poverty: Special Causes and Policy Reforms*, Greenwood Press, New York.
Salant, Priscilla, 1983, 'Farm Women: Contribution to Farm and Family', *Agricultural Economics*, Research Report No. 140, Economic Development Division, Economic Research Service, US Department of Agriculture.
Saupe, William E., 1990, *Status of Wisconsin Farming, 1990*, University of Wisconsin-Madison.
Schultz, T.W., 1975, 'The Value of the Ability to Deal with Disequilibria', *Journal of Economic Literature*, 13, pp. 827–46.
Singh, I., Squire, L. and Strauss, J. (eds), 1986, *Agricultural Household Models*, Johns Hopkins University Press, Baltimore, MD.
Tokle, J.G. and Huffman, Wallace E., 1991, 'Local Economic Conditions and Wage Labor Decisions of Farm and Rural Nonfarm Couples', *American Journal of Agricultural Economics*, 73, (3).
US Department of Agriculture, Economic Research Service, 1990, *Agricultural Income and Finance Situation and Outlook Report*, AFO-39, December.
US Department of Commerce, Bureau of the Census, 1982, *1978 Census of Agriculture, Special Report, Part 6, 1979 Farm Finance Survey*, Report AC 78-SR-6.

US Department of Commerce, Bureau of the Census, 1985, *Poverty in the United States*, Series P-60, No. 158.

DISCUSSION OPENING – HIROYUKI NISHIMURA*

Dr Jensen is able to show increased trends in off-farm and non-farm income earned by women in the United States, using empirical results derived by herself and by others. The effects have been favourable both in terms of increasing and in stabilizing total household income. This phenomenon has also been observed in Japan, as well as in other countries, though it can occur in different forms and the policy implications which follow can have different impacts on agricultural structure.

I would like to begin by raising a question concerning increased off-farm employment. Are the impacts of dependency on non-farming jobs taken by members of farm households favourable when viewed in a long-term perspective? Dr Jensen's view seems optimistic. I think the phenomenon is not always favourable, either for the farming business or for family relations in households. Usually, it is difficult to achieve harmonious and complementary relationships between farming and non-farm employment. Reliance on supplementary income does not, in itself, provide the incentive to increase productivity or efficiency in agriculture. Much depends on the type of labour which is diverted towards non-farm employment.

In Japan, for example, it is widely recognized that farm operators (normally husbands) have tended to take off-farm jobs, and their spouses, with aged family members, perform an increasing amount of farmwork. This appears to contrast to the situation in the United States. Japanese women usually have to put in more physical work and cope with the additional problems of acquiring knowledge of rapidly changing technology of farming. To me this appears to be a short-term expedient rather than a long-term foundation for successful farm business operation.

It is also important to consider the case of developing countries. Commonly, the family income derived from small-scale agriculture is not sufficient to maintain large families, and they come to depend on different sources of income from a number of jobs. They are not specialized and do not operate their farms on an efficient scale. In order to achieve higher productivity, and reduce production costs, farm structure must alter. Off-farm and non-farm income is a vital supplement when farm technology is backward, but it is again short-term in relation to the need to improve the farming base.

I found it extremely interesting to listen to a paper dealing with the particular conditions of the United States, but I do want to emphasize that its lessons may be country-specific and that the general theme needs to be explored on a case-by-case basis. In short, the influence of non-farm activities in the economy of rural areas can influence agriculture in complex, and not always helpful, ways.

*Kyoto University, Japan.

STEVEN HAGGBLADE AND CARL LIEDHOLM*

*Agriculture, Rural Labour Markets
and the Evolution of the Rural Non-farm Economy* [1]

INTRODUCTION

The size of the rural non-farm economy depends primarily on agricultural demand. As farm income grows, it generates spillover growth in the rural non-farm economy, since rising farm income increases rural purchases of non-farm goods and services. The well-known debates on agricultural growth linkages revolve around how powerful these demand linkages are (Mellor, 1975; Johnston and Kilby, 1975; Bell and Hazell, 1980). Yet agriculture affects the supply of non-farm goods and services as well. Operating primarily through the labour market, these supply-side linkages have been largely overlooked in the growth linkage discussions. This is unfortunate, because a focus on the labour market alters conclusions about the magnitude of farm–non-farm linkages. It also highlights how agriculture affects not only the size but also the composition of the rural non-farm economy.

This paper explores the relationship between agricultural growth, the rural labour market, and the size and composition of rural non-farm activity. It begins by reviewing what is known about the rural non-farm economy in developing countries, followed by a review of empirical evidence on the relationship between agriculture, labour markets and the transformation of the rural non-farm economy. The paper then introduces a simple price-endogenous model that projects the non-farm employment, wage and income effects of alternative forms of agricultural growth. The model highlights the labour market interactions that contribute importantly to a shifting composition of rural non-farm activity.

PROFILE OF THE RURAL NON-FARM ECONOMY

Static profile

Non-farm activities form an important and integral part of the rural economies of developing countries. They provide 20 to 45 per cent of full-time employment and 30 to 50 per cent of rural household income (Chuta and Liedholm,

*Bodija Associates and Michigan State University, USA, respectively.

1979; Haggblade and Hazell, 1989; Liedholm and Kilby, 1989). Amid wide variation, the composition of rural non-farm employment typically includes one-third manufacturing and one-third commerce, with services, mining and construction making up the remainder (Chuta and Liedholm, 1979). Most non-farm enterprises are quite small. Self-employed, one-person firms predominate. Unlike the case of the formal wage labour force, women constitute 40 per cent or more of those engaged; frequently they account for the majority of the rural non-farm entrepreneurs. Because of extremely low capital requirements and seasonal demand, most businesses operate with excess capacity (Liedholm and Mead, 1987).

Dynamic profile

Employment data, the only indicator routinely available, suggest that rural non-farm activity has generally increased across continents and over time (Anderson, 1982; Chuta and Liedholm, 1979; Haggblade and Hazell, 1989; Liedholm, 1990), yet employment growth can signal good news or bad. In prosperous regions, where rising wages and buoyant demand stimulate growth in increasingly productive non-farm activity, non-farm employment growth signals prosperity. But in stagnant rural regions, a surge in non-farm employment may reflect the bad news that population growth is forcing non-farm activities to act as a sponge, soaking up excess workers in marginal, low-paying jobs (Shand, 1986). Differences in wage rates and the composition of non-farm activity help in interpreting the employment data to distinguish between the two.

In prosperous regions, employment growth concentrates increasingly in rural towns and in full-time enterprises with hired employees. The composition of activity also changes, with a decline in very labour-intensive, often household-based activities and an increase in higher-investment, higher-productivity enterprises. Transport, food preparation, repair and other services normally grow, while household manufacturing industries decline. A great deal of churning accompanies this aggregate growth; 10 per cent or more of total enterprises disappear each year, while other, new firms emerge. Among the deceased, one-person firms predominate (Liedholm, 1990).

Women typically bear the brunt of this adjustment. They predominate in weaving, basket making, pottery and many of the household-based activities that generally decline. While many growing non-farm services – milling, food preparation and many domestic services – normally employ women, the necessary capital investment in mechanical milling, transport, some food processing and manufacturing can form an intimidating barrier preventing them from participation in this transformation and growth. Although rural transformation offers improved opportunities, for non-farm labourers and for the rural poor in general, women's access to the larger, full-time, higher-investment and higher-productivity non-farm businesses is not assured. Access to investment funds and education combine with child rearing and other household obligations to constrain women as they try to respond to new opportunities.

TRANSFORMATION OF THE RURAL NON-FARM ECONOMY

A complex interaction of forces drives the evolution of the rural non-farm economy. On the demand side, growth in agricultural income, changes in urban and foreign preferences, and income transfers from urban areas all influence the growth and composition of demand for non-farm goods and services. On the supply side, natural resource availability, technological change, the supply of investment capital as well as physical and institutional infrastructure all influence the magnitude and shape of the rural non-farm economy.

Agriculture, however, plays a central role in this process. As the principal source of rural income, agriculture generates the principal source of demand for rurally produced consumer and intermediate goods. Through the rural labour market, agriculture also affects the supply side of the rural non-farm economy. As farm production and income grow, they generate increased demand, not only for more production inputs, but also for rurally produced consumer goods. Recent estimates suggest that agricultural growth multipliers generally lie in the range of 1.3 to 1.8, which means that every dollar of technologically induced agricultural income generates an additional 30 to 80 cents in rural non-farm income (Haggblade and Hazell, 1989). Irrigated rice regions in Asia growing high-yielding varieties (HYVs) generate the largest multipliers, while traditional smallholder regions in Africa and Latin America produce the smallest. About two-thirds of the total agricultural growth multipliers stem from the consumption linkages, with the production linkages providing the remainder.

Rapid agricultural growth also affects the composition of rural non-farm activity in two important ways. First, where agricultural income growth outpaces population, rising per capita agricultural income leads to consumption diversification into a broader array of non-foods, many of which are produced in rural areas. Second, on the supply side of the rural non-farm economy, agricultural growth affects the rural wage and hence the opportunity cost of labour available for non-farm activities. This induces a movement away from many low-return non-farm activities towards those that are more remunerative. In contrast, in regions where agricultural growth lags and employment prospects in agriculture cannot keep pace with population growth, low-return non-farm activities proliferate, with no increase in wage rates. In these cases, the rural non-farm economy becomes an employer of last resort, a sponge, absorbing by default labour force increments unemployed in agriculture. Whether buoyant or lagging, agriculture plays a key role in the structural transformation of the rural non-farm economy.

Recent evidence from Bangladesh describes this combined effect of agricultural growth on the composition of rural non-farm activity (Table 1). Employment in services, the highest-return non-farm activities, increase dramatically in prosperous agricultural regions. In contrast, villagers reduce time spent in low-return cottage industries, in earth hauling and in petty trading. Within cottage industry and trading, the doubling and tripling of labour returns suggests a considerable shift in the composition of activity.

Labour market interactions are of major significance. Green revolution farm technology has typically increased demand for farm labour. In its early

TABLE 1   *Differences in the size and composition of rural non-farm activity in agriculturally developed and under-developed[a] regions of Bangladesh, 1982*

	Income per hour in under-developed regions (taka/hour)	Per cent by which agriculturally developed regions exceed under-developed areas		
		Income/hour[b]	Employment, hours per week	Income per household
Agriculture	5.14	29	8	40
Non-agriculture				
Services	11.41	4	30	35
Cottage industry	4.35	90	−81	−63
Wage labour [c]	2.82	6	−41	−38
Trade	2.30	195	−28	113
Total non-agriculture	4.35	59	−29	12

Notes: [a]Agriculturally developed and under-developed regions are distinguished by a number of criteria: access to irrigation, use of modern rice varieties, and fertilizer consumption, among others. In the agriculturally developed regions, modern varieties cover 60 per cent of cropped area, compared with only 5 per cent in the under-developed areas.
[b]Calculations based on Hossain (1988), Tables 48 and 64.
[c]Non-farm wage labour includes earth hauling, construction, transport and 'other' employment.

Source: Hossain (1988, pp. 95, 120).

phases, biological innovations increase labour demand by between 20 and 40 per cent (Jayasuriya and Shand, 1986; Lipton, 1989). In contrast, the mechanical technologies normally lower the demand for agricultural labour. Village-level studies reveal declines ranging from 6 per cent in India (Sisler and Coleman, 1979) to 8 per cent in Sierra Leone (Byerlee, Eicher, Liedholm and Spencer, 1977) and 26, 33 and 34 per cent in Thailand, the Philippines and Indonesia, respectively (Jayasuriya and Shand, 1986). Normally, mechanical innovations, especially in threshing and soil preparation, arrive after the biological ones. Induced by rising rural wages, they reduce initial gains in farm labour demand.

Labour supply, in the short run, depends on households' willingness to forego leisure. In the medium and long run, it depends on population growth and ease of migration. Most household studies indicate short-run household labour supply as being inelastic, in the range of 0.1 to 0.26 (Singh, Squire and Strauss, 1986), yet, over time, aggregate estimates point to a growing rural labour force in all regions, spurred importantly by the growth of population (Anderson and Leiserson, 1978).

Trends in the rural wage rate reveal the relative strength of these supply and demand forces in the rural labour market. Real wages have increased in some areas following the introduction of biological innovations in farm technology, for example in the Punjab region of India (Chanda, 1986), Thailand and Malaysia (Lipton, 1989), yet, in countries with similar new farm technology –

Indonesia, the Philippines and Mexico – real rural wages declined or stagnated, indicating that increases in agricultural demand were insufficient to offset increases in the rural labour supply. In countries with mechanical innovations or with stagnant agricultural sectors, such as most of Sub-Saharan Africa, real rural wages have frequently declined (Griffin, 1989).

Changing rural wage rates signal a shifting opportunity cost of labour in rural non-farm activity. They raise costs of non-farm production but at the same time offer prospects of higher-productivity employment for landless and poor households who have only their labour to sell. Changing wage rates affect the rate of non-farm output growth as well as the composition of rural non-farm activity. A formal model of the farm–non-farm rural economy – one that includes a labour market – allows us to trace these different effects more clearly.

MODELLING LABOUR MARKET
LINKAGES AND THE RURAL NON-FARM ECONOMY

Virtually all earlier work has modelled rural non-farm activity as a purely demand-driven spin-off of agricultural income growth. Normally, analysts have not embellished the supply side of the rural non-farm economy; they simply assume non-farm output supply to be perfectly elastic. Implicitly, this assumption requires excess capacity in fixed non-farm inputs as well as a perfectly elastic supply of non-farm labour.

This model begins to build up the supply side of the rural non-farm economy by adding a labour market to the standard demand linkage models of rural non-farm growth. Modelling the classic demand linkages allows estimates of the impact of agricultural growth on the size of the rural non-farm economy. The addition of a labour market enables the tracking of changing wage rates and employment and hence offers a window onto the shifting composition of rural non-farm activity.

The model (see appendix) compares two sources of growth in rural non-farm activity: (a) technological change in agriculture; and (b) population growth. The first raises farm income, thereby increasing demand for rural non-farm output and simultaneously raising demand for non-farm labour (Figure 1, Panel 1). To the extent that new agricultural technology requires additional labour, labour demand and wage rates will rise even further. Note that, where labour supply is upward-sloping, the inclusion of the labour market dampens non-farm income and output response from N_1' to N_1. Population growth, on the other hand, increases labour supply, lowers wage rates, spurs demand for labour and thereby increases rural non-farm employment (Figure 1, Panel 2). By contrasting the changes resulting from these two driving forces, the model examines analytically the characteristics of non-farm activity in stagnant and growing agricultural regions.

Within prosperous agricultural regions, the model considers three forms of technological progress: labour-neutral, labour-using and labour-saving. Figure 1 depicts labour-using technological change, the most common experience in the green revolution. Labour-neutral change would differ only in that the

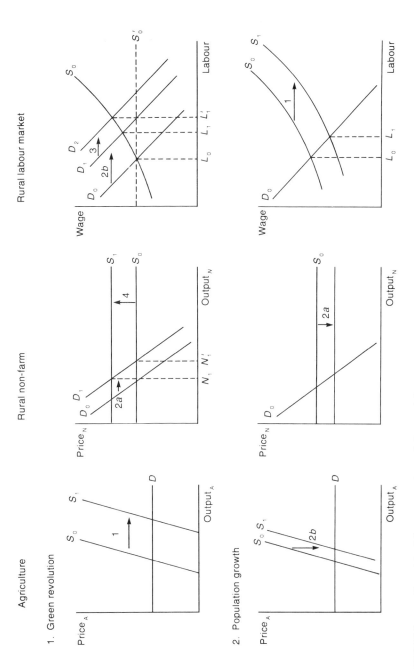

FIGURE 1 *Graphical interpretation of the model*

labour demand shift 2b would not occur, and hence wage increases and the cost-push inflation in the rural non-farm supply curve would diminish. Labour-saving technological change would further dampen wage increases and hence lead to the largest non-farm output response.

The three technological options can be thought of, respectively, as investment in irrigation infrastructure that allows expansion and replication of existing agricultural technology, introduction of high-yielding packages of seed and fertilizer, and mechanization. Because the biological packages are perfectly divisible and normally labour-using, many associate them with employment-oriented, small farmer growth strategies. Mechanization, which displaces labour, is normally associated with large-farmer growth, what Johnston, Kilby and Mellor call a bimodal agricultural growth strategy.

The model presented here is a slightly embellished verson of one developed by Haggblade, Hammer and Hazell (1991). It includes two sectors, one tradable and one non-tradable. For simplicity, this application assumes all agricultural commodities are tradable outside the rural region. Given the predominance of foodgrains and cash crops in much of the Third World, this assumption is not unreasonable. In contrast, the model assumes non-farm activity to be non-tradable. This likewise does not depart dramatically from reality, since non-farm income typically accounts for over 80 per cent of incremental non-tradable income (Haggblade and Hazell, 1989).

The model takes the price of agricultural tradables as fixed outside the rural region, invoking the standard assumption that imports will stabilize prices at border cost plus transport. In contrast, because non-farm goods and services are not tradable, the model must determine their price endogenously. The model incorporates a simple neo-classical rural labour market. Rural households supply labour in response to the real wage rate and population pressure. Farms and non-farm businesses demand labour as a function of the nominal wage and technology. In response to shifting labour supply and demand, the rural wage rate adjusts until the labour market clears. A single rural wage prevails in both farm and non-farm activity.

Although the model accommodates any production function technology, the following experiments adopt very simple assumptions. In both agriculture and non-farm activity, Leontief technology governs the demand for intermediates. Non-farm businesses enjoy excess capacity in fixed inputs. For agriculture, land and technology constrain output supply, making it inelastic. The two exogenous shifters in this system, agricultural technology and population growth, trace out changes in the model's four endogenous variables: the price of non-farm output (P_n), the nominal rural wage rate (w), the rural inflation rate (I) and rural income (Y). Since the formal mathematics have been developed elsewhere (Haggblade, Hammer and Hazell, 1991), they are not discussed in this paper. The present model differs from the original version in two ways: it introduces population as a determinant of labour supply and it considers alternative forms of technological change in agriculture.

The data for the following experiments represent a stylized Asian rice-growing economy. The production parameters are drawn from the Muda River region of Malaysia. Data from a wide range of sources scale the stylized economy as follows. Foodgrains account for 25 per cent of both income and employment. New agricultural technology increases output by 80 per cent among adopting farmers and increases their foodgrain income by 50 per cent. Farmers accounting for 50 per cent of the cropped area adopt the improved technology. Labour-using technology increases labour demand in foodgrains by 20 per cent (low) to 40 per cent (high), while labour-saving technology reduces foodgrain labour demand by 20 per cent.

The series of experiments summarized in Table 2 suggest four principal conclusions about the relationship between agriculture and evolution of the rural non-farm economy:

(1) *Rising wage rates dampen non-farm income growth.* If the rural labour supply is perfectly elastic, a demand injection from any new agricultural technology will stimulate the same increases in rural non-farm income and employment. The rural wage rate will not rise, even in the face of increasing labour demand by both farm and non-farm businesses (Table 2, Experiment 1). With no cost-push inflation in non-farm supply, spin-off growth in non-farm activity is highest in this setting.

But unlimited supplies of labour rarely occur, and, where labour supply comes only at increasing wage rates, labour-using technology will generate the smallest increase in rural non-farm income.[2] In the stylized rice-growing region described in Table 2, mechanization, or similar labour-saving farm technology, raises non-farm income by an amount equal to 3 per cent of total rural income. Yet labour-using biological innovations raise non-farm, income by only 1.1 – 1.7 per cent, one-third to one-half as much (Table 2, Experiment 2).

The smaller income multipliers result because, when agricultural technology increases the demand for labour, it raises the rural wage rate. This raises the cost of production in non-farm activity and hence the price of rural non-farm output. At the higher price, rural households demand fewer non-farm goods and services. The dampened output response lowers rural non-farm income. Of course, smaller income multipliers do not represent unambiguously bad news. The opposite side of a dampened non-farm income growth is higher wage rates and consequently improved living standards for labour-selling households, typically the very poor. Note that the rural wage rises 6.6 per cent under labour-using agricultural technology and only 1.7 per cent when increased farm output results from introduction of labour-saving technology. Total rural employment also increases most with labour-using agricultural change, growing by 6.6 per cent compared to 1.7 per cent in response to labour-saving technology (Table 2, Experiment 2).

So the pure labour market effect suggests a trade-off between employment and growth in alternative agricultural development strategies. Models that consider only demand linkages ignore this tension. To the extent that small-farmer growth strategies are synonymous with labour-using technological change, the labour market effects suggest that a small farmer focus may lower income growth in return for greater equity and employment. Of course, consumption patterns, savings rates and investment propensities may also differ among large and small farmers. So the conventional wisdom in favour of a small farmer focus (Mellor, 1975; Johnston and Kilby, 1975) cannot be over-

TABLE 2   *Modelling the impact of the green revolution and population growth on the rural non-farm economy in a stylized Asian rice-growing economy*

	Initial change in agriculture		Rural wage	Rural non-farm		Total rural	
				Resulting change (% of regional totals)			
				Real income per capita[a]	Employment	Real income per capita[a]	Employment
	Income	Employment					
Labour supply elasticity infinite							
1 Green revolution (improved agricultural technology)							
(a) labour-saving	6.3	-2.5	0.0	3.7	5.2	10.0	2.7
(b) labour-neutral	6.3	0.0	0.1	3.7	5.2	10.0	5.2
(c) labour-using, low	6.3	2.5	0.1	3.7	5.2	10.0	7.6
(d) labour-using, high	6.3	5.0	0.1	3.7	5.1	10.0	10.1
Labour supply elasticity = 1							
2 Green revolution (improved agricultural technology)							
(a) labour-saving	6.3	-2.5	1.7	3.0	4.3	9.3	1.7
(b) labour-neutral	6.3	0.0	3.3	2.4	3.5	8.7	3.3
(c) labour-using, low	6.3	2.5	5.0	1.7	2.7	8.0	5.0
(d) labour-using, high	6.3	5.0	6.6	1.1	1.9	7.4	6.6
3 The sponge (population growth with stagnant agriculture)							
population growth, 6.0%	0	0	-3.9	-4.7	1.9	-4.4	2.1
4 Green revolution plus population growth							
(a) slow population growth, 1.8% for 4 yrs = 7.4%	6.3	5.0	1.8	-0.4	4.3	1.9	9.2
(b) rapid population growth, 2.8% for 4 yrs = 11.7%	6.3	5.0	-1.0	-3.8	5.7	-1.3	10.7

Notes: [a]Real income includes a deduction for inflation in the price of non-farm goods and services. Using the small country assumption, however, the price of agricultural tradables in the rural region remains unchanged. Note that the per capita adjustment only affects experiments 3 and 4.

turned on the wage-dampening effects alone. A final pronouncement will require simultaneous comparison of demand, investment and labour market linkages, an important excursion that ventures beyond the scope of the current paper.

(2)   *The composition of rural non-farm activity changes most following labour-using technological change in agriculture.* Rural wage rates rise most in the face of growing labour demand in agriculture – 6.6 per cent compared to 1.7 per cent with labour-saving farm technology (Experiment 2). This jump in the opportunity cost of non-farm labour signals a sizable shift in the composition of rural non-farm activity. Evidence from Table 1 and elsewhere suggests the shift involves an increase in high-value services, trade and a decline in low-productivity non-farm activity, often very labour-intensive manufacturing, and most prominently female-dominated cottage industries.

In contrast, where population pressure outpaces agricultural output growth, returns to farming labour decline. In these settings, the rural non-farm economy operates as a sponge, absorbing labour force into increasingly low-paying activities. This scenario plays out frequently in South Asia and Sub-Saharan Africa, where observers lament growing rural non-farm employment as a signal of diminished opportunities. Experiment 3 in Table 2 describes this situation: wage rates and per capita income decline, while non-farm employment increases in increasingly unrewarding activity.

(3) *Employment data alone can mislead.* Because of this, employment data can be dangerously misleading if considered by themselves. As Table 2 indicates, rural non-farm employment grows at the same rate, 1.9 per cent, in both Experiments 2(d) and 3. Yet trends in rural welfare differ dramatically in the two settings. Where new technology makes agricultural advance possible, the rising non-farm employment brings with it rising income and rising returns to labour and a shift to increasingly remunerative activities. The poor benefit especially as the labour they sell brings increasing remuneration.

To avoid misinterpreting employment data, students of the non-farm economy must track changes in rural wage rates together with the employment figures. Activity breakdowns of the employment data, if sufficiently detailed, can likewise signal shifts in the composition of non-farm activity and enable diagnosis of employment growth as a harbinger of opportunity or malaise.

(4) *The race between population and technology.* The last panel of Table 2 measures the impact of population growth together with new agricultural technology. Panel 4(b) indicates that population growth of 2.8 per cent per year, over four years, will nullify the wage and income gains resulting from typical new foodgrain technology. This result closely matches the common empirical finding of stagnant or declining real wages in areas where rapid population growth accompanies the green revolution.

CONCLUSIONS

Labour market linkages between agriculture and rural non-farm enterprises highlight the potential trade-off between employment and growth in alternative agricultural development strategies. Because labour-using agricultural technology raises wage rates, it dampens non-farm output supply response and reduces income gains as well. Thus the pure labour market effects suggest that an employment-oriented small farmer strategy will lead to lower growth than labour-saving farm technology, except where labour supply is perfectly elastic. Of course, since consumption and investment patterns may also differ between small and large farmers, this result does not constitute the final word on the small versus large farm debate.

Agriculture affects not only the size but also the composition of rural non-farm activity. Through the labour market and the rising opportunity cost of non-farm labour, agricultural growth fosters a shifting composition of non-farm activity. Although much of the literature on structural transformation highlights changing sectoral shares, this review suggests that intrasectoral shifts, especially within manufacturing, may be equally important in assessing rural welfare.

Women are especially vulnerable. They predominate in the declining, household-based activities and at the same time enjoy opportunities in the growing, high-return market segments. Although they have the most to gain from a shift to higher-return non-farm activities, institutional rigidities on many occasions make this difficult. To facilitate transformation of the rural non-farm economy, policy makers will need to pay particular attention to opportunities and constraints facing women, both in agriculture and off the farm.

NOTES

[1] The helpful comments of Carl Eicher, Peter Hazell, John Staatz and John Strauss on an earlier draft are gratefully acknowledged.

[2] These experiments compare technological options for raising farm output. All raise foodgrain income by 50 per cent, equivalent to a 6.3 per cent increase in rural income. This green revolution income injection assumes that foodgrains constitute 25 per cent of rural income and 50 per cent of all foodgrain output shifts from traditional to improved varieties. Thus 0.5 x 0.25 x 0.5 = 6.25 per cent. The technologies differ only in that some demand more labour in agriculture, while others demand less.

REFERENCES

Anderson, D., 1982, 'Small industry in developing countries: a discussion of the issues', *World Development*, 10, pp. 913–48.

Anderson, D. and Leiserson, M., 1978, *Rural Enterprises and Non-farm Employment*, World Bank, Washington, DC.

Bell, C. and Hazell, P., 1980, 'Measuring the indirect effects of an agricultural investment project on its surrounding region', *American Journal of Agricultural Economics*, 62, pp. 75–86.

Byerlee, D., Eicher, Carl, Liedholm, C. and Spencer, D., 1977, 'Rural Employment in Tropical Africa: Summary of Findings', *African Rural Economy Paper No. 20*, Michigan State University, East Lansing, Michigan.

Chanda, G. K., 1986, 'The off-farm economic structure of agriculturally growing regions: a study of the Indian Punjab', in R. Shand (ed.), *Off-farm Employment in the Development of Rural Asia*, Australian National University, Canberra.

Chuta, E. and Liedholm, C., 1979, 'Rural non-farm employment: a review of the state of the art', M.S.U. Rural Development Paper No. 4, Michigan State University, East Lansing, Michigan.

Griffin, K., 1989, *Alternative Strategies for Economic Development*, St. Martin's Press, New York.

Haggblade, S. and Hazell, P., 1989, 'Agricultural technology and farm–non-farm linkages', *Agricultural Economics*, 3, pp. 345–64.

Haggblade, S., Hammer J., and Hazell, P., 1991, 'Modeling Agricultural Growth Multipliers', *American Journal of Agricultural Economics*, May.

Hossain, M., 1988, *Nature and Impact of the Green Revolution in Bangladesh*, Research Report No. 67, International Food Policy Research Institute, Washington, DC.

Jayasuriya, S. and Shand, K., 1986, 'Technical change and labour absorption in Asian agriculture: some emerging trends', *World Development*, 14, pp. 415–28.

Johnston, B. and Kilby, P., 1975, *Agriculture and Structural Transformation: Economic Strategies in Late Developing Countries*, Oxford University Press, London.

Liedholm, C., 1990, 'The dynamics of small-scale industry in Africa and the role of policy', *Gemini Working Paper No. 2*, Development Alternatives Inc., Washington DC.

Liedholm, C. and Kilby P., 1989, 'The role of nonfarm activities in the rural economy', in J. Williamson and V. Panchamukhi (eds), *The Balance Between Industry and Agriculture in Economic Development*, St Martins Press, New York.

Liedholm, C. and Mead, D., 1987, 'Small-scale industries in developing countries: empirical evidence and policy implications', *MSU International Development Paper No. 9*, Michigan State University, East Lansing, Michigan.

Lipton, M., 1989, *New Seeds and Poor People*, Johns Hopkins Press, Baltimore, MD.

Mellor, J., 1975, *The New Economics of Growth: A Strategy for India and the Developing World*, Cornell University Press, Ithaca, New York.

Shand, R., 1986, *Off-farm Employment in the Development of Rural Asia*, Australian National University, Canberra.

Singh, I., Squire, L. and Strauss, J., (eds), 1986, *Agricultural Household Models*, Johns Hopkins Press, Baltimore, MD.

Sisler, D. and Coleman, D. R., 1979, 'Poor rural households, technical change, and income distribution in developing countries: insights from Asia', *AERS – 79/13*, Cornell University, Ithaca, New York.

APPENDIX

A labour market linkages model

Six equations summarize the formal model:

$$T(P_t, P_n, w, \theta) = H_t(P_t, P_n, Y) + D_{tt}(P_t, P_n, w, \theta) + D_{tn}(P_t, P_n, Y) + G_t + V_t + X_t \quad (1)$$

$$N(P_t, P_n, w) = H_n(P_t, P_n, Y) + D_{nt}, P_t, w, \theta) + D_{nn}(P_t, P_n, Y) + G_n + V_n \quad (2)$$

$$L_s(\bar{w}, Z) = L_{dt}(P_t, P_n, w, \theta) + L_{dn}(P_t, P_n, Y) \quad (3)$$

$$\bar{w} = w/I \quad (4)$$

$$I = P_{nb} P_{t(1-b)} \quad (5)$$

$$Y = \pi_t(P_t, P_n, w) + \pi_n(P_t, P_n, w) + w \cdot L_s \quad (6)$$

The first two set supply equal to demand in agricultural tradables (*T*) and non-tradable non-farm activities (*N*). Supply of both depends on input and output prices, that is on the price of non-tradables (P_n), tradables (P_t) and the wage rate (*w*). In addition, the supply of tradables is influenced by a technology shift parameter (θ). Through its effect input demand in tradables (D_{tt}, D_{nt} and L_{dt}), θ offers the flexibility to model a wide array of new technology, including neutral, input-using or input-saving technical change.

The demand for tradables and non-tradables depends on household consumpion (H_i) of each, intermediate input requirements (D_{ij}) and exogenous government (G_i) and investment (V_i) demand for each sector's output. In addition, because it can be imported or exported, tradable demand includes net exports (X_t) from the region. Household consumption (H_i) depends on relative commodity prices $(P_n$ and $P_t)$ as well as household income (Y).

A full-employment, neo-classical labour market clears through equation (3), which sets labour supply (L_s) equal to sum of labour demanded in each sector $(L_{dt}$ and $L_{dn})$. Labour demand depends on nominal input and output prices, while supply is a function of the real wage rate $(\bar{w})$ and population (Z). The inflation rate (I) is defined in equation (5). Finally, equation 6 defines regional income as the sum of profits (π_n, π_t) and wages $(w \cdot L_s)$.

When solved, the model traces changes in four endogenous variables $(P_n, w, I$ and $Y)$ in response to exogenous changes in agricultural technology (θ) and population (Z). Using (^) to represent percentage changes, the model's solution becomes:

$$\hat{E} = A^{-1} [B \cdot d\theta + C \cdot \hat{Z}]$$

where $\hat{E}$ is a 4×1 column vector representing percentage changes in the four endogenous variables, A^{-1} is a 4×4 matrix of multipliers, and B and C and 4×1 column vectors containing shift parameters for each exogenous variable. Because it is additive, the model can solve for any combination of exogenous shifts, or it can isolate the effect of any single exogenous shock.

The full model mathematics have been presented elsewhere (Haggblade, Hazell and Hammer, 1991) for A and B. For the population vector in C as well as the parameter restrictions associated with alternative forms of technical change in agriculture (θ), a technical Appendix is available on request from the authors.

DISCUSSION OPENING – DANIEL A. SUMNER*

No area of economics research is more important than the study of rural labour markets. It remains true that, for most people on earth, understanding rural labour markets is required to understand the factors that determine their well-being and economic prospects. It is particularly important to incorporate labour market considerations into the analysis of the linkages between farm and non-farm rural sectors, just as labour market considerations are important to improve our analysis of rural and non-farm rural sectors of the economy.

Haggblade and Liedholm have provided some examples of the importance of the labour market characterisitics to an accurate understanding of the effects of technical change and population growth. They have emphasized that to ignore labour market linkages is to risk misunderstanding the consequences of factors that drive economic development in rural economies. On these points, they are surely right and have provided a useful reminder to analysts and policy officials alike.

In these comments, I want to clarify and extend some points made by the authors. My remarks can be summarized in three categories: (1) clarification of income effects of labour-using and labour-saving agricultural investments;

*US Department of Agriculture and North Carolina State University, USA.

(2) the importance of labour market relationships in understanding distribution impacts of agricultural investments; and (3) the importance of farm–non-farm labour market linkages in rural areas of developed countries.

First, I want to highlight a feature of the example simulation presented in Table 2 of the paper. In characterizing the four improvements in technology as labour-saving or labour-using, the authors emphasize the different impacts on wage rates, employment and income. They point out a trade-off between total rural employment and income, particularly in the relevant case of an upward-sloping supply curve of labour. But notice that the technologies labelled as labour-saving or labour-using differ fundamentally. The labour-saving technology uses less of all resources. It is really a pure gift to the economy. For the same 6.3 per cent increase in agricultural income the labour-saving technology uses 2.5 per cent less labour and no more of any resources. The high labour-using technology uses 5 per cent more labour and no less of any resource. Obviously, an economy gains more income from a pure improvement in technology, which is like manna from heaven, than from one that uses more of the scarce economic resources of the economy. It would be interesting to compare technologies that differed by being either a free lunch or not. The authors suggest mechanization as an example of labour-saving technology and, in that case, the capital-using characteristics are evident.

Clearly, this short paper by Haggblade and Liedholm is not attempting a fully realistic explanation of the consequences of technological change, but it is important not to make too much of the income trade-offs inherent in labour-saving versus labour-using technologies. One contribution of the model is to show, in a simple example, that even technologies which themselves reduce the use of labour can benefit rural employees (as seen by the wage rate gain of 1.7 per cent even if on-farm employment is reduced).

When a technological adoption increases the demand for labour in a rural area, the added labour comes from three sources: more hours per worker, more labour force participation by locals, and migration into the area. Leaving aside inward migration, the consequence of the demand increase in the rural labour market is reduced leisure and less time spent on other household activities. Therefore, because leisure is a 'good' and household hours are also productive, the gain in well-being is less than the gain in measured income. Only if the supply curve of labour were vertical would the higher earnings be a pure gain to the workers.

Second, Haggblade and Liedholm allude to the income distribution effects of agricultural investments and I want to reinforce their comments with an example from rural India. These comments are based on a very preliminary examination of data from a village-level survey at the Institute for Crop Research in the Semi-Arid Tropics in Hyderabad by Rolf Mueller and myself.

Some have suggested that policies to encourage individual investments in ground water irrigation in central India would be regressive because the larger farms have more access to capital. This means that the large farms where the the necessary wells have been dug would profit most. However, irrigation does increase the demand for labour. Our preliminary examination of wage and employment data indicates that the incomes of the poorest segment of the population – the landless labourers and the small farmers who also work off their farms – gain substantially from irrigation. So, in addition to profit for the larger farms, there are benefits for the very poor. To ignore the labour market consequences in this case is to get the distributional effects of the technology wrong. If irrigation was to be left out of the policy agenda, ignoring the labour market effects would also mean that we would be getting the policy wrong. The paper by Otsuka, in this volume, provides massive evidence from several Asian economies illustrating the importance of labour markets in understanding the distributional impacts of technology.

Finally, the issues of rural labour markets raised by Haggblade and Liedholm apply not only to developing economies but also to relatively wealthy countries. However, there are major differences in rural economies between rich and poor countries. These include the share of the economy accounted for by farming and, perhaps, the level of integration between rural and urban markets. In the United States, for example, less than 10 per cent of rural employment is in farming. Further, farms in the developed countries not only sell on the world market but also buy more of their inputs from off the farm and from outside the local rural economy. The highly developed information and distribution systems of rich countries mean that goods and services flow more easily between rural and urban sectors.

These basic economic differences suggest some differences in the way we should think about farm and non-farm rural labour market interactions in rich and poor economies. With few exceptions, technology adoption on farms is a relatively minor issue for rural labour markets in wealthy countries. The small share of the labour force employed on farms, and the larger share of farm and household inputs acquired from the urban sector, means that on-farm changes have less effect on the local rural economy. In the United States, for example, there is growing recognition that rural development policy is not a farm policy question. For an elaboration of this point, see a large body of work from the Economic Research Service of the US Department of Agriculture (Parker and Whitener, 1989) and a paper presented at the International Forum on Rural Development (Sumner, 1991a).

In both wealthy and poor countries, one feature of the farm–non-farm labour market connection is the tendency for farmers and farm family members to work off the farm. In the case of rich countries, almost all off-farm work is at non-farm jobs, whereas off-farm work on another farm is more common in poor countries.

The linkage between non-farm local labour market conditions and the opportunities for off-farm work deserves more attention. In the United States, at most 60 per cent of farm operators cite farming as their primary occupation, and some of those are people who have retired from a non-farm job and earn most of their income from non-farm sources. (US Department of Commerce, Bureau of the Census, 1989). To date, most of the research on off-farm work by farmers tends to treat it as a choice to work off the farm made by farmers, rather than a choice to work part-time as a self-employed farmer by someone whose primary occupation is in the non-farm labour market (Parker and Whitener, 1989; Sumner, 1991b). For this research we need more survey data that include both farm and non-farm households and information on human capital, farm characteristics and time allocation.

Off-farm work is also important in poor rural labour markets and, while not dealt with explicitly by Haggblade and Liedholm, is consistent with their model. When they refer to a labour-saving technology, it may well be an innovation that allows a farmer to satisfy the demand for labour on the farm by working only part-time on the farm and, thus, be available to work off the farm in the non-farm labour market. It may also be the spouse of the farmer who seeks off-farm employment to supplement the farm family income.

The analysis of Haggblade and Liedholm has considered an important set of issues. Agricultural economists often spend so much effort on commodity market or farm management issues that they tend to leave aside the labour market. These examples of the way labour market conditions affect the consequences of technologies and how different technologies affect labour market outcomes are useful reminders. The interactions between the farm and non-farm sectors are complex and important, and deserve continuing research effort.

REFERENCES

Parker, T.S. and Whitener, LA., 1989, 'Farmers and Their Search for Off-Farm Employment', *Rural Development Perspectives*, Economic Research Service, US Department of Agriculture, vol. 5, issue 2, February.

Sumner, Daniel A., 1991a, 'Rural Revitalization: Facts and Notions from the United States', *International Forum on Rural Development*, Tokyo, Japan, 30 August.

Sumner, Daniel A., 1991b, 'Useful Directions for Research on Multiple Job Holding Among Farm Families', *Multiple Job Holding Among Farm Families in North America*, Iowa State University Press, Ames.

US Department of Commerce, Bureau of the Census, 1989, *Census on Agriculture: 1987*, Washington, DC.

RYOHEI KADA*

Capital Formation of the Farm Household and Resource Allocation in Agricultural–Rural Sectors: An Analysis of Sustainability of Japanese Agriculture

INTRODUCTION

The terminology of 'sustainability' is quite new to Japanese agriculture. The main reason for the lack of the concept of sustainable development is the fact that rice-based Japanese agriculture has developed and continued for nearly two thousand years under the temperate, monsoon climate. Needless to say, rice is produced in paddy fields, which necessarily contain a certain depth and amount of water during the growing period. Hence rice production has possessed an ecologically and biologically sustainable system, which has enabled farmers to produce rice in paddy fields year after year, for a considerable time.

However, such a sustainable system has been gradually changing since the early 1960s, when Japan entered a highly industrial economic development stage. Part-time farming has been widespread among rice farmers, and rice production has also changed from one of relatively low input and low output, to one of high input and high output. In other words, Japanese agriculture has rapidly increased its productivity by means of intensified capital inputs, which has at the same time caused such problems as deterioration of soil fertility, soil and ground water pollution, and other conservation problems. Recently, Japanese people have become more and more concerned about these issues, as global environment problems such as global warming and desertification have become serious global issues.

The main purpose of this paper is three-fold. First, we outline the development and major characteristics of capital formation among farm households in Japan. Second, we examine the impact of such capital investment on income and employment patterns, and on the sustainability of land use systems. Finally, using these findings we discuss the future perspectives for long-term development and research needs.

*Faculty of Agriculture, Kyoto University, Japan.

CHANGE IN EMPLOYMENT-INCOME
STRUCTURE OF JAPANESE AGRICULTURE

One of the major factors restricting the development of Japanese agriculture has been the limited land resources for agricultural purposes. About 66 per cent of the total land area in Japan is mountainous. Of the remaining 34 per cent, 20 per cent is given over to urban use, which leaves the agricultural land area in Japan at about only 14 per cent.

When compared to the major agricultural producing countries, the percentage of arable land in Japan is one of the lowest. In particular, the land given over to meadows and pasture in Japan is only a tiny percentage of the total land area and illustrates the restrictions placed on dairy and livestock production in this country. The land use figures of Japan basically show the differing nature of Japanese agriculture – the large percentage of forested area, and limited arable and pasture land, make for a highly intensive farming structure.

An overview of the general trend in the farm population and in the number of farms in Japan is necessary for the purpose of this section. Before the Second World War, the Japanese farm population was stable at around 35 million. Since the war the number has declined, from 37 million in 1950 to 20 million in 1985. The share of the farm population in the total population has gradually decreased, from around 45 per cent just after the war to 16.4 per cent in 1985. The rate of decline is, however, rather slow compared with that in the United States. The decline in the number of farms was even slower than the decline in the farm population.

Like the farm population, the number of farms in prewar days was stable at around 5.3 to 5.5 million, but, in the postwar period, the number of farms declined gradually, from 6.2 million in 1950 to 4.2 million in 1988. Since the total area of cultivated land in Japan has not changed very much throughout these periods, the average size of holding has been very stable at around one hectare, although the present trend shows a slight increase in average size. It should be noted, however, that the composition of the working population has changed significantly. Also changed is the inner structure of farm families, a change due especially to off-farm employment and income. In other words, the increase of part-time farm families is the most noticeable postwar change in Japanese agricultural and rural structure.

Part-time farming *per se* is not a recent phenomenon in rural Japan. Almost 30 per cent of farm households were classified as part-time farm households even in prewar days. Opportunities for rural people in those days were to work as craftsmen or local merchants, or to run small cottage industries. After the Second World War, however, the number of full-time farm households decreased quite drastically, from about 2 million in 1960 to only 0.06 million in 1990. At the same time, the absolute number and ratio of Type II part-time farm households (where off-farm income is greater than farm income) increased from 1.9 million in 1960 to 3.0 million in 1990.

The ratio of off-farm income to total family income has steadily increased, from 12.2 per cent in 1921 to 67.8 per cent in 1975, with an especially remarkable increase after 1960. The share of off-farm income exceeded that of farm income in 1963 for the first time. Moreover, it should be noted that per capita disposable income of the farm population exceeded that of the non-farm population in 1972 for the first time in Japanese history. The trend thereafter has been a further widening of the gap between the two.

The variety of off-farm jobs, in the postwar period especially, has widened from rural or agriculture-related employment to include urban manufacturing or service-related jobs. This widening of job opportunities was made possible through decentralization of industry and concomitant development of transport. Now off-farm jobs are available not only for the young, but also for middle-aged and elderly members of the farm family. In addition, the nature of off-farm jobs has changed from casual, part-time to permanent (or at least long-term) employment.

Increased off-farm employment by farm family members means a reduction in the working population engaged in the family's agricultural work. The number of the working population in agriculture fell dramatically during the period 1950–85, decreasing from 16 million to 5 million. Also changed was the composition of the working population. The decrease of young workers in agriculture was quite drastic between 1960–65, when the Japanese economy expanded more than 10 per cent per annum. Thus, combined with the changes in the number of farm households mentioned above, the working population per farm household decreased from 2.6 persons in 1950 to 1.6 persons by 1975.

TECHNOLOGICAL CHANGE
WITH INTENSIFIED CAPITAL FORMATION

Technological change has been a remarkable feature in postwar Japanese agriculture. Just after the war the pressing need to increase the food supply was achieved mainly by the development and diffusion of yield-increasing techniques such as a new rice bedding system, breeding of more high-yielding varieties, and the rationalization of fertilizer inputs. After the late 1950s, however, as Japan entered a labour-scarce economy, labour was replaced by

TABLE 1   *Agricultural capital formation in Japan (nominal; unit:1 billion yen)*

Item Year	1960	1965	1970	1975	1980	1985	1988
1. Agricultural fixed capital	240.4	523.9	1091.7	2453.5	3674.4	3670.9	3929.5
Land improvement	99.6	212.4	406.8	872.6	1765.9	1763.3	2063.9
Agr'l buildings	30.2	82.5	213.7	427.1	635.2	544.9	517.5
Farm machinery	84.1	166.0	376.2	968.5	1045.3	1146.0	1122.8
Plants & trees	12.3	30.4	44.2	83.4	76.9	80.4	72.1
Livestock	14.2	32.6	50.8	101.9	151.1	136.3	153.2
2. Changes in inventory	7.5	27.6	3.4	20.1	15.5	−20.6	8.2
Crops	−3.3	0.0	−0.3	0.3	−52.6	−15.8	−63.3
Livestock	10.8	−26.3	3.0	13.8	63.0	23.1	53.3
Farm equipment	0.0	−1.3	0.7	−12.0	5.1	−27.9	18.2
3. Total capital formation	247.9	496.3	1095.1	2445.6	3689.9	3650.3	3937.7

machinery in the work of land tillage, and later, in the late 1960s to 1970s, the substitution was further intensified by machines doing the work of harvesting and rice transplanting (see Table 1 on the change in farm investment).

Farm mechanization has increased even among part-time farm families, where the labour shortage in the family is keenly felt. In other words, the pattern of part-time farming has been sustained by the development of labour-saving technologies. Historically, Japanese agricultural development has been achieved by means of land-saving or labour-intensive technology. Now, however, more machinery is used instead of labour, and land intensity is dropping.

In terms of land intensity, winter crops have been abandoned as the trend towards part-time farming has increased. Part-time farm families have specialized in single cropping of rice, which combined with decreasing per capita consumption of rice, brought about serious rice surplus problems in the late 1960s.

CHANGES IN THE NATURE OF
RURAL AREAS AND COMMUNITIES

Starting in the early 1960s, the high economic growth of Japan has changed not only urban–metropolitan areas, but also rural areas, under the wave of continuous industrialization and modernization. Technological progress within the agricultural sector is characterized by modern farm mechanization and other heavy capital investments in farming systems. This has certainly lessened the burden of human labour inputs, but, in the process, more and more young people have left rural areas and most agricultural labour has been taken over by elderly people and women.

This exodus of the rural labour force has, on the one hand, endangered the maintenance and management of rural community functions. There are basically two categories of rural areas involved in this process. In the relatively flat rural areas close to urban centres, the majority of farm households became more dependent upon non-agricultural incomes and employment. Furthermore, more and more non-agricultural city people have moved to the rural areas to live. The mixture of agricultural and non-agricultural households in the boundaries of rural areas is very common in this flat zone. On the other hand, in more remote and mountainous areas, such rural exodus has caused what may be called 'depopulation' problems, where many communities have been deserted and have lost major social and economic functions, especially during the 1960s and early 1970s.

However, when we look at the essential nature and function of rural villages, most of the rural communities still maintain their institutional and cultural functions to maintain their agricultural land, farm production and rural life. It is a surprising fact that the last three decades of high economic growth have undoubtedly changed the physical and material life of rural areas, but have not changed the institutional systems and social relationships within the village. The reasons for this are important for our understanding of present-day rural Japan in a comparative sense.

Let us next examine the changing nature of farmers and farm households. In the past, most farmers were dependent upon solely small-scale, paddy-based agriculture. But in the long-term process of Japanese industrialization and modernization, such a homogeneous structure of farms has given way to a more heterogeneous structure. In brief, we can say that there exist three basic types of farm households in contemporary rural Japan. The first category is full-time farmers, who believe agriculture is the most important family business to be continued over generations. Although the number of these farms is very small and still shrinking in the overall agricultural population, they are 'real farmers', undertaking modern, capital-intensive types of agriculture, and are thus considered to play an important role in Japanese agricultural production. The problem is, however, that they do not possess much political power in decision-making processes because of their small numbers.

The second group is part-time farming households, which comprise the majority of rural households. Their heavy dependence upon outside incomes and time allocation to non-agricultural activity has tended to shift their concerns towards things urban and modern. But the important point is that, when it comes to the issues of agricultural land use and rural community problems, their attitudes are still very conservative and they try to maintain the traditional functions of the community. This is largely due to their way of thinking about the possession and succession of assets and status holding within rural communities.

The third group is farm households where only elderly people are living, maintaining more or less subsistence agriculture. Statistically speaking, about 10 per cent of the total of farm households are classified in this category, but the category is not important in agricultural production or political power in rural areas. However, as the age structure of rural areas shifts to one in which the elderly are increasing, this category of farmers will also increase in the future. Another point which must not be forgotten is that there are more and more middle-aged and retired people who are returning to the rural areas, taking up again the farm households which they had left in their younger days.

One of the difficulties facing rural communities today is this heterogeneity of farming populations, particularly when it comes to land consolidation projects and land-use planning in general. Rationalization of land use by means of increased farm size has been constrained by the difficulty of creating a consensus among villagers. Interestingly enough, decision making in rural communities in Japan is not by majority vote, nor is it by election. Rather, their rule might be expressed as unanimous decision making, after protracted discussion among all rural community members. It may appear that this is old-fashioned, but it was actually quite wise of villagers to have such a well defined and sustainable community function. It has been their biggest concern not to split the communities among villagers, which in the past meant disastrous consequences in their lives in poorly resourced and over-populated rural Japan.

Another reason for this unanimous decision-making process is that resource conservation, including farmland, water and community roads, is only made possible by the participation of all the members. The unit of agriculture in Japan is not really farming by individual families. It is rather more a system of the 'community farm', where the whole village maintains its territory and functions cooperatively; for example, cropping patterns are in effect decided upon as a group, according to unspoken social conventions.

Probably the nature of Japanese rural areas can be identified by the following three key ideas: (1) continuity of rural community functions, (2) adjustability to outside changes, and (3) egalitarian principles and unanimous decision-making processes. For hundreds of years, rural community people have cooperatively maintained their agricultural production and rural life. Egalitarian principles are the basis of their attitudes even in modern times. Illustrations of this can be easily found in such examples as the maintenance of water use and drainage systems, common rural roads they use in their daily life, and other communal activities such as agriculture-related festivals and religious activities. Irrespective of their farm size, economic well-being, or any other indicator of social class, each farm household has equal opportunities and equal political rights.

Another example of this is the method of setting aside paddy fields, where in Japan quotas were almost equally distributed using the same percentage of reduction among rice growers. Certainly, by doing this, these communities have sacrificed to a great degree the efficiency of rice production from a narrowly defined economic point of view. But they have enjoyed a relatively harmonious, long-standing community life instead.

CHANGES IN AGRICULTURAL PRODUCTION AND LAND USE

Agricultural production in Japan is currently facing over-supply problems in several major farm products, including rice and mandarin oranges, while low self-sufficiency commodities such as wheat, soybeans and feed-grains emphasize the need for diversification. The decline in the growth of products such as rice is countered by the growth of livestock products, from 15 per cent of total agricultural output in 1960 to 27 per cent in 1987. Japan's area for cultivation is small in comparison with her competitors, thus the increase in livestock is connected with her need for a highly intensive farming system, and increases the need for her to import feed.

The pattern of land use in Japan has also been changing constantly. The most noticeable change has been the overall reduction in Japan's agricultural land. Total agricultural land use has been reduced by 8 per cent, from 5796 thousand hectares in 1970 to 5317 thousand hectares in 1988. This reduction is basically a result of the difference between agricultural land expanded and land ruined. The total area of cultivated land expanded was 23.7 thousand hectares in 1988, and the total area of cultivated land ruined in the same year was 46.6 thousand hectares, which includes 40.4 thousand hectares for artificially transformed land. In Japan, there has been a strong external demand for agricultural land, and though the Japanese government has endeavoured to conserve the superior farmland, overall figures have decreased. It is remarkable that the upland field area has increased by 2 per cent, while the paddy field area has decreased by 15 per cent during the same period.

In 1988, paddy fields constituted 54 per cent, continuing to dominate the total agricultural land area. Of the 46 per cent of upland field area, 24 per cent was normal upland field, 12 per cent meadows and 10 per cent land under permanent field. Although the importance of meadow land to the Japanese economy is growing, it still only constitutes 12 per cent of Japan's total agricultural land area.

A few words should be said about the Rice Crop Diversion Policy which was initiated in 1971, and the Paddy Field Reorganization Policy which was introduced in 1978. In recent years, the Japanese government has endeavoured to combat the surplus situation in rice production through diversification. The total diversification area in 1988 was 817 thousand hectares, which was 15.4 per cent of the total cultivated area. The paddy use diversion from rice is concentrated upon vegetables, soybeans, forage crops, wheat and barley. For example, in 1988, 817 thousand hectares diverted from rice included 120 thousand hectares for vegetable production, 135 thousand hectares for forage crops, 134 thousand hectares for wheat and barley and 96 thousand hectares for soybeans. The fact that forage crops constitute the largest area of diversification from rice helps the shift towards livestock production. Diversification is also being promoted by the increase in multi-purpose paddy-fields, which can produce a variety of other crops as well as rice.

It is not surprising to note that the actual planted area has been reduced by 12 per cent during the period from 1970 to 1987. The major crops reduced are obviously rice, of which the area decreased from 2923 thousand hectares in 1970 to 2146 thousand hectares in 1987, a reduction of 26.6 per cent, while wheat and barley have also been reduced on a large scale – total reduction is around 20 per cent. The question of farm size and productivity is obviously dictated by this intensive agricultural system. The paddy field area reached its peak in 1965 at 0.71 ha per farm household, and by 1988 had been reduced to 0.61 ha.

On the other hand, the growth of livestock and greenhouse production is still continuing at a steady pace. Dairy cattle have increased 8.4 times, from 3.4 head per farm household in 1965 to 28.6 head in 1988; beef cattle have increased 7.8 times, from 1.3 head in 1965 to 10.2 head in 1988; hogs have increased 35.8 times, from 5.7 head in 1965 to 203.9 head in 1988; layers have increased from 27 head in 1965 to 1356 head in 1988, which is 50.2 times the earlier number; chickens have increased 28.6 times, from 892 in 1965 to 25 500 in 1988. The number of greenhouses per farm household has also undergone a tremendous increase (see Table 2 on these farm size changes).

Before the Second World War, higher productivity was achieved by a number of methods. These included improved rice varieties, use of animal power and improved fertilization. The small farm management system was also improved through a heavier input of both fertilizer and labour. Following the war, however, it was brought about by labour-saving techniques, mainly through mechanized farming. The difference in productivity by cultivation size basically shows that, the smaller the area of agricultural production, the longer are the hours worked per unit area, and the higher the cost of farm production. For example, farmers with agricultural areas of less than 0.3 ha. worked 82 hours

TABLE 2   *Change in farm size*

	1965	1988
Paddy field (ha)	0.71	0.61
Dairy cattle (head)	3.4	28.6
Beef cattle (head)	1.3	10.2
Hog (head)	5.7	203.9
Layer (head)	27.0	1 356.0
Chicken (head)	892.0	25 500.0

per 0.1 ha in 1982. Those farmers with more than 5 ha worked only 32 hours per 0.1 ha.

Under the monsoon climatic conditions of high summer temperatures and high humidity, Japan has high productivity of vegetation growth. The average production of organic matter of Japanese forests, for example, is 13 to 23 tonnes (dry matter basis) per hectare, which is often as much as tropical forests producing 10 to 35 tonnes, as contrasted to Mediterranean Europe where average forest production is between 2.5 and 15 tonnes. Such high production capacity has enabled Japan to produce rice as the most productive, population-supportive crop for nearly 2000 years. In paddy fields, soil tends to possess the following characteristics, as rice is continuously planted over long periods. First, since the paddy field is covered by water during summer, decomposition of organic matter in the soil is very slight even at high temperatures. During winter, too, the soil becomes dry, but decomposition is minimal owing to cold temperatures. As a whole, decomposition of organic matter can be minimized in the paddy field. In addition, rice straws and roots are retained, which aids soil fertility to a certain degree. As a result, rice production tends to be relatively easily stabilized over long periods of time.

Secondly, ferrous ions deoxidize to ferric ions in a watered paddy field. The acid condition of soil becomes neutral, which makes rice plants grow well. Thirdly, phosphorous ions separated from ferrous compounds become available for rice plants in that condition. Furthermore, many minerals dissolved in water are supplied to the paddy fields during the irrigation period. These minerals can be conserved and used as nutrients for growing plants since water does not leak out through the base soil of the field. Finally, soil erosion caused by heavy rainfall is minimized when terraced paddy fields are located stepwise on the slope of a mountain.

For all these reasons, Japanese farmers have endeavoured to create paddy fields, as much as possible, out of the total agricultural land area. In areas where winter is not very cold, paddy fields have been utilized to produce barley and other upland crops in winter. This system of crop rotation has long been established in various parts of Japan. Generally speaking, since the average farm size in Japan is very small, farmers have tried to produce as much as possible per unit of land, and adopted a crop rotation system. In this system, livestock and human manure, as well as residues of leaves, are widely used to supplement soil fertility. This established a self-contained, recycling system of paddy land use.

However, when Japan entered a high economic growth period, Japanese agriculture started to change very dramatically. More and more farm household members took up non-farm employment, and the agricultural population started to decline very rapidly. As a result, many rice farmers became part-time farmers, with only rice being produced in a monoculture system. The remaining farmers tended to specialize in vegetable or livestock. Among rice farmers, there was heavy use of machinery, chemicals, and chemical fertilizer. The intensity of land use became very much lower. In livestock farming, feeds were mostly purchased to increase the number of livestock.

These changes have not only broken down the crop rotation system, but also created agricultural pollution problems, such as an accumulation of intensive livestock manure. On the other hand, soil fertility on crop farms gradually deteriorated owing to the heavy dependence upon chemicals and chemical fertilizer.

These days Japanese people are eating more and more livestock and dairy products, less and less rice. Demand-side change has necessarily reduced the acreage of paddy production. The Japanese government has encouraged rice farmers to divert to other crops, with government subsidies. However, these crops diverted from rice are not as profitable. Some paddy fields have become idle, without any production. Naturally, the self-sufficiency rate of food production in Japan has rapidly decreased.

CONCLUDING REMARKS : A TIME OF CHANGE

Since the high economic growth period, the nature of Japanese agriculture has shifted towards more and more energy-intensive and resource-exploitative techniques. As a result of increased use of chemical fertilizers and agricultural chemicals, yield has been increased and stabilized, and quality has been improved. At the same time, however, a number of problems have occurred in this transformation process, such as over-production of rice and other crops, deterioration of soil fertility, and dangers for food safety. These issues are not only raised by the agricultural side, but also by consumer groups and the general public in Japan. There is also an increasing concern among Japanese consumers to seek safety of agricultural products, both domestically produced and imported.

It is a time of change in the system of agricultural production and land use. More and more efforts should be devoted to recreating the viability and sustainability of Japanese agriculture, not simply by encouraging farmers to seek more intensive farming.

DISCUSSION OPENING – YANG BOO CHOE*

The paper presented by Professor Kada, which focuses attention on rural labour market conditions in Japan, is particularly interesting since it deals with a country in which there has been massive industrial development impinging on a very traditional, small-scale, agricultural sector.

The sustainability issues in Japanese agriculture include both ecological and socio-political concerns. Traditional paddy farming has been sustainable over many centuries in Japan but, since the 1960s, considerable intensification has occurred in rice production. Increasingly, Kada emphasized, the Japanese, too, are becoming more aware of the issues of sustainable agriculture. My particular interest, however, was in the socio-political area, where the author noted the aging structure of the agricultural labour force, the very limited entry of the young into farming, and the depopulation of remoter areas. This has been accompanied by an increasing role for part-time farming.

The question is whether the latter is sustainable over the longer term. There appears to be an extremely difficult policy problem in Japan associated with part-time farmers who receive most of their income from off-farm sources. In most Japanese farm households, the younger generation tends to work in urban areas, leaving agricultural work to older people and to females. It is not easy to squeeze the latter out of agriculture because their skills and work opportunities are relatively poorer. Furthermore, it appears that the key part-time 'farmer' often tends to over-invest in farm machinery in order to be able to cope with peak workloads. The result must be that the sector as a whole becomes less than responsive to market signals. Change has taken place, but there is little indication that it has been aided by any real policy response. Specifically, I wonder whether the concept of 'part-time farming' is still relevant to the needs of the rural sector, or whether there should be greater recognition of the need to encourage the development of the 'pluriactive household', as has been occurring in Europe.

*Rural Economics Institute, Korea.

MAHABUB HOSSAIN*

Section Summary

The main theme of the Conference has been sustainable agricultural development. In his Presidential Address, John W. Longworth has drawn our attention
to the economic and social dimensions of the sustainability issue, emphasizing the critical need for appropriate human capital formation so that scientists
can identify real problems and tackle broader socio-economic issues for agricultural development. V.S. Vyas has reminded us of the interdependence of
agrarian structure, environmental sustenance and poverty and has argued that
poverty alleviation could be a step towards an environmentally safe world.
The structure of this section is important in this context, since agricultural
production is organized mostly around the household and poverty is concentrated among small farmers. The papers which were presented in this section
focused strongly on the organization of the farm household, though they also
had a wider dimension. One is concerned with alternative forms of organization,
while some of them look outward, regarding the household not only as a
farming entity but as a base for other forms of economic activity, either within
the confines of the household or in the economy at large.

The analysis of household behaviour began with a detailed case study, by
Anthony Ikpi, of the time budgets of Nigerian farmers, indicating how a rural
household allocates family labour among three interrelated sectors – farming,
non-farm commercial activity and non-monetarized home production activity.
He argued that acceptance of an exogenous intervention, such as improved
crop technology, can depend both on its financial implications and on its
impact on the previously established time allocation equilibrium. It was widely
agreed that the case study approach provided valuable insights into a number
of aspects of low income household behaviour, not only in respect of technology adoption but also in relation to such key issues as willingness to borrow
and choice of family size.

The second plenary session paper, by Günther Schmitt, adopted a more
theoretical approach to the question as to why the family farm is the dominant
form of agricultural organization. His main focus was the developed world,
though it appeared clear that the underlying methodology is of general applicability. Despite the appearance of innovations which favour large-scale farming with hired labour, it was argued that there are many economic and social
advantages which make family farms the most competitive institution among
various forms of agricultural organization. The 'farm' should be viewed as a

*Bangladesh Institute of Development Studies, Bangladesh.

'household', rather than simply as a 'firm', with analysis of all aspects of behaviour and choice being integrated within a comprehensive framework. There was little dissent from this view, though it was emphasized that the measurement of key elements within the theory, notably transaction and supervision costs, could pose severe difficulty in applied work.

Though he was primarily concerned with cooperative forms of organization in Israel, Yoav Kislev echoed much of the argument of Schmitt by saying that family organization of production minimizes transaction costs through 'specialization by identity' and harmony and trust that comes with it. The specific nature of control of the biological process which characterizes agriculture, and the time-dependent nature of the production process, make control of labour more difficult in agriculture than in manufacturing. This provides the family farm with comparative advantage over large firms dependent on hired labour. Kislev modified this stance to the extent that, for realization of economies of scale, cooperatives could be preferable institutional forms. However, he did point to the many difficulties which had recently been experienced in Israel.

In the paper occupying the remainder of the session, Helen Jensen provided a mass of information on the importance of off-farm income, earned by women and other household members, in the United States. Though seemingly rather different, the two papers provoked considerable reaction. Reliance on off-farm income was viewed by some as an impediment to the efficiency of farm production *per se*, and it was argued that long-term sustainability of rural areas required the evolution of farming systems which would favour the basic production base.

Linkages between the farm and non-farm sectors were further considered in the final session, beginning with a paper by Steven Haggblade and Carl Liedholm dealing with the somewhat neglected question of labour market effects. This was set in the context of Third World economies where the development process changes the resource base of the farm household and the productivity of labour in different components of household economic activities. Rural households constantly adjust to these changes. The relative importance of farm and non-farm activities, as well as their composition, are determined by the adjustment process, and attention was drawn to the role of labour market linkages within that context. This paper was paired with a paper by Ryohei Kada, who drew attention both to the rise in environmental concern in Japan and to the importance of the steady increase in the share of off-farm income in rural household income. The two papers again provided an opportunity for discussion of what, in the European jargon, is known as the 'pluriactive' household both in terms of its adjustment behaviour and with respect to the analytical and policy problems which it poses.

Chairpersons: I.J.Singh, G.H.Peters, I.Kajii.
Rapporteurs: M.Rukini, D.Shah, T.Veeman.
Floor discussion: M. Collinson, J. Ntsangi, R.Spitze, F.G.Mack, J.C.Umeh, J.M.Boussard, G.Thijssen, G.Schmitt, P. Ollila, T.Dams, K.Daubner, D.Pick, V.Abiad, C.L.Zhang, A.Brun.

SECTION VII

Smallholder Agriculture and Changes in Development

IFTIKAR AHMED*

Will biotechnology alleviate poverty?

POVERTY CHARACTERISTICS AND TRENDS

The greatest numbers of the poor, including the very poorest, live almost entirely in rural areas. Therefore it is not surprising that agriculture is the main source of income of the *world's* poor (World Bank, 1990) and agricultural growth has a decisive influence on the evolution of poverty (Singh and Tabatabai, 1990). The rural poor belong to wage labour or marginal farmer households, poverty resulting as much from low returns as from unemployment and under-employment. Their number has increased from 767 million in 1970 to 850 million in 1985 (Singh and Tabatabai, 1990.).

World food crop production grew half a per cent faster than growth in population over the last two decades. Despite the positive margin, the absolute number of under-nourished in the Third World increased from 460 million in 1969–71 to 512 million in 1983–5 (Singh and Tabatabai, 1990). This increase cannot be attributed to variability of food production (ILO, 1990). The demographic projections suggest that demand for food could grow by 3 to 4 per cent per annum in the coming years (Pronk, 1991).

Hunger results from the inability of poor countries, poor families and poor individuals to purchase sufficient quantities of food from available food supplies. Biotechnology could make a contribution to poverty alleviation if it is accompanied by widespread gains in the poor's purchasing power by improving labour absorption without sacrificing growth.

The Third World agricultural labour force is projected to increase at 0.8 per cent per annum until the year 2000 (Singer, 1990). Furthermore, a yield plateau has been reached for major crops, particularly for rice (Barker, 1989). The green revolution rice remained unchanged around the yield potential of the IR-8 rice variety released in 1984 (Lipton and Longhurst, 1989). Therefore future yield gains critically depend on what biotechnology has to offer.

BIOTECHNOLOGY AND POVERTY-RELATED ISSUES

Biotechnology consists of a cluster of commercial techniques which use living organisms to make or modify a product, including techniques for improving the characteristics of economically important plants and animals and for

*International Labour Office, Geneva, Switzerland.

developing micro-organisms which act on the environment. This paper also
deals with micropropagation (tissue culture) techniques often labelled 'second
generation' biotechnologies.

Given the time-frame for the release of many of these biotechnologies for
crops of importance to the Third World (Table 1), forecasting of the probable
impact on poverty alleviation is based on *ex ante* assessment to help influence
research in a pro-poor direction. *A priori* deductive reasoning is used to match
specific biotechnology breakthroughs in industrialized countries with the prevailing socio-economic context of developing countries to assess likely impact
of new technologies on Third World poverty. This is supplemented by hypothesis testing when possible. The paper analyses both the pro- and anti-
poor features of current and future biotechnology developments. The poverty-
related issues are described below.

Will *transgenic* plants (containing a foreign gene) and genetically-engineered microbes be potentially more scale-neutral at the farm level than the
green revolution and mechanical innovations? Will the cost-reducing or output-
enhancing potential of biotechnologies be more beneficial to small farmers?[1]
Do resource-saving agricultural biotechnologies depress GDP and reduce aggregate employment? Would the newly emerging biotechnologies reduce previously uncontrollable production variances in the agricultural sector?[2] This is
of concern to the risk-averse, resource-poor small farm sector.

TABLE 1   *Availability of new biotechnologies for selected Third World crops, 1989*[a]

Crop	New diagnostics[b]	Rapid propagation systems[c]	Transformation systems[d]	Regeneration systems[e]	Time-frame for commercial applications (years)
Banana/Plantain	+	+	–	+	5–10
Cassava	+	+	–	–	5–10
Cocoa	+	–	–	–	>10
Coconut	+	–	–	–	>10
Coffee	+	+	–	+	5–10
Oilpalm	+	+	–	–	>10
Potato	+	+	+	+	0–5
Rape-seed	+	+	+	+	0–5
Rice	+	+	+	+	0–5
Wheat	+	+	–	–	>10

Notes: [a]Source: Persley (1989, p. 23, Table 3.1).
[b]Availability of new diagnostics for pests or diseases based on the use of monoclonal antibodies or DNA probes.
[c]Availability of rapid propagation systems to allow the multiplication of new varieties.
[d]Availability of transformation systems to enable new genetic information to be inserted into single plant cells.
[e]Availability of regeneration systems to enable single cells to be regenerated into whole plants, after transformation.

The paper deals with changes in labour intensity in agriculture, structure and stability of rural employment, and the impact of biotechnology on the rural labour market. The magnitude and skill composition of jobs created and the non-farm employment generated through linkages to laboratories and to marketing and crop processing are analysed. Can biotechnologies be specifically designed and deliberately released to alleviate rural poverty? Several biotechnology developments have affected the international division of labour through disruptions in global trading patterns. What categories of rural workers and producers are affected? Finally, are all external biotechnology developments anti-poor? The related issue of improving Third World countries' access to pro-poor, but patented, biotechnologies is also taken up.

CROSSING THE YIELD FRONTIER

One traditional biotechnology in China, 'Shan Yu 63', which resists the rice blast disease, increased output by 4.7 million tonnes (valued at 1.9 billion yuan) in 1987 and saved 100 million yuan on chemical pesticides, producing a total return of 2 billion yuan (Yuanliang, 1989). A return of 15.7 yuan is obtained from 1 yuan of investment in biofertilizers for wheat. Breakthroughs in cellular engineering included important cereal crops in addition to non-food crops. Genetic material from Sui Yan wheat straw created engineered

TABLE 2 *China: output gains from cellular engineering on major crops[a]*

Crop	Type of cellular engineering	Crop variety	Sown area (ha) in 1988	Increase in yield
Wheat	Chromosomal engineering	Nos 4, 5, 6 Xiao Yan	2 000 000	900 000 tonnes
Wheat	Culture of pollen haploid cells[b]	No. 1 Jing Hua	70 000	15–20%
Rice	Pollen haploid	Xin Xiu, Wan Gen 959, etc.	170 000[c]	About 10%
Rice	Pollen haploid	Nos 8, 9 Zhong Hua	70 000[c]	15–20%
Rice	Marker rescue	No. 1 Hu Yu	3 000[c]	15%
Potatoes	Tissue culture		70 000[d]	Over 50%
Sugar-cane	Tissue culture		4 000	Over 50%
Tobacco	Pollen haploid		10 000	Over 50%
Banana	Tissue culture		100 000 test-tube seedlings	Over 50%

Notes: [a]Source: Yuanliang (1989).
[b]Pollen and ovules have half the number of chromosomes present in all other tissues of a plant. By chemical treatment this number can be doubled so that plants generated from these cells have two sets of identical chromosomes and therefore identical genes. These homozygous plants are very useful in searching for mutants and for breeding.
[c]1985.
[d]1984.

wheat to resist drought, hot wind and diseases (Table 2). The total area sown (in ten provinces of China) is 2 million hectares, increasing yields by 900 000 tonnes. The area under pollen haploid rice, wheat and tobacco was 466 700 hectares between 1981 and 1985.

The new biotechnology in potatoes doubled land productivity, labour intensity and profitability and increased labour productivity by 24 per cent in Kenya (Table 3). Its contribution (value added) to national income is twice that of the traditional technology. The relative efficiency (value added as a proportion of gross output) is higher for the adopters. Nitrogen fixing biotechnologies could increase per hectare maize yields by 0.5 tonnes, that is, 26 per cent on 5.2 million hectares on small farms in Mexico. While national output would increase by 21 per cent, Mexican farmers' income would in-

TABLE 3   *Biotechnology and farm size: potato and tea in Kenya, 1987[a]*

Key Indicators	Potato farms (Number = 33)			Tea farms/estates (Number = 39)
	Biotechnology (BT)	Traditional technology (TT)	Relationship to farm size	Relationship to farm size (biotechnology only)
Labour productivity (gross output/ha) in shillings)	33,210	16,382	Inverse for BT and TT	Inverse
Labour intensity (work-days/ha)	301	144	BT: unclear TT: inverse	Unclear
Labour productivity (kg. work-day)	124	100	Positive for both BT and TT	Positive (sh/work-day)
Labour's factor share (wages as % of value added)	27	23	Positive for both BT and TT	Positive
Capital use sh/w-day sh/ha	3 867	3 426	Inverse for both BT and TT	Positive Positive
Intermediate inputs (sh/ha)	3,553	3,008	Inverse for both BT and TT	Inverse
Value added as a proportion of gross output (%)	89	82	BT: positive TT: inverse	Positive
Profitability (gross output minus operating costs in sh/ha)	20,816	9,916	Inverse for both BT and TT	Positive
Income ratio[b]	8	4	—	3 (small farms[c]) 2 (large estates[d])

Notes: [a] Source: calculated from data in Mureithi and Makau (1991).
[b] Ratio of income of the 30 per cent of richer farmers to income of the 70 per cent of poorer farmers.
[c] Up to 3 ha.
[d] Over 20 ha.

crease by 55 per cent (Gilliland, 1988). This limited evidence offers hope for overcoming the present yield barrier with biotechnology.

BOOSTING PURCHASING POWER

The probable impact of biotechnology on the level and pattern of rural employment is analysed below.

Labour absorption in agriculture

Advanced biotechnologies may lead to a saving in labour use for chemical means of plant protection (Ahmed, 1991). Micropropagation in Mexico need not cause labour displacement in citrus cultivation as this would be compensated by more intensive labour use in weeding, pruning, irrigation and harvesting. In Kenya the doubling of labour intensity per unit of land was due to more labour being needed for ridging before cultivating micropropagated potatoes and in Malawi for nursery and planting operations.

Seasonality and structure of employment

Through biotechnologies applied to crops, animal feed and milk production, an indirect and steady source of employment could be created by linkages to juice processing (Mexico), poultry production (Nigeria), coffee, henequen, tequila and dairy industries (Mexico) and the tea industry (Kenya). Underemployment in the south-eastern region of Mexico could be reduced by applying advanced plan biotechnologies (APB)[3] to create and widen crop varieties and prolong the growing season, and by increasing harvests (Ahmed, 1991). Agriculture labour released by APB is absorbed in new and sideline activities, with a change in social organization of delivering these services in China (Yuanliang, 1989). Similarly, increased labour use in Malawi and Kenya was due to structural adjustments in new farm practices for APB.

Employment linkages

The chemicals used in rural areas are not produced there. Over 40 per cent of the fertilizers are imported by developing countries from industrialized countries. Fertilizers are among the most capital-intensive products (Johnston and Kilby, 1975). A large plant can cost between US$300 and 700 million (Doyle, 1985). Reducing fertilizer use should not cause labour displacement.

Biotechnology use requires blending of workers with 'low-tech' skills engaged in traditional agricultural work with 'highly skilled' technicians involved in the generation of advanced biotechnologies (Ahmed, 1991). In Mexico, scientists and technicians produced plantlets cultivated by agricultural labour. In addition to the employment created for the traditional workforce in agriculture, 933 scientists were employed by the Tea Research Foundation of Kenya in 1986 with an income of K704 371 (Mureithi and Makau, 1991). About 512 plant scientists work on cellular engineering in China (Yuanliang, 1989). In Nepal, scientific personnel produce 8 000 to 10 000 potato plantlets per day through micropropagation which are transferred to sandbeds by semi-skilled workers (Rajbhandari, 1988).

Women dominate the micropropagation laboratories in the Philippines (Halos, 1991) and Mexico (Eastmond *et al.*, 1989). They constitute 80 per cent, 74 per cent and 85 per cent of the Philippine Society for Microbiology, Cell/Molecular Biology and Biotechnology Societies, respectively. These were considered low-paid jobs concerned with basic science with limited linkage to industry. Moreover, work in tissue culture laboratories is tedious, requiring patience and perseverance.

Because of the resource-saving character of biotechnological innovations an input–output simulation exercise reveals a series of inter-industry repercussions throughout the economy, the cumulative impact of which may be to depress GDP and aggregate employment (Lee and Tank, 1991).

RURAL LABOUR MARKETS

Wage labour

APB could increase the demand for hired labour (in Mexico and Kenya), boost wages, improve labour's factor share and reduce rural–urban wage differentials. Gross earnings from APB in Kenya compare favourably with wage incomes in a modern-sector job, important for dampening the pace of rural–urban migration.

Displacing female wage labour: worsening poverty

The green revolution (GR) relied on manual labour for weeding, which had the following characteristics (Ahmed, 1991): (1) weeding is one of the most labour-intensive of all agricultural operations; (2) there is a significant increase in the demand for hired labour in weeding (doubled in Sri Lanka); (3) weeding labour doubled or tripled over that of the pre-GR crops (as, for example, in Bangladesh and the Philippines); (4) small GR farmers recorded much higher labour intensity in weeding than larger farmers; and (5) women constituted between 72 and 82 per cent of such labour input. The genetically engineered plants will substitute chemical herbicides for manual weeding, massively displacing women. Therefore genetic engineering will not only introduce a new fixed cost for farmers by forcing them to purchase the herbicide genetically tied to the seed supplied by the same company, but will also strike a colossal blow at the poor.

ENHANCING CROPPING INTENSITY

Multiple cropping, facilitated by early-maturing GR varieties, contributed to greater labour use. Micropropagated potatoes could similarly improve cropping intensity. Since potatoes in most Third World climates take only 40–90 days to grow (compared to 150 days in temperate climates), it can easily be accommodated with current cropping patterns.

Thirty poor countries can already micropropagate potatoes, a major source of food for poor families in Africa and Asia. Indeed, micropropagation has made potatoes the second biggest crop after rice in Vietnam and quadrupled production in China over 30 years (*The Economist*, 13 October 1990). In Vietnam, micropropagation has increased potato yields from 200 tonnes to 8000 tonnes per year on 450 hectares during 1980–4 (Uyen and Zaag, 1985). Yield has increased from 8 tonnes to 18 tonnes per hectare in Nepal (Rajbhandari, 1988). The following reasons make micropropagation a very attractive option: (1) year-round production of plantlets is possible; in one rural valley of Vietnam each family produces up to 150 000 plantlets per year (Walgate, 1990); (2) it saves costs and reduces difficulties of physical transportation of tubers to the fields for planting; (3) by generating plantlets from tissues, a substantial volume of tubers spared from planting can now be eaten; (4) disease-free planting material could reduce production variances from disease, potato being vulnerable to 268 diseases, and late blight can wipe out more than 50 per cent of crop (Manandhar *et al.*, 1988); and (5) while yields benefit landowners, increased cropping intensity also benefits the landless by increasing hired labour demand.

WILL SMALL FARMERS BENEFIT?

As with the GR diffusion process, large farmers are pioneering adoption of APB in Kenya. Economic inducements exist for all categories of farms for adoption of advanced biotechnologies. The Chinese experience provides proof of APB profitability. In Nigeria, escalating costs of vegetative sources of animal feed and the lower relative price of single cell protein (SCP) serves as an inducement. More than half the small-scale growers in Mexico were willing to adopt APB-based disease-free planting material. Although biotechnological innovations constitute variable costs, collecting information on the technology represents a fixed cost. This is an important reason for the continuation of a bias in favour of large farmers (Kinnucan *et al.*, 1989).

Small farmers in Asia adopted the GR technology only after large ones had applied it and raised yields. While large farmers obtained 'innovators' rent', food prices had fallen by the time poorer, later adopters were ready to sell (Lipton and Longhurst, 1989). Similar experience may be repeated for biotechnology, although with a reduced lag. The farmers in non-GR areas of Asia, often the poorest, gained nothing from the GR. They lost when extra output from GR areas depressed the returns from their meagre output, for example when the extra GR sales from Punjab (wheat) or Central Luzon (rice) pulled down farm-gate prices in impoverished Madhya Pradesh (India) and

Mindanao (Philippines) respectively (Lipton and Longhurst, 1989). Through its potential for less favoured areas, biotechnology may help redress this disparity.

REDUCTION IN FARMING COSTS: A BOON FOR WHOM?

Nitrogen fertilizers account for 75 per cent of agricultural production costs in Brazil (InterAmerican Development Bank, 1988) and 60 per cent of the energy costs of wheat production in India. Fertilizers and pesticides constitute over 80 per cent of costs of production of GR rice in Thailand and wheat in Europe (*The Economist*, 1987). Biotechnology breakthroughs in nitrogen fixation would certainly reduce costs of farming.

In Mexico, micropropagated flowering plants can be produced at half the cost of the imported ones, even after 60 per cent profit. Similar comparative advantages exist for micropropagated *tequilina* (Eastmond *et al.*, 1989). Commercial micropropagation of orchids is insufficient, requiring imports into the Philippines, although prices charged locally are still too high for small farms (Zamora and Barba, 1990). Imported certified potato seed in Nepal resuls in 40 to 60 per cent higher production cost for micropropagation (Manandhar *et al.*, 1988). Domestic micropropagation could reduce costs from Nepalese Rupees 1.5 to about Rupees 0.30–0.50 per plantlet (Rajbhandari, 1988). Cost of producing one pot of micropropagated potato is two-thirds the cost of one seed tuber in Vietnam (Uyen, 1991).

Biotechnology's protection of crops from insects is more economic compared to chemical alternatives. One species accounts for 40 per cent of all soybean crop losses caused by insects in Brazil. An insect-destroying virus introduced on 11 000 hectares during 1983–4 led to about 75 per cent savings in the cost of protecting soybeans, as compared to the cost of chemicals (InterAmerican Development Bank, 1988).

DESIGNING BIOTECHNOLOGIES FOR POVERTY ALLEVIATION

This section demonstrates how exactly biotechnologies could be designed by scientists to launch a planned assault to solve location-specific constraints responsible for impoverishment. The orange leaf rust disease ravaging coffee cultivation in Mexico has threatened the survival of the small growers. Chemical means of control is beyond their reach. Application of APB to supply disease-free or disease-resistant plant materials will not only save but expand the employment opportunities of these producers and of the large body of hired labour, and generate indirect employment through its linkage to the coffee industry and to the micropropagation laboratories and nurseries (Eastmond and Robert, 1991).

The plant *tequila agave* takes nine years to grow in Mexico to a mature stage before it can be utilized. Some 12 million plantlets are required to replenish existing stocks.[4] Some 6000 small contract growers who supply the large agro-industrial companies with the raw material will stand to benefit from rapid micropropagation (Eastmond *et al.*, 1989). Linkages to the tequila drink industry will result in additional and more stable employment.

Combating malnutrition

The application of SCP could help alleviate protein malnutrition in general and boost animal protein intake by the Nigerian protein-deficient population (Okereke, 1991).[5] The economic climate is favourable for its acceptance in Nigeria because: (1) the income elasticity of demand for poultry products is higher than that for beef; (2) the supply–demand projections reveal an excess demand for poultry products; (3) relative prices of other sources of poultry feed are higher and increasing; and (4) the ban on the import of poultry products and feed provides the protection needed.

Between 1979 and 1987, the import of soybeans and protein meal as sources of animal feed increased by 433 per cent to 516 per cent in Venezuela (Martel, 1990). The abundant supply of natural gas in Venezuela could easily be harnessed to produce SCP. Cuba has already established 13 SCP plants based on cane-molasses (InterAmerican Development Bank, 1988).

Milk for the thirsty

In Mexico the bovine somatotropin (BST) technology could reduce the daily deficit of 12.5 million litres of milk and make it more accessible to the population (37 per cent of whom currently consume only 14.5 per cent of the available milk supply). It increases milk production in cows by 10–25 per cent. This is like having extra milk without extra feed. Employment would increase in the production and processing of milk and the feed industry, all of which are concentrated in a few hands (Otero, 1991a). BST also holds prospects for Pakistan. Despite having three and a half times as much pasture as Wisconsin and over one and a half times as many dairy cows, Pakistan produces only a quarter as much milk. Pakistan's cows are only 15 per cent as efficient as Wisconsin's. Pakistan spends $30 million on milk imports each year (*The Economist*, 13 January 1990). Most astonishingly, milk produced per day could be five times higher for BST-treated cows in Zimbabwe (Kirk, 1990).

Returns over variable cost could be 26 per cent for dairy farmers using BST (Otero, 1991a). With assumptions of milk prices and costs, if BST can increase milk production by 15 per cent, a farm with 500 cows could make an extra $82 000 profit per year in the United States (*New Scientist*, 24 March 1988). There should now be less concern for consumer safety since the US Food and Drug Administration and a team of US doctors have announced that BST causes no changes in milk composition of any practical importance to consumers (*Chicago Tribune*, 1990 and *International Herald Tribune*, 1990).[6]

Four major manufacturers of BST argue that its cost of less than one dollar a day per cow would make it scale-neutral (Schneider, 1989). This would be facilitated by the expanding global (approaching $1 billion annually) and international ($100–$500 million annually) markets (Schneider, 1989, UNDP, 1989).

GENETIC ENGINEERING FOR THE POOR

Some of the fragmented and widely scattered information on primarily single gene-based genetic engineering breakthroughs assembled reveals some trends (Ahmed, 1991). The private sector corporations dominate genetic engineering research and their eyes are on agronomic traits and on crops which promote markets for their seeds and/or agrochemicals. These also concern crops of importance to industrialized countries as it is difficult to police patency infringements in Third World countries. Developing countries generally do not have patent laws. The private industry cannot recover revenue through royalties and licences.

The private sector accounts for two-thirds of total global funding (US$4 billion) of biotechnology research, and large chemical multinational companies spent 50 per cent of the total R&D budget on biotechnology. They spent $10 billion over the last decade to buy up seed companies for marketing their own biotechnology products (James and Persley, 1989). After 2001, 75 per cent of all major seed will be based on biotechnology (*McGraw Hill Biotechnology*, 1989). The cost of seeds as a proportion of total cost of wheat in Europe could rise from 20 per cent to 50 per cent at about that time (*The Economist*, 1987).

The possible socio-economic impact of the *transgenic* plants and microbes could involve the following: (1) Pest and disease resistance and drought tolerance will reduce output variance, important for risk-averse farmers; together with breakthroughs for nitrogen fixation this will reduce farmers' costs of production; further research by Cornell University Boyce Thompson Institute for Plant Research which has discovered a bacterium that can fix nitrogen without depending on the plants for energy should be encouraged (*Genetic Engineering News*, 1989); (2) production of less thirsty crops will increase labour absorption through area expansion and multiple cropping; (3) lower labour requirements in pest and disease control may be made up by overall increases in labour use in other new operations; (4) herbicide resistance will directly displace labour for weeding, particularly for the vulnerable groups; (5) prolonged shelf life of freshly harvested agricultural produce will help the poor faced with inadequate marketing infrastructure; (6) genetic engineering breakthroughs in (1) and (2) above will help compensate for inadequacies of extension services and delivery failures; (7) genetically engineered microbes may benefit the small farmers if these spill over to the poor neighbours' plots and fix nitrogen there or protect the crops from pests and diseases there; a rough comparison of chemicals with microbial controls shows the clear economic and safety advantages, although costs need to be further reduced and effective means of dispersing the microbes would make their use feasible for smallholders (Bunders, 1990); (8) the major obstacles to the Third World countries and poor farmers' access to socially beneficial biotechnologies are the legal and financial barriers associated with the proprietary rights over these technologies through patents; moreover, increasing research partnership between industry and the academics tends to diminish Third World countries' access to technology previously available freely as a public good; (9) Third World countries could gain from increased wool production through genetically-engineered pasture crop (saves on grazing land) and from cost reduction through biotechnologies which substitute for mechanical wool harvesting technologies.

USING THIRD WORLD BIOTECHNOLOGY CAPABILITIES

The 'second-generation' biotechnologies are within the scientific and financial reach of Third World countries. A fully equipped laboratory might cost US$250 000 (Lipton and Longhurst, 1989). More astonishingly, a tiny micropropagation facility in a farming household costs only US$354 to produce 200 000 plantlets annually in Vietnam (Ministry of Agriculture and Food, Vietnam, 1987).

Micropropagation is already applied to potatoes in 30 poor countries. Singapore and Brazil produce coffee plantlets on a large scale (*Biotechnology and Development Monitor*, 1990). Micropropagation capabilities were noted for Malawi, Nepal, Vietnam and Kenya. Unfortunately, this capacity used for non-food crops meets the needs of the commercial large-farm sector in the Philippines, Mexico and India (Mani, 1990; Zamora and Barba, 1990; Eastmond *et al.*, 1989).

Tissue culture in Japan could produce 3 billion rice seedlings from a single seed in about six months. It saves seeds (releasing more grains to feed the hungry) and supplies vastly more seedlings to be planted by the countless unemployed hands in the densely populated areas of the Third World. In labour-scarce Japan robots are being sought to meet the intensive labour demands for root separation of seedlings grown by the new technique (UNIDO, 1989). On the other hand, tissue culture in California could turn sour for Third World grape producers if they are denied access to it (*Scientific American*, 1991).

THE NEW INTERNATIONAL DIVISION
OF LABOUR : WHAT HAPPENS TO THE POOR?

By the end of the decade, biotechnology will affect developing country exports worth US$66 billion (Table 4). These countries may lose annually $10 billion of their export income (Kumar, 1988) with serious repercussions on the international division of labour. While 90 per cent of the sugar internationally traded came from developing countries in 1975, it declined to about 67 per cent in 1981 (Otero, 1991a). Sugar imports by developed countries declined from 70 per cent to 57 per cent during the same period. World consumption of high fructose corn syrup (HFCS) accounted for only 1 per cent of total sweeteners in 1975, but rose to 6 per cent in 1985 (Wald, 1989). A total of 34 soft drink manufacturers in the US have switched to the immobilized enzyme technology. In consequence, sugar exports from the Philippines declined from $624 million in 1980 to $246 million in 1984. In the Caribbean the decline of sugar export to the USA was similar. This was accompanied by a crash in sugar prices from US cents to 63.20 per kg. in 1980 to US cents 8.36 in 1985 (Panchamukhi and Kumar, 1988). The livelihood of over 50 million workers engaged in the sugar industry of Third World countries were affected by the decline in exports (ibid.)

Replacement of vanilla flavour by biotechnology products threatens 70 000 small farmers in Madagascar, which could lose US$50 million of its annual export earnings (Mushita, 1989). Comoros will be similarly affected (Junne, 1990). A Californian company commercially produces vanilla plantlets through tissue culture for the lucrative annual flavouring market worth US$200 million (*New Scientist*, 1991). The lower cost of tissue culture compared to the traditional vanilla extract makes it more profitable. Cacao, the second most important agricultural commodity in the Third World, faces similar threat. Africa accounts for nearly 60 per cent of the world production of cacoa. Small cocoa producers in Cameroon, Ghana and Côte d'Ivoire will be affected by the biotechnology developments in the Swiss-based company, Nestlé (Hobbelink, 1989). Patent applications have been made by Kao Corporation of Japan for genetically engineered enzymes for making cocoa butter substitutes (Svarstad, 1988).

Biotechnology research in Germany could produce a substitute for coffee (Otero, 1991b). Third World countries account for almost the entire world's coffee exports (US$10 to 50 billion each year). Apart from the adverse impact on the balance of payments of Colombia, Burundi, Uganda, Rwanda and Ethiopia, the livelihood and jobs of 500 000 small producers in Rwanda and another 650 000 in Indonesia would be threatened (*Biotechnology and Development Monitor*, 1990).

The next oil crisis: who are the victims?

Biotechnology converting plant oils into structural lipids or tailored fats will affect the market shares of 11 vegetative oil crops traded by the Third World (Ruivenkamp, 1991). Biotechnology will dramatically increase the market for castor, palm and groundnut oil and reduce eight others (Kumar, 1988). While coconut is the source of only 2 per cent of the world's oils and fats market, the Philippines alone supplies 80 per cent of it. Decline in its export would affect 15 million Filipinos, who are poorer than the rest of the farming population (Halos, 1991).

IS AGRARIAN REFORM STILL IMPORTANT?

Inverse relationships between farm size and productivity, both under the traditional and the new biotechnology, for potatoes and tea are observed in Kenya (Table 3) and Malawi. Small farms as compared to large ones, make a larger contribution to national income, extract higher levels of profit and demonstrate

TABLE 4   *Biotechnology's impact on Third World exports by type of biotechnology and time-frame*

Time-frame for routine use	Micropropagation techniques		Transgenic plants	
	Value of exports (US$ billions)	Exports affected (number of developing countries)	Value of exports (US$ billions)	Exports affected (number of developing countries)
Up to 1995	20.9	Coffee (28), Bananas/plantains (16), rice (6), rubber (5), tobacco (2), vanilla (2), cassava (1), potatoes (1)	6.4	Rubber (5), tobacco (2), maize (1), potatoes (1), tomatoes (0)
1995–2000	21.2	Sugar-cane/sugar-beet (16), cocoa (15), tea (4), soyabeans (3), oil palm (3), wheat (3), maize (1), sunflower (1), pineapple (0), sorghum (0), barley (0), sweet potatoes (0), yams (0)	17.5	Sugar-beet (16), cotton (15), bananas/plantains (16), rice (6), soyabeans (3), cassava (1), sunflower (1), barley (0), rape-seed (0), sweet potatoes (0), yams (0)
Year 2000 and beyond	3.4	Cotton (15), coconut (10), rape-seed (0), millet (0)	21.7	Coffee (27), sugar-cane (16), cocoa (15), coconut (10), tea (4), oil-palm (3), wheat and flour (3), pineapple (0), sorghum (0), millet (0)

Source: UNCTAD (1991).

stronger linkages to agricultural input suppliers. A small farm-based development strategy would increase output and prevent worsening income inequality without sacrificing employment.

With the application of APB, labour's factor share increases in Malawi and Kenya. This share is already high in Mexico. However, the past trend of increasing social differentiation will be accentuated in Mexico through limited mobility across occupational class structures (Eastmond and Robert, 1991) and in Kenya (increase in the concentration ratio), unless essential agrarian reforms are adopted. The advent of biotechnology into an agrarian system in which land is unequally distributed will tend to reinforce the existing inequality (Table 3). The Kenyan evidence with biotechnology represents a close parallel to the experience of inequality created by the green revolution in Asia, although it is likely to be less acute.

NOTES

[1]Farmers are assumed to select on the cost curve the least-cost combination of inputs to produce a given level of output. Technological change should reduce the use of farm inputs at any relative price so that those scarce resources can be reallocated to other uses.

[2]Insect- and disease-resistant biotechnologies counter some of the stress and pathological losses associated with disease and insect infestations. The new biotechnologies could reduce the previously uncontrollable fluctuations in production.

[3]APB will be used henceforth to signify micropropagation techniques.

[4]The plant is used to produce high-quality spirits, particularly tequila, an important Mexican export.

[5]SCP are dried cells of micro-organisms which are grown in large-scale culture systems for use as protein sources in human and animal feeds. Micro-organisms can develop between 100 and 1000 times more quickly than a plant or an animal.

[6]It was feared that BST could transfer from the milk into the blood, producing hormonal and allergic effects. It was also considered as an unnecessary and cruel way of squeezing out more milk from the cow to pour into overflowing milk lakes in the industrialized countries.

BIBLIOGRAPHY

Ahmed, I. ed., 1991, *Biotechnology: A hope or a threat?*, Macmillan, London.

Barker, R., 1989, 'Socioeconomic impact of modern biotechnology on world trade and economic development in the developing countries', *Biotechnology study project papers: Summaries of commissioned papers*, ISNAR, The Hague.

Bifani, P., 1991, 'New biotechnologies for rural development', in I. Ahmed, *Biotechnology*.

Biotechnology and Development Monitor, 1990, Amsterdam, no. 4, September.

Bunders, F.G. (ed.), 1990, *Biotechnology for small-scale farmers in developing countries: Analysis and assessment procedures*, VU University Press, Amsterdam.

Chicago Tribune, Wednesday 22 August, 1990.

Chipita, C. and Mhango, M.W., 1991, 'Biotechnology and labour absorption in Malawi agriculture' in I. Ahmed, *Biotechnology*.

Doyle, J., 1985, *Altered harvest: Agriculture, genetics and the fate of the world's food supply*, Viking, New York.

Eastmond, A. and Robert, M.L., 1991, 'Advanced plant biotechnology in Mexico: A hope for the neglected?' in I. Ahmed, *Biotechnology*.

Eastmond, A., Gonzalez, R.L., Saldana, H.L. and Robert, M.L., 1989, *Towards the application and commercialisation of plant biotechnology in Mexico*, Universidad Autonama de Yucatán, Mexico, October, unpublished draft.

Genetic Engineering News, February 1989.
Gilliland, M., 1988, 'A study of nitrogen fixing biotechnologies for corn in Mexico', *Environment*, April.
Halos, S.C., 1991, 'Biotechnology trends: A threat to Philippine agriculture?', in I. Ahmed, *Biotechnology*.
Hobbelink, H., 1989, 'Agricultural biotechnology and the Third World: International context, impact and policy options', in *Development-related research: The role of the Netherlands*, University of Groningen, Groningen.
Hobbelink, H., 1991, *Biotechnology and the future of world agriculture*, Zed Books, London.
ILO, 1990, *Structural adjustment and its socioeconomic effects in rural areas*, Advisory Committee on Rural Development, Eleventh Session, ILO, Geneva.
InterAmerican Development Bank, 1988, *Economic and social progress in Latin America*, 1988 Report, Washington, DC.
International Herald Tribune, 25–26 August, 1990.
James, C. and Persley, G., 1989, 'Private/public sector collaboration in biotechnology for developing countries in World Bank/ISNAR/AIDAB/ACIAR', *Biotechnology study project papers: Summaries of commissioned papers*.
Johnston, B.F. and Kilby, P., 1975, *Agriculture and structural transformation: Economic strategies in late-developing countries*, Oxford University Press, London.
Junne, G., 1990, 'The impact of biotechnology on international commodity trade', paper presented at the International Seminar on the *Economic and Sociocultural Implications of Biotechnologies*, Vézelay, France, 29–31 October, organized by UNESCO, Paris.
Kinnucan, H., Molnar, J.J. and Hatch, U., 1989, 'Theories of technical change in agriculture with implications for biotechnologies', in J.J. Molnar and H. Kinnucan (eds), *Biotechnology and the New Agricultural Revolution*, Westview Press, Boulder, Co.
Kirk, R., 1990, 'BST trials on indigenous cattle', in *The Farmer* (Harare, Zimbabwe), 24 May 1990.
Kumar, N., 1988, 'Biotechnology revolution and the Third World: An Overview', in *Biotechnology Revolution and the Third World*, Research and Information System for the Non-Aligned and other Developing Countries, New Delhi.
Lee, H.H. and Tank, F.E., 1991, 'A conceptual framework for biotechnology assessment', in I. Ahmed, *Biotechnology*.
Lipton, M.A. and Longhurst, R., 1989, *New Seeds and Poor People*, Unwin Hyman, London.
Manandhar, A., Rajbhandari, S., Joshi, P. and Rajbhandari, S.B., 1988, 'Micropropagation of potato cultivars and their field performance', in *Proceedings of national conference on science and technology*, Royal Nepal Academy of Science and Technology, Kathmandu.
Mani, S., 1990, 'Biotechnology research in India: Implications for Indian public sector enterprises', *Economic and Political Weekly*, Bombay, 25 August.
Martel, A., 1990, 'Possible impacts of biotechnology on Venezuela's agroindustry', paper presented at the International Seminar on the *Economic and Sociocultural Implications of Biotechnologies*, Vézelay, France, 29–31 October, organized by UNESCO, Paris.
McGraw Hill Biotechnology, no. 9, 3 April 1989.
Ministry of Agriculture and Food, Vietnam, 1987, *Potato research and development in Vietnam*, The International Potato Centre (CIP), Southeast Asia and Pacific Regional Offices, IRRI, Manila, Philippines.
Molnar, J.J. and Kinnucan, H., 1989, 'Introduction: The biotechnology revolution', in J.J. Molnar and H. Kinnucan (eds), *Biotechnology and the New Agricultural Revolution*.
Mureithi, L.P. and Makau, B.F., 1991, 'Biotechnology and farm size in Kenya', in I. Ahmed, *Biotechnology*.
Mushita, A.T., 1989, 'The impact of biotechnology in developing countries', in *Development*, Society for International Development, Rome.
New Scientist, 14 March,1988.
New Scientist, 24 March 1988.
New Scientist, 5 March 1991.
Okereke, G.U., 1991, 'Biotechnology to combat malnutrition in Nigeria', in I. Ahmed, *Biotechnology*.
Organisation for Economic Cooperation and Development (OECD), 1989, *Biotechnology: Economic and wider impacts*, OECD, Paris.

Otero, G., 1991a, 'The differential impact of biotechnology: The Mimeo-United States contrast', in I. Ahmed, *Biotechnology*.

Otero, G., 1991b, 'The coming revolution of biotechnology: A critique of Buttel', in *Sociological Forum*, vol. 6, no. 3, September.

Panchamukhi, V.R. and Kumar, N., 1988, 'Impact on commodity exports', in *Biotechnology Revolution and the Third World*, Research and Information System for the Non-Aligned and other Developing Countries, New Delhi.

Persley, G. (ed.), 1989, *Agricultural biotechnology opportunities for international development: Synthesis report*, World Bank/ISNAR/AIDAB/ACIAR, World Bank, Washington DC, May.

Pronk, John, 1991, speech delivered at the public debate on *Biotechnology and farmer's rights: Opportunities and threats for small-scale farmers in developing countries*, Amsterdam 8–9 April, organized by the Department of Biotechnology & Security, Vrije Universiteit and Netherlands Organization for International Development Cooperation, Amsterdam.

Rajbhandari, S.B., 1988, 'Plant tissue culture method and its potential' in *Proceedings of national conference on science and technology*, Royal Nepal Academy of Science and Technology, Kathmandu.

Research and Information System for the Non-aligned and other Developing Countries (RIS), 1988, *Biotechnology revolution and the Third World*.

Ruivenkamp, G., 1991, 'Can we avert an oil crisis?' in I. Ahmed, *Biotechnology*.

Schneider, K., 1989, '5 big chains bar milk produced with aid of BST', in the *New York Times*, 24 August.

Scientific American, 1991, 'Through the Grapevine', March.

Singer, H.W., 1990, *Research at the World Employment Programme*, ILO, Geneva draft.

Singh, A. and Tabatabai, H., 1990, 'Facing the crisis: Third World agriculture in the 1980s', *International Labour Review*, vol. 129, no. 4.

Svarstad, H., (1988) *Biotechnology and the international division of labour*, Institute of Sociology, University of Oslo, Oslo.

The Economist, 'Seed firms: Fruit Machines', London, 15–21 August 1987.

The Economist, 'The slow march of technology', London, 13 January 1990.

The Economist, 'Let the sky rain potatoes', London, 13 October 1990.

UNDP, 1989, *Programme Advisory Note: Plant biotechnology including tissue culture and cell culture*, UNDP, New York, July.

UNIDO, 1989, *Genetic engineering and biotechnology monitor*, Vienna, issue no. 26, December.

Uyen, Nguen Van, 1991, 'Plant biotechnology and farmer's right in Vietnam', paper presented at the public debate in *Biotechnology and farmer's rights: Opportunities and threats for small-scale farmers in developing countries*, Amsterdam 8–9 April, organized by the Department of Biotechnology & Security, Vrije Universiteit and Netherlands Organization for the International Development Cooperation, Amsterdam.

Uyen, N.V. and Zaag, P.V., 1985, 'Potato production using tissue culture in Vietnam: The status after four years', *American Potato Journal*, vol. 62.

UNCTAD, 1991, 'Trade and development aspects and implications of new and emerging technologies', in *The Case of biotechnology*, TD/B/C.6/154 UNCTAD, Geneva.

Wald, S., 1989, 'The biotechnological revolution', in *OECD Observer*, Paris, no. 156, February/March.

Walgate, R., 1990, *Miracle or Menace? Biotechnology and the Third World*, Panos Publications, London.

World Bank, 1990, *World Development Report 1990*, Oxford University Press, Oxford.

Yuanliang, M., 1989, '*Modern plant biotechnology and structure of rural employment in China*, ILO, Geneva, draft.

Zamora, A.B. and Barba, R.C., 1990, 'Status of tissue culture activities and the prospects of their commercialisation in the Philippines', *Australian Journal of Biotechnology*, vol. 4, no. 1, January.

DISCUSSION OPENING – MARCO FERRONI*

I enjoyed reading Dr Ahmed's review of some of the key issues relating to biotechnology and poverty alleviation. He discusses yield enhancement, employment and labour market effects, scale neutrality, the challenge of meeting the needs of the poor, the problem of private sector control over world genetic resources, and the replacement of traditional export crops by synthetic substitutes. I did feel, however, that a clear answer to the basic question of whether biotechnology will alleviate poverty was lacking. Dr Ahmed certainly implies that the potential exists for biotechnology to do much good, though this is qualified. There are considerable uncertainties regarding the pace of application and others relating to distributional consequences. Furthermore, there are some potential applications (notably the production of synthetic substitutes) which are decidedly anti-poor, or at least against the interests of developing countries. If this is a correct interpretation of the paper I would declare myself in agreement with its main thrust. To substantiate this view, I would like to discuss four points not covered (or only partially covered) by Dr Ahmed.

In setting out to promote pro-poor technological change (regardless of whether or not it has a biotechnology component), decision makers and their advisers need to:

(1)   identify the poor and their technology needs as consumers, labourers, and producers, bearing in mind that in this context the focus needs to be on absolute poverty;
(2)   reflect on what we know about the distribution of benefits from technological change in agriculture;
(3)   establish criteria for using technology and associated delivery systems to help the poor; and
(4)   address the institutional, managerial and legal challenges which may hamper the generation and adoption of agricultural technology capable of helping the poor.

First, agricultural technology can help alleviate poverty if it leads to a reduction in the real price of food staples, an increase in employment and earnings, and an increase in the level and stability of yields. Thus the patterns of consumption, agricultural employment and farm production of the poor must be analysed. Location-specific work on these topics will yield information on the crops and livestock activities which may be candidates for productivity improvement. An issue which should be addressed is whether the poor, in a given context, should be helped as farm producers, as workers or as consumers. It is as well to remember that investments geared to the development of yield-enhancing technology for resource-poor smallholders may have opportunity costs. There may be a trade-off between equity and efficiency in the allocation of resources to agricultural research. Other approaches to poverty

*World Bank, Washington and Switzerland Directorate of Development Co-operation and Humanitarian Aid.

reduction, including public works, migration away from marginal areas, and better health and education services, may be more cost-effective.

Regarding the distribution of benefits from technological change many factors have to be considered (Binswanger and Von Braun, 1991). Under conditions of inelastic demand, consumers gain from technological change which engenders an expansion in the supply of commodities they use. However, consumers in the Third World do not benefit when products are mainly exported, unless added production for export causes favourable general equilibrium effects. Producers gain from technical change when demand is elastic, which occurs when new markets (including export markets) are opened up. When demand is inelastic (the usual condition for domestically consumed food) they may gain or lose, depending on how fast costs decline relative to prices received. Thus, if producers are net buyers of food (a frequently encountered condition among smallholders), their gains as consumers may outweigh their gains or losses as producers. Labourers gain if technical change leads to a net increase in the demand for hired labour, or to an increase in employment due to growth linkages.

This simple framework is useful for an initial assessment of the required provision of pro-poor agricultural technology. Since consumers stand a good chance of gaining from technical change, as long as it occurs in the right commodities, the focus must be on producers and labourers. Technology for resource-poor smallholders should be *input-extensive* (stress-resistant varieties, which do not require ancillary chemical inputs, are a case in point). It should focus on 'orphan crops' (a term used in the paper by Collinson at this Conference), particularly in Africa (cassava, yams, plantains, coarse grains) and on the less well endowed agro-climatological areas. It should facilitate diversification to enable producers to maintain their share of the gains from innovation and to raise the demand for labour, and it should be accompanied by programmes to speed up adoption (agricultural extension, credit, infrastructure). Land improvement programmes can play an important role in raising the demand for labour, as well as enhancing the future productive base.

Biotechnology would appear to offer a number of potentially useful avenues which meet the pro-poor condition. Possibilities (and in some cases realities) mentioned by Dr Ahmed include tissue culture to produce disease-free planting material, genetic engineering for pathogen and pest resistance, and genetic manipulation to raise cold tolerance or, in the potato, to reduce high temperature susceptibility. The main benefit, however, is likely to be capability to raise the efficiency of conventional plant-breeding programmes. That has been mentioned on a number of occasions at this Conference.

This brings me to my last point about institutional, managerial and legal challenges which I would like to address from the point of view of needs in Sub-Saharan Africa. Caveats about the possible trade-off between equity and efficiency notwithstanding, there is a strong consensus among analysts that 'significant reorientations towards neglected areas and economically weak groups are needed to bring major social and economic gains to the rapidly growing numbers of poor Africans' (Gnaegy and Anderson, 1991). However, this has not crystallized into a shared vision and effective long-term programmes and alliances between donors, international centres and national governments, to raise the dismally low performance of African research institutions. Facing the legal challenges of patenting technical advances in developed countries also requires strong institutions working in favour of the developing world.

REFERENCES

Binswanger, H.P., and Von Braun J., 1991, 'Technological Change and the Commercialization in Agriculture: The Effect on the Poor', *World Bank Research Observer*, 6/1, January.
Gnaegy, S. and Anderson J.R., 1991, 'Agricultural Technology in Sub-Saharan Africa (A Workshop on Research Issues)', *World Bank Discussion Papers*, 126, World Bank, Washington, DC.

ROBERT W. HERDT*

Agricultural Biotechnology and the Poor in Developing Countries

INTRODUCTION

The poor can be helped by improving the conditions or environment in which they live, lowering the costs of what they consume, and increasing their incomes. In the developing world, if agricultural biotechnology can be directed towards such changes it can benefit the poor. Directing biotechnology requires a clear identification of the poor, where they live, what they consume, how they generate their income, how biotechnology could improve their living conditions, how it could reduce the costs of what they consume, and how it might increase their income. I briefly discuss identifying the poor and their needs, and then concentrate on the potentials of biotechnology.

THE POOR

Poor people are everywhere, from the streets of New York, London and Calcutta to the rural outback of Texas, Nigeria and Brazil. Statistics about the poor are inadequate and the definitional basis for those statistics are questionable, but, still, we know that poverty is an overwhelming problem in the developing world. 'More than one billion people in the developing world are living in poverty. *World Development Report 1990* estimates that this is the number of people who are struggling to survive on less than $370 a year' (World Bank, 1990). A total of 42 nations with a population of 5.8 billion people were in the lowest income group with per capita incomes of $545 or less in 1988. Many of those are in Africa, but the most populous are in Asia, including China and India. Population growth rates are slower in most countries of Asia than in most of Africa and per capita income growth rate faster, but Asia is nonetheless expected to have about 800 million poor in the year 2000, compared to 260 million in Sub-Saharan Africa and about 100 million in the rest of the developing world.

The poor lack assets (wealth), have low incomes, low consumption levels, and live in primitive conditions. Within poor countries the poor live in the worst conditions and have the lowest levels of health. The obvious potential of biotechnology for improving human health suggests that may be a priority

*Director for Agricultural Sciences, The Rockefeller Foundation, USA.

area for its application in the developing world, but one may contrast that with the great potential for improving health through the application of known, low-cost, public health measures in most developing countries. Even in the area of health research, there is a continuing need to increase emphasis on diseases that are important in developing countries (Commission on Health Research for Development, 1990). Although the complex issues related to health and biotechnology for developing countries are recognized, space limitations preclude further discussion of that issue.

Analysts point out that the application of biotechnology in the industrialized world will create substitutes for agricultural goods now produced in the developing world, undercut export markets and thereby reduce developing country incomes. Trade policies have similar effects. As with health, the value of these points is recognized, but they are not discussed: this paper concentrates on the way in which biotechnology should be directed to make positive contributions to the poor in the developing world. The most obvious need of the poor is for higher income: how can biotechnology be used to that end? The needs of the poor would be better met if the things they consumed were available at lower real costs: how can biotechnology be used to reduce the costs of what the poor now consume and would consume if their real incomes were higher?

Increasing Incomes for the Poor

Low incomes are the result of either low productivity by the self-employed or, among those employed by others, low wages, generally in combination with inadequate duration of employment. The poor self-employed work with very few of their own capital resources, generating income essentially from their labour input. If biotechnology is to raise the productivity of the poor self-employed, their enterprises must be identified, the factors that limit their labour productivity must be identified and some strategy to overcome such limitations must be followed. Strategies which require the self-employed individual to make capital investments are biased against the poor, by definition. It may be possible to offset such biases by policy or institutional means, but those go beyond the scope of directing biotechnology, which is the focus of this paper.

Self-employment enterprises include agricultural, hunting, gathering, handicraft, service, and other types of activities. Crop and animal enterprises are important among these, but agriculturalists probably over-estimate their importance for total income. For example, a comprehensive, careful study of a Philippine village, in the middle of the most intensive rice-growing area of that country, showed that 42 per cent of average family income came from rice, the balance from other agricultural enterprises (14 per cent), labour earnings (28 per cent) and transfer payments (15 per cent) (Hayami, 1978).

Most poor do not own land and they are employed by others (in the Philippine village study cited above the landless received 61 per cent as labour earnings and 33 per cent as transfer payments). If biotechnology is to raise the income of the poor, it must increase either wage rates or employment opportunities, or both. That is, it should increase the demand for labour and should not create substitutes for labour. Biotechnology can increase the demand for labour by increasing the productivity of labour-intensive enterprises and by creating new labour-intensive enterprises. Examples include making crop production possible on land that previously was not usable, or making it possible to grow crops in a season when previously it was not possible.

Reducing consumption costs of the poor

Among the developing world's poor, the products of agriculture – food, fodder, fibre and fuel – generally absorb the largest fraction of total income. No global data exist, but in countries as diverse as Indonesia, Côte d'Ivoire and Peru about 70 per cent of poor households' spending is for food (World Bank, 1990). Many poor suffer from malnutrition and, although some malnutrition is observed among groups at all levels of income, the rate of food consumption generally rises sharply as incomes rise above the lowest levels. Technology holds out the promise of increasing agricultural production and reducing the real price of basic agricultural commodities in the developing world by increasing the efficiency with which they are produced, if it can be directed and focused on appropriate problems.

Cereals, roots, tubers and legumes contribute 60 per cent of the food protein and over 60 per cent of the energy of developing world people, with rice alone contributing over 25 per cent (FAO 1980). This is in sharp contrast to North America and Europe, where cereals, roots, tubers and legumes contribute only 30 per cent. Comparisons within countries show that consumption patterns of the poor differ considerably from those of the wealthy. It follows that, if one is to direct biotechnology towards food consumed by the poor, those commodities must be clearly identified. This is the first task of those who would use biotechnology to benefit the poor.

THE IMPACT OF TECHNOLOGICAL CHANGE

Research on the effect of technical change in agriculture has established that, when technological improvements are made in the production of a basic food crop such as rice, for which demand is price inelastic, the largest fraction of benefits accrues to consumers, assuming product prices are relatively free to respond to markets. If that downward adjustment of price is frustrated by government policy, then the benefits of the technological change may be absorbed by producers. This is true in high-income countries (Cochrane, 1958) as well as low-income countries (Hayami and Kikuchi, 1981). The impact on producers who also consume some of what they produce, which is the case in most developing countries, has also been analysed and is reviewed below (Hayami and Herdt, 1977).

Disaggregated effects of technological improvement

A technological improvement is a decline in the amount of input per unit of output and hence in unit cost of production, all other things held constant. One of the most important factors determining whether farmers benefit from a new technology is whether they adopt it. When widely adopted over the long run, technical improvement affects the earnings of all land and capital used in agriculture and hence affects both adopters and nonadopters. Technical change also affects non-farmers, both rural and urban, and labourers who work for farmers.

Table 1 illustrates the impact of a technological improvement in rice production that is suitable for one agro-ecology but not for a second. The impact on individuals depends on their relationship to rice production and consumption. Seven groups are defined. Group 1 consists of rice consumers who do not produce rice. They are not affected by farm prices received or rice production costs, but, because the price of rice they consume falls with a technological improvement, their welfare improves. Groups 2 and 4 include producers located in the agro-ecology for which the technology is suitable and who adopt the technology. After the technological improvement they receive lower prices for the rice they sell but also have lower costs of production. Whether they gain depends on the relative change in costs and prices, and cannot be determined *a priori*. Net buyers gain because they buy rice at lower prices, but because some changes increase benefits while others reduce them, the net impact on groups 3 and 6 are also indeterminate. Groups 5 and 7 lose from the technical change because the price of rice they sell declines and they get no benefits from the technology because they either cannot (7) or do not (5) adopt. One may divide each of the groups in Table 1 into a poor and non-poor component, thereby focusing more sharply on the issue of the poor, but using Table 1 or an expanson of it requires much empirical work (Binswanger, 1980).

TABLE 1   *Impact of a technological improvement in rice production applicable in agro-ecology 1 but not 2, which leads to a fall in market price*

Group	Impact on price received	Impact on rice production costs	Impact on rice consumption costs	Net impact on welfare
1. Consumers, non-producers	0	0	−	+
Producers in agro-ecology 1 (technology suitable)				
2. Net buyers who adopt	−	−	−	?
3. Net buyers, non-adopters	−	0	−	?
4. Net sellers who adopt	−	−	0	?
5. Net sellers, non-adopter	−	0	0	−
Producers in agro-ecology 2 (technology not suitable)				
6. Net buyers	−	0	−	?
7. Net sellers	−	0	0	−

Note: The effect on variables in column headings are decrease (−), increase (+), no effect (0), or indeterminate (?).

Robert W. Herdt

EMPIRICAL STUDIES OF TECHNICAL CHANGE IN RICE

As indicated above, adoption is necessary to obtain benefits as a producer. There was concern in the 1960s and 1970s about whether farmers with limited resources would adopt new technology (for example, Falcon, 1970). Since then, many empirical studies have shown that, although adoption of semi-dwarf rice varieties in the first one or two years after their introduction was concentrated on larger farms, adoption thereafter spread generally throughout the farming population in agro-ecologies for which the new technologies were biologically well adapted (Ruttan, 1977; Barker and Herdt, 1985; Herdt and Capule, 1983).

A recent thorough review of the equity effects of new technology by one of its early critics concluded that the major defect of the 'green revolution' rices is that they were not equally well suited to all areas (Lipton and Longhurst, 1989). A new set of studies comparing the impact in areas where new rices were widely adopted with areas where they were not shows that the new varieties increased the demand for labour in the adopting areas and that interregional migration, both seasonal and permanent, occurred, with the consequence that wage rates tended to rise in the non-adopting areas as well as in the adopting areas (David and Otsuma, 1989; Hossain, 1988; Jateliksono, 1987). That research also shows that irrigation and other physical factors are much more important than farm size and tenure in determining adoption of semi-dwarf varieties and that returns to land tend to rise in the adopting areas and fall in the non-adopting areas.

The effect of a technical change on individuals, whether poor or non-poor, will depend on the magnitude of changes in production costs, product prices, input use and input prices. The net effect of changes will differ individual by individual, depending on consumption patterns and the resources and enterprises each controls. In addition to the attributes noted in Table 1, some individuals may work as labourers, others may receive the residual earnings from rice land, and others may supply inputs required by the new technology. All these attributes will affect the impact of a technical change. The aggregate effect on the 'poor' in a country will also depend on the number of people in each group. Thus it is impossible to predict the size and magnitude of impact on the poor from a technological change in agriculture, whether it arises from biotechnology or from some other source, although it is clear that consumers will gain and non-adopting farmers will lose when market prices are allowed to reflect the cost-reducing effect of the technological improvement.

Decision makers in any particular country who wish to direct biotechnology at the needs of the poor in their country will need to operationalize the above general comments by empirical analyses that identify:

(1) the health problems of the poor,
(2) major commodities consumed by the poor,
(3) enterprises controlled by the poor,
(4) constraints that limit productivity of those enterprises, and
(5) constraints that can be addressed by biotechnology.

THE POTENTIALS OF BIOTECHNOLOGY

To help direct biotechnology it is important to have an understanding of the basics of biotechnology (see NRC, 1987; Walgate, 1990; Messer and Heywood, 1990). Too few social scientists or economists have been willing to gain such an appreciation. Biotechnology does not directly affect such agriculturally important factors as the physical composition of soil, the strength of metals, the access to financial assets, the distribution of land, the regulations governing international trade, or the phases of the moon! It does offer a number of new ways to modify the genetic compositions of living organisms – plants, animals and micro-organisms. All of its potentials and limitations derive from that capacity.

Mankind has been changing the genetic code of the living organisms it uses since the beginning of time, at first simply by choosing to use some plants and animals rather than others, more recently by deliberately breeding plants and animals. In agriculture, biotechnology can be directed towards creating genetically improved crops, genetically improved animals, or genetically modified micro-organisms that produce something used in crop or animal production. In order to keep the discussion manageable, I concentrate on crops biotechnology. Two sets of techniques are important: recombinant DNA and tissue culture.

Recombinant DNA

Recombinant DNA, which (theoretically, at any rate) enables the transfer and activation of genes from any living organism to any other living organism, lies behind the explosion of scientific and commercial interest in biotechnology. Related biotechnology techniques offer great potential, but recombinant DNA or genetic engineering is the key: 'The essence of genetic engineering is the ability to identify a particular gene – one that encodes a desired trait in an organism – isolate the gene, study its function and regulation, modify the gene, and reintroduce it into its natural host or other organism' (NRC, 1987).

Identify a particular gene This apparently simple statement is rather misleading. Genes cannot be physically observed, even with the most powerful of electron microscopes, because they are segments of DNA strands. Chromosomes, which are composed of huge amounts of DNA, can be observed; rice has 12, human beings have 46, but being able to see chromosomes does little to help identify a gene. Molecular biology has developed laboratory techniques, the main one being gel electrophoresis, to distinguish reproducibly the presence or absence of fragments of DNA with varying molecular weights. DNA is prepared by treating cells with restriction enzymes which cut the DNA of each chromosome into many, many pieces. Some fragments contain genes or fragments of genes and many other fragments which either have no function or have unknown functions.

One that encodes a desired trait Traits are just what they sound like: observable or measurable characteristics that make an organism what it is.

Simple human examples are hair colour, number of fingers, or the capacity for 'perfect pitch'. By observing the DNA of progeny of individuals with and without a particular trait, it may be possible to identify the fragment of DNA associated with the trait of interest. Biologically, what a gene does is to provide the instructions or code required to produce a given protein. Each gene encodes a different protein. Proteins are extraordinarily versatile molecules that make up most of a cell's structure and ensure that cells do whatever it is they are intended to do.

Study its function and regulation   Some useful traits are conferred by single genes, such as the toxin produced by *Bacillus thuringiensis*, which can kill insect pests. Many important traits are controlled by several genes, however, which makes working with such multi-gene traits orders of magnitude more complex than with single-gene traits. And to make the challenges even greater. Some of the most desirable properties of plants, such as drought-tolerance, high yield or hardiness, are the result of many traits working together, just as a good sportsman has good eyesight, endurance, strength, balance, timing and 'heart'.

Re-introduce the gene into another organism   Before it can be transferred to another organism, a gene has to be cloned, or reproduced in large numbers. It is introduced into the DNA of a bacterium, the bacterium makes many 'copies' of the gene and itself multiplies millions of times, thereby multiplying the DNA and making it available in large quantities. The DNA is then introduced into cells of the organism which is to be changed, either by infection with the vector *Agrobacterium tumaefaciens*, which has the natural capacity to incorporate its DNA into plants, by shooting small particles through the cell wall using an explosive charge, by temporarily perforating the cell wall with electrical charges so the DNA can permeate it, or by treating the cell with a chemical substance for the same purpose. Millions of cells are treated and a small fraction take up the foreign DNA. Then, if the target organism is a plant, the cell is manipulated in such a way as to grow into a whole plant. If everything happens as planned, the transformed plant has the desired new trait.

DNA mapping   The ability to cut DNA into segments and clone genes has made it possible to determine whether specific pre-identified segments of DNA exist in a given organism. If one has determined that the gene for a given trait falls on a given segment of DNA, then knowing whether that segment is present in a plant indicates whether the gene is present. Restriction fragment length polymorphism (RFLP) mapping and DNA probes make use of this capacity. DNA probes of plant pathogens can be used to detect the presence or absence of such pathogens on seeds, and to re-assure importing countries that seeds its plant-breeders may wish to import are pathogen-free.

Tissue culture

Tissue culture is the process of keeping tissue alive independent from the organism from which it was obtained, say in a test-tube. Plant tissue culture is a key part of plant biotechnology: it can be used to produce genetically modified plants itself, and is a necessary step in obtaining recombinant plants. Tissue culture is a generic term: many different tissues/organs can be cultured, including pollen cells (pollen culture), cells from the growing tip of a plant (meristem culture) and portions of ordinary leaves (somatic cell culture). The capacity to produce a whole plant from a cell differs for various plants and plant organs. Tissue culture is generally conducted in glass containers, that is *in vitro*.

Somaclonal variation   Until ten or so years ago, the theory of genetics suggested that, if a single cell were induced to produce many cells, and some of those cells were induced to grow into whole plants, those plants would all be genetically identical. In fact, many are not, and the resulting variation is known as somaclonal variation. Some of the plants that result from the variation may have characteristics that are more desirable than others, so, in effect, somaclonal variation is one technique by which tissue culture can be used to induce genetic variation, much as the crossing operation in plant breeding generates variation.

Protoplast fusion   As described in elementary biology, a single cell 'grows' by dividing, first into two, then into four and so on. (In this process, the DNA replicates so that the new cells each have identical DNA to the cell from which they were formed.) Almost the opposite can be made to happen: cells can be 'fused' together so that their respective chromosomes mix and, if the fused cell can be induced to produce a plant, it will have all or some genes from each 'parent cell'. If one begins with cells of, say, tomato and potato, the resulting plant would be expected to have characteristics of each crop. Plant cells have cell walls, so fusing them together is difficult, but removing or degrading their cell walls, and thereby making them into protoplasts, makes fusion possible. The technique works best with related species.

In vitro selection   Somaclonal variation and protoplast fusion both generate many genetically modified cells relatively easily. Taking a large number of cells to the full plant stage is laborious and time consuming, and may be inefficient if most of the resulting plants do not have any especially desired characteristic. Subjecting the original millions of cells to carefully designed selection pressure *in vitro* can kill a large fraction of them, thereby reducing the number that have to be carried to maturity. If the cells that survive the selection pressure have the special characteristic that is being sought, then a significant savings in time and resources may be effected.

Micropropagation   The clump of cells at the growing tip of a plant (the apical meristem) generally grows quite rapidly into a seedling that can mature into a complete plant. This process may occur so rapidly that plant diseases or viruses (pathogens) present in the 'mother plant' are not in the meristem tissue and hence the plants produced by micropropagation are pathogen-free. Pathogens infect most plants grown by conventional means, although their effect is generally to depress productivity rather than kill the plants; pathogen-free plants have enhanced productivity.

Interspecific crossing and embryo rescue   A species is composed of individuals that are capable of sexually reproducing, so, by definition, plants of different species cannot generally be successfully crossed. Plants in the same genus may be thought of as close relatives, and plants in the same family are somewhat more distant relatives. Attempts to use the pollen of close or distant relatives to pollinate a plant in a different species ('interspecific crosses') sometimes result in the formation of an embryo, but in nature these generally abort. Embryo rescue techniques can be used to remove the embryo, place it in culture solution and coax it into living and growing. Like DNA transformation, interspecific crossing is a way to get new genes into a plant.

Limitations and costs

Each of the techniques mentioned has been implemented in plants, and some have produced new genetic compositions superior to previous ones. Many of the techniques require carefully controlled light, temperature and nutrient environments, and most can be routinely carried out only on a limited number of species. Even where they have been demonstrated, some cannot be routinely replicated. Hence a good deal more development is necessary before plant biotechnology is as predictable as car repairs (and most car owners know that process is fairly unreliable)!

Four realities about biotechnology should be recognized by those seeking to direct it to the needs of the poor: (1) biotechnology research is a risky investment; (2) the pay-off to plant biotechnology depends on adequate 'conventional' plant-breeding capacity; (3) recombinant DNA research requires a relatively large initial investment; and (4) the environmental safety of biotechnology is being questioned by a number of observers, slowing its broad-scale testing in the industrialized countries.

Risky investment   Research is an investment: it requires a period of years before yielding a pay-off. It is risky: the size of the pay-off is not known with certainty when the investment is made. There is considerable private investment in biotechnology in the industrialized countries, some for agricultural but more for medical applications. Private firms or individuals invest in research only if there is a way for them to capture a return on their investment and the expected value of the return compensates for the risk. Biotechnology provides a basis for distinguishing between the genetic composition of individual plants or animals; that is, with the appropriate knowledge base, it is now possible to trace whether a particular gene is present in a particular individual. This capacity provides a practical basis for granting property rights to those who discover, construct, isolate and clone genes, and this prospect has induced investors to provide venture capital to biotechnologists in the United States and other countries where such property rights are granted.

In countries where property rights for genetically engineered plants or animals do not exist, or where most farmers are subsistence producers so that the market for agricultural inputs is too small to attract private firms, it is unlikely that private biotechnology research and development will be undertaken. In such cases, there may be a rationale for publicly funded biotechnology research, but it must have a higher social rate of return than alternatives, if it is to be economically justified.

Complementary to other research In order to produce a 'better' crop variety, a specific desired trait or set of traits must be identified, a source of the trait must be found, that source must be combined with an existing variety and the new combination be shown to 'breed true', or be reproduced in the subsequent generations.

Genetic engineering does the work of combining and hence substitutes for the crossing step of conventional breeding. DNA markers can substitute for physical, biological or chemical markers, and hence determine the presence or absence of a gene (if it has been previously identified and cloned) but nothing in biotechnology can substitute for *determining* the important traits, *testing* crops containing new genes under farmers' conditions, *multiplying* seeds, and *distributing* them. This amounts to saying that a country must have good plant-breeding capacity for a crop in order to make biotechnology applied to that crop a useful investment. A country incapable of generating a continuous stream of genetic technological innovations that have been adopted by farmers is likely to have too weak a crop research base to make investment in plant biotechnology worthwhile.

Financial requirement Recombinant DNA research requires materials and equipment more sophisticated and costly than most crop and animal breeders use. To build the capacity for recombinant research, institutions will have to invest at a different order of magnitude than traditionally. They will have to go beyond the 'few simple sheds' and fields that, together with hard work, are the essence of an effective plant breeding programme (Jennings, Coffman and Kauffman, 1979). Tissue culture applications have more modest investment requirements.

Table 2 shows my rough estimate of the cost and time necessary to demonstrate certain relatively well defined plant biotechnology goals on crops which have not heretofore been subject to each technique. Additional time and money would be required to make each process routine and to make it possible with varieties different from the one on which it was developed. The guesses derive from the Rockefeller Foundation's experience with rice. Additional time and resources will be needed to get the genetically engineered crops into the hands of farmers. There is considerable imprecision associated with the estimates.

The Rockefeller Foundation has expended roughly US$35 million to develop the techniques of biotechnology that have enabled scientists to demonstrate genetic engineering of rice, *Oryza sativa*. We hope that in two to five

years, and with an additional expenditure of US$15 to 25 million, a geneti-
cally engineered rice will be available that incorporates a new kind of resistance
to rice tungro virus, one of the major diseases of rice. Assuming it will be
possible to put that genetically engineered rice into farmers' hands in developing Asia by the year 2000, the additional rice produced in the decade
between 2000 and 2010 would have a present value of about US$4.0 billion.
Success with other traits will add to the benefits. That economic benefit will
be distributed among poor and non-poor, consumers and producers, urban and
rural, as discussed in the first section of this paper.

TABLE 2   *Rough estimate of resource requirements for biotechnology*
research

Technique	Cost ($ mill)	Time (yrs)
Disease-free propagation of perennials	0.5–1.0	3–6
RFLP map to assist breeding	1.0–5.0	3–8
Protoplast fusion, regeneration	0.5–15	2–15
Genetic engineering for virus resistance	40–60	8–15
Genetic engineering for nitrogen fixation	100–500	20–40

Genetically engineered resistance to tungro virus, and all other traits for
which the Foundation is supporting research, will not require farmers to use
additional inputs other than seeds (which in rice are self-reproducing, except
for 'true hybrids'). For tungro, genetic resistance is the only available control
strategy, other than pesticides that are sometimes used to kill the insect which
transmits the virus. To date, resistance incorporated by conventional plant
breeding has not been durable; the incorporation of DNA that encodes for the
protein coat of the virus particle itself is expected to be extremely durable on
theoretical grounds. I believe that this and other strategies usable through
biotechnology have the potential to provide the poor with more benefits than
input using agricultural technologies.

Environmental Bio-safety   There is considerable apprehension in some quar-
ters about the potential environmental problems that may be associated with
the introduction of genetically engineered organisms into the environment,
especially genetically engineered micro-organisms. Less danger is perceived
from genetically engineered plants, but potential dangers analogous to the
ecological nightmares caused by the introduction of plants or animals from
one continent to another are possible: rabbits to Australia, kudzu to the United
States, kariba weed in Africa, and many others. On the other hand, many see
biotechnology as holding the promise of replacing chemical pesticides with
genetic means, thereby eliminating the need for farmers to handle dangerous
chemicals and contributing positively to the environment.

The issues of environmental bio-safety are among the most complex accompanying biotechnology (Walgate, 1990). Some organizations are calling for 'a general slowdown in the rate at which organisms are engineered, tougher regulations, control on an international scale, and prohibitions against release of genetically engineered organisms in certain locations' (Mellon, 1988). Ecologists are cautious. While they agree that the capability of genetic engineering to make precise changes increases confidence that unintended changes in the genetic make-up of organisms will not occur, 'precise genetic characterization does not ensure that all ecologically important aspects of the phenotype can be predicted for the environments into which an organism will be introduced' (Tiede, *et al*, 1989).

A panel commissioned by the National Academy of Sciences to advise the Agency for International Development of the United States put its bio-safety recommendations at the top of the list of what USAID should do to assist developing countries to benefit from biotechnology (NRC, 1990). The potential environmental dangers of possible bio-engineered crops will have to be weighed against the potential benefits, along with the other costs associated with their use. If biotechnology can be directed specifically towards problems of the poor, then increased equity may be added to its other benefits.

DIRECTING RICE BIOTECHNOLOGY RESEARCH

Some vocal critics of biological research imply that biologists are unaware of the potential value to farmers of plants that are drought-tolerant, resistant to soil problems, or that give high yield with low inputs, and are too reluctant to turn their energies to such problems. But that over-simplifies the challenges of bio-engineering. It is extremely difficult to determine what genes are involved in producing traits like drought resistance, high yield and resistance to toxic soils conditions, to say nothing of identifying the genes that give such traits.

A more productive approach for those interested in equity would be to assist biological scientists to identify the major problems on crops produced or consumed by the poor. The Rockefeller Foundation decided to focus its biotechnology programme on rice in part because rice is the most important food crop for the poor in many poor countries. Having decided to focus attention on rice, the Foundation drew up a list of the possible rice production challenges at an early stage of the programme. A quantitative analysis was undertaken to help in directing the Foundation's programme among the challenges listed, and a modified cost–benefit approach was used to set priorities for the Foundation's support of applied rice biotechnology research (Herdt, 1991). It essentially consisted of:

(1) quantitatively estimating, for all possible challenges, the expected benefits to society from solving each;
(2) weighing the benefits by their contributions to environmental and equity goals; and
(3) evaluating, for each challenge, the likely effectiveness of biotechnological as compared to conventional approaches.

The result was an ordered list of priority traits for attention. The list was publicized among participants and potential participants in the programme and outside (Herdt, 1987). The analysis gave added weight to strategies or traits that were especially beneficial to the poor and to the environment. The Foundation has used the priorities list to help guide programme development and funding decisions, and has been pleased with the response of the research community in developing projects on challenges of high priority. I believe that such clearly stated priorities need to be developed for individual rice-growing countries and for issues beyond rice biotechnology.

REFERENCES

Barker, Randolph, and Herdt, Robert W., 1985, *The Rice Economy of Asia*, Resources for the Future, Washington DC, p. 324.

Binswanger, H.P., 1980, 'Income Distribution Effects of Technical Change: Some Analytical Issues', *South East Asian Economic Review*, vol. 1, pp. 179–218.

Cochrane, Willard W., 1958, *Farm Prices: Myth and Reality*, University of Minnesota Press, Minneapolis.

Commission on Health Research for Development, 1990, *Health Research Essential Link to Equity in Development*, Oxford University Press, New York, p. 136.

David, C.C. and Otsuka, Keijiro, 1989, *The Modern Seed-Fertilizer Technology and Adoption of Labor-Saving Technologies: The Philippine Case*, International Rice Research Institute, Department of Agricultural Economics Paper No. 89–15., Los Baños, Philippines.

Falcon, Walter P., 1970, 'The Green Revolution: Generations of Problems', *The American Journal of Agricultural Economics*, 52,(3).

FAO, 1980, 'Food Balance Sheets and Per Caput Food Supplies 1961–65 average and 1967 to 1977', Food and Agriculture Organization of the United Nations, Rome.

Hayami, Yujiro, 1978, *Anatomy of a Peasant Economy*, International Rice Research Institute, Los Baños, Philippines.

Hayami, Yujiro and Herdt, R. W., 1977, 'Market Price Effects of Technical Change on Income Distribution in Semi-Subsistence Agriculture', *American Journal of Agricultural Economics*, 59,(2), May, pp. 245–56.

Hayami, Yujiro and Kikuchi, Masao, 1981, *Asian Village Economy at the Crossroads. An Economic Approach to Institutional Change*, Tokyo University Press, Tokyo.

Herdt, R.W., 1991, 'Research Priorities for Rice Biotechnology', in G.S. Khush and G.H. Toenniessen (eds), *Rice Biotechnology*, CAB International, Wallingford.

Herdt, R.W. and Capule, C. 1983, *Adoption, Spread, and Production Impact of Modern Rice Varieties in Asia*, International Rice Research Institute, Los Baños, Philippines.

Hossain, Mahabub, 1988, *Nature and Impact of the Green Revolution in Bangladesh*, International Food Policy Research Institute, Research Report 67, Washington, DC.

Jatcliksono, Tumari, 1987, *Equity Achievement in the Indonesia Rice Economy*, Gadjah Mada University Press, Yogyakarta, Indonesia.

Jennings P.R., Coffman, W.R. and Kauffman, H.E., 1979, *Rice Improvement*, International Rice Research Institute, Los Banos, Philippines, p. 186.

Lipton, Michael, with Longhurst, Richard, 1989, *New Seeds and Poor People*, Johns Hopkins Press, Baltimore, p. 473.

Mellon, M., 1988, *Biotechnology and the Environment*, National Wildlife Federation, Washington, DC, p. 64.

Messer, E. and Heywood, P., 1990, 'Trying Technology', *Food Policy*, August, pp. 336–345.

NRC, 1990, *Plant Biotechnology Research for Developing Countries*, National Research Council, National Academy Press, Washington, DC.

NRC, 1987, *Agricultural Biotechnology, Strategies for National Competitiveness*, Committee on a National Strategy for Biotechnology in Agriculture, Board on Agriculture, National Research Council, National Academy Press, Washington, DC.

Ruttan, V.W., 1977, 'The Green Revolution: Seven Generalizations', *International Development Review*, 19, pp. 16–23.
Tiede, J.M. *et al.*, 1989, 'The Planned Introduction of Genetically Engineered Organisms: Ecological Considerations and Recommendations', *Ecology*, April 1989, vol. 70, pp. 297–315.
Walgate, R., 1990, *Miracle or Menace?*, The Panos Institute, Washington, DC, p. 199.
World Bank, 1990, *World Development Report 1990*, Oxford University Press for the World Bank, Washington DC, p. 260.

DISCUSSION OPENING – ANTONIO L. LEDESMA*

Coming as I do from the non-governmental organization (NGO) community to this IAAE Conference is akin to a quasar's light threading its way across a vast distance and finally arriving as a minuscule presence in an enormous gathering. I was quite reluctant to come to Tokyo, since I felt some degree of bias against what I felt might be a meeting of scientists linked, in some way, to the interests of the powerful rather than to those of the poor. However, Professor Dams assured me that the IAAE is 'NGO-friendly'. It has certainly proved to be so. I have heard so much about sustainability, equity, ethics, poverty alleviation and participation, from so many speakers, that I feel there is congruence between the values of the typical NGO worker and those who make up your membership.

I do have bias since there is considerable debate relating to the question of whether positive benefits of biotechnology can ever be directed towards aiding the poor. In that debate, fear is often expressed that biotechnology has the capacity to displace traditional agricultural commodities on a massive scale, that it is likely to accentuate inequality, and perhaps increase the vulnerability and dependence of farmers through concentration in the power of transnational agribusiness. Prior experience adds to those concerns. The green revolution, in my view, brought dependence on the use of chemical inputs at the farm level, while the use of powerful materials provided hazards to public health and to the environment. The Bhopal pesticides tragedy, and other examples of chemical spillage, are stark reminders of what can occur.

Against that background of old apprehensions, I listened to Dr Herdt with considerable interest. I also remembered some of his earlier work, references to which he quotes, in which there was obvious concern with the central issue of the distribution of the benefits of technical change. I can best summarize his paper by saying that it attempts both to de-mystify the scientific techniques utilized in crop technology, and to set out the research agenda, priorities and basic aims of the Rockefeller Foundation. He did, however, begin with a short but very striking description of the conditions facing the poor, and he ended with an outline of an evaluation system which stresses both equity considerations and environmental safety. Thus his sensitivity to the needs of the poor is very much in evidence. It is a significant paper, very much in the tradition of the founders of the IAAE, who put the first objective of the Association as being 'to foster the application of the science of agricultural

*Centre for the Development of Human Resources in Rural Asia, Manila, Philippines.

economics in the improvement of the economic and social conditions of rural people and their associated communities'.

Despite all of that, however, I still retain my own worries. I cannot criticize Dr Herdt for having neglected them, since he has much to cover in a short paper. Those of us who work in NGOs are still concerned about the potentially damaging effects of the aggressive marketing of biotechnology-based products in monopoly or near-monopoly conditions, and about the privatization of biotechnology through patent protection, thereby restricting access by the potential Third World user. We are also concerned about the extent to which it is possible for Third World organizations, working among the poor, to secure relevant information and to ensure that it can be used to promote greater participation by potential, and needy, beneficiaries. This Conference, through the efforts of Dr Herdt and others, has provided me with a great deal of useful information on the latter count, though it has not really addressed the issues of control of use.

KEIJIRO OTSUKA*

Green Revolution, Agrarian Structure and Income Distribution in Asia

INTRODUCTION

The rice sector in developing countries of Asia has achieved remarkable productivity growth for the last two decades as a result of the adoption of high-yielding modern rice varieties (MVs) – a phenomenon popularly known as the green revolution. As of now, about 50 per cent of the rice area is planted to MVs. The productivity effect of MVs, however, is constrained by production environments, particularly by the availability of irrigation water (Barker and Herdt, 1985). As a result, a significant productivity gap has emerged between favourable and unfavourable rice-growing areas. The critics argue that the green revolution has by-passed unfavourable areas, where farm populations are generally poor, and worsened the regional income distribution (for example, Lipton and Longhurst, 1989).

However, the impact of the MV adoption on the regional income distribution depends not only on the direct effects of MVs on productivity and factor demands, but also on the indirect effects on factor markets. Since the adoption of MVs increases labour demand, particularly for hired labour, wage rates will increase in favourable areas in the short run. In the longer run, higher wages in favourable areas may induce interregional migration from unfavourable to favourable areas leading to the equalization of regional wages. Furthermore, farm size in favourable areas may decline through land and tenancy market adjustments relative to unfavourable areas, as the demand for land increases with the MV adoption. Such factor market adjustments may alleviate the regional income inequality of farming households. Also important may be the availability of profitable alternative crops and lucrative non-farm employment opportunities in unfavourable areas.

The green revolution literature focused on the productivity impact of MVs in favourable areas without due attention to the factor market adjustments and the resulting income distribution between favourable and unfavourable areas. In order to explore the distributional consequences of the green revolution in Asia, the International Rice Research Institute (IRRI) organized the international collaborative project on the direct and indirect impacts of MV technology on income distribution in the Philippines, Indonesia, Thailand, Bangladesh, India and Nepal. Two sets of village surveys were conducted, for 1985–7 in the Philippines and for 1987–8 in other countries. The first covers 40 to 70

*Tokyo Metropolitan University, Japan.

villages across wide areas, and collects village-level information on technology adoption, rice production, factor prices and agrarian structure represented by farm size, tenure distribution and the incidence of landless labourer households. The second is an intensive survey of farmer and landless labourer households in several selected villages in representative production environments in each country, and aims to obtain detailed information on rice production and household annual income, including income from non-rice and non-farm sources. For the sake of cross-country comparability, essentially the same sets of survey questionnaire have been used.

This paper reports the major findings of this project. First, I summarize the adoption patterns of MVs and the productivity differential across production environments, using the extensive survey. Second, also using the extensive survey, I examine the difference in the agrarian structure across production environments. Third, I compare annual household income per capita by production environment and by status of households (owner-cultivator, tenant, and landless agricultural labourer), based on the intensive survey. Finally, I discuss the policy implications of our findings for future rice research and alleviation of rural poverty.

MV ADOPTION AND REGIONAL PRODUCTIVITY DIFFERENTIAL

It is widely believed from the green revolution literature that larger farmers adopt MVs more rapidly than smaller farmers and that tenants, particularly share tenants, are slower to adopt MVs than owner-cultivators. It is consistently found in our country studies, however, that those socio-economic factors did not significantly affect the adoption of MVs (for example, David and Otsuka, 1991). This may be because the dynamic process of MV adoption had largely ended by the late 1980s and because share tenancy and small farms are not generally inefficient in Asia, as argued by Hayami and Otsuka (1991). In contrast, favourable physical production environments, particularly the presence and the quality of irrigation as well as the rain-fed condition free from flood and drought problems, were decisive factors affecting MV adoption. It is also statistically found that productivity of rice production, as measured by rice yield, is primarily affected by the production environments and the adoption of MVs.

Table 1 reports the adoption rate of MVs during the wet season across highly irrigated areas, favourable shallow rain-fed areas without or with very low rates of irrigation, and unfavourable rain-fed areas prone to drought and flood. It is clear at a glance that the adoption rate of MVs is positively associated with the production environments. In Indonesia and the Philippines, where rainfall is relatively plentiful during the wet season and the problem of drainage is not generally serious, the rate of MV adoption is particularly high, not only in irrigated areas but also in favourable rain-fed areas. It is low in Bangladesh, where the production environments are harsh because of the climatic and topographical conditions prone to flood and drought. In unfavourable rain-fed areas, the adoption rate is low in Nepal and the Philippines mainly because of the drought problems in hilly environments and

TABLE 1 *Adoption rate of MVs and paddy yield by country and environment[a]*

Country	Irrigated	Favourable rain-fed	Unfavourable rain-fed
Bangladesh[b]			
MV (%)	31	20	13
Yield (t/ha)	3.6	2.6	2.5
India[c]			
MV(%)	100	66	—
Yield (t/ha)	5.8	3.5	—
Indonesia[d]			
MV (%)	90	81	0
Yield (t/ha)	5.5	4.1	1.7
Nepal[e]			
MV (%)	74	46	24
Yield (t/ha)	3.0	2.5	1.8
Philippines[f]			
MV (%)	97	99	42
Yield (t/ha)	3.6	3.3	2.6
Thailand[g]			
MV (%)	71	11	0
Yield (t/ha)	3.8	2.2	1.8

Notes: [a]The data pertain to the wet season between 1985 and 1988.
[b]Source: Hossain *et al.* (1991). Irrigated area refers to highly irrigated area, favourable rain-fed area to low irrigated area, and unfavourable rain-fed area to drought- and flood-prone areas.
[c]Source: Ramasamy *et al.* (1991). Favourable rain-fed area refers to rain-fed area with supplementary tank irrigation. The data pertain to Tamil Nadu.
[d]Source: Sudaryanto *et al.* (1991). Unfavourable rain-fed area refers to saline water area in South Kalimanthan.
[e]Source: Thapa and Upadhyaya (1991). Unfavourable rain-fed area refers to drought-prone area mostly located in the hills.
[f]Source: David *et al.* (1991). Unfavourable rain-fed area refers to both flood- and drought-prone areas.
[g]Source: Isvilanonda and Wattnutchariya (1991). Owing to the distinctly different patterns of rice production among major rice-growing regions, only data from Central Plain are shown. Unfavourable rain-fed area refers to deep-water area growing floating rice.

even zero in Indonesia (South Kalimanthan) and Thailand (Central Plain), where flooding is a major problem. Thus, MVs favour favourable areas, even though MVs recently developed are more resistant to drought and submergence problems than the 'first-generation' MVs.

Reflecting partly the differential adoption of MVs and partly the differences in the physical production environments, rice yields are significantly higher in more favourable areas (Table 1). The average yields range from 3 to 6 tons per hectare in irrigated areas, 2 to 4 tons in favourable rain-fed areas and 1.7 to 2.6 tons in unfavourable areas. Yields in irrigated areas are particularly high in India (Tamil Nadu) and Indonesia (mostly Java) partly because of the high adoption rate of MVs and partly because of the well-maintained gravity irrigation system. In Nepal, yield in irrigated areas is lowest among the six countries because popular MVs are not IRRI-type semi-dwarf, fertilizer-responsive varieties but 'Masuli', which is more tolerant to drought but less high-yielding than IRRI-type MVs.

In our extensive survey, we collected recall data on yields of traditional varieties (TVs) before MVs were introduced. According to the farmers' recall, yields of TVs have been essentially unchanged over time and were largely similar across production environments. Therefore the adoption of MVs undoubtedly created the significant yield differential across production environments. Though unreported here, our country studies also found that MVs were conducive to multiple cropping of rice because of the shorter growth duration and non-photo period sensitivity. Thus the statistical evidence strongly supports the popular view that the green revolution has created the large productivity gap between favourable and unfavourable areas.

GREEN REVOLUTION AND AGRARIAN STRUCTURE

Our country studies found that MVs as well as irrigation significantly increased the labour use per hectare by increasing labour requirements for crop care, harvesting and threshing. Particularly pronounced was an increase in the hired labour use, which in some cases led to the absolute decline in family labour use. Because of the increased rice cropping intensity with the adoption of MVs, the labour use per year further increased. Therefore we expect that the wage rate in favourable areas will tend to increase faster than in unfavourable areas, unless interregional migration from unfavourable to favourable areas took place. Unfortunately, however, we failed to collect reliable information on migration in the past from farmers' recall. Yet, studies in the Philippines (David *et al.* 1991; Otsuka *et al.*, 1991a) and India (Ramasamy *et al.*, 1991) found according to the population census statistics, that the growth rate of village population during 1970, when MVs were rapidly diffused, was positively and significantly associated with MV adoption. Casual observation also suggests that a large number of landless agricultural labourers, who are geographically more mobile than farmers, migrated from unfavourable to favourable areas, permanently as well as seasonally. The increase in the labour–land ratio in favourable areas must have increased the demand for land. Also the demand for land would have increased in favourable areas with the adoption of MVs. The increased demand for land may have triggered changes in farm size and tenure structure, which, in turn, may have affected the income distribution.

Table 2 compares farm size, ratio of tenanted areas and ratio of landless agricultural labourer households across production environments. Farm size differs vastly among countries, reflecting the difference in rural population and the endowment of cultivable land. More important for our analysis is that farm size is found to be largely similar among irrigated, favourable rain-fed, and unfavourable rainfed areas in each country, even though it is, in general, slightly larger in unfavourable rain-fed areas. Such observation indicates that farm size did not change substantially with changes in rice technology. In some countries, farm size did not adjust, partly because land reform law prohibits the transfer of cultivation right. For example, in the Philippines, land reform was effectively implemented in favourable rice growing areas (Otsuka, 1991), which prevented the farm size adjustment to a considerable extent. The market for land is known to be inactive in Asian countries, which might also have been reflected in the lack of the farm size adjustment. In any case, the fact that farm size is not significantly smaller in more favourable production environments indicates that rice income of farming households in favourable areas increased relative to unfavourable areas owing to the differential adoption of MVs.

Income of farmers depends not only on farm size but also on their tenure status. We may expect that the larger number of farmers in favourable areas are tenant cultivators, given the inactive land market and the larger demand for land. Data in Table 2, however, do not support such expectations. Except in the Philippines and Nepal, the ratio of tenanted area is not substantially higher in more favourable areas. Moreover, owner-cultivation tends to dominate, except in the Philippines, where large rice haciendas prevailed before land reform was implemented from 1972. Therefore there is no strong indication that the tenancy market adjusts to changes in rice technology. Being owner-cultivators, the majority of farmers in favourable areas seem to have benefited from the gain from technological change in rice farming.

However, remarkable difference in the agrarian structure is observed across production environments with respect to the incidence of landless agricultural labourer households. In India (Tamil Nadu), as many as a half of households in the irrigated villages are landless labourer households, whereas about one-fifth are labourer households in favourable rain-fed villages. In the Philippines, the ratio varies from 31 per cent in irrigated villages to 14 per cent in unfavourable rain-fed villages. While the ratio of labourer households is generally low in the Central Plain of Thailand, it is highest in irrigated villages. Since landless labourers are geographically mobile, these observations strongly indicate that inter-regional migration from unfavourable to favourable areas took place, corresponding to increased labour employment opportunities in favourable areas.

Such interregional migration has contributed significantly to the equalization of wage rates across production environments. According to the estimation results of wage determination functions, which include the technology factors (the rate of MV adoption and irrigation ratio) and some socio-economic

TABLE 2   *Farm size, ratio of tenanted area and ratio of landless labourer households, by country and environment*

Country	Irrigated	Favourable rain-fed	Unfavourable rain-fed
Farm size (ha)			
Bangladesh	0.5	0.6	0.6
India	1.4	1.8	–
Indonesia	0.5	0.5	2.3
Nepal	1.5	1.8	1.0
Philippines	1.7	1.7	1.6
Thailand	3.9	5.4	5.3
Ratio of tenanted area (%)			
Bangladesh	25	26	18
India	22	22	–
Indonesia	23	14	27
Nepal	31	29	10
Philippines	78	81	57
Thailand	47	42	51
Ratio of labourer households (%)			
India	49	21	–
Indonesia	26	11	8
Nepal	18	19	2
Philippines	31	18	14
Thailand	12	7	9

Notes: As for Table 1.

factors as explanatory variables, no significant regional difference in wage rates between favourable and unfavourable areas is found in Indonesia (Sudaryanto *et al*, 1991), Nepal (Upadhyaya *et al*, 1990; Thapa and Upadhyaya, 1991) or the Philippines (Otsuka *et al*, 1991a; David *et al*, 1991). The significant difference is found in Bangladesh (Hossain *et al*, 1991), India (Ramasamy *et al.*, 1991) and Thailand (Isvilanonda and Wattanutchariya, 1991), but the rate of difference between favourable and unfavourable areas is of the order of only 10 per cent – 15 per cent. Such difference may well be explained by the cost of migration rather than the lack of interregional labour market adjustments.

Not only migrants, but also those who have remained in unfavourable areas would have benefited from the MV adoption in favourable areas because the wage rate should have increased in unfavourable areas owing to out-migration of working population. In this way the income gain from the technological change in favourable areas seems to have been shared by landless labourers at large. Therefore, as far as the well-being of poor landless labourers is concerned, the impact of the green revolution does not seem to be as inequitable as is generally thought.

THE GREEN REVOLUTION AND
REGIONAL INCOME DISTRIBUTION

While agricultural wage rates are found to be largely equalized across production environments, some studies found significant regional difference in land rents (Isvilanonda and Wattanutchariya, 1991; Sudaryanto *et al*, 1991). Further, the analysis of the recall data of transacted land prices, collected by the intensive household survey in South Sumatra in Indonesia by Jatileksono (1991), demonstrates that the adoption of the series of improved MVs significantly and substantially increased land prices. Also our country studies consistently found, through the factor share analysis based on the intensive survey data, that the residual return to land, which is estimated by subtracting actual and imputed costs of current, labour and capital inputs from the gross value of production, is far larger in favourable areas than in unfavourable areas. In fact, it is common to observe that the factor share of the residual return is 40–50 percent in irrigated areas, whereas its share is typically less than 30 per cent in unfavourable rain-fed areas. Since the value of production is significantly higher in favourable areas because of the higher rice yields, the absolute difference in the estimated return to land per hectare was more than three-fold between the most and the least favourable environments. The estimated return to land in unfavourable areas is found to be intermediate.

Thus a major income inequalizer associated with the differential adoption of the modern rice technology is identified to be land income. There is, however, a tendency for owner-farmers, who have captured the increased returns to land, to allocate less time to farm and non-farm activities, presumably because of the income effect on the preference for leisure. Moreover, we found that income from non-rice and non-farm sources is generally important, even though our survey villages are typical rice-dependent villages commonly observed in Asia. In fact, the share of non-rice income exceeds 50 per cent in most cases, particularly in unfavourable areas. It appears that, depending on the availability of alternative crops and non-farm employment opportunities, farmers and landless labourers in unfavourable areas allocate more resources to non-rice activities in accordance with the regional comparative advantage of rice production and non-rice activities. To what extent the total household income remains different across favourable and unfavourable areas, and among owner-cultivators, tenants and agricultural labourers, is an important empirical question.

Table 3 compares annual per capita household income among owner-cultivator, tenant and landless labourer households across production environments by tenure status. The data are based on the intensive survey of the relatively small number of villages. The income is expressed in US dollars, using the prevailing official exchange rates in survey years of each country. Since survey years are somewhat different, and the official exchange rates do not necessarily reflect the difference in the purchasing power of local currencies, caution must be exercised in the comparison of the income levels across countries. In this paper, I focus on the difference in per capita income across regions and by tenure status in each country.

The major findings with respect to the inter-village comparison may be summarized as follows. First, per-capita income of landless labourer households tends to be similar across production environments, even though the income data of the landless are not reported in all countries. This finding supports the hypothesis that labour income tends to be equalized across production environments, owing to the inter-regional labour market adjustments. Second, per-capita income of owner-cultivator households is substantially larger in more favourable areas. A major exception is Indonesia, where income from non-farm jobs is high in favourable rain-fed villages and income from non-rice crops is high in unfavourable villages located in upland areas. The

TABLE 3   *Comparison of annual income per capita among members of owner, tenant and labourer households (US $) by country and environment[a]*

Country	Irrigated	Favourable rain-fed	Unfavourable rain-fed
India[b]			
Owner	339	151	n.a.
Labourer	162	114	n.a.
Indonesia[c]			
Owner	149	116	110[d]
Tenant	114	90	134[d]
Nepal[e]			
Owner	267	110	n.a.
Tenant	99	81	n.a.
Labourer	66	54	n.a.
Philippines[f]			
Owner	302	202	71
Tenant	232	152	81
Labourer	129	116	70
Thailand[g]			
Owner	681	245	313
Tenant	614	232	384
Labourer	244	n.a.	n.a.

Notes: [a] Sources and definition of environments are the same as Table 1, except for Indonesia.
[b] The data pertain to two villages in Tamil Nadu.
[c] Source: Jatileksono (1991). The data pertain to six villages in Lampung, South Sumatra.
[d] Upland area.
[e] The data pertain to two villages in the western Tarai region.
[f] The data pertain to two villages in Central Luzon and three villages in Panay island.
[g] The data pertain to three villages in the Central Plain.

detailed income decomposition analysis demonstrates that a major part of the difference in the income of owner-cultivators across production environments can be explained by the difference in land income in rice production. Third, the regional difference in income of tenant households is much smaller than in the case of owner-cultivator households, but larger than in the case of labourer households. Since land rent corresponds to the return to land, tenants receive essentially the return to labour. Therefore we expect that their income tends to be equalized so long as wage rates are equalized across production environments, unless the land rent is distorted. This is the case in the Philippines, where the majority of tenants in favourable areas are land reform beneficiaries and leasehold rents and amortization fees are regulated at very low levels (Otsuka *et al*, 1991b). Thus tenants in favourable areas received part of the increased returns to land, which otherwise would have accrued to the wealthy landlord class. Income of tenant households in irrigated areas is also substantially higher than in other areas in Thailand. This may be explained by the sluggish adjustment of land rents in the irrigated village, where MVs were introduced from the early 1980s.

With respect to the intra-village income distribution among owner, tenant, and labourer households, the following points can be made. First, unless farm size is very small (as in Indonesia) or land rents are lower than the rental value of land (as in the Philippines and Thailand), income of owner-cultivators is much higher than that of tenant and labourer households in irrigated areas. Note that per capita income of tenant households in Nepal is substantially lower than that of owner households, partly because the family size of tenant households is 50 per cent larger owing to the larger number of children. Second, the income difference among the three types of household in favourable rain-fed areas is much smaller than in irrigated areas. Third, the income difference is almost nil in an unfavourable production environment. These intra-village income differences can be consistently explained by the difference in land income in rice farming.

CONCLUDING REMARKS

We found in this study that, while labour income tends to be equalized across production environments because of interregional labour market adjustments, significant regional difference in land income has emerged owing to the differential adoption of the modern rice technology. We also found that the poorest in rural areas of Asia are landless labourers, whose livelihood depends on labour earnings. Owing to the increased labour demand associated with the MV adoption and the subsequent interregional labour market adjustments, the landless, even in unfavourable areas, seem to have benefited from the green revolution. It is therefore misleading to argue that the green revolution bypassed unfavourable areas, where no technological change took place. If large social weight ought to be given to the fate of those poor landless in the evaluation of the distributional consequences of the green revolution, it seems fair to conclude that the distributional impact of the green revolution is not as inequitable as is generally thought. We admit, however, that the green revolution brought about larger economic gains to wealthier owner-cultivators as well as landowners in favourable areas, thereby aggravating the income distribution in rice-growing rural areas.

One may be tempted to conclude that larger research efforts should be directed to the development of MVs suitable for unfavourable areas so as to attain more equitable distribution of income. Scientifically, however, it is more difficult to develop such varieties. Moreover, unfavourable environments are highly heterogeneous, so that a single superior rice variety, even if successfully bred, can be diffused only in limited areas. In other words, there is a trade-off between efficiency and equity in research resource allocation. In evaluating the equity impact of the green revolution,we must recognize that MVs developed for favourable areas benefited the rural poor at large, in addition to the fact that rice is a staple food for both the urban and rural poor.

Since the rural poor are typically the landless, the policy effort should be directed to enhancing the labour demand to alleviate rural poverty. Since human capital measured by schooling, is generally found to be an important determinant of labour earnings from non-farm sources in our country studies, the investment in human capital should be strengthened. Also important is land tenure policy. As Otsuka (1990) argues, the tenancy regulations in some Asian countries created the labourer class by inducing tenant eviction and by restricting the tenancy transactions. If the purpose of land reform is to enhance social equity, it must be so designed as to transfer wealth from landowners to landless labourers, particularly in favourable areas, through land taxation and other means.

REFERENCES

Barker, R. and Herdt, R.W., *The Rice Economy of Asia*, Resources for the Future, Washington DC.

David, C.C., Cordova, V.G. and Otsuka, K., 1991,'Differential Adoption of Modern Rice Technology across Production Environments: The Philippine Case', IRRI, Los Baños.

David, C.C. and Otsuka, K., 1991, 'The Modern Seed-Fertilizer Technology and Adoption of Labor-Saving Technologies: The Philippine Case', *Australian Journal of Agricultural Economics*, August, 1991.

Hayami, Y. and Otsuka, K., 1991, *The Economics of Contract Choice: An Agrarian Perspective*, Clarendon Press, Oxford.

Hossain, M. *et al.*, 1991, 'Differential Adoption of Modern Rice Technology across Production Environments: The Bangladesh Case", IRRI, Los Baños.

Isvilanonda, S. and Wattanutchariya, S., 1991, 'Differential Adoption of Modern Rice Technology across Production Environments: A Case Study of Thai Rice Villages', IRRI, Los Baños.

Jatileksono, T., 1991, 'Differential Adoption of Modern Rice Technology across Production Environments: The Case of Lampung in Indonesia', IRRI, Los Baños.

Lipton, M. and Longhurst, R., 1989, *New Seeds and Poor People*, Unwin Hyman, London.

Otsuka, K., 1990, 'Land Tenure and Rural Poverty in Asia', paper presented at the Finalization Workshop on Rural Poverty in Asia, Asian Development Bank.

Otsuka, K., 1991, 'Determinants and Consequences of Land Reform Implementation in the Philippines', *Journal of Development Economics*, 35(2) April.

Otsuka, K., Cordova, V.A. and David, C.C., 1991a, 'Modern Rice Technology and Regional Wage Differential in the Philippines', *Agricultural Economics* 4(314).

Otsuka, K., Cordova, V.G. and David, C.C., 1991b, 'Green Revolution, Land Reform, and Household Income Distribution in the Philippines', *Economic Development and Culture Change*, (forthcoming).

Ramasamy, C., Kandaswamy, A. and Paramasivam, P., 1991, 'Differential Impact of Modern Rice Technology across Production Environments: The Tamil Nadu Case', IRRI, Los Baños.

Sudaryanto, T., *et al.*, 1991, 'Differential Adoption of Modern Rice Technology across Production Environments: The Indonesian Case', IRRI, Los Baños.

Thapa, G. and Upadhyaya, H., 1991, 'Differential Adoption of Modern Rice Technology across Production Environments: The Nepalese Case', IRRI, Los Baños.

Upadhyaya, H., Otsuka, K. and David, C.C., 1990, 'Differential Adoption of Modern Rice Technology and Regional Wage Differential in Nepal', *Journal of Development Studies*, 26(3) April.

DISCUSSION OPENING – DIBYO PRABOWO*

Over the past two decades there have been fundamental and far-reaching changes in the Asian rice economy. Many of these have been associated with the so-called 'green revolution', which has led to a rapid increase in the use of both modern high-yielding varieties and fertilizer. Dr Otsuka provides an analytical description of these changes. He summarizes the adoption of modern varieties and the productivity differential across production environments. In addition, he comments on differences in agrarian structure and associated variation in annual household incomes. Dr Otsuka has reviewed a large body of research into the nature and impact of new rice technology in order to address the major policy issues confronting planners and researchers. Also, in drawing conclusions, he uses material from an intensive collaborative study which was recently completed.

The theme of Dr Otsuka's paper is to challenge critics who argue that the green revolution has by-passed unfavourable areas, where farm populations are generally poor, and worsened the regional income distribution. I feel that these assertions need to be considered rather more carefully and need further elaboration.

Basically, the view put forward is that increased labour demand associated with use of modern varieties, and subsequent inter-regional labour market adjustment, results in the landless obtaining benefit from the green revolution. Using lessons from Indonesia, let me describe briefly the long-term effects of the green revolution on employment and income. Because of the increased number of rice harvests, higher yields and the more even pattern of cultivation over the year, employment in rice production rose, as did the absolute income of farmers and farm hands. On closer inspection, however, the situation is found to be rather more complex and has produced a surprising outcome.

In Indonesia, the higher costs of seed-grain for the more expensive new varieties, together with fertilizer and pesticide costs, have led to an increase of the *tebasan* system, with crops being sold prior to harvest. At the same time, the introduction of rice mills eliminating the task of pounding rice by hand, as well as the use of the sickle instead of a less efficient knife (especially by female labour), has tended to increase the supply of labour looking for em-

*Gadjahmada University, Indonesia.

ployment. As a result, harvesters now appear to earn a smaller hourly wage in the tebasan than in owner harvesting, but this disadvantage has clearly been more than offset by the additional work generated by the increased overall yield. In spite of a relative drop in wages, the harvesters appear to be better off in aggregate. Farmers' income has risen because of higher yields. Increased income for the total group involved in rice production, but with greater income inequality, is thus the paradoxical outcome of the green revolution.

NIU RUOFENG AND CHEN JIYUAN*

'Small farmers' in China and their Development[1]

INTRODUCTION

The population in mainland China almost doubled between 1952 and 1989, but total grain production rose about 2.5 times. The period 1979–89, in particular, was one of the most successful decades, since total grain production increased by 2.7 per cent per year (Niu, Guo *et al.*, 1991). As a result, per capita food supplies improved considerably and availability of most farm produce increased by a wide margin.

The biggest contribution of the Chinese agricultural sector has been to solve the food and clothing problem for over one billion people, despite limited land resources. At the same time, the agricultural sector has also provided a huge amount of funds to the state so that industrialization of the country could be initiated and developed despite starting from a very low level of economic development (for example, per capita national income was around US$ 60–70 in the early 1950s). As a result of these massive transfers of capital, economic development in China has been biased against agriculture. The cumulative growth rate of industry was 16 times greater than agriculture between 1952 and 1989. Therefore, although there has been considerable overall beneficial progress in the agricultural sector, it is still characterized by wide fluctuations in output from year to year; by a relatively weak resource base; by having insufficient reserves to withstand adverse conditions; and by a long-standing shortage of many farm inputs. This situation is unlikely to change in the near future.

Several radical transformations of the agricultural management system have occurred in China since 1949. These changes have been from independent management of privately owned land by individual farmers after the agrarian reforms, to mutual aid groups and then cooperatives, from cooperatives to communes, and finally the abandonment of the peoples' communes, with unified collective management and the introduction of the family contract responsibility system. Each of these drastic changes was basically achieved within a relatively short period of time. Each change was designed to liberate the productive forces of farmers and to promote agricultural productivity.

*Professors Niu and Chen are Vice-Presidents of the Chinese Association of Agricultural Economists. Niu Ruofeng is ex-Director of the Institute of Agricultural Economics, Chinese Academy of Agricultural Sciences, and Chen Jiyuan is Director of the Institute for Rural Development, Chinese Academy of Social Sciences, China–Mainland.

Unfortunately, the rush for quick results and the forced application of a single pattern of management everywhere, without regard to local circumstances, often had detrimental results, not only on the productive capacity of agriculture but on the level of uncertainty faced by farmers.

THE CHINESE FARM HOUSEHOLD
AND ITS CHANGING POSITION IN RETROSPECT

Generally speaking, a small farmer is a worker who is engaged in independent agricultural production based on the family household as the management unit. The distinguishing features of small farmers in China are their enormous numbers, wide dispersion over the country, and strong survival ability over a dozen centuries despite numerous wars and social chaos.

The small farm as traditional management form in China

The Warring States (475–221 BC) marked the transition of the slave society to the feudal society when the necessary material conditions for the development of individual labour units became available owing to the popularization of the use of iron farm tools and draft cattle. Surplus goods increased. Slave owners were replaced by landlords, and slaves became peasants. Slavish corvée was abolished and replaced by land rent. Historians agree that the small farmer emerged in about 216 BC, during the Qin Dynasty, when the feudal social system and feudal landlord ownership were established. Small-scale private land ownership accompanied this change (Li, 1990; Liu Xiaoduo, 1990).

The landlord economy, based on feudal land ownership, occupied the predominant position for about two thousand years, both during the feudal era and during the century of semi-feudal, semi-colonial society which followed. During this long period, the small farm economy was always subordinate to the landlord economy. The small peasants included people who owned small private plots and tenant peasants dependent on land rented from landlords. Both types of small farmer engaged in agricultural production in their own right. Hence the emergence and existence of small peasants in Old China was connected with feudal land ownership as well as small private land ownership. This was an outstanding characteristic of small peasants in the past.

The distribution of land was very inequitable in Old China up to the last year before the founding of the People's Republic. Landlords and rich peasants accounted for less than 10 per cent of total rural population, but they owned 70–80 per cent of the land and the lion's share of farm animals and tools. On the other hand, over 90 per cent of rural labourers (including poor peasants, farmhands and middle peasants) possessed only 20–30 per cent of the land. In addition to farmhands and tenant peasants, a great number of small peasants were exploited by the landlords because they had to rent land or take usurious loans from the latter. Small farmers paid about 35 million tons of grain as rent in kind and paid enormous interest debts to the landlords each year (Niu, Guo *et al.*, 1991). Millions of peasants lived in hunger and cold, and were distressed by famine in years of disasters. Therefore, for generations, the great mass of Chinese peasants had long cherished the wish that the available land should be redistributed.

'Land to the tiller' and agrarian reform

Hong Xiuquan (1814–64) and Sun Yat-sen (1866–1925) were two well known advocates of the abolition of the feudal agrarian system and stood for the principle of 'land to the tiller'. It also became the basic policy of the Communist Party of China (CPC) durlng the period of the people's democratic revolution (1919–49). Agrarian reforms were carried out first in the old liberated areas according to 'Outline Land Law of China' in the 1940s, and then, in the newly liberated areas, according to the 'Agrarian Reform Law of the People's Republic of China'. The land reform was completed in mainland China between 1950 and 1952. This reform abolished the feudal land system, confiscated the redundant land of landlords, and distributed it among landless peasants, or to those peasants with insufficient land, in a gratuitous way. The redistributed land acquired new owners and thus the control over this resource shifted to the tillers. As a result, about 300 million peasants in mainland China obtained 750 million mu (1 ha = 15 mu) and other means of production. In addition, the new owners were freed from all debts in terms of rent and usurious borrowings.

Small private farms, therefore, became almost the only management form after the agrarian reform. Land was distributed in an egalitarian manner. Per capita and per farm household land availability for the country was 2.5 mu and 12.4 mu, respectively. A middle peasant household owned on average 15 to 30 mu. Some of the new small peasants rapidly became middle peasants. These, along with peasants who previously owned their land, took a dominant position in terms of number of households and land size. However, the commercial sales of these small-scale farmers remained very limited. According to surveys conducted by the State Statistical Bureau in 1955, of 16 000 farm households in 25 provinces, the ratio of commercial sales to total output averaged only 25.7 per cent in 1954 – among poor peasants the ratio was 22.1 per cent, middle peasants 25.2 per cent, and rich peasants 43.1 per cent (Niu, Guo et al., 1991).

The state allowed farm produce to be freely purchased and sold and prices to fluctuate on the market until the state monopoly purchasing and marketing system was introduced in November 1953. Farm households, being fully independent decision makers in production management, were also independent marketers of farm products in the first years of the People's Republic.

Agriculture, which had been destroyed during the protracted warfare prior to 1949, recovered and was quickly revitalized. Total output rose by 14 per cent annually in the early 1950s.

From individual farm household to mutual aid groups and elementary agricultural producers' cooperatives

The government encouraged small farmers to create mutual aid groups (MAGs) and elementary agricultural producers' cooperatives (EAPCs) to overcome some of the difficulties faced by individual farmholders, such as lack of farm implements and draught animals, and lack of capacity to resist natural disasters.

MAGs were based on private land ownership. Farm households helped each other with manpower, farm tools and draught cattle in fieldwork, but they could decide what to produce on their own land, as they remained independent management units. There were both seasonal and permanent MAGs in various regions in the early 1950s, the average size being six to seven households.

A small number of EAPCs appeared in some areas in the early 1950s. The land, draught animals and other basic means of production were put into collective use and controlled by unified management, but they were still owned by the EAPC participants. The revenue was distributed according to both the land shares and man-days of labour contributed. The average size of EAPCs was about 20 to 30 households to cultivate 200–300 mu of land. Farm households in EAPCs became semi-independent management units. Apart from the land placed under the control of the EAPCs, farmers often retained sizable land plots for family-run undertakings. They also engaged in diversified sideline occupations. Thus farm households were free to undertake a considerable portion of individually managed activities. This was a distinctive feature of EAPCs. Farmers welcomed both MAGs and EAPCs and agricultural production developed at a relatively high rate.

From elementary to advanced cooperatives

In September 1953, the CPC Central Committee made public its general policy for a step-by-step socialist transformation of agriculture, handicrafts and capitalist industry and commerce. After considering all the relevant factors, including the direction of agricultural development, the need to accumulate funds to finance the nation's industrialization, and the availability of material supplies relative to demand, the CPC made an historical choice to carry out the cooperative transformation of agriculture to establish a collective ownership system in the agriculture sector. By means of this system the government could, on the one hand, put the rural economy under the control of centralized plans and, on the other hand, extract huge funds from agriculture for industrial development by introducing the monopolized purchasing and marketing system which enforced a great price difference between industrial and agricultural goods.

Urged on by the above general line of argument and the desire to meet industrialization targets, the government began to impose the state monopoly for purchasing and marketing of major farm products. Free marketing was replaced by state-planned purchasing and marketing. At the same time, the socialist transformation of agriculture was conducted and, by the end of 1956, the cooperative transformation of agriculture had on the whole, been realized. A total of 740 000 advanced agricultural producers' cooperatives (AAPCs) were set up nation-wide, covering over 90 per cent of all farm households. The average size of AAPCs was 150 to 200 households, with 2 000 to 2 500 mu of land.

AAPCs were characterized by a fully collective management system and two-level accounting (at the cooperative and team levels). The land, draught animals and large farm implements were collectively owned. The private ownership of the cooperative members over these means of production was abandoned. They also lost production and marketing decision-making power. Their income came only from the collectives according to their man-days of labour, and rights to income based on their share of the assets assigned to the cooperative were abolished. Despite this, the farmers were permitted to use small plots of land for personal needs, to own small farm tools, and to engage in some sideline occupations. However, these private activities played a very limited role in the total agricultural production of the country.

The reorganization of EAPCs into AAPCs reached a high tide in the summer of 1955 and the growth rate of agricultural production began to slow down, although the total output continued to increase.

From advanced cooperatives to people's communes

The people's communes were created country-wide in the summer and autumn of 1958. This change was part of a newly-set strategy of 'leap forward to catch up with and surpass the industrial powers'. It took place before the agricultural producers' cooperatives had time to consolidate and to develop the productive forces available. The AAPCs were merged into 26 000 people's communes. A commune had on average 4615 farm households and 55–60 thousand mu of arable land. Under this system, all means of production were owned by the collective, and the unified collective management of these large productive units was integrated with government administration.

People's communes, being larger in size and higher in degree of public ownership, were unable to generate new productive forces because they, like the AAPCs before them, were merely combinations of human and animal power with the chief means of production being collective labour. Later, a number of communes were readjusted in size, but in general they remained unchanged for more than 20 years. The communes functioned under a system of 'three-level ownership by the commune, the production brigade and the production team, with the production team as the basic accounting unit'.

Under the commune system, sometimes the family plots and livestock kept for personal needs became owned by the production team. At other times, the farmers could not sell privately produced commodities because the free markets had been completely closed down. Occasionally, even piecemeal trade was forbidden, on the grounds that it was necessary to 'cut the tails of capitalism'. Farm family households, being regarded only as blood-tied living units, played almost no role as production or marketing management units. Farmers became solely labourers in collectives, as they were totally divorced from the means of production and basic management activities.

People's communes exercised a high degree of unified management, which led to an overly concentrated labour management system and an egalitarian distribution of collective income. The production teams did not really have any important decision-making power, although they were the basic accounting units. For example, the teams were required to arrange their production activities in accordance with the plans and orders issued by the government authorities and sell their products to the state at prices set by the government. Almost any farm activities other than grain production were regarded as not being honest jobs. There were no economic incentives to stimulate the enthusiasm of farmers. In general, agricultural production grew slowly during the 20 years of communes, resulting in stagnation of the rural economy, per capita output and consumption levels. However, agricultural production dropped sharply during the first three years of the commune era, at the end of the 1950s and beginning of the 1960s. Farmers suffered most severely in these years. Despite some recovery in the 1962–5 period, the Cultural Revolution (1966–77) was also a very harsh time for farm families in China. There were over 100 million people in rural areas who were under-fed and who did not have the necessary clothing when the rural reforms began at the end of 1978.

From the people's commune to the family contract responsibility system

The first step in reforming the commune system was carried out after the Third Plenary Session of the CPC Eleventh Central Committee held in December 1978. A new dual-level management system, combining both unified and individual management with the family contract responsibility system as its main characteristic, replaced the unitary collective/commune management system in most regions of China.

Land and other collectively owned means of production were assigned to every household through contracts under which the farmers were required to pay for the resources used. The farm household, therefore, regained its previous status of being a productive management unit. Farmers, being themselves managers as well as workers, could integrate directly the land and other means of production. They also became, to some extent, independent commodity marketers because they obtained decision-making power over the marketing of a significant part of their output. The small farmer has taken a dominant place in Chinese agriculture since the responsibility system (also known as 'dabaogan') was introduced. The reform has eliminated major disadvantages of the people's communes. In particular, the troublesome problems associated with the effective supervision of farm labour and the distribution of income according to work done (which created major disincentives under the commune system) have both been resolved.

The state monopoly and obligatory purchasing and marketing of major agricultural products were abolished step by step, beginning in 1985, to encourage the development of the commodity economy. Free markets have arisen as the times require, both in urban and rural areas, to break the government monopoly. Farmers now have more choices in regard to selling their products.

China's agriculture reached its highest average annual growth rate of 8.4 per cent in the period 1979–84, thanks to the reform of the management system and a series of other beneficial factors. Total grain output grew by one-third between 1979 and 1984, reaching 407.31 million tonnes in 1984. The output of most other commodities also increased considerably. For example, cotton production increased 280 per cent, exceeding 6.25 million tonnes in 1984. The per capita availability of major farm products rose significantly, creating a greatly improved situation with respect to foods, the supply of which had been precarious for a long time. 'Dabaogan' as a form of small farm management, was widely accepted, mainly because it was able to promote agricultural development and hence increase the effective food supply and thus provide a new chance for the industrialization of the nation. However, the new system was not deliberately designed by the government. Rather, it evolved from necessity (and is still evolving).

RECONSTRUCTION OF FARM HOUSEHOLD MANAGEMENT AND DUAL-LEVEL MANAGEMENT

Step by step reforms since 1978

The re-emergence of the farm households as small family farms based on the family contract system after 1978 occurred in several steps and has not been introduced as a unified pattern over the whole country. In particular, there have been major differences in the way the system has evolved in agricultural and pastoral areas.

In agricultural areas, the step-by-step reconstruction of farm households generally occurred as follows: from 'fixing output quotas on a team basis and with remuneration to each household linked with output' to 'fixing output quotas for each household', to the system of 'contracting land to each house-hold for productive management'. It is the last stage which has the Chinese name 'dabaogan'.

In September 1979, the Central Committee of the CPC adopted the 'Reso-lution on Some Problems Concerning Accelerated Development of Agriculture', calling for reforms within the framework of the collective economy based on the people's communes. The first reform was to give more decision-making power to the production teams: 'Especially, it is necessary to strengthen the responsibility system and realize the income distribution according to work' by means of 'fixing output quotas on a team basis and linking remunerations with output'. That is, a kind of team production responsibility system was introduced. At the same time, the party encouraged farmers to develop family-run side occupations to increase individual income and enliven the rural economy.

A year later, in September 1980, another Central Committee document was issued on further consolidating and perfecting the production responsibility system. The document emphasized: 'Our general direction is to develop the collective economy', but affirmed officially that 'the practice of fixing output quotas for each household, or contracting land to each household for production management is allowed... in the remote mountain and poverty-stricken regions' because it 'is a necessary measure to solve the food and clothing problem... There is no need in other regions to fix output quotas for each household.' Instead, the main efforts should aim to 'carry out the specialized contract responsibility system on the basis of the production brigade' to further develop and consolidate the collective economy.

The system of 'fixing output quotas for each household' was not entirely new. It was a kind of responsibility system which had been used previously in some areas in the 1962–5 period. The main points were that fixed output quotas and fixed amounts of farm work were established for each household. The household was required to deliver all major products to the collective. The collective economic organization calculated the remunerations (work points) of the households according to their output; that is, the collective still controlled the distribution of income, but households which exceeded their quotas were rewarded accordingly.

The famous No. 1 Document of the Central Committee of the CPC, in 1982, announced that the practice of 'contracting land to each household for productive management is also a kind of responsibility system'. This new responsibility system soon spread rapidly over the whole country. By the end of 1982, 99 per cent of the production teams had assigned 95.7 per cent of collective-owned land to households. The majority of rural collective economic organizations which controlled the communes were spontaneously disbanded. A great quantity of collective property was distributed to the farm households. The rural township administrative authorities were re-established as the lowest level of formal state administration (replacing the communes). Little attention was paid to the creation of new cooperatives to take charge of the community assets such as irrigation infrastructure and so on, and consequently some areas now face undesirable consequences of the rush to abandon communes.

'Contracting land to households for productive management' clearly differs significantly from the responsibility system as originally understood in China. After the land was distributed and assigned to each household, the farmers, operating their farms independently according to the contracts, could satisfy their commitments to the new collective organizations by paying in cash once their delivery quota had been met. They own their proceeds as well as their products. There is no longer any linkage between output quotas and remunerations. The collective economic organizations do not interfere in the production and management of their members.

In the pastoral areas, the basic experiences of the agricultural areas were repeated, but with a one- to two-year time-lag. However, the livestock farmers usually also practise a double-contract system. Both the previously collectively owned animals and grassland have been assigned to the farm households. The process of implementing this double-contract system usually occurred in three steps, spread over about two years:

(1) The previously collective-owned animals were assigned to each house-

hold. Livestock were usually assigned according to the existing size of families and the number of available animals, but the method of assignment varied considerably from place to place. The ratio of family members to animals ranged from 1:3 to 1:2, or simply reflected the division of the number of available animals by the total population of the village. The households were then responsible for the livestock to the collective. The households were required to pay the collective for any losses, but had the right to share any gains according to a ratio agreed in the contract (4:6 or 3:7, for example). The livestock remained the property of the collective.

(2)   The livestock were valued and assigned to the households under a contract system. The household became responsible for losses or profits which had to be shared with the collective. In addition, the households had to pay levies/taxes in the form of cash payments to the collective once a year.

(3)   The livestock were valued and then sold to the herdsman households on a discounted basis. The ownership of the animals shifted to the herdsmen for individual production. The households were free to keep all profits (and losses) but had to pay agreed fees/levies/taxes each year.

In most pastoral areas, once the ownership of livestock changed, the pasture land was also assigned (under contract) to the households. In some places, households were required to make a payment for the grassland assigned to the households, with the users being required to hand over RMB¥ 0.2–0.3 per mu to the collective. But the sum involved was negligible and differs little from free usage.

There is now almost no collectively owned livestock in the pastoral area. Public ownership of livestock exists only on state farms, and with a small number of breeding stock at artificial insemination stations set up by some villages. However, not all the pasture land has been contracted to the households and, even when it has been assigned, it is often still grazed 'in common'.

Characteristics of the reconstructed farm households as 'small farmers'

As indicated earlier, a large number of farmers faced major food and clothing problems when the first stage of the reform of the agricultural management system was started in 1978. Land was the essential resource upon which the farmers' livelihood depended. Therefore land assignment on a per capita basis was the most feasible and simple way of distributing the land which was acceptable to the majority of farmers.[2] The farm household was reconstructed as the basic production and management unit after the reforms. These 'small farmer' households exhibit some important distinctive features:

(1)   Apart from the economically developed areas and suburbs of large and medium cities, most farmer (herdsmen) households throughout China still mainly use human and animal power in agricultural production, supplemented by machinery. These traditional factors of production play

an essential role. This is especially true in the less developed western part and pastoral areas of China, where traditional farming dominates. Nomadic animal husbandry is quite common in Inner Mongolia, Tibet and Qinghai.

(2) In general, the farms are too small in scale and their land parcels are too scattered geographically. A survey was made by the Research Centre for Rural Development under the State Council in 1987. The survey covered 27 576 households in 208 villages. On average, there were 4.8 persons per household, with 2.7 full- and part-time workers and 8.35 mu of contracted land per farm. This later statistic can be compared to 12.4 mu per farm household after the land reform in the early 1950s. At that time, every male worker cultivated 5 mu. On average, the present management scale of a farm household is one-third smaller than was the case 40 years ago. Besides this, the land contracted to each household is commonly divided into 8 or 9 pieces, which are too small and scattered. Table 1 presents data for 1986.

TABLE 1   *Distribution of land per household in China, 1986*

Area of contract land per household (mu)	As percentage of total number of households (%)
Less than 10	70.6
10 – 20	19.6
20 – 30	5.6
30 – 50	2.9
50 – 100	1.3
More than 100	0.2

Source: CAAS (1989)

The per capita land availability varies from region to region; so does the scale of land management of farm households. In general, the size is small in the southern paddy rice regions, it is bigger in the dry farming zone of North China, and even larger in the north-east. It is even more extensive in the pastoral areas, where both livestock and grassland are under contract. For example, according to the survey in Inner Mongolia by the Institute of Agricultural Economics of the Chinese Academy of Agricultural Sciences, the average herdsman household in Gangacha Township of Ongniud Qi (that is county) had 300 mu of grassland, or 68 mu per capita, keeping 10 sheep units in 1989.

(3) Most products are for own consumption, indicating a low percentage of commercial sales. By and large, over two-thirds of grain produced by the present farm households are for their own consumption, and only about 30 per cent marketed. As demonstrated by the data in Table 2, this

percentage has remained remarkably constant since 1978. However, the commodity sale ratio is very high in cash crop production. As a result, the overall sales to total production ratio for the agricultural sector amounts to just over 50 per cent, which is much higher than in 1978 (Table 2).

TABLE 2   *Dynamics of commodity ratios, selected years (percentage)*

	1978	1984	1985	1987	1988	1989
Grain	30.48	30.80	28.39	30.08	30.44	29.78
Agricultural Sector	39.94	44.80	46.42	50.67	53.36	51.81

(4) The savings capacity is low, and farm households are financially weak, with little in reserve in case of natural calamities. Surveys show that about 80 per cent of farmers' net income is used for consumption, indicating a savings ratio near to 20 per cent, but, since the net income is low, the absolute amount saved is very small. It is well known that small farmers are also vulnerable to market risks.

Dual-level management versus individual small farmers

Following the rural reforms of the early 1980s, China's agricultural economy is based on collectively-owned land which is managed by the small farmers under contract to the collectives. The resources available are therefore subject to two levels of management: the unified and basic levels.

There are many functions to be performed at the unified or community management level by a collective or cooperative economic organization. For instance:

— management of the collectively owned land, issuing land contracts and developing contract norms, supervising the implementation of the contracts, and solving the issues associated with renewing the land contracts or the re-adjustment of land contracts;
— initiating and managing investment to improve and expand the collectively owned infrastructure by such activities as land reclamation, the building and maintenance of irrigation and drainage facilities, and arbitration in case of economic disputes about the use of these facilities;
— organizing the purchase and supply of major means of production and marketing of farm products;
— management and use of collectively owned large-size farm machinery and the provision of services for major farm operations, such as tractor ploughing and sowing, and mechanized harvesting;

— the provision of improved seeds and artifical insemination (AI) facilities (or stud stock) for livestock breeding, technical training, extension and consultation among farmers;
— initiation of measures to fight droughts or floods, to protect crops and animals from pests and diseases;
— coordinating the interests among households of the village, between the village and other economic entities, and between the latter and administrative institutions;
— assisting the government to achieve its purchase plans of agricultural products and providing relevant consultations to the farmers;
— undertaking village affairs concurrently if the village authorities fail to function well; and
— having discussions on behalf of the farmers with the government and other economic, social and administrative institutions.

These functions were all previously performed by the commune administration. Nowadays, village cooperatives have been set up to undertake these tasks. However, as stated earlier, these agricultural cooperatives do not interfere in the farmers' production and management decisions beyond the administration of the contracts.

The reconstructed farm households established by the 1980s reforms are, therefore, members of agricultural cooperative organizations based on the collective-owned land. They are not individual small farmers in the same sense as was the case in the early 1950s. They do not own their land. In this respect, the system of 'dabaogan' can be said to have created an agricultural system in which the basic management units are modified 'small farmers'.

Small farmers in China must now operate under two sets of economic rules. That is, they must respond to (comply with) their contracts and, at the same time, they can operate in their own best interests in free markets. The two sets of rules differ in their nature, although in practical operation the farmers manage all the resources available to them as a single entity. The contract land belongs to the collective. The relationship between the farmers and the agricultural cooperatives managing the contracts for the collective is that of contractor and service institutions. The contractor must comply with regulations set out in the contracts, including payments of land use fees and other deductions to the collective/cooperative. When the farmers operate on their own behalf in the free market, they need not have any regard to the collective/cooperative, except for the services they wish voluntarily to obtain from these organizations, including coordination and management of the social infrastructure. The farmers may use the contract land, but they do not own it. The land is not to be sold, leased, mortgaged, left uncultivated, or destroyed. In addition, under the contract, farmers accept the state guiding plans and sell the contracted amounts of their products to the government at fixed prices according to quotas. The decision-making power of the farmers in regard to crop production and management, therefore, is far from being completely free. The existence of the dual-level management system implies that the farmers are not fully independent. The planned economy still requires obligatory tasks to be undertaken by the farmers.

It should be recognized that this 'dual-level' system of management is not a perfect organizational arrangement. It has given the best results in the eastern coastal regions and suburbs of large and middle-size cities. In these regions, the collective economic organizations were relatively strong and their industrial and other non-agricultural enterprises not only remained but also have grown considerably following the implementation of the 'dabaogan' system. In the central zone, however, the 'dual-level' system has succeeded only in about 30 per cent of the villages which have had good organization and strong local cadres. About 40 per cent of villages in the central zone of China remain weak at the unified level, practicing scattered individual operations. The remaining villages in the central zone (30 per cent of the total), being in a difficult situation, have had to distribute all collective property for consumption and there is now no effective unified level management. The collective economy in the western zone was poor even under the commune system. However, it has been eroded further by the improper distribution of resources among farmers. The majority of villages have no collectively run undertakings, while the farm households are faced with a series of difficulties at the basic level. Naturally, exceptions can be found. For example, the 'dual-level' management system works well in Gansu Province, maybe better even than in some economically strong regions.

The lack of a unified management structure to provide and to maintain critically important social infrastructure and to create non-agricultural employment in many parts of the country has been the basis for major criticism of the current organizational structure of Chinese agriculture. This problem must be addressed by future reforms.

Better results from new options

While the introduction of the dual-level management system in the early 1980s has attracted criticism, it has also had many beneficial effects. In particular, it has provided Chinese small farmers with many new options and, hence, opportunities to improve the living standard of their families.

As part of the management structure reforms, the government removed the restriction that a farmer should only cultivate the land. A new economic policy has now emerged which stresses all-round development and comprehensive management in rural areas to stimulate the socialist commodity economy. The farmers now have more options. Apart from crop cultivation, they may raise animals for sale if they want to. They can also seek employment in secondary or tertiary industries.

The net per capita income of the farmers in China increased by 450 per cent in the decade from 1979 to 1989, and the composition of their income has changed significantly (Table 3). In the past, the major source of their income was agricultural production for the collective economy, whereas at present it is the family management. Non-farm income on average in 1989 constituted 38.2 per cent for the whole country (it differs greatly from region to region), indicating a remarkable decline of the agricultural share.

The dramatic changes in the farmers' net income level (along with its composition) in the decade of reform could not have occurred without the economic reforms which have created the dual-level management system and given more choices to the farmers.

TABLE 3   *Dynamics of the net income of farm households in rural China, selected years*

	1978	1980	1985	1988	1989
Net per capita					
income of farmers (¥)	133.57	191.33	397.6	544.94	601.51
Source of income (%)					
From collective economy	66.3	56.6	8.4	9.1	9.4
From economic combinations	–	–	0.9	0.7	0.6
From family management	26.8	32.7	81.1	83.2	82.2
Other non-borrowed income	6.9	10.7	6.9	7.0	7.8
Share of sources (%)					
Agricultural	85.0	72.8	66.34	63.4	61.79
Non-agricultural	7.0	8.8	21.7	27.3	28.03
Non-productive	8.0	13.0	11.96	9.3	10.18

Source: *Chinese Statistical Yearbook*, 1985, 1990.

CHANGING ROLE OF
'SMALL FARMERS' AND THEIR NON-FARMING TENDENCY

Differentiation of farmers and specialization

Under the previous system of people's communes, farmers all over the country were commune members, who were engaged in collective production at almost the same level. The initial economic status of most reconstructed small farmers (or herdsman households) was similar following the restructuring. However, with the passage of time, the gap between farmers is widening, owing to their different responses to the development of the commodity economy and the new opportunities the changes in policy have made available.

Considering the data on farm household annual per capita net income, farmers can now be divided into three levels: rich, well-to-do and poor. There is a most encouraging trend for the number of rich households to grow while the number of poor households is declining. The farm households with per capita net income over ¥1 000 accounted for 9.4 per cent of the total in 1989, and those over ¥500 made up more than a half of the total farm households in 1989. The households with less than ¥200 were reduced from 82.6 per cent in 1978 to 4.7 per cent of the total in 1989 (Niu, Guo et al. 1991). The absolute number of farm households below the ¥200 level was 57 885 in 1989, or 6.6 per cent, of the rural population. These households are mainly located in the economically undeveloped regions of the south-west and north-west of China, regions which are disadvantaged by severe natural and ecological conditions. Only a small number of these households live in counties of the central and eastern zones.

As far as the employment structure and economic classification of the farmers are concerned, they have now been split up into farm managers, rural workers, farmworkers, individual workers, industrialists or businessmen, enterprise owners, entrepreneurs and various administrators. They belong to different interest groups and social strata with varied demands and wishes. Nevertherless, the farm managers, as a working stratum, still represent the majority of the rural population. As explained earlier, they are self-employers engaged in crop or livestock production on the collective-owned farmland or pasture land which they farm under contracts. The only or main source of their livelihood depends on agricultural income. The most viable among them are big contractors or specialized households, although some larger farmers have failed.

The Research Centre for Rural Development under the State Council made a survey of 26 666 grain-producing farmers in various regions of the country in 1987. The results show quite good performance by 235 big producers. Their commodity ratio was nearly six times that of common households, and other indicators such as per mu yields, net income from one mu and one work-day were higher by 37.48 per cent, 35.96 per cent and 50.59 per cent, respectively (see CAAS, 1989).[3]

There are also a number of big households specializing in production of pigs, poultry and fish. They are often very specialized and highly effective.

Enlarging the management scale and the associated surplus labour problem

From the late 1980s, there have been frequent appeals for an enlargement of the management scale. It is argued that the present structure suffers from major scale diseconomies, made worse because the area available to each worker is split up into a large number of small scattered plots. This situation leads to low labour productivity.

Given the current combinations of manual and draught labour as the chief production factors, supplemented by machinery, it has been estimated that the desirable size of a farm could be enlarged by one-third to one-half. A worker could cultivate 7–8 mu of paddy field in the south, 12–15 mu of land in the northern dry farming region, 10 mu in cash crop producing areas, and 50–100 mu in the north-east. In general terms, this would mean that the average size of contract farms should be increased by a half, from 8 mu to 12 mu per household, or from 5 mu to 7.5 mu per worker. If such a change occurred, then each worker would be able to provide 625 kg of commodity grain, which is enough to satisfy 1.5 non-agricultural persons. The result would be that about one-third of all farm households (or 100 million workers) would need to be transferred out of grain production. It will probably take at least 15 to 20 years to create other jobs for such a large contingent of farmworkers. Such a change needs to be an evolutionary process in economic development and it will not be easy to achieve.

It should be noted that 100 million surplus rural labourers have been left in the countryside, despite the outflow of huge funds from the agricultural sector to fund the national industrialization programme since 1949. The problem of providing more worthwhile employment for these people has become a serious challenge in the overall economic development of the nation. In order to enlarge the management scale in agriculture, there is need to promote the secondary and tertiary industries in rural areas. Only in such a way can the enormous surplus labour in agricultural production be transferred to other sectors. This depends mainly on a proper development of rural industrial enterprises, rather than a continued concentration on urban industrialization based on state-run enterprises, because the capacity of these urban enterprises to absorb the surplus rural labour is virtually nil.

Chinese farmers are reluctant to leave their land, even when seeking employment in non-agricultural industries. Scarce land is the last guarantee for their livelihood when it is difficult for them, in their tens of millions, to get jobs in the town. The comparative advantages of crop farming are clearly low; nevertheless, most farmers do not want to give up their contract plots.

The Ministry of Agriculture carried out a case study on 3200 farm households in 28 provinces, autonomous regions and central municipalities in 1988. The data obtained show that 72.4 per cent of farmers were in favour of keeping the current situation unchanged, 26.6 per cent considered the size of contract land appropriate, 50 per cent expressed their wishes that their contract land plots should not be re-adjusted for 15 years, and 15.6 per cent hoped the land would be re-adjusted and allocated principally to the skilled farmers. Some 21.3 per cent wanted to take on more contract land, while only 6.2 per cent were ready to re-contract out their land plots to others. Many farmers supported re-adjustment of land plots and/or their concentration only for the sake of removing the inconveniences caused by the scattered location of their fields.

Some 60 per cent of agricultural labour in the suburbs of Shanghai Municipality has already been shifted to the secondary and tertiary industries, yet most farm households do not want to alter the present situation of land contract conditions. According to a survey of 4015 households, 52.6 per cent of them refused to re-contract their food fields and responsibility fields, 38.4 per cent agreed to re-contract with others their responsibility fields, on condition that the food fields were to be retained, 7.3 per cent were ready to give up all contract land plots, and only 1.7 per cent would like to enlarge their farmland size and to become bigger specialized households.

Similar to the outskirts of Shanghai, southern Jiangsu is also characterized by a developed economy and two crops (wheat and rice) a year. Over 55 per cent of agricultural labour has been shifted to the secondary and tertiary industries. The majority of farm households are engaged in agriculture in a part-time way, but remain reluctant to give up their contract land plots. For example, there are 60 big grain-producing households in Rongnan village in Wuxi County, with on average only one farm worker to cultivate 1.5 mu of food field and 7.5 mu of responsibility field. Rongnan village 'compensates' (subsidizes) agriculture from its industrial enterprises. On average the subsidy is ¥125 per mu. As a result, these big grain producers receive around ¥1000 as 'compensation'. Their own income from crop production comes to approximately ¥1000. Thus the total income for a farm worker is about ¥2000, which exceeds that of industrial workers of the village by 70 per cent (see CAAS, 1989). Besides, the main field operations are done by the mechanical services of the collective. The farmworkers really only perform some auxiliary work. These facts explain why they do not want to give up their contract land.[4]

It is expected that the general picture of the predominance of 'small farmers' in China will not change greatly over a long period of time. The practical choice is the coexistence of the various management forms discussed in Appendix C (See also CAAS, 1989).

Farmers' behaviour and non-farming tendency

Near-sightedness in farmers' production behaviour has been a widespread phenomenon in rural China since the reforms. The farm households are reluctant to invest in cropland and do not seem to care about increasing land fertility. For example, they want to harvest the highest possible yields only for some years by the application of large amounts of chemical fertilizers, instead of following the time-honoured sustainable methods based on farm manure. The herdsmen, for example, have unduly expanded their livestock numbers, which has resulted in the degradation of pasture. These undesirable changes in farmer and herdsmen attitudes are another major basis for criticism by those who oppose the reforms which have occurred since 1978.

The basic conservative and sustainable production methods of the past are being abandoned both by the collective/cooperatives and by the small farmers in the course of pursuit of higher income. The non-farming tendency is getting the upper hand. One obvious indicator of this problem is the willingness of village collectives and individual farmers to invest in non-agricultural activities, rather than in agriculture, even if these investments are in the cities. The total investment in collective-run non-agricultural enterprises grows year after year, whereas the amount invested by collectives in agriculture declines proportionally, from 30.1 per cent in 1983 to 9.4 per cent in 1988 (Zhao, 1990). The funds used in agriculture as fixed assets by the collective economic organizations totalled ¥4.25 billion in 1987, but those in industry, construction and so on were ¥25.09 billion. The 'compensation' (or subsidy) to agriculture from the industrial enterprises accounted for around 30 per cent of their net profit in 1978, and dropped to 5 per cent in 1984 and 4.5 per cent in 1988. In 1988, 53.3 per cent, of the net profit from non-agricultural enterprises was re-invested in these enterprises (Niu, Guo *et al.* 1991). Likewise, the share of farm households' savings re-invested in agriculture decreased from 60 per cent to 40 per cent. The amount invested in agricultural assets constituted only 12.2 per cent of the fixed assets of the farm households, while the sum invested in house construction was as high as 70 per cent (Zhao, 1990).

The development of part-time farming has encouraged the development of the farmers' non-farming tendency. This is especially the case in economically developed eastern and southern coastal areas and suburbs of large and middle cities, where two-thirds of the farmers are part-time farmers. These farmers refer to agriculture as 'morning–noon–evening business' or 'Sunday agriculture'. Most of the full-time agricultural workers in these districts are aged or female.

The basic causes of the above phenomena seem to be rising production costs and the unfavourable terms of trade faced by agriculture, which have led to agricultural incomes falling behind incomes in other industries. During the period 1979–85, the prices farmers had to pay for industrial products used as inputs in agriculture increased 50 per cent faster than the state purchasing prices of agricultural products. Remarkable rises in production costs were widely reported. For example, according to a survey in Shaanxi Province, the production costs of six grain crops in 1985–8 were ¥60.3 per mu, compared to ¥45.3 in 1979–83, an increase of one-third; the production cost of 100 kg of wheat rose by Y12.68 from 1984 to 1988, but the purchase price was up by only ¥2.98 (Liu Yang, 1990).

Income differentials shifted sharply against agriculture. For example, the survey in Shaanxi Province just referred to showed that, in 1985, the average agricultural product value of a farmer was ¥1176 per year. If this is given an index value of 100, then the corresponding indices for secondary and tertiary output values were 433 and 533, respectively, with industry being 533, commerce 449 and transportation 353. Another sample survey of 67 households in Jiangling County, Hubei Province, showed that yearly per worker income for a grain producer was ¥1661, while the income in construction was ¥2307, in transport ¥3202, and in trade ¥3840. This Hubei Province survey also showed that, within the agricultural sector, grain producers' incomes were one-third to one-half lower than the incomes of growers of cash crop (Niu, 1989). A much larger, national survey undertaken by the State Council in 1986 covered 30 000 households. The results of this survey also demonstrated a significant distinction among sectors in terms of daily income per workday. It ranged from ¥4 in the case of crop producers, to ¥8.4 for produce processors, ¥8.6 for workers in commerce and catering trade, and up to ¥15 for employees in transport and processing enterprises (see CAAS, 1989). Another important finding of the State Council survey was that the three broad groups of workers engaged in agricultural production, rural enterprises, and state-run industry enjoyed very different average incomes. The ratios of average income for these three groups of workers were 1:6:12, which is a remarkable range of income levels.

It is clear that most farm households which have become rich have done so through non-farming activities. In particular, the owners of the relatively large private enterprises have become people of wealth in rural areas. Thus non-farming activities have become very attractive to many farmers and have given them a false impression that farming has no prospects. Many people claim that this is the major reason for grain production stagnating after 1984 and the rate of increase in the output of many other agricultural products slowing down. This 'loss of enthusiasm for farming' has been attributed to the reforms of the early 1980s. The critics of the reforms claim that this move away from farming is a serious threat to the nation's food supply.

PROSPECTS FOR 'SMALL FARMERS' IN CHINA

Considering the major constraints on the scarce agricultural resources of China and the extraordinary demand pressures generated by a steadily growing population with rising incomes, we may conclude that the small farmers of China have some reasons to be more optimistic than pessimistic. To be sure, the weaknesses of the small farmer (such as small management scale and the inherent diseconomy, his relative lack of resistance to natural calamities and market risks, limitations in using modern machinery and advanced science and technologies) will become more apparent as the economy develops further. Nevertheless, as we noted earlier, China will not be able to change the predominance of the small farmer for a very long period of time. In view of the overall needs of socio-economic progress we must let the small farms survive and develop to a certain extent. In fact, they are viable and indomitable. They will survive and continue to develop because of the geographical features and dispersion inherent in agriculture. The production of living plants and animals requires the attention and care of the producers. This kind of production requires the decentralization of micro-decision making to the grass-root level (to the farm households as at present in China). We should recognize that small farmers operating under the 'dabaogan' system are well suited to the development of productive forces, and there is no reason to change the present system.

Naturally, the existence and development of the small farmers in the future should be assisted by ensuring favourable inside and outside conditions such as a stable policy in regard to the family responsibility system; the consolidation and perfection of the 'dual-level' management system integrating 'unified and individual management'; the intensification of socialized services provided by the state, local organizations and cooperatives; a relatively beneficial market environment and reasonable price system; and a proper ratio of the government (both national and local) budget for capital investment in agriculture. In short, the future of small farmers in China depends upon the implementation of the policy for a coordinated development of industry and agriculture. Evidence shows that the Chinese Government is making efforts to adopt such an approach. Whatever the case, we have some reasons to be optimistic about agricultural development in China.

NOTES

[1]This paper has been developed as part of a collaborative research project between Chinese and Australian agricultural economists. The project, entitled 'Economic Aspects of Raw Wool Production and Marketing in China', is being funded by the Australian Centre for International Agricultural Research. The authors wish to thank John Longworth, Colin Brown and Greg Williamson of the Department of Agriculture at The University of Queensland, Australia, for comments and suggestions on earlier drafts.

[2]For some details on how the land was allocated see Appendix A.

[3]For further discussion of the success of larger-scale farms, see Appendix B.

[4]Actually, under the prevailing conditions in the southern Jiangsu, a farm worker could cultivate 22.5 mu of paddy rice. If a farm household consists of 4 workers, 90 mu would be more effective economically. The size of the current farm households is 12–14 mu of paddy

fleld, or 6–7 mu per worker. This has resulted in irrational use of both labour and machinery. The utilization ratios are 60 per cent and 20 per cent, respectively.

REFERENCES

CAAS, 1989, *Nongye Shidu Guimo Jingying Wenti (Problems on the Optimal Management Scale in Agriculture)*, Nongye Chubanshe, Beijing.
Chinese Statistical Yearbook (CSY), 1985, 1990, Zhongguo Tongji Chubanshe.
Li Fangwei, 1990, 'The Land Systems in Chinese History', *Nongye Jingji Wenti*, no. 7.
Liu Xiaoduo, 1990, 'The Evolution of Land Systems and the Development of Productive Forces', *Nongye Jingji Wenti*, no. 11.
Liu Yang, 1990, 'The Grain Price Reform on the Condition of Rising Costs of Agricultural Production' *Nongye Jingji Wenti*, no. 2.
Liu Yunzi, 1990, *Nongye Zuzhi yu Guimo: Lilun, Shijian, Bijiao Yanjiu (Organization and Scale in Agriculture: Theory, Practice and Comparisons)*, Zhongguo Jihua Chubanshe.
Niu Ruofeng, 1989, 'China's Grain Production Toward 2000', in J.W. Longworth (ed.), *China's Rural Development Miracle; with International Comparisons*, University of Queensland Press, Queensland.
Niu Ruofeng, Guo Wei *et al.*, 1991, *A Study on the Relatlonship between Industry and Agriculture in China*, Institute of Agricultural Economics, Chinese Academy of Agricultural Sciences (CAAS), Beijing.
Ye Sifa and Min Zengfu, 1990, 'Running the Family Farm is an Effective Way to Make the Farmer Rich and to Increase Grain Production', *Nongcun Fazhan yu Guanli*, no. 3.
Zhao Donghuan, 1990, 'Analysis on the Problems in Macroeconomic Readjustment in Agriculture and their Solutions' *Nongye Jingji Wenti*, no. 6.

APPENDIX A

Some notes on land allocation procedures

The basic principle of land allocation was 'an equal plot to everybody'; that is, distribution of land on a per capita basis. In recent years, over 20 provinces have introduced the 'two-field system', namely a food field and a responsibility field. We shall illustrate this system with a case study in Pingdu County, Shandong Province.

The collective economic organizations were not disbanded in this county in the first stage of reform, and the reform of the economic system went quite smoothly. When the 'dabaogan' system was implemented in 1982–3, the contract farmland was divided into 'two big plots' and 'two small plots'. 'Two big plots' referred to a food field for own consumption and a responsibility plot contracted with households, according to the number of family members, for several years; 'two small plots' included a 'free plot' and a 'mobile plot'. The former accounted for 5 per cent of the total arable land in the village and every family had it on a per capita basis; this plot could be used for any crop production, and no payment was required. The latter made up 10 per cent of the total; the plot was contracted to the bid winners on a yearly basis. At the same time, they had six 'unified activities' as follows: (1) crop planting plans; (2) management of large-size farm machinery; (3) planning, construction and management of water conservancy projects; (4) mechanized field operations, such as ploughing, sowing, irrigation and harvesting; (5) breeding improved crop seeds and technical services; and (6) industrial and sideline production.

The above system was a form of 'an equal plot to everybody'. It was replaced by the 'two-field system' in 1986–7 in this county in an experimental way. The new system spread quickly and stood firm in 90 per cent of villages in 1988, leaving 3 per cent of land as 'mobile plot'. The 'food field' was 0.7–0.8 mu per capita, which was given for free use and accounted for 31.8–36.4 per cent of village farmland. The 'responsibility field' amounted to 63.6–68.2 per cent and was put out to contractors for five years, at 1.4–1.5 mu per capita. The contracts can be prolonged when they expire. The collective organizations should deal with all issues in the case of transfer of the responsibility plots. Every responsibility field was classified in all villages according to fertility (9 grades in 3 levels) and the land plots were registered in files to record the changes of land fertility. The fertility grades were given on the spot, in a democratic way. The farmers were free to bid for the land plots. The land contract payments ranged from ¥30–70 per mu, or 30–40 per cent of the total income per mu. Except for these payments, the farmers do not observe any other commitments, even deductions (see CAAS, 1989).

Many villages in Pingdu County issue their land bids on a large rectangle basis to volunteers, who should create a coordinated group of several farmers in crop cultivation. Such measures have also being taken in other places because of the advantages in production planning, irrigation and mechanized operations.

The basic conditions for the introduction of this 'two-field' system' are relatively strong collective economic power (for example, with developed industrial enterprises or sideline occupations) and the existence of a group of big specialized cropping farm households demanding the shift of land in the village.

Occasionally, the land lease contract system has been introduced in some places.

APPENDIX B

Examples of the Success of Larger-scale farmers

Some big crop producers in the land-scarce yet economically developed regions have begun to run a number of small-sized family farms as a single enterprise, using the available family labour and purchased machinery. Some of these ventures have been extremely successful. For example, in Qianfeng Township (Jinhu County, Jiangsu Province) there were six such ventures in 1985, but this increased to 38 in 1987. At that time, these 38 big producers had by contract about 1573 mu of farmland and waste beachland. They employed 191 persons in total (of which 94 were full-time workers) or 41.4 mu per participating household and 16.7 mu per worker. They had 230 pieces of various machinery and auxiliary tools (including tractors) and other fixed assets valued at ¥346 thousand, or ¥3680.85 per worker. This figure is not small in China. These 38 large farms produced 811.8 tons of grain, and commodity sales were 684.5 tons. This ratio surpassed remarkably the average level of the township (Ye and Min, 1990).

	38 family farms	Percentage of the township average level
Grain commodity ratio (%)	83.8	127.3
Per mu yields (kg)	516*	110.1
Net income from one mu of grain (¥)	259.7	123.5
Net income per worker (¥)	2 523	260
Input–output ratio in grain production	1:2.63	142.2

Note: *Including the newly reclaimed waste beachland; the yield of original existing farmland was 719 kg per mu.

Source: Ye and Min (1990).

In Jinhu County, the per household farmland availability is 50 mu, and where land is more plentiful 60 mu, which is more in line with adequate scale in current conditions. The scale of the above family farms seems to enable full use of labour and machinery.

The average scale of the contract farmers is larger in the less developed agricultural areas with more land resources. The big specialized grain producers have taken more than 50 mu of farmland. The 409 000 farm households in the counties under the jurisdiction of Jiamusi City (Heilongjiang Province) are an example. The farm households with 51–150 mu of contract farmland accounted for nearly 25 per cent in 1988; big households with 151–300 mu made up 1.03 per cent. There are several big specialized grain producers with more than 450 mu of land. An analysis was made of 59 households (from 212 households in the four counties of Fujin, Baoqing, Suibin and Huachuan in Jiamusi City). The results suggest that the most effective are 18 big specialized households with over 50 mu of land per worker.

APPENDIX C

Possible Management Forms which can Co-exist with 'Small Farmers'

(1)   *Big specialized households:* with more contract land, combining the traditional planting techniques and mechanized operations. Their land plots may be contracted in from other villages.

(2) *Family farms:* engaged in production of one or several crops, mainly mechanized, and managed in a manner as if they are industrial enterprises.

(3) *Cooperative farms:* organized by farmers on a voluntary basis. Farmland is managed in a unified way; field operations are assigned to households or workers.

(4) *Collective specialized teams, groups or farms:* may be sub-divided into three types:
  — specialized teams or groups make contracts and cultivate the collective-owned land, assign field operations to households or workers (mainly according to the number of workers) and distribute the income on the basis of the contract targets;
  — mechanized farms: mechanized teams or groups make contracts and fulfil all field operations with farm machinery, which is more progressive technically; and
  — collective farms: the collective management system of the original communes or production brigades has been retained. There are unified plans, resource allocation, accounting and income distribution within the teams. Some sign contracts with individual households in doing field operations, some record work points in production, which was typical under the commune system, and some have introduced wage systems similar to industrial enterprises.

(5) *Agricultural workshops (specialized teams attached to industrial enterprises) on condition of agro-industrial integration:* usually composed of skilled farmers under the auspices of rural industrial enterprises. The farmers are responsible for planting the contract land of the employees of these enterprises. The specialized teams are thus attached to them and the farmers share all the advantages, as do the employees.

DISCUSSION OPENING – AKIMI FUJIMOTO*

Niu and Chen discuss the process of small farm development in China, with the focus on various changes and problems which emerged under the economic reforms after 1979. There were two major changes in the last decade: the establishment of the family contract responsibility system, and the introduction of free markets for farm commodities. Farmers quickly responded to new opportunities and increased agricultural production and incomes. The Chinese experience in policy and institutional changes, as well as farmers' responses to them, present a most interesting case for the study of small farm development, which is a vital issue for many countries in Asia.

One of the most serious problems in small farm development is the difficulty of achieving a balance between promotion of growth and equity. The Chinese case demonstrates that the heavy emphasis given to equity under the commune system was responsible for the stagnation of production and continuing poverty among farmers, whereas new options provided under the

*Tokyo University of Agriculture, Japan.

economic reforms resulted in an overall improvement. However, there has also been increased differentiation among farmers, who used to have almost equal resources available to them. Farmers appear now to be operating under two economic rules and their decision making in crop production and management is not free. It would be useful if the actual operation of small farms could be clarified, especially with reference to the choice of product and entry into free markets. How can farmers function in free markets if the use of land is controlled by the contract?

Whatever the actual operation may be, the important implications of the Chinese experience are that the relationship between equity and growth depends upon the nature of farmers' responses. As long as growth is determined by individual performance, it can be expected that it will be accompanied by some degree of inequity. The question is how to tolerate the differences in production capabilities among individual farmers in order to achieve growth, while maintaining equity. This is a continuing problem in the history of economic development, demonstrated once again by the Chinese experience.

Actually, agricultural production rose rapidly in the early years of economic reforms, but stagnated after 1985. More detailed analysis of this phenomenon would be useful in identifying the mechanism of agricultural development in China's golden decade. Some economists argue that production incentives provided under the responsibility system and free markets did not last long, and there emerged a need for the introduction of a price determination system, in the market, for the promotion of further growth.

Niu and Chen's main arguments concerning small farm development in China involve the need for creation of off-farm employment opportunities and the enlargement of the scale of farm operation. These are typical arguments in the theory of small farm development. Not only in China, but also in many countries of Asia, farms are generally of small size and their level of income is low, presenting a most urgent agenda for political action. The major policy approach has been the promotion of rural industrialization and the creation of off-farm employment opportunities. In short, an increase in total farm household income is sought through the increase in off-farm earnings. Because farmers do not easily give up their land, this can result in there being increasing numbers of part-time farmers. In fact, this has been the main feature of development in Japan and seems to be taking place in other parts of Southeast Asia. It is also happening in China, and the authors seem to support such a trend, which may be termed the 'part-time farming approach'.

While the necessity to identify the kind of farmers who have been active in obtaining off-farm employment, and the particular factors that made them so successful is recogized, the fundamental question in agriculture is how farm size can be enlarged if farmers do not give up their land. Certainly, part-time farming may be one way of maintaining rural society and agricultural production in the Asian setting, but it is not necessarily the ideal form of farm management. Since increased production costs and decline of output were also noted by the authors, it is clear that the merits of part-time farming need further exploration.

It is my experience that most farmers in Southeast Asia and Japan are losing their enthusiasm for farming, or at least they do not want to see their children become farmers. This causes serious questions to be raised about the sustainability of the sector and of stability in the supply of food and other agricultural products. We can hardly expect incentives to be generated in agriculture when there is so much emphasis on the need to secure external sources of income. I would therefore like to emphasize the need to consider the possibility of adopting a 'full-time farming approach' in agricultural development. It is most important to establish viable farmers who obtain sufficient income from farming alone, and regard it as their main professional activity.

It is certain that all the problems of small farm size, low output prices, high input costs and low farmer capability could be limiting factors in the 'full-time farming approach'. There will probably be a need to expand the physical size of farms to some extent, and that can be attained only by reduction in numbers. In this sense, the availability of off-farm employment opportunities has its importance. However, we need to conceive of the scale of farming as the size of the farming business, and it is this which should be expanded, rather than the physical sizes of farms. We must seek ways in which physically small farms can also be very productive and obtain a high level of direct income. More intensive systems with appropriate technology, for instance, should be worked out in order to secure the sustainability of small but viable enterprises, capable of maintaining the supply of agricultural products. I strongly feel that agricultural economists should devote greater efforts to a study of the endogenous development process of the small farm sector itself, rather than arguing for an increase in off-farm income.

For the 'full-time farming approach' to be successful, some reduction in the number of farmers is probably inevitable. The selection of farmers who would remain would be a problem, but this should be determined by the preferences and capabilities of farmers themselves. In the creation of off-farm employment, emphasis should be placed upon industries which are related to agriculture, in order to establish a sound regional economy. Is the Chinese attempt to create more off-farm jobs regarded as an evolutionary step for the creation of full-time farmers, or is it a goal in itself? It seems to me that, under the controlled system, it should be much easier to implement the full-time farming approach, which will present significant lessons for other countries in Asia.

RICHARD L. MEYER AND GEETHA NAGARAJAN*

An Assessment of the Role of Informal Finance in the Development Process

INTRODUCTION

For more than two decades, formal financial markets have been emphasized by policy makers in their attempt to direct more credit into rural areas in pursuit of production and income goals. Informal finance, and especially the role of moneylenders, has often been considered to be exploitative and a hindrance to modernizing agriculture. Policies have often been employed, therefore, with an explicit objective of increasing formal finance and diminishing the role of informal finance.

In fact, the supply-leading approach to formal finance has largely failed (Adams, Graham and Von Pischke, 1984). Loan quotas and targets set for commercial banks have not led to sustained agricultural lending. Interest rate controls and special small farmer credit projects have not resulted in large amounts of lending to small farmers, because larger farmers have received most of the cheap loans. Agricultural development banks and cooperatives specifically established to lend to agriculture have often failed or are saddled with huge non-performing assets. Ironically, in areas where formal finance temporarily displaced informal finance, the latter is now re-emerging and is once again the main source of financial services for most rural firms and households.

Researchers and policy makers have rediscovered informal finance. But, unlike earlier times, when informal finance was considered exploitative, today the predominant view is that informal finance may play a useful role in developing countries. It may be the only viable financial source in the short term, and may have certain informational and cost advantages, so it will continue to exist even when the formal financial system has become better developed.

Analysing and understanding informal finance presents a challenge, for two reasons: first, informal finance consists of a complex set of individual and group financial arrangements and contracts; second, unlike the use of bank-firm theory for formal finance, there is no standard framework for studying informal finance. Frequently, therefore, widely divergent conclusions and policy recommendations are presented about the role of informal finance in the development process.

*Ohio State University, USA.

The purposes of this paper are briefly to describe the wide variety of arrangements and contracts that exist in informal finance, to discuss some of the approaches used to study it, and to summarize the policy alternatives that emerge from recent studies. The paper is divided into three parts. The first describes informal finance, the second briefly discusses theories and frameworks used in informal finance studies, and the third discusses policies.

INFORMAL FINANCE IN THE FINANCIAL SYSTEM

No standard definition exists for informal finance.[1] Table 1 presents examples of several financial arrangements found in financial systems in developing countries. These arrangements could be arranged along a continuum from most formal (a bank) to least formal (family and friends) but for simplicity's sake they are classified into formal, semi-formal and informal.

The formal financial system has three components. The first is the central bank and other regulatory and prudential institutions that establish the financial rules of the game for most formal institutions, and sometimes for semi-formal and informal finance. The second is a wide variety of financial institutions that provide financial services directly to savers and borrowers. They have been the focus of most supply-leading financial policies. The third component represents capital markets, but they generally play an insignificant role in low-income countries.

Semi-formal finance includes those financial arrangements that fall between formal and informal. Usually they are not regulated and supervised by the same regulatory authorities as formal finance, but they may have some official sanctions, such as being licensed and audited by a cooperative registrar. They may also have formal written statutes and operating procedures. Some are actively encouraged, assisted and subsidized by NGOs (non-governmental organizations) and PVOs (private voluntary organizations), or governments and donor agencies. Owing to the failure of formal finance in many countries, semi-formal financial arrangements have recently become the focus of increased government and donor attention.

Informal finance includes a heterogeneous set of individual and group financial arrangements. Most fall outside the scope of government support and regulation, although some countries have usury and other laws intended to cover them. Some types of informal finance are autonomous, while others emerge as a reaction to the repression of formal finance (Chandavarkar, 1986). Some have strong links to formal finance, such as the input supplier who borrows from formal institutions and sells inputs on credit, while others operate completely outside the formal system.

Informal finance is often loosely defined to include both the financial arrangements classified here as semi-formal and informal. Several attributes can be identified for informal finance:

(1) *Heterogeneity.* Informal finance includes a wide variety of institutional forms and, within any one type, a variety of financial contracts between savers and borrowers can be found. The frequency with which these

TABLE 1   *Important components of the rural financial system*

Formal	Semi-formal	Informal
Regional and national central banks, Treasury, and other regulatory bodies	Agricultural cooperatives	Communal clubs
	Credit unions	Mutual aid associations
	Banques populaires	Moneylenders
	Integrated dev. projects	Moneykeepers/mobile bankers
Financial intermediaries	Village banks	Input suppliers
Commercial banks	Self-help groups	Store-owners/merchants
Development banks	Savings clubs	Trader-lenders
Savings banks		Farmer-lenders
Postal savings		Friends
Cooperative banks		Neighbours
Unit rural banks		Relatives/family
Finance corporations		
Capital markets		

financial arrangements are found varies widely: Africa and Asia seem to have a greater variety of informal finance arrangements than Latin America, for example.

(2) *Services.* Informal finance offers both loan and savings unlike supply-leading finance which often ignores savings. Many types of informal finance also offer marketing and services.

(3) *Specialization.* Some informal financial arrangements serve a broad clientele, but informational problems often encourage specialization. For example, moneykeepers, moneylenders and trader-lenders often provide only one type of financial service and it is limited to those clients for which the supplier has good information. This specialization may provide efficient services, but also limits access for new clients.

(4) *Access.* Informal finance is used by the rich and the poor but often it is the only source for the poor, while it is an alternative source for the rich.

(5) *Collateral.* As with formal finance, some informal financial arrangements may require collateral for loans (for example credit unions, pawnshops) but frequently informal finance has developed effective collateral substitutes through interlinked contracts, peer monitoring and group lending.

(6) *Interest rates and transaction costs.* Moneylenders have criticized for charging exploitative interest rates on loans, but other types of informal lenders, such as farmers, friends and relatives, often charge no interest at all. Some credit unions have experienced difficulty because they have not charged rates high enough to cover costs and inflation. What tends to distinguish informal finance is the relatively low transaction costs for savers and borrowers because of close proximity and a minimum of formal procedures.

(7) *Growth and decline.* It is difficult conclusively to determine the trends of informal finance. In countries as diverse as Brazil and the Philippines, it appears that informal finance has increased with the decline of formal

financial institutions. Credit unions are expanding in Togo and the Cameroon, but declining in Ecuador and other countries with high inflation. The large informal sector observed in many countries implies that informal finance will also be large.

(8) *Contribution to development.* Clearly, informal finance is already contributing to development because it is often the only source of financial services for the poor, the landless, and those living in isolated regions. The question frequently asked is how the positive benefits of informal finance can be expanded and improved so its contributions can be increased.

Researchers face a challenging task in conducting studies of, and evaluating, policy alternatives for informal finance. There are no well-defined theoretical and methodological tools to guide research in the way that bank–firm theory serves as a standard framework for studying formal finance. The following section highlights some of the issues in developing an ideal framework to examine informal finance.

THEORIES AND FRAMEWORK TO STUDY INFORMAL FINANCE

There are a number of unique features of financial market transactions compared to transactions in other factor and product markets that create special research problems. A wide variety of contractual arrangements exist in financial markets in which a loan is exchanged in the current period for a promise to repay in the future. The fulfilment of the promise, however, is influenced by exogenous and endogenous risks that affect borrowers. While exogenous risks are due to random shocks, endogenous risks arise because of borrower incentives to default and because of asymmetric information, so that lenders cannot perfectly screen borrowers and enforce contracts (Hoff and Stiglitz, 1990). Therefore rural financial markets tend to be highly segmented, owing to variations in financial technologies used by lenders for screening and enforcement (Esguerra and Meyer, 1992). Furthermore, since rural financial markets generally operate in environments where contingent markets are incomplete or absent, financial contracts must incorporate insurance features.

Although informal financial markets are faced with these inherent problems, they have a comparative advantage over formal financial institutions in their ability to internalize the externalities caused by information problems and exogenous risks. Complex interlinked loan contracts are often designed to mitigate or accommodate exogenous and endogenous risks in a given institutional environment. Peer monitoring and transactions among participants with established social and familial relations are effectively used to reduce information problems. Therefore moral hazard and adverse selection problems, which can occur under asymmetric information, are not as prominent in informal as in formal financial markets.[2] Informal financial markets effectively use collateral substitutes such as factor and product market linkages and third-party guarantees to overcome rigid collateral requirements which restrict access in formal financial markets.[3] Furthermore, the engendering of reciprocal obligations and risk-sharing contracts observed in informal finance function as insurance substitutes (Platteau and Abraham, 1987). Interest rates are often more flexible in informal than in formal finance, so financial contracts can adjust more easily to differences in costs and risks. The flexibility of informal financial contracts and the heterogeneity of the financial agents make it difficult to model informal finance quantitatively. A robust framework is required for analysis, however, in order to understand the structure of informal finance, the nature of financial contracts and the efficiency of financial transactions.

Several theories and economic frameworks used to study formal, semi-formal and informal finance are outlined in Table 2. Bank–firm, contract and insurance theories have been effectively used to examine the supply and demand sides of formal financial institutions. The frameworks of the new institutional economics and the new household economics have used club theory in addition to bank–firm and insurance theories to study semi-formal finance. The normative nature of the new household economics guides rational choice and decision making of borrowers under risk, uncertainty and costless markets (Singh, Squire and Strauss, 1986). On the other hand, the new institutional economics incorporates transaction and information costs that arise within a given institutional environment to explain the contractual relations between lenders and borrowers (Eggertsson, 1990; Floro and Yotopoulos, 1991).

The theoretical tools developed for formal and semi-formal finance are used to examine contractual relations between anonymous lenders and borrowers. They are not appropriate, however, to analyse informal finance, where the identity and reputation of the contracting parties influence the multifaceted transactions conducted among heterogeneous participants. Empirical studies of informal finance, therefore, have often relied on hypothetical conjectures

TABLE 2   *Theories and Frameworks to Study Rural Financial Markets*

Formal	Semi-formal	Informal
	Theories	
Bank–firm theory under:	Bank–firm theory under:	Game theory
(a) asymmetric, incomplete and imperfect information	(a) lender-dominated institutions	
(b) risks and uncertainty	(b) borrower-dominated institutions	
Insurance theory	Insurance theory	Insurance theory
Contract theory	Club theory under:	Contract theory
	(a) exclusive and non-exclusive contracts	
	(b) queuing models	
	Frameworks	
Supply side: profit maximization	New institutional economics	New institutional economics
Demand side: new household economics	New household economics	New household economics
		Neo-institutional economics

based on artificial paradigms or have simply limited themselves to descriptive analysis. Recent studies have attempted to evaluate more rigorously informal contractual arrangements and market structure.[4] The new institutional economics, the new household economics and neo-institutional economics have been employed to analyse the behaviour of informal financial market participants and to explain complex informal contractual arrangements. For instance, the bulk of the literature on interlinked credit contracts follows the new institutional economics framework using the theories of games, contracts and insurance under the axioms of micro-economics.[5] These models explain the existence of heterogeneous interlinked contractual arrangements as a means to economize on transaction and information costs. While models based on game theory consider interlinked contracts between lenders and borrowers as a repetitive game that incorporates reputation and pre-commitments in terms of reciprocity, the models based on insurance theory explain interlinked contracts as insurance substitutes.[6] These models are theoretically appealing, but they have limited empirical application because of data problems. The empirical regularities of informal contractual arrangements are better explained by the positive nature of a neo-institutional economics framework that integrates both the institutional and household economics frameworks (Eggertsson, 1990).

The multiplicity of frameworks identified above might suggest the availability of well developed methodological tools to study informal finance. The segmentation of informal financial markets and the heterogeneity among the contracts, however, inhibit the development of a single model that will capture the details of most of the transactions. The basic structure of informal financial markets is often subject to controversies, so there is a lack of consensus about how to study them, especially in developing countries. This has led to a proliferation of conflicting propositions based on divergent models. The contradicting explanations provided to rationalize high interest rates charged in informal credit markets is one example.[7] Consequently, these conflicting opinions have confused policy makers regarding the role of informal finance in the development process.[8]

It can be concluded that the several theories and frameworks used to study informal finance are as heterogeneous and controversial as the informal financial markets themselves. Therefore research on informal markets is confronted with unresolved issues such as (1) What is the structure of informal financial markets? Are they competitive, monopolistic or contestable? (2) How can the terms and conditions of differentiated informal financial contracts from heterogeneous lenders in a given institutional environment be explained by a plausible framework that incorporates costs and risks? These issues need to be addressed if we are to evaluate adequately the role of informal finance in the development process and to suggest appropriate policies.

POLICY ALTERNATIVES

Several policy alternatives have been suggested for informal finance in developing countries. Some implicitly propose controlling or reducing informal transactions because they are rooted in the old exploitation school of thought.

Others follow the newer approach in which informal finance is perceived as providing valuable services. The difficulty in generalizing about informal finance is due to its heterogeneity and to the lack of a consistent framework for analysis.

The purpose of this section is to summarize five policy alternatives that are found in the developing country literature. It is important to understand the nature of these recommendations, even though there is little consensus about when, where and for which type of informal finance they apply.[9]

(1) *Benign neglect* Those who are most frustrated about the negative experience of two decades of supply-leading formal finance argue that the worst thing to do would be to meddle with informal finance. They argue that past failures demonstrate the difficulty of effectively developing a financial system through government efforts. Attempts to intervene in informal financial markets could result in the same negative consequences. Efforts to 'use' informal finance as a development strategy may damage or destroy its integrity. Part of its rationale derives precisely from its informality and immunity from official regulation (Chandavarkar, 1986).

(2) *Regulate and supervise* Although informal finance may be ubiquitous in developing countries and successfully operate in environments where formal finance fails, a number of problems exist which lead some analysts to argue that social welfare would probably increase with some degree of government regulation and supervision.[10] The problems identified include the perception that interlinked credit transactions are exploitative, that participants in group savings arrangements run the risk of losing their savings, and that the self-regulation of cooperatives and credit unions is too weak to prevent abuses. Besides the possibility of systematic abuse, there are also the normal financial problems such as excessive portfolio concentration of localized institutions operating in areas with high covariance in income.

The regulations that are proposed to avoid or reduce these problems range from the simple establishment of minimum conditions for obtaining operating licences, to more complex rules about portfolio diversification, lending to insiders, capital requirements, bookkeeping and accounting procedures, and contract terms and conditions. Reporting requirements and supervisory procedures are also recommended.

Some successful attempts to regulate informal finance exist, such as with the licensing of pawnshops in the Philippines and Sri Lanka. There is much scepticism about regulation, however, because of the regulatory failure that has frequently occurred in formal finance. The first priority, therefore, would seem to be to improve regulation and supervision of formal finance before trying to extend it to the greater complexity of informal finance. Another problem is that governments may implicitly incur additional unwanted liabilities if they attempt to license and regulate informal agents and transactions.

(3) *Imitate informal finance* Since informal finance has prospered in many environments where formal finance has failed, there is an obvious implication that informal finance must be doing something correct that should be imitated

by formal finance. This alternative implies discovering the technologies and procedures that make informal finance work, and transferring them to formal finance.

To a degree, imitation is already occurring. One of the important comparative advantages of informal finance is access to information useful for screening participants and peer monitoring (Burkett,1989). For example, members of savings clubs and ROSCAs tend to know each other; otherwise they will not undertake group activities. Likewise, trader-lenders and input suppliers accumulate information about potential borrowers through their other transactions. The highly touted Grameen Bank in Bangladesh uses group lending techniques similar to those employed by informal groups (Hossain, 1988). The borrowing groups use their informational advantages to screen and monitor their members so that bank operating costs and loan losses are reduced. Other techniques, such as mobile banks and collection agents, have been employed in some rural credit programmes to reduce costs by bringing financial institutions closer to their client. These innovations can reduce transaction costs for both borrower and lender.

Apart from these examples, it is difficult to imagine many ways in which formal finance can imitate informal techniques. The comparative advantage of formal financial institutions is that they conduct large-scale, low-margin transactions in which formal bureaucratic procedures substitute for personal relations and first-hand knowledge. This discourages them from servicing clients for which information and transaction costs are large relative to loan size.

(4) *Link formal with informal*   The idea behind this alternative is that financial services can be expanded to the poor most efficiently by linking formal with informal finance. This approach is being actively promoted in Indonesia, where self-help groups are being formed to mobilize group savings that are deposited in banks as guarantees for loans to the group. Many governments and donors are now channelling funds for on-lending through credit unions and informal groups rather than through agricultural or commercial banks. The hope is that these groups will be able to lend and recover the funds more efficiently than formal institutions, many of which have failed when trying to serve a rural clientele.

One problem with using informal finance as a conduit for funds is group dynamics. The credit union experience in many Latin American countries illustrates the problem. When they were first organized and operated with their own savings in the 1950s and 1960s, many were healthy, self-sustaining institutions. In the 1970s, however, many began to lend borrowed funds and this changed their operations from being saver-dominated to borrower-dominated (Poyo, 1985). Savings mobilization was ignored, interest rates were not properly adjusted for costs and inflation, and loan recovery declined. There is not yet any clear understanding about the appropriate incentive structure to use so that external funds do not destroy these groups.

(5) *Graduate informal into formal*   Many advocates of credit for the poor believe that formal institutions, even under liberalized financial conditions,

will never lend significant amounts of funds to groups currently without access. Therefore it is argued that NGOs and PVOs that attempt to serve marginal groups must eventually evolve into financial institutions (Boomgard, 1989). Rather than pursue the popular strategy of trying to graduate poor clients from specialized lending agencies to commercial banking facilities, the entire programme needs to graduate.

Some attempts to formalize or scale up informal financial arrangements are already taking place. The Grameen Bank became a formal bank after several years operating as a programme linked to existing banks. Credit union development programmes are being developed in Africa, based on the widespread incidence of moneykeepers and savings groups. Credit union central liquidity funds are being created in the Philippines and Portugal to link individual credit unions together as a way to spread risk and allocate funds among deficit and surplus units. These developments are relatively simple, however, compared with the large task implied by trying to convert many of the existing NGOs and PVOs into financial intermediaries.

CONCLUSIONS

Informal finance consists of a variety of individual and group financial arrangements and contracts. Many types of informal finance can be found in most developing countries. Often they provide the only financial services available to the poor, to small farmers, to the landless, and to other groups that lack access to formal finance. Although informal finance is ubiquitous and provides important financial services for many people, its role in the development process is subject to debate. Many researchers and policy makers still characterize it as being exploitative of people who have no choice but to rely on informal finance. Others believe that informal finance makes positive contributions to development precisely because it provides services not provided by formal finance.

One of the explanations for the divergent views about informal finance is the lack of a clear set of theories and framework for use in studying it. The study of formal finance has been able successfully to adopt the tools of bank-firm theory to examine the behaviour of formal financial institutions in developing countries. The complexity of informal finance makes it impossible to use a single framework to study its characteristics and performance. It is difficult, therefore, to arrive at a commonly accepted set of theoretical propositions to explain why informal finance exists, how financial terms and conditions are formed, and whether or not social welfare will be improved by some form of government intervention.

There is an urgent need more fully to understand informal finance. The collapse of the rural formal financial system in many developing countries is causing concern about the possible impact on production, technological change and rural incomes. The sluggish response of formal finance to recent attempts to liberalize and privatize rural financial markets has raised doubts about the ability of formal financial institutions to serve effectively the needs of agriculture, but at the same time informal finance is often not regarded as an effective substitute.

Policy alternatives ranging from benign neglect to actively promoting formal–informal financial linkages have been proposed in the research that has been completed to date. Although these recommendations may have relevance for specific cases, there is no general understanding about the appropriate set of policies to use towards informal finance. Much more work must be done to conceptualize and empirically study informal financial arrangements, their strengths and their weaknesses. Fortunately, there is generally a more positive attitude towards informal finance today than existed several years ago. This situation makes it easier to conduct the large amount of research that is needed so that more appropriate policies can be developed to ensure the most positive impact of informal finance on development.

NOTES

[1] Examples can be found in the literature where attempts have been made to establish definitions, at least for the purposes of individual studies. Recent examples include Bouman (1989), Chandavarkar (1986), Germidies, Kessler and Meghir (1991), Ghate (1988), Lamberte and Balbosa (1988).

[2] Bell (1988) claims that informal financial markets face problems of moral hazard due to asymmetric information rather than adverse selection, while Siamwalla *et al.* (1990) assume symmetric information among informal credit market participants. Nonetheless, they argue that information problems are better managed in informal than in formal finance.

[3] Transactions are said to be interlinked when the contracting parties trade in more than one market on the condition that terms and conditions of all trades are jointly determined (Bell, 1988).

[4] See, for example, Bardhan (1989), Basu (1989), Bell (1988), Bhaduri (1983), Mitra (1983).

[5] See Bardhan (1989), Bell (1988), also Quibria (1987) for a state-of-the-art review on interlinked contracts.

[6] See Roth (1985) for a game theory model that incorporates reputation, and Binswanger and Rosenzweig (1986) for a model based on insurance theory.

[7] While some perceive high interest rates as surplus extraction by exploitative lenders (Basu, 1989; Bhaduri, 1983), others justify it as a risk premium required to compensate for the high risks faced by lenders (Bottomley, 1963).

[8] The emphasis on usury laws in the Philippines in the 1960s, and the subsequent policy to relax them and use informal lenders as conduits for formal finance in the 1980s, are due in part to conflicting evaluations of informal finance.

[9] Insightful discussions about policy alternatives for informal finance are found in Adams and Ghate (1992); Chandavarkar (1986); Ghate (1988).

[10] See, for example, Floro and Yotopoulos (1991).

REFERENCES

Adams, Dale W. and Ghate, P.D., 1992, 'Where to From Here in Informal Finance?' in Dale W. Adams and Del Fitchett (eds), *Informal Finance in Low-Income Countries*, Westview Press, Boulder, Co.

Adams, Dale W., Graham, Douglas H. and Von Pischke, J.D. (eds), 1984, *Undermining Rural Development with Cheap Credit*, Westview Press, Boulder, Co.

Bardhan, P., 1989, 'Alternative Approaches to the Theory of Institutions in Economic Development', in Pranab Bardhan (ed.), *The Economic Theory of Agrarian Institutions*, Clarendon Press, Oxford.

Basu, K., 1989, 'Rural Credit Markets: The Structure of Interest Rates, Exploitation and Efficiency', in Pranab Bardhan (ed.), *The Economic Theory of Agrarian Institutions*, Clarendon Press, Oxford.

Bell, Clive, 1988, 'Credit Markets and Interlinked Transactions', in H. Chenery, and T.N. Srinivasan (eds), *Handbook of Development Economics*, vol. 1, North Holland, Amsterdam.

Bhaduri, A., 1983, 'On the Formation of Usurious Interest Rates in Backward Agriculture', *Cambridge Journal of Economics*, vol. 1, no. 4.

Binswanger, Hans and Rosenzweig, Mark R., 1986, 'Behavioural and Material Determinants of Production Relations in Agriculture', *Journal of Development Studies*, April.

Boomgard, James J., 1989, 'A.I.D. Microenterprise Stocktaking: Synthesis Report', *A.I.D. Special Study No. 65*, A.I.D., Washington, DC.

Bottomley, Anthony, 1963, 'The Premium for Risk as a Determinant of Interest Rates in Underdeveloped Rural Areas', *Quarterly Journal of Economics*, vol. 72, no.4.

Bouman, F. J. A., 1989, *Small , Short and Unsecured: Informal Rural Finance in India*, Oxford University Press, Oxford.

Burkett, P., 1989, 'Group Lending Programs and Rural Finance in Developing Countries', *Savings and Development*, vol. XIII, no. 4.

Chandavarkar, Anand G., 1986, 'The Informal Financial Sector in Developing Countries: Analysis, Evidence and Policy Implications', IMF paper DM/86/13, Washington, DC.

Eggertsson, T., 1990, *Economic Behaviour and Institutions*, Cambridge University Press, Cambridge.

Esguerra, E. F. and Meyer, Richard L., 1992, 'Collateral Substitutes in Rural Informal Financial Markets: Evidence from an Agricultural Rice Economy' in Dale W. Adams and Del Fitchett (eds), *Informal Finance in Low-Income Countries*, Westview Press, Boulder, Co.

Floro, Sagrario L. and Yotopoulos, Pan A., 1991, *Informal Credit Markets and the New Institutional Economics: The Case of Philippine Agriculture*, Westview Press, Boulder, Co.

Germidies, Dimitri, Kessler, Denis and Meghir, Rachel, 1991, *Financial Systems and Development: What Role for the Formal and Informal Financial Sector?* OECD, Paris.

Ghate, P. B., 1988, 'Informal Credit Markets in Asian Developing Countries', *Asian Development Review*, vol. 6, February.

Hoff, Karla, and Stiglitz, Joseph E., 1990, 'Introduction: Imperfect Information and Rural Credit Markets: Puzzles and Policy Perspectives', *The World Bank Economic Review*, vol. 4, no. 3.

Hossain, M., 1988, 'Credit for Alleviation of Rural Poverty: The Grameen Bank in Bangladesh', *Research Report No. 65*, International Food Policy Research Institute, Washington, DC.

Lamberte, Mario B. and Balbosa, Joven Z., 1988, 'Informal Savings and Credit Institutions in the Urban Areas: The Case of Cooperative Credit Unions', *Working Paper Series No. 88–06*, Philippine Institute for Development Studies, Manila.

Mitra, Pradeep K., 1983, 'A Theory of Interlinked Rural Transactions', *Journal of Public Economics*, vol. 20.

Platteau, Jean-Philippe and Abraham, Anita, 1987, 'An Inquiry into Quasi-Credit Contracts: The Role of Reciprocal Credit and Interlinked Deals in Small Scale Fishing Communities', *The Journal of Development Studies*, vol. 23.

Poyo, J., 1985, 'Development Without Dependency: Financial Repression and Deposit Mobilization Among the Rural Credit Unions in Honduras', Unpublished PhD. dissertation, Syracuse University.

Quibria, M.G., 1987, *The Enigma of Interlinked Rural Transactions: An Introductory Survey of Literature*, Asian Development Bank, Manila.

Roth, A., 1985, *Game-Theoretic Models of Bargaining*, Cambridge University Press, Cambridge.

Siamwalla, A. *et al.*, 1990, 'The Thai Rural Credit System: Public Subsidies, Private Information, and Segmented Markets', *The World Bank Economic Review*, vol. 4, no. 3.

Singh, Inderjit, Squire, Lyn and Strauss, John (eds), 1986, *Agricultural Household Models : Extensions, Applications and Policy*, Johns Hopkins Press, Baltimore, MD.

DISCUSSION OPENING – CRISTINA DAVID*

I agree with the authors that further theoretical and empirical research is required to develop a full understanding of the operation of formal and informal markets in rural sectors, where interlinking of credit to other markets dominates. Such studies require understanding not only of the credit side of the transaction but the nature of the product, land and labour markets themselves. Nonetheless, I think that sufficient evidence is now available to derive some policy implications for the improvement of rural credit markets. In the Philippines case, for example, none of the five possible options appears to provide the correct approach. What is needed is policy reform in the formal financial markets as well as product and input markets, to promote the development of the formal financial markets and lower further the cost of credit in the informal credit market. Although informal credit markets may be efficient, the whole range of services provided by the formal financial markets is not as yet available to the vast majority of the rural sector.

Two sets of reforms are suggested:

(1) Financial markets:
  (a) Interest rates should be allowed to reflect true lending cost.
  (b) Regulations restricting entry to the banking sector, such as limitation on branch banking, should be removed.
  (c) Development of formal financial sectors in rural areas is restricted as a means of maintaining monopoly profits, and steps are needed to foster competition.
(2) Other markets:
  (a) Land reform measures, at least in the Philippine case, need modification. Potentially, land reform provides assets which can be used as collateral in the formal financial market, though this advantage has not materialized. Land transfers are prohibited and hence land rights and titles are not acceptable collateral. This appears less inhibiting in the case of informal financial markets, but it remains a significant problem affecting more formal financing.
  (b) Prohibition of share tenancy should be removed. Landless labourers are prevented from having access to cultivation rights, and therefore cannot take advantage of informal credit provision extended by landlords for use in production or consumption. Landless households do now have access to consumption credit, though only if they become permanent labour or attached workers receiving lower average daily wages than casual labourers.
  (c) Greater public expenditure is needed on market infrastructure, down to the level of postal and telecommunication services, to lower the costs of operation of formal financial markets and increase competition.

*International Rice Research Institute, Philippines.

(d) Increased public support for irrigation and agricultural research would reduce the inhibition on borrowing associated with risk, and increase credit demand by fostering improved repayment capacity.

(e) Macro-economic reform is needed to correct policies which artificially lower agricultural incentives. This would also improve repayment capacity and help generate demand, which could probably then be met at lower cost.

K.A.S. MURSHID*

Informal Credit Markets in Bangladesh Agriculture: Bane or Boon?

INTRODUCTION

Agricultural growth occurs through capital accumulation and technical change. The modern HYV technology, which has been the vehicle of growth in agriculture since the late 1960s, is capital-intensive in the sense that it increases the share of capital inputs in total output. It also represents technical change in that yields and factor productivities (including that of labour) are increased (Bhalla, Alagh and Sharma, 1984). Mundlak (1988) points out that capital accumulation leads to the employment of capital-intensive techniques and to a switch to modern technology and that 'it is impossible to increase the relative importance of the modern techniques without capital accumulation'.[1] Thus policies that extract resources away from agriculture will impact negatively on agricultural performance, while the opposite is true for policies that facilitate flow of resources to agriculture.

Governments have often sought to influence the rate and pace of HYV diffusion through subsidised credit programmes. The traditional view is that credit programmes help to overcome the capital constraint inhibiting adoption of HYV by LDC farmers. It also serves to offset the disincentive effects of an overvalued exchange rate and price controls. A further objective is to stem the flow of 'expoitative' informal credit which is deemed anti-developmental and inadequate.[2]

The traditional view has been criticized on the ground that subsidized credit results in rationing, is highly biased in favour of rich farmers, serves to discourage savings mobilization, and generally leads to a misallocation of resources (Von Pischke, Adams and Donald, 1983). It is also argued that credit programmes are not essential for HYV adoption as the technology involved is highly divisible. The way out is through credit at 'market' rates. Since the bank rate is thought to be a poor indicator of the cost of credit, attention has inevitably been re-focused on the informal credit market, which, despite the bad press, appears to be engaged in legitimate economic roles, and which could point to a more relevant rate of interest. In this paper we explore empirical evidence from Bangladesh to attempt an assessment of the role of informal credit in agricultural growth and technological transformation.

There is a large literature on the link between formal credit and agricultural growth, but few studies have focused on the role of informal credit systems.

*Bangladesh Institute of Development Studies, Bangladesh.

The 'interlocked market' literature (including markets for labour, land and credit) provides a theoretical approach for understanding certain forms of surplus appropriation, exploitation and persistence of under-development in traditional agriculture, but empirical validation has been scarce.[3]

DATA AND METHODOLOGY

This paper draws heavily on two sources of information: a BIDS study on the nature, structure and role of informal credit markets in Bangladesh (Murshid and Rahman, 1991) and a joint Open University–BIDS study on foodgrain market participation by rural households and traders (Crow and Murshid, 1990). The latter study generated detailed data on informal financing of trade and participation of different types of rural households in informal credit markets. Prolonged field stay enabled us to generate excellent panel data on a sensitive subject.

Our data relate to two types of rural settings: (a) an agriculturally backward area where a single paddy crop is cultivated under traditional rain-fed conditions, where there are few other economic activities present, and where the potential for technical change is very limited, owing to a weak irrigation potential;[4] (b) an agriculturally advanced area, where HYV cultivation has made great strides under controlled irrigation. The type and incidence of different forms of informal credit in each of these settings is reviewed in terms of the likely impact on agricultural performance.

We are concerned mainly with the demand side of informal credit and we have implicitly assumed that supply will change only sluggishly and that, over a period of a year and a half, supply changes will reflect seasonal change rather than a change in trend.

Area Description

The evidence produced here relates to two very different geo-ecological areas of the country (Table 1). These areas are characterized by considerable differences in cropping patterns, social organization and the level of technological development. Our backward area, situated in the south of the country, on the confluence of the Meghna River and the Bay of Bengal, is predominantly single-cropped, with the traditional Aman crop cultivation comprising the major agricultural activity. Sharecropping is extensive and constitutes the dominant tenurial arrangement. The ownership of the 'char' lands is often vested with absentee landlords resident in the towns, but who maintain close control over their estates. These landlords are often very powerful, frequently laying claim to many hundreds of acres, and backing up these claims by keeping on call armed retainers known as 'lathials'.[5] This area is normally said to be in net deficit in foodgrain production, but periodically generates a surplus. Agricultural growth has stagnated, with crop output registering a growth rate of 0.3 per cent in recent years.[6]

TABLE 1   *Socio-economic characteristics of the two study areas*

Indicator	Backward	Advanced
Population per cultd. acre	3.43	4.10
Land cultivated per household	1.84	1.23
Land owned per household	0.79	1.20
Area sharecropped (%)	66.0	30.1
Cropping intensity	0.89	1.78
Literacy Rate[1]	25.9	60.7

Note: [1]Persons over 12 years of age reporting ever going to school.

The advanced area to the north of the country is situated in the very old alluvium of the Barind Tract, unlike the still-active chars. Intensive irrigation development, beginning in the early 1980s, has led to widespread diffusion of modern cultivation practices and technologies. Far-reaching changes in cropping patterns and productivity have occurred. HYV diffusion is almost universal, with double cropping the norm and triple cropping not rare. The dominant tenurial mode is one of owner cultivation, with fixed rent contracts more popular than share rents. The area is generally endowed with a good communication network and is well served by both primary and assembly markets for agricultural produce. It has been designated as a surplus area by the government, especially for purposes of grain procurement by the official food distribution agency. This has been one of the fastest growing areas in the country, with output expanding at over 6 per cent per annum in recent years.[7]

Table 1 clearly brings out some sharp differences between the two areas. Land cultivated per household is greater in the backward area, while the actual ownership situation reveals that households own more land in the advanced area. Sharecropping is much more prevalent in the former (66 per cent of area) compared to 30 per cent in the latter. The low cropping intensity and the literacy rate also confirms the relative backwardness of the backward area.

THE DIVERSITY OF INFORMAL CREDIT

There is a great variety of informal credit arrangements in operation in rural Bangladesh, with considerable regional variation in terms of incidence and type. Most of the major forms reported in the literature were found in our study areas.[8]

Dhaner Upore (DU)

Dhaner Upore is a cash-for-kind loan or advance sale, taken three to five months before harvest, to be repaid in paddy. Implicit interest rates are very high or, alternatively, if this is viewed as a sale rather than a credit transaction, the price received is well below the market price at harvest. Repayment rates were typically 7 maunds (260 kg) per Tk. 1000 borrowed. Default leads to recalculation of liability and roll-over of the loan. First, the volume of paddy to be repaid is converted to cash at the ongoing market rate and then it is reconverted to paddy at the lending rate, with repayment period extended up to the next harvest. A variant of this system was also frequently observed, where a cash loan is taken two to four weeks before harvest, the principal is repaid in cash at harvest but the interest payment is in the form of paddy. Both these systems are widespread, but appear to be especially popular in poor backward regions.[9]

Land mortgage (LM)

This is also widely practised, particularly when credit requirement is large or when alternative sources have been exhausted. The most popular form of LM found in the study areas is transfer of user rights in land in exchange for cash, with the stipulation that such rights will revert back to the owner once repayment is completed. LM is associated with greater distress or urgency than DU. A variant of this system is sometimes encountered in which the land reverts back to owner control automatically at the end of a stipulated period.

Cash credit with positive interest (CCPI)

This is widely found all over the country, particularly among non-farm and labour households. Interest rates can be as high as 25 per cent per month. Collateral requirements and repayment period are sometimes stipulated, but this is less common. When collateral is taken, interest rates are usually lower.

Interest free credit (IFC)

In this case, amounts involved tend to be small, with transacting parties usually being from the same class, often involved in a web of mutual obligations. Such transactions have a specific, clear local name (for example, karja or hawlat).

Higher than or highest market price (HMP)

These are paddy loans repayable in cash at harvest. Two variants are available, depending on the timing of the transaction. If the loan is taken just after harvest, usually for the whole season (that is until the next harvest), repayment is at the highest price attained over the course of the season (HMP2). Alternatively, for a loan taken later on, when prices being to rise before reaching their pre-harvest peaks, repayment is negotiated at a price that is fixed at a level that is well above the prevailing market price (HMP1). These forms of credit have been reported from all over the country, but seem to be especially popular in areas of grain surplus. Sometimes repayment is in the form of labour (RL), particularly when the borrower is poor and finds it difficult to service the loan.

The relative importance of different credit forms found in our study areas is indicated in Tables 2 and 3, which show that in the deficit, backward area, formal credit contributes very little to credit needs, while in the surplus, advanced area it is the single most important source of credit. A number of informal credit arrangements have been discussed. Of these, the dominant forms are IFC and HMP in the advanced area, and DU, LM and HMP in the backward area. It will be observed that IFC is unimportant in our backward

TABLE 2 *Structure of credit transactions in advanced and backward areas*

Area	Borrower households(%)	Transactions Nos.	Formal (%)	Informal (%)					
				DU	HMP	LM	CCPI	IFC	RL
ADV	52	57	33.3	3.5	28.1	–	10.5	24.6	–
BACK	88	111	1.8	63.1	11.7	9.0	2.7	8.1	4.5

Notes: ADV: Agriculturally advanced area with substantial grain surplus and widespread use of modern irrigation. BACK: backward, mono-cropped area, no irrigation and with sizable grain deficit. See text for definitions of informal loan types.

TABLE 3 *Volume of Credit by Type*

Area	Credit Per Household	Formal	Informal					
			DU	LM	CCPI	IFC	HMP	RL
ADV	2 003	49 355 (19)	2 000 (2)	– –	1 892 (6)	26 750 (8)	20 150 (16)	– –
BACK	10 093	7 000 (1)	115 475 (70)	72 700 (10)	2 000 (3)	7 700 (9)	43 360 (13)	4 080 (5)

Notes: See notes to Table 1. Figures are in Taka ($1US=36 Taka). Figures in brackets are number of transactions.

area, where per household credit is found to be considerably higher than in the advanced area (Table 3).

Cash requirements are much higher for HYV cultivation, especially under irrigation, relative to cultivation under rain-fed conditions. In the case of the latter, the major elements of cost are labour and draught power, while for irrigated HYV paddy, water and fertilizer costs are very important in addition. A recent estimate suggests that the cash cost of cultivation with the modern technology is more than three times that under rain-fed conditions.[10]

The role of IFM in agriculture will have to be assesed in terms of the use of credit (a) to finance capital investment, and (b) to meet working capital needs. The implicit assumption here is that modern HYV agriculture is associated with greater credit demand to cover increased cash costs of cultivation.[11]

INFORMAL CREDIT IN A BACKWARD AREA

In terms of the number of transactions and amounts borrowed by rural house-holds from alternative sources of informal finance, our backward area stands in sharp contrast to our advanced area. Table 4 shows the monthly deviation of total outstanding credit from the mean, over a period of 18 months. Total credit reaches a peak in October, tapering off quickly after the harvest in December. A clear bunching of credit from May to August and October to

TABLE 4   *Deviation of total monthly outstanding credit from mean*

Month	Backward area	Advanced area
Dec 1987	–	0.48
Jan	0.85	0.45
Feb	0.59	1.10
Mar	0.59	1.40
Apr	0.52	1.22
May	1.04	0.65
Jun	0.92	1.55
Jul	0.83	–
Aug	1.10	1.67
Sep	0.97	1.64
Oct	2.43	1.43
Nov	2.36	1.52
Dec	1.68	0.67
Jan 1988	0.58	0.98
Feb	0.62	0.67
Mar	0.47	0.70
Apr	0.55	0.65
May	0.88	0.23

December, is at once apparent. The first period corresponds to the sowing–transplanting period of Aman paddy (the major paddy crop) and the second period coincides with the pre-harvest lean season.

The seasonal movement of total outstanding credit, and its concentration during the transplanting period and the pre-harvest lean period, suggests that both input costs and consumption are important motives behind these transactions. Table 5 shows that over half of the credit volume is reported to have been taken for purchase of agricultural inputs. On the other hand, consumption accounted for 13 per cent of credit.

Three important forms of informal credit were identified in our backward area: a cash-for-kind loan (DU), paddy loans (kind-for-cash or HMP) and land mortgages (LM).

TABLE 5 *Allocation of informal credit by rural households (percentages)*

Area	Farm inputs	Draught animals	Consumption	Trade	Other	Total
ADV	8.3	3.2	19.5	13.3	55.7	100
BACK	51.2	7.4	13.1	1.2	27.1	100

Dhaner Upore

DU is the most popular form of informal credit in the backward area, a high proportion of which (around 70 per cent) is invested in agriculture. The bulk of DU was found to be financed by large paddy traders in the nearby market town. The relatively large operational holdings and extensive sharecropping means that labour demands are highly concentrated in time. The credit demands so generated are largely met by traders, often lending through local agents. This suggests a flow of resources to agriculture to enable cultivation, indicating a positive role for credit in agricultural growth. There is no doubt that, in the short run, DU enables many cultivators to harvest their crop. The longer-run (dynamic) implications, however, tell a different story: technological possibilities are limited here by scarce irrigation potential; also the incidence of default is high and, combined with rigorous credit terms, causes a net outflow of resources away from agriculture.

The implicit rate of interest, and the 'loss' incurred by growers taking DU, are substantial.[12] Under 'normal' circumstances, growers are able to sustain this loss to reproduce the system. Default in terms of not being able to meet obligations on time is not rare, and leads to rescheduling of the debt and roll-over of the credit amount (Table 6). Thus the volume of paddy due at harvest will be valued at the roll-over price and then re-invested at the 7 maunds per 1000 rate until the next harvest.[13] An initial debt equivalent to 7 maunds of paddy is likely to involve repayment of 12.25 maunds, if rolled over to the next season. Of our 25 sample households, 50 per cent reported at least one roll over case. Of these, half reported asset sale and/or land mortgage to enable repayment.

The source of credit is primarily large and medium grain traders, motivated by two factors: acquisition of cheap grain, and utilization of idle finance during the lean trading period. Risk is minimized through a complex and well-organized network of agents and subordinates, so that loans are hardly ever written off. The rules are not totally rigid, with some scope for renegotiation, for example of the roll-over price.

TABLE 6   *Roll-over cases reported in the backward area*

Case	Original loan(Tk)	Roll-over price (Tk/Md)	Harvest price (Tk/Md)	Asset sale (Tk)	Land mortgage (Tk)
1	25 000	260	210	–	30 000
2	11 000	260	210	–	–
3	7 000	250	210	3 500	–
4	1 000	250	210	4 000	–
5	3 000	250	210	–	–
6	2 500	250	210	–	1 500
7	700	270	210	3 100	–
8	3 500	250	210	1 500	–
9	400	250	210	–	–
10	4 000	260	210	–	–
11	7 000	260	210	–	–
12	500	232	210	–	–
13	8 500	250	210	4 000	9 500

Notes: Tk=Taka; Md=Maund.

To recapitulate briefly, DU is the dominant credit form in our backward area, and finances both production and consumption. The terms are hard, but serviceable under 'normal' conditions. In the event of a shock, rapid de-accumulation occurs through asset sale and land mortgage.

Higher than or highest market price

HMP is a form of paddy loan which has been reported from all over Bangladesh, but is especially favoured in grain surplus area. Individual producers with surpluses, even in a deficit area, may prefer to lend in paddy than in cash. A few cases were reported from the backward area, but none was related to agricultural production. There are nevertheless implications of this type of loans for agriculture. Since this is the major form of informal credit in the advanced (surplus) area, we will postpone this discussion until the next section.

Land mortgage

Land mortgage is the last but one step removed from outright land sale, and is an attempt to forestall what often becomes inevitable. The use of the land mortgaged out rarely reverts back to the owner, as repayment is difficult. Eleven of our respondents (44 per cent) reported mortgaging out some land in the last five years and only one case of repossession was encountered.

INFORMAL CREDIT IN AN ADVANCED AREA

The seasonal dispersal of credit in the advanced area is distinctly different, with peaks occuring in February to April and June to November, presumably reflecting the different cropping pattern there. The first period precedes the HYV Boro crop, grown in the dry season under controlled irrigation, and harvested in May–June. The second period corresponds to the Aman crop which is transplanted in June–July and harvested in November–December.

Table 3 indicates that credit per household is much smaller, with informal credit playing a much more modest role. Of the various forms of informal credit, HMP is the most popular. Table 5 shows that only a small part of informal credit feeds directly into the agricultural sector, while loans for consumption and trading are much more important. On the other hand, formal credit sources were found to be important (57 per cent) and do appear to support agriculture. Out of 19 transactions, 14 were for cultivation; similarly 79 per cent of formal credit went to finance agriculture. The informal market, on the other hand, accounted for 100 per cent of the consumption loans, and 100 per cent of the loans for land purchase and house repair. Use of high-cost informal credit is clearly not preferred.

As already noted, HMP is an important form of informal credit here. Its significance for agriculture is not as a source of credit but as an avenue for surplus investment. This aspect is rarely discussed in the literature, and deserves closer examination, especially as there are distinct implications for growers' returns.

Surplus peasants will normally be in a position to play the market and will attempt to derive a higher return through grain storage and sale at the pre-harvest high price period. HMP has opened up new opportunities to the surplus cultivator that allows him a higher return compared to what the market can afford. Around 20 per cent of all households in the advanced area reported lending on HMP. Of those owning 2.5 acres or more of land (that is most of the surplus producers) 66 per cent were HMP lenders, and 78 per cent of the grain outflow from these households was accounted for by HMP. A comparison of returns to growers under HMP with returns from speculative participation in the paddy market was attempted (Table 7). The advantage of HMP to the surplus grower is clearly brought out.

TABLE 7   *Relative returns to HMP*

Respondent no.	HMP type	Returns from HMP	Speculative returns
1	HMP1	19.5	9.5
2	HMP1	20.2	4.6
3	HMP1	17.6	9.5
4	HMP2	8.9	2.1
5	HMP2	6.4	−3.6
6	HMP2	17.6	9.8
7	HMP2	20.9	9.8
8	HMP2	15.0	6.7
9	HMP2	11.7	−4.0
10	HMP2	6.4	3.5
11	HMP2	20.0	11.8
12	HMP2	24.3	6.9
13	HMP2	8.9	1.4
14	HMP2	20.8	5.4
Average (simple)		15.6	5.2

Notes: Returns from HMP are actual, while speculative returns are hypothetical figures, based on the assumption that the paddy was stocked for sale. Returns are net of marketing costs and adjusted for paddy weight loss, which is likely to have been incurred if the paddy was stored. Prices used to estimate speculative returns were expected prices, which in the event turned out to be higher than actual prices.

CONCLUDING OBSERVATIONS

In our backward area, informal credit sources were extremely important, but were found to be dominated by trading capital, which appeared to be injecting resources into agriculture. While the immediate impact of this resource flow is beneficial for production, the adverse terms result in a net outflow of resources from agriculture in the longer run. The incidence of default, roll-over and asset liquidation all point in this direction.

By contrast, informal credit sources play a much smaller role in financing agriculture in the advanced area. There is a large and expanding credit demand for HMP, particularly from the processing sector, suggesting a flow of resources out of agriculture. The overall rural sector stands to benefit in two ways: forward linkages tend to be facilitated, as the surplus finds its way into linked activities like processing (a trend that is unlikely to stop just here) and, more to the point, the return to (surplus) growers is boosted enormously, which should go some way towards raising demand and prices of paddy.

Informal credit systems are not homogeneous. While the main forms are to be found all over the country, some forms tend to be dominant in certain areas, suggesting that the role of credit will vary. In a stagnant, traditional context, informal credit appears to be regressive. Where there is potential for technological change, the particular form of credit that becomes popular could be supportive of such change, helping to promote both agriculture and non-farm activity.

NOTES

[1]Capital is broadly defined to include both physical and non-physical components (for example, education).

[2]See Rosegrant and Siamwalla (1988, p. 219).

[3]See, for example, Bardhan (1980) for an excellent review of the literature on interlocked markets. See also Rahman, (1979).

[4]This area is part of a newly emerged 'char' land which has been settled only in the last 50 years. It lacks sufficient surface water for irrigation development and the ground water is excessively saline.

[5]The historical sequence of settlement in the Chars, which is not untypical of newly emerged land arising from riverine action, established sharecropping under the control of absentee landlords. Some details of this historical background is provided in Crow and Murshid (1990).

[6]See Parthasarathy and Choudhury (1989).

[7]Ibid.

[8]See Murshid and Rahman (1991) and Maloney and Ahmed (1988) for a review of different forms.

[9]The author has led a number of 'rapid rural appraisals' in recent months and has found these types of credit arrangements to be particularly popular in backward areas of, for example, Mymensingh and Rangpur, which are known to be famine-prone.

[10]See Government of Bangladesh (1991).

[11]See Khandker (1987).

[12]Calculation of the rate of interest is made difficult by the fact that the terms of repayment (usually 7 maunds per Tk. 1000) are not very sensitive to the time-period involved, suggesting limited alternative use for credit. The implicit DU price was Tk. 143 or 70 per cent of the actual harvest price. The interest element for the season implied is 40 per cent.

[13]Roll-over price is the price at which the value of outstanding paddy debt is calculated by the lender. This is usually the price achieved by most surplus farmers or absentee landlords when they sell their paddy (Crow and Murshid, 1990).

REFERENCES

Bardhan, P.K., 1980, 'Interlocking Factor Markets and Agrarian Development: A Review of Issues', *Oxford Economic Papers*, vol.32, no. 1.

Bhalla, G.S., Alagh, Y.K. and Sharma, R.K., 1984, *Foodgrains Growth – District Wise Study*, Jawaharlal University, New Delhi (mimeo)

Crow, B. and Murshid, K.A.S., 1990, 'The Finance of Forced and Free Markets: Merchants' Capital in Bangladesh Grain Markets', paper for AEA and Union for Radical Political Economics session on *Reassessing the Role of Finance in Development*, ASSA Meetings, Washington, DC.

Government of Bangladesh (GOB), 1991, *Task Force Report on Food Policy*, Ministry of Planning, Dhaka.

Khandker, S., 1987, 'Effect of Institutional Credit on Agricultural Output, Investment, Employ-
    ment and Wage in India', *Journal of Development Studies*, October.
Maloney, C. and Ahmed, A.B.S., 1988, *Rural Savings and Credit in Bangladesh*, University Press,
    Dhaka.
Mundlak, Y., 1988, 'Capital Accumulation, the Choice of Techniques and Agricultural Output',
    in J.W. Mellor, and R. Ahmed (eds), *Agricultural Price Policies for Developing Countries*,
    Johns Hopkins Press, London and Baltimore, MD.
Murshid, K.A.S. and Rahman, A., 1991, 'Rural Informal Financial Markets in Bangladesh: An
    Overview', *BIDS Research Report*, Dhaka.
Parthasarathy, G. and Choudhury, A.U., 1989, 'Growth Performance of Cereal Production
    Since the Middle of 1970s and Regional Variations', in *Bangladesh Agricultural Sector Re-
    view*, UNDP, Dhaka.
Rahman, A., 1979, 'Using Capital and Credit Relations in Bangladesh Agriculture: Some
    Implications for Capital Formation and Capitalist Growth', *Bangladesh Development Stud-
    ies*, vol. 7.
Rosegrant, M.W. and Siamwalla, A., 1988, 'Government Credit Programs: Justification, Ben-
    efits and Costs', in J.W. Mellor and R. Ahmed (eds) *Agricultural Price Policies*.
Von Pischke, J.D., Adams, D.W. and Donald, G., 1983, *Rural Financial Markets in Developing
    Countries*, Johns Hopkins Press, Baltimore, MD.

DISCUSSION OPENING – GERSHON FEDER*

Dr Murshid's paper describes credit transactions in two areas of rural Bangladesh, attempting an assessment of the role of informal credit in agricultural
growth and technological transformation. One of the observations made is
that, in the more backward area, formal credit is not as prevalent as in the
more advanced area. While this is apparently a supply-side problem due to
political or bureaucratic decisions, an explanation would be useful to readers.
Secondly, the separation between formal and informal credit in the analysis
would only be legitimate if there was evidence that fungibility of the two
types of credit across uses (for example, between consumption and production) was not feasible. A discussion on this issue would help, as typically
credit is fungible.

The focus of the discussion is on the impact of credit on agricultural
growth; the indicator being used is whether there is a net total inflow or
outflow of resources from agriculture. The implied conclusion of the paper is
that, in the longer term, informal credit in a stagnant area is regressive and
harmful to agriculture, as it leads to the outflow of resources from the sector.
The real issue of interest is rather whether the informal credit market increases overall efficiency in the rural economy. This issue is not taken up
explicitly, although the various indications in the paper (and common sense)
suggest it does increase overall welfare. That is, it facilitates both consumption and input use during the period between planting and harvest. If, for some
reason, the activity of traders–lenders providing the funding during this interim period were to be prohibited, the most likely outcome would be a lower
level of consumption and production and, over the long term, less wealth
accumulation. The point that, when a natural calamity takes place, it triggers a
loss of wealth through roll-over, on harsher terms, is not crucial in the discus-

*World Bank, Washington, USA.

sion of the efficiency impact of informal credit. This is because the most likely outcome in the absence of such credit would be distress sales of land which are another, and more permanent, loss of wealth.

MAHABUB HOSSAIN*

Section Summary

In the previous section of the conference programme attention focused on the resilience of the small farm sector. The final section was devoted to a number of closely related questions. It began with another discussion of biotechnology within the specific context of the small farm sector, where poverty may be one of the key underlying features. From that beginning the section looked closely at the characteristic features of the sector in a number of Asian countries, and in more detail at mainland China. Discussion then turned to the key issue of credit.

The potential of modern biotechnology for improving the well-being of the small farmer and promoting sustainable agricultural development through use of varieties resistant to pests and diseases and to abiotic stresses was discussed in the papers by Iftikar Ahmed and Robert Herdt. A key point which emerged was that there are encouraging prospects of reducing farmers' dependence on agro-chemicals and raising yields in unfavourable production environments which have been by-passed by the green revolution technology. To many at the Conference that was a central issue. The green revolution has frequently been criticized for a lack of specific focus on the 'poor', and the question now is whether another 'biological revolution' will not only be production enhancing but also be directed towards more specific equity goals. Dr Ahmed's paper contains a vast amount of detailed evidence relating to the nature of biotechnological developments, drawn from numerous specific situations. In some of those, newer technology could be 'small farmer-friendly', though he was able to quote disturbing evidence of its potential to replace labour and to remove markets through technical substitution. He was criticized for not having 'answered' the question posed in his title, and for not having paid sufficient attention to the institutional arrangements needed to direct research and subsequently assist poor farmers in their efforts to obtain knowledge of, and apply, new techniques. Nevertheless, what he did amply demonstrate was the sheer complexity of the task of understanding the many interrelationships which emerge between technical developments and their detailed economic effects.

A similar theme was heard from Robert Herdt. Within his paper there was a mass of information relating to the scientific nature of biotechnology, though, before considering what might be done, Dr Herdt laid great stress on what he called the 'four realities' about the new developments. Furthermore, writing

*Bangladesh Institute of Development Studies, Bangladesh.

670

from his position within an organization which has fostered so much research, he was able to speak with authority about both the need to set priorities and the difficulty of doing so.

The papers stimulated a lively discussion concerning role of both governments and non-government organizations in less developed countries in more fully exploiting the potential created by biotechnology research than appears to have been the case with green revolution technology.

Against that background it was particularly interesting to read the first invited paper by Keijiro Otsuka, which dealt specifically with green revolution effects in regions studied in an International Rice Research Institute project, supported by the Rockefeller Foundation. The paper was concerned with the income distribution effects of adoption of the first wave of new technology across rural households and agro-ecological regions, by assessing the impact of diffusion on the operation of land and labour markets. It was suggested that the poor have gained not only from lower foodgrain prices due to increases in supply, but also through the operation of labour and tenancy markets. Labour income tends to be equalized across production environments owing to rural–rural migration, and employment for the low-income group increases as the result of substitution of family labour for leisure in relatively high-income households. Significant regional differences in land income have, however, emerged following differential adoption of improved varieties across agro-ecological zones.

Though they provoked a sharp reaction from the discussion opener the findings suggest that the major problems of poverty alleviation are not to be found in the nature of technology *per se,* but in the promotion of 'human capital formation' and in reform of agrarian structures. The reception of the paper from the floor suggested that the full results of the study will provoke a major re-opening of debate.

The accompanying invited paper by Niu Ruofeng and Chen Jiyuan provided the opportunity for a detailed examination of the small farm sector in mainland China in relation both to farm production and to rural non-farm activities. The importance of the latter increased substantially after 1979, when farmers were given relatively more freedom in the management of family resources. It appears that, when the gap in labour productivity between rural and urban areas increases with rapid development of urban manufacturing and services, the small farmer responds not only by introducing labour-saving technology in farming, but also through rapid capital formation and allocation of labour in non-farm enterprises.

The final invited paper session was devoted to the key issue of credit. One of the critical needs of the small farmer in sharing the benefits from technological process is access to agricultural credit. But the experience of the less developed countries shows that public sector institutions have had limited success in reaching the small farmer with credit and in recovering the loans for sustaining credit operations. The small farmer still depends largely on high-cost informal markets for loans. Richard Meyer and Geetha Nagarajan, presenting a paper at short notice, provided some innovative theoretical thinking concerning mechanisms to incorporate informal credit markets into the development process in a more satisfactory way.

In more applied work, the diversity of credit transactions in informal markets was described in K.A.S. Murshid's case study in Bangladesh, which indicated how informal credit transactions change with economic progress. The session was one of the few occasions at the Conference in which credit was explicitly discussed. It provoked a lively floor debate on the relationships between the formal and informal systems, on ways of improving both, and on the contributions which improvement could make in the development process. It was also suggested that there is a need for better understanding of household decision making with respect to the willingness to borrow and the choice of credit source. In short, the type of household analysis so much in evidence in the papers of Section VI, as well as that dealing with inter-linked markets, could have particular relevance to the future discussion of a central issue for agricultural economists.

Chairpersons: Mahabub Hossain, Sjarisuddin Baharsjha, Glenn Johnson.
Rapporteurs: Chaur Syan Lee, J.C. Umeh, Peter Calkins.
Floor discussion: G.T. Jones, D. Belshaw, J.C. Umeh, T. Engelhardt, R. Kada, J. Strasma, P. Dixit, P.B. Hazell, J. Groenewald, J.M. Boussard, M. Petit.

CSABA CSÁKI*
PRESIDENT ELECT

*Synoptic view****

We have come to the end of the 21st International Conference of Agricultural Economists. We have behind us eight busy, eventful days, full of rich impressions and experiences. We have been able to take part in an excellent, well-organized programme for which Professor Stanton and his team, and the Japanese organizing committèe headed by Professors Yamada, Kawano and Imamura deserve our praise and gratitude, and we thank them for their thorough and painstaking work.

The 21st Conference is another important milestone in the history of the IAAE. Our members, and all of us present here today, can be justly proud of our Association's rich traditions. At the initiative of Leonard Elmhirst an international organization of agricultural economists was founded in 1929 as one of the first in the field of economics, providing an organized frame for international contacts. In the following decades the IAAE has won general recognition and prestige in international scholarly life. Its conferences are highly important events, acknowledged all over the world. In the 1960s the accelerated development of the various disciplines and the intensification of international scientific relations created a new situation in many respects for the Association. Regional associations of agricultural economists were formed one after another. The IAAE provided substantial help for the establishment of each of them. A whole series of international scholarly societies were also set up in related disciplines, including general economics and the agricultural sciences. The 21st Conference has shown beyond doubt that the IAAE has successfully adapted to new conditions, and has been able to work together with other societies and face the situation created by the new possibilities and forms of scientific cooperation. Our Tokyo meeting sets a new record for the number of participants. The membership of IAAE is expected to be higher than ever before in the 1990–3 period and will exceed 2000 for the first time in its 52-year history.

I am pleased to report that 1413 participants registered for the 21st Conference. The number of accompanying persons exceeds 200. The genuinely

*Budapest University of Economic Sciences, Hungary.
**In this traditional overview of the 1991 Conference the papers mentioned are to be found in this volume, or in IAAE *Occasional Papers, No 6* (the Contributed Papers volume). No further references are necessary. The author received valuable support in preparing the final version of the manuscript from Glenn Johnson, Michel Petit, Bernard Stanton, George Peters, Ferenc Fekete, Mária Sebestyén, Csaba Forgács and Bruce Greenshields.

global character of our congress is very gratifying; the delegates represent 61 countries, showing that the world organization of agricultural economists is present on all continents and in all the important countries of the world. I would like to express my special thanks to the host country, and the supporting foundations and institutions, for their help, which has been indispensable in ensuring that all regions of the world are appropriately represented at our event.

It is now becoming a tradition to evaluate our meetings in the spirit of our founding president. Leonard Elmhirst offers the following recipe for the holding of a successful conference:

A Recipe:
Friendshlp, Tolerance; Gravity;
Humour; Thought; Play
Mix Well and use freely while you are here.

I believe that we have followed our founder's advice with success in Tokyo. Old friends have been able to greet each other again and new friendships have been formed in a relaxed and cheerful atmosphere. There can be no doubt that we have enjoyed ourselves here in Tokyo Shinjuku, in the colourful bustle of this ultramodern yet attractive and human corner in the Land of the Rising Sun.

We will all take home with us the memory of pleasant suppers, real Japanese sushi and tempura, and informal conversations and excursions. However, it is not the principal task of the President-elect on this occasion to evoke pleasant memories and experiences before we set out for home. Our conference is first and foremost a scholarly event in which professional problems, papers and discussions, the 'thoughts' as Elmhirst called them, constitute the most important element. I am convinced that in the final analysis this is what will determine the opinion each of us forms of the meeting.

Our scientific programme was on a large, we could confidently say grand, scale:

— in the 7 blocks of the plenary and invited sections a total of 40 lectures were given, together with the Elmhirst lecture and the Presidential Address;
— those who became speakers in the contributed paper section had to prove their worth in very stiff competition: out of the 380 abstracts submitted, only 45 authors were invited to present their papers;
— the lecturers represented 45 countries and international organizations;
— the poster section has become a popular element in the programme; the majority of over 300 posters presented the latest achievements of authors representing the younger generation in our field, displaying a very high standard; it would appear that there has been no decline in the popularity of the Discussion Groups. Many were organized in the new framework of the mini-symposium. These sessions, for the most part held late in the evening, successfully competed with the enticements of the big city. It has been evident again that there is a demand among our members for the opportunity for more informal debates.

The last lecture was concluded only a few minutes ago. The professional experiences and impressions are still very fresh. In rushing from lecture to lecture, I have made every effort to prepare for the traditional task of the President-elect, to evaluate the contents of the programme. Obviously, I have not been able to make a full and thorough study of all the lectures and related discussions, and of the activity of the discussion groups. I shall therefore attempt only to sum up the content of the core of the scientific programme. I would like to share with participants my personal opinions and first impressions. I must therefore ask you from the outset to overlook any errors, oversights or possible imbalances, unfortunately the inevitable concomitants of such rapid and subjective summings up.

SOME GENERAL REMARKS

First of all, I note a few general observations on the scientific programme as a whole:

(1) In keeping with its traditions, agricultural economics appeared basically as an applied science. The overwhelming majority of papers dealt with a concrete problem of agriculture and sought a concrete solution. The rules of the IAAE state our most important goal as follows: 'To foster the application of the science of agricultural economics in the improvement of the economic and social conditions of rural people and their associated communities.' I believe that this goal has been fully attained at the 21st Conference. The real problems of agriculture have been examined. Our programme has been imbued by the effort to achieve a more efficient agricultural economy, offering more for the farming population. A multitude of concrete proposals and ideas that could be put to direct use in practice on both macro and micro levels have been expounded. It is my conviction that the touchstone of the social sciences is usefulness, although that is not always an advantage among general economists. It is probably precisely this concrete nature and the results that can be put to direct use in real life which explain the favourable opinion which Professor W. Leontief held of agricultural economics. In a particularly interesting report of the president presented to the American Economics Association in the early 1970s, the Nobel-laureate professor summed up the achievements of the different branches of economics and reached the conclusion that especially good results have been obtained in agricultural economics. It would appear, on the basis of this meeting, that this concreteness and the effort to solve practical problems continues to be the strong side of agricultural economics. Our lecturers acted in the spirit of Professor Glenn Johnson: they undertook tasks they felt themselves capable of solving.

(2) Owing to the characteristic features of the sector, agricultural economics is a multidisciplinary science. Agricultural production is one of the most complex spheres of economic activity, since the production technology is generally accompanied by special social relations. At this meeting the

interdisciplinary character of agricultural economics has found better expression than on earlier occasions. This year special emphasis has been given, above all, to technological and ecological aspects. It was encouraging to see agricultural economists think in terms of systems theory and to observe their familiarity with environmental and techno- logical topics.

(3) The congress has highlighted a great variety of problems, areas of inves- tigation, approaches, methods of research and analysis. This has always been a characteristic of our discipline. The two decisive main trends in agricultural economics were also clearly and strikingly present on this occasion too. Again a major role was played by the Anglo-Saxon tradition based on neo-classical economics and quantitative methods, though the Continental European school has provoked attention with its precise formulation of problems and its greater sensitivity to social consequences. On this occasion, however, the difference between the two schools was far from being as striking as that perceived by Michel Petit at the 1985 conference. The demand for methodological perfection and the efforts to check hypotheses by statistical methods at all costs have become less marked. The approach that stresses methodology for its own sake has diminished. At this conference concrete problems quite clearly constituted the essence; however, most authors have used a wide variety of sophisti- cated methodology in their analysis. Greater scope has been given to research adapted to the direct service and requirements of agricultural development and at the same time authors were more pragmatic in methodology and ready to accept compromises.

(4) The descriptive and the analytical approaches continue to be clearly distinct. There were also considerable differences in the papers in the level of abstraction and the depth of analysis. It would appear that in some regions, including my own, Central Eastern Europe, the analytical approach is not sufficiently deep-rooted. There are many reasons for this. Allow me to mention what is perhaps the most important of these. The over-proliferation of ideology and politics has never favoured the social sciences, and agricultural economics is no exception. Eastern Europe and the Soviet Union are now emerging from a period in which the conditions did not always exist for an objective analysis of social and economic problems. In many respects, agricultural economics was downgraded to a descriptive science and those in power expected it to serve day-to-day political interests. The change has now begun in Eastern European and Soviet agricultural economics and we are witnessing rapid renewal. I believe that the IAAE should also undertake a bigger role in this.

(5) The complex and mutual interdependency of the problems and issues of concern to agricultural economists found strong expression at the con- gress. The fact that the phenomena of agricultural development are closely interrelated in time and space, and that ecological, biological, economic and social aspects all form part of the problems examined by agricultural economics, could be felt in the overwhelming majority of the papers. Most authors attempted to formulate conclusions based on a thorough

understanding of decisive interrelationships. The papers dealing with narrower questions, for the most part, also presented experiences that can be generalized and reached conclusions that can be applied in other situations, too. Michel Petit distinguishes macro-economic, market, technological and ecological–environmental components in agricultural economics. Looking back at our programme we can say that all of these have been included in the debates over the past week.

(6) The problem of hunger and poverty continues to be of great concern to agricultural economists. To a certain extent the problem of poverty appeared in a new context at this conference. Our Elmhirst Memorial Lecturer, V.S. Vyas, brought much new insight into the subject. More generally, authors no longer interpreted poverty as a natural disaster or simply as the product of colonial oppression and exploitation but as an unequivocal development problem. In this spirit they approached policies leading to the elimination of hunger through an understanding of the mechanism that gives rise to it. As K. Parikh observed, hunger is a phenomenon which is closely interrelated with the mechanism of the world food system. Its elimination is not simply a question of production or finance. The roots of hunger are deeply embedded in the world food system and it cannot be expected that its mechanism will spontaneously eliminate hunger. Without the combined efforts of the developed countries, and international redistribution of access to income producing resources, it is not probable that hunger will disappear from the world in the foreseeable future. Efforts to eliminate hunger and poverty must also be one of the mainsprings of the reform of agricultural policy and of national restructuring of society at large in the countries affected.

TOWARDS A SUSTAINABLE AGRICULTURE

Focal points of our programme have clearly been determined by the main themes of our Conference. The debate over sustainability has dominated our discussions, having been based on the thought-provoking Presidential Address and Elmhirst Lecture and on a number of outstanding papers.

In the development of agricultural production technologies, the availability of cheap energy sources on a mass scale made major growth in agriculture possible, especially under conditions where the food needs of growing populations rose very sharply from year to year. It was easy to ignore the possible environmental harm this was causing. However, the growing number of ever more serious cases of environmental pollution is increasingly obliging us to reflect. It is not easy to draw some kind of global balance of the harmful environmental effects of agricultural production: even for individual countries only fragmentary data are available. We were able to learn of new cases at our congress. There can be no doubt that the increased use of artificial fertilizers has a serious impact on water management systems through nutrient enrichment and pollution of water with nitrates. Phosphate fertilizers represent a special danger to the quality of water in lakes. More than 70 per cent of the nitrogen entering surface waters originates in agricultural sources. We still do not have a precise knowledge of the biological effects of herbicides and plant-protecting agents. However, two things appear to be more or less certain. One is that plant-protecting agents and herbicides accumulating in the human organism over a longer period of time can cause dangerous harm to health. The other is that the resistance of crops can be lowered through the use of these substances and there can be a change in the composition of flora and fauna, especially of insects.

It took a relatively long while from the recognition of the harmful environmental effects of modern agricultural production until the conditions ripened for a change in attitude and a shift towards a sustainable agriculture came within reach. It would appear that the first stage of debate on the interactions of agriculture and the natural environment has come to a close. No one any longer questions the significance of the environmental and ecological aspects of agricultural production. The debates over concepts and interpretations have also died down. Sustainable development which meets the needs of the present without compromising the ability of future generations to meet their own needs has become a goal in agriculture too. It is not by chance that the IAAE chose this problem as one of the main themes for the Tokyo meeting.

A sustainable agricultural development strategy strives over the long term to achieve harmony between the production activity of man in agriculture and the human and natural environment. It involves three things: a long-term perspective, equity between generations and understanding dynamic phenomena. Achieving sustainable agricultural development is a complex task which is only possible with harmonization between economic and ecological considerations. The papers presented at our congress examined the problems of achieving this goal. The human and research aspects of sustainability have been strongly emphasized, and it has been pointed out that sustainability represents a new challenge for research, education and administration.

From the conference discussions it appears that one of the most important conditions for sustainable development is the transformation of production technologies. Efforts are needed to achieve biologically more efficient solutions, more soundly based on concrete ecological endowments, and on understanding biological and natural processes. The possible paths of technological development which are now taking shape involve a great deal of uncertainty and their analysis also requires special expertise. There can be no doubt, though, that the first signs of this 'second industrial' or, if you prefer, 'biological' revolution within agriculture, are already very striking.

Present technologies can be divided into roughly three groups.

(1) The first consists of present industrialized agriculture (considered to be conventional) guided by the principle of short-term agricultural rationality which maximizes profit and necessarily leaves out of account the external impacts of its operation and those that will be felt by future generations.
(2) The second comprises organic agricultural technologies which stress the preservation of 'wholeness' of natural processes and strive to respect the ecosystems found in nature as well as the 'wholeness' of social structures and communities.

(3) The third group is made up of applications of the most advanced technology, involving strict control of biological systems with the aim of achieving precise predictability and uniform quality. In their logic these methods essentially follow the present industrial methods but they use the latest findings of biological and genetic research for their implementation.

Among the future technological alternatives the congress discussed mainly the economic and social impact of wider use of the methods of biotechnology. It appears that they have much to offer both for developed and under-developed countries, though the potential remains far from fully explored! Other directions of technological development, including the question of organic or biological agriculture featured only in the programme of the Poster Section.

WORLD AGRICULTURAL TRADE IN THE PROCESS OF CHANGE

International cooperation has been the other main theme of our conference. The present system of international agricultural trade has long been the subject of criticism and debate in numerous international and scholarly forums. Exchange of views on this topic continued within the framework of our discussions.

The basis of the world food economy system today is a market mechanism which is being increasingly distorted by agricultural policy interventions applied independently of each other in different countries or groups of countries. The goals and instruments of national agricultural policy differ very widely in their nature and methods. However, intervention in the agricultural economy is often characterized by protectionism.

Naturally, different countries do not have the same possibilities. The policy of the developed countries is generally effective and usually offensive. The production stimulated by unrealistically high producers' and consumers' prices in the developed countries, coupled with the protection of the domestic market, results in the accumulation of vast surpluses. In the final analysis, these surpluses flow out onto the international markets at prices set unrealistically low, often covering only the storage costs. Since there is a limit to absorption capacity, this becomes one of the most important factors contributing to extremely depressed world agricultural market prices. The export of the vast, state-subsidized food surpluses of the developed countries undermines the market positions of those, otherwise efficiently producing, countries which are not willing or able to take part in price competition financed from the state budget. The world market for agricultural produce has increasingly become a competition of state budgets rather than a competition of agricultural producers. There are both losers and winners in this situation. The smaller, poorer countries are obviously on the losing side, while the rich importers are on the winning side. While it is not easy to find one's bearings in this situation, one thing that is quite clear is that, with the present practice, mankind is renouncing a very substantial part of the comparative advantages inherent in the rational regional location of agricultural production.

The protection of national agricultural markets and the support of agricultural producers created an especially serious situation on agricultural world markets in the 1980s. As the conference has shown, the main characteristics of the present contradictory situation can be summed up as follows:

— The world agricultural market has become destabilized and world market prices continue to decline.
— The support of agricultural producers' and export prices, through protectionism consumes vast sums in the countries concerned. In 1990, the total costs of agricultural support were US$300 billion in the 24 OECD countries (A.de Zeeuw).
— The situation of the traditional agricultural exporting countries is becoming increasingly serious. Their rapid loss of position can be observed to be a consequence of price dumping by the protectionist countries which is achieved with the help of export subsidies.
— Agricultural trade is increasingly becoming the source of conflict between countries, an issue stressed by de Zeeuw and analysed very fully in contributed papers. The agricultural trade dispute and the subsidies war have been going on for years between the USA and the European Economic Community. Relations between the USA and Japan are also marked by sharp contradictions in the area of agricultural trade (for example, over Japanese beef import policy). The Cairns group has been set up to represent the joint interests of the agricultural exporting countries.
— The support of the agricultural sector and export subsidies are also the source of growing tensions within the countries concerned. The contradiction within the European Community in connection with the Common Agricultural Policy is well known, but the agricultural budget of the US administration is also a constant subject of debate.

This situation is increasingly untenable and demand is increasing for fundamental changes both in the international pattern of trade in farm products and in agricultural policies of individual countries. The idea and demand for liberalization of trade in agricultural produce have long been on the agenda. Many researchers have studied the possible consequences and effects of agricultural liberalization. Some of their findings were presented here. We are all closely watching the Uruguay Round of talks and we listened with interest to de Zeeuw and Warley and McClatchy. On the whole, a modest optimism emerged at the conference. I sincerely hope that these expectations will be confirmed in the near future, with successful conclusion of the GATT negotiations.

REFORM IN NATIONAL AGRICULTURAL POLICIES

The conference has made it clear once again that the reform in national agricultural policies is a major precondition for the restructuring of international agricultural trade. Today policy reform is in progress in practically all countries of the world. Of course, as the papers also suggested, the motivations for, and depth of, the changes differ from country to country and region to region.

Changes in agricultural policies of developed countries

A number of papers dealt with the agricultural policy of the developed countries, above all of the EC and the USA. They showed that the most important lesson from the crisis in international agricultural markets in the 1980s is precisely that only changes affecting the majority of countries together can bring a solution. The key thus lies in the coordination of national agricultural policies. The experiences of recent years also suggest that international trade policy efforts will be ineffective if they are not accompanied by concerted reform of national policies.

Decoupling, a new concept that appeared recently in agricultural economics, indicates the new direction in the transformation of national agricultural policies in developed countries. The essence is the shift to trade-neutral policy measures; in other words, preserving the parity of agricultural incomes without simultaneous provision of incentives and support for production. A variety of different interpretations can still be encountered regarding the details of this concept. However, it is obvious that the developed countries are seeking new instruments to replace price supports, export subsidies, guaranteed prices, trade quotas and tariffs. A debate is being conducted on solutions that will ensure the attainment of the social, income and environmental protection goals of the present agricultural policies without exerting an unfavourable influence on the world market for agricultural produce or preventing the assertion of comparative advantages in that market. Several speakers pointed out that agricultural policy reforms directed at trade are often linked with measures to conserve the natural environment or are implemented in a form that also fosters environmental protection.

Structural reforms in the developing countries

Many countries in the developing world wish to pursue a new general development strategy. Comprehensive agricultural policy reforms are also being implemented as a part of this. The congress had the opportunity to learn about the experiences of a wide range of these reforms (in Africa, Chile, Morocco, Cameroon, Brazil and Ethiopia for example).

The majority of developing countries are characterized by a price and support policy that is unfavourable to agriculture. In the African countries examined by A. Valdés in his study, for example, the direct support rate for agriculture is minus 20 per cent, while the non-agricultural sector enjoys the benefits of positive support of 36 per cent. Price interventions unfavourable to agriculture lowered the agricultural GDP by around 28 per cent, while at the same time the positive effect of price subsidies provided for agricultural inputs was equivalent to 8 per cent of agricultural GDP. From papers presented, the frame of agricultural policy reform in developing countries consists of four main elements:

(1) policy reforms to improve the economic environment for agriculture;
(2) strengthening the public sector to support technology development and transfer, education in rural areas, and infrastructure projects supportive of agriculture;
(3) encouraging opportunities for increased participation in the economy by historically disadvantaged small farmers and landless workers; and
(4) promoting improved natural resources management.

Although, as we have heard from our speakers, some reforms introduced after suitable preparation have generally brought favourable results, opinions continue to differ sharply on some specific issues. Views are particularly divided over the role of the free market and the desirable manner and extent of state intervention. The criticisms expressed in connection with the operation of the market and the references to the extreme impacts of the market are based on unequivocal, concrete facts. U. Koester rightly notes that: 'There is much support for the hypothesis that markets are not efficient in directing agricultural development.' At the same time, it seems to me, in the light of successful reforms and especially the experiences of the Eastern European countries, that negative observations concerning the free market are often over-simplified. It is reasonable to speak of market failures only in the context of existing and operating markets.

What is needed first of all in the countries concerned is markets and it is only after they have been created that it is worth debating the necessary or desirable extent of state market interventions. The reservations or negative experiences of agricultural economists in the developed countries concerning the free market or the market in general arise from centuries of developed market economy practice. However, the history of the former socialist countries is a warning that nothing causes greater harm to the development of agriculture than the lack of a working market. Therefore it is not helpful to try to alarm countries striving to create an elementary market, right from the start, with the possible adverse features of a well-established, operating market.

Restructuring of agriculture in the socialist countries

The first economic reforms were carried out in the so-called socialist countries as early as the late 1960s. At the end of the 1970s, Chinese agricultural reform aroused attention. As our discussions suggested, the changes that have occurred in China still attract considerable interest among us. The events of the second half of 1989 and later developments turned the spotlight on Eastern Europe and the Soviet states.

The first wave of agricultural reforms in the socialist countries sought ways of making the system more efficient within the framework of the socialist economy. The aim of the Chinese agricultural reform and the earlier Eastern European reform attempts was to make the system more efficient and the changes did not affect the basic elements of socialism. The political ferment that arose in Eastern Europe from economic difficulties brought a spectacular change of direction. Declining economic performance, the lag behind the mainstream of world development and the lack of success of the attempts at economic and political reform questioned the very foundations of the post-Stalinist economic and political system. Changes now under way in most of these countries have gone beyond the bounds of reform in the traditional sense and quite clearly point in the direction of the total rejection of the socialist-communist economic and political system.

In examining the restructuring of agriculture in Eastern Europe in the early 1990s, we must therefore clearly understand that the changes now under way constitute a process reaching far deeper than the reforms of earlier years. What has begun to emerge is a new agricultural structure based on private ownership and the market economy. This restructuring has already advanced beyond the first steps in the countries of Central Europe and similar changes can be expected in the Soviet Union. The main characteristics of the new system, and the critical points for the future, can only be traced in general outline as yet. Nevertheless, it is already quite clear that the most important tasks for the creation of the new agricultural structure are the following:

— restructuring of ownership of the means of production, including the land, and creation of marketable landed property;
— creation of a farming structure based on private ownership;
— creation of a mechanism of market regulation;
— establishment of the physical, organizational and institutional conditions for the operation of the market;
— shaping a new type of role for the government in agriculture; and
— creation of the legislative framework required for all this.

One of the biggest restructurings of agriculture in history is now taking place in Eastern Europe and the Soviet Union. This transformation involves far more than replacing central planning with market regulation. It is an unprecedented attempt to create a democratic market economy by peaceful means from the autocratic one-party system and the planned economy. It is regrettable that the agricultural aspects of this process were not discussed in their full complexity, since events of historical importance taking place during the days of the conference prevented our Soviet speakers from delivering their papers in person. One is to be included in the Proceedings volume. We are grateful to Dr Karen Brooks and Professors Fekete, Forgacs and Hunek for enabling us to have a very informative discussion of current developments.

FAMILY FARMS AND SMALL-SCALE PRODUCTION

The micro-sphere came to the fore again and was given a new profile in our discussions. It was particularly welcome since such topics can too easily become pushed into the background, either consciously or simply for a lack of interest. It is an extremely gratifying development and one to the credit of those who planned the programme that the spotlight has been turned once again onto the problems of farms, and above all onto the present condition and future of family farms. It would appear that these are universal issues which are of concern alike to agricultural economists in the developed, the developing and the former socialist countries now undergoing transformation. It is not farm management that has been considered at our congress! It was the micro-economy which returned to the programme. A certain aristocratism, and an excessive macro-orientation, has often caused this area to be underservedly neglected. On this occasion we have had a rich selection of papers relating to part-time and small-scale farming. It was only in connection with cooperatives that I felt something to be missing. We devoted relatively little attention to them. The seminar to be held in Israel in March 1992 offers an excellent opportunity for detailed examination of this subject. I am convinced that the experiences of agricultural cooperatives are of exceptional importance for all countries. The transformation of the Eastern European kolkhoz-type cooperatives is perhaps one of the most exciting issues of micro-level agricultural economics at the present time. The paper by Brooks and Braverman provided an interesting overview, suggesting that the change might require more time and effort than is often envisaged.

GLOBAL AGRICULTURAL DILEMMAS
ON THE THRESHOLD OF THE TWENTY-FIRST CENTURY

The IAAE congresses are always a good opportunity for an overview of the situation of the world food economy and the identification of its critical points for the future. Here in Tokyo the papers presented by N. Alexandratos and K. Parikh set the keynote for the discussion on this theme, although in the final analysis practically every paper contributed in some way to shaping the global picture.

The progress that has occurred in recent developments in world food production is one of the greatest achievements of mankind. Only a century ago, hunger and malnutrition were found in practically all areas of the world. The situation has now improved substantially. The per capita daily calorie consumption in the developing countries now stands at 2440 calories, in contrast with the 1950 of the early 1960s. And this progress has been achieved in a period when the population of the developing countries has grown by 1.6 billion; that is, by 77 per cent (N. Alexandratos data). As a result of this change, according to estimates by the World Bank in 1990, the proportion of those who suffer hunger and malnutrition has significantly declined.

However, the discussion at this congress indicates that this global picture inspiring optimism for the future hides very serious contradictions:

(1) Hunger has not disappeared from the earth. A considerable part of mankind still does not receive sufficient nourishment. According to FAO estimates, in 1986–7 the daily calorie consumption of 54 per cent of the population of the developing countries (not including China), a total of 1471 million people, was less than 2200 calories per day. The fact that

the daily calorie consumption of such a substantial part of the population of the developing countries remains closer to the 2000 calorie per person per day level than to 3000 calories remains a cause for concern. In a considerable part of the world, agricultural production is having increasing difficulty in keeping pace with the growth of the population and the growth rate of agricultural production is steadily declining.

(2) At the same time, vast food surpluses have accumulated in many countries and a number of countries in the world, including also developing countries, are struggling to cope with the problems of over-production of food and how to sell the surpluses. The over-production faced by the EC and the USA are well known. However, it is also worth mentioning that foodgrain stocks of 21 million tons accumulated in India in 1991, while at the same time there was no appreciable decline in the number of those suffering from hunger (K. Parikh).

(3) Protectionism has become the general characteristic of agricultural trade. The equilibrium of the world food market has been lastingly upset. Supply regularly exceeds the actual demand and as a result there is a steady decline in the relative world market prices for agricultural produce.

(4) Food has to be produced for an increasing number of people per unit of land. The area of arable land per capita fell rapidly, from 0.24 to 0.15 hectares, between 1950 and 1986. The forecast for the year 2000 is 0.12 hectares per capita. The stagnation in the area of land under cultivation is accompanied by a relatively rapid deterioration in the quality of the soil.

(5) In recent years the development of agricultural production has been characterized by the replacement on an ever greater scale of human labour and natural biological processes by the use of energy of a non-agricultural origin. The utilization of energy and artificial fertilizers and of the different plant-protecting agents and herbicides is now increasingly decisive in the growth of agricultural production. All this is undoubtedly further aggravating the contradictions between agriculture and the natural environment, heightening the ecological and biological concerns regarding modern agricultural production.

All these phenomena clearly show the symptoms of the sickness of the global food economy. Reflection on all this raises a host of questions:

— If the development is as dynamic and spectacular as a number of speakers at this congress also noted, how is it possible that, despite this, hunger has not disappeared, since the quantities required to eliminate undernourishment appear to be marginal compared to the scale of the development?

— The rapid development of production could easily have provided the possibility of eliminating hunger. The fact that signs of over-production – large stocks and relatively low prices – appear simultaneously with hunger gives cause for thought. If there is lasting over-production, how can there be hunger at the same time?

— Not only is there over-production, there is also growing, or at least stubborn, protectionism. How is it possible that, although attempts are repeated practically every year to dispose of the almost unsaleable, vast stocks of food (such as in the countries of the European Common Market), protection has nevertheless not declined, not to mention disappeared?

— If such a combined pressure arising from the growth of the world population and over-production weighs on the arable land of the world, why is it that over-production does not stop? Why do we waste the resources that will have to play such a big role with the anticipated population growth? What is the secret of the vicious circle of over-production, protectionism and hunger?

— If the environmental problems arising from the energy shortage and the present technologies are as serious as a number of speakers have shown them to be, why has all this not acted more strongly in the direction of technological change?

— If the present technologies already clearly reveal the seeds of present and future conflict, why are the new technologies about which we have heard so much here in Tokyo being introduced so sporadically, slowly and with such delay?

We have received a satisfactory reply to only some of these questions here at the conference. Answering the majority of them remains a further task for agricultural economists and not least of all for policy makers.

CONCLUSIONS

I have reached the end of my review. I am afraid that I risk exhausting your patience, although there are still many other issues which could be raised. I have made no mention of research, consultancy, the achievements of the international research institutes, the experiences of rural credit, the effectiveness of product marketing and efficiency of production systems. These were all valuable elements in our scientific programme of the congress but, unfortunately, for lack of time, I have not been able to refer to them.

In conclusion, I would like to stress the words of Dennis Britton, President in 1981, who said, 'the IAAE is neither a pressure group nor an action group. We have not come ...to pass resolutions, nor to organize some dramatic piece of world-wide collective activity.' The agricultural economists of the world once again came together, here in Tokyo, as independent, autonomous experts, to debate the problems of world agriculture. As our conference has again demonstrated, none of the many possible forms of scholarly contact can replace personal face-to-face encounters and the thought-provoking experience of participation in scholarly debate.

We have debated major themes, and more detailed questions, of world agriculture. There can be no doubt that as we approach the twenty-first century, agriculture throughout the world faces new challenges: the problem of hunger must be solved once and for all; a new, globally more harmonious, order must be established for national agricultural policies and for international trade of agricultural products; and real, substantive, progress must be made in the direction of sustainable agriculture,

These are very big tasks for the agricultural economics profession and for our Association. Their accomplishment requires even more mutual understanding than in the past, with international cooperation and a world less marked by tension than in the past. At the end of the 1934 conference, George Warren said much in one final sentence 'I think this has been an excellent conference .... as a stimulus to science and to friendship – two things badly needed in this sick world.'

I think it would be difficult to add anything to that. Ladies and gentlemen! Sayonara!

INDEX

Abbot, P. C., 62
Abraham, A., 648
Achutuni, R., 451
Adams, D. W., 644, 657
Adams, R. M.
  on climate change, 450–57, 476
adaptive management strategy, 399–400
Adelman, I., 51, 52
adjustment policy, 306–8
  agricultural development and, 211–22,
      225–6, 234–7
  Cameroonian case study of, 265–78,
      280
  case study of E. European
      decollectivization and, 254–62,
      263–4
  Chilean case study of, 242–51, 252–3
  Moroccan case study of, 281–9, 290–
      91
  structural model of, 232–7
Aereboe, F., 511
African Development Bank, 284
age structure of population, 63
aggregate measure of support (AMS), 131
agrarian structure see labour market;
      land
agricultural markets, 91
  analysis of efficiency of, 66–80, 115
  internationalization of, 63–4
  model of effects of liberalization of,
      136–9
  Policy Analysis Matrix for evaluation
      of, 166–80
  prospects for GATT agreement on,
      119–28, 129–39, 148–50
Ahearn, M. C., 536
  on resource allocation by households,
      515–18
Ahluwalia, M. S., 102

Ahmad, Y. J., 394
Ahmed, I., 319
  on biotechnology and poverty, 573–
      86, 670
Ahmed, R., 72
air pollution, 462–4
  costs of control of, 469–70
  effects of, 465–9, 472–3
Alagh, Y. K., 657
Alauddin, M., 104, 110
Alcamo, J., 463, 465, 470
Alexandratos, N.
  on world agriculture in the next
      century, 44–53, 115
Allue, M., 70, 245
Alston, J. M., 198
Altieri, M. A., 406
Alves, E.
  on government and agricultural
      development, 197–207, 307–8
Andean Development Bank, 360
Anderson, D., 543, 545
Anderson, J. R., 106, 590
Anderson, K., 48, 50, 138–9
anti-sense technology, 314–15
Argentina, 42
  plant breeders' rights legislation in,
      330, 331–2, 333, 334, 335, 339
Aribisala, T. S. B., 481
Armah, P. W., 84
Armbruster, W. J.
  on efficiency of agricultural markets,
      77–80
Armstrong, C., 431
Arnason, R., 432
Arrow, K. J., 67
Asian Development Bank, 12
Australia, 137, 347, 378
autarkic policies, 69

Bajracharya, D., 415
balance of payments problems, 208, 211
Balassa, B., 217
Ballabh, V.
    on management of forest resources,
        435–44, 475
Bangladesh, 10, 38, 544
    Green Revolution in, 608, 612
    informal credit market in, 651, 657–
        67, 668–9
banking system *see* credit market
Banskota, M., 412
Barahona, P., 246
Barba, R. C., 580, 583
Bardhan, P., 256
Barichello, R. R.
    on concepts and strategies for
        sustainable development, 402–4,
        474
Barkai, H., 522
Barker, R., 104, 110, 363, 379, 573, 596,
        616
    on biotechnology, 328–9
Barton, J. H., 363, 364
Batie, S. S., 18, 19, 452
    on concepts and strategies for
        sustainable development, 391–
        401, 474
Bator, F., 94
Beachy, R. N., 364
Becker, G., 503, 512, 515
Beghin, J.
    on private versus public sector in
        biotechnology, 383–5
Belgium, 210
Bell, C., 542
Bellman, K., 467
Ben Porath, Y., 520
Berkes, F., 437, 439
Bertram, R. B., 362
Bhaduri, A., 12
Bhalla, G. S., 657
Binswanger, H. P., 82, 206, 297, 590,
        595
biotechnology, 311–13, 386–7
    costs of, 600–602
    current techniques of, 313–17, 377–9,
        597–600
    employment effects of, 577–8, 583–4
    environmental safety in, 602–3
    future of, 317
    impact on developed countries of,
        321–7

    poverty alleviation by, 573–86, 589–
        91, 592–604, 605–6
    private versus public sector in, 371–
        81, 383–5
    relevance to developing countries of,
        317–21
    research scientist's perspective on,
        328–9
    role of CGIAR in, 356–65, 368–70
Bjørndal, T.
    on management of fisheries, 422–33,
        475
Blandford, D., 137
    on agricultural and trade policy
        reform, 148–50
Boadway, R. W., 344
Bonnen, J. T.
    on financing agricultural research,
        343–52
Boomgard, J. J., 652
Botswana, 49
Bouanani, M.
    on adjustment policy in Morocco,
        281–9, 306
Boussard, J. M., 531
    on private versus public sector in
        agriculture, 90–91
bovine somatotropin, 581–2
Bradnock, W. T.
    on plant breeders' rights, 340–42
Brandao, A. S., 236, 306–8
Brander, J. A., 197
Braverman, A.
    on decollectivization, 254–62, 308
Brazil, 14, 42, 137, 295, 580, 583
    inflation and government debt in,
        227–37, 238–40
Breton, A., 344
Brewster, J. M., 520
Britton, D., 686
Bromley, D. W., 436, 437
Brooks, K.
    on decollectivization, 254–62, 308
Brown, L. R. 481
Brundtland Report, 18, 391, 405
budget deficits
    adjustment policy and, 219–21
    inflation and, 230–32
buffer stocks, 39–40
Bulgaria
    agricultural reform in, 256–7, 259,
        261
Bunders, F. G., 582

bureaucratic behaviour, 198
Burfisher, M., 226
Burkett, P., 651
Burkino Faso, 226
Busch, L., 371, 373–4
Buttel, F., 342
Byerlee, D., 545

Cairns Group, 121, 126, 130–31, 134, 136
Calkins, P. H., 408
Cameroon
    effects of trade liberalization on, 185–8, 190
    price controls in, 270, 276–7
    state intervention in, 266–9, 270, 271–2, 275–6
    structural adjustment policy in, 265–6, 273–8, 280
    trade policy in, 269–70, 272–3, 276
Canada, 123, 137, 342
capital, financial *see* credit market
capital, human, 18–25, 62, 202
capital formation in households, 558–67
Capule, C., 596
Carter, C. A., 343
Carvalho, J. L., 236
Cassava Biotechnology Network, 358–9
Cautley, E., 538
Cavallo, D., 297
Centro Internacional de Agricultura Tropical
    (CIAT), 358–9, 360
Centro Internacional de la Papa (CIP), 359
Cernea, M. M., 436, 437
Chadwick, M. J., 465
Chalfant, J. A., 522
Chambers, R., 408
Chanda, G. K., 545
Chandavarkar, A. G., 645, 650
Chandrashekar, H.
    on Green Revolution in India, 102–12, 116
Chen, H. H.
    on financing agricultural research, 353–4
Chen, J., 23
    on small farmers in China, 619–37, 671
Chile, 70, 210
    agricultural reforms in, 241–51, 252–3

plant breeders' rights legislation in, 330, 331, 333, 334, 336, 338, 339
China, 13, 409, 619
    agricultural reform in, 23, 137, 625–32
    biotechnology in, 363, 575–6, 577, 578, 579
    history of agriculture in, 620–25
    small farmers in, 620, 622, 624, 625–37, 641–3
Choe, Y. B.
    on sustainability of Japanese agriculture, 567
Choksi, A. M., 71
Christy, F. T., 422
Chuta, E., 542, 543
Clark, C. E., 427
climate
    changes in, 450–57, 458–61
    yield instability and, 105–6
Coase, R., 509
Cobb, J. B., 95
Cobia, D., 531
Cochrane, W. W., 594
coevolution framework, 97–9
Coeymans, J. E., 246, 297
Coffman, W. R., 601
Cohen, J. I., 357, 358, 359
Colby, M. E., 92, 391, 394
Coleman, D. R., 545
collective farming, 13, 154–6, 158, 622–5
    compared with family farms, 520–29, 530–32
    transition to market economy and, 74–5, 156–7, 160–61, 254–62, 263–4
Collinson, M. P.
    on the role of CGIAR, 355–66
Colman, D., 115–16, 192–3
command economies
    transition to market economy by, 72–5, 79–80, 152–6, 159–65, 682–3
    collective farming and, 74–5, 156–7, 160–61, 254–62, 263–4
common resource management
    fisheries, 422–33
    forests, 435–44, 445–8
communications and transport, 204, 407–8, 413, 415–16
communist system *see* command economies
competition, imperfect, 181–91, 192
Comte, A., 92, 96

Conable, B. B., 397
conceptual pluralism, 96–7
Condon, T., 181
conference overview, 673–87
constant margin price support, 140
Consultative Group on International
    Agricultural Research (CGIAR),
    328, 355–6
  biotechnology in, 356–65, 368–70
  future of, 365–6
consumers
  needs of, 63, 325
  subsidy to, 35–7
Contini, E.
  on government and agricultural
    development, 197–207, 307–8
Convention on the Law of the Sea, 429
cooperative farms *see* collective farms
Corbo, V., 300
Corden, W. M., 66
Corn Laws, 199
Costa Rica, 437
Côte d'Ivoire, 219, 226, 295
Crawford, E., 225
credit market, 204–6, 236, 526–7
  informal, 644–53, 655–6
    Bangladesh case study of, 657–67,
      668–9
Crocker, T. D., 454, 455–6
Crosson, R. R., 452
Crow, B., 658
Csáki, C.
  conference overview by, 673–87
Cuba, 581
Czechoslovakia, agricultural reform in,
    257, 259–60, 261–2

Dafu, Y., 409
Daly, H. E., 95, 393
Dareshwar, A., 51
Dasgupta, P. S., 429
Datta, S., 517
David, C. C., 596, 608, 610, 612
  on informal finance, 655–6
Davis, J. G., 343, 344, 347
debt, government
  inflation and, 227–37, 238–40
de Castro, J. P. R., 343
Decker, W. L., 451
DeDatta, S. K., 355
de Faro, C.
  on government and agricultural
    development, 197–207, 307–8

deficiency payment systems, 68
de Janvry, A., 48, 49, 52
demand aspect of sustainability, 411–13
deMelo, J., 181
demography *see* population
DeRose, D., 71
Desai, G. M., 13
Deshpande, R. S., 102, 105, 106, 107
Deshpande, S. H., 38
Dev, M. S., 38
Devarajan, S., 181, 182, 185–8, 190
de Zeeuw, A., 680
  on GATT agricultural agreement,
    119–25, 192
Dias, G. L. S.
  on inflation and government debt in
    Brazil, 227–37, 307
diet *see* nutrition
disease resistance, biotechnology and,
    313–14
distribution of income, Green Revolu-
    tion and, 613–15
distributional problems, sustainability
    and, 22
Dodds, J. H., 359, 360, 364, 365
Domenech, R., 297
Donald, J. W. G., 657
Doyle, J. J., 359, 577
Dryzek, J., 399
Du, Y., 23
dual pricing policies, 37, 41
Dudek, D. J., 452
Dutch disease, 219

Easterling, W., 452
Eastmond, A., 578, 580, 581, 583, 586
Eckert, R. D., 425
ecological science, 392–5
economic growth *see* growth
education, agricultural, 19–20, 359–60
  sustainability concepts in, 21–4
education policies, 201–3
Edwards, G. W., 343, 347
efficiency of markets, analysis of, 66–
    80, 115
egalitarianism, 203
Eggertsson, T., 648, 649
Egypt, 49
Eicher, C. K., 349, 481, 545
Eisner, R., 503, 512
Elgin, M., 206
Elmhirst, L. K., 3, 673, 674
El-Osta, H., 536

Emel'ianov, A.
  on Soviet agricultural reform, 152–7
employment *see* labour market
environmental problems, 4, 64, 103,
    107, 208, 209
  air pollution, 462–70, 472–3
  attitudes to, 394–5, 396, 397
  biotechnology and, 602–3
  climate change, 450–57, 458–61
  fishery management and, 422–33
  forestry management and, 435–44,
    445–8
  poverty and, 9–11
  sustainability concepts and, 21, 22–4
Epraums, J. J., 450
equity issues
  biotechnology and, 319–20
  intergenerational, 92–6
erosion, 421
Esguerra, E. F., 647
ethical issues
  biotechnology and, 324–5
  plant breeders' rights and, 341–2
European Community
  agricultural policy in, 68, 69, 295
    prospects for reform of, 120–23,
      127, 130–31, 133, 135, 137, 140
European Investment Bank, 284
Evans, P., 230
Evenson, R. E., 335, 343, 347, 349, 377,
    380
  on role of CGIAR, 368–70
exchange rates, 68, 208–9
  adjustment policies and, 215–17, 225
  trade and, 69–70, 74
exclusive economic zones (EEZS), 425
export crops, 218–19
export licensing, 236
export subsidies, 132
Extended Fisheries Jurisdiction (EFJ),
    429

Faber, M., 396
Falcon, W. P., 596
family farming *see* households
famines, 38–40
Fanelli, J. M., 237
Fardoust, S., 51
Farmer, G., 450
Fast, I., 508
Feder, G.
  on informal credit market in
    Bangladesh, 668–9

Feeny, D., 437
Fekete, F.
  on Hungarian agricultural reform,
    158–64
Ferroni, M.
  on biotechnology and poverty, 589–91
fertilizer, 106, 209, 318, 577, 580
  subsidies for, 35, 41
finance *see* credit market
Finkelshtain, I., 522
Finland, 473
Fishcer, G., 31, 451, 452
Fishel, W. L., 343
fisheries, 422–5
  bio-economic modelling of, 425–7
  inefficiencies in exploitation of, 427–
    8
  management of, 428–9
  migration issues and, 430–33
Fishlow, A., 198
fixed external reference price (FERP),
    140, 142–3
Flaaten, O., 431
Flanders, M. J., 51
Floro, S. L., 648
Fontaine, A., 242
Food and Agricultural Organization
    (FAO), 9, 29, 30, 48, 49, 51, 446,
    594
food security, 115, 295
  biotechnology and, 319
  buffer stocks for, 39–40
  infrastructure development and, 37–8,
    41
  insurance scheme for, 40
  poverty and, 29–30, 41
  price policy options for, 30–33, 41–2
  subsidies for, 34–7, 41
  trade policy options for, 33–4, 39, 41–
    2, 70, 71
forestry, 412, 414–15, 420, 453
  effects of air pollution on, 465–9, 473
  management of, 435–44, 445–8
Forgács, C.
  on Hungarian agricultural reform,
    158–64
Fraley, R. T., 356
Franc zone, 217, 275
France
  plant breeders' rights legislation in,
    330–31, 332, 333, 335
fraud, 68
Freebairn, J. W., 343, 347

Frenkel, R., 237
Frisvold, G., 209
Fujimoto, A.
  on small farmers in China, 641–3
future generations, rights of, 92–6
future outlook for world agriculture, 44–
  53, 62–5, 115

Gadgil, M., 439
game theory, 197, 430–31, 649
Gardner, B., 137
General Agreement on Tariffs and Trade
  (GATT), 67, 341, 380
  prospects for agricultural agreement
    in, 119–28, 129–39, 144, 148–50,
    181, 192
general equilibrium model, effects of
  trade liberalization in, 181–91
generations, future, rights of, 92–6
genetic engineering *see* biotechnology
German Democratic Republic, 74, 75
Germany, 68, 584
Ghana, 84, 217, 295
Gilliland, M., 577
Gittinger, J. P., 494
Gnaegy, S., 590
Goeller, H. E., 23
Goldin, I., 53, 137, 138, 236
Goldstein, M., 300
government (state, public sector)
  adjustment policy and, 211–22
  agricultural development role of, 197–
    207, 208–10
  agricultural research and, 344–7, 349
  agriculture and, 81, 86–91
    see also collective farming
  biotechnology and, 320–21, 371–81,
    383–5
  cost of subsidies to, 35
  debt of, and inflation, 227–37, 238–40
  informal finance market and, 650
  intervention in markets by, 66–76, 77–
    80
  legal structures and, 85–6, 204
  policy evaluation for, 166–80
  poverty alleviation and, 7–8, 15
  price intervantions by, 292, 293–6
  trade policy guidelines for, 298–303
Graham, D. H., 230, 644
Grameen Bank, 651, 652
Gramlich, E. M., 346
Green, J., 376
Green Revolution (1960s), 102–3, 110–

  14, 116, 578, 579, 607–10, 617–18
  agrarian structure and, 610–13
  cost problems of, 107–8, 113–14
  income distribution and, 613–15
  instability of, 103–7, 113
  watershed development programmes
    and, 108, 110
Greene, J., 71
greenhouse effect, 450–57, 458–61
Griffin, K., 546
Griliches, Z., 521
Grotius, H., 433
group activities, 84–5
  see also collective farming
growth, economic, 391
  agricultural development and, 38, 44,
    49, 51–2
  poverty alleviation and, 7
  sustainability and, 21–2, 23, 392–5
Guha, R., 439
Guillaumont, P.
  on adjustment policy, 211–22
Guinea, 217
Guiterrez, M. B., 334
Gunter, L., 505
Guo, W., 619, 620, 621, 632, 635

Haggblade, S.
  on rural non-farm economy, 542–54,
    569
Hahn, F., 91
Hallam, J. A.
  on climate change, 458–61
Halos, S. C., 578, 584
Hammer, J., 548
Hansen, J., 450
Harford, J., 346, 349
Hartmans, E. H., 481
Hassan, R. M.
  on agricultural trade and pricing
    policies, 303–5
  on structural adjustment in Morocoo,
    290–91
Hathaway, D. E., 135
Hayami, Y., 12, 49, 50, 295, 324, 520,
  593, 594, 608
Hayek, F. A., 76
Hazell, P. B. R., 83, 102, 105, 106, 542,
  543, 544, 548
Heady, E. O., 531
Heal, G. M., 429
Healy, R., 443

Hedley, D. D.
  on plant breeders' rights, 340–42
Helpman, E., 181, 197
Herdt, R. W., 355, 607
  on biotechnology and poverty, 592–604, 670
Hernández-Estrada, J.
  on government and agricultural development, 208–10
Hertel, T. W.
  on effects of trade liberalization, 181–91, 192
Heywood, P., 597
high yielding varieties (HYV), 13, 15, 106, 657
Hillman, J. S., 145
Hirschman, A. O., 230
Hirway, I., 8
Hobbelink, H., 584
Hoff, K., 647
Hojman, E., 250
Honma, M., 295
Hopper, W. D., 355
Horne, M. E., 357, 358, 359
Hossain, M., 568–9, 596, 612, 651, 670–72
Hotelling, H., 94
Houck, J. P.
  on sustainability paradigm, 100–101
Houghton, J. T., 450
Houllier, F., 467
households (families, small farms), 62, 648, 683–4
  capital formation in, 558–67
  compared with cooperatives and collective farms, 520–29, 530–32
  contribution to local economy of members of, 533–40, 541
  production by, 503, 511–12
  theory of resource allocation by, 502–13, 515–18
  time allocation in, 503, 505
    Nigerian case study of, 481–99, 500–501
Howarth, R. B., 93, 94, 95
Huffman, W. E., 335, 369, 505, 516, 517, 535, 536, 538
human capital formation, 18–25, 62, 202
Human Development Index, 5
Hunek, T.
  on Hungarian agricultural reform, 164–5

Hungary
  agricultural reform in, 158–65, 257, 259, 261
Hurtado, H., 246
Hurwicz, L., 197
Hwa, E.-C., 49
hydropower, 412, 413

Idachaba, F. S., 343, 348, 398
Ikpi, A.
  on household time allocation in Nigeria, 481–99, 568
imperfect competition
  effects of trade liberalization in model with, 181–91, 192
import substitution, 47–8
incentive structures, 87, 293, 296–7
imcome (wages), 243, 270
  distribution of, Green Revolution and, 613–15
  policy of restraint on, 229, 234, 236
  see also poverty
India
  agricultural productivity in, 210
  agricultural research in, 349
  biotechnology in, 583
  fertilizer subsidies in, 35, 41
  forest management in, 439–41, 446–7
  Green Revolution in, 102–14, 116, 610, 611, 612
  infrastructure development in, 38
  mountain area development in, 409–10
  poverty in, 5, 6, 7–8, 10
  problems of agricultural growth in, 38
  rural non-farm employment in, 555–6
  sharecropping in, 12
individual economic agents, problems facing, 82–4
individual transferable quotas (ITQs), 432–3
Indonesia
  Green Revolution in, 608, 610, 612, 613, 617–18
industrial development, 49, 51–2, 199, 200
  see also non-farm employment
inflation, 205–6
  adjustment policies and, 217, 225
  government debt and, 227–37, 238–40
informal markets, 217
  finance and credit from, 644–53, 655–6

Bangladesh case study of, 657–67,
    668–9
infrastructure, 204, 407–8, 413, 415–16
    food security and, 37–8
Ingram, M. J., 450
input subsidies, 35, 41, 107
institutional economics, 81–2, 509–11,
    513, 516, 648–9
insurance scheme for food security, 40
intellectual property rights, 380, 383–4
    plant breeders and, 330–39, 340–42,
        363–4
interest rates, 206, 236, 242, 646, 648,
    660
intergenerational equity, 92–6
intermediate technology, 202
internal trade, 71–2, 74–5
International Agricultural Trade Re-
    search Consortium (IATRC), 131,
    135
International Development Association,
    444
International Food Policy Research
    Institute (IFPRI), 108
International Institute of Tropical
    Agriculture (IITA), 328, 359, 361
International Laboratory for Research on
    Animal Diseases (ILRAD), 359
International Monetary Fund (IMF)
    adjustment policy and, 232, 265, 275
    cereal facility of, 39, 40
International Rice Research Institute,
    358, 607
international trade *see* trade
intervention stores, 91
irrigation, 9, 35, 415
    see also Green Revolution
Israel, 210
    cooperative farm system in, 522–9,
        531–2
Isvilanonda, S., 612,613

Jabara, C. L., 62
James, C., 378, 582
Japan, 137, 541
    analysis of sustainability of agricul-
        ture in, 558–67
    biotechnology in, 583, 584
    fisheries in, 423, 424
    opposition to agricultural reform in,
        119, 120, 121, 123, 126, 130–31,
        295
Jarvis, L., 245

Jateliksono, T., 596, 613
Javier, E., 357
Jayasuriya, S., 545
Jayne, T. S., 225
Jenkins, G. J., 450
Jennings, P. R., 601
Jensen, H. C.
    on contribution of women and
        household members to rural
        economy, 533–40, 569
Jensen, M. C., 523
Jodha, N. S., 6, 405, 406, 407, 412, 413
Johnson, D. G., 30
Johnson, S. R., 534
Johnston, B., 542, 549, 577
joint farming *see* collective farming
Joly, P.-B., 335
Jones, R. W., 71
Jones, V. K., 451
Josserand, H., 226
Judd, M. A., 343
Junne, G., 584

Kada, R.
    analysis of sustainability of Japanese
        agriculture by, 558–66, 569
Kaiser, H., 451, 452
Kaitala, V. T., 431
Kane, S., 451
Kauffman, H. E., 601
Kenya, 576, 577, 578, 579, 584, 586
Kettunen, L.
    on air pollution, 472–3
Khan, A. R., 412
Khan, M., 300
kibbutz system, 522–9, 531–2
Kikuchi, M., 594
Kilby, P., 542, 543, 549, 577
King, P. G., 392, 393
Kinnucan, H., 579
Kirk, R., 581
Kirkpatrick, C. H., 181
Kislev, Y., 349, 377
    on family farms compared to
        cooperatives and collectives,
        520–29, 569
Klaassen, G., 470
Klaus, V., 74
Kloppenberg, J. R., 336
Knerr, B.
    on forest management, 445–8
Knudsen, O., 137, 138, 201, 207
Koester, U.

on efficiency of agricultural markets, 66–76, 115, 682
Konandreas, P., 105
Korea, South, 49, 120, 121, 295
Kramer, R. A.
    on management of forest resources, 435–44, 475
Krueger, A. O., 71, 76, 198, 292, 293
Krugman, P., 181, 197
Kumar, N., 320, 583, 584
Kuylenstierna, J., 465

labour market
    Chilean case study of, 242–3, 250–51
    effect of biotechnology on, 577–8, 583–4
    functioning of, 12, 15
    migration and, 8, 202, 610–12
    see also non-farm employment
Lacy, W. B., 371, 373–4
Lamb, P. J., 451
Lancaster, K. J., 455
Lanclos, K.
    on effects of trade liberalization, 181–91, 192
land, 8–9
    access to, and poverty, 11–16
    collectivization of, in China, 621
    degradation of *see* environmental problems
    denationalization of, in communist states, 156–7, 159–61, 254, 256–60, 263–4, 638–9
    impact of Green Revolution on, 610–13
    markets for, 12, 15, 245
    rights to, 204, 245, 254, 256–60, 420–21
Lange, M., 505, 535
Larue B., 137
Lasch, C., 92
Lashof, D. A., 450
Lau, L., 49
Lawrence, D., 517
Lawrence, R. Z., 129
lease of land, 12, 15
leasehold forestry, 414–15, 420
Ledesma, A. L.
    on biotechnology and poverty, 605–6
Lee, H. H., 578
Lee, J. E., 515
Lee, N., 181
legal structures, 85–6, 204

legislation on plant breeders' intellectual property rights, 330–39, 340–42
Leipert, Ch., 465, 469
Leiserson, M., 545
Lent, R.
    on adjustment policy, 225–6, 306
Leontief, W., 675
Levhari, D., 431
Li, F., 620
Liddle, M. J., 407
Liedholm, C.
    on rural non-farm economy, 542–54, 569
Lindert, P. H., 225
Lindner, R. K.
    on private versus public sector in biotechnology, 371–81
Linnemann, H., 30
Lipton, M. A., 5, 408, 545, 573, 579–80, 583, 596, 607
Liu, X., 620
Liu, Y., 23, 636
loan market *see* credit market
Locay, L., 502
Lomax, E., 209
Longhurst, R., 573, 579–80, 583, 596, 607
Longworth, J. W., 568
    on human capital formation, 18–25
Lopes, M., 236
Lopez, R. E., 522
Lough, J. M., 450
Lucas, R. E., 202
Lutz, E., 394
Lyons, D. C., 63

McClatchy, D.
    on agricultural and trade policy reform, 129–45, 192
McIntire, J., 69
McIver, R., 23
McNamara, K. T., 505
McRae, D., 429, 430
macro-economic adjustment *see* adjustment policy
Madagascar, 584
    adjustment policy in, 218, 220
    forest management in, 441–2, 447
Mahe, L. P.
    on prospects for GATT agricultural agreement, 125–8
Makau, B. F., 578
Malawi, 577, 584, 586

Malaysia, 295
Mali, 218
Manabe, S., 450
Manandhar, A., 579, 580
Mani, S., 583
Mansfield, E., 384
Manstetten, R., 396
marginal lands, sustainable development
    strategies for, 405–18, 419–21
Markandya, A., 94, 95
market price support (MPS), 133
market system, transition to *see* com-
    mand economies
marketing, 88, 271–2
markets *see* agricultural; credit; labour;
    land
Markusen, J. R., 184
Martens, L.
    on family farming compared with
        cooperative farming, 530–32
Martin, F., 225
Marx, K., 507
Mastenbroek, C., 335
Matlon, P., 50
Matoussi, M. S., 516
Matthews, R., 509
Mead, D., 543
Meadows, D. H., 450
Mehra, S., 102, 106
Mehta, M., 412
Meier, G. M., 76
Meilke, K. D., 137
Mellon, M., 603
Mellor, J. W., 13, 47, 51, 52, 394, 405,
    414, 542, 549
Mendelsohn, R., 443
Mendonca de Barros, J. R., 230
mercantilism, 62
Mercer, E., 442
Merlo, M.
    on India's Green Revolution, 112–14
Messenberg, R. P., 234
Messer, E., 597
Mexico, 137, 209, 210
    biotechnology in, 576–81 *passim*, 583
Meyer, R.L.
    on decollectivization in E. Europe,
        263–4
    on inflation and government debt in
        Brazil, 238–40
    on informal finance, 644–53, 671
Michaely, M., 71
migration, 8, 202, 610–12

of fish, 430–33
Mikesell, R. F., 95
Miner, W. M., 135
mining, 412
Mirman, L. J., 431
Mirowski, P., 393
Missaien, M., 226
Monke, E.
    on evaluation of policy choice, 166–
        77, 193
Moock, J., 406
moral hazard, 526, 527
moral issues *see* ethical issues
Morande, F., 243
Morocco
    agricultural sector in, 282–3, 284–8
    economy of, 281–2
    price and marketing policies in, 285,
        286–7
    public investment programmes in,
        281–2, 285, 287–8
    structural adjustment in, 283–9
mountain areas, strategies for sustain-
    able development in, 405–18, 419–
    21
Muchnik de Rubinstein, E., 70
    on agricultural reform in Chile, 241–
        51, 307
Mueesen, R. L., 357, 361, 364
multinational companies, 325–6
Mundlak, Y., 246, 297, 521, 657
Munro, B., 508
Munro, G. R., 427, 429, 430–31
Mureithi, L. P., 578
Murphy, P. W., 335, 338
Murshid, K. A. S.
    on informal credit market in Bangla-
        desh, 657–67, 672
Musgrave, R. A., 350
Mushita, A. T., 584
Myers, N., 439

Nadkarni, M. V., 102, 105, 106, 107, 108
Nagarajan, G.
    on informal finance, 644–53, 671
Nagy, J. G., 20
Nakajima, C., 493
Narayana, N. S. S., 35
national income accounting, 21
Nelson, J. M., 281, 512
Nelson, R. H., 92
Nelson, R. R., 371, 375–6, 378

neo-classical economics, 81, 93, 96, 181, 393
Nepal
    biotechnology in, 578, 579, 580
    Green Revolution in, 608, 610, 612
    sustainable development strategies for, 411, 412, 414
Netherlands, 209, 360
networks in biotechnology, 358–9
New Zealand, 68, 137
Newby, H., 531
Niehans, J., 511
Nielsen, A. H., 531
Niger, 49, 218
Nigeria, 42, 361, 577, 579, 581
    household time allocation in, 481–99, 500–501
Nilsson, S.
    on air pollution, 462–70, 476
Ninan, K. N.
    on Green Revolution in India, 102–12, 116
Nishimura, H.
    on women's and household members' contribution to rural economy, 541
Niu, R., 23
    on small farmers in Cina, 619–37, 671
Nixon, F. I., 181
Nodine, L., 336, 338
non-farm employment 415, 536–8, 542–52, 554–7
    Chinese case study of, 632–6, 642
    Japanese case study of, 559–60
    Nigerian case study of, 505, 515
Norgaard, R. B., 393, 394, 395
    on sustainability paradigm, 92–9, 116
North, D., 85
Norton, G. W., 343
Norway, 137, 431
Ntangsi, J.
    on structural adjustment in Cameroon, 265–78, 306–7
Nugent, J. B., 516, 517
nutrition
    biotechnology and, 314, 318, 581
    problems of, 45–9, 51, 62, 594, 684–5

Oakerson, R. J., 435
off-farm employment *see* non-farm employment
Okereke, G. U., 581
Okigbo, P., 481

Olayide, S. O., 481
Olivera, J. H., 227
Olnoch-Kosura, W.
    on household time-allocation, 500–501
Olson, M., 84, 198, 525, 531
Olstrom, E., 436, 437
Onchan, T.
    on sustainable development strategies in less favoured areas, 419–21
Ordover, J. A., 384
Organization for Economic Cooperation and Development (OECD), 130, 138, 148–50, 324, 371, 379
Otero, G., 581, 583, 584
O'Toole, J. C., 320
Otsuka, K., 12, 520, 596
    on Green Revolution, 607–16, 671
overview of conference, 673–87
Oxley, A., 129

Paarlberg, R., 130
Pakistan, 10, 42, 210, 410–11, 581
Palutikof, J. P., 450
Panagariya, A., 295
Panchamukhi, V. R., 320, 584
Papageorgiou, D., 71
Papua New Guinea, 219
parastatals, 72
Pardey, P. G., 343, 356
Pareto-efficiency, 66
Parikh, K. S., 52, 677
    on food security, 29–40, 115
Parker, T. S., 556
Parry, M. L., 452
Partap, T., 410
Pastore, A. C., 230
patents *see* intellectual property rights
Paz, L. J., 343
Peacock, W. J.
    on biotechnology, 311–17, 386
Pearce, D. W., 94, 95
Pearson, S.
    on evaluation of policy choice, 166–77, 193
Peart, R. M., 451
Perrin, R. K.
    on private versus public sector in biotechnology, 383–5
Persley, G. J., 363, 364, 372, 373, 379, 582
Peru, 209–10, 423

pest resistance, genetic engineering and, 313–14
Peterson, W., 521
Petit, M., 68
Philip, T. R.
 on food security, 41–2
Philippines, 6, 14, 578, 580, 583, 584, 593
 Green Revolution in, 608, 611, 612, 615
Pigou, A. C., 344
Pingali, P., 51
Pinstrup-Anderson, P., 380
plant breeders, intellectual property rights of, 330–39, 340–42, 363–4
Platais, K. W.
 on role of CGIAR, 355–66
Platteau, J.-P., 648
Plucknett, D., 329, 357, 358, 359, 363
Poland, 74
 agricultural reform in, 137, 254, 260, 261
Policy Analysis Matrix (PAM), 166–80, 193
Pollak, R., 502, 509, 510, 511, 512, 515
pollution *see* environmental problems
Popper, K., 96
population
 growth of, 10, 46, 210, 412
  control of, 413
 structural changes in, 63, 274
poverty, rural, 3–5, 534
 agrarian structure and, 11–16
 approaches to alleviation of, 7–8
  biotechnology and, 573–86, 589–91, 592–604, 605–6
 characteristics of, 6–7
 environmental degradation and, 9–11
 extent of, 5–6, 573, 592
 food insecurity and, 29–30, 41
 nutrition and, 45, 594, 684–5
 recent improvement of, 6
Poyo, J., 651
Prabowo, D.
 on Green Revolution, 617–18
Pradhan, P., 415
Pray, C. E., 350
 on plant breeders' rights legislation, 330–39
Price, M. F., 436
price(s), 88
 adjustment policies and, 213, 215–16, 218–19, 225, 236, 246

agricultural productivity and, 63
E. European reforms and, 255, 260–62, 263
food security and, 29–33, 41–2
incentives and, 293, 296–7
interventions in, 292, 293–6
policy reform of, 296–8, 300, 301–2
subsidies to, 35–7
suport and stabilization of, 140–43
see also inflation
principal-agent problem, 508, 509–11
processing of food, 63, 326–7
producer subsidies, 35–7
Producer Subsidy Equivalents (PSEs), 125
production, household, 503, 511–12
productivity, agricultural, 19, 209–10
 adjustment policy and, 213–14, 250
 biotechnology and, 319, 322–3
 prospects for growth in, 49–51, 62–3
Pronk, J., 573
Proops, R., 396
property *see* intellectual property; land
protectionism, 68, 295–6
 prospects for GATT agreement on reduction of, 119–28
public expenditure, 220–21
public finance theory: on agricultural research spillovers, 343–52
Putman, J. D., 380

Qinfa, L., 409
Quiggin, J., 517
Quiroz, J., 246

Radner, R., 508
Rahman, A., 658
Raj, K. N., 447
Rajbhandari, S. B., 578, 579, 580
Ramasamy, C., 610, 612
Ramos, R. S., 227, 228
Rana, R. S., 410
Rao, C. H. H., 102, 105, 106, 107
 on biotechnology for developing countries, 317–21
Rao, V. K. R. V., 102, 107
ration shops, 36–7, 41
Reardon, T., 226
recombinant DNA technology *see* biotechnology
regional trade agreements, 124
Reiger, H. C., 407
rent seeking, 198

Repetto, R., 394, 448
research, agricultural, 22, 24, 203–4
    financing of, 343–52, 353–4
    plant breeders' rights and, 334–7, 341
    see also biotechnology
resource allocation in households, 502–13, 515–18
resource management strategies, 394
Reutlinger, S., 47
Reynolds, R., 68, 300
Rezende, G. C., 53, 236
Riebe, K., 509
Robert, M. L., 580, 586
Robinson, S., 181
Roca, W. M., 360
Rockefeller Foundation, 601–2, 603–4
Rodrik, D., 181, 182, 184, 185–8, 190
Rogers, G., 10
Romania
    agricultural reform in, 256, 257–9, 261, 262, 263
Rooseboom, J., 356
Rosenzweig, C., 451
Rosenzweig, M. R., 82
Ross, R. L., 87
Rozenwurcel, G., 237
Ruivenkamp, G., 320, 584
Runge, C. F., 145
Russia *see* Soviet Union
Ruttan, V. W., 49, 324, 343, 349, 350, 399, 596

Sadoulet, E., 48, 49, 52
Saith, A., 5
Salant, P., 538
Sallnäs, O., 467
Sanders, J. H., 20
Sandrey, R., 68, 300
Sarris, A.
    on evaluation of policy choice, 177–80
Saupe, W. E., 538
scale economies, 521
Schafer, H., 71
Schiff, M., 292, 293, 295, 297, 298
Schmid, A. A., 377–8
Schmitt, G. H., 531
    on theory of resource allocation by farm households, 502–13, 568
Schneider, K., 581–2
Schott, J. J., 129
Schuh, G. E., 62, 68, 343, 392, 394
Schultz, T. W., 115, 202, 522, 525, 533, 535

Schultze, C. L., 129
Schweikhardt, D. B.
    on financing agricultural research, 343–52
Scotchmer, S., 376, 385
Scott, A. D., 344, 422, 427
security *see* food security
Seddon, D., 282
selective subsidies, 36–7
self-employment, 593
self-sufficiency, 71
Selowsky, M., 47
Sen, S. R., 102, 106
Sen Index of Poverty, 5
Senegal, 225
Serafy, S. E., 394
Sexton, R. J., 522
Shand, K., 545
Shand, R., 543
sharecropping, 12, 516–17, 520
Sharma, L. R., 410
Sharma, N. P., 443
Sharma, P., 413
Sharma, R. K., 657
Sharples, J. A., 139
Shaw, R., 469
Shetty, S. L., 108
Shugart, H. H., 452
Shyamsundar, P., 442
Siamwalla, A., 69, 302
Siebeck, W. E., 363, 364
Simonis, U. E., 465, 469
Singapore, 583
Singer, H. W., 49, 573
Singh, A., 573
Singh, I., 515, 522, 535, 545, 648
Singh, K., 439, 440
Sisler, D., 545
size of farms, optimal, 507–8, 521, 633
Slesinger, D. P., 538
Sloping Agricultural Land Technology, 416
small farming *see* households
Smith, A., 88
Smith, L. D.
    on private versus public sector agriculture, 81–90, 115
Smith, R. J., 439
social hardship argument, 73
social security, 68
socialist economics, 243
    see also command economy
soil erosion, 421

Sonka, S. T., 451
South Korea, 49, 120, 121, 295
Soviet Union
    agricultural reform in, 137, 152–7, 257
    collective farming in, 13, 154–7
    fisheries in, 431
    food imports by, 30, 152
    sanctions against, 68
Spain, 209
Spencer, D., 545
spillovers, agricultural research and,
        343–52, 353–4
Squire, L., 515, 522, 535, 545, 648
Srinivasan, T. N., 29, 35
stabilization policy *see* adjustment
        policy
stabilizing price support, 140–43
Stallman, J. I., 377–8
state *see* government
Stern, N., 72, 75
Stifel, L. D., 481
Stiglitz, J. E., 67, 71, 197, 198, 204, 205,
        509, 647
Stoeckel, A., 130
Stoevener, H., 386–7, 474–6
Strauss, J., 515, 522, 535, 545, 648
structural adjustment *see* adjustment
        policy
structural economics, 227–37
subsidiarity principle, 242
subsidies, 107, 163
    credit, 205–6
    food security and, 34–7, 41
    prospects for GATT agreement on
        reduction of, 119–28, 132
Sudaryanto, T., 612
Sumner, D. A., 517
    on rural non-farm economy, 554–7
suply aspects of sustainability, 406–11
Suryanarayana, M. H., 35
Sushma, S., 5
sustainability, 677–9
    analysis of Japanese agriculture in
        terms of, 558–67
    basic concepts of, 20–24, 92–9, 100–
        101, 116, 391–401, 402–4
    strategies in less favoured areas for,
        405–18, 419–21
Svarstad, H., 584
Swaminathan, M. S., 9, 481
Sweden, 137, 295

Tabatabai, H., 573

Tacio, H. D., 416
Tank, F. E., 578
Tanzania, 217
taxation, 245, 270, 304–5
    adjustment policy and, 219–21, 234,
        236
    of exports, 71
    finance of agricultural research by,
        350
Taylor, D., 393
Taylor, J., 52
technological developments in agricul-
        ture, 13, 14–15, 64, 201–3, 210,
        250, 560–61
    time allocation and, 494–6
    see also biotechnology; Green
        Revolution
technology transfer, 338–9, 359–60
Tejada, M., 359, 360
Thailand, 420–21, 580, 610, 612
Thapa, G., 612
Tharakan, M., 447
Thompson, R. L.
    on world agriculture in next century,
        62–5
Thomson, A. M.
    on private versus public sector
        agriculture, 81–90, 115
Thursby, M.
    on effects of trade liberalization, 181–
        91, 192
Tiede, J. M., 603
time allocation, 503, 505
    Nigerian case study of, 481–99, 500–
        501
Timmer, C. P., 517
Tinbergen, J., 455
Tirpak, D. A., 450
Tisdell, C., 104, 110
tissue culture *see* biotechnology
Tiwari, S. C., 410
Toenniessen, G. H., 358
Tokle, J. G., 536, 538
Toledo, J. E. C., 232
tourism, 412–13, 419–20
trade, 679–80
    adjustment policy and, 218–19, 234,
        248–9
    agricultural development and, 62, 63–
        4
    effects of liberalization on, 181–91,
        192, 296–8

food security and, 33–4, 39, 41–2, 70,
71
government intervention in, 69–75
policy guidelines for, 298–303
prospect for GATT agriculture
agreement on, 119–28, 129–39,
144, 148–50, 181, 192
Trainer, T., 405
training *see* education
transaction costs, 73, 509–11, 516–17,
520–21
transfer payments, 68
transport, 204, 407–8, 413, 415–16
'trickle down', 7
Trigo, E. J., 350
Troll, C., 408
Tullock, G., 198
Tunisia, 516
Tweeten, L., 281
Tyers, R., 138–9
Tyner, W. E.
on adjustment policy in Morocco,
281–9, 306

Underwood, D. A., 392, 393
unemployment, 68, 250
United Kingdom, 199
plant breeders' rights legislation in,
330, 331, 332, 333, 335, 338, 339
United Nations, 4–5, 6, 18, 51, 429
United States of America
agricultural markets in, 68, 78
agricultural reform in, 137
agricultural research in, 349, 356,
372–4
case study of women's and household
members' contribution to rural
economy in, 533–40, 541, 556
effects of climate change in, 451–3
environmental problems in, 399, 400
farm households in, 515–18, 521
fisheries in, 424, 425
GATT agricultural agreement and,
121, 123, 126–7, 130–31, 134,
136
plant breeders' rights legislation in,
330, 331, 332–3, 335–6
Upadhyaya, H., 612
Uruguay Round, 119–28, 129–39, 144,
148–50, 341, 380
USSR *see* Soviet Union
Uyen, N. V., 579, 580

Valdés, A., 69, 71, 105, 139, 246
on agricultural trade and pricing
policies, 292–303
Van der Meer, C. L. J.
on biotechnology, 321–7
Van Panchayats, 439–41, 446–7
Veeman, T. S., 20
Venables, A. J., 184
Venezian, E., 336
Venezuela, 581
Verma, L. R., 410
Vietnam, 579, 580, 583
Vlastuin, C., 517
voluntary group activities, 84–5
Von Braun, J., 590
Von Pischke, J. D., 644, 657
Vyas, V. S., 3–4, 568, 677
on agrarian structure, 11–13
on land resources, 8–11
on rural poverty, 4–8
on sustainability, 13–16

Wachter, M. L., 515
wages *see* income
Wald, S., 583
Walgate, R., 579, 597, 603
Walker, M., 197
Walters, C. J., 399
Warley, T. K.
on agricultural and trade policy
reform, 129–45, 192
Warren, G., 687
Washington Consensus, 232
water quality, 64
watershed development programmes,
108, 110
Wattanutchariya, S., 612, 613
Weatherald, R. J., 450
Weber, M. T., 225
Weinberg, A. M., 23
Weiss, A., 205
welfare economics, 68, 81
Wenpu, L., 409
Whalley, J., 33
White, G., 397
White, T. K., 48
Whiteman, P. T. S., 406
Whitener, L. A., 556
Widawsky, D. A., 320
Wigley, T. M. L., 450
Wildasin, D. E., 344
Wildavsky, A. B., 391
Williamson, G. J., 23

Williamson, O., 509
Wolf, E. C., 481
women, 6
    employment of, 63, 107, 251, 543,
        552, 578
        Nigerian case study of, 483–92, 501
        US case study of, 533–40, 541
Woodwell, G. M., 453
work *see* labour market
World Bank
    and adjustment policy, 232, 265, 275,
        281, 283, 284–5
    on agricultural reform, 130, 181
    on economic and agricultural growth,
        49, 51
    and forest management, 442, 443
    on poverty, 5–6, 7, 11, 29, 45, 573,
        592, 594
    on trade policy, 62
World Commission on Environment and
    Development (WCED), 10, 18
World Resources Institute, 437, 443
Wright, B., 375

Yadav, R. P.
    on sustainable development strategies
        for less favoured areas, 405–18,
        475
Yamada, S., 322
Yee, Y. L., 22
yield augmentation, 13, 15, 318, 319,
    575–7
    see also biotechnology; Green
        Revolution
Yotopoulos, P. A., 49, 648
Yuanling, M., 575, 577

Zaag, P. V., 579
Zambia, 295
Zamora, A. B., 580, 583
Zhang, C., 23
Zhao, D., 635
Zhu, M., 393
Zietz, J., 139
Zimbabwe, 581
Zusman, P., 522